Asymptotic Statistics

This book is an introduction to the field of asymptotic statistics. The treatment is both practical and mathematically rigorous. In addition to most of the standard topics of an asymptotics course, including likelihood inference, M-estimation, asymptotic efficiency, U-statistics, and rank procedures, the book also presents recent research topics such as semiparametric models, the bootstrap, and empirical processes and their applications.

One of the unifying themes is the approximation by limit experiments. This entails mainly the local approximation of the classical i.i.d. set-up with smooth parameters by location experiments involving a single, normally distributed observation. Thus, even the standard subjects of asymptotic statistics are presented in a novel way.

Suitable as a text for a graduate or Master's level statistics course, this book also gives researchers in statistics, probability, and their applications an overview of the latest research in asymptotic statistics.

A.W. van der Vaart is Professor of Statistics in the Department of Mathematics and Computer Science at the Vrije Universiteit, Amsterdam.

Asymptotic Statistics

A.W. VAN DER VAART

CAMBRIDGE
UNIVERSITY PRESS

University Printing House, Cambridge CB2 8BS, United Kingdom

One Liberty Plaza, 20th Floor, New York, NY 10006, USA

477 Williamstown Road, Port Melbourne, VIC 3207, Australia

314–321, 3rd Floor, Plot 3, Splendor Forum, Jasola District Centre,
New Delhi – 110025, India

103 Penang Road, #05–06/07, Visioncrest Commercial, Singapore 238467

Cambridge University Press is part of the University of Cambridge.

It furthers the University's mission by disseminating knowledge in the pursuit of
education, learning, and research at the highest international levels of excellence.

www.cambridge.org
Information on this title: www.cambridge.org/9780521784504
DOI: 10.1017/CBO9780511802256

First published 1998
First paperback edition 2000
8th printing 2007

A catalogue record for this publication is available from the British Library

Library of Congress Cataloguing in Publication Data
Vaart, A. W. van der
Asymptotic statistics / A. W. van der Vaart.
p. cm. – (Cambridge series in statistical and probabilistic
mathematics)
Includes bibliographical references.
1. Mathematical statistical – Asymptotic theory. I. Title.
II. Series: Cambridge series on statistical and probabilistic mathematics
CA2276. V22 1998
519.5–dc21   98-15176

ISBN 978-0-521-49603-2 Hardback
ISBN 978-0-521-78450-4 Paperback

To Maryse and Marianne

Contents

vii

Preface

This book grew out of courses that I gave at various places, including a graduate course in the Statistics Department of Texas A&M University, Master's level courses for mathematics students specializing in statistics at the Vrije Universiteit Amsterdam, a course in the DEA program (graduate level) of Université de Paris-sud, and courses in the Dutch AIO-netwerk (graduate level).

The mathematical level is mixed. Some parts I have used for second year courses for mathematics students (but they find it tough), other parts I would only recommend for a graduate program. The text is written both for students who know about the technical details of measure theory and probability, but little about statistics, and vice versa. This requires brief explanations of statistical methodology, for instance of what a rank test or the bootstrap is about, and there are similar excursions to introduce mathematical details. Familiarity with (higher-dimensional) calculus is necessary in all of the manuscript. Metric and normed spaces are briefly introduced in Chapter 18, when these concepts become necessary for Chapters 19, 20, 21 and 22, but I do not expect that this would be enough as a first introduction. For Chapter 25 basic knowledge of Hilbert spaces is extremely helpful, although the bare essentials are summarized at the beginning. Measure theory is implicitly assumed in the whole manuscript but can at most places be avoided by skipping proofs, by ignoring the word "measurable" or with a bit of handwaving. Because we deal mostly with i.i.d. observations, the simplest limit theorems from probability theory suffice. These are derived in Chapter 2, but prior exposure is helpful.

Sections, results or proofs that are preceded by asterisks are either of secondary importance or are out of line with the natural order of the chapters. As the chart in Figure 0.1 shows, many of the chapters are independent from one another, and the book can be used for several different courses.

A unifying theme is approximation by a limit experiment. The full theory is not developed (another writing project is on its way), but the material is limited to the "weak topology" on experiments, which in 90% of the book is exemplified by the case of smooth parameters of the distribution of i.i.d. observations. For this situation the theory can be developed by relatively simple, direct arguments. Limit experiments are used to explain efficiency properties, but also why certain procedures asymptotically take a certain form.

A second major theme is the application of results on abstract empirical processes. These already have benefits for deriving the usual theorems on M-estimators for Euclidean parameters but are indispensable if discussing more involved situations, such as M-estimators with nuisance parameters, chi-square statistics with data-dependent cells, or semiparametric models. The general theory is summarized in about 30 pages, and it is the applications

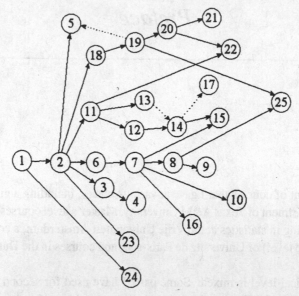

Figure 0.1. Dependence chart. A solid arrow means that a chapter is a prerequisite for a next chapter. A dotted arrow means a natural continuation. Vertical or horizontal position has no independent meaning.

that we focus on. In a sense, it would have been better to place this material (Chapters 18 and 19) earlier in the book, but instead we start with material of more direct statistical relevance and of a less abstract character. A drawback is that a few (starred) proofs point ahead to later chapters.

Almost every chapter ends with a "Notes" section. These are meant to give a rough historical sketch, and to provide entries in the literature for further reading. They certainly do not give sufficient credit to the original contributions by many authors and are not meant to serve as references in this way.

Mathematical statistics obtains its relevance from applications. The subjects of this book have been chosen accordingly. On the other hand, this is a mathematician's book in that we have made some effort to present results in a nice way, without the (unnecessary) lists of "regularity conditions" that are sometimes found in statistics books. Occasionally, this means that the accompanying proof must be more involved. If this means that an idea could go lost, then an informal argument precedes the statement of a result.

This does not mean that I have strived after the greatest possible generality. A simple, clean presentation was the main aim.

Leiden, September 1997
A.W. van der Vaart

Notation

A^*	adjoint operator
$\mathbb{B}^*$	dual space
$C_b(T), UC(T), C(T)$	(bounded, uniformly) continuous functions on T
$\ell^\infty(T)$	bounded functions on T
$\mathcal{L}_r(Q), L_r(Q)$	measurable functions whose rth powers are Q-integrable
$\|f\|_{Q,r}$	norm of $L_r(Q)$
$\|z\|_\infty, \|z\|_T$	uniform norm
lin	linear span
$\mathbb{C}, \mathbb{N}, \mathbb{Q}, \mathbb{R}, \mathbb{Z}$	number fields and sets
$EX, E^*X, \operatorname{var} X, \operatorname{sd} X, \operatorname{Cov} X$	(outer) expectation, variance, standard deviation, covariance (matrix) of X
$\mathbb{P}_n, \mathbb{G}_n$	empirical measure and process
$\mathbb{G}_P$	P-Brownian bridge
$N(\mu, \Sigma), t_n, \chi_n^2$	normal, t and chisquare distribution
$z_\alpha, \chi_{n,\alpha}^2, t_{n,\alpha}$	upper α-quantiles of normal, chisquare and t distributions
$\ll$	absolutely continuous
$\triangleleft, \triangleleft \triangleright$	contiguous, mutually contiguous
$\underset{\sim}{<}$	smaller than up to a constant
$\rightsquigarrow$	convergence in distribution
$\xrightarrow{P}$	convergence in probability
$\xrightarrow{as}$	convergence almost surely
$N(\varepsilon, T, d), N_{[]}(\varepsilon, T, d)$	covering and bracketing number
$J(\varepsilon, T, d), J_{[]}(\varepsilon, T, d)$	entropy integral
$o_P(1), O_P(1)$	stochastic order symbols

1

Introduction

Why asymptotic statistics? The use of asymptotic approximations is two-fold. First, they enable us to find approximate tests and confidence regions. Second, approximations can be used theoretically to study the quality (efficiency) of statistical procedures.

1.1   Approximate Statistical Procedures

To carry out a statistical test, we need to know the critical value for the test statistic. In most cases this means that we must know the distribution of the test statistic under the null hypothesis. Sometimes this is known exactly, but more often only approximations are available. This may be because the distribution of the statistic is analytically intractable, or perhaps the postulated statistical model is considered only an approximation of the true underlying distributions. In both cases the use of an approximate critical value may be fully satisfactory for practical purposes.

Consider for instance the classical t-test for location. Given a sample of independent observations $X_1, \ldots, X_n$, we wish to test a null hypothesis concerning the mean $\mu = \mathrm{E}X$. The t-test is based on the quotient of the sample mean $\overline{X}_n$ and the sample standard deviation S_n. If the observations arise from a normal distribution with mean μ_0, then the distribution of $\sqrt{n}(\overline{X}_n - \mu_0)/S_n$ is known exactly: It is a t-distribution with $n - 1$ degrees of freedom. However, we may have doubts regarding the normality, or we might even believe in a completely different model. If the number of observations is not too small, this does not matter too much. Then we may act as if $\sqrt{n}(\overline{X}_n - \mu_0)/S_n$ possesses a standard normal distribution. The theoretical justification is the limiting result, as $n \to \infty$,

$$\sup_x \left| P_\mu \left(\frac{\sqrt{n}(\overline{X}_n - \mu)}{S_n} \le x \right) - \Phi(x) \right| \to 0,$$

provided the variables X_i have a finite second moment. This variation on the central limit theorem is proved in the next chapter. A "large sample" level α test is to reject $H_0 : \mu = \mu_0$ if $\left| \sqrt{n}(\overline{X}_n - \mu_0)/S_n \right|$ exceeds the upper $\alpha/2$ quantile of the standard normal distribution. Table 1.1 gives the significance level of this test if the observations are either normally or exponentially distributed, and $\alpha = 0.05$. For $n \ge 20$ the approximation is quite reasonable in the normal case. If the underlying distribution is exponential, then the approximation is less satisfactory, because of the skewness of the exponential distribution.

Table 1.1. *Level of the test with critical region*
$\left| \sqrt{n}(\overline{X}_n - \mu_0)/S_n \right| > 1.96$ *if the observations
are sampled from the normal or
exponential distribution.*

n	Normal	Exponential[a]
5	0.122	0.19
10	0.082	0.14
15	0.070	0.11
20	0.065	0.10
25	0.062	0.09
50	0.056	0.07
100	0.053	0.06

[a] The third column gives approximations based on 10,000
simulations.

In many ways the t-test is an uninteresting example. There are many other reasonable
test statistics for the same problem. Often their null distributions are difficult to calculate.
An asymptotic result similar to the one for the t-statistic would make them practically
applicable at least for large sample sizes. Thus, one aim of asymptotic statistics is to derive
the asymptotic distribution of many types of statistics.

There are similar benefits when obtaining confidence intervals. For instance, the given
approximation result asserts that $\sqrt{n}(\overline{X}_n - \mu)/S_n$ is approximately standard normally distributed if μ is the true mean, whatever its value. This means that, with probability approximately $1 - 2\alpha$,

$$-z_\alpha \leq \frac{\sqrt{n}(\overline{X}_n - \mu)}{S_n} \leq z_\alpha.$$

This can be rewritten as the confidence statement $\mu = \overline{X}_n \pm z_\alpha S_n/\sqrt{n}$ in the usual manner.
For large n its confidence level should be close to $1 - 2\alpha$.

As another example, consider maximum likelihood estimators $\hat{\theta}_n$ based on a sample of
size n from a density p_θ. A major result in asymptotic statistics is that in many situations
$\sqrt{n}(\hat{\theta}_n - \theta)$ is asymptotically normally distributed with zero mean and covariance matrix the
inverse of the Fisher information matrix I_θ. If Z is k-variate normally distributed with mean
zero and nonsingular covariance matrix Σ, then the quadratic form $Z^T \Sigma^{-1} Z$ possesses a
chi-square distribution with k degrees of freedom. Thus, acting as if $\sqrt{n}(\hat{\theta}_n - \theta)$ possesses
an $N_k(0, I_\theta^{-1})$ distribution, we find that the ellipsoid

$$\left\{ \theta : (\theta - \hat{\theta}_n)^T I_{\hat{\theta}_n} (\theta - \hat{\theta}_n) \leq \frac{\chi^2_{k,\alpha}}{n} \right\}$$

is an approximate $1 - \alpha$ confidence region, if $\chi^2_{k,\alpha}$ is the appropriate critical value from the
chi-square distribution. A closely related alternative is the region based on inverting the
likelihood ratio test, which is also based on an asymptotic approximation.

1.2 Asymptotic Optimality Theory

For a relatively small number of statistical problems there exists an exact, optimal solution.
For instance, the Neyman-Pearson theory leads to optimal (uniformly most powerful) tests

in certain exponential family models; the Rao-Blackwell theory allows us to conclude that certain estimators are of minimum variance among the unbiased estimators. An important and fairly general result is the Cramér-Rao bound for the variance of unbiased estimators, but it is often not sharp.

If exact optimality theory does not give results, be it because the problem is untractable or because there exist no "optimal" procedures, then asymptotic optimality theory may help. For instance, to compare two tests we might compare approximations to their power functions. To compare estimators, we might compare asymptotic variances rather than exact variances. A major result in this area is that for smooth parametric models maximum likelihood estimators are asymptotically optimal. This roughly means the following. First, maximum likelihood estimators are asymptotically consistent: The sequence of estimators converges in probability to the true value of the parameter. Second, the rate at which maximum likelihood estimators converge to the true value is the fastest possible, typically $1/\sqrt{n}$. Third, their asymptotic variance, the variance of the limit distribution of $\sqrt{n}(\hat{\theta}_n - \theta)$, is minimal; in fact, maximum likelihood estimators "asymptotically attain" the Cramér-Rao bound. Thus asymptotics justify the use of the maximum likelihood method in certain situations. It is of interest here that, even though the method of maximum likelihood often leads to reasonable estimators and has great intuitive appeal, in general it does not lead to best estimators for finite samples. Thus the use of an asymptotic criterion simplifies optimality theory considerably.

By taking limits we can gain much insight in the structure of statistical experiments. It turns out that not only estimators and test statistics are asymptotically normally distributed, but often also the whole sequence of statistical models converges to a model with a normal observation. Our good understanding of the latter "canonical experiment" translates directly into understanding other experiments asymptotically. The mathematical beauty of this theory is an added benefit of asymptotic statistics. Though we shall be mostly concerned with normal limiting theory, this theory applies equally well to other situations.

1.3 Limitations

Although asymptotics is both practically useful and of theoretical importance, it should not be taken for more than what it is: approximations. Clearly, a theorem that can be interpreted as saying that a statistical procedure works fine for $n \to \infty$ is of no use if the number of available observations is $n = 5$.

In fact, strictly speaking, most asymptotic results that are currently available are logically useless. This is because most asymptotic results are limit results, rather than approximations consisting of an approximating formula plus an accurate error bound. For instance, to estimate a value a, we consider it to be the 25th element $a = a_{25}$ in a sequence $a_1, a_2, \ldots,$ and next take $\lim_{n \to \infty} a_n$ as an approximation. The accuracy of this procedure depends crucially on the choice of the sequence in which a_{25} is embedded, and it seems impossible to defend the procedure from a logical point of view. This is why there is good asymptotics and bad asymptotics and why two types of asymptotics sometimes lead to conflicting claims.

Fortunately, many limit results of statistics do give reasonable answers. Because it may be theoretically very hard to ascertain that approximation errors are small, one often takes recourse to simulation studies to judge the accuracy of a certain approximation.

Just as care is needed if using asymptotic results for approximations, results on asymptotic optimality must be judged in the right manner. One pitfall is that even though a certain procedure, such as maximum likelihood, is asymptotically optimal, there may be many other procedures that are asymptotically optimal as well. For finite samples these may behave differently and possibly better. Then so-called higher-order asymptotics, which yield better approximations, may be fruitful. See e.g., [7], [52] and [114]. Although we occasionally touch on this subject, we shall mostly be concerned with what is known as "first-order asymptotics."

1.4   The Index n

In all of the following n is an index that tends to infinity, and *asymptotics* means taking limits as $n \to \infty$. In most situations n is the number of observations, so that usually asymptotics is equivalent to "large-sample theory." However, certain abstract results are pure limit theorems that have nothing to do with individual observations. In that case n just plays the role of the index that goes to infinity.

1.5   Notation

A symbol index is given on page xv.

For brevity we often use operator notation for evaluation of expectations and have special symbols for the empirical measure and process.

For P a measure on a measurable space $(\mathcal{X}, \mathcal{B})$ and $f : \mathcal{X} \mapsto \mathbb{R}^k$ a measurable function, Pf denotes the integral $\int f \, dP$; equivalently, the expectation $\mathrm{E}_P f(X_1)$ for X_1 a random variable distributed according to P. When applied to the empirical measure $\mathbb{P}_n$ of a sample $X_1, \ldots, X_n$, the discrete uniform measure on the sample values, this yields

$$\mathbb{P}_n f = \frac{1}{n} \sum_{i=1}^{n} f(X_i).$$

This formula can also be viewed as simply an abbreviation for the average on the right. The empirical process $\mathbb{G}_n f$ is the centered and scaled version of the empirical measure, defined by

$$\mathbb{G}_n f = \sqrt{n}(\mathbb{P}_n f - Pf) = \frac{1}{\sqrt{n}} \sum_{i=1}^{n} \big(f(X_i) - \mathrm{E}_P f(X_i)\big).$$

This is studied in detail in Chapter 19, but is used as an abbreviation throughout the book.

2

Stochastic Convergence

This chapter provides a review of basic modes of convergence of sequences of stochastic vectors, in particular convergence in distribution and in probability.

2.1 Basic Theory

A *random vector* in $\mathbb{R}^k$ is a vector $X = (X_1, \ldots, X_k)$ of real random variables.[†] The *distribution function* of X is the map $x \mapsto P(X \le x)$.

A sequence of random vectors X_n is said to *converge in distribution* to a random vector X if

$$P(X_n \le x) \to P(X \le x),$$

for every x at which the limit distribution function $x \mapsto P(X \le x)$ is continuous. Alternative names are *weak convergence* and *convergence in law*. As the last name suggests, the convergence only depends on the induced laws of the vectors and not on the probability spaces on which they are defined. Weak convergence is denoted by $X_n \rightsquigarrow X$; if X has distribution L, or a distribution with a standard code, such as $N(0, 1)$, then also by $X_n \rightsquigarrow L$ or $X_n \rightsquigarrow N(0, 1)$.

Let $d(x, y)$ be a distance function on $\mathbb{R}^k$ that generates the usual topology. For instance, the Euclidean distance

$$d(x, y) = \|x - y\| = \left(\sum_{i=1}^{k} (x_i - y_i)^2 \right)^{1/2}.$$

A sequence of random variables X_n is said to *converge in probability* to X if for all $\varepsilon > 0$

$$P\big(d(X_n, X) > \varepsilon\big) \to 0.$$

This is denoted by $X_n \xrightarrow{P} X$. In this notation convergence in probability is the same as $d(X_n, X) \xrightarrow{P} 0$.

[†] More formally it is a Borel measurable map from some probability space in $\mathbb{R}^k$. Throughout it is implicitly understood that variables X, $g(X)$, and so forth of which we compute expectations or probabilities are measurable maps on some probability space.

As we shall see, convergence in probability is stronger than convergence in distribution. An even stronger mode of convergence is almost-sure convergence. The sequence X_n is said to *converge almost surely* to X if $d(X_n, X) \to 0$ with probability one:

$$P\big(\lim d(X_n, X) = 0\big) = 1.$$

This is denoted by $X_n \overset{as}{\to} X$. Note that convergence in probability and convergence almost surely only make sense if each of X_n and X are defined on the same probability space. For convergence in distribution this is not necessary.

2.1 Example (Classical limit theorems). Let $\overline{Y}_n$ be the average of the first n of a sequence of independent, identically distributed random vectors $Y_1, Y_2, \ldots$. If $E\|Y_1\| < \infty$, then $\overline{Y}_n \overset{as}{\to} EY_1$ by the *strong law of large numbers*. Under the stronger assumption that $E\|Y_1\|^2 < \infty$, the *central limit theorem* asserts that $\sqrt{n}(\overline{Y}_n - EY_1) \rightsquigarrow N(0, \operatorname{Cov} Y_1)$. The central limit theorem plays an important role in this manuscript. It is proved later in this chapter, first for the case of real variables, and next it is extended to random vectors. The strong law of large numbers appears to be of less interest in statistics. Usually the *weak law of large numbers*, according to which $\overline{Y}_n \overset{P}{\to} EY_1$, suffices. This is proved later in this chapter. □

The portmanteau lemma gives a number of equivalent descriptions of weak convergence. Most of the characterizations are only useful in proofs. The last one also has intuitive value.

2.2 Lemma (Portmanteau). *For any random vectors X_n and X the following statements are equivalent.*
 (i) $P(X_n \le x) \to P(X \le x)$ *for all continuity points of $x \mapsto P(X \le x)$;*
 (ii) $Ef(X_n) \to Ef(X)$ *for all bounded, continuous functions f;*
 (iii) $Ef(X_n) \to Ef(X)$ *for all bounded, Lipschitz[†] functions f;*
 (iv) $\liminf Ef(X_n) \ge Ef(X)$ *for all nonnegative, continuous functions f;*
 (v) $\liminf P(X_n \in G) \ge P(X \in G)$ *for every open set G;*
 (vi) $\limsup P(X_n \in F) \le P(X \in F)$ *for every closed set F;*
 (vii) $P(X_n \in B) \to P(X \in B)$ *for all Borel sets B with $P(X \in \delta B) = 0$, where $\delta B = \overline{B} - \mathring{B}$ is the boundary of B.*

Proof. (i) $\Rightarrow$ (ii). Assume first that the distribution function of X is continuous. Then condition (i) implies that $P(X_n \in I) \to P(X \in I)$ for every rectangle I. Choose a sufficiently large, compact rectangle I with $P(X \notin I) < \varepsilon$. A continuous function f is uniformly continuous on the compact set I. Thus there exists a partition $I = \cup_j I_j$ into finitely many rectangles I_j such that f varies at most ε on every I_j. Take a point x_j from each I_j and define $f_\varepsilon = \sum_j f(x_j) 1_{I_j}$. Then $|f - f_\varepsilon| < \varepsilon$ on I, whence if f takes its values in $[-1, 1]$,

$$\big|Ef(X_n) - Ef_\varepsilon(X_n)\big| \le \varepsilon + P(X_n \notin I),$$
$$\big|Ef(X) - Ef_\varepsilon(X)\big| \le \varepsilon + P(X \notin I) < 2\varepsilon.$$

[†] A function is called *Lipschitz* if there exists a number L such that $|f(x) - f(y)| \le L d(x, y)$, for every x and y. The least such number L is denoted $\|f\|_{\text{lip}}$.

For sufficiently large n, the right side of the first equation is smaller than 2ε as well. We combine this with

$$\left|\mathrm{E}f_\varepsilon(X_n) - \mathrm{E}f_\varepsilon(X)\right| \le \sum_j \left|\mathrm{P}(X_n \in I_j) - \mathrm{P}(X \in I_j)\right| \left|f(x_j)\right| \to 0.$$

Together with the triangle inequality the three displays show that $\left|\mathrm{E}f(X_n) - \mathrm{E}f(X)\right|$ is bounded by 5ε eventually. This being true for every $\varepsilon > 0$ implies (ii).

Call a set B a *continuity set* if its boundary δB satisfies $\mathrm{P}(X \in \delta B) = 0$. The preceding argument is valid for a general X provided all rectangles I are chosen equal to continuity sets. This is possible, because the collection of discontinuity sets is sparse. Given any collection of pairwise disjoint measurable sets, at most countably many sets can have positive probability. Otherwise the probability of their union would be infinite. Therefore, given any collection of sets $\{B_\alpha : \alpha \in A\}$ with pairwise disjoint boundaries, all except at most countably many sets are continuity sets. In particular, for each j at most countably many sets of the form $\{x : x_j \le \alpha\}$ are not continuity sets. Conclude that there exist dense subsets $Q_1, \dots, Q_k$ of $\mathbb{R}$ such that each rectangle with corners in the set $Q_1 \times \cdots \times Q_k$ is a continuity set. We can choose all rectangles I inside this set.

(iii) $\Rightarrow$ (v). For every open set G there exists a sequence of Lipschitz functions with $0 \le f_m \uparrow 1_G$. For instance $f_m(x) = (md(x, G^c)) \wedge 1$. For every fixed m,

$$\liminf_{n \to \infty} \mathrm{P}(X_n \in G) \ge \liminf_{n \to \infty} \mathrm{E}f_m(X_n) = \mathrm{E}f_m(X).$$

As $m \to \infty$ the right side increases to $\mathrm{P}(X \in G)$ by the monotone convergence theorem.

(v) $\Leftrightarrow$ (vi). Because a set is open if and only if its complement is closed, this follows by taking complements.

(v) + (vi) $\Rightarrow$ (vii). Let $\mathring{B}$ and $\overline{B}$ denote the interior and the closure of a set, respectively. By (v)

$$\mathrm{P}(X \in \mathring{B}) \le \liminf \mathrm{P}(X_n \in \mathring{B}) \le \limsup \mathrm{P}(X_n \in \overline{B}) \le \mathrm{P}(X \in \overline{B}),$$

by (vi). If $\mathrm{P}(X \in \delta B) = 0$, then left and right side are equal, whence all inequalities are equalities. The probability $\mathrm{P}(X \in B)$ and the limit $\lim \mathrm{P}(X_n \in B)$ are between the expressions on left and right and hence equal to the common value.

(vii) $\Rightarrow$ (i). Every cell $(-\infty, x]$ such that x is a continuity point of $x \mapsto \mathrm{P}(X \le x)$ is a continuity set.

The equivalence (ii) $\Leftrightarrow$ (iv) is left as an exercise. ∎

The continuous-mapping theorem is a simple result, but it is extremely useful. If the sequence of random vectors X_n converges to X and g is continuous, then $g(X_n)$ converges to $g(X)$. This is true for each of the three modes of stochastic convergence.

2.3 **Theorem (Continuous mapping).** *Let $g : \mathbb{R}^k \mapsto \mathbb{R}^m$ be continuous at every point of a set C such that $\mathrm{P}(X \in C) = 1$.*

(i) If $X_n \rightsquigarrow X$, then $g(X_n) \rightsquigarrow g(X)$;

(ii) If $X_n \overset{\mathrm{P}}{\to} X$, then $g(X_n) \overset{\mathrm{P}}{\to} g(X)$;

(iii) If $X_n \overset{\mathrm{as}}{\to} X$, then $g(X_n) \overset{\mathrm{as}}{\to} g(X)$.

Proof. (i). The event $\{g(X_n) \in F\}$ is identical to the event $\{X_n \in g^{-1}(F)\}$. For every closed set F,

$$g^{-1}(F) \subset \overline{g^{-1}(F)} \subset g^{-1}(F) \cup C^c.$$

To see the second inclusion, take x in the closure of $g^{-1}(F)$. Thus, there exists a sequence x_m with $x_m \to x$ and $g(x_m) \in F$ for every F. If $x \in C$, then $g(x_m) \to g(x)$, which is in F because F is closed; otherwise $x \in C^c$. By the portmanteau lemma,

$$\limsup P\big(g(X_n) \in F\big) \leq \limsup P\big(X_n \in \overline{g^{-1}(F)}\big) \leq P\big(X \in \overline{g^{-1}(F)}\big).$$

Because $P(X \in C^c) = 0$, the probability on the right is $P\big(X \in g^{-1}(F)\big) = P\big(g(X) \in F\big)$. Apply the portmanteau lemma again, in the opposite direction, to conclude that $g(X_n) \rightsquigarrow g(X)$.

(ii). Fix arbitrary $\varepsilon > 0$. For each $\delta > 0$ let B_δ be the set of x for which there exists y with $d(x, y) < \delta$, but $d\big(g(x), g(y)\big) > \varepsilon$. If $X \notin B_\delta$ and $d\big(g(X_n), g(X)\big) > \varepsilon$, then $d(X_n, X) \geq \delta$. Consequently,

$$P\Big(d\big(g(X_n), g(X)\big) > \varepsilon\Big) \leq P(X \in B_\delta) + P\big(d(X_n, X) \geq \delta\big).$$

The second term on the right converges to zero as $n \to \infty$ for every fixed $\delta > 0$. Because $B_\delta \cap C \downarrow \emptyset$ by continuity of g, the first term converges to zero as $\delta \downarrow 0$.

Assertion (iii) is trivial. ∎

Any random vector X is *tight*: For every $\varepsilon > 0$ there exists a constant M such that $P(\|X\| > M) < \varepsilon$. A set of random vectors $\{X_\alpha : \alpha \in A\}$ is called *uniformly tight* if M can be chosen the same for every X_α: For every $\varepsilon > 0$ there exists a constant M such that

$$\sup_\alpha P\big(\|X_\alpha\| > M\big) < \varepsilon.$$

Thus, there exists a compact set to which all X_α give probability "almost" one. Another name for uniformly tight is *bounded in probability*. It is not hard to see that every weakly converging sequence X_n is uniformly tight. More surprisingly, the converse of this statement is almost true: According to Prohorov's theorem, every uniformly tight sequence contains a weakly converging subsequence. Prohorov's theorem generalizes the Heine-Borel theorem from deterministic sequences X_n to random vectors.

2.4 Theorem (Prohorov's theorem). *Let X_n be random vectors in $\mathbb{R}^k$.*
  (i) *If $X_n \rightsquigarrow X$ for some X, then $\{X_n : n \in \mathbb{N}\}$ is uniformly tight;*
  (ii) *If X_n is uniformly tight, then there exists a subsequence with $X_{n_j} \rightsquigarrow X$ as $j \to \infty$, for some X.*

Proof. (i). Fix a number M such that $P\big(\|X\| \geq M\big) < \varepsilon$. By the portmanteau lemma $P\big(\|X_n\| \geq M\big)$ exceeds $P\big(\|X\| \geq M\big)$ arbitrarily little for sufficiently large n. Thus there exists N such that $P\big(\|X_n\| \geq M\big) < 2\varepsilon$, for all $n \geq N$. Because each of the finitely many variables X_n with $n < N$ is tight, the value of M can be increased, if necessary, to ensure that $P\big(\|X_n\| \geq M\big) < 2\varepsilon$ for every n.

(ii). By Helly's lemma (described subsequently), there exists a subsequence F_{n_j} of the sequence of cumulative distribution functions $F_n(x) = P(X_n \le x)$ that converges weakly to a possibly "defective" distribution function F. It suffices to show that F is a proper distribution function: $F(x) \to 0, 1$ if $x_i \to -\infty$ for some i, or $x \to \infty$. By the uniform tightness, there exists M such that $F_n(M) > 1 - \varepsilon$ for all n. By making M larger, if necessary, it can be ensured that M is a continuity point of F. Then $F(M) = \lim F_{n_j}(M) \ge 1 - \varepsilon$. Conclude that $F(x) \to 1$ as $x \to \infty$. That the limits at $-\infty$ are zero can be seen in a similar manner. ∎

The crux of the proof of Prohorov's theorem is Helly's lemma. This asserts that any given sequence of distribution functions contains a subsequence that converges weakly to a possibly defective distribution function. A *defective distribution function* is a function that has all the properties of a cumulative distribution function with the exception that it has limits less than 1 at ∞ and/or greater than 0 at $-\infty$.

2.5 Lemma (Helly's lemma). *Each given sequence F_n of cumulative distribution functions on $\mathbb{R}^k$ possesses a subsequence F_{n_j} with the property that $F_{n_j}(x) \to F(x)$ at each continuity point x of a possibly defective distribution function F.*

Proof. Let $\mathbb{Q}^k = \{q_1, q_2, \ldots\}$ be the vectors with rational coordinates, ordered in an arbitrary manner. Because the sequence $F_n(q_1)$ is contained in the interval $[0, 1]$, it has a converging subsequence. Call the indexing subsequence $\{n_j^1\}_{j=1}^\infty$ and the limit $G(q_1)$. Next, extract a further subsequence $\{n_j^2\} \subset \{n_j^1\}$ along which $F_n(q_2)$ converges to a limit $G(q_2)$, a further subsequence $\{n_j^3\} \subset \{n_j^2\}$ along which $F_n(q_3)$ converges to a limit $G(q_3), \ldots$, and so forth. The "tail" of the diagonal sequence $n_j := n_j^j$ belongs to every sequence n_j^i. Hence $F_{n_j}(q_i) \to G(q_i)$ for every $i = 1, 2, \ldots$. Because each F_n is nondecreasing, $G(q) \le G(q')$ if $q \le q'$. Define

$$F(x) = \inf_{q > x} G(q).$$

Then F is nondecreasing. It is also right-continuous at every point x, because for every $\varepsilon > 0$ there exists $q > x$ with $G(q) - F(x) < \varepsilon$, which implies $F(y) - F(x) < \varepsilon$ for every $x \le y \le q$. Continuity of F at x implies, for every $\varepsilon > 0$, the existence of $q < x < q'$ such that $G(q') - G(q) < \varepsilon$. By monotonicity, we have $G(q) \le F(x) \le G(q')$, and

$$G(q) = \lim F_{n_j}(q) \le \liminf F_{n_j}(x) \le \lim F_{n_j}(q') = G(q').$$

Conclude that $\left| \liminf F_{n_j}(x) - F(x) \right| < \varepsilon$. Because this is true for every $\varepsilon > 0$ and the same result can be obtained for the lim sup, it follows that $F_{n_j}(x) \to F(x)$ at every continuity point of F.

In the higher-dimensional case, it must still be shown that the expressions defining masses of cells are nonnegative. For instance, for $k = 2$, F is a (defective) distribution function only if $F(b) + F(a) - F(a_1, b_2) - F(a_2, b_1) \ge 0$ for every $a \le b$. In the case that the four corners $a, b, (a_1, b_2)$, and (a_2, b_1) of the cell are continuity points; this is immediate from the convergence of F_{n_j} to F and the fact that each F_n is a distribution function. Next, for general cells the property follows by right continuity. ∎

2.6 *Example (Markov's inequality).* A sequence X_n of random variables with $\mathrm{E}|X_n|^p = O(1)$ for some $p > 0$ is uniformly tight. This follows because by *Markov's inequality*

$$\mathrm{P}(|X_n| > M) \leq \frac{\mathrm{E}|X_n|^p}{M^p}$$

The right side can be made arbitrarily small, uniformly in n, by choosing sufficiently large M.

Because $\mathrm{E}X_n^2 = \mathrm{var}\, X_n + (\mathrm{E}X_n)^2$, an alternative sufficient condition for uniform tightness is $\mathrm{E}X_n = O(1)$ and $\mathrm{var}\, X_n = O(1)$. This cannot be reversed. $\quad\square$

Consider some of the relationships among the three modes of convergence. Convergence in distribution is weaker than convergence in probability, which is in turn weaker than almost-sure convergence, except if the limit is constant.

2.7 *Theorem. Let X_n, X and Y_n be random vectors. Then*
  (i) $X_n \overset{\text{as}}{\to} X$ *implies* $X_n \overset{\text{P}}{\to} X$;
  (ii) $X_n \overset{\text{P}}{\to} X$ *implies* $X_n \rightsquigarrow X$;
  (iii) $X_n \overset{\text{P}}{\to} c$ *for a constant c if and only if $X_n \rightsquigarrow c$*;
  (iv) *if $X_n \rightsquigarrow X$ and $d(X_n, Y_n) \overset{\text{P}}{\to} 0$, then $Y_n \rightsquigarrow X$*;
  (v) *if $X_n \rightsquigarrow X$ and $Y_n \overset{\text{P}}{\to} c$ for a constant c, then $(X_n, Y_n) \rightsquigarrow (X, c)$*;
  (vi) *if $X_n \overset{\text{P}}{\to} X$ and $Y_n \overset{\text{P}}{\to} Y$, then $(X_n, Y_n) \overset{\text{P}}{\to} (X, Y)$.*

Proof. (i). The sequence of sets $A_n = \cup_{m \geq n}\{d(X_m, X) > \varepsilon\}$ is decreasing for every $\varepsilon > 0$ and decreases to the empty set if $X_n(\omega) \to X(\omega)$ for every ω. If $X_n \overset{\text{as}}{\to} X$, then $\mathrm{P}(d(X_n, X) > \varepsilon) \leq \mathrm{P}(A_n) \to 0$.

(iv). For every f with range $[0, 1]$ and Lipschitz norm at most 1 and every $\varepsilon > 0$,

$$\left|\mathrm{E}f(X_n) - \mathrm{E}f(Y_n)\right| \leq \varepsilon \mathrm{E}1\{d(X_n, Y_n) \leq \varepsilon\} + 2\mathrm{E}1\{d(X_n, Y_n) > \varepsilon\}.$$

The second term on the right converges to zero as $n \to \infty$. The first term can be made arbitrarily small by choice of ε. Conclude that the sequences $\mathrm{E}f(X_n)$ and $\mathrm{E}f(Y_n)$ have the same limit. The result follows from the portmanteau lemma.

(ii). Because $d(X_n, X) \overset{\text{P}}{\to} 0$ and trivially $X \rightsquigarrow X$, it follows that $X_n \rightsquigarrow X$ by (iv).

(iii). The "only if" part is a special case of (ii). For the converse let ball(c, ε) be the open ball of radius ε around c. Then $\mathrm{P}(d(X_n, c) \geq \varepsilon) = \mathrm{P}(X_n \in \mathrm{ball}(c, \varepsilon)^c)$. If $X_n \rightsquigarrow c$, then the lim sup of the last probability is bounded by $\mathrm{P}(c \in \mathrm{ball}(c, \varepsilon)^c) = 0$, by the portmanteau lemma.

(v). First note that $d\big((X_n, Y_n), (X_n, c)\big) = d(Y_n, c) \overset{\text{P}}{\to} 0$. Thus, according to (iv), it suffices to show that $(X_n, c) \rightsquigarrow (X, c)$. For every continuous, bounded function $(x, y) \mapsto f(x, y)$, the function $x \mapsto f(x, c)$ is continuous and bounded. Thus $\mathrm{E}f(X_n, c) \to \mathrm{E}f(X, c)$ if $X_n \rightsquigarrow X$.

(vi). This follows from $d\big((x_1, y_1), (x_2, y_2)\big) \leq d(x_1, x_2) + d(y_1, y_2)$. ∎

According to the last assertion of the lemma, convergence in probability of a sequence of vectors $X_n = (X_{n,1}, \ldots, X_{n,k})$ is equivalent to convergence of every one of the sequences of components $X_{n,i}$ separately. The analogous statement for convergence in distribution is false: Convergence in distribution of the sequence X_n is stronger than convergence of every one of the sequences of components $X_{n,i}$. The point is that the distribution of the components $X_{n,i}$ separately does not determine their joint distribution: They might be independent or dependent in many ways. We speak of *joint convergence* in distribution versus *marginal convergence* .

Assertion (v) of the lemma has some useful consequences. If $X_n \rightsquigarrow X$ and $Y_n \rightsquigarrow c$, then $(X_n, Y_n) \rightsquigarrow (X, c)$. Consequently, by the continuous mapping theorem, $g(X_n, Y_n) \rightsquigarrow g(X, c)$ for every map g that is continuous at every point in the set $\mathbb{R}^k \times \{c\}$ in which the vector (X, c) takes its values. Thus, for every g such that

$$\lim_{x \to x_0, y \to c} g(x, y) = g(x_0, c), \qquad \text{for every } x_0.$$

Some particular applications of this principle are known as Slutsky's lemma.

2.8 Lemma (Slutsky). *Let X_n, X and Y_n be random vectors or variables. If $X_n \rightsquigarrow X$ and $Y_n \rightsquigarrow c$ for a constant c, then*
  (i) $X_n + Y_n \rightsquigarrow X + c$;
  (ii) $Y_n X_n \rightsquigarrow cX$;
  (iii) $Y_n^{-1} X_n \rightsquigarrow c^{-1} X$ *provided $c \neq 0$.*

In (i) the "constant" c must be a vector of the same dimension as X, and in (ii) it is probably initially understood to be a scalar. However, (ii) is also true if every Y_n and c are matrices (which can be identified with vectors, for instance by aligning rows, to give a meaning to the convergence $Y_n \rightsquigarrow c$), simply because matrix multiplication $(x, y) \mapsto yx$ is a continuous operation. Even (iii) is valid for matrices Y_n and c and vectors X_n provided $c \neq 0$ is understood as c being invertible, because taking an inverse is also continuous.

2.9 Example (t-statistic). Let $Y_1, Y_2, \ldots$ be independent, identically distributed random variables with $EY_1 = 0$ and $EY_1^2 < \infty$. Then the *t*-statistic $\sqrt{n}\bar{Y}_n/S_n$, where $S_n^2 = (n-1)^{-1}\sum_{i=1}^n (Y_i - \bar{Y}_n)^2$ is the sample variance, is asymptotically standard normal.

To see this, first note that by two applications of the weak law of large numbers and the continuous-mapping theorem for convergence in probability

$$S_n^2 = \frac{n}{n-1}\left(\frac{1}{n}\sum_{i=1}^n Y_i^2 - \bar{Y}_n^2\right) \xrightarrow{P} 1\left(EY_1^2 - (EY_1)^2\right) = \operatorname{var} Y_1.$$

Again by the continuous-mapping theorem, S_n converges in probability to sd Y_1. By the central limit theorem $\sqrt{n}\bar{Y}_n$ converges in law to the $N(0, \operatorname{var} Y_1)$ distribution. Finally, Slutsky's lemma gives that the sequence of *t*-statistics converges in distribution to $N(0, \operatorname{var} Y_1)/$ sd Y_1 = $N(0, 1)$. $\square$

2.10 Example (Confidence intervals). Let T_n and S_n be sequences of estimators satisfying

$$\sqrt{n}(T_n - \theta) \rightsquigarrow N(0, \sigma^2), \qquad S_n^2 \xrightarrow{P} \sigma^2,$$

for certain parameters θ and σ^2 depending on the underlying distribution, for every distribution in the model. Then $\theta = T_n \pm S_n/\sqrt{n}\, z_\alpha$ is a confidence interval for θ of asymptotic level $1 - 2\alpha$. More precisely, we have that the probability that θ is contained in $[T_n - S_n/\sqrt{n} \, z_\alpha, T_n + S_n/\sqrt{n} \, z_\alpha]$ converges to $1 - 2\alpha$.

This is a consequence of the fact that the sequence $\sqrt{n}(T_n - \theta)/S_n$ is asymptotically standard normally distributed. $\quad\square$

If the limit variable X has a continuous distribution function, then weak convergence $X_n \rightsquigarrow X$ implies $P(X_n \leq x) \to P(X \leq x)$ for every x. The convergence is then even uniform in x.

2.11 *Lemma. Suppose that $X_n \rightsquigarrow X$ for a random vector X with a continuous distribution function. Then $\sup_x |P(X_n \leq x) - P(X \leq x)| \to 0$.*

Proof. Let F_n and F be the distribution functions of X_n and X. First consider the one-dimensional case. Fix $k \in \mathbb{N}$. By the continuity of F there exist points $-\infty = x_0 < x_1 < \cdots < x_k = \infty$ with $F(x_i) = i/k$. By monotonicity, we have, for $x_{i-1} \leq x \leq x_i$,

$$F_n(x) - F(x) \leq F_n(x_i) - F(x_{i-1}) = F_n(x_i) - F(x_i) + 1/k$$
$$\geq F_n(x_{i-1}) - F(x_i) = F_n(x_{i-1}) - F(x_{i-1}) - 1/k.$$

Thus $|F_n(x) - F(x)|$ is bounded above by $\sup_i |F_n(x_i) - F(x_i)| + 1/k$, for every x. The latter, finite supremum converges to zero as $n \to \infty$, for each fixed k. Because k is arbitrary, the result follows.

In the higher-dimensional case, we follow a similar argument but use hyperrectangles, rather than intervals. We can construct the rectangles by intersecting the k partitions obtained by subdividing each coordinate separately as before. $\quad\blacksquare$

2.2 Stochastic o and O Symbols

It is convenient to have short expressions for terms that converge in probability to zero or are uniformly tight. The notation $o_P(1)$ ("small oh-P-one") is short for a sequence of random vectors that converges to zero in probability. The expression $O_P(1)$ ("big oh-P-one") denotes a sequence that is bounded in probability. More generally, for a given sequence of random variables R_n,

$$X_n = o_P(R_n) \quad \text{means} \quad X_n = Y_n R_n \quad \text{and} \quad Y_n \overset{P}{\to} 0;$$
$$X_n = O_P(R_n) \quad \text{means} \quad X_n = Y_n R_n \quad \text{and} \quad Y_n = O_P(1).$$

This expresses that the sequence X_n converges in probability to zero or is bounded in probability at the "rate" R_n. For deterministic sequences X_n and R_n, the stochastic "oh" symbols reduce to the usual o and O from calculus.

There are many rules of calculus with o and O symbols, which we apply without comment. For instance,

$$o_P(1) + o_P(1) = o_P(1)$$
$$o_P(1) + O_P(1) = O_P(1)$$
$$O_P(1)o_P(1) = o_P(1)$$

$$\left(1 + o_P(1)\right)^{-1} = O_P(1)$$
$$o_P(R_n) = R_n o_P(1)$$
$$O_P(R_n) = R_n O_P(1)$$
$$o_P\left(O_P(1)\right) = o_P(1).$$

To see the validity of these rules it suffices to restate them in terms of explicitly named vectors, where each $o_P(1)$ and $O_P(1)$ should be replaced by a different sequence of vectors that converges to zero or is bounded in probability. In this way the first rule says: If $X_n \xrightarrow{P} 0$ and $Y_n \xrightarrow{P} 0$, then $Z_n = X_n + Y_n \xrightarrow{P} 0$. This is an example of the continuous-mapping theorem. The third rule is short for the following: If X_n is bounded in probability and $Y_n \xrightarrow{P} 0$, then $X_n Y_n \xrightarrow{P} 0$. If X_n would also converge in distribution, then this would be statement (ii) of Slutsky's lemma (with $c = 0$). But by Prohorov's theorem, X_n converges in distribution "along subsequences" if it is bounded in probability, so that the third rule can still be deduced from Slutsky's lemma by "arguing along subsequences."

Note that both rules are in fact implications and should be read from left to right, even though they are stated with the help of the equality sign. Similarly, although it is true that $o_P(1) + o_P(1) = 2o_P(1)$, writing down this rule does not reflect understanding of the o_P symbol.

Two more complicated rules are given by the following lemma.

2.12   Lemma. *Let R be a function defined on domain in $\mathbb{R}^k$ such that $R(0) = 0$. Let X_n be a sequence of random vectors with values in the domain of R that converges in probability to zero. Then, for every $p > 0$,*
  (i) if $R(h) = o\left(\|h\|^p\right)$ as $h \to 0$, then $R(X_n) = o_P\left(\|X_n\|^p\right)$;
  (ii) if $R(h) = O\left(\|h\|^p\right)$ as $h \to 0$, then $R(X_n) = O_P\left(\|X_n\|^p\right)$.

Proof.   Define $g(h)$ as $g(h) = R(h)/\|h\|^p$ for $h \neq 0$ and $g(0) = 0$. Then $R(X_n) = g(X_n)\|X_n\|^p$.

(i) Because the function g is continuous at zero by assumption, $g(X_n) \xrightarrow{P} g(0) = 0$ by the continuous-mapping theorem.

(ii) By assumption there exist M and $\delta > 0$ such that $\left|g(h)\right| \leq M$ whenever $\|h\| \leq \delta$. Thus $P\left(|g(X_n)| > M\right) \leq P\left(\|X_n\| > \delta\right) \to 0$, and the sequence $g(X_n)$ is tight. ∎

*2.3   Characteristic Functions

It is sometimes possible to show convergence in distribution of a sequence of random vectors directly from the definition. In other cases "transforms" of probability measures may help. The basic idea is that it suffices to show characterization (ii) of the portmanteau lemma for a small subset of functions f only.

The most important transform is the *characteristic function*

$$t \mapsto \mathrm{E}e^{it^T X}, \qquad t \in \mathbb{R}^k.$$

Each of the functions $x \mapsto e^{it^T x}$ is continuous and bounded. Thus, by the portmanteau lemma, $\mathrm{E}e^{it^T X_n} \to \mathrm{E}e^{it^T X}$ for every t if $X_n \rightsquigarrow X$. By Lévy's continuity theorem the converse is also true: Pointwise convergence of characteristic functions is equivalent to weak convergence.

2.13 Theorem (Lévy's continuity theorem). *Let X_n and X be random vectors in $\mathbb{R}^k$. Then $X_n \rightsquigarrow X$ if and only if $\mathrm{E} e^{it^T X_n} \to \mathrm{E} e^{it^T X}$ for every $t \in \mathbb{R}^k$. Moreover, if $\mathrm{E} e^{it^T X_n}$ converges pointwise to a function $\phi(t)$ that is continuous at zero, then ϕ is the characteristic function of a random vector X and $X_n \rightsquigarrow X$.*

Proof. If $X_n \rightsquigarrow X$, then $\mathrm{E} h(X_n) \to \mathrm{E} h(X)$ for every bounded continuous function h, in particular for the functions $h(x) = e^{it^T x}$. This gives one direction of the first statement.

For the proof of the last statement, suppose first that we already know that the sequence X_n is uniformly tight. Then, according to Prohorov's theorem, every subsequence has a further subsequence that converges in distribution to some vector Y. By the preceding paragraph, the characteristic function of Y is the limit of the characteristic functions of the converging subsequence. By assumption, this limit is the function $\phi(t)$. Conclude that every weak limit point Y of a converging subsequence possesses characteristic function ϕ. Because a characteristic function uniquely determines a distribution (see Lemma 2.15), it follows that the sequence X_n has only one weak limit point. It can be checked that a uniformly tight sequence with a unique limit point converges to this limit point, and the proof is complete.

The uniform tightness of the sequence X_n can be derived from the continuity of ϕ at zero. Because marginal tightness implies joint tightness, it may be assumed without loss of generality that X_n is one-dimensional. For every x and $\delta > 0$,

$$1\{|\delta x| > 2\} \leq 2\left(1 - \frac{\sin \delta x}{\delta x}\right) = \frac{1}{\delta} \int_{-\delta}^{\delta} (1 - \cos tx) \, dt.$$

Replace x by X_n, take expectations, and use Fubini's theorem to obtain that

$$\mathrm{P}\left(|X_n| > \frac{2}{\delta}\right) \leq \frac{1}{\delta} \int_{-\delta}^{\delta} \mathrm{Re}(1 - \mathrm{E} e^{it X_n}) \, dt.$$

By assumption, the integrand in the right side converges pointwise to $\mathrm{Re}(1 - \phi(t))$. By the dominated-convergence theorem, the whole expression converges to

$$\frac{1}{\delta} \int_{-\delta}^{\delta} \mathrm{Re}(1 - \phi(t)) \, dt.$$

Because ϕ is continuous at zero, there exists for every $\varepsilon > 0$ a $\delta > 0$ such that $|1 - \phi(t)| < \varepsilon$ for $|t| < \delta$. For this δ the integral is bounded by 2ε. Conclude that $\mathrm{P}(|X_n| > 2/\delta) \leq 2\varepsilon$ for sufficiently large n, whence the sequence X_n is uniformly tight. ∎

2.14 Example (Normal distribution). The characteristic function of the $N_k(\mu, \Sigma)$ distribution is the function

$$t \mapsto e^{it^T \mu - \frac{1}{2} t^T \Sigma t}.$$

Indeed, if X is $N_k(0, I)$ distributed and $\Sigma^{1/2}$ is a symmetric square root of Σ (hence $\Sigma = (\Sigma^{1/2})^2$), then $\Sigma^{1/2} X + \mu$ possesses the given normal distribution and

$$\mathrm{E} e^{z^T (\Sigma^{1/2} X + \mu)} = e^{z^T \mu} \int e^{(\Sigma^{1/2} z)^T x - \frac{1}{2} x^T x} \, dx \, \frac{1}{(2\pi)^{k/2}} = e^{z^T \mu + \frac{1}{2} z^T \Sigma z}.$$

For real-valued z, the last equality follows easily by completing the square in the exponent. Evaluating the integral for complex z, such as $z = it$, requires some skill in complex function theory. One method, which avoids further calculations, is to show that both the left- and righthand sides of the preceding display are analytic functions of z. For the right side this is obvious; for the left side we can justify differentiation under the expectation sign by the dominated-convergence theorem. Because the two sides agree on the real axis, they must agree on the complex plane by uniqueness of analytic continuation. $\square$

2.15 Lemma. *Random vectors X and Y in $\mathbb{R}^k$ are equal in distribution if and only if $\mathrm{E}e^{it^T X} = \mathrm{E}e^{it^T Y}$ for every $t \in \mathbb{R}^k$.*

Proof. By Fubini's theorem and calculations as in the preceding example, for every $\sigma > 0$ and $y \in \mathbb{R}^k$,

$$\int e^{-it^T y} e^{-\frac{1}{2}t^T t\sigma^2} \mathrm{E}e^{it^T X} \, dt = \mathrm{E} \int e^{it^T (X-y)} e^{-\frac{1}{2}t^T t\sigma^2} \, dt$$

$$= \frac{(2\pi)^{k/2}}{\sigma^k} \mathrm{E}e^{-\frac{1}{2}(X-y)^T (X-y)/\sigma^2}.$$

By the convolution formula for densities, the righthand side is $(2\pi)^k$ times the density $p_{X+\sigma Z}(y)$ of the sum of X and σZ for a standard normal vector Z that is independent of X. Conclude that if X and Y have the same characteristic function, then the vectors $X + \sigma Z$ and $Y + \sigma Z$ have the same density and hence are equal in distribution for every $\sigma > 0$. By Slutsky's lemma $X + \sigma Z \rightsquigarrow X$ as $\sigma \downarrow 0$, and similarly for Y. Thus X and Y are equal in distribution. $\blacksquare$

The characteristic function of a sum of independent variables equals the product of the characteristic functions of the individual variables. This observation, combined with Lévy's theorem, yields simple proofs of both the law of large numbers and the central limit theorem.

2.16 Proposition (Weak law of large numbers). *Let $Y_1, \ldots, Y_n$ be i.i.d. random variables with characteristic function ϕ. Then $\overline{Y}_n \overset{\mathrm{P}}{\to} \mu$ for a real number μ if and only if ϕ is differentiable at zero with $i\mu = \phi'(0)$.*

Proof. We only prove that differentiability is sufficient. For the converse, see, for example, [127, p. 52]. Because $\phi(0) = 1$, differentiability of ϕ at zero means that $\phi(t) = 1 + t\phi'(0) + o(t)$ as $t \to 0$. Thus, by Fubini's theorem, for each fixed t and $n \to \infty$,

$$\mathrm{E}e^{it\overline{Y}_n} = \phi^n\left(\frac{t}{n}\right) = \left(1 + \frac{t}{n}i\mu + o\left(\frac{1}{n}\right)\right)^n \to e^{it\mu}.$$

The right side is the characteristic function of the constant variable μ. By Lévy's theorem, $\overline{Y}_n$ converges in distribution to μ. Convergence in distribution to a constant is the same as convergence in probability. $\blacksquare$

A sufficient but not necessary condition for $\phi(t) = \mathrm{E}e^{itY}$ to be differentiable at zero is that $\mathrm{E}|Y| < \infty$. In that case the dominated convergence theorem allows differentiation under the expectation sign, and we obtain

$$\phi'(t) = \frac{d}{dt} \mathrm{E} e^{itY} = \mathrm{E} i Y e^{itY}.$$

In particular, the derivative at zero is $\phi'(0) = i\mathrm{E}Y$ and hence $\overline{Y}_n \overset{\mathrm{P}}{\to} \mathrm{E}Y_1$.

If $\mathrm{E}Y^2 < \infty$, then the Taylor expansion can be carried a step further and we can obtain a version of the central limit theorem.

2.17   Proposition (Central limit theorem). *Let $Y_1, \dots, Y_n$ be i.i.d. random variables with $\mathrm{E}Y_i = 0$ and $\mathrm{E}Y_i^2 = 1$. Then the sequence $\sqrt{n}\overline{Y}_n$ converges in distribution to the standard normal distribution.*

Proof.   A second differentiation under the expectation sign shows that $\phi''(0) = i^2 \mathrm{E}Y^2$. Because $\phi'(0) = i\mathrm{E}Y = 0$, we obtain

$$\mathrm{E} e^{it\sqrt{n}\overline{Y}_n} = \phi^n\left(\frac{t}{\sqrt{n}}\right) = \left(1 - \frac{1}{2}\frac{t^2}{n}\mathrm{E}Y^2 + o\left(\frac{1}{n}\right)\right)^n \to e^{-\frac{1}{2}t^2 \mathrm{E}Y^2}.$$

The right side is the characteristic function of the normal distribution with mean zero and variance $\mathrm{E}Y^2$. The proposition follows from Lévy's continuity theorem. $\blacksquare$

The characteristic function $t \mapsto \mathrm{E} e^{it^T X}$ of a vector X is determined by the set of all characteristic functions $u \mapsto \mathrm{E} e^{iu(t^T X)}$ of linear combinations $t^T X$ of the components of X. Therefore, Lévy's continuity theorem implies that weak convergence of vectors is equivalent to weak convergence of linear combinations:

$$X_n \rightsquigarrow X \quad \text{if and only if} \quad t^T X_n \rightsquigarrow t^T X \quad \text{for all} \quad t \in \mathbb{R}^k.$$

This is known as the *Cramér-Wold device*. It allows to reduce higher-dimensional problems to the one-dimensional case.

2.18   Example (Multivariate central limit theorem). Let $Y_1, Y_2, \dots$ be i.i.d. random vectors in $\mathbb{R}^k$ with mean vector $\mu = \mathrm{E}Y_1$ and covariance matrix $\Sigma = \mathrm{E}(Y_1 - \mu)(Y_1 - \mu)^T$. Then

$$\frac{1}{\sqrt{n}}\sum_{i=1}^n (Y_i - \mu) = \sqrt{n}(\overline{Y}_n - \mu) \rightsquigarrow N_k(0, \Sigma).$$

(The sum is taken coordinatewise.)   By the Cramér-Wold device, this can be proved by finding the limit distribution of the sequences of real variables

$$t^T\left(\frac{1}{\sqrt{n}}\sum_{i=1}^n (Y_i - \mu)\right) = \frac{1}{\sqrt{n}}\sum_{i=1}^n (t^T Y_i - t^T \mu).$$

Because the random variables $t^T Y_1 - t^T \mu, t^T Y_2 - t^T \mu, \dots$ are i.i.d. with zero mean and variance $t^T \Sigma t$, this sequence is asymptotically $N_1(0, t^T \Sigma t)$-distributed by the univariate central limit theorem. This is exactly the distribution of $t^T X$ if X possesses an $N_k(0, \Sigma)$ distribution. $\square$

*2.4 Almost-Sure Representations

Convergence in distribution certainly does not imply convergence in probability or almost surely. However, the following theorem shows that a given sequence $X_n \rightsquigarrow X$ can always be replaced by a sequence $\tilde{X}_n \rightsquigarrow \tilde{X}$ that is, marginally, equal in distribution and converges almost surely. This construction is sometimes useful and has been put to good use by some authors, but we do not use it in this book.

2.19 Theorem (Almost-sure representations). *Suppose that the sequence of random vectors X_n converges in distribution to a random vector X_0. Then there exists a probability space $(\tilde{\Omega}, \tilde{\mathcal{U}}, \tilde{P})$ and random vectors $\tilde{X}_n$ defined on it such that $\tilde{X}_n$ is equal in distribution to X_n for every $n \geq 0$ and $\tilde{X}_n \to \tilde{X}_0$ almost surely.*

Proof. For random variables we can simply define $\tilde{X}_n = F_n^{-1}(U)$ for F_n the distribution function of X_n and U an arbitrary random variable with the uniform distribution on $[0, 1]$. (The "quantile transformation," see Section 21.1.) The simplest known construction for higher-dimensional vectors is more complicated. See, for example, Theorem 1.10.4 in [146], or [41]. ∎

*2.5 Convergence of Moments

By the portmanteau lemma, weak convergence $X_n \rightsquigarrow X$ implies that $\mathrm{E}f(X_n) \to \mathrm{E}f(X)$ for every continuous, bounded function f. The condition that f be bounded is not superfluous: It is not difficult to find examples of a sequence $X_n \rightsquigarrow X$ and an unbounded, continuous function f for which the convergence fails. In particular, in general convergence in distribution does not imply convergence $\mathrm{E}X_n^p \to \mathrm{E}X^p$ of moments. However, in many situations such convergence occurs, but it requires more effort to prove it.

A sequence of random variables Y_n is called *asymptotically uniformly integrable* if

$$\lim_{M \to \infty} \limsup_{n \to \infty} \mathrm{E}|Y_n| 1\{|Y_n| > M\} = 0.$$

Uniform integrability is the missing link between convergence in distribution and convergence of moments.

2.20 Theorem. *Let $f : \mathbb{R}^k \mapsto \mathbb{R}$ be measurable and continuous at every point in a set C. Let $X_n \rightsquigarrow X$ where X takes its values in C. Then $\mathrm{E}f(X_n) \to \mathrm{E}f(X)$ if the sequence of random variables $f(X_n)$ is asymptotically uniformly integrable.*

Proof. We give the proof only in the most interesting direction. (See, for example, [146] (p. 69) for the other direction.) Suppose that $Y_n = f(X_n)$ is asymptotically uniformly integrable. Then we show that $\mathrm{E}Y_n \to \mathrm{E}Y$ for $Y = f(X)$. Assume without loss of generality that Y_n is nonnegative; otherwise argue the positive and negative parts separately. By the continuous mapping theorem, $Y_n \rightsquigarrow Y$. By the triangle inequality,

$$|\mathrm{E}Y_n - \mathrm{E}Y| \leq |\mathrm{E}Y_n - \mathrm{E}Y_n \wedge M| + |\mathrm{E}Y_n \wedge M - \mathrm{E}Y \wedge M| + |\mathrm{E}Y \wedge M - \mathrm{E}Y|.$$

Because the function $y \mapsto y \wedge M$ is continuous and bounded on $[0, \infty)$, it follows that the middle term on the right converges to zero as $n \to \infty$. The first term is bounded above by

$EY_n 1\{Y_n > M\}$, and converges to zero as $n \to \infty$ followed by $M \to \infty$, by the uniform integrability. By the portmanteau lemma (iv), the third term is bounded by the lim inf as $n \to \infty$ of the first and hence converges to zero as $M \uparrow \infty$. ∎

2.21   Example. Suppose X_n is a sequence of random variables such that $X_n \rightsquigarrow X$ and $\limsup E|X_n|^p < \infty$ for some p. Then all moments of order strictly less than p converge also: $EX_n^k \to EX^k$ for every $k < p$.

By the preceding theorem, it suffices to prove that the sequence X_n^k is asymptotically uniformly integrable. By Markov's inequality

$$E|X_n|^k 1\{|X_n|^k \geq M\} \leq M^{1-p/k} E|X_n|^p.$$

The limit superior, as $n \to \infty$ followed by $M \to \infty$, of the right side is zero if $k < p$. □

The moment function $p \mapsto EX^p$ can be considered a transform of probability distributions, just as can the characteristic function. In general, it is not a true transform in that it does determine a distribution uniquely only under additional assumptions. If a limit distribution is uniquely determined by its moments, this transform can still be used to establish weak convergence.

2.22   Theorem. *Let X_n and X be random variables such that $EX_n^p \to EX^p < \infty$ for every $p \in \mathbb{N}$. If the distribution of X is uniquely determined by its moments, then $X_n \rightsquigarrow X$.*

Proof.   Because $EX_n^2 = O(1)$, the sequence X_n is uniformly tight, by Markov's inequality. By Prohorov's theorem, each subsequence has a further subsequence that converges weakly to a limit Y. By the preceding example the moments of Y are the limits of the moments of the subsequence. Thus the moments of Y are identical to the moments of X. Because, by assumption, there is only one distribution with this set of moments, X and Y are equal in distribution. Conclude that every subsequence of X_n has a further subsequence that converges in distribution to X. This implies that the whole sequence converges to X. ∎

2.23   Example. The normal distribution is uniquely determined by its moments. (See, for example, [123] or [133, p. 293].) Thus $EX_n^p \to 0$ for odd p and $EX_n^p \to (p-1)(p-3)\cdots 1$ for even p implies that $X_n \rightsquigarrow N(0, 1)$. The converse is false. □

*2.6   Convergence-Determining Classes

A class $\mathcal{F}$ of functions $f : \mathbb{R}^k \to \mathbb{R}$ is called *convergence-determining* if for every sequence of random vectors X_n the convergence $X_n \rightsquigarrow X$ is equivalent to $Ef(X_n) \to Ef(X)$ for every $f \in \mathcal{F}$. By definition the set of all bounded continuous functions is convergence-determining, but so is the smaller set of all differentiable functions, and many other classes. The set of all indicator functions $1_{(-\infty,t]}$ would be convergence-determining if we would restrict the definition to limits X with continuous distribution functions. We shall have occasion to use the following results. (For proofs see Corollary 1.4.5 and Theorem 1.12.2, for example, in [146].)

2.24   Lemma. On $\mathbb{R}^k = \mathbb{R}^l \times \mathbb{R}^m$ the set of functions $(x, y) \mapsto f(x)g(y)$ with f and g ranging over all bounded, continuous functions on $\mathbb{R}^l$ and $\mathbb{R}^m$, respectively, is convergence-determining.

2.25   Lemma. There exists a countable set of continuous functions $f : \mathbb{R}^k \mapsto [0, 1]$ that is convergence-determining and, moreover, $X_n \rightsquigarrow X$ implies that $\mathrm{E}f(X_n) \to \mathrm{E}f(X)$ uniformly in $f \in \mathcal{F}$.

*2.7   Law of the Iterated Logarithm

The law of the iterated logarithm is an intriguing result but appears to be of less interest to statisticians. It can be viewed as a refinement of the strong law of large numbers. If $Y_1, Y_2, \ldots$ are i.i.d. random variables with mean zero, then $Y_1 + \cdots + Y_n = o(n)$ almost surely by the strong law. The law of the iterated logarithm improves this order to $O(\sqrt{n \log \log n})$, and even gives the proportionality constant.

2.26   Proposition (Law of the iterated logarithm). Let $Y_1, Y_2, \ldots$ be i.i.d. random variables with mean zero and variance 1. Then

$$\limsup_{n \to \infty} \frac{Y_1 + \cdots + Y_n}{\sqrt{n \log \log n}} = \sqrt{2}, \quad a.s.$$

Conversely, if this statement holds for both Y_i and $-Y_i$, then the variables have mean zero and variance 1.

The law of the iterated logarithm gives an interesting illustration of the difference between almost sure and distributional statements. Under the conditions of the proposition, the sequence $n^{-1/2}(Y_1 + \cdots + Y_n)$ is asymptotically normally distributed by the central limit theorem. The limiting normal distribution is spread out over the whole real line. Apparently division by the factor $\sqrt{\log \log n}$ is exactly right to keep $n^{-1/2}(Y_1 + \cdots + Y_n)$ within a compact interval, eventually.

A simple application of Slutsky's lemma gives

$$Z_n := \frac{Y_1 + \cdots + Y_n}{\sqrt{n \log \log n}} \overset{\mathrm{P}}{\to} 0.$$

Thus Z_n is with high probability contained in the interval $(-\varepsilon, \varepsilon)$ eventually, for any $\varepsilon > 0$. This appears to contradict the law of the iterated logarithm, which asserts that Z_n reaches the interval $(\sqrt{2} - \varepsilon, \sqrt{2} + \varepsilon)$ infinitely often with probability one. The explanation is that the set of ω such that $Z_n(\omega)$ is in $(-\varepsilon, \varepsilon)$ or $(\sqrt{2} - \varepsilon, \sqrt{2} + \varepsilon)$ fluctuates with n. The convergence in probability shows that at any advanced time a very large fraction of ω have $Z_n(\omega) \in (-\varepsilon, \varepsilon)$. The law of the iterated logarithm shows that for each particular ω the sequence $Z_n(\omega)$ drops in and out of the interval $(\sqrt{2} - \varepsilon, \sqrt{2} + \varepsilon)$ infinitely often (and hence out of $(-\varepsilon, \varepsilon)$).

The implications for statistics can be illustrated by considering confidence statements. If μ and 1 are the true mean and variance of the sample $Y_1, Y_2, \ldots$, then the probability that

$$\bar{Y}_n - \frac{2}{\sqrt{n}} \le \mu \le \bar{Y}_n + \frac{2}{\sqrt{n}}$$

converges to $\Phi(2) - \Phi(-2) \approx 95\%$. Thus the given interval is an asymptotic confidence interval of level approximately 95%. (The confidence level is exactly $\Phi(2) - \Phi(-2)$ if the observations are normally distributed. This may be assumed in the following; the accuracy of the approximation is not an issue in this discussion.) The point $\mu = 0$ is contained in the interval if and only if the variable Z_n satisfies

$$|Z_n| \leq \frac{2}{\sqrt{\log \log n}}.$$

Assume that $\mu = 0$ is the true value of the mean, and consider the following argument. By the law of the iterated logarithm, we can be sure that Z_n hits the interval $(\sqrt{2} - \varepsilon, \sqrt{2} + \varepsilon)$ infinitely often. The expression $2/\sqrt{\log \log n}$ is close to zero for large n. Thus we can be sure that the true value $\mu = 0$ is outside the confidence interval infinitely often.

How can we solve the paradox that the usual confidence interval is wrong infinitely often? There appears to be a conceptual problem if it is imagined that a statistician collects data in a sequential manner, computing a confidence interval for every n. However, although the frequentist interpretation of a confidence interval is open to the usual criticism, the paradox does not seem to rise within the frequentist framework. In fact, from a frequentist point of view the curious conclusion is reasonable. Imagine 100 statisticians, all of whom set 95% confidence intervals in the usual manner. They all receive one observation per day and update their confidence intervals daily. Then every day about five of them should have a false interval. It is only fair that as the days go by all of them take turns in being unlucky, and that the same five do not have it wrong all the time. This, indeed, happens according to the law of the iterated logarithm.

The paradox may be partly caused by the feeling that with a growing number of observations, the confidence intervals should become better. In contrast, the usual approach leads to errors with certainty. However, this is only true if the usual approach is applied naively in a sequential set-up. In practice one would do a genuine sequential analysis (including the use of a stopping rule) or change the confidence level with n.

There is also another reason that the law of the iterated logarithm is of little practical consequence. The argument in the preceding paragraphs is based on the assumption that $2/\sqrt{\log \log n}$ is close to zero and is nonsensical if this quantity is larger than $\sqrt{2}$. Thus the argument requires at least $n \geq 1619$, a respectable number of observations.

*2.8   Lindeberg-Feller Theorem

Central limit theorems are theorems concerning convergence in distribution of sums of random variables. There are versions for dependent observations and nonnormal limit distributions. The Lindeberg-Feller theorem is the simplest extension of the classical central limit theorem and is applicable to independent observations with finite variances.

2.27   Proposition (Lindeberg-Feller central limit theorem). *For each n let $Y_{n,1}, \ldots,$ Y_{n,k_n} be independent random vectors with finite variances such that*

$$\sum_{i=1}^{k_n} \mathrm{E}\|Y_{n,i}\|^2 1\{\|Y_{n,i}\| > \varepsilon\} \to 0, \qquad \text{every } \varepsilon > 0,$$

$$\sum_{i=1}^{k_n} \mathrm{Cov}\, Y_{n,i} \to \Sigma.$$

Then the sequence $\sum_{i=1}^{k_n}(Y_{n,i} - \mathrm{E}Y_{n,i})$ converges in distribution to a normal $N(0, \Sigma)$ distribution.

A result of this type is necessary to treat the asymptotics of, for instance, regression problems with fixed covariates. We illustrate this by the linear regression model. The application is straightforward but notationally a bit involved. Therefore, at other places in the manuscript we find it more convenient to assume that the covariates are a random sample, so that the ordinary central limit theorem applies.

2.28   Example (Linear regression). In the linear regression problem, we observe a vector $Y = X\beta + e$ for a known $(n \times p)$ matrix X of full rank, and an (unobserved) error vector e with i.i.d. components with mean zero and variance σ^2. The least squares estimator of β is

$$\hat{\beta} = (X^T X)^{-1} X^T Y.$$

This estimator is unbiased and has covariance matrix $\sigma^2 (X^T X)^{-1}$. If the error vector e is normally distributed, then $\hat{\beta}$ is exactly normally distributed. Under reasonable conditions on the design matrix, the least squares estimator is asymptotically normally distributed for a large range of error distributions. Here we fix p and let n tend to infinity.

This follows from the representation

$$(X^T X)^{1/2}(\hat{\beta} - \beta) = (X^T X)^{-1/2} X^T e = \sum_{i=1}^{n} a_{ni} e_i,$$

where $a_{n1}, \ldots, a_{nn}$ are the columns of the $(p \times n)$ matrix $(X^T X)^{-1/2} X^T =: A$. This sequence is asymptotically normal if the vectors $a_{n1}e_1, \ldots, a_{nn}e_n$ satisfy the Lindeberg conditions. The norming matrix $(X^T X)^{1/2}$ has been chosen to ensure that the vectors in the display have covariance matrix $\sigma^2 I$ for every n. The remaining condition is

$$\sum_{i=1}^{n} \|a_{ni}\|^2 \mathrm{E} e_i^2 1\{\|a_{ni}\| \|e_i\| > \varepsilon\} \to 0.$$

This can be simplified to other conditions in several ways. Because $\sum \|a_{ni}\|^2 = \mathrm{trace}(AA^T) = p$, it suffices that $\max \mathrm{E} e_i^2 1\{\|a_{ni}\| \|e_i\| > \varepsilon\} \to 0$, which is equivalent to

$$\max_{1 \le i \le n} \|a_{ni}\| \to 0.$$

Alternatively, the expectation $\mathrm{E}e^2 1\{a|e| > \varepsilon\}$ can be bounded by $\varepsilon^{-k} \mathrm{E}|e|^{k+2} a^k$ and a second set of sufficient conditions is

$$\sum_{i=1}^{n} \|a_{ni}\|^k \to 0; \qquad \mathrm{E}|e_1|^k < \infty, \qquad (k > 2).$$

Both sets of conditions are reasonable. Consider for instance the simple linear regression model $Y_i = \beta_0 + \beta_1 x_i + e_i$. Then

$$(X^T X)^{-1/2} X^T = \frac{1}{\sqrt{n}} \left(\begin{matrix} 1 & \overline{x} \\ \overline{x} & \overline{x^2} \end{matrix} \right)^{-1/2} \left(\begin{matrix} 1 & 1 & \cdots & 1 \\ x_1 & x_2 & \cdots & x_n \end{matrix} \right).$$

It is reasonable to assume that the sequences $\overline{x}$ and $\overline{x^2}$ are bounded. Then the first matrix on the right behaves like a fixed matrix, and the conditions for asymptotic normality simplify to

$$\max_{1 \le i \le n} |x_i| = o(n^{1/2}); \quad \text{or} \quad n^{1-k/2}\overline{|x|^k} \to 0, \qquad E|e_1|^k < \infty.$$

Every reasonable design satisfies these conditions. □

*2.9   Convergence in Total Variation

A sequence of random variables converges in *total variation* to a variable X if

$$\sup_B \left|P(X_n \in B) - P(X \in B)\right| \to 0,$$

where the supremum is taken over all measurable sets B. In view of the portmanteau lemma, this type of convergence is stronger than convergence in distribution. Not only is it required that the sequence $P(X_n \in B)$ converges for every Borel set B, the convergence must also be uniform in B. Such strong convergence occurs less frequently and is often more than necessary, whence the concept is less useful.

A simple sufficient condition for convergence in total variation is pointwise convergence of densities. If X_n and X have densities p_n and p with respect to a measure μ, then

$$\sup_B \left|P(X_n \in B) - P(X \in B)\right| = \frac{1}{2} \int |p_n - p|\, d\mu.$$

Thus, convergence in total variation can be established by convergence theorems for integrals from measure theory. The following proposition, which should be compared with the monotone and dominated convergence theorems, is most appropriate.

2.29   Proposition.   *Suppose that f_n and f are arbitrary measurable functions such that $f_n \to f$ μ-almost everywhere (or in μ-measure) and $\limsup \int |f_n|^p\, d\mu \le \int |f|^p\, d\mu < \infty$, for some $p \ge 1$ and measure μ. Then $\int |f_n - f|^p\, d\mu \to 0$.*

Proof.   By the inequality $(a + b)^p \le 2^p a^p + 2^p b^p$, valid for every $a, b \ge 0$, and the assumption, $0 \le 2^p|f_n|^p + 2^p|f|^p - |f_n - f|^p \to 2^{p+1}|f|^p$ almost everywhere. By Fatou's lemma,

$$\int 2^{p+1}|f|^p\, d\mu \le \liminf \int \left(2^p|f_n|^p + 2^p|f|^p - |f_n - f|^p\right) d\mu$$

$$\le 2^{p+1} \int |f|^p\, d\mu - \limsup \int |f_n - f|^p\, d\mu,$$

by assumption. The proposition follows. ■

2.30   Corollary (Scheffé).   *Let X_n and X be random vectors with densities p_n and p with respect to a measure μ. If $p_n \to p$ μ-almost everywhere, then the sequence X_n converges to X in total variation.*

The central limit theorem is usually formulated in terms of convergence in distribution. Often it is valid in terms of the total variation distance, in the sense that

$$\sup_B \left| P(Y_1 + \cdots + Y_n \in B) - \int_B \frac{1}{\sqrt{n}\sigma\sqrt{2\pi}} e^{-\frac{1}{2}(x-n\mu)^2/n\sigma^2} \, dx \right| \to 0.$$

Here μ and σ^2 are mean and variance of the Y_i, and the supremum is taken over all Borel sets. An integrable characteristic function, in addition to a finite second moment, suffices.

2.31 *Theorem (Central limit theorem in total variation). Let $Y_1, Y_2, \ldots$ be i.i.d. random variables with finite second moment and characteristic function ϕ such that $\int |\phi(t)|^\nu \, dt < \infty$ for some $\nu \geq 1$. Then $Y_1 + \cdots + Y_n$ satisfies the central limit theorem in total variation.*

Proof. It can be assumed without loss of generality that $EY_1 = 0$ and $\operatorname{var} Y_1 = 1$. By the inversion formula for characteristic functions (see [47, p. 509]), the density p_n of $Y_1 + \cdots + Y_n/\sqrt{n}$ can be written

$$p_n(x) = \frac{1}{2\pi} \int e^{-itx} \phi\left(\frac{t}{\sqrt{n}}\right)^n dt.$$

By the central limit theorem and Lévy's continuity theorem, the integrand converges to $e^{-itx} \exp(-\frac{1}{2}t^2)$. It will be shown that the integral converges to

$$\frac{1}{2\pi} \int e^{-itx} e^{-\frac{1}{2}t^2} \, dt = \frac{e^{-\frac{1}{2}x^2}}{\sqrt{2\pi}}.$$

Then an application of Scheffé's theorem concludes the proof.

The integral can be split into two parts. First, for every $\varepsilon > 0$,

$$\int_{|t|>\varepsilon\sqrt{n}} \left| e^{-itx} \phi\left(\frac{t}{\sqrt{n}}\right)^n \right| dt \leq \sqrt{n} \sup_{|t|>\varepsilon} |\phi(t)|^{n-\nu} \int |\phi(t)|^\nu \, dt.$$

Here $\sup_{|t|>\varepsilon} |\phi(t)| < 1$ by the Riemann-Lebesgue lemma and because ϕ is the characteristic function of a nonlattice distribution (e.g., [47, pp. 501, 513]). Thus, the first part of the integral converges to zero geometrically fast.

Second, a Taylor expansion yields that $\phi(t) = 1 - \frac{1}{2}t^2 + o(t^2)$ as $t \to 0$, so that there exists $\varepsilon > 0$ such that $|\phi(t)| \leq 1 - t^2/4$ for every $|t| < \varepsilon$. It follows that

$$\left| e^{-itx} \phi\left(\frac{t}{\sqrt{n}}\right)^n \right| 1\{|t| \leq \varepsilon\sqrt{n}\} \leq \left(1 - \frac{t^2}{4n}\right)^n \leq e^{-t^2/4}.$$

The proof can be concluded by applying the dominated convergence theorem to the remaining part of the integral. ∎

Notes

The results of this chapter can be found in many introductions to probability theory. A standard reference for weak convergence theory is the first chapter of [11]. Another very readable introduction is [41]. The theory of this chapter is extended to random elements with values in general metric spaces in Chapter 18.

PROBLEMS

1. If X_n possesses a t-distribution with n degrees of freedom, then $X_n \rightsquigarrow N(0, 1)$ as $n \to \infty$. Show this.

2. Does it follow immediately from the result of the previous exercise that $EX_n^p \to EN(0, 1)^p$ for every $p \in \mathbb{N}$? Is this true?

3. If $X_n \rightsquigarrow N(0, 1)$ and $Y_n \overset{P}{\to} \sigma$, then $X_n Y_n \rightsquigarrow N(0, \sigma^2)$. Show this.

4. In what sense is a chi-square distribution with n degrees of freedom approximately a normal distribution?

5. Find an example of sequences such that $X_n \rightsquigarrow X$ and $Y_n \rightsquigarrow Y$, but the joint sequence (X_n, Y_n) does not converge in law.

6. If X_n and Y_n are independent random vectors for every n, then $X_n \rightsquigarrow X$ and $Y_n \rightsquigarrow Y$ imply that $(X_n, Y_n) \rightsquigarrow (X, Y)$, where X and Y are independent. Show this.

7. If every X_n and X possess discrete distributions supported on the integers, then $X_n \rightsquigarrow X$ if and only if $P(X_n = x) \to P(X = x)$ for every integer x. Show this.

8. If $P(X_n = i/n) = 1/n$ for every $i = 1, 2, \ldots, n$, then $X_n \rightsquigarrow X$, but there exist Borel sets with $P(X_n \in B) = 1$ for every n, but $P(X \in B) = 0$. Show this.

9. If $P(X_n = x_n) = 1$ for numbers x_n and $x_n \to x$, then $X_n \rightsquigarrow x$. Prove this
   (i) by considering distributions functions
   (ii) by using Theorem 2.7.

10. State the rule $o_P(1) + O_P(1) = O_P(1)$ in terms of random vectors and show its validity.

11. In what sense is it true that $o_P(1) = O_P(1)$? Is it true that $O_P(1) = o_P(1)$?

12. The rules given by Lemma 2.12 are not simple plug-in rules.
   (i) Give an example of a function R with $R(h) = o(\|h\|)$ as $h \to 0$ and a sequence of random variables X_n such that $R(X_n)$ is not equal to $o_P(X_n)$.
   (ii) Given an example of a function R such $R(h) = O(\|h\|)$ as $h \to 0$ and a sequence of random variables X_n such that $X_n = O_P(1)$ but $R(X_n)$ is not equal to $O_P(X_n)$.

13. Find an example of a sequence of random variables such that $X_n \rightsquigarrow 0$, but $EX_n \to \infty$.

14. Find an example of a sequence of random variables such that $X_n \overset{P}{\to} 0$, but X_n does not converge almost surely.

15. Let $X_1, \ldots, X_n$ be i.i.d. with density $f_{\lambda,a}(x) = \lambda e^{-\lambda(x-a)} 1\{x \geq a\}$. Calculate the maximum likelihood estimator of $(\hat{\lambda}_n, \hat{a}_n)$ of (λ, a) and show that $(\hat{\lambda}_n, \hat{a}_n) \overset{P}{\to} (\lambda, a)$.

16. Let $X_1, \ldots, X_n$ be i.i.d. standard normal variables. Show that the vector $U = (X_1, \ldots, X_n)/N$, where $N^2 = \sum_{i=1}^n X_i^2$, is uniformly distributed over the unit sphere S^{n-1} in $\mathbb{R}^n$, in the sense that U and OU are identically distributed for every orthogonal transformation O of $\mathbb{R}^n$.

17. For each n, let U_n be uniformly distributed over the unit sphere S^{n-1} in $\mathbb{R}^n$. Show that the vectors $\sqrt{n}(U_{n,1}, U_{n,2})$ converge in distribution to a pair of independent standard normal variables.

18. If $\sqrt{n}(T_n - \theta)$ converges in distribution, then T_n converges in probability to θ. Show this.

19. If $EX_n \to \mu$ and var $X_n \to 0$, then $X_n \overset{P}{\to} \mu$. Show this.

20. If $\sum_{n=1}^\infty P(|X_n| > \varepsilon) < \infty$ for every $\varepsilon > 0$, then X_n converges almost surely to zero. Show this.

21. Use characteristic functions to show that binomial$(n, \lambda/n) \rightsquigarrow$ Poisson(λ). Why does the central limit theorem not hold?

22. If $X_1, \ldots, X_n$ are i.i.d. standard Cauchy, then $\overline{X}_n$ is standard Cauchy.
   (i) Show this by using characteristic functions
   (ii) Why does the weak law not hold?

23. Let $X_1, \ldots, X_n$ be i.i.d. with finite fourth moment. Find constants a, b, and c_n such that the sequence $c_n(\overline{X}_n - a, \overline{X_n^2} - b)$ converges in distribution, and determine the limit law. Here $\overline{X}_n$ and $\overline{X_n^2}$ are the averages of the X_i and the X_i^2, respectively.

3

Delta Method

The delta method consists of using a Taylor expansion to approximate a random vector of the form $\phi(T_n)$ by the polynomial $\phi(\theta) + \phi'(\theta)(T_n - \theta) + \cdots$ in $T_n - \theta$. It is a simple but useful method to deduce the limit law of $\phi(T_n) - \phi(\theta)$ from that of $T_n - \theta$. Applications include the nonrobustness of the chi-square test for normal variances and variance stabilizing transformations.

3.1 Basic Result

Suppose an estimator T_n for a parameter θ is available, but the quantity of interest is $\phi(\theta)$ for some known function ϕ. A natural estimator is $\phi(T_n)$. How do the asymptotic properties of $\phi(T_n)$ follow from those of T_n?

A first result is an immediate consequence of the continuous-mapping theorem. If the sequence T_n converges in probability to θ and ϕ is continuous at θ, then $\phi(T_n)$ converges in probability to $\phi(\theta)$.

Of greater interest is a similar question concerning limit distributions. In particular, if $\sqrt{n}(T_n - \theta)$ converges weakly to a limit distribution, is the same true for $\sqrt{n}\big(\phi(T_n) - \phi(\theta)\big)$? If ϕ is differentiable, then the answer is affirmative. Informally, we have

$$\sqrt{n}\big(\phi(T_n) - \phi(\theta)\big) \approx \phi'(\theta)\,\sqrt{n}(T_n - \theta).$$

If $\sqrt{n}(T_n - \theta) \rightsquigarrow T$ for some variable T, then we expect that $\sqrt{n}\big(\phi(T_n) - \phi(\theta)\big) \rightsquigarrow \phi'(\theta)\,T$. In particular, if $\sqrt{n}(T_n - \theta)$ is asymptotically normal $N(0, \sigma^2)$, then we expect that $\sqrt{n}\big(\phi(T_n) - \phi(\theta)\big)$ is asymptotically normal $N\big(0, \phi'(\theta)^2\sigma^2\big)$. This is proved in greater generality in the following theorem.

In the preceding paragraph it is silently understood that T_n is real-valued, but we are more interested in considering statistics $\phi(T_n)$ that are formed out of several more basic statistics. Consider the situation that $T_n = (T_{n,1}, \ldots, T_{n,k})$ is vector-valued, and that $\phi : \mathbb{R}^k \mapsto \mathbb{R}^m$ is a given function defined at least on a neighbourhood of θ. Recall that ϕ is *differentiable* at θ if there exists a linear map (matrix) $\phi'_\theta : \mathbb{R}^k \mapsto \mathbb{R}^m$ such that

$$\phi(\theta + h) - \phi(\theta) = \phi'_\theta(h) + o\big(\|h\|\big), \qquad h \to 0.$$

All the expressions in this equation are vectors of length m, and $\|h\|$ is the Euclidean norm. The linear map $h \mapsto \phi'_\theta(h)$ is sometimes called a "total derivative," as opposed to partial derivatives. A sufficient condition for ϕ to be (totally) differentiable is that all partial derivatives $\partial \phi_j(x)/\partial x_i$ exist for x in a neighborhood of θ and are continuous at θ. (Just existence of the partial derivatives is not enough.) In any case, the total derivative is found from the partial derivatives. If ϕ is differentiable, then it is partially differentiable, and the derivative map $h \mapsto \phi'_\theta(h)$ is matrix multiplication by the matrix

$$\phi'_\theta = \begin{pmatrix} \frac{\partial \phi_1}{\partial x_1}(\theta) & \cdots & \frac{\partial \phi_1}{\partial x_k}(\theta) \\ \vdots & & \vdots \\ \frac{\partial \phi_m}{\partial x_1}(\theta) & \cdots & \frac{\partial \phi_m}{\partial x_k}(\theta) \end{pmatrix}.$$

If the dependence of the derivative ϕ'_θ on θ is continuous, then ϕ is called *continuously differentiable*.

It is better to think of a derivative as a linear approximation $h \mapsto \phi'_\theta(h)$ to the function $h \mapsto \phi(\theta + h) - \phi(\theta)$ than as a set of partial derivatives. Thus the derivative at a point θ is a linear map. If the range space of ϕ is the real line (so that the derivative is a horizontal vector), then the derivative is also called the *gradient* of the function.

Note that what is usually called the derivative of a function $\phi : \mathbb{R} \mapsto \mathbb{R}$ does not completely correspond to the present derivative. The derivative at a point, usually written $\phi'(\theta)$, is written here as ϕ'_θ. Although $\phi'(\theta)$ is a number, the second object ϕ'_θ is identified with the map $h \mapsto \phi'_\theta(h) = \phi'(\theta) h$. Thus in the present terminology the usual derivative function $\theta \mapsto \phi'(\theta)$ is a map from $\mathbb{R}$ into the set of linear maps from $\mathbb{R} \mapsto \mathbb{R}$, not a map from $\mathbb{R} \mapsto \mathbb{R}$. Graphically the "affine" approximation $h \mapsto \phi(\theta) + \phi'_\theta(h)$ is the tangent to the function ϕ at θ.

3.1 Theorem. *Let $\phi : \mathbb{D}_\phi \subset \mathbb{R}^k \mapsto \mathbb{R}^m$ be a map defined on a subset of $\mathbb{R}^k$ and differentiable at θ. Let T_n be random vectors taking their values in the domain of ϕ. If $r_n(T_n - \theta) \rightsquigarrow T$ for numbers $r_n \to \infty$, then $r_n(\phi(T_n) - \phi(\theta)) \rightsquigarrow \phi'_\theta(T)$. Moreover, the difference between $r_n(\phi(T_n) - \phi(\theta))$ and $\phi'_\theta(r_n(T_n - \theta))$ converges to zero in probability.*

Proof. Because the sequence $r_n(T_n - \theta)$ converges in distribution, it is uniformly tight and $T_n - \theta$ converges to zero in probability. By the differentiability of ϕ the remainder function $R(h) = \phi(\theta + h) - \phi(\theta) - \phi'_\theta(h)$ satisfies $R(h) = o(\|h\|)$ as $h \to 0$. Lemma 2.12 allows to replace the fixed h by a random sequence and gives

$$\phi(T_n) - \phi(\theta) - \phi'_\theta(T_n - \theta) \equiv R(T_n - \theta) = o_P(\|T_n - \theta\|).$$

Multiply this left and right with r_n, and note that $o_P(r_n \|T_n - \theta\|) = o_P(1)$ by tightness of the sequence $r_n(T_n - \theta)$. This yields the last statement of the theorem. Because matrix multiplication is continuous, $\phi'_\theta(r_n(T_n - \theta)) \rightsquigarrow \phi'_\theta(T)$ by the continuous-mapping theorem. Apply Slutsky's lemma to conclude that the sequence $r_n(\phi(T_n) - \phi(\theta))$ has the same weak limit. ∎

A common situation is that $\sqrt{n}(T_n - \theta)$ converges to a multivariate normal distribution $N_k(\mu, \Sigma)$. Then the conclusion of the theorem is that the sequence $\sqrt{n}(\phi(T_n) - \phi(\theta))$ converges in law to the $N_m(\phi'_\theta \mu, \phi'_\theta \Sigma (\phi'_\theta)^T)$ distribution.

3.2 Example (Sample variance). The sample variance of n observations $X_1, \ldots, X_n$ is defined as $S^2 = n^{-1} \sum_{i=1}^n (X_i - \overline{X})^2$ and can be written as $\phi(\overline{X}, \overline{X^2})$ for the function

$\phi(x, y) = y - x^2$. (For simplicity of notation, we divide by n rather than $n-1$.) Suppose that S^2 is based on a sample from a distribution with finite first to fourth moments $\alpha_1, \alpha_2, \alpha_3, \alpha_4$. By the multivariate central limit theorem,

$$\sqrt{n}\left(\begin{pmatrix} \overline{X} \\ \overline{X^2} \end{pmatrix} - \begin{pmatrix} \alpha_1 \\ \alpha_2 \end{pmatrix}\right) \rightsquigarrow N_2\left(\begin{pmatrix} 0 \\ 0 \end{pmatrix}, \begin{pmatrix} \alpha_2 - \alpha_1^2 & \alpha_3 - \alpha_1\alpha_2 \\ \alpha_3 - \alpha_1\alpha_2 & \alpha_4 - \alpha_2^2 \end{pmatrix}\right).$$

The map ϕ is differentiable at the point $\theta = (\alpha_1, \alpha_2)^T$, with derivative $\phi'_{(\alpha_1,\alpha_2)} = (-2\alpha_1, 1)$. Thus if the vector $(T_1, T_2)'$ possesses the normal distribution in the last display, then

$$\sqrt{n}\big(\phi(\overline{X}, \overline{X^2}) - \phi(\alpha_1, \alpha_2)\big) \rightsquigarrow -2\alpha_1 T_1 + T_2.$$

The latter variable is normally distributed with zero mean and a variance that can be expressed in $\alpha_1, \ldots, \alpha_4$. In case $\alpha_1 = 0$, this variance is simply $\alpha_4 - \alpha_2^2$. The general case can be reduced to this case, because S^2 does not change if the observations X_i are replaced by the centered variables $Y_i = X_i - \alpha_1$. Write $\mu_k = \mathrm{E}Y_i^k$ for the *central moments* of the X_i. Noting that $S^2 = \phi(\overline{Y}, \overline{Y^2})$ and that $\phi(\mu_1, \mu_2) = \mu_2$ is the variance of the original observations, we obtain

$$\sqrt{n}\big(S^2 - \mu_2\big) \rightsquigarrow N\big(0, \mu_4 - \mu_2^2\big).$$

In view of Slutsky's lemma, the same result is valid for the unbiased version $n/(n-1)S^2$ of the sample variance, because $\sqrt{n}\big(n/(n-1) - 1\big) \to 0$. □

3.3 *Example (Level of the chi-square test).* As an application of the preceding example, consider the chi-square test for testing variance. Normal theory prescribes to reject the null hypothesis $H_0 : \mu_2 \leq 1$ for values of nS^2 exceeding the upper α point $\chi^2_{n,\alpha}$ of the χ^2_{n-1} distribution. If the observations are sampled from a normal distribution, then the test has exactly level α. Is this still approximately the case if the underlying distribution is not normal? Unfortunately, the answer is negative.

For large values of n, this can be seen with the help of the preceding result. The central limit theorem and the preceding example yield the two statements

$$\frac{\chi^2_{n-1} - (n-1)}{\sqrt{2n-2}} \rightsquigarrow N(0, 1), \qquad \sqrt{n}\left(\frac{S^2}{\mu_2} - 1\right) \rightsquigarrow N(0, \kappa + 2),$$

where $\kappa = \mu_4/\mu_2^2 - 3$ is the *kurtosis* of the underlying distribution. The first statement implies that $\big(\chi^2_{n,\alpha} - (n-1)\big)/\sqrt{2n-2}$ converges to the upper α point z_α of the standard normal distribution. Thus the level of the chi-square test satisfies

$$\mathrm{P}_{\mu_2=1}\big(nS^2 > \chi^2_{n,\alpha}\big) = \mathrm{P}\left(\sqrt{n}\left(\frac{S^2}{\mu_2} - 1\right) > \frac{\chi^2_{n,\alpha} - n}{\sqrt{n}}\right) \to 1 - \Phi\left(\frac{z_\alpha\sqrt{2}}{\sqrt{\kappa+2}}\right).$$

The asymptotic level reduces to $1 - \Phi(z_\alpha) = \alpha$ if and only if the kurtosis of the underlying distribution is 0. This is the case for normal distributions. On the other hand, heavy-tailed distributions have a much larger kurtosis. If the kurtosis of the underlying distribution is "close to" infinity, then the asymptotic level is close to $1 - \Phi(0) = 1/2$. We conclude that the level of the chi-square test is nonrobust against departures of normality that affect the value of the kurtosis. At least this is true if the critical values of the test are taken from the chi-square distribution with $(n-1)$ degrees of freedom. If, instead, we would use a

Table 3.1. *Level of the test that rejects
if nS^2/μ_2 exceeds the 0.95 quantile
of the χ^2_{19} distribution.*

Law	Level
Laplace	0.12
$0.95\, N(0, 1) + 0.05\, N(0, 9)$	0.12

Note: Approximations based on simulation of
10,000 samples.

normal approximation to the distribution of $\sqrt{n}(S^2/\mu_2 - 1)$ the problem would not arise, provided the asymptotic variance $\kappa + 2$ is estimated accurately. Table 3.1 gives the level for two distributions with slightly heavier tails than the normal distribution. □

In the preceding example the asymptotic distribution of $\sqrt{n}(S^2 - \sigma^2)$ was obtained by the delta method. Actually, it can also and more easily be derived by a direct expansion. Write

$$\sqrt{n}(S^2 - \sigma^2) = \sqrt{n}\left(\frac{1}{n}\sum_{i=1}^{n}(X_i - \mu)^2 - \sigma^2\right) - \sqrt{n}(\overline{X} - \mu)^2.$$

The second term converges to zero in probability; the first term is asymptotically normal by the central limit theorem. The whole expression is asymptotically normal by Slutsky's lemma.

Thus it is not always a good idea to apply general theorems. However, in many examples the delta method is a good way to package the mechanics of Taylor expansions in a transparent way.

3.4   Example. Consider the joint limit distribution of the sample variance S^2 and the t-statistic $\overline{X}/S$. Again for the limit distribution it does not make a difference whether we use a factor n or $n - 1$ to standardize S^2. For simplicity we use n. Then $(S^2, \overline{X}/S)$ can be written as $\phi(\overline{X}, \overline{X^2})$ for the map $\phi : \mathbb{R}^2 \mapsto \mathbb{R}^2$ given by

$$\phi(x, y) = \left(y - x^2, \frac{x}{(y - x^2)^{1/2}}\right).$$

The joint limit distribution of $\sqrt{n}(\overline{X} - \alpha_1, \overline{X^2} - \alpha_2)$ is derived in the preceding example. The map ϕ is differentiable at $\theta = (\alpha_1, \alpha_2)$ provided $\sigma^2 = \alpha_2 - \alpha_1^2$ is positive, with derivative

$$\phi'_{(\alpha_1, \alpha_2)} = \begin{pmatrix} -2\alpha_1 & 1 \\ \frac{\alpha_1^2}{(\alpha_2 - \alpha_1^2)^{3/2}} + \frac{1}{(\alpha_2 - \alpha_1^2)^{1/2}} & \frac{-\alpha_1}{2(\alpha_2 - \alpha_1^2)^{3/2}} \end{pmatrix}.$$

It follows that the sequence $\sqrt{n}(S^2 - \sigma^2, \overline{X}/S - \alpha_1/\sigma)$ is asymptotically bivariate normally distributed, with zero mean and covariance matrix,

$$\phi'_{(\alpha_1, \alpha_2)} \begin{pmatrix} \alpha_2 - \alpha_1^2 & \alpha_3 - \alpha_1\alpha_2 \\ \alpha_3 - \alpha_1\alpha_2 & \alpha_4 - \alpha_2^2 \end{pmatrix} (\phi'_{(\alpha_1, \alpha_2)})^T.$$

It is easy but uninteresting to compute this explicitly. □

3.5 *Example (Skewness).* The sample *skewness* of a sample $X_1, \ldots, X_n$ is defined as

$$l_n = \frac{n^{-1}\sum_{i=1}^{n}(X_i - \overline{X})^3}{\left(n^{-1}\sum_{i=1}^{n}(X_i - \overline{X})^2\right)^{3/2}}.$$

Not surprisingly it converges in probability to the skewness of the underlying distribution, defined as the quotient $\lambda = \mu_3/\sigma^3$ of the third central moment and the third power of the standard deviation of one observation. The skewness of a symmetric distribution, such as the normal distribution, equals zero, and the sample skewness may be used to test this aspect of normality of the underlying distribution. For large samples a critical value may be determined from the normal approximation for the sample skewness.

The sample skewness can be written as $\phi(\overline{X}, \overline{X^2}, \overline{X^3})$ for the function ϕ given by

$$\phi(a, b, c) = \frac{c - 3ab + 2a^3}{(b - a^2)^{3/2}}.$$

The sequence $\sqrt{n}(\overline{X} - \alpha_1, \overline{X^2} - \alpha_2, \overline{X^3} - \alpha_3)$ is asymptotically mean-zero normal by the central limit theorem, provided EX_1^6 is finite. The value $\phi(\alpha_1, \alpha_2, \alpha_3)$ is exactly the population skewness. The function ϕ is differentiable at the point $(\alpha_1, \alpha_2, \alpha_3)$ and application of the delta method is straightforward. We can save work by noting that the sample skewness is location and scale invariant. With $Y_i = (X_i - \alpha_1)/\sigma$, the skewness can also be written as $\phi(\overline{Y}, \overline{Y^2}, \overline{Y^3})$. With $\lambda = \mu_3/\sigma^3$ denoting the skewness of the underlying distribution, the Ys satisfy

$$\sqrt{n}\begin{pmatrix} \overline{Y} \\ \overline{Y^2} - 1 \\ \overline{Y^3} - \lambda \end{pmatrix} \rightsquigarrow N\left(0, \begin{pmatrix} 1 & \lambda & \kappa + 3 \\ \lambda & \kappa + 2 & \mu_5/\sigma^5 - \lambda \\ \kappa + 3 & \mu_5/\sigma^5 - \lambda & \mu_6/\sigma^6 - \lambda^2 \end{pmatrix}\right).$$

The derivative of ϕ at the point $(0, 1, \lambda)$ equals $(-3, -3\lambda/2, 1)$. Hence, if T possesses the normal distribution in the display, then $\sqrt{n}(l_n - \lambda)$ is asymptotically normal distributed with mean zero and variance equal to $\text{var}(-3T_1 - 3\lambda T_2/2 + T_3)$. If the underlying distribution is normal, then $\lambda = \mu_5 = 0$, $\kappa = 0$ and $\mu_6/\sigma^6 = 15$. In that case the sample skewness is asymptotically $N(0, 6)$-distributed.

An approximate level α test for normality based on the sample skewness could be to reject normality if $\sqrt{n}|l_n| > \sqrt{6}\,z_{\alpha/2}$. Table 3.2 gives the level of this test for different values of n. $\square$

Table 3.2. *Level of the test that rejects if $\sqrt{n}|l_n|/\sqrt{6}$ exceeds the 0.975 quantile of the normal distribution, in the case that the observations are normally distributed.*

n	Level
10	0.02
20	0.03
30	0.03
50	0.05

Note: Approximations based on simulation of 10,000 samples.

3.2   Variance-Stabilizing Transformations

Given a sequence of statistics T_n with $\sqrt{n}(T_n - \theta) \overset{\theta}{\rightsquigarrow} N(0, \sigma^2(\theta))$ for a range of values of θ, asymptotic confidence intervals for θ are given by

$$\left(T_n - z_\alpha \frac{\sigma(\theta)}{\sqrt{n}}, T_n + z_\alpha \frac{\sigma(\theta)}{\sqrt{n}} \right).$$

These are asymptotically of level $1 - 2\alpha$ in that the probability that θ is covered by the interval converges to $1 - 2\alpha$ for every θ. Unfortunately, as stated previously, these intervals are useless, because of their dependence on the unknown θ. One solution is to replace the unknown standard deviations $\sigma(\theta)$ by estimators. If the sequence of estimators is chosen consistent, then the resulting confidence interval still has asymptotic level $1 - 2\alpha$. Another approach is to use a variance-stabilizing transformation, which often leads to a better approximation.

The idea is that no problem arises if the asymptotic variances $\sigma^2(\theta)$ are independent of θ. Although this fortunate situation is rare, it is often possible to transform the parameter into a different parameter $\eta = \phi(\theta)$, for which this idea can be applied. The natural estimator for η is $\phi(T_n)$. If ϕ is differentiable, then

$$\sqrt{n}\big(\phi(T_n) - \phi(\theta)\big) \overset{\theta}{\rightsquigarrow} N\big(0, \phi'(\theta)^2 \sigma^2(\theta)\big).$$

For ϕ chosen such that $\phi'(\theta)\sigma(\theta) \equiv 1$, the asymptotic variance is constant and finding an asymptotic confidence interval for $\eta = \phi(\theta)$ is easy. The solution

$$\phi(\theta) = \int \frac{1}{\sigma(\theta)} d\theta$$

is a *variance stabililizing transformation*. If it is well defined, then it is automatically monotone, so that a confidence interval for η can be transformed back into a confidence interval for θ.

3.6 *Example (Correlation).* Let $(X_1, Y_1), \ldots, (X_n, Y_n)$ be a sample from a bivariate normal distribution with correlation coefficient ρ. The *sample correlation coefficient* is defined as

$$r_n = \frac{\sum_{i=1}^n (X_i - \bar{X})(Y_i - \bar{Y})}{\left\{ \sum_{i=1}^n (X_i - \bar{X})^2 \sum (Y_i - \bar{Y})^2 \right\}^{1/2}}.$$

With the help of the delta method, it is possible to derive that $\sqrt{n}(r_n - \rho)$ is asymptotically zero-mean normal, with variance depending on the (mixed) third and fourth moments of (X, Y). This is true for general underlying distributions, provided the fourth moments exist. Under the normality assumption the asymptotic variance can be expressed in the correlation of X and Y. Tedious algebra gives

$$\sqrt{n}\,(r_n - \rho) \rightsquigarrow N\big(0, (1 - \rho^2)^2\big).$$

It does not work very well to base an asymptotic confidence interval directly on this result.

Table 3.3. *Coverage probability of the asymptotic 95%
confidence interval for the correlation coefficient, for two
values of n and five different values of the true correlation ρ.*

n	$\rho = 0$	$\rho = 0.2$	$\rho = 0.4$	$\rho = 0.6$	$\rho = 0.8$
15	0.92	0.92	0.92	0.93	0.92
25	0.93	0.94	0.94	0.94	0.94

Note: Approximations based on simulation of 10,000 samples.

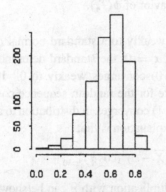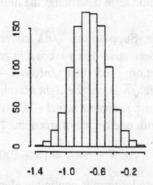

Figure 3.1. Histogram of 1000 sample correlation coefficients, based on 1000 independent samples of the the bivariate normal distribution with correlation 0.6, and histogram of the arctanh of these values.

The transformation

$$\phi(\rho) = \int \frac{1}{1-\rho^2} d\rho = \frac{1}{2} \log \frac{1+\rho}{1-\rho} = \text{arctanh } \rho$$

is variance stabilizing. Thus, the sequence $\sqrt{n}(\text{arctanh } r_n - \text{arctanh } \rho)$ converges to a standard normal distribution for every ρ. This leads to the asymptotic confidence interval for the correlation coefficient ρ given by

$$\left(\tanh(\text{arctanh } r_n - z_\alpha/\sqrt{n}), \tanh(\text{arctanh } r_n + z_\alpha/\sqrt{n})\right).$$

Table 3.3 gives an indication of the accuracy of this interval. Besides stabilizing the variance the arctanh transformation has the benefit of symmetrizing the distribution of the sample correlation coefficient (which is perhaps of greater importance), as can be seen in Figure 5.3. □

*3.3 Higher-Order Expansions

To package a simple idea in a theorem has the danger of obscuring the idea. The delta method is based on a Taylor expansion of order one. Sometimes a problem cannot be exactly forced into the framework described by the theorem, but the principle of a Taylor expansion is still valid.

In the one-dimensional case, a Taylor expansion applied to a statistic T_n has the form

$$\phi(T_n) = \phi(\theta) + (T_n - \theta)\phi'(\theta) + \tfrac{1}{2}(T_n - \theta)^2\phi''(\theta) + \cdots.$$

Usually the linear term $(T_n - \theta)\phi'(\theta)$ is of higher order than the remainder, and thus determines the order at which $\phi(T_n) - \phi(\theta)$ converges to zero: the same order as $T_n - \theta$. Then the approach of the preceding section gives the limit distribution of $\phi(T_n) - \phi(\theta)$. If $\phi'(\theta) = 0$, this approach is still valid but not of much interest, because the resulting limit distribution is degenerate at zero. Then it is more informative to multiply the difference $\phi(T_n) - \phi(\theta)$ by a higher rate and obtain a nondegenerate limit distribution. Looking at the Taylor expansion, we see that the linear term disappears if $\phi'(\theta) = 0$, and we expect that the quadratic term determines the limit behavior of $\phi(T_n)$.

3.7   Example. Suppose that $\sqrt{n}\overline{X}$ converges weakly to a standard normal distribution. Because the derivative of $x \mapsto \cos x$ is zero at $x = 0$, the standard delta method of the preceding section yields that $\sqrt{n}(\cos \overline{X} - \cos 0)$ converges weakly to 0. It should be concluded that $\sqrt{n}$ is not the right norming rate for the random sequence $\cos \overline{X} - 1$. A more informative statement is that $-2n(\cos \overline{X} - 1)$ converges in distribution to a chi-square distribution with one degree of freedom. The explanation is that

$$\cos \overline{X} - \cos 0 = (\overline{X} - 0)0 + \tfrac{1}{2}(\overline{X} - 0)^2(\cos x)''_{|x=0} + \cdots.$$

That the remainder term is negligible after multiplication with n can be shown along the same lines as the proof of Theorem 3.1. The sequence $n\overline{X}^2$ converges in law to a χ_1^2 distribution by the continuous-mapping theorem; the sequence $-2n(\cos \overline{X} - 1)$ has the same limit, by Slutsky's lemma. $\quad\square$

A more complicated situation arises if the statistic T_n is higher-dimensional with coordinates of different orders of magnitude. For instance, for a real-valued function ϕ,

$$\phi(T_n) - \phi(\theta) = \sum_{i=1}^{k} \frac{\partial \phi}{\partial x_i}(\theta)(T_{n,i} - \theta_i)$$

$$+ \frac{1}{2}\sum_{i=1}^{k}\sum_{j=1}^{k} \frac{\partial^2 \phi}{\partial x_i \partial x_j}(\theta)(T_{n,i} - \theta_i)(T_{n,j} - \theta_j) + \cdots.$$

If the sequences $T_{n,i} - \theta_i$ are of different order, then it may happen, for instance, that the linear part involving $T_{n,i} - \theta_i$ is of the same order as the quadratic part involving $(T_{n,j} - \theta_j)^2$. Thus, it is necessary to determine carefully the rate of all terms in the expansion, and to rearrange these in decreasing order of magnitude, before neglecting the "remainder."

*3.4   Uniform Delta Method

Sometimes we wish to prove the asymptotic normality of a sequence $\sqrt{n}\big(\phi(T_n) - \phi(\theta_n)\big)$ for centering vectors θ_n changing with n, rather than a fixed vector. If $\sqrt{n}(\theta_n - \theta) \to h$ for certain vectors θ and h, then this can be handled easily by decomposing

$$\sqrt{n}\big(\phi(T_n) - \phi(\theta_n)\big) = \sqrt{n}\big(\phi(T_n) - \phi(\theta)\big) - \sqrt{n}\big(\phi(\theta_n) - \phi(\theta)\big).$$

Several applications of Slutsky's lemma and the delta method yield as limit in law the vector $\phi'_\theta(T + h) - \phi'_\theta(h) = \phi'_\theta(T)$, if T is the limit in distribution of $\sqrt{n}(T_n - \theta_n)$. For $\theta_n \to \theta$ at a slower rate, this argument does not work. However, the same result is true under a slightly stronger differentiability assumption on ϕ.

3.8 Theorem. *Let $\phi : \mathbb{R}^k \mapsto \mathbb{R}^m$ be a map defined and continuously differentiable in a neighborhood of θ. Let T_n be random vectors taking their values in the domain of ϕ. If $r_n(T_n - \theta_n) \rightsquigarrow T$ for vectors $\theta_n \to \theta$ and numbers $r_n \to \infty$, then $r_n(\phi(T_n) - \phi(\theta_n)) \rightsquigarrow \phi'_\theta(T)$. Moreover, the difference between $r_n(\phi(T_n) - \phi(\theta_n))$ and $\phi'_\theta(r_n(T_n - \theta_n))$ converges to zero in probability.*

Proof. It suffices to prove the last assertion. Because convergence in probability to zero of vectors is equivalent to convergence to zero of the components separately, it is no loss of generality to assume that ϕ is real-valued. For $0 \le t \le 1$ and fixed h, define $g_n(t) = \phi(\theta_n + th)$. For sufficiently large n and sufficiently small h, both θ_n and $\theta_n + h$ are in a ball around θ inside the neighborhood on which ϕ is differentiable. Then $g_n : [0, 1] \mapsto \mathbb{R}$ is continuously differentiable with derivative $g'_n(t) = \phi'_{\theta_n + th}(h)$. By the mean-value theorem, $g_n(1) - g_n(0) = g'_n(\xi)$ for some $0 \le \xi \le 1$. In other words

$$R_n(h) := \phi(\theta_n + h) - \phi(\theta_n) - \phi'_\theta(h) = \phi'_{\theta_n + \xi h}(h) - \phi'_\theta(h).$$

By the continuity of the map $\theta \mapsto \phi'_\theta$, there exists for every $\varepsilon > 0$ a $\delta > 0$ such that $\left\| \phi'_\zeta(h) - \phi'_\theta(h) \right\| < \varepsilon \|h\|$ for every $\|\zeta - \theta\| < \delta$ and every h. For sufficiently large n and $\|h\| < \delta/2$, the vectors $\theta_n + \xi h$ are within distance δ of θ, so that the norm $\left\| R_n(h) \right\|$ of the right side of the preceding display is bounded by $\varepsilon \|h\|$. Thus, for any $\eta > 0$,

$$P\left(r_n \left\| R_n(T_n - \theta_n) \right\| > \eta \right) \le P\left(\|T_n - \theta_n\| \ge \frac{\delta}{2} \right) + P\left(r_n \|T_n - \theta_n\| \varepsilon > \eta \right).$$

The first term converges to zero as $n \to \infty$. The second term can be made arbitrarily small by choosing ε small. ∎

*3.5 Moments

So far we have discussed the stability of convergence in distribution under transformations. We can pose the same problem regarding moments: Can an expansion for the moments of $\phi(T_n) - \phi(\theta)$ be derived from a similar expansion for the moments of $T_n - \theta$? In principle the answer is affirmative, but unlike in the distributional case, in which a simple derivative of ϕ is enough, global regularity conditions on ϕ are needed to argue that the remainder terms are negligible.

One possible approach is to apply the distributional delta method first, thus yielding the qualitative asymptotic behavior. Next, the convergence of the moments of $\phi(T_n) - \phi(\theta)$ (or a remainder term) is a matter of uniform integrability, in view of Lemma 2.20. If ϕ is uniformly Lipschitz, then this uniform integrability follows from the corresponding uniform integrability of $T_n - \theta$. If ϕ has an unbounded derivative, then the connection between moments of $\phi(T_n) - \phi(\theta)$ and $T_n - \theta$ is harder to make, in general.

Notes

The Delta method belongs to the folklore of statistics. It is not entirely trivial; proofs are sometimes based on the mean-value theorem and then require continuous differentiability in a neighborhood. A generalization to functions on infinite-dimensional spaces is discussed in Chapter 20.

PROBLEMS

1. Find the joint limit distribution of $\left(\sqrt{n}(\overline{X} - \mu), \sqrt{n}(S^2 - \sigma^2)\right)$ if $\overline{X}$ and S^2 are based on a sample of size n from a distribution with finite fourth moment. Under what condition on the underlying distribution are $\sqrt{n}(\overline{X} - \mu)$ and $\sqrt{n}(S^2 - \sigma^2)$ asymptotically independent?

2. Find the asymptotic distribution of $\sqrt{n}(r - \rho)$ if r is the correlation coefficient of a sample of n bivariate vectors with finite fourth moments. (This is quite a bit of work. It helps to assume that the mean and the variance are equal to 0 and 1, respectively.)

3. Investigate the asymptotic robustness of the level of the t-test for testing the mean that rejects $H_0 : \mu \le 0$ if $\sqrt{n}\overline{X}/S$ is larger than the upper α quantile of the t_{n-1} distribution.

4. Find the limit distribution of the sample kurtosis $k_n = n^{-1}\sum_{i=1}^{n}(X_i - \overline{X})^4/S^4 - 3$, and design an asymptotic level α test for normality based on k_n. (Warning: At least 500 observations are needed to make the normal approximation work in this case.)

5. Design an asymptotic level α test for normality based on the sample skewness and kurtosis jointly.

6. Let $X_1, \ldots, X_n$ be i.i.d. with expectation μ and variance 1. Find constants such that $a_n(\overline{X}_n^2 - b_n)$ converges in distribution if $\mu = 0$ or $\mu \ne 0$.

7. Let $X_1, \ldots, X_n$ be a random sample from the Poisson distribution with mean θ. Find a variance stabilizing transformation for the sample mean, and construct a confidence interval for θ based on this.

8. Let $X_1, \ldots, X_n$ be i.i.d. with expectation 1 and finite variance. Find the limit distribution of $\sqrt{n}(\overline{X}_n^{-1} - 1)$. If the random variables are sampled from a density f that is bounded and strictly positive in a neighborhood of zero, show that $E|\overline{X}_n^{-1}| = \infty$ for every n. (The density of $\overline{X}_n$ is bounded away from zero in a neighborhood of zero for every n.)

4

Moment Estimators

The method of moments determines estimators by comparing sample and theoretical moments. Moment estimators are useful for their simplicity, although not always optimal. Maximum likelihood estimators for full exponential families are moment estimators, and their asymptotic normality can be proved by treating them as such.

4.1 Method of Moments

Let $X_1, \ldots, X_n$ be a sample from a distribution P_θ that depends on a parameter θ, ranging over some set Θ. The method of moments consists of estimating θ by the solution of a system of equations

$$\frac{1}{n} \sum_{i=1}^{n} f_j(X_i) = E_\theta f_j(X), \quad j = 1, \ldots, k,$$

for given functions $f_1, \ldots, f_k$. Thus the parameter is chosen such that the sample moments (on the left side) match the theoretical moments. If the parameter is k-dimensional one usually tries to match k moments in this manner. The choices $f_j(x) = x^j$ lead to the method of moments in its simplest form.

Moment estimators are not necessarily the best estimators, but under reasonable conditions they have convergence rate $\sqrt{n}$ and are asymptotically normal. This is a consequence of the delta method. Write the given functions in the vector notation $f = (f_1, \ldots, f_k)$, and let $e : \Theta \mapsto \mathbb{R}^k$ be the vector-valued expectation $e(\theta) = P_\theta f$. Then the moment estimator $\hat{\theta}_n$ solves the system of equations

$$\mathbb{P}_n f \equiv \frac{1}{n} \sum_{i=1}^{n} f(X_i) = e(\theta) \equiv P_\theta f.$$

For existence of the moment estimator, it is necessary that the vector $\mathbb{P}_n f$ be in the range of the function e. If e is one-to-one, then the moment estimator is uniquely determined as $\hat{\theta}_n = e^{-1}(\mathbb{P}_n f)$ and

$$\sqrt{n}(\hat{\theta}_n - \theta_0) = \sqrt{n}\big(e^{-1}(\mathbb{P}_n f) - e^{-1}(P_{\theta_0} f)\big).$$

If $\mathbb{P}_n f$ is asymptotically normal and e^{-1} is differentiable, then the right side is asymptotically normal by the delta method.

35

The derivative of e^{-1} at $e(\theta_0)$ is the inverse $e'^{-1}_{\theta_0}$ of the derivative of e at θ_0. Because the function e^{-1} is often not explicit, it is convenient to ascertain its differentiability from the differentiability of e. This is possible by the inverse function theorem. According to this theorem a map that is (continuously) differentiable throughout an open set with nonsingular derivatives is locally one-to-one, is of full rank, and has a differentiable inverse. Thus we obtain the following theorem.

4.1   Theorem. *Suppose that $e(\theta) = P_\theta f$ is one-to-one on an open set $\Theta \subset \mathbb{R}^k$ and continuously differentiable at θ_0 with nonsingular derivative e'_{θ_0}. Moreover, assume that $P_{\theta_0} \|f\|^2 < \infty$. Then moment estimators $\hat\theta_n$ exist with probability tending to one and satisfy*

$$\sqrt{n}(\hat\theta_n - \theta_0) \overset{\theta_0}{\rightsquigarrow} N\left(0, e'^{-1}_{\theta_0} P_{\theta_0} f f^T \left(e'^{-1}_{\theta_0}\right)^T\right).$$

Proof. Continuous differentiability at θ_0 presumes differentiability in a neighborhood and the continuity of $\theta \mapsto e'_\theta$ and nonsingularity of e'_{θ_0} imply nonsingularity in a neighborhood. Therefore, by the inverse function theorem there exist open neighborhoods U of θ_0 and V of $P_{\theta_0} f$ such that $e : U \mapsto V$ is a differentiable bijection with a differentiable inverse $e^{-1} : V \mapsto U$. Moment estimators $\hat\theta_n = e^{-1}(\mathbb{P}_n f)$ exist as soon as $\mathbb{P}_n f \in V$, which happens with probability tending to 1 by the law of large numbers.

The central limit theorem guarantees asymptotic normality of the sequence $\sqrt{n}(\mathbb{P}_n f - P_{\theta_0} f)$. Next use Theorem 3.1 on the display preceding the statement of the theorem. ∎

For completeness, the following two lemmas constitute, if combined, a proof of the inverse function theorem. If necessary the preceding theorem can be strengthened somewhat by applying the lemmas directly. Furthermore, the first lemma can be easily generalized to infinite-dimensional parameters, such as used in the semiparametric models discussed in Chapter 25.

4.2   Lemma. *Let $\Theta \subset \mathbb{R}^k$ be arbitrary and let $e : \Theta \mapsto \mathbb{R}^k$ be one-to-one and differentiable at a point θ_0 with a nonsingular derivative. Then the inverse e^{-1} (defined on the range of e) is differentiable at $e(\theta_0)$ provided it is continuous at $e(\theta_0)$.*

Proof. Write $\eta = e(\theta_0)$ and $\Delta h = e^{-1}(\eta + h) - e^{-1}(\eta)$. Because e^{-1} is continuous at η, we have that $\Delta h \mapsto 0$ as $h \mapsto 0$. Thus

$$\eta + h = e\, e^{-1}(\eta + h) = e(\Delta h + \theta_0) = e(\theta_0) + e'_{\theta_0}(\Delta h) + o(\|\Delta h\|),$$

as $h \mapsto 0$, where the last step follows from differentiability of e. The displayed equation can be rewritten as $e'_{\theta_0}(\Delta h) = h + o(\|\Delta h\|)$. By continuity of the inverse of e'_{θ_0}, this implies that

$$\Delta h = e'^{-1}_{\theta_0}(h) + o(\|\Delta h\|).$$

In particular, $\|\Delta h\|(1 + o(1)) \leq \left\|e'^{-1}_{\theta_0}(h)\right\| = O(\|h\|)$. Insert this in the displayed equation to obtain the desired result that $\Delta h = e'^{-1}_{\theta_0}(h) + o(\|h\|)$. ∎

4.3   Lemma. *Let $e : \Theta \mapsto \mathbb{R}^k$ be defined and differentiable in a neighborhood of a point θ_0 and continuously differentiable at θ_0 with a nonsingular derivative. Then e maps every* *sufficiently small open neighborhood U of θ_0 onto an open set V and $e^{-1} : V \mapsto U$ is well defined and continuous.*

Proof. By assumption, $e'_\theta \to A^{-1} := e'_{\theta_0}$ as $\theta \mapsto \theta_0$. Thus $\|I - Ae'_\theta\| \leq \frac{1}{2}$ for every θ in a sufficiently small neighborhood U of θ_0. Fix an arbitrary point $\eta_1 = e(\theta_1)$ from $V = e(U)$ (where $\theta_1 \in U$). Next find an $\varepsilon > 0$ such that $\overline{\text{ball}}(\theta_1, \varepsilon) \subset U$, and fix an arbitrary point η with $\|\eta - \eta_1\| < \delta := \frac{1}{2}\|A\|^{-1}\varepsilon$. It will be shown that $\eta = e(\theta)$ for some point $\theta \in \overline{\text{ball}}(\theta_1, \varepsilon)$. Hence every $\eta \in \text{ball}(\eta_1, \delta)$ has an original in $\overline{\text{ball}}(\theta_1, \varepsilon)$. If e is one-to-one on U, so that the original is unique, then it follows that V is open and that e^{-1} is continuous at η_1.

Define a function $\phi(\theta) = \theta + A(\eta - e(\theta))$. Because the norm of the derivative $\phi'_\theta = I - Ae'_\theta$ is bounded by $\frac{1}{2}$ throughout U, the map ϕ is a contraction on U. Furthermore, if $\|\theta - \theta_1\| \leq \varepsilon$,

$$\|\phi(\theta) - \theta_1\| \leq \|\phi(\theta) - \phi(\theta_1)\| + \|\phi(\theta_1) - \theta_1\| \leq \frac{1}{2}\|\theta - \theta_1\| + \|A\| \, \|\eta - \eta_1\| < \varepsilon.$$

Consequently, ϕ maps $\overline{\text{ball}}(\theta_1, \varepsilon)$ into itself. Because ϕ is a contraction, it has a fixed point $\theta \in \overline{\text{ball}}(\theta_1, \varepsilon)$: a point with $\phi(\theta) = \theta$. By definition of ϕ this satisfies $e(\theta) = \eta$.

Any other $\tilde{\theta}$ with $e(\tilde{\theta}) = \eta$ is also a fixed point of ϕ. In that case the difference $\tilde{\theta} - \theta = \phi(\tilde{\theta}) - \phi(\theta)$ has norm bounded by $\frac{1}{2}\|\tilde{\theta} - \theta\|$. This can only happen if $\tilde{\theta} = \theta$. Hence e is one-to-one throughout U. ∎

4.4   Example. Let $X_1, \ldots, X_n$ be a random sample from the beta-distribution: The common density is equal to

$$x \mapsto \frac{\Gamma(\alpha + \beta)}{\Gamma(\alpha)\Gamma(\beta)} x^{\alpha-1}(1 - x)^{\beta-1} 1_{0 < x < 1}.$$

The moment estimator for (α, β) is the solution of the system of equations

$$\overline{X}_n = E_{\alpha,\beta} X_1 = \frac{\alpha}{\alpha + \beta},$$

$$\overline{X^2_n} = E_{\alpha,\beta} X^2_1 = \frac{(\alpha + 1)\alpha}{(\alpha + \beta + 1)(\alpha + \beta)}.$$

The righthand side is a smooth and regular function of (α, β), and the equations can be solved explicitly. Hence, the moment estimators exist and are asymptotically normal. □

*4.2   Exponential Families

Maximum likelihood estimators in full exponential families are moment estimators. This can be exploited to show their asymptotic normality. Actually, as shown in Chapter 5, maximum likelihood estimators in smoothly parametrized models are asymptotically normal in great generality. Therefore the present section is included for the benefit of the simple proof, rather than as an explanation of the limit properties.

Let $X_1, \ldots, X_n$ be a sample from the k-dimensional *exponential family* with density

$$p_\theta(x) = c(\theta) h(x) e^{\theta^T t(x)}.$$

Thus h and $t = (t_1, \ldots, t_k)$ are known functions on the sample space, and the family is given in its natural parametrization. The parameter set Θ must be contained in the *natural parameter space* for the family. This is the set of θ for which p_θ can define a probability density. If μ is the dominating measure, then this is the right side in

$$\Theta \subset \left\{ \theta \in \mathbb{R}^k : c(\theta)^{-1} \equiv \int h(x)\, e^{\theta^T t(x)}\, d\mu(x) < \infty \right\}.$$

It is a standard result (and not hard to see) that the natural parameter space is convex. It is usually open, in which case the family is called "regular." In any case, we assume that the true parameter is an inner point of Θ. Another standard result concerns the smoothness of the function $\theta \mapsto c(\theta)$, or rather of its inverse. (For a proof of the following lemma, see [100, p. 59] or [17, p. 39].)

4.5 Lemma. *The function* $\theta \mapsto \int h(x)\, e^{\theta^T t(x)}\, d\mu(x)$ *is analytic on the set* $\{\theta \in \mathbb{C}^k : \mathrm{Re}\,\theta \in \overset{\circ}{\Theta}\}$. *Its derivatives can be found by differentiating (repeatedly) under the integral sign:*

$$\frac{\partial^p \int h(x)\, e^{\theta^T t(x)}\, d\mu(x)}{\partial \theta_1^{i_1} \cdots \partial \theta_k^{i_k}} = \int h(x)\, t_1(x)^{i_1} \cdots t_k(x)^{i_k}\, e^{\theta^T t(x)}\, d\mu(x),$$

for any natural numbers p and $i_1 + \cdots + i_k = p$.

The lemma implies that the log likelihood $\ell_\theta(x) = \log p_\theta(x)$ can be differentiated (infinitely often) with respect to θ. The vector of partial derivatives (the score function) satisfies

$$\dot{\ell}_\theta(x) = \frac{\dot{c}}{c}(\theta) + t(x) = t(x) - E_\theta t(X).$$

Here the second equality is an example of the general rule that score functions have zero means. It can formally be established by differentiating the identity $\int p_\theta\, d\mu \equiv 1$ under the integral sign: Combine the lemma and the Leibniz rule to see that

$$\frac{\partial}{\partial \theta_i} \int p_\theta\, d\mu = \int \frac{\partial c(\theta)}{\partial \theta_i}\, h(x)\, e^{\theta^T t(x)}\, d\mu(x) + \int c(\theta)\, h(x)\, t_i(x)\, e^{\theta^T t(x)}\, d\mu(x).$$

The left side is zero and the equation can be rewritten as $0 = \dot{c}/c(\theta) + E_\theta t(X)$.

It follows that the likelihood equations $\sum \dot{\ell}_\theta(X_i) = 0$ reduce to the system of k equations

$$\frac{1}{n} \sum_{i=1}^n t(X_i) = E_\theta t(X).$$

Thus, the maximum likelihood estimators are moment estimators. Their asymptotic properties depend on the function $e(\theta) = E_\theta t(X)$, which is very well behaved on the interior of the natural parameter set. By differentiating $E_\theta t(X)$ under the expectation sign (which is justified by the lemma), we see that its derivative matrices are given by

$$e'_\theta = \mathrm{Cov}_\theta\, t(X).$$

The exponential family is said to be of *full rank* if no linear combination $\sum_{j=1}^k \lambda_j t_j(X)$ is constant with probability 1; equivalently, if the covariance matrix of $t(X)$ is nonsingular. In view of the preceding display, this ensures that the derivative e'_θ is strictly positive-definite throughout the interior of the natural parameter set. Then e is one-to-one, so that there exists at most one solution to the moment equations. (Cf. Problem 4.6.) In view of the expression for $\dot{\ell}_\theta$, the matrix $-n e'_\theta$ is the second-derivative matrix (Hessian) of the log likelihood $\sum_{i=1}^n \ell_\theta(X_i)$. Thus, a solution to the moment equations must be a point of maximum of the log likelihood.

A solution can be shown to exist (within the natural parameter space) with probability 1 if the exponential family is "regular," or more generally "steep" (see [17]); it is then a point of absolute maximum of the likelihood. If the true parameter is in the interior of the parameter set, then a (unique) solution $\hat{\theta}_n$ exists with probability tending to 1 as $n \mapsto \infty$, in any case, by Theorem 4.1. Moreover, this theorem shows that the sequence $\sqrt{n}(\hat{\theta}_n - \theta_0)$ is asymptotically normal with covariance matrix

$$e'_{\theta_0}{}^{-1} \operatorname{Cov}_{\theta_0} t(X) \left(e'_{\theta_0}{}^{-1}\right)^T = \left(\operatorname{Cov}_{\theta_0} t(X)\right)^{-1}.$$

So far we have considered an exponential family in standard form. Many examples arise in the form

$$p_\theta(x) = d(\theta) h(x) e^{Q(\theta)^T t(x)}, \tag{4.6}$$

where $Q = (Q_1, \ldots, Q_k)$ is a vector-valued function. If Q is one-to-one and a maximum likelihood estimator $\hat{\theta}_n$ exists, then by the invariance of maximum likelihood estimators under transformations, $Q(\hat{\theta}_n)$ is the maximum likelihood estimator for the natural parameter $Q(\theta)$ as considered before. If the range of Q contains an open ball around $Q(\theta_0)$, then the preceding discussion shows that the sequence $\sqrt{n}\big(Q(\hat{\theta}_n) - Q(\theta_0)\big)$ is asymptotically normal. It requires another application of the delta method to obtain the limit distribution of $\sqrt{n}(\hat{\theta}_n - \theta_0)$. As is typical of maximum likelihood estimators, the asymptotic covariance matrix is the inverse of the *Fisher information matrix*

$$I_\theta = \mathrm{E}_\theta \dot{\ell}_\theta(X) \dot{\ell}_\theta(X)^T.$$

4.6   Theorem. *Let $\Theta \subset \mathbb{R}^k$ be open and let $Q : \Theta \mapsto \mathbb{R}^k$ be one-to-one and continuously differentiable throughout Θ with nonsingular derivatives. Let the (exponential) family of densities p_θ be given by (4.6) and be of full rank. Then the likelihood equations have a unique solution $\hat{\theta}_n$ with probability tending to 1 and $\sqrt{n}(\hat{\theta}_n - \theta) \overset{\theta}{\rightsquigarrow} N(0, I_\theta^{-1})$ for every θ.*

Proof.   According to the inverse function theorem, the range of Q is open and the inverse map Q^{-1} is differentiable throughout this range. Thus, as discussed previously, the delta method ensures the asymptotic normality. It suffices to calculate the asymptotic covariance matrix. By the preceding discussion this is equal to

$$Q'_\theta{}^{-1} \left(\operatorname{Cov}_\theta t(X)\right)^{-1} \left(Q'_\theta{}^{-1}\right)^T.$$

By direct calculation, the score function for the model is equal to $\dot{\ell}_\theta(x) = \dot{d}/d(\theta) + (Q'_\theta)^T t(x)$. As before, the score function has mean zero, so that this can be rewritten as $\dot{\ell}_\theta(x) = (Q'_\theta)^T \big(t(x) - \mathrm{E}_\theta t(X)\big)$. Thus, the Fisher information matrix equals $I_\theta = (Q'_\theta)^T \operatorname{Cov}_\theta t(X) Q'_\theta$. This is the inverse of the asymptotic covariance matrix given in the preceding display. ∎

Not all exponential families satisfy the conditions of the theorem. For instance, the normal $N(\theta, \theta^2)$ family is an example of a "curved exponential family." The map $Q(\theta) = (\theta^{-2}, \theta^{-1})$ (with $t(x) = (-x^2/2, x)$) does not fill up the natural parameter space of the normal location-scale family but only traces out a one-dimensional curve. In such cases the result of the theorem may still hold. In fact, the result is true for most models with "smooth parametrizations," as is seen in Chapter 5. However, the "easy" proof of this section is not valid.

PROBLEMS

1. Let $X_1, \ldots, X_n$ be a sample from the uniform distribution on $[-\theta, \theta]$. Find the moment estimator of θ based on $\overline{X^2}$. Is it asymptotically normal? Can you think of an estimator for θ that converges faster to the parameter?

2. Let $X_1, \ldots, X_n$ be a sample from a density p_θ and f a function such that $e(\theta) = E_\theta f(X)$ is differentiable with $e'(\theta) = E_\theta \dot\ell_\theta(X) f(X)$ for $\ell_\theta = \log p_\theta$.

   (i) Show that the asymptotic variance of the moment estimator based on f equals $\text{var}_\theta(f)/\text{cov}_\theta(f, \dot\ell_\theta)^2$.

   (ii) Show that this is bigger than I_θ^{-1} with equality for all θ if and only if the moment estimator is the maximum likelihood estimator.

   (iii) Show that the latter happens only for exponential family members.

3. To what extent does the result of Theorem 4.1 require that the observations are i.i.d.?

4. Let the observations be a sample of size n from the $N(\mu, \sigma^2)$ distribution. Calculate the Fisher information matrix for the parameter $\theta = (\mu, \sigma^2)$ and its inverse. Check directly that the maximum likelihood estimator is asymptotically normal with zero mean and covariance matrix I_θ^{-1}.

5. Establish the formula $e'_\theta = \text{Cov}_\theta\, t(X)$ by differentiating $e(\theta) = E_\theta t(X)$ under the integral sign. (Differentiating under the integral sign is justified by Lemma 4.5, because $E_\theta t(X)$ is the first derivative of $c(\theta)^{-1}$.)

6. Suppose a function $e : \Theta \mapsto \mathbb{R}^k$ is defined and continuously differentiable on a convex subset $\Theta \subset \mathbb{R}^k$ with strictly positive-definite derivative matrix. Then e has at most one zero in Θ. (Consider the function $g(\lambda) = (\theta_1 - \theta_2)^T e(\lambda\theta_1 + (1 - \lambda)\theta_2)$ for given $\theta_1 \neq \theta_2$ and $0 \leq \lambda \leq 1$. If $g(0) = g(1) = 0$, then there exists a point λ_0 with $g'(\lambda_0) = 0$ by the mean-value theorem.)

5

M- and Z-Estimators

This chapter gives an introduction to the consistency and asymptotic normality of M-estimators and Z-estimators. Maximum likelihood estimators are treated as a special case.

5.1 Introduction

Suppose that we are interested in a parameter (or "functional") θ attached to the distribution of observations $X_1, \ldots, X_n$. A popular method for finding an estimator $\hat{\theta}_n = \hat{\theta}_n(X_1, \ldots, X_n)$ is to maximize a criterion function of the type

$$\theta \mapsto M_n(\theta) = \frac{1}{n} \sum_{i=1}^{n} m_\theta(X_i). \tag{5.1}$$

Here $m_\theta: \mathcal{X} \mapsto \overline{\mathbb{R}}$ are known functions. An estimator maximizing $M_n(\theta)$ over Θ is called an *M-estimator*. In this chapter we investigate the asymptotic behavior of sequences of M-estimators.

Often the maximizing value is sought by setting a derivative (or the set of partial derivatives in the multidimensional case) equal to zero. Therefore, the name *M-estimator* is also used for estimators satisfying systems of equations of the type

$$\Psi_n(\theta) = \frac{1}{n} \sum_{i=1}^{n} \psi_\theta(X_i) = 0. \tag{5.2}$$

Here ψ_θ are known vector-valued maps. For instance, if θ is k-dimensional, then ψ_θ typically has k coordinate functions $\psi_\theta = (\psi_{\theta,1}, \ldots, \psi_{\theta,k})$, and (5.2) is shorthand for the system of equations

$$\sum_{i=1}^{n} \psi_{\theta,j}(X_i) = 0, \qquad j = 1, 2, \ldots, k.$$

Even though in many examples $\psi_{\theta,j}$ is the jth partial derivative of some function m_θ, this is irrelevant for the following. Equations, such as (5.2), defining an estimator are called *estimating equations* and need not correspond to a maximization problem. In the latter case it is probably better to call the corresponding estimators *Z-estimators* (for zero), but the use of the name *M-estimator* is widespread.

Sometimes the maximum of the criterion function M_n is not taken or the estimating equation does not have an exact solution. Then it is natural to use as estimator a value that almost maximizes the criterion function or is a near zero. This yields approximate M-estimators or Z-estimators. Estimators that are sufficiently close to being a point of maximum or a zero often have the same asymptotic behavior.

An operator notation for taking expectations simplifies the formulas in this chapter. We write P for the marginal law of the observations $X_1, \ldots, X_n$, which we assume to be identically distributed. Furthermore, we write Pf for the expectation $\mathrm{E} f(X) = \int f \, dP$ and abbreviate the average $n^{-1} \sum_{i=1}^n f(X_i)$ to $\mathbb{P}_n f$. Thus $\mathbb{P}_n$ is the *empirical distribution*: the (random) discrete distribution that puts mass $1/n$ at every of the observations $X_1, \ldots, X_n$. The criterion functions now take the forms

$$M_n(\theta) = \mathbb{P}_n m_\theta, \quad \text{and} \quad \Psi_n(\theta) = \mathbb{P}_n \psi_\theta.$$

We also abbreviate the centered sums $n^{-1/2} \sum_{i=1}^n \big(f(X_i) - Pf\big)$ to $\mathbb{G}_n f$, the *empirical process* at f.

5.3 *Example (Maximum likelihood estimators).* Suppose $X_1, \ldots, X_n$ have a common density p_θ. Then the *maximum likelihood estimator* maximizes the likelihood $\prod_{i=1}^n p_\theta(X_i)$, or equivalently the log likelihood

$$\theta \mapsto \sum_{i=1}^n \log p_\theta(X_i).$$

Thus, a maximum likelihood estimator is an M-estimator as in (5.1) with $m_\theta = \log p_\theta$. If the density is partially differentiable with respect to θ for each fixed x, then the maximum likelihood estimator also solves an equation of type (5.2), with ψ_θ equal to the vector of partial derivatives $\dot{\ell}_{\theta,j} = \partial / \partial \theta_j \log p_\theta$. The vector-valued function $\dot{\ell}_\theta$ is known as the *score function* of the model.

The definition (5.1) of an M-estimator may apply in cases where (5.2) does not. For instance, if $X_1, \ldots, X_n$ are i.i.d. according to the uniform distribution on $[0, \theta]$, then it makes sense to maximize the log likelihood

$$\theta \mapsto \sum_{i=1}^n \big(\log 1_{[0,\theta]}(X_i) - \log \theta\big).$$

(Define $\log 0 = -\infty$.) However, this function is not smooth in θ and there exists no natural version of (5.2). Thus, in this example the definition as the location of a maximum is more fundamental than the definition as a zero. □

5.4 *Example (Location estimators).* Let $X_1, \ldots, X_n$ be a random sample of real-valued observations and suppose we want to estimate the location of their distribution. "Location" is a vague term; it could be made precise by defining it as the mean or median, or the center of symmetry of the distribution if this happens to be symmetric. Two examples of location estimators are the sample mean and the sample median. Both are Z-estimators, because they solve the equations

$$\sum_{i=1}^n (X_i - \theta) = 0; \quad \text{and} \quad \sum_{i=1}^n \text{sign}(X_i - \theta) = 0,$$

respectively.[†] Both estimating equations involve functions of the form $\psi(x - \theta)$ for a function ψ that is monotone and odd around zero. It seems reasonable to study estimators that solve a general equation of the type

$$\sum_{i=1}^{n} \psi(X_i - \theta) = 0.$$

We can consider a Z-estimator defined by this equation a "location" estimator, because it has the desirable property of location equivariance. If the observations X_i are shifted by a fixed amount α, then so is the estimate: $\hat{\theta} + \alpha$ solves $\sum_{i=1}^{n} \psi(X_i + \alpha - \theta) = 0$ if $\hat{\theta}$ solves the original equation.

Popular examples are the *Huber estimators* corresponding to the functions

$$\psi(x) = [x]_{-k}^{k} := \begin{cases} -k & \text{if } x \leq -k, \\ x & \text{if } |x| \leq k, \\ k & \text{if } x \geq k. \end{cases}$$

The Huber estimators were motivated by studies in robust statistics concerning the influence of extreme data points on the estimate. The exact values of the largest and smallest observations have very little influence on the value of the median, but a proportional influence on the mean. Therefore, the sample mean is considered nonrobust against outliers. If the extreme observations are thought to be rather unreliable, it is certainly an advantage to limit their influence on the estimate, but the median may be too successful in this respect. Depending on the value of k, the Huber estimators behave more like the mean (large k) or more like the median (small k) and thus bridge the gap between the nonrobust mean and very robust median.

Another example are the *quantiles*. A pth sample quantile is roughly a point θ such that pn observations are less than θ and $(1 - p)n$ observations are greater than θ. The precise definition has to take into account that the value pn may not be an integer. One possibility is to call a *pth sample quantile* any $\hat{\theta}$ that solves the inequalities

$$-1 < \sum_{i=1}^{n} \big((1 - p)1\{X_i < \theta\} - p1\{X_i > \theta\} \big) < 1. \tag{5.5}$$

This is an approximate M-estimator for $\psi(x) = 1 - p, 0, -p$ if $x < 0, x = 0$, or $x > 0$, respectively. The "approximate" refers to the inequalities: It is required that the value of the estimating equation be inside the interval $(-1, 1)$, rather than exactly zero. This may seem a rather wide tolerance interval for a zero. However, all solutions turn out to have the same asymptotic behavior. In any case, except for special combinations of p and n, there is no hope of finding an exact zero, because the criterion function is discontinuous with jumps at the observations. (See Figure 5.1.) If no observations are tied, then all jumps are of size one and at least one solution $\hat{\theta}$ to the inequalities exists. If tied observations are present, it may be necessary to increase the interval $(-1, 1)$ to ensure the existence of solutions. Note that the present ψ function is monotone, as in the previous examples, but not symmetric about zero (for $p \neq 1/2$).

[†] The *sign-function* is defined as $\text{sign}(x) = -1, 0, 1$ if $x < 0, x = 0$ or $x > 0$, respectively. Also x^+ means $x \vee 0 = \max(x, 0)$. For the median we assume that there are no tied observations (in the middle).

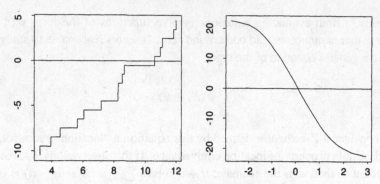

Figure 5.1. The functions $\theta \mapsto \Psi_n(\theta)$ for the 80% quantile and the Huber estimator for samples of size 15 from the gamma(8,1) and standard normal distribution, respectively.

All the estimators considered so far can also be defined as a solution of a maximization problem. Mean, median, Huber estimators, and quantiles minimize $\sum_{i=1}^n m(X_i - \theta)$ for m equal to x^2, $|x|$, $x^2 1_{|x| \le k} + (2k|x| - k^2) 1_{|x| > k}$ and $(1-p)x^- + px^+$, respectively. $\square$

5.2 Consistency

If the estimator $\hat{\theta}_n$ is used to estimate the parameter θ, then it is certainly desirable that the sequence $\hat{\theta}_n$ converges in probability to θ. If this is the case for every possible value of the parameter, then the sequence of estimators is called *asymptotically consistent*. For instance, the sample mean $\overline{X}_n$ is asymptotically consistent for the population mean EX (provided the population mean exists). This follows from the law of large numbers. Not surprisingly this extends to many other sample characteristics. For instance, the sample median is consistent for the population median, whenever this is well defined. What can be said about M-estimators in general? We shall assume that the set of possible parameters is a metric space, and write d for the metric. Then we wish to prove that $d(\hat{\theta}_n, \theta_0) \xrightarrow{P} 0$ for some value θ_0, which depends on the underlying distribution of the observations.

Suppose that the M-estimator $\hat{\theta}_n$ maximizes the random criterion function

$$\theta \mapsto M_n(\theta).$$

Clearly, the "asymptotic value" of $\hat{\theta}_n$ depends on the asymptotic behavior of the functions M_n. Under suitable normalization there typically exists a deterministic "asymptotic criterion function" $\theta \mapsto M(\theta)$ such that

$$M_n(\theta) \xrightarrow{P} M(\theta), \qquad \text{every } \theta. \tag{5.6}$$

For instance, if $M_n(\theta)$ is an average of the form $\mathbb{P}_n m_\theta$ as in (5.1), then the law of large numbers gives this result with $M(\theta) = Pm_\theta$, provided this expectation exists.

It seems reasonable to expect that the maximizer $\hat{\theta}_n$ of M_n converges to the maximizing value θ_0 of M. This is what we wish to prove in this section, and we say that $\hat{\theta}_n$ is (asymptotically) consistent for θ_0. However, the convergence (5.6) is too weak to ensure

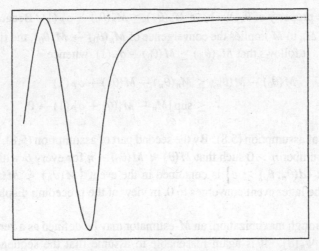

Figure 5.2. Example of a function whose point of maximum is not well separated.

the convergence of $\hat{\theta}_n$. Because the value $\hat{\theta}_n$ depends on the whole function $\theta \mapsto M_n(\theta)$, an appropriate form of "functional convergence" of M_n to M is needed, strengthening the pointwise convergence (5.6). There are several possibilities. In this section we first discuss an approach based on uniform convergence of the criterion functions. Admittedly, the assumption of uniform convergence is too strong for some applications and it is sometimes not easy to verify, but the approach illustrates the general idea.

Given an arbitrary random function $\theta \mapsto M_n(\theta)$, consider estimators $\hat{\theta}_n$ that *nearly maximize* M_n, that is,

$$M_n(\hat{\theta}_n) \geq \sup_\theta M_n(\theta) - o_P(1).$$

Then certainly $M_n(\hat{\theta}_n) \geq M_n(\theta_0) - o_P(1)$, which turns out to be enough to ensure consistency. It is assumed that the sequence M_n converges to a nonrandom map $M : \Theta \mapsto \overline{\mathbb{R}}$. Condition (5.8) of the following theorem requires that this map attains its maximum at a unique point θ_0, and only parameters close to θ_0 may yield a value of $M(\theta)$ close to the maximum value $M(\theta_0)$. Thus, θ_0 should be a *well-separated* point of maximum of M. Figure 5.2 shows a function that does not satisfy this requirement.

5.7 Theorem. *Let M_n be random functions and let M be a fixed function of θ such that for every $\varepsilon > 0$[†]*

$$\sup_{\theta \in \Theta} \left| M_n(\theta) - M(\theta) \right| \xrightarrow{\mathrm{P}} 0,$$
$$\sup_{\theta \,:\, d(\theta, \theta_0) \geq \varepsilon} M(\theta) < M(\theta_0). \tag{5.8}$$

Then any sequence of estimators $\hat{\theta}_n$ with $M_n(\hat{\theta}_n) \geq M_n(\theta_0) - o_P(1)$ converges in probability to θ_0.

[†] Some of the expressions in this display may be nonmeasurable. Then the probability statements are understood in terms of outer measure.

Proof. By the property of $\hat{\theta}_n$, we have $M_n(\hat{\theta}_n) \geq M_n(\theta_0) - o_P(1)$. Because the uniform convergence of M_n to M implies the convergence of $M_n(\theta_0) \xrightarrow{P} M(\theta_0)$, the right side equals $M(\theta_0) - o_P(1)$. It follows that $M_n(\hat{\theta}_n) \geq M(\theta_0) - o_P(1)$, whence

$$M(\theta_0) - M(\hat{\theta}_n) \leq M_n(\hat{\theta}_n) - M(\hat{\theta}_n) + o_P(1)$$
$$\leq \sup_{\theta} |M_n - M|(\theta) + o_P(1) \xrightarrow{P} 0.$$

by the first part of assumption (5.8). By the second part of assumption (5.8), there exists for every $\varepsilon > 0$ a number $\eta > 0$ such that $M(\theta) < M(\theta_0) - \eta$ for every θ with $d(\theta, \theta_0) \geq \varepsilon$. Thus, the event $\{d(\hat{\theta}_n, \theta_0) \geq \varepsilon\}$ is contained in the event $\{M(\hat{\theta}_n) < M(\theta_0) - \eta\}$. The probability of the latter event converges to 0, in view of the preceding display. ∎

Instead of through maximization, an M-estimator may be defined as a zero of a criterion function $\theta \mapsto \Psi_n(\theta)$. It is again reasonable to assume that the sequence of criterion functions converges to a fixed limit:

$$\Psi_n(\theta) \xrightarrow{P} \Psi(\theta).$$

Then it may be expected that a sequence of (approximate) zeros of Ψ_n converges in probability to a zero of Ψ. This is true under similar restrictions as in the case of maximizing M estimators. In fact, this can be deduced from the preceding theorem by noting that a zero of Ψ_n maximizes the function $\theta \mapsto -\|\Psi_n(\theta)\|$.

5.9   Theorem. *Let Ψ_n be random vector-valued functions and let Ψ be a fixed vector-valued function of θ such that for every $\varepsilon > 0$*

$$\sup_{\theta \in \Theta} \|\Psi_n(\theta) - \Psi(\theta)\| \xrightarrow{P} 0,$$
$$\inf_{\theta : d(\theta, \theta_0) \geq \varepsilon} \|\Psi(\theta)\| > 0 = \|\Psi(\theta_0)\|.$$

Then any sequence of estimators $\hat{\theta}_n$ such that $\Psi_n(\hat{\theta}_n) = o_P(1)$ converges in probability to θ_0.

Proof. This follows from the preceding theorem, on applying it to the functions $M_n(\theta) = -\|\Psi_n(\theta)\|$ and $M(\theta) = -\|\Psi(\theta)\|$. ∎

The conditions of both theorems consist of a stochastic and a deterministic part. The deterministic condition can be verified by drawing a picture of the graph of the function. A helpful general observation is that, for a compact set Θ and continuous function M or Ψ, uniqueness of θ_0 as a maximizer or zero implies the condition. (See Problem 5.27.)

For $M_n(\theta)$ or $\Psi_n(\theta)$ equal to averages as in (5.1) or (5.2) the uniform convergence required by the stochastic condition is equivalent to the set of functions $\{m_\theta : \theta \in \Theta\}$ or $\{\psi_{\theta, j} : \theta \in \Theta, j = 1, \ldots, k\}$ being *Glivenko-Cantelli*. Glivenko-Cantelli classes of functions are discussed in Chapter 19. One simple set of sufficient conditions is that Θ be compact, that the functions $\theta \mapsto m_\theta(x)$ or $\theta \mapsto \psi_\theta(x)$ are continuous for every x, and that they are dominated by an integrable function.

Uniform convergence of the criterion functions as in the preceding theorems is much stronger than needed for consistency. The following lemma is one of the many possibilities to replace the uniformity by other assumptions.

5.10   Lemma. *Let Θ be a subset of the real line and let Ψ_n be random functions and Ψ a fixed function of θ such that $\Psi_n(\theta) \to \Psi(\theta)$ in probability for every θ. Assume that each map $\theta \mapsto \Psi_n(\theta)$ is continuous and has exactly one zero $\hat\theta_n$, or is nondecreasing with $\Psi_n(\hat\theta_n) = o_P(1)$. Let θ_0 be a point such that $\Psi(\theta_0 - \varepsilon) < 0 < \Psi(\theta_0 + \varepsilon)$ for every $\varepsilon > 0$. Then $\hat\theta_n \overset{P}{\to} \theta_0$.*

Proof.   If the map $\theta \mapsto \Psi_n(\theta)$ is continuous and has a unique zero at $\hat\theta_n$, then

$$P\big(\Psi_n(\theta_0 - \varepsilon) < 0, \Psi_n(\theta_0 + \varepsilon) > 0\big) \le P(\theta_0 - \varepsilon < \hat\theta_n < \theta_0 + \varepsilon).$$

The left side converges to one, because $\Psi_n(\theta_0 \pm \varepsilon) \to \Psi(\theta_0 \pm \varepsilon)$ in probability. Thus the right side converges to one as well, and $\hat\theta_n$ is consistent.

If the map $\theta \mapsto \Psi_n(\theta)$ is nondecreasing and $\hat\theta_n$ is a zero, then the same argument is valid. More generally, if $\theta \mapsto \Psi_n(\theta)$ is nondecreasing, then $\Psi_n(\theta_0 - \varepsilon) < -\eta$ and $\hat\theta_n \le \theta_0 - \varepsilon$ imply $\Psi_n(\hat\theta_n) < -\eta$, which has probability tending to zero for every $\eta > 0$ if $\hat\theta_n$ is a near zero. This and a similar argument applied to the right tail shows that, for every $\varepsilon, \eta > 0$,

$$P\big(\Psi_n(\theta_0 - \varepsilon) < -\eta, \Psi_n(\theta_0 + \varepsilon) > \eta\big) \le P(\theta_0 - \varepsilon < \hat\theta_n < \theta_0 + \varepsilon) + o(1).$$

For 2η equal to the smallest of the numbers $-\Psi(\theta_0 - \varepsilon)$ and $\Psi(\theta_0 + \varepsilon)$ the left side still converges to one.   ∎

5.11   Example (Median).   The sample median $\hat\theta_n$ is a (near) zero of the map $\theta \mapsto \Psi_n(\theta) = n^{-1}\sum_{i=1}^n \text{sign}(X_i - \theta)$. By the law of large numbers,

$$\Psi_n(\theta) \overset{P}{\to} \Psi(\theta) = E\,\text{sign}(X - \theta) = P(X > \theta) - P(X < \theta),$$

for every fixed θ. Thus, we expect that the sample median converges in probability to a point θ_0 such that $P(X > \theta_0) = P(X < \theta_0)$: a population median.

This can be proved rigorously by applying Theorem 5.7 or 5.9. However, even though the conditions of the theorems are satisfied, they are not entirely trivial to verify. (The uniform convergence of Ψ_n to Ψ is proved essentially in Theorem 19.1) In this case it is easier to apply Lemma 5.10. Because the functions $\theta \mapsto \Psi_n(\theta)$ are nonincreasing, it follows that $\hat\theta_n \overset{P}{\to} \theta_0$ provided that $\Psi(\theta_0 - \varepsilon) > 0 > \Psi(\theta_0 + \varepsilon)$ for every $\varepsilon > 0$. This is the case if the population median is unique: $P(X < \theta_0 - \varepsilon) < \frac{1}{2} < P(X < \theta_0 + \varepsilon)$ for all $\varepsilon > 0$.   □

5.2.1   Wald's Consistency Proof

Consider the situation that, for a random sample of variables $X_1, \ldots, X_n$,

$$M_n(\theta) = \mathbb{P}_n m_\theta = \frac{1}{n}\sum_{i=1}^n m_\theta(X_i), \qquad M(\theta) = P m_\theta.$$

In this subsection we consider an alternative set of conditions under which the maximizer $\hat\theta_n$ of the process M_n converges in probability to a point of maximum θ_0 of the function M. This "classical" approach to consistency was taken by Wald in 1949 for maximum likelihood estimators. It works best if the parameter set Θ is compact. If not, then the argument must be complemented by a proof that the estimators are in a compact set eventually or be applied to a suitable compactification of the parameter set.

Assume that the map $\theta \mapsto m_\theta(x)$ is upper-semicontinuous for almost all x: For every θ

$$\limsup_{\theta_n \to \theta} m_{\theta_n}(x) \leq m_\theta(x), \qquad \text{a.s..} \tag{5.12}$$

(The exceptional set of x may depend on θ.) Furthermore, assume that for every sufficiently small ball $U \subset \Theta$ the function $x \mapsto \sup_{\theta \in U} m_\theta(x)$ is measurable and satisfies

$$P \sup_{\theta \in U} m_\theta < \infty. \tag{5.13}$$

Typically, the map $\theta \mapsto Pm_\theta$ has a unique global maximum at a point θ_0, but we shall allow multiple points of maximum, and write Θ_0 for the set $\{\theta_0 \in \Theta : Pm_{\theta_0} = \sup_\theta Pm_\theta\}$ of all points at which M attains its global maximum. The set Θ_0 is assumed not empty. The maps $m_\theta : \mathcal{X} \mapsto \overline{\mathbb{R}}$ are allowed to take the value $-\infty$, but the following theorem assumes implicitly that at least Pm_{θ_0} is finite.

5.14   Theorem. *Let $\theta \mapsto m_\theta(x)$ be upper-semicontinuous for almost all x and let (5.13) be satisfied. Then for any estimators $\hat{\theta}_n$ such that $M_n(\hat{\theta}_n) \geq M_n(\theta_0) - o_P(1)$ for some $\theta_0 \in \Theta_0$, for every $\varepsilon > 0$ and every compact set $K \subset \Theta$,*

$$\mathrm{P}\big(d(\hat{\theta}_n, \Theta_0) \geq \varepsilon \wedge \hat{\theta}_n \in K\big) \to 0.$$

Proof.   If the function $\theta \mapsto Pm_\theta$ is identically $-\infty$, then $\Theta_0 = \Theta$, and there is nothing to prove. Hence, we may assume that there exists $\theta_0 \in \Theta_0$ such that $Pm_{\theta_0} > -\infty$, whence $P|m_{\theta_0}| < \infty$ by (5.13).

Fix some θ and let $U_l \downarrow \theta$ be a decreasing sequence of open balls around θ of diameter converging to zero. Write $m_U(x)$ for $\sup_{\theta \in U} m_\theta(x)$. The sequence m_{U_l} is decreasing and greater than m_θ for every l. Combination with (5.12) yields that $m_{U_l} \downarrow m_\theta$ almost surely. In view of (5.13), we can apply the monotone convergence theorem and obtain that $Pm_{U_l} \downarrow Pm_\theta$ (which may be $-\infty$).

For $\theta \notin \Theta_0$, we have $Pm_\theta < Pm_{\theta_0}$. Combine this with the preceding paragraph to see that for every $\theta \notin \Theta_0$ there exists an open ball U_θ around θ with $Pm_{U_\theta} < Pm_{\theta_0}$. The set $B = \{\theta \in K : d(\theta, \Theta_0) \geq \varepsilon\}$ is compact and is covered by the balls $\{U_\theta : \theta \in B\}$. Let $U_{\theta_1}, \ldots, U_{\theta_p}$ be a finite subcover. Then, by the law of large numbers,

$$\sup_{\theta \in B} \mathbb{P}_n m_\theta \leq \sup_{j=1,\ldots,p} \mathbb{P}_n m_{U_{\theta_j}} \overset{\text{as}}{\to} \sup_j Pm_{U_{\theta_j}} < Pm_{\theta_0}.$$

If $\hat{\theta}_n \in B$, then $\sup_{\theta \in B} \mathbb{P}_n m_\theta$ is at least $\mathbb{P}_n m_{\hat{\theta}_n}$, which by definition of $\hat{\theta}_n$ is at least $\mathbb{P}_n m_{\theta_0} - o_P(1) = Pm_{\theta_0} - o_P(1)$, by the law of large numbers. Thus

$$\{\hat{\theta}_n \in B\} \subset \left\{\sup_{\theta \in B} \mathbb{P}_n m_\theta \geq Pm_{\theta_0} - o_P(1)\right\}.$$

In view of the preceding display the probability of the event on the right side converges to zero as $n \to \infty$.   ∎

Even in simple examples, condition (5.13) can be restrictive. One possibility for relaxation is to divide the n observations in groups of approximately the same size. Then (5.13) may be replaced by, for some k and every $k \leq l < 2k$,

$$P^l \sup_{\theta \in U} \sum_{i=1}^{l} m_\theta(x_i) < \infty. \tag{5.15}$$

Surprisingly enough, this simple device may help. For instance, under condition (5.13) the preceding theorem does not apply to yield the asymptotic consistency of the maximum likelihood estimator of (μ, σ) based on a random sample from the $N(\mu, \sigma^2)$ distribution (unless we restrict the parameter set for σ), but under the relaxed condition it does (with $k = 2$). (See Problem 5.25.) The proof of the theorem under (5.15) remains almost the same. Divide the n observations in groups of k observations and, possibly, a remainder group of l observations; next, apply the law of large numbers to the approximately n/k group sums.

5.16 *Example (Cauchy likelihood).* The maximum likelihood estimator for θ based on a random sample from the Cauchy distribution with location θ maximizes the map $\theta \mapsto \mathbb{P}_n m_\theta$ for

$$m_\theta(x) = -\log\bigl(1 + (x - \theta)^2\bigr).$$

The natural parameter set $\mathbb{R}$ is not compact, but we can enlarge it to the extended real line, provided that we can define m_θ in a reasonable way for $\theta = \pm\infty$. To have the best chance of satisfying (5.13), we opt for the minimal extension, which in order to satisfy (5.12) is

$$m_{-\infty}(x) = \limsup_{\theta \mapsto -\infty} m_\theta(x) = -\infty; \qquad m_\infty(x) = \limsup_{\theta \mapsto \infty} m_\theta(x) = -\infty.$$

These infinite values should not worry us: They are permitted in the preceding theorem. Moreover, because we maximize $\theta \mapsto \mathbb{P}_n m_\theta$, they ensure that the estimator $\hat{\theta}_n$ never takes the values $\pm\infty$, which is excellent.

We apply Wald's theorem with $\Theta = \overline{\mathbb{R}}$, equipped with, for instance, the metric $d(\theta_1, \theta_2) = |\text{arctg}\, \theta_1 - \text{arctg}\, \theta_2|$. Because the functions $\theta \mapsto m_\theta(x)$ are continuous and nonpositive, the conditions are trivially satisfied. Thus, taking $K = \overline{\mathbb{R}}$, we obtain that $d(\hat{\theta}_n, \Theta_0) \overset{\text{P}}{\to} 0$. This conclusion is valid for any underlying distribution P of the observations for which the set Θ_0 is nonempty, because so far we have used the Cauchy likelihood only to motivate m_θ.

To conclude that the maximum likelihood estimator in a Cauchy location model is consistent, it suffices to show that $\Theta_0 = \{\theta_0\}$ if P is the Cauchy distribution with center θ_0. This follows most easily from the identifiability of this model, as discussed in Lemma 5.35. □

5.17 *Example (Current status data).* Suppose that a "death" that occurs at time T is only observed to have taken place or not at a known "check-up time" C. We model the observations as a random sample $X_1, \ldots, X_n$ from the distribution of $X = \bigl(C, 1\{T \leq C\}\bigr)$, where T and C are independent random variables with completely unknown distribution functions F and G, respectively. The purpose is to estimate the "survival distribution" $1 - F$.

If G has a density g with respect to Lebesgue measure λ, then $X = (C, \Delta)$ has a density

$$p_F(c, \delta) = \bigl(\delta F(c) + (1 - \delta)(1 - F)(c)\bigr)g(c)$$

with respect to the product of λ and counting measure on the set $\{0, 1\}$. A maximum likelihood estimator for F can be defined as the distribution function $\hat{F}$ that maximizes the likelihood

$$F \mapsto \prod_{i=1}^{n} (\Delta_i F(C_i) + (1 - \Delta_i)(1 - F)(C_i))$$

over all distribution functions on $[0, \infty)$. Because this only involves the numbers $F(C_1)$, $\ldots, F(C_n)$, the maximizer of this expression is not unique, but some thought shows that there is a unique maximizer $\hat{F}$ that concentrates on (a subset of) the observation times $C_1, \ldots, C_n$. This is commonly used as an estimator.

We can show the consistency of this estimator by Wald's theorem. By its definition $\hat{F}$ maximizes the function $F \mapsto \mathbb{P}_n \log p_F$, but the consistency proof proceeds in a smoother way by setting

$$m_F = \log \frac{p_F}{p_{(F+F_0)/2}} = \log \frac{2p_F}{p_F + p_{F_0}}.$$

Because the likelihood is bigger at $\hat{F}$ than it is at $\frac{1}{2}\hat{F} + \frac{1}{2}F_0$, it follows that $\mathbb{P}_n m_{\hat{F}} \geq 0 = \mathbb{P}_n m_{F_0}$. (It is not claimed that $\hat{F}$ maximizes $F \mapsto \mathbb{P}_n m_F$; this is not true.)

Condition (5.13) is satisfied trivially, because $m_F \leq \log 2$ for every F. We can equip the set of all distribution functions with the topology of weak convergence. If we restrict the parameter set to distributions on a compact interval $[0, \tau]$, then the parameter set is compact by Prohorov's theorem.[†] The map $F \mapsto m_F(c, \delta)$ is continuous at F, relative to the weak topology, for every (c, δ) such that c is a continuity point of F. Under the assumption that G has a density, this includes almost every (c, δ), for every given F. Thus, Theorem 5.14 shows that $\hat{F}_n$ converges under F_0 in probability to the set $\mathcal{F}_0$ of all distribution functions that maximize the map $F \mapsto P_{F_0} m_F$, provided $F_0 \in \mathcal{F}_0$. This set always contains F_0, but it does not necessarily reduce to this single point. For instance, if the density g is zero on an interval $[a, b]$, then we receive no information concerning deaths inside the interval $[a, b]$, and there can be no hope that $\hat{F}_n$ converges to F_0 on $[a, b]$. In that case, F_0 is not "identifiable" on the interval $[a, b]$.

We shall show that $\mathcal{F}_0$ is the set of all F such that $F = F_0$ almost everywhere according to G. Thus, the sequence $\hat{F}_n$ is consistent for F_0 "on the set of time points that have a positive probability of occurring."

Because $p_F = p_{F_0}$ under P_{F_0} if and only if $F = F_0$ almost everywhere according to G, it suffices to prove that, for every pair of probability densities p and p_0, $P_0 \log 2p/(p+p_0) \leq 0$ with equality if and only if $p = p_0$ almost surely under P_0. If $P_0(p = 0) > 0$, then $\log 2p/(p + p_0) = -\infty$ with positive probability and hence, because the function is bounded above, $P_0 \log 2p/(p + p_0) = -\infty$. Thus we may assume that $P_0(p = 0) = 0$. Then, with $f(u) = -u \log(\frac{1}{2} + \frac{1}{2}u)$,

$$P_0 \log \frac{2p}{(p + p_0)} = Pf\left(\frac{p_0}{p}\right) \leq f\left(P\frac{p_0}{p}\right) = f(1) = 0,$$

[†] Alternatively, consider all probability distributions on the compactification $[0, \infty]$ again equipped with the weak topology.

by Jensen's inequality and the concavity of f, with equality only if $p_0/p = 1$ almost surely under P, and then also under P_0. This completes the proof. $\square$

5.3 Asymptotic Normality

Suppose a sequence of estimators $\hat{\theta}_n$ is consistent for a parameter θ that ranges over an open subset of a Euclidean space. The next question of interest concerns the order at which the discrepancy $\hat{\theta}_n - \theta$ converges to zero. The answer depends on the specific situation, but for estimators based on n replications of an experiment the order is often $n^{-1/2}$. Then multiplication with the inverse of this rate creates a proper balance, and the sequence $\sqrt{n}(\hat{\theta}_n - \theta)$ converges in distribution, most often to a normal distribution. This is interesting from a theoretical point of view. It also makes it possible to obtain approximate confidence sets. In this section we derive the asymptotic normality of M-estimators.

We can use a characterization of M-estimators either by maximization or by solving estimating equations. Consider the second possibility. Let $X_1, \ldots, X_n$ be a sample from some distribution P, and let a random and a "true" criterion function be of the form:

$$\Psi_n(\theta) \equiv \frac{1}{n} \sum_{i=1}^{n} \psi_\theta(X_i) = \mathbb{P}_n \psi_\theta, \qquad \Psi(\theta) = P\psi_\theta.$$

Assume that the estimator $\hat{\theta}_n$ is a zero of Ψ_n and converges in probability to a zero θ_0 of Ψ. Because $\hat{\theta}_n \to \theta_0$, it makes sense to expand $\Psi_n(\hat{\theta}_n)$ in a Taylor series around θ_0. Assume for simplicity that θ is one-dimensional. Then

$$0 = \Psi_n(\hat{\theta}_n) = \Psi_n(\theta_0) + (\hat{\theta}_n - \theta_0)\dot{\Psi}_n(\theta_0) + \frac{1}{2}(\hat{\theta}_n - \theta_0)^2 \ddot{\Psi}_n(\tilde{\theta}_n),$$

where $\tilde{\theta}_n$ is a point between $\hat{\theta}_n$ and θ_0. This can be rewritten as

$$\sqrt{n}(\hat{\theta}_n - \theta_0) = \frac{-\sqrt{n}\Psi_n(\theta_0)}{\dot{\Psi}_n(\theta_0) + \frac{1}{2}(\hat{\theta}_n - \theta_0)\ddot{\Psi}_n(\tilde{\theta}_n)}. \qquad (5.18)$$

If $P\psi_{\theta_0}^2$ is finite, then the numerator $-\sqrt{n}\Psi_n(\theta_0) = -n^{-1/2}\sum \psi_{\theta_0}(X_i)$ is asymptotically normal by the central limit theorem. The asymptotic mean and variance are $P\psi_{\theta_0} = \Psi(\theta_0) = 0$ and $P\psi_{\theta_0}^2$, respectively. Next consider the denominator. The first term $\dot{\Psi}_n(\theta_0)$ is an average and can be analyzed by the law of large numbers: $\dot{\Psi}_n(\theta_0) \xrightarrow{P} P\dot{\psi}_{\theta_0}$, provided the expectation exists. The second term in the denominator is a product of $\hat{\theta}_n - \theta = o_P(1)$ and $\ddot{\Psi}_n(\tilde{\theta}_n)$ and converges in probability to zero under the reasonable condition that $\ddot{\Psi}_n(\tilde{\theta}_n)$ (which is also an average) is $O_P(1)$. Together with Slutsky's lemma, these observations yield

$$\sqrt{n}(\hat{\theta}_n - \theta_0) \rightsquigarrow N\left(0, \frac{P\psi_{\theta_0}^2}{\left(P\dot{\psi}_{\theta_0}\right)^2}\right). \qquad (5.19)$$

The preceding derivation can be made rigorous by imposing appropriate conditions, often called "regularity conditions." The only real challenge is to show that $\ddot{\Psi}_n(\tilde{\theta}_n) = O_P(1)$ (see Problem 5.20 or section 5.6).

The derivation can be extended to higher-dimensional parameters. For a k-dimensional parameter, we use k estimating equations. Then the criterion functions are maps $\Psi_n : \mathbb{R}^k \mapsto$

$\mathbb{R}^k$ and the derivatives $\dot{\Psi}_n(\theta_0)$ are $(k \times k)$-matrices that converge to the $(k \times k)$ matrix $P\dot{\psi}_{\theta_0}$ with entries $P\partial/\partial\theta_j\psi_{\theta_0,i}$. The final statement becomes

$$\sqrt{n}(\hat{\theta}_n - \theta_0) \rightsquigarrow N_k\left(0, \left(P\dot{\psi}_{\theta_0}\right)^{-1}P\psi_{\theta_0}\psi_{\theta_0}^T\left(P\dot{\psi}_{\theta_0}^T\right)^{-1}\right). \tag{5.20}$$

Here the invertibility of the matrix $P\dot{\psi}_{\theta_0}$ is a condition.

In the preceding derivation it is implicitly understood that the function $\theta \mapsto \psi_\theta(x)$ possesses two continuous derivatives with respect to the parameter, for every x. This is true in many examples but fails, for instance, for the function $\psi_\theta(x) = \text{sign}(x-\theta)$, which yields the median. Nevertheless, the median is asymptotically normal. That such a simple, but important, example cannot be treated by the preceding approach has motivated much effort to derive the asymptotic normality of M-estimators by more refined methods. One result is the following theorem, which assumes less than one derivative (a Lipschitz condition) instead of two derivatives.

5.21   Theorem. *For each θ in an open subset of Euclidean space, let $x \mapsto \psi_\theta(x)$ be a measurable vector-valued function such that, for every θ_1 and θ_2 in a neighborhood of θ_0 and a measurable function $\dot{\psi}$ with $P\dot{\psi}^2 < \infty$,*

$$\left\|\psi_{\theta_1}(x) - \psi_{\theta_2}(x)\right\| \leq \dot{\psi}(x)\|\theta_1 - \theta_2\|.$$

Assume that $P\|\psi_{\theta_0}\|^2 < \infty$ and that the map $\theta \mapsto P\psi_\theta$ is differentiable at a zero θ_0, with nonsingular derivative matrix V_{θ_0}. If $\mathbb{P}_n\psi_{\hat{\theta}_n} = o_P(n^{-1/2})$, and $\hat{\theta}_n \xrightarrow{P} \theta_0$, then

$$\sqrt{n}(\hat{\theta}_n - \theta_0) = -V_{\theta_0}^{-1}\frac{1}{\sqrt{n}}\sum_{i=1}^n \psi_{\theta_0}(X_i) + o_P(1),$$

In particular, the sequence $\sqrt{n}(\hat{\theta}_n - \theta_0)$ is asymptotically normal with mean zero and covariance matrix $V_{\theta_0}^{-1}P\psi_{\theta_0}\psi_{\theta_0}^T(V_{\theta_0}^{-1})^T$.

Proof. For a fixed measurable function f, we abbreviate $\sqrt{n}(\mathbb{P}_n - P)f$ to $\mathbb{G}_n f$, the empirical process evaluated at f. The consistency of $\hat{\theta}_n$ and the Lipschitz condition on the maps $\theta \mapsto \psi_\theta$ imply that

$$\mathbb{G}_n\psi_{\hat{\theta}_n} - \mathbb{G}_n\psi_{\theta_0} \xrightarrow{P} 0. \tag{5.22}$$

For a nonrandom sequence $\hat{\theta}_n$ this is immediate from the fact that the means of these variables are zero, while the variances are bounded by $P\|\psi_{\theta_n} - \psi_{\theta_0}\|^2 \leq P\dot{\psi}^2\|\theta_n - \theta_0\|^2$ and hence converge to zero. A proof for estimators $\hat{\theta}_n$ under the present mild conditions takes more effort. The appropriate tools are developed in Chapter 19. In Example 19.7 it is seen that the functions ψ_θ form a Donsker class. Next, (5.22) follows from Lemma 19.24. Here we accept the convergence as a fact and give the remainder of the proof.

By the definitions of $\hat{\theta}_n$ and θ_0, we can rewrite $\mathbb{G}_n\psi_{\hat{\theta}_n}$ as $\sqrt{n}P(\psi_{\theta_0} - \psi_{\hat{\theta}_n}) + o_P(1)$. Combining this with the delta method (or Lemma 2.12) and the differentiability of the map $\theta \mapsto P\psi_\theta$, we find that

$$\sqrt{n}V_{\theta_0}(\theta_0 - \hat{\theta}_n) + \sqrt{n}\, o_P\left(\|\hat{\theta}_n - \theta_0\|\right) = \mathbb{G}_n\psi_{\theta_0} + o_P(1).$$

In particular, by the invertibility of the matrix V_{θ_0},

$$\sqrt{n}\|\hat{\theta}_n - \theta_0\| \leq \|V_{\theta_0}^{-1}\|\|\sqrt{n}\|V_{\theta_0}(\hat{\theta}_n - \theta_0)\| = O_P(1) + o_P\big(\sqrt{n}\|\hat{\theta}_n - \theta_0\|\big).$$

This implies that $\hat{\theta}_n$ is $\sqrt{n}$-consistent: The left side is bounded in probability. Inserting this in the previous display, we obtain that $\sqrt{n}V_{\theta_0}(\hat{\theta}_n - \theta_0) = -\mathbb{G}_n\psi_{\theta_0} + o_P(1)$. We conclude the proof by taking the inverse $V_{\theta_0}^{-1}$ left and right. Because matrix multiplication is a continous map, the inverse of the remainder term still converges to zero in probability. ∎

The preceding theorem is a reasonable compromise between simplicity and general applicability, but, unfortunately, it does not cover the sample median. Because the function $\theta \mapsto \text{sign}(x - \theta)$ is not Lipschitz, the Lipschitz condition is apparently still stronger than necessary. Inspection of the proof shows that it is used only to ensure (5.22). It is seen in Lemma 19.24, that (5.22) can be ascertained under the weaker conditions that the collection of functions $x \mapsto \psi_\theta(x)$ are a "Donsker class" and that the map $\theta \mapsto \psi_\theta$ is continuous in probability. The functions $\text{sign}(x - \theta)$ do satisfy these conditions, but a proof and the definition of a Donsker class are deferred to Chapter 19.

If the functions $\theta \mapsto \psi_\theta(x)$ are continuously differentiable, then the natural candidate for $\dot{\psi}(x)$ is $\sup_\theta \|\dot{\psi}_\theta\|$, with the supremum taken over a neighborhood of θ_0. Then the main condition is that the partial derivatives are "locally dominated" by a square-integrable function: There should exist a square-integrable function $\dot{\psi}$ with $\|\dot{\psi}_\theta\| \leq \dot{\psi}$ for every θ close to θ_0. If $\theta \mapsto \dot{\psi}_\theta(x)$ is also continuous at θ_0, then the dominated-convergence theorem readily yields that $V_{\theta_0} = P\dot{\psi}_{\theta_0}$.

The properties of M estimators can typically be obtained under milder conditions by using their characterization as maximizers. The following theorem is in the same spirit as the preceding one but does cover the median. It concerns M-estimators defined as maximizers of a criterion function $\theta \mapsto \mathbb{P}_n m_\theta$, which are assumed to be consistent for a point of maximum θ_0 of the function $\theta \mapsto Pm_\theta$. If the latter function is twice continuously differentiable at θ_0, then, of course, it allows a two-term Taylor expansion of the form

$$Pm_\theta = Pm_{\theta_0} + \tfrac{1}{2}(\theta - \theta_0)^T V_{\theta_0}(\theta - \theta_0) + o\big(\|\theta - \theta_0\|^2\big).$$

It is this expansion rather than the differentiability that is needed in the following theorem.

5.23   Theorem. *For each θ in an open subset of Euclidean space let $x \mapsto m_\theta(x)$ be a measurable function such that $\theta \mapsto m_\theta(x)$ is differentiable at θ_0 for P-almost every x[†] with derivative $\dot{m}_{\theta_0}(x)$ and such that, for every θ_1 and θ_2 in a neighborhood of θ_0 and a measurable function $\dot{m}$ with $P\dot{m}^2 < \infty$*

$$\big|m_{\theta_1}(x) - m_{\theta_2}(x)\big| \leq \dot{m}(x)\,\|\theta_1 - \theta_2\|.$$

Furthermore, assume that the map $\theta \mapsto Pm_\theta$ admits a second-order Taylor expansion at a point of maximum θ_0 with nonsingular symmetric second derivative matrix V_{θ_0}. If $\mathbb{P}_n m_{\hat{\theta}_n} \geq \sup_\theta \mathbb{P}_n m_\theta - o_P(n^{-1})$ and $\hat{\theta}_n \xrightarrow{P} \theta_0$, then

$$\sqrt{n}(\hat{\theta}_n - \theta_0) = -V_{\theta_0}^{-1}\frac{1}{\sqrt{n}}\sum_{i=1}^{n}\dot{m}_{\theta_0}(X_i) + o_P(1).$$

[†] Alternatively, it suffices that $\theta \mapsto m_\theta$ is differentiable at θ_0 in P-probability.

In particular, the sequence $\sqrt{n}(\hat{\theta}_n - \theta_0)$ is asymptotically normal with mean zero and covariance matrix $V_{\theta_0}^{-1} P\dot{m}_{\theta_0}\dot{m}_{\theta_0}^T V_{\theta_0}^{-1}$.

*Proof.   The Lipschitz property and the differentiability of the maps $\theta \mapsto m_\theta$ imply that, for every random sequence $\tilde{h}_n$ that is bounded in probability,

$$\mathbb{G}_n\left[\sqrt{n}\left(m_{\theta_0 + \tilde{h}_n/\sqrt{n}} - m_{\theta_0}\right) - \tilde{h}_n^T \dot{m}_{\theta_0}\right] \xrightarrow{P} 0.$$

For nonrandom sequences $\tilde{h}_n$ this follows, because the variables have zero means, and variances that converge to zero, by the dominated convergence theorem. For general sequences $\tilde{h}_n$ this follows from Lemma 19.31.

A second fact that we need and that is proved subsequently is the $\sqrt{n}$-consistency of the sequence $\hat{\theta}_n$. By Corollary 5.53, the Lipschitz condition, and the twice differentiability of the map $\theta \mapsto Pm_\theta$, the sequence $\sqrt{n}(\hat{\theta}_n - \theta)$ is bounded in probability.

The remainder of the proof is self-contained. In view of the twice differentiability of the map $\theta \mapsto Pm_\theta$, the preceding display can be rewritten as

$$n\mathbb{P}_n\left(m_{\theta_0 + \tilde{h}_n/\sqrt{n}} - m_{\theta_0}\right) = \tfrac{1}{2}\tilde{h}_n^T V_{\theta_0}\tilde{h}_n + \tilde{h}_n^T \mathbb{G}_n \dot{m}_{\theta_0} + o_P(1).$$

Because the sequence $\hat{\theta}_n$ is $\sqrt{n}$-consistent, this is valid both for $\tilde{h}_n$ equal to $\hat{h}_n = \sqrt{n}(\hat{\theta}_n - \theta_0)$ and for $\tilde{h}_n = -V_{\theta_0}^{-1}\mathbb{G}_n\dot{m}_{\theta_0}$. After simple algebra in the second case, we obtain the equations

$$n\mathbb{P}_n\left(m_{\theta_0 + \hat{h}_n/\sqrt{n}} - m_{\theta_0}\right) = \frac{1}{2}\hat{h}_n^T V_{\theta_0}\hat{h}_n + \hat{h}_n^T \mathbb{G}_n \dot{m}_{\theta_0} + o_P(1),$$

$$n\mathbb{P}_n\left(m_{\theta_0 - V_{\theta_0}^{-1}\mathbb{G}_n\dot{m}_{\theta_0}/\sqrt{n}} - m_{\theta_0}\right) = -\frac{1}{2}\mathbb{G}_n\dot{m}_{\theta_0}^T V_{\theta_0}^{-1}\mathbb{G}_n\dot{m}_{\theta_0} + o_P(1).$$

By the definition of $\hat{\theta}_n$, the left side of the first equation is larger than the left side of the second equation (up to $o_P(1)$) and hence the same relation is true for the right sides. Take the difference, complete the square, and conclude that

$$\frac{1}{2}\left(\hat{h}_n + V_{\theta_0}^{-1}\mathbb{G}_n\dot{m}_{\theta_0}\right)^T V_{\theta_0}\left(\hat{h}_n + V_{\theta_0}^{-1}\mathbb{G}_n\dot{m}_{\theta_0}\right) + o_P(1) \geq 0.$$

Because the matrix V_{θ_0} is strictly negative-definite, the quadratic form must converge to zero in probability. The same must be true for $\|\hat{h}_n + V_{\theta_0}^{-1}\mathbb{G}_n\dot{m}_{\theta_0}\|$.   ∎

The assertions of the preceding theorems must be in agreement with each other and also with the informal derivation leading to (5.20). If $\theta \mapsto m_\theta(x)$ is differentiable, then a maximizer of $\theta \mapsto \mathbb{P}_n m_\theta$ typically solves $\mathbb{P}_n \psi_\theta = 0$ for $\psi_\theta = \dot{m}_\theta$. Then the theorems and (5.20) are in agreement provided that

$$\dot{V}_\theta = \frac{\partial^2}{\partial\theta^2}Pm_\theta = \frac{\partial}{\partial\theta}P\psi_\theta = P\dot{\psi}_\theta = P\dot{m}_\theta.$$

This involves changing the order of differentiation (with respect to θ) and integration (with respect to x), and is usually permitted. However, for instance, the second derivative of Pm_θ may exist without $\theta \mapsto m_\theta(x)$ being differentiable for all x, as is seen in the following example.

5.24   *Example (Median).* The sample median maximizes the criterion function $\theta \mapsto -\sum_{i=1}^n |X_i - \theta|$. Assume that the distribution function F of the observations is differentiable

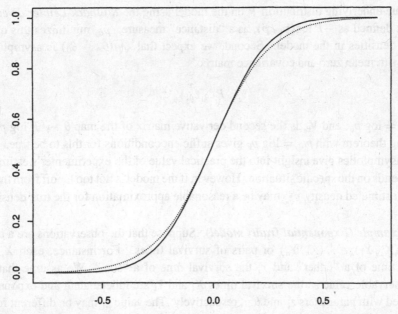

Figure 5.3. The distribution function of the sample median (dotted curve) and its normal approximation for a sample of size 25 from the Laplace distribution.

at its median θ_0 with positive derivative $f(\theta_0)$. Then the sample median is asymptotically normal.

This follows from Theorem 5.23 applied with $m_\theta(x) = |x - \theta| - |x|$. As a consequence of the triangle inequality, this function satisfies the Lipschitz condition with $\dot{m}(x) \equiv 1$. Furthermore, the map $\theta \mapsto m_\theta(x)$ is differentiable at θ_0 except if $x = \theta_0$, with $\dot{m}_{\theta_0}(x) = -\text{sign}(x - \theta_0)$. By partial integration,

$$Pm_\theta = \theta F(0) + \int_{(0,\theta]} (\theta - 2x) \, dF(x) - \theta\big(1 - F(\theta)\big) = 2 \int_0^\theta F(x) \, dx - \theta.$$

If F is sufficiently regular around θ_0, then Pm_θ is twice differentiable with first derivative $2F(\theta) - 1$ (which vanishes at θ_0) and second derivative $2f(\theta)$. More generally, under the minimal condition that F is differentiable at θ_0, the function Pm_θ has a Taylor expansion $Pm_{\theta_0} + \frac{1}{2}(\theta - \theta_0)^2 2f(\theta_0) + o(|\theta - \theta_0|^2)$, so that we set $V_{\theta_0} = 2f(\theta_0)$. Because $P\dot{m}_{\theta_0}^2 = \text{E}1 = 1$, the asymptotic variance of the median is $1/\big(2f(\theta_0)\big)^2$. Figure 5.3 gives an impression of the accuracy of the approximation. $\square$

5.25 Example (Misspecified model). Suppose an experimenter postulates a model $\{p_\theta : \theta \in \Theta\}$ for a sample of observations $X_1, \ldots, X_n$. However, the model is misspecified in that the true underlying distribution does not belong to the model. The experimenter decides to use the postulated model anyway, and obtains an estimate $\hat{\theta}_n$ from maximizing the likelihood $\sum \log p_\theta(X_i)$. What is the asymptotic behaviour of $\hat{\theta}_n$?

At first sight, it might appear that $\hat{\theta}_n$ would behave erratically due to the use of the wrong model. However, this is not the case. First, we expect that $\hat{\theta}_n$ is asymptotically consistent for a value θ_0 that maximizes the function $\theta \mapsto P \log p_\theta$, where the expectation is taken under the true underlying distribution P. The density p_{θ_0} can be viewed as the "projection" of the true underlying distribution P on the model using the *Kullback-Leibler divergence*, which is defined as $-P \log(p_\theta/p)$, as a "distance" measure: p_{θ_0} minimizes this quantity over all densities in the model. Second, we expect that $\sqrt{n}(\hat{\theta}_n - \theta_0)$ is asymptotically normal with mean zero and covariance matrix

$$V_{\theta_0}^{-1} P \dot{\ell}_{\theta_0} \dot{\ell}_{\theta_0}^T V_{\theta_0}^{-1}.$$

Here $\ell_\theta = \log p_\theta$, and V_{θ_0} is the second derivative matrix of the map $\theta \mapsto P \log p_\theta$. The preceding theorem with $m_\theta = \log p_\theta$ gives sufficient conditions for this to be true.

The asymptotics give insight into the practical value of the experimenter's estimate $\hat{\theta}_n$. This depends on the specific situation. However, if the model is not too far off from the truth, then the estimated density $p_{\hat{\theta}_n}$ may be a reasonable approximation for the true density. □

5.26 Example (Exponential frailty model). Suppose that the observations are a random sample $(X_1, Y_1), \ldots, (X_n, Y_n)$ of pairs of survival times. For instance, each X_i is the survival time of a "father" and Y_i the survival time of a "son." We assume that given an unobservable value z_i, the survival times X_i and Y_i are independent and exponentially distributed with parameters z_i and θz_i, respectively. The value z_i may be different for each observation. The problem is to estimate the ratio θ of the parameters.

To fit this example into the i.i.d. set-up of this chapter, we assume that the values $z_1, \ldots, z_n$ are realizations of a random sample $Z_1, \ldots, Z_n$ from some given distribution (that we do not have to know or parametrize).

One approach is based on the sufficiency of the variable $X_i + \theta Y_i$ for z_i in the case that θ is known. Given $Z_i = z$, this "statistic" possesses the gamma-distribution with shape parameter 2 and scale parameter z. Corresponding to this, the conditional density of an observation (X, Y) factorizes, for a given z, as $h_\theta(x, y) g_\theta(x + \theta y \mid z)$, for $g_\theta(s \mid z) = z^2 s e^{-zs}$ the gamma-density and

$$h_\theta(x, y) = \frac{\theta}{x + \theta y}.$$

Because the density of $X_i + \theta Y_i$ depends on the unobservable value z_i, we might wish to discard the factor $g_\theta(s \mid z)$ from the likelihood and use the factor $h_\theta(x, y)$ only. Unfortunately, this "conditional likelihood" does not behave as an ordinary likelihood, in that the corresponding "conditional likelihood equation," based on the function $\dot{h}_\theta/h_\theta(x, y) = \partial/\partial\theta \log h_\theta(x, y)$, does not have mean zero under θ. The bias can be corrected by conditioning on the sufficient statistic. Let

$$\psi_\theta(X, Y) = 2\theta \frac{\dot{h}_\theta}{h_\theta}(X, Y) - 2\theta \mathrm{E}_\theta\left(\frac{\dot{h}_\theta}{h_\theta}(X, Y) \mid X + \theta Y\right) = \frac{X - \theta Y}{X + \theta Y}.$$

Next define an estimator $\hat{\theta}_n$ as the solution of $\mathbb{P}_n \psi_\theta = 0$.

This works fairly nicely. Because the function $\theta \mapsto \psi_\theta(x, y)$ is continuous, and decreases strictly from 1 to -1 on $(0, \infty)$ for every $x, y > 0$, the equation $\mathbb{P}_n \psi_\theta = 0$ has a unique solution. The sequence of solutions $\hat{\theta}_n$ can be seen to be consistent by Lemma 5.10. By straightforward calculation, as $\theta \to \theta_0$,

$$P_{\theta_0} \psi_\theta = -\frac{\theta + \theta_0}{\theta - \theta_0} - \frac{2\theta\theta_0}{(\theta - \theta_0)^2} \log \frac{\theta_0}{\theta} = \frac{1}{3\theta_0}(\theta_0 - \theta) + o(\theta_0 - \theta).$$

Hence the zero of $\theta \mapsto P_{\theta_0} \psi_\theta$ is taken uniquely at $\theta = \theta_0$. Next, the sequence $\sqrt{n}(\hat{\theta}_n - \theta_0)$ can be shown to be asymptotically normal by Theorem 5.21. In fact, the functions $\dot{\psi}_\theta(x, y)$ are uniformly bounded in x, $y > 0$ and θ ranging over compacta in $(0, \infty)$, so that, by the mean value theorem, the function $\dot{\psi}$ in this theorem may be taken equal to a constant.

On the other hand, although this estimator is easy to compute, it can be shown that it is not asymptotically optimal. In Chapter 25 on semiparametric models, we discuss estimators with a smaller asymptotic variance. $\square$

5.27 *Example (Nonlinear least squares).* Suppose that we observe a random sample $(X_1, Y_1), \ldots, (X_n, Y_n)$ from the distribution of a vector (X, Y) that follows the regression model

$$Y = f_{\theta_0}(X) + e, \qquad E(e \mid X) = 0.$$

Here f_θ is a parametric family of regression functions, for instance $f_\theta(x) = \theta_1 + \theta_2 e^{\theta_3 x}$, and we aim at estimating the unknown vector θ. (We assume that the independent variables are a random sample in order to fit the example in our i.i.d. notation, but the analysis could be carried out conditionally as well.) The least squares estimator that minimizes

$$\theta \mapsto \sum_{i=1}^n \left(Y_i - f_\theta(X_i)\right)^2$$

is an M-estimator for $m_\theta(x, y) = \left(y - f_\theta(x)\right)^2$ (or rather minus this function). It should be expected to converge to the minimizer of the limit criterion function

$$\theta \mapsto P m_\theta = P(f_{\theta_0} - f_\theta)^2 + E e^2.$$

Thus the least squares estimator should be consistent if θ_0 is identifiable from the model, in the sense that $\theta \neq \theta_0$ implies that $f_\theta(X) \neq f_{\theta_0}(X)$ with positive probability.

For sufficiently regular regression models, we have

$$P m_\theta \approx P\left((\theta - \theta_0)^T \dot{f}_{\theta_0}\right)^2 + E e^2.$$

This suggests that the conditions of Theorem 5.23 are satisfied with $V_{\theta_0} = 2P \dot{f}_{\theta_0} \dot{f}_{\theta_0}^T$ and $\dot{m}_{\theta_0}(x, y) = -2\left(y - f_{\theta_0}(x)\right)\dot{f}_{\theta_0}(x)$. If e and X are independent, then this leads to the asymptotic covariance matrix $V_{\theta_0}^{-1} 2E e^2$. $\square$

Besides giving the asymptotic normality of $\sqrt{n}(\hat{\theta}_n - \theta_0)$, the preceding theorems give an asymptotic representation

$$\hat{\theta}_n = \theta_0 + \frac{1}{n} \sum_{i=1}^n V_{\theta_0}^{-1} \psi_{\theta_0}(X_i) + o_P\left(\frac{1}{\sqrt{n}}\right).$$

If we neglect the remainder term,[†] then this means that $\hat{\theta}_n - \theta_0$ behaves as the average of the variables $V_{\theta_0}^{-1} \psi_{\theta_0}(X_i)$. Then the (asymptotic) "influence" of the nth observation on the

---

[†] To make the following derivation rigorous, more information concerning the remainder term would be necessary.

value of $\hat{\theta}_n$ can be computed as

$$\hat{\theta}_n(X_1, \ldots, X_n) - \hat{\theta}_{n-1}(X_1, \ldots, X_{n-1}) \approx \frac{1}{n} V_{\theta_0}^{-1} \psi_{\theta_0}(X_n) - \frac{1}{n(n-1)} \sum_{i=1}^{n-1} V_{\theta_0}^{-1} \psi_{\theta_0}(X_i)$$

$$= \frac{1}{n} V_{\theta_0}^{-1} \psi_{\theta_0}(X_n) + o_P\left(\frac{1}{n}\right).$$

Because the "influence" of an extra observation x is proportional to $V_\theta^{-1} \psi_\theta(x)$, the function $x \mapsto V_\theta^{-1} \psi_\theta(x)$ is called the *asymptotic influence function* of the estimator $\hat{\theta}_n$. Influence functions can be defined for many other estimators as well, but the method of Z-estimation is particularly convenient to obtain estimators with given influence functions. Because V_{θ_0} is a constant (matrix), any shape of influence function can be obtained by simply choosing the right functions ψ_θ.

For the purpose of robust estimation, perhaps the most important aim is to bound the influence of each individual observation. Thus, a Z-estimator is called *B-robust* if the function ψ_θ is bounded.

5.28   *Example (Robust regression).* Consider a random sample of observations (X_1, Y_1), $\ldots, (X_n, Y_n)$ following the linear regression model

$$Y_i = \theta_0^T X_i + e_i,$$

for i.i.d. errors $e_1, \ldots, e_n$ that are independent of $X_1, \ldots, X_n$. The classical estimator for the regression parameter θ is the least squares estimator, which minimizes $\sum_{i=1}^n (Y_i - \theta^T X_i)^2$. Outlying values of X_i ("leverage points") or extreme values of (X_i, Y_i) jointly ("influence points") can have an arbitrarily large influence on the value of the least-squares estimator, which therefore is nonrobust. As in the case of location estimators, a more robust estimator for θ can be obtained by replacing the square by a function $m(x)$ that grows less rapidly as $x \to \infty$, for instance $m(x) = |x|$ or $m(x)$ equal to the primitive function of Huber's ψ. Usually, minimizing an expression of the type $\sum_{i=1}^n m(Y_i - \theta X_i)$ is equivalent to solving a system of equations

$$\sum_{i=1}^n \psi(Y_i - \theta^T X_i) X_i = 0.$$

Because $E\psi(Y - \theta_0^T X)X = E\psi(e)EX$, we can expect the resulting estimator to be consistent provided $E\psi(e) = 0$. Furthermore, we should expect that, for $V_{\theta_0} = E\psi'(e)XX^T$,

$$\sqrt{n}(\hat{\theta}_n - \theta_0) = \frac{1}{\sqrt{n}} V_{\theta_0}^{-1} \sum_{i=1}^n \psi(Y_i - \theta_0^T X_i) X_i + o_P(1).$$

Consequently, even for a bounded function ψ, the influence function $(x, y) \mapsto V_\theta^{-1} \psi(y - \theta^T x)x$ may be unbounded, and an extreme value of an X_i may still have an arbitrarily large influence on the estimate (asymptotically). Thus, the estimators obtained in this way are protected against influence points but may still suffer from leverage points and hence are only partly robust. To obtain fully robust estimators, we can change the estimating equations to

$$\sum_{i=1}^{n} \psi\big((Y_i - \theta^T X_i) v(X_i)\big) w(X_i) = 0.$$

Here we protect against leverage points by choosing w bounded. For more flexibility we have also allowed a weighting factor $v(X_i)$ inside ψ. The choices $\psi(x) = x$, $v(x) = 1$ and $w(x) = x$ correspond to the (nonrobust) least-squares estimator.

The solution $\hat{\theta}_n$ of our final estimating equation should be expected to be consistent for the solution of

$$0 = \mathrm{E}\psi\big((Y - \theta^T X) v(X)\big) w(X) = \mathrm{E}\psi\Big((e + \theta_0^T X - \theta^T X) v(X)\Big) w(X).$$

If the function ψ is odd and the error symmetric, then the true value θ_0 will be a solution whenever e is symmetric about zero, because then $\mathrm{E}\psi(e\sigma) = 0$ for every σ.

Precise conditions for the asymptotic normality of $\sqrt{n}(\hat{\theta}_n - \theta_0)$ can be obtained from Theorems 5.21 and 5.9. The verification of the conditions of Theorem 5.21, which are "local" in nature, is relatively easy, and, if necessary, the Lipschitz condition can be relaxed by using results on empirical processes introduced in Chapter 19 directly. Perhaps proving the consistency of $\hat{\theta}_n$ is harder. The biggest technical problem may be to show that $\hat{\theta}_n = O_P(1)$, so it would help if θ could a priori be restricted to a bounded set. On the other hand, for bounded functions ψ, the case of most interest in the present context, the functions $(x, y) \mapsto \psi\big((y - \theta^T x) v(x)\big) w(x)$ readily form a Glivenko-Cantelli class when θ ranges freely, so that verification of the strong uniqueness of θ_0 as a zero becomes the main challenge when applying Theorem 5.9. This leads to a combination of conditions on ψ, v, w, and the distributions of e and X. $\square$

5.29 *Example (Optimal robust estimators).* Every sufficiently regular function ψ defines a location estimator $\hat{\theta}_n$ through the equation $\sum_{i=1}^{n} \psi(X_i - \theta) = 0$. In order to choose among the different estimators, we could compare their asymptotic variances and use the one with the smallest variance under the postulated (or estimated) distribution P of the observations. On the other hand, if we also wish to guard against extreme obervations, then we should find a balance between robustness and asymptotic variance. One possibility is to use the estimator with the smallest asymptotic variance at the postulated, ideal distribution P under the side condition that its influence function be uniformly bounded by some constant c. In this example we show that for P the normal distribution, this leads to the Huber estimator.

The Z-estimator is consistent for the solution θ_0 of the equation $P\psi(\cdot - \theta) = \mathrm{E}\psi(X_1 - \theta) = 0$. Suppose that we fix an underlying, ideal P whose "location" θ_0 is zero. Then the problem is to find ψ that minimizes the asymptotic variance $P\psi^2/(P\psi')^2$ under the two side conditions, for a given constant c,

$$\sup_x \left| \frac{\psi(x)}{P\psi'} \right| \leq c, \quad \text{and} \quad P\psi = 0.$$

The problem is homogeneous in ψ, and hence we may assume that $P\psi' = 1$ without loss of generality. Next, minimization of $P\psi^2$ under the side conditions $P\psi = 0$, $P\psi' = 1$ and $\|\psi\|_\infty \leq c$ can be achieved by using Lagrange multipliers, as in problem 14.6 This leads to minimizing

$$P\psi^2 + \lambda P\psi + \mu(P\psi' - 1) = P\Big(\psi^2 + \psi\big(\lambda + \mu(p'/p)\big) - \mu\Big)$$

for fixed "multipliers" λ and μ under the side condition $\|\psi\|_\infty \le c$ with respect to ψ. This expectation is minimized by minimizing the integrand pointwise, for every fixed x. Thus the minimizing ψ has the property that, for every x separately, $y = \psi(x)$ minimizes the parabola $y^2 + \lambda y + \mu y (p'/p)(x)$ over $y \in [-c, c]$. This readily gives the solution, with $[y]_c^d$ the value y truncated to the interval $[c, d]$,

$$\psi(x) = \left[-\frac{1}{2}\lambda - \frac{1}{2}\mu \frac{p'}{p}(x) \right]_{-c}^c.$$

The constants λ and μ can be solved from the side conditions $P\psi = 0$ and $P\psi' = 1$. The normal distribution $P = \Phi$ has location score function $p'/p(x) = -x$, and by symmetry it follows that $\lambda = 0$ in this case. Then the optimal ψ reduces to Huber's ψ function. $\square$

*5.4 Estimated Parameters

In many situations, the estimating equations for the parameters of interest contain preliminary estimates for "nuisance parameters." For example, many robust location estimators are defined as the solutions of equations of the type

$$\sum_{i=1}^n \psi\left(\frac{X_i - \theta}{\hat{\sigma}} \right) = 0. \tag{5.30}$$

Here $\hat{\sigma}$ is an initial (robust) estimator of scale, which is meant to stabilize the robustness of the location estimator. For instance, the "cut-off" parameter k in Huber's ψ-function determines the amount of robustness of Huber's estimator, but the effect of a particular choice of k on bounding the influence of outlying observations is relative to the range of the observations. If the observations are concentrated in the interval $[-k, k]$, then Huber's ψ yields nothing else but the sample mean, if all observations are outside $[-k, k]$, we get the median. Scaling the observations to a standard scale gives a clear meaning to the value of k. The use of the *median absolute deviation from the median* (see. section 21.3) is often recommended for this purpose.

If the scale estimator is itself a Z-estimator, then we can treat the pair $(\hat{\theta}, \hat{\sigma})$ as a Z-estimator for a system of equations, and next apply the preceding theorems. More generally, we can apply the following result. In this subsection we allow a condition in terms of Donsker classes, which are discussed in Chapter 19. The proof of the following theorem follows the same steps as the proof of Theorem 5.21.

5.31 Theorem. *For each θ in an open subset of $\mathbb{R}^k$ and each η in a metric space, let $x \mapsto \psi_{\theta,\eta}(x)$ be an $\mathbb{R}^k$-valued measurable function such that the class of functions $\{\psi_{\theta,\eta} : \|\theta - \theta_0\| < \delta, d(\eta, \eta_o) < \delta\}$ is Donsker for some $\delta > 0$, and such that $P\|\psi_{\theta,\eta} - \psi_{\theta_0,\eta_0}\|^2 \to 0$ as $(\theta, \eta) \to (\theta_0, \eta_0)$. Assume that $P\psi_{\theta_0,\eta_0} = 0$, and that the maps $\theta \mapsto P\psi_{\theta,\eta}$ are differentiable at θ_0, uniformly in η in a neighborhood of η_0 with nonsingular derivative matrices $V_{\theta_0,\eta}$ such that $V_{\theta_0,\eta} \to V_{\theta_0,\eta_0}$. If $\sqrt{n}\,\mathbb{P}_n \psi_{\hat{\theta}_n, \hat{\eta}_n} = o_P(1)$ and $(\hat{\theta}_n, \hat{\eta}_n) \xrightarrow{P} (\theta_0, \eta_0)$, then*

$$\sqrt{n}(\hat{\theta}_n - \theta_0) = -V_{\theta_0,\eta_0}^{-1} \sqrt{n}\, P\psi_{\theta_0, \hat{\eta}_n} - V_{\theta_0,\eta_0}^{-1} \mathbb{G}_n \psi_{\theta_0,\eta_0} + o_P(1 + \sqrt{n}\|P\psi_{\theta_0, \hat{\eta}_n}\|).$$

Under the conditions of this theorem, the limiting distribution of the sequence $\sqrt{n}(\hat{\theta}_n - \theta_0)$ depends on the estimator $\hat{\eta}_n$ through the "drift" term $\sqrt{n} P \psi_{\theta_0, \hat{\eta}_n}$. In general, this gives a contribution to the limiting distribution, and $\hat{\eta}_n$ must be chosen with care. If $\hat{\eta}_n$ is $\sqrt{n}$-consistent and the map $\eta \mapsto P \psi_{\theta_0, \eta}$ is differentiable, then the drift term can be analyzed using the delta-method.

It may happen that the drift term is zero. If the parameters θ and η are "orthogonal" in this sense, then the auxiliary estimators $\hat{\eta}_n$ may converge at an arbitrarily slow rate and affect the limit distribution of $\hat{\theta}_n$ only through their limiting value η_0.

5.32 *Example (Symmetric location).* Suppose that the distribution of the observations is symmetric about θ_0. Let $x \mapsto \psi(x)$ be an antisymmetric function, and consider the Z-estimators that solve equation (5.30). Because $P\psi\big((X - \theta_0)/\sigma\big) = 0$ for every σ, by the symmetry of P and the antisymmetry of ψ, the "drift term" due to $\hat{\eta}$ in the preceding theorem is identically zero. The estimator $\hat{\theta}_n$ has the same limiting distribution whether we use an arbitrary consistent estimator of a "true scale" σ_0 or σ_0 itself. □

5.33 *Example (Robust regression).* In the linear regression model considered in Example 5.28, suppose that we choose the weight functions v and w dependent on the data and solve the robust estimator $\hat{\theta}_n$ of the regression parameters from

$$0 = \frac{1}{n} \sum_{i=1}^{n} \psi\big((Y_i - \theta^T X_i)\hat{v}_n(X_i)\big)\hat{w}_n(X_i).$$

This corresponds to defining a nuisance parameter $\eta = (v, w)$ and setting $\psi_{\theta, v, w}(x, y) = \psi\big((y - \theta^T x)v(x)\big)w(x)$. If the functions $\psi_{\theta, v, w}$ run through a Donsker class (and they easily do), and are continuous in (θ, v, w), and the map $\theta \mapsto P\psi_{\theta, v, w}$ is differentiable at θ_0 uniformly in (v, w), then the preceding theorem applies. If $\mathrm{E}\psi(e\sigma) = 0$ for every σ, then $P\psi_{\theta_0, v, w} = 0$ for any v and w, and the limit distribution of $\sqrt{n}(\hat{\theta}_n - \theta_0)$ is the same, whether we use the random weight functions $(\hat{v}_n, \hat{w}_n)$ or their limit (v_0, w_0) (assuming that this exists).

The purpose of using random weight functions could be, besides stabilizing the robustness, to improve the asymptotic efficiency of $\hat{\theta}_n$. The limit (v_0, w_0) typically is not the same for every underlying distribution P, and the estimators $(\hat{v}_n, \hat{w}_n)$ can be chosen in such a way that the asymptotic variance is minimal. □

5.5 Maximum Likelihood Estimators

Maximum likelihood estimators are examples of M-estimators. In this section we specialize the consistency and the asymptotic normality results of the preceding sections to this important special case. Our approach reverses the historical order. Maximum likelihood estimators were shown to be asymptotically normal first by Fisher in the 1920s and rigorously by Cramér, among others, in the 1940s. General M-estimators were not introduced and studied systematically until the 1960s, when they became essential in the development of robust estimators.

If $X_1, \ldots, X_n$ are a random sample from a density p_θ, then the maximum likelihood estimator $\hat{\theta}_n$ maximizes the function $\theta \mapsto \sum \log p_\theta(X_i)$, or equivalently, the function

$$M_n(\theta) = \frac{1}{n}\sum_{i=1}^{n} \log \frac{p_\theta}{p_{\theta_0}}(X_i) = \mathbb{P}_n \log \frac{p_\theta}{p_{\theta_0}}.$$

(Subtraction of the "constant" $\sum \log p_{\theta_0}(X_i)$ turns out to be mathematically convenient.) If we agree that $\log 0 = -\infty$, then this expression is with probability 1 well defined if p_{θ_0} is the true density. The asymptotic function corresponding to M_n is[†]

$$M(\theta) = \mathrm{E}_{\theta_0} \log \frac{p_\theta}{p_{\theta_0}}(X) = P_{\theta_0} \log \frac{p_\theta}{p_{\theta_0}}.$$

The number $-M(\theta)$ is called the *Kullback-Leibler divergence* of p_θ and p_{θ_0}; it is often considered a measure of "distance" between p_θ and p_{θ_0}, although it does not have the properties of a mathematical distance. Based on the results of the previous sections, we may expect the maximum likelihood estimator to converge to a point of maximum of $M(\theta)$. Is the true value θ_0 always a point of maximum? The answer is affirmative, and, moreover, the true value is a unique point of maximum if the true measure is *identifiable*:

$$P_\theta \neq P_{\theta_0}, \qquad \text{every } \theta \neq \theta_0. \tag{5.34}$$

This requires that the model for the observations is not the same under the parameters θ and θ_0. Identifiability is a natural and even a necessary condition: If the parameter is not identifiable, then consistent estimators cannot exist.

5.35 Lemma. *Let $\{p_\theta : \theta \in \Theta\}$ be a collection of subprobability densities such that (5.34) holds and such that P_{θ_0} is a probability measure. Then $M(\theta) = P_{\theta_0} \log p_\theta/p_{\theta_0}$ attains its maximum uniquely at θ_0.*

Proof. First note that $M(\theta_0) = P_{\theta_0} \log 1 = 0$. Hence we wish to show that $M(\theta)$ is strictly negative for $\theta \neq \theta_0$.

Because $\log x \leq 2(\sqrt{x} - 1)$ for every $x \geq 0$, we have, writing μ for the dominating measure,

$$P_{\theta_0} \log \frac{p_\theta}{p_{\theta_0}} \leq 2 P_{\theta_0}\left(\sqrt{\frac{p_\theta}{p_{\theta_0}}} - 1\right) = 2\int \sqrt{p_\theta p_{\theta_0}}\, d\mu - 2$$

$$\leq -\int (\sqrt{p_\theta} - \sqrt{p_{\theta_0}})^2\, d\mu.$$

(The last inequality is an equality if $\int p_\theta\, d\mu = 1$.) This is always nonpositive, and is zero only if p_θ and p_{θ_0} are equal. By assumption the latter happens only if $\theta = \theta_0$. ∎

Thus, under conditions such as in section 5.2 and identifiability, the sequence of maximum likelihood estimators is consistent for the true parameter.

[†] Presently we take the expectation P_{θ_0} under the parameter θ_0, whereas the derivation in section 5.3 is valid for a generic underlying probability structure and does not conceptually require that the set of parameters θ indexes a set of underlying distributions.

This conclusion is derived from viewing the maximum likelihood estimator as an M-estimator for $m_\theta = \log p_\theta$. Sometimes it is technically advantageous to use a different starting point. For instance, consider the function

$$m_\theta = \log \frac{p_\theta + p_{\theta_0}}{2p_{\theta_0}}.$$

By the concavity of the logarithm, the maximum likelihood estimator $\hat{\theta}$ satisfies

$$\mathbb{P}_n m_{\hat{\theta}} \geq \mathbb{P}_n \frac{1}{2} \log \frac{p_{\hat{\theta}}}{p_{\theta_0}} + \mathbb{P}_n \frac{1}{2} \log 1 \geq 0 = \mathbb{P}_n m_{\theta_0}.$$

Even though $\hat{\theta}$ does not maximize $\theta \mapsto \mathbb{P}_n m_\theta$, this inequality can be used as the starting point for a consistency proof, since Theorem 5.7 requires that $M_n(\hat{\theta}) \geq M_n(\theta_0) - o_P(1)$ only. The true parameter is still identifiable from this criterion function, because, by the preceding lemma, $P_{\theta_0} m_\theta = 0$ implies that $(p_\theta + p_{\theta_0})/2 = p_{\theta_0}$, or $p_\theta = p_{\theta_0}$. A technical advantage is that $m_\theta \geq \log(1/2)$. For another variation, see Example 5.17.

Consider asymptotic normality. The maximum likelihood estimator solves the likelihood equations

$$\frac{\partial}{\partial \theta} \sum_{i=1}^{n} \log p_\theta(X_i) = 0.$$

Hence it is a Z-estimator for ψ_θ equal to the *score function* $\dot{\ell}_\theta = \partial/\partial\theta \log p_\theta$ of the model. In view of the results of section 5.3, we expect that the sequence $\sqrt{n}(\hat{\theta}_n - \theta)$ is, under θ, asymptotically normal with mean zero and covariance matrix

$$(P_\theta \ddot{\ell}_\theta)^{-1} P_\theta \dot{\ell}_\theta \dot{\ell}_\theta^T (P_\theta \ddot{\ell}_\theta^T)^{-1}. \tag{5.36}$$

Under regularity conditions, this reduces to the inverse of the Fisher information matrix

$$I_\theta = P_\theta \dot{\ell}_\theta \dot{\ell}_\theta^T.$$

To see this in the case of a one-dimensional parameter, differentiate the identity $\int p_\theta \, d\mu \equiv 1$ twice with respect to θ. Assuming that the order of differentiation and integration can be reversed, we obtain $\int \dot{p}_\theta \, d\mu \equiv \int \ddot{p}_\theta \, d\mu \equiv 0$. Together with the identities

$$\dot{\ell}_\theta = \frac{\dot{p}_\theta}{p_\theta}; \qquad \ddot{\ell}_\theta = \frac{\ddot{p}_\theta}{p_\theta} - \left(\frac{\dot{p}_\theta}{p_\theta}\right)^2,$$

this implies that $P_\theta \dot{\ell}_\theta = 0$ (scores have mean zero), and $P_\theta \ddot{\ell}_\theta = -I_\theta$ (the curvature of the likelihood is equal to minus the Fisher information). Consequently, (5.36) reduces to I_θ^{-1}. The higher-dimensional case follows in the same way, in which we should interpret the identities $P_\theta \dot{\ell}_\theta = 0$ and $P_\theta \ddot{\ell}_\theta = -I_\theta$ as a vector and a matrix identity, respectively.

We conclude that maximum likelihood estimators typically satisfy

$$\sqrt{n}(\hat{\theta}_n - \theta) \overset{\theta}{\rightsquigarrow} N(0, I_\theta^{-1}).$$

This is a very important result, as it implies that maximum likelihood estimators are asymptotically optimal. The convergence in distribution means roughly that the maximum likelihood estimator $\hat{\theta}_n$ is $N(\theta, (nI_\theta)^{-1})$-distributed for every θ, for large n. Hence, it is asymptotically unbiased and asymptotically of variance $(nI_\theta)^{-1}$. According to the Cramér-Rao theorem, the variance of an unbiased estimator is at least $(nI_\theta)^{-1}$. Thus, we could infer that the maximum likelihood estimator is asymptotically uniformly minimum-variance unbiased, and in this sense optimal. We write "could" because the preceding reasoning is informal and unsatisfying. The asymptotic normality does not warrant any conclusion about the convergence of the moments $E_\theta \hat{\theta}_n$ and $var_\theta \hat{\theta}_n$; we have not introduced an asymptotic version of the Cramér-Rao theorem; and the Cramér-Rao bound does not make any assertion concerning asymptotic normality. Moreover, the unbiasedness required by the Cramér-Rao theorem is restrictive and can be relaxed considerably in the asymptotic situation.

However, the message that maximum likelihood estimators are *asymptotically efficient* is correct. We give a precise discussion in Chapter 8. The justification through asymptotics appears to be the only general justification of the method of maximum likelihood. In some form, this result was found by Fisher in the 1920s, but a better and more general insight was only obtained in the period from 1950 through 1970 through the work of Le Cam and others.

In the preceding informal derivations and discussion, it is implicitly understood that the density p_θ possesses at least two derivatives with respect to the parameter. Although this can be relaxed considerably, a certain amount of smoothness of the dependence $\theta \mapsto p_\theta$ is essential for the asymptotic normality. Compare the behavior of the maximum likelihood estimators in the case of uniformly distributed observations: They are neither asymptotically normal nor asymptotically optimal.

5.37   Example (Uniform distribution). Let $X_1, \ldots, X_n$ be a sample from the uniform distribution on $[0, \theta]$. Then the maximum likelihood estimator is the maximum $X_{(n)}$ of the observations. Because the variance of $X_{(n)}$ is of the order $O(n^{-2})$, we expect that a suitable norming rate in this case is not $\sqrt{n}$, but n. Indeed, for each $x < 0$

$$P_\theta\big(n(X_{(n)} - \theta) \leq x\big) = P_\theta\left(X_1 \leq \theta + \frac{x}{n}\right)^n = \left(\frac{\theta + x/n}{\theta}\right)^n \to e^{x/\theta}.$$

Thus, the sequence $-n(X_{(n)} - \theta)$ converges in distribution to an exponential distribution with mean θ. Consequently, the sequence $\sqrt{n}(X_{(n)} - \theta)$ converges to zero in probability.

Note that most of the informal operations in the preceding introduction are illegal or not even defined for the uniform distribution, starting with the definition of the likelihood equations. The informal conclusion that the maximum likelihood estimator is asymptotically optimal is also wrong in this case; see section 9.4. $\square$

We conclude this section with a theorem that establishes the asymptotic normality of maximum likelihood estimators rigorously. Clearly, the asymptotic normality follows from Theorem 5.23 applied to $m_\theta = \log p_\theta$, or from Theorem 5.21 applied with $\psi_\theta = \dot{\ell}_\theta$ equal to the score function of the model. The following result is a minor variation on the first theorem. Its conditions somehow also ensure the relationship $P_\theta \ddot{\ell}_\theta = -I_\theta$ and the twice-differentiability of the map $\theta \mapsto P_{\theta_0} \log p_\theta$, even though the existence of second derivatives is not part of the assumptions. This remarkable phenomenon results from the trivial fact that square roots of probability densities have squares that integrate to 1. To exploit this, we require the differentiability of the maps $\theta \mapsto \sqrt{p_\theta}$, rather than of the maps $\theta \mapsto \log p_\theta$. A statistical model $(P_\theta : \theta \in \Theta)$ is called *differentiable in quadratic mean* if there exists a measurable vector-valued function $\dot{\ell}_{\theta_0}$ such that, as $\theta \to \theta_0$,

$$\int \left[\sqrt{p_\theta} - \sqrt{p_{\theta_0}} - \frac{1}{2}(\theta - \theta_0)^T \dot{\ell}_{\theta_0} \sqrt{p_{\theta_0}} \right]^2 d\mu = o\big(\|\theta - \theta_0\|^2 \big), \qquad (5.38)$$

This property also plays an important role in asymptotic optimality theory. A discussion, including simple conditions for its validity, is given in Chapter 7. It should be noted that

$$\frac{\partial}{\partial \theta} \sqrt{p_\theta} = \frac{1}{2\sqrt{p_\theta}} \frac{\partial}{\partial \theta} p_\theta = \frac{1}{2} \left(\frac{\partial}{\partial \theta} \log p_\theta \right) \sqrt{p_\theta}.$$

Thus, the function $\dot{\ell}_{\theta_0}$ in the integral really is the score function of the model (as the notation suggests), and the expression $I_{\theta_0} = P_{\theta_0} \dot{\ell}_{\theta_0} \dot{\ell}_{\theta_0}^T$ defines the Fisher information matrix. However, condition (5.38) does not require existence of $\partial/\partial\theta \, p_\theta(x)$ for every x.

5.39   Theorem. *Suppose that the model $(P_\theta : \theta \in \Theta)$ is differentiable in quadratic mean at an inner point θ_0 of $\Theta \subset \mathbb{R}^k$. Furthermore, suppose that there exists a measurable function $\dot{\ell}$ with $P_{\theta_0} \dot{\ell}^2 < \infty$ such that, for every θ_1 and θ_2 in a neighborhood of θ_0,*

$$\big| \log p_{\theta_1}(x) - \log p_{\theta_2}(x) \big| \leq \dot{\ell}(x) \, \|\theta_1 - \theta_2\|.$$

If the Fisher information matrix I_{θ_0} is nonsingular and $\hat{\theta}_n$ is consistent, then

$$\sqrt{n}(\hat{\theta}_n - \theta_0) = I_{\theta_0}^{-1} \frac{1}{\sqrt{n}} \sum_{i=1}^{n} \dot{\ell}_{\theta_0}(X_i) + o_{P_{\theta_0}}(1).$$

In particular, the sequence $\sqrt{n}(\hat{\theta}_n - \theta_0)$ is asymptotically normal with mean zero and covariance matrix $I_{\theta_0}^{-1}$.

*Proof.   This theorem is a corollary of Theorem 5.23. We shall show that the conditions of the latter theorem are satisfied for $m_\theta = \log p_\theta$ and $V_{\theta_0} = -I_{\theta_0}$.

Fix an arbitrary converging sequence of vectors $h_n \to h$, and set

$$W_n = 2\left(\sqrt{\frac{p_{\theta_0 + h_n/\sqrt{n}}}{p_{\theta_0}}} - 1 \right).$$

By the differentiability in quadratic mean, the sequence $\sqrt{n} W_n$ converges in $L_2(P_{\theta_0})$ to the function $h^T \dot{\ell}_{\theta_0}$. In particular, it converges in probability, whence by a delta method

$$\sqrt{n}\big(\log p_{\theta_0 + h_n/\sqrt{n}} - \log p_{\theta_0} \big) = 2\sqrt{n} \log\big(1 + \tfrac{1}{2} W_n \big) \xrightarrow{P} h^T \dot{\ell}_{\theta_0}.$$

In view of the Lipschitz condition on the map $\theta \mapsto \log p_\theta$, we can apply the dominated-convergence theorem to strengthen this to convergence in $L_2(P_{\theta_0})$. This shows that the map $\theta \mapsto \log p_\theta$ is differentiable in probability, as required in Theorem 5.23. (The preceding argument considers only sequences θ_n of the special form $\theta_0 + h_n/\sqrt{n}$ approaching θ_0. Because h_n can be any converging sequence and $\sqrt{n+1}/\sqrt{n} \to 1$, these sequences are actually not so special. By re-indexing the result can be seen to be true for any $\theta_n \to \theta_0$.)

Next, by computing means (which are zero) and variances, we see that

$$\mathbb{G}_n \Big[\sqrt{n}\big(\log p_{\theta_0 + h_n/\sqrt{n}} - \log p_{\theta_0} \big) - h^T \dot{\ell}_{\theta_0} \Big] \xrightarrow{P} 0.$$

Equating this result to the expansion given by Theorem 7.2, we see that

$$n P_{\theta_0}\left(\log p_{\theta_0 + h_n/\sqrt{n}} - \log p_{\theta_0}\right) \to -\tfrac{1}{2}h^T I_{\theta_0} h.$$

Hence the map $\theta \mapsto P_{\theta_0} \log p_\theta$ is twice-differentiable with second derivative matrix $-I_{\theta_0}$, or at least permits the corresponding Taylor expansion of order 2. ∎

5.40 *Example (Binary regression).* Suppose that we observe a random sample $(X_1, Y_1), \ldots, (X_n, Y_n)$ consisting of k-dimensional vectors of "covariates" X_i, and 0-1 "response variables" Y_i, following the model

$$P_\theta(Y_i = 1 \mid X_i = x) = \Psi(\theta^T x).$$

Here $\Psi: \mathbb{R} \mapsto [0, 1]$ is a known continuously differentiable, monotone function. The choices $\Psi(\theta) = 1/(1 + e^{-\theta})$ (the logistic distribution function) and $\Psi = \Phi$ (the normal distribution function) correspond to the *logit model* and *probit model*, respectively. The maximum likelihood estimator $\hat{\theta}_n$ maximizes the (conditional) likelihood function

$$\theta \mapsto \prod_{i=1}^n p_\theta(Y_i \mid X_i) := \prod_{i=1}^n \Psi(\theta^T X_i)^{Y_i}\left(1 - \Psi(\theta^T X_i)\right)^{1-Y_i}.$$

The consistency and asymptotic normality of $\hat{\theta}_n$ can be proved, for instance, by combining Theorems 5.7 and 5.39. (Alternatively, we may follow the classical approach given in section 5.6. The latter is particularly attractive for the logit model, for which the log likelihood is strictly concave in θ, so that the point of maximum is unique.) For identifiability of θ we must assume that the distribution of the X_i is not concentrated on a $(k-1)$-dimensional affine subspace of $\mathbb{R}^k$. For simplicity we assume that the range of X_i is bounded.

The consistency can be proved by applying Theorem 5.7 with $m_\theta = \log(p_\theta + p_{\theta_0})/2$. Because p_{θ_0} is bounded away from 0 (and ∞), the function m_θ is somewhat better behaved than the function $\log p_\theta$.

By Lemma 5.35, the parameter θ is identifiable from the density p_θ. We can redo the proof to see that, with $\lesssim$ meaning "less than up to a constant,"

$$P_{\theta_0}(m_\theta - m_{\theta_0}) \lesssim -\int \left(\left(\tfrac{1}{2}(p_\theta + p_{\theta_0})\right)^{1/2} - p_{\theta_0}^{1/2}\right)^2 d\mu$$

$$\lesssim -\mathrm{E}\left(\Psi(\theta^T X) - \Psi(\theta_0^T X)\right)^2.$$

This shows that θ_0 is the unique point of maximum of $\theta \mapsto P_{\theta_0} m_\theta$. Furthermore, if $P_{\theta_0} m_{\theta_k} \to P_{\theta_0} m_{\theta_0}$, then $\theta_k^T X \xrightarrow{\mathrm{P}} \theta_0^T X$. If the sequence θ_k is also bounded, then $\mathrm{E}\left((\theta_k - \theta_0)^T X\right)^2 \to 0$, whence $\theta_k \mapsto \theta_0$ by the nonsingularity of the matrix $\mathrm{E} X X^T$. On the other hand, $\|\theta_k\|$ cannot have a diverging subsequence, because in that case $\theta_k^T/\|\theta_k\| X \xrightarrow{\mathrm{P}} 0$ and hence $\theta_k/\|\theta_k\| \to 0$ by the same argument. This verifies condition (5.8).

Checking the uniform convergence to zero of $\sup_\theta |\mathbb{P}_n m_\theta - P m_\theta|$ is not trivial, but it becomes an easy exercise if we employ the Glivenki-Cantelli theorem, as discussed in Chapter 19. The functions $x \mapsto \Psi(\theta^T x)$ form a VC-class, and the functions m_θ take the form $m_\theta(x, y) = \phi\left(\Psi(\theta^T x), y, \Psi(\theta_0^T x)\right)$, where the function $\phi(\gamma, y, \eta)$ is Lipschitz in its first argument with Lipschitz constant bounded above by $1/\eta + 1/(1-\eta)$. This is enough to ensure that the functions m_θ form a Donsker class and hence certainly a Glivenko-Cantelli class, in view of Example 19.20.

The asymptotic normality of $\sqrt{n}(\hat\theta_n - \theta)$ is now a consequence of Theorem 5.39. The score function

$$\dot\ell_\theta(y \mid x) = \frac{y - \Psi(\theta^T x)}{\Psi(\theta^T x)(1 - \Psi)(\theta^T x)} \Psi'(\theta^T x)x$$

is uniformly bounded in x, y and θ ranging over compacta, and continuous in θ for every x and y. The Fisher information matrix

$$I_\theta = \mathrm{E} \frac{\Psi'(\theta^T X)^2}{\Psi(\theta^T X)(1 - \Psi)(\theta^T X)} X X^T$$

is continuous in θ, and is bounded below by a multiple of $\mathrm{E} X X^T$ and hence is nonsingular. The differentiability in quadratic mean follows by calculus, or by Lemma 7.6. $\quad\square$

*5.6 Classical Conditions

In this section we discuss the "classical conditions" for asymptotic normality of M-estimators. These conditions were formulated in the 1930s and 1940s to make the informal derivations of the asymptotic normality of maximum likelihood estimators, for instance by Fisher, mathematically rigorous. Although Theorem 5.23 requires less than a first derivative of the criterion function, the "classical conditions" require existence of third derivatives. It is clear that the classical conditions are too stringent, but they are still of interest, because they are simple, lead to simple proofs, and nevertheless apply to many examples. The classical conditions also ensure existence of Z-estimators and have a little to say about their consistency.

We describe the classical approach for general Z-estimators and vector-valued parameters. The higher-dimensional case requires more skill in calculus and matrix algebra than is necessary for the one-dimensional case. When simplified to dimension one the arguments do not go much beyond making the informal derivation leading from (5.18) to (5.19) rigorous.

Let the observations $X_1, \ldots, X_n$ be a sample from a distribution P, and consider the estimating equations

$$\Psi_n(\theta) = \frac{1}{n} \sum_{i=1}^{n} \psi_\theta(X_i) = \mathbb{P}_n \psi_\theta, \qquad \Psi(\theta) = P \psi_\theta.$$

The estimator $\hat\theta_n$ is a zero of Ψ_n, and the true value θ_0 a zero of Ψ. The essential condition of the following theorem is that the second-order partial derivatives of $\psi_\theta(x)$ with respect to θ exist for every x and satisfy

$$\left| \frac{\partial^2 \psi_{\theta,h}(x)}{\partial\theta_i\theta_j} \right| \le \ddot\psi(x),$$

for some integrable measurable function $\ddot\psi$. This should be true at least for every θ in a neighborhood of θ_0.

5.41   Theorem.   *For each θ in an open subset of Euclidean space, let $\theta \mapsto \psi_\theta(x)$ be twice continuously differentiable for every x. Suppose that $P\psi_{\theta_0} = 0$, that $P\|\psi_{\theta_0}\|^2 < \infty$ and that the matrix $P\dot{\psi}_{\theta_0}$ exists and is nonsingular. Assume that the second-order partial derivatives are dominated by a fixed integrable function $\ddot{\psi}(x)$ for every θ in a neighborhood of θ_0. Then every consistent estimator sequence $\hat{\theta}_n$ such that $\Psi_n(\hat{\theta}_n) = 0$ for every n satisfies*

$$\sqrt{n}(\hat{\theta}_n - \theta_0) = -\left(P\dot{\psi}_{\theta_0}\right)^{-1} \frac{1}{\sqrt{n}} \sum_{i=1}^{n} \psi_{\theta_0}(X_i) + o_P(1).$$

In particular, the sequence $\sqrt{n}(\hat{\theta}_n - \theta_0)$ is asymptotically normal with mean zero and covariance matrix $(P\dot{\psi}_{\theta_0})^{-1} P\psi_{\theta_0}\psi_{\theta_0}^T (P\dot{\psi}_{\theta_0})^{-1}$.

Proof.   By Taylor's theorem there exist (random) vectors $\tilde{\theta}_n$ on the line segment between θ_0 and $\hat{\theta}_n$ (possibly different for each coordinate of the function Ψ_n) such that

$$0 = \Psi_n(\hat{\theta}_n) = \Psi_n(\theta_0) + \dot{\Psi}_n(\theta_0)(\hat{\theta}_n - \theta_0) + \frac{1}{2}(\hat{\theta}_n - \theta_0)^T \ddot{\Psi}_n(\tilde{\theta}_n)(\hat{\theta}_n - \theta_0).$$

The first term on the right $\Psi_n(\theta_0)$ is an average of the i.i.d. random vectors $\psi_{\theta_0}(X_i)$, which have mean $P\psi_{\theta_0} = 0$. By the central limit theorem, the sequence $\sqrt{n}\Psi_n(\theta_0)$ converges in distribution to a multivariate normal distribution with mean 0 and covariance matrix $P\psi_{\theta_0}\psi_{\theta_0}^T$. The derivative $\dot{\Psi}_n(\theta_0)$ in the second term is an average also. By the law of large numbers it converges in probability to the matrix $V = P\dot{\psi}_{\theta_0}$. The second derivative $\ddot{\Psi}_n(\tilde{\theta}_n)$ is a k-vector of $(k \times k)$ matrices depending on the second-order derivatives $\ddot{\psi}_\theta$. By assumption, there exists a ball B around θ_0 such that $\ddot{\psi}_\theta$ is dominated by $\|\ddot{\psi}\|$ for every $\theta \in B$. The probability of the event $\{\hat{\theta}_n \in B\}$ tends to 1. On this event

$$\left\|\ddot{\Psi}_n(\tilde{\theta}_n)\right\| = \left\|\frac{1}{n}\sum_{i=1}^{n}\ddot{\psi}_{\tilde{\theta}_n}(X_i)\right\| \le \frac{1}{n}\sum_{i=1}^{n}\left\|\ddot{\psi}(X_i)\right\|.$$

This is bounded in probability by the law of large numbers. Combination of these facts allows us to rewrite the preceding display as

$$-\Psi_n(\theta_0) = \left(V + o_P(1) + \tfrac{1}{2}(\hat{\theta}_n - \theta_0)O_P(1)\right)(\hat{\theta}_n - \theta_0) = (V + o_P(1))(\hat{\theta}_n - \theta_0),$$

because the sequence $(\hat{\theta}_n - \theta_0)O_P(1) = o_P(1)O_P(1)$ converges to 0 in probability if $\hat{\theta}_n$ is consistent for θ_0. The probability that the matrix $V_{\theta_0} + o_P(1)$ is invertible tends to 1. Multiply the preceding equation by $\sqrt{n}$ and apply $\left(V + o_P(1)\right)^{-1}$ left and right to complete the proof.   ∎

In the preceding sections, the existence and consistency of solutions $\hat{\theta}_n$ of the estimating equations is assumed from the start. The present smoothness conditions actually ensure the existence of solutions. (Again the conditions could be significantly relaxed, as shown in the next proof.) Moreover, provided there exists a consistent estimator sequence at all, it is always possible to select a consistent sequence of solutions.

5.42 Theorem.   *Under the conditions of the preceding theorem, the probability that the equation $\mathbb{P}_n\psi_\theta = 0$ has at least one root tends to 1, as $n \to \infty$, and there exists a sequence of roots $\hat{\theta}_n$ such that $\hat{\theta}_n \to \theta_0$ in probability. If $\psi_\theta = \dot{m}_\theta$ is the gradient of some function*

m_θ and θ_0 is a point of local maximum of $\theta \mapsto Pm_\theta$, then the sequence $\hat{\theta}_n$ can be chosen to be local maxima of the maps $\theta \mapsto \mathbb{P}_n m_\theta$.

Proof. Integrate the Taylor expansion of $\theta \mapsto \psi_\theta(x)$ with respect to x to find that, for points $\tilde{\theta} = \tilde{\theta}(x)$ on the line segment between θ_0 and θ (possibly different for each coordinate of the function $\theta \mapsto P\psi_\theta$),

$$P\psi_\theta = P\psi_{\theta_0} + P\dot{\psi}_{\theta_0}(\theta - \theta_0) + \tfrac{1}{2}(\theta - \theta_0)^T P\ddot{\psi}_{\tilde{\theta}}(\theta - \theta_0).$$

By the domination condition, $\|P\ddot{\psi}_{\tilde{\theta}}\|$ is bounded by $P\|\ddot{\psi}\| < \infty$ if θ is sufficiently close to θ_0. Thus, the map $\Psi(\theta) = P\psi_\theta$ is differentiable at θ_0. By the same argument Ψ is differentiable throughout a small neighborhood of θ_0, and by a similar expansion (but now to first order) the derivative $P\dot{\psi}_\theta$ can be seen to be continuous throughout this neighborhood. Because $P\dot{\psi}_{\theta_0}$ is nonsingular by assumption, we can make the neighborhood still smaller, if necessary, to ensure that the derivative of Ψ is nonsingular throughout the neighborhood. Then, by the inverse function theorem, there exists, for every sufficiently small $\delta > 0$, an open neighborhood G_δ of θ_0 such that the map $\Psi : \overline{G}_\delta \mapsto \overline{\text{ball}}(0, \delta)$ is a homeomorphism. The diameter of $\overline{G}_\delta$ is bounded by a multiple of δ, by the mean-value theorem and the fact that the norms of the derivatives $(P\dot{\psi}_\theta)^{-1}$ of the inverse Ψ^{-1} are bounded.

Combining the preceding Taylor expansion with a similar expansion for the sample version $\Psi_n(\theta) = \mathbb{P}_n \psi_\theta$, we see

$$\sup_{\theta \in \overline{G}_\delta} \|\Psi_n(\theta) - \Psi(\theta)\| \leq o_P(1) + \delta o_P(1) + \delta^2 O_P(1),$$

where the $o_P(1)$ terms and the $O_p(1)$ term result from the law of large numbers, and are uniform in small δ. Because $P\big(o_P(1) + \delta o_P(1) > \tfrac{1}{2}\delta\big) \to 0$ for every $\delta > 0$, there exists $\delta_n \downarrow 0$ such that $\text{P}\big(o_P(1) + \delta_n o_P(1) > \tfrac{1}{2}\delta_n\big) \to 0$. If $K_{n,\delta}$ is the event where the left side of the preceding display is bounded above by δ, then $\text{P}(K_{n,\delta_n}) \to 1$ as $n \to \infty$.

On the event $K_{n,\delta}$ the map $\theta \mapsto \theta - \Psi_n \circ \Psi^{-1}(\theta)$ maps $\overline{\text{ball}}(0, \delta)$ into itself, by the definitions of $\overline{G}_\delta$ and $K_{n,\delta}$. Because the map is also continuous, it possesses a fixed-point in $\overline{\text{ball}}(0, \delta)$, by Brouwer's fixed point theorem. This yields a zero of Ψ_n in the set $\overline{G}_\delta$, whence the first assertion of the theorem.

For the final assertion, first note that the Hessian $P\dot{\psi}_{\theta_0}$ of $\theta \mapsto Pm_\theta$ at θ_0 is negative-definite, by assumption. A Taylor expansion as in the proof of Theorem 5.41 shows that $\mathbb{P}_n \dot{\psi}_{\hat{\theta}_n} - \mathbb{P}_n \dot{\psi}_{\theta_0} \overset{P}{\to} 0$ for every $\hat{\theta}_n \overset{P}{\to} \theta_0$. Hence the Hessian $\mathbb{P}_n \dot{\psi}_{\hat{\theta}_n}$ of $\theta \mapsto \mathbb{P}_n m_\theta$ at any consistent zero $\hat{\theta}_n$ converges in probability to the negative-definite matrix $P\dot{\psi}_{\theta_0}$ and is negative-definite with probability tending to 1. $\blacksquare$

The assertion of the theorem that there exists a consistent sequence of roots of the estimating equations is easily misunderstood. It does not guarantee the existence of an asymptotically consistent sequence of estimators. The only claim is that a clairvoyant statistician (with preknowledge of θ_0) can choose a consistent sequence of roots. In reality, it may be impossible to choose the right solutions based only on the data (and knowledge of the model). In this sense the preceding theorem, a standard result in the literature, looks better than it is.

The situation is not as bad as it seems. One interesting situation is if the solution of the estimating equation is unique for every n. Then our solutions must be the same as those of the clairvoyant statistician and hence the sequence of solutions is consistent.

In general, the deficit can be repaired with the help of a preliminary sequence of estimators $\tilde{\theta}_n$. If the sequence $\tilde{\theta}_n$ is consistent, then it works to choose the root $\hat{\theta}_n$ of $\mathbb{P}_n \psi_\theta = 0$ that is closest to $\tilde{\theta}_n$. Because $\|\hat{\theta}_n - \tilde{\theta}_n\|$ is smaller than the distance $\|\hat{\theta}_n^c - \tilde{\theta}_n\|$ between the clairvoyant sequence $\hat{\theta}_n^c$ and $\tilde{\theta}_n$, both distances converge to zero in probability. Thus the sequence of closest roots is consistent.

The assertion of the theorem can also be used in a negative direction. The point θ_0 in the theorem is required to be a zero of $\theta \mapsto P\psi_\theta$, but, apart from that, it may be arbitrary. Thus, the theorem implies at the same time that a malicious statistician can always choose a sequence of roots $\hat{\theta}_n$ that converges to any given zero. These may include other points besides the "true" value of θ. Furthermore, inspection of the proof shows that the sequence of roots can also be chosen to jump back and forth between two (or more) zeros. If the function $\theta \mapsto P\psi_\theta$ has multiple roots, we must exercise care. We can be sure that certain roots of $\theta \mapsto \mathbb{P}_n \psi_\theta$ are bad estimators.

Part of the problem here is caused by using estimating equations, rather than maximization to find estimators, which blurs the distinction between points of absolute maximum, local maximum, and even minimum. In the light of the results on consistency in section 5.2, we may expect the location of the point of absolute maximum of $\theta \mapsto \mathbb{P}_n m_\theta$ to converge to a point of absolute maximum of $\theta \mapsto P m_\theta$. As long as this is unique, the absolute maximizers of the criterion function are typically consistent.

5.43 Example (Weibull distribution). Let $X_1, \ldots, X_n$ be a sample from the Weibull distribution with density

$$p_{\theta,\sigma}(x) = \frac{\theta}{\sigma} x^{\theta-1} e^{-x^\theta/\sigma}, \qquad x > 0, \ \theta > 0, \ \sigma > 0.$$

(Then $\sigma^{1/\theta}$ is a scale parameter.) The score function is given by the partial derivatives of the log density with respect to θ and σ:

$$\dot{\ell}_{\theta,\sigma}(x) = \left(\frac{1}{\theta} + \log x - \frac{x^\theta}{\sigma} \log x, \ w - \frac{1}{\sigma} + \frac{x^\theta}{\sigma^2} \right).$$

The likelihood equations $\sum \dot{\ell}_{\theta,\sigma}(x_i) = 0$ reduce to

$$\sigma = \frac{1}{n} \sum_{i=1}^n x_i^\theta; \qquad \frac{1}{\theta} + \frac{1}{n} \sum_{i=1}^n \log x_i - \frac{\sum_{i=1}^n x_i^\theta \log x_i}{\sum_{i=1}^n x_i^\theta} = 0.$$

The second equation is strictly decreasing in θ, from ∞ at $\theta = 0$ to $\overline{\log x} - \log x_{(n)}$ at $\theta = \infty$. Hence a solution exists, and is unique, unless all x_i are equal. Provided the higher-order derivatives of the score function exist and can be dominated, the sequence of maximum likelihood estimators $(\hat{\theta}_n, \hat{\sigma}_n)$ is asymptotically normal by Theorems 5.41 and 5.42. There exist four different third-order derivatives, given by

$$\frac{\partial^3 \ell_{\theta,\sigma}(x)}{\partial \theta^3} = \frac{2}{\theta^3} - \frac{x^\theta}{\sigma} \log^3 x$$

$$\frac{\partial^3 \ell_{\theta,\sigma}(x)}{\partial \theta^2 \partial \sigma} = \frac{x^\theta}{\sigma^2} \log^2 x$$

$$\frac{\partial^3 \ell_{\theta,\sigma}(x)}{\partial \theta \partial \sigma^2} = -\frac{2x^\theta}{\sigma^3} \log x$$

$$\frac{\partial^3 \ell_{\theta,\sigma}(x)}{\partial \sigma^3} = -\frac{2}{\sigma^3} + \frac{6x^\theta}{\sigma^4}.$$

For θ and σ ranging over sufficiently small neighborhoods of θ_0 and σ_0, these functions are dominated by a function of the form

$$M(x) = A(1 + x^B)(1 + |\log x| + \cdots + |\log x|^3),$$

for sufficiently large A and B. Because the Weibull distribution has an exponentially small tail, the mixed moment $\mathrm{E}_{\theta_0,\sigma_0} X^p |\log X|^q$ is finite for every $p, q \geq 0$. Thus, all moments of $\dot{\ell}_\theta$ and $\ddot{\ell}_\theta$ exist and M is integrable. $\square$

*5.7 One-Step Estimators

The method of Z-estimation as discussed so far has two disadvantages. First, it may be hard to find the roots of the estimating equations. Second, for the roots to be consistent, the estimating equation needs to behave well throughout the parameter set. For instance, existence of a second root close to the boundary of the parameter set may cause trouble. The *one-step method* overcomes these problems by building on and improving a preliminary estimator $\tilde{\theta}_n$.

The idea is to solve the estimator from a linear approximation to the original estimating equation $\Psi_n(\theta) = 0$. Given a preliminary estimator $\tilde{\theta}_n$, the one-step estimator is the solution (in θ) to

$$\Psi_n(\tilde{\theta}_n) + \dot{\Psi}_n(\tilde{\theta}_n)(\theta - \tilde{\theta}_n) = 0.$$

This corresponds to replacing $\Psi_n(\theta)$ by its tangent at $\tilde{\theta}_n$, and is known as the method of *Newton-Rhapson* in numerical analysis. The solution $\theta = \hat{\theta}_n$ is

$$\hat{\theta}_n = \tilde{\theta}_n - \dot{\Psi}_n(\tilde{\theta}_n)^{-1} \Psi_n(\tilde{\theta}_n).$$

In numerical analysis this procedure is iterated a number of times, taking $\hat{\theta}_n$ as the new preliminary guess, and so on. Provided that the starting point $\tilde{\theta}_n$ is well chosen, the sequence of solutions converges to a root of Ψ_n. Our interest here goes in a different direction. We suppose that the preliminary estimator $\tilde{\theta}_n$ is already within range $n^{-1/2}$ of the true value of θ. Then, as we shall see, just one iteration of the Newton-Rhapson scheme produces an estimator $\hat{\theta}_n$ that is as good as the Z-estimator defined by Ψ_n. In fact, it is better in that its consistency is guaranteed, whereas the true Z-estimator may be inconsistent or not uniquely defined.

In this way consistency and asymptotic normality are effectively separated, which is useful because these two aims require different properties of the estimating equations. Good initial estimators can be constructed by ad-hoc methods and take care of consistency. Next, these initial estimators can be improved by the one-step method. Thus, for instance, the good properties of maximum likelihood estimation can be retained, even in cases in which the consistency fails.

In this section we impose the following condition on the random criterion functions Ψ_n. For every constant M and a given nonsingular matrix $\dot{\Psi}_0$,

$$\sup_{\sqrt{n}\|\theta-\theta_0\|<M} \left\| \sqrt{n}\big(\Psi_n(\theta) - \Psi_n(\theta_0)\big) - \dot{\Psi}_0 \sqrt{n}(\theta - \theta_0) \right\| \xrightarrow{P} 0. \qquad (5.44)$$

Condition (5.44) suggests that Ψ_n is differentiable at θ_0, with derivative tending to $\dot{\Psi}_0$, but this is not an assumption. We do not require that a derivative $\dot{\Psi}_n$ exists, and introduce a further refinement of the Newton-Rhapson scheme by replacing $\dot{\Psi}_n(\tilde{\theta}_n)$ by arbitrary estimators. Given nonsingular, random matrices $\dot{\Psi}_{n,0}$ that converge in probability to $\dot{\Psi}_0$ define the *one-step estimator*

$$\hat{\theta}_n = \tilde{\theta}_n - \dot{\Psi}_{n,0}^{-1}\Psi_n(\tilde{\theta}_n).$$

Call an estimator sequence $\tilde{\theta}_n$ $\sqrt{n}$-*consistent* if the sequence $\sqrt{n}(\tilde{\theta}_n - \theta_0)$ is uniformly tight. The interpretation is that $\tilde{\theta}_n$ already determines the value θ_0 within $n^{-1/2}$-range.

5.45   Theorem (One-step estimation). *Let $\sqrt{n}\Psi_n(\theta_0) \rightsquigarrow Z$ and let (5.44) hold. Then the one-step estimator $\hat{\theta}_n$, for a given $\sqrt{n}$-consistent estimator sequence $\tilde{\theta}_n$ and estimators $\dot{\Psi}_{n,0} \xrightarrow{P} \dot{\Psi}_0$, satisfies*

$$\sqrt{n}(\hat{\theta}_n - \theta_0) = -\dot{\Psi}_0^{-1}\sqrt{n}\Psi_n(\theta_0) + o_P(1).$$

5.46   Addendum. *For $\Psi_n(\theta) = \mathbb{P}_n\psi_\theta$ condition (5.44) is satisfied under the conditions of Theorem 5.21 with $\dot{\Psi}_0 = V_{\theta_0}$, and under the conditions of Theorem 5.41 with $\dot{\Psi}_0 = P\dot{\psi}_{\theta_0}$.*

Proof.   The standardized estimator $\dot{\Psi}_{n,0}\sqrt{n}(\hat{\theta}_n - \theta_0)$ equals

$$\dot{\Psi}_{n,0}\sqrt{n}(\tilde{\theta}_n - \theta_0) - \sqrt{n}\big(\Psi_n(\tilde{\theta}_n) - \Psi_n(\theta_0)\big) - \dot{\Psi}_{n,0}^{-1}\sqrt{n}\Psi_n(\theta_0).$$

By (5.44) the second term can be replaced by $-\dot{\Psi}_0\sqrt{n}(\tilde{\theta}_n - \theta_0) + o_P(1)$. Thus the expression can be rewritten as

$$(\dot{\Psi}_{n,0} - \dot{\Psi}_0)\sqrt{n}(\tilde{\theta}_n - \theta_0) - \sqrt{n}\Psi_n(\theta_0) + o_P(1).$$

The first term converges to zero in probability, and the theorem follows after application of Slutsky's lemma.

For a proof of the addendum, see the proofs of the corresponding theorems.   ∎

If the sequence $\sqrt{n}(\hat{\theta}_n - \theta_0)$ converges in distribution, then it is certainly uniformly tight. Consequently, a sequence of one-step estimators is $\sqrt{n}$-consistent and can itself be used as preliminary estimator for a second iteration of the modified Newton-Rhapson algorithm. Presumably, this would give a value closer to a root of Ψ_n. However, the limit distribution of this "two-step estimator" is the same, so that repeated iteration does not give asymptotic improvement. In practice a multistep method may nevertheless give better results.

We close this section with a discussion of the *discretization* trick. This device is mostly of theoretical value and has been introduced to relax condition (5.44) to the following. For every *nonrandom* sequence $\theta_n = \theta_0 + O(n^{-1/2})$,

$$\left\| \sqrt{n}\big(\Psi_n(\theta_n) - \Psi_n(\theta_0)\big) - \dot{\Psi}_0\sqrt{n}(\theta_n - \theta_0) \right\| \xrightarrow{P} 0. \tag{5.47}$$

This new condition is less stringent and much easier to check. It is sufficiently strong if the preliminary estimators $\tilde{\theta}_n$ are *discretized* on grids of mesh width $n^{-1/2}$. For instance, $\tilde{\theta}_n$ is suitably discretized if all its realizations are points of the grid $n^{-1/2}\mathbb{Z}^k$ (consisting of the points $n^{-1/2}(i_1, \ldots, i_k)$ for integers $i_1, \ldots, i_k$). This is easy to achieve, but perhaps unnatural. Any preliminary estimator sequence $\tilde{\theta}_n$ can be discretized by replacing its values by the closest points of the grid. Because this changes each coordinate by at most $n^{-1/2}$, $\sqrt{n}$-consistency of $\tilde{\theta}_n$ is retained by discretization.

Define a one-step estimator $\hat{\theta}_n$ as before, but now use a discretized version of the preliminary estimator.

5.48 Theorem (Discretized one-step estimation). *Let $\sqrt{n}\Psi_n(\theta_0) \rightsquigarrow Z$ and let (5.47) hold. Then the one-step estimator $\hat{\theta}_n$, for a given $\sqrt{n}$-consistent, discretized estimator sequence $\tilde{\theta}_n$ and estimators $\dot{\Psi}_{n,0} \overset{P}{\to} \dot{\Psi}_0$, satisfies*

$$\sqrt{n}(\hat{\theta}_n - \theta_0) = -\dot{\Psi}_0^{-1}\sqrt{n}\Psi_n(\theta_0) + o_P(1).$$

5.49 Addendum. *For $\Psi_n(\theta) = \mathbb{P}_n\psi_\theta$ and $\mathbb{P}_n$ the empirical measure of a random sample from a density p_θ that is differentiable in quadratic mean (5.38), condition (5.47), is satisfied, with $\dot{\Psi}_0 = -P_{\theta_0}\psi_{\theta_0}\dot{\ell}_{\theta_0}^T$, if, as $\theta \to \theta_0$,*

$$\int \left[\psi_\theta\sqrt{p_\theta} - \psi_{\theta_0}\sqrt{p_{\theta_0}}\right]^2 d\mu \to 0.$$

Proof. The arguments of the previous proof apply, except that it must be shown that

$$R(\tilde{\theta}_n) := \sqrt{n}\left(\Psi_n(\tilde{\theta}_n) - \Psi_n(\theta_0)\right) - \dot{\Psi}_0^{-1}\sqrt{n}(\tilde{\theta}_n - \theta_0)$$

converges to zero in probability. Fix $\varepsilon > 0$. By the $\sqrt{n}$-consistency, there exists M with $P(\sqrt{n}\|\tilde{\theta}_n - \theta_0\| > M) < \varepsilon$. If $\sqrt{n}\|\tilde{\theta}_n - \theta_0\| \leq M$, then $\tilde{\theta}_n$ equals one of the values in the set $S_n = \{\theta \in n^{-1/2}\mathbb{Z}^k : \|\theta - \theta_0\| \leq n^{-1/2}M\}$. For each M and n there are only finitely many elements in this set. Moreover, for fixed M the number of elements is bounded independently of n. Thus

$$P(\|R(\tilde{\theta}_n)\| > \varepsilon) \leq \varepsilon + \sum_{\theta_n \in S_n} P(\|R(\theta_n)\| > \varepsilon \wedge \tilde{\theta}_n = \theta_n)$$

$$\leq \varepsilon + \sum_{\theta_n \in S_n} P(\|R(\theta_n)\| > \varepsilon).$$

The maximum of the terms in the sum corresponds to a sequence of nonrandom vectors θ_n with $\theta_n = \theta_0 + O(n^{-1/2})$. It converges to zero by (5.47). Because the number of terms in the sum is bounded independently of n, the sum converges to zero.

For a proof of the addendum, see proposition A.10 in [139]. ∎

If the score function $\dot{\ell}_\theta$ of the model also satisfies the conditions of the addendum, then the estimators $\dot{\Psi}_{n,0} = -P_{\tilde{\theta}_n}\psi_{\tilde{\theta}_n}\dot{\ell}_{\tilde{\theta}_n}^T$ are consistent for $\dot{\Psi}_0$. This shows that discretized one-step estimation can be carried through under very mild regularity conditions. Note that the addendum requires only continuity of $\theta \mapsto \psi_\theta$, whereas (5.47) appears to require differentiability.

5.50 Example (Cauchy distribution). Suppose $X_1, \ldots, X_n$ are a sample from the Cauchy location family $p_\theta(x) = \pi^{-1}\left(1 + (x - \theta)^2\right)^{-1}$. Then the score function is given by

$$\dot{\ell}_\theta(x) = \frac{2(x - \theta)}{1 + (x - \theta)^2}.$$

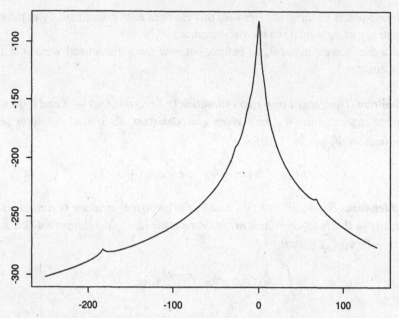

Figure 5.4. Cauchy log likelihood function of a sample of 25 observations, showing three local maxima. The value of the absolute maximum is well-separated from the other maxima, and its location is close to the true value zero of the parameter.

This function behaves like $1/x$ for $x \to \pm\infty$ and is bounded in between. The second moment of $\dot{\ell}_\theta(X_1)$ therefore exists, unlike the moments of the distribution itself. Because the sample mean possesses the same (Cauchy) distribution as a single observation X_1, the sample mean is a very inefficient estimator. Instead we could use the median, or another M-estimator. However, the asymptotically best estimator should be based on maximum likelihood. We have

$$\dot{\ell}_\theta(x) = \frac{4(x - \theta)\big((x - \theta)^2 - 3\big)}{\big(1 + (x - \theta)^2\big)^3}.$$

The tails of this function are of the order $1/x^3$, and the function is bounded in between. These bounds are uniform in θ varying over a compact interval. Thus the conditions of Theorems 5.41 and 5.42 are satisfied. Since the consistency follows from Example 5.16, the sequence of maximum likelihood estimators is asymptotically normal.

The Cauchy likelihood estimator has gained a bad reputation, because the likelihood equation $\sum \dot{\ell}_\theta(X_i) = 0$ typically has several roots. The number of roots behaves asymptotically as two times a Poisson$(1/\pi)$ variable plus 1. (See [126].) Therefore, the one-step (or possibly multi-step method) is often recommended, with, for instance, the median as the initial estimator. Perhaps a better solution is not to use the likelihood equations, but to determine the maximum likelihood estimator by, for instance, visual inspection of a graph of the likelihood function, as in Figure 5.4. This is particularly appropriate because the difficulty of multiple roots does not occur in the two parameter location-scale model. In the model with density $p_\theta(x/\sigma)/\sigma$, the maximum likelihood estimator for (θ, σ) is unique. (See [25].)   □

5.51   *Example (Mixtures).* Let f and g be given, positive probability densities on the real line. Consider estimating the parameter $\theta = (\mu, \nu, \sigma, \tau, p)$ based on a random sample from the mixture density

$$x \mapsto pf\left(\frac{x-\mu}{\sigma}\right)\frac{1}{\sigma} + (1-p)g\left(\frac{x-v}{\tau}\right)\frac{1}{\tau}.$$

If f and g are sufficiently regular, then this is a smooth five-dimensional parametric model, and the standard theory should apply. Unfortunately, the supremum of the likelihood over the natural parameter space is ∞, and there exists no maximum likelihood estimator. This is seen, for instance, from the fact that the likelihood is bigger than

$$pf\left(\frac{x_1-\mu}{\sigma}\right)\frac{1}{\sigma}\prod_{i=2}^{n}(1-p)g\left(\frac{x_i-v}{\tau}\right)\frac{1}{\tau}.$$

If we set $\mu = x_1$ and next maximize over $\sigma > 0$, then we obtain the value ∞ whenever $p > 0$, irrespective of the values of v and τ.

A one-step estimator appears reasonable in this example. In view of the smoothness of the likelihood, the general theory yields the asymptotic efficiency of a one-step estimator if started with an initial $\sqrt{n}$-consistent estimator. Moment estimators could be appropriate initial estimators. $\square$

*5.8 Rates of Convergence

In this section we discuss some results that give the rate of convergence of M-estimators. These results are useful as intermediate steps in deriving a limit distribution, but also of interest on their own. Applications include both classical estimators of "regular" parameters and estimators that converge at a slower than $\sqrt{n}$-rate. The main result is simple enough, but its conditions include a maximal inequality, for which results such as in Chapter 19 are needed.

Let $\mathbb{P}_n$ be the empirical distribution of a random sample of size n from a distribution P, and, for every θ in a metric space Θ, let $x \mapsto m_\theta(x)$ be a measurable function. Let $\hat{\theta}_n$ (nearly) maximize the criterion function $\theta \mapsto \mathbb{P}_n m_\theta$.

The criterion function may be viewed as the sum of the deterministic map $\theta \mapsto Pm_\theta$ and the random fluctuations $\theta \mapsto \mathbb{P}_n m_\theta - Pm_\theta$. The rate of convergence of $\hat{\theta}_n$ depends on the combined behavior of these maps. If the deterministic map changes rapidly as θ moves away from the point of maximum and the random fluctuations are small, then $\hat{\theta}_n$ has a high rate of convergence. For convenience of notation we measure the fluctuations in terms of the empirical process $\mathbb{G}_n m_\theta = \sqrt{n}(\mathbb{P}_n m_\theta - Pm_\theta)$.

5.52 Theorem (Rate of convergence). *Assume that for fixed constants C and $\alpha > \beta$, for every n, and for every sufficiently small $\delta > 0$,*

$$\sup_{\frac{\delta}{2}\leq d(\theta,\theta_0)<\delta} P(m_\theta - m_{\theta_0}) \leq -C\delta^\alpha,$$

$$E^* \sup_{d(\theta,\theta_0)<\delta}\left|\mathbb{G}_n(m_\theta - m_{\theta_0})\right| \leq C\delta^\beta.$$

If the sequence $\hat{\theta}_n$ satisfies $\mathbb{P}_n m_{\hat{\theta}_n} \geq \mathbb{P}_n m_{\theta_0} - O_P\left(n^{\alpha/(2\beta-2\alpha)}\right)$ and converges in outer probability to θ_0, then $n^{1/(2\alpha-2\beta)}d(\hat{\theta}_n,\theta_0) = O_P^*(1)$.

Proof. Set $r_n = n^{1/(2\alpha - 2\beta)}$ and suppose that $\hat{\theta}_n$ maximizes the map $\theta \mapsto \mathbb{P}_n m_\theta$ up to a variable $R_n = O_P(r_n^{-\alpha})$.

For each n, the parameter space minus the point θ_0 can be partitioned into the "shells" $S_{j,n} = \{\theta : 2^{j-1} < r_n d(\theta, \theta_0) \leq 2^j\}$, with j ranging over the integers. If $r_n d(\hat{\theta}_n, \theta_0)$ is larger than 2^M for a given integer M, then $\hat{\theta}_n$ is in one of the shells $S_{j,n}$ with $j \geq M$. In that case the supremum of the map $\theta \mapsto \mathbb{P}_n m_\theta - \mathbb{P}_n m_{\theta_0}$ over this shell is at least $-R_n$ by the property of $\hat{\theta}_n$. Conclude that, for every $\varepsilon > 0$,

$$\mathrm{P}^*\left(r_n d(\hat{\theta}_n, \theta_0) > 2^M\right) \leq \sum_{\substack{j \geq M \\ 2^j \leq \varepsilon r_n}} \mathrm{P}^*\left(\sup_{\theta \in S_{j,n}} (\mathbb{P}_n m_\theta - \mathbb{P}_n m_{\theta_0}) \geq -\frac{K}{r_n^\alpha}\right)$$
$$+ \mathrm{P}^*\left(2d(\hat{\theta}_n, \theta_0) \geq \varepsilon\right) + \mathrm{P}\left(r_n^\alpha R_n \geq K\right).$$

If the sequence $\hat{\theta}_n$ is consistent for θ_0, then the second probability on the right converges to 0 as $n \to \infty$, for every fixed $\varepsilon > 0$. The third probability on the right can be made arbitrarily small by choice of K, uniformly in n. Choose $\varepsilon > 0$ small enough to ensure that the conditions of the theorem hold for every $\delta \leq \varepsilon$. Then for every j involved in the sum, we have

$$\sup_{\theta \in S_{j,n}} P(m_\theta - m_{\theta_0}) \leq -C \frac{2^{(j-1)\alpha}}{r_n^\alpha}.$$

For $\frac{1}{2} C 2^{(M-1)\alpha} \geq K$, the series can be bounded in terms of the empirical process $\mathbb{G}_n$ by

$$\sum_{\substack{j \geq M \\ 2^j \leq \varepsilon r_n}} \mathrm{P}^*\left(\left\|\mathbb{G}_n(m_\theta - m_{\theta_0})\right\|_{S_{j,n}} \geq C\sqrt{n}\frac{2^{(j-1)\alpha}}{2r_n^\alpha}\right) \leq \sum_{j \geq M} \frac{(2^j/r_n)^\beta 2r_n^\alpha}{\sqrt{n} 2^{(j-1)\alpha}},$$

by Markov's inequality and the definition of r_n. The right side converges to zero for every $M = M_n \to \infty$. ∎

Consider the special case that the parameter θ is a Euclidean vector. If the map $\theta \mapsto P m_\theta$ is twice-differentiable at the point of maximum θ_0, then its first derivative at θ_0 vanishes and a Taylor expansion of the limit criterion function takes the form

$$P(m_\theta - m_{\theta_0}) = \frac{1}{2}(\theta - \theta_0)^T V (\theta - \theta_0) + o(\|\theta - \theta_0\|^2).$$

Then the first condition of the theorem holds with $\alpha = 2$ provided that the second-derivative matrix V is nonsingular.

The second condition of the theorem is a *maximal inequality* and is harder to verify. In "regular" cases it is valid with $\beta = 1$ and the theorem yields the "usual" rate of convergence $\sqrt{n}$. The theorem also applies to nonstandard situations and yields, for instance, the rate $n^{1/3}$ if $\alpha = 2$ and $\beta = \frac{1}{2}$. Lemmas 19.34, 19.36 and 19.38 and corollary 19.35 are examples of maximal inequalities that can be appropriate for the present purpose. They give bounds in terms of the entropies of the classes of functions $\{m_\theta - m_{\theta_0} : d(\theta, \theta_0) < \delta\}$.

A Lipschitz condition on the maps $\theta \mapsto m_\theta$ is one possibility to obtain simple estimates on these entropies and is applicable in many applications. The result of the following corollary is used earlier in this chapter.

5.53   Corollary. *For each θ in an open subset of Euclidean space let $x \mapsto m_\theta(x)$ be a measurable function such that, for every θ_1 and θ_2 in a neighborhood of θ_0 and a measurable function $\dot{m}$ such that $P\dot{m}^2 < \infty$,*

$$\left| m_{\theta_1}(x) - m_{\theta_2}(x) \right| \le \dot{m}(x) \, \|\theta_1 - \theta_2\|.$$

Furthermore, suppose that the map $\theta \mapsto Pm_\theta$ admits a second-order Taylor expansion at the point of maximum θ_0 with nonsingular second derivative. If $\mathbb{P}_n m_{\hat{\theta}_n} \ge \mathbb{P}_n m_{\theta_0} - O_P(n^{-1})$, then $\sqrt{n}(\hat{\theta}_n - \theta_0) = O_P(1)$, provided that $\hat{\theta}_n \overset{P}{\to} \theta_0$.

Proof.   By assumption, the first condition of Theorem 5.52 is valid with $\alpha = 2$. To see that the second one is valid with $\beta = 1$, we apply Corollary 19.35 to the class of functions $\mathcal{F} = \{m_\theta - m_{\theta_0} \colon \|\theta - \theta_0\| < \delta\}$. This class has envelope function $F = \dot{m}\delta$, whence

$$E^* \sup_{\|\theta - \theta_0\| < \delta} |\mathbb{G}_n(m_\theta - m_{\theta_0})| \lesssim \int_0^{\|\dot{m}\|_{P,2}\delta} \sqrt{\log N_{[]}(\varepsilon, \mathcal{F}, L_2(P))}\, d\varepsilon.$$

The bracketing entropy of the class $\mathcal{F}$ is estimated in Example 19.7. Inserting the upper bound obtained there into the integral, we obtain that the preceding display is bounded above by a multiple of

$$\int_0^{\|\dot{m}\|_{P,2}\delta} \sqrt{\log\left(\frac{\delta}{\varepsilon}\right)}\, d\varepsilon.$$

Change the variables in the integral to see that this is a multiple of δ.   ∎

Rates of convergence different from $\sqrt{n}$ are quite common for M-estimators of infinite-dimensional parameters and may also be obtained through the application of Theorem 5.52. See Chapters 24 and 25 for examples. Rates slower than $\sqrt{n}$ may also arise for fairly simple parametric estimates.

5.54   Example (Modal interval). Suppose that we define an estimator $\hat{\theta}_n$ of location as the center of an interval of length 2 that contains the largest possible fraction of the observations. This is an M-estimator for the functions $m_\theta = 1_{[\theta-1,\theta+1]}$.

For many underlying distributions the first condition of Theorem 5.52 holds with $\alpha = 2$. It suffices that the map $\theta \mapsto Pm_\theta = P[\theta - 1, \theta + 1]$ is twice-differentiable and has a proper maximum at some point θ_0. Using the maximal inequality Corollary 19.35 (or Lemma 19.38), we can show that the second condition is valid with $\beta = \frac{1}{2}$. Indeed, the bracketing entropy of the intervals in the real line is of the order δ/ε^2, and the envelope function of the class of functions $1_{[\theta-1,\theta+1]} - 1_{[\theta_0-1,\theta_0+1]}$ as θ ranges over $(\theta_0 - \delta, \theta_0 + \delta)$ is bounded by $1_{[\theta_0-1-\delta,\theta_0-1+\delta]} + 1_{[\theta_0+1-\delta,\theta_0+1+\delta]}$, whose squared L_2-norm is bounded by $\|p\|_\infty 2\delta$.

Thus Theorem 5.52 applies with $\alpha = 2$ and $\beta = \frac{1}{2}$ and yields the rate of convergence $n^{1/3}$. The resulting location estimator is very robust against outliers. However, in view of its slow convergence rate, one should have good reasons to use it.

The use of an interval of length 2 is somewhat awkward. Every other fixed length would give the same result. More interestingly, we can also replace the fixed-length interval by the smallest interval that contains a fixed fraction, for instance $1/2$, of the observations. This still yields a rate of convergence of $n^{1/3}$. The intuitive reason for this is that the length of a "shorth" settles down by a $\sqrt{n}$-rate and hence its randomness is asymptotically negligible relative to its center. $\square$

The preceding theorem requires the consistency of $\hat{\theta}_n$ as a condition. This consistency is implied if the other conditions are valid for every $\delta > 0$, not just for small values of δ. This can be seen from the proof or the more general theorem in the next section. Because the conditions are not natural for large values of δ, it is usually better to argue the consistency by other means.

5.8.1   *Nuisance Parameters*

In Chapter 25 we need an extension of Theorem 5.52 that allows for a "smoothing" or "nuisance" parameter. We also take the opportunity to insert a number of other refinements, which are sometimes useful.

Let $x \mapsto m_{\theta,\eta}(x)$ be measurable functions indexed by parameters (θ, η), and consider estimators $\hat{\theta}_n$ contained in a set Θ_n that, for a given $\hat{\eta}_n$ contained in a set H_n, maximize the map

$$\theta \mapsto \mathbb{P}_n m_{\theta,\hat{\eta}_n}.$$

The sets Θ_n and H_n need not be metric spaces, but instead we measure the discrepancies between $\hat{\theta}_n$ and θ_0, and $\hat{\eta}_n$ and a limiting value η_0, by nonnegative functions $\theta \mapsto d_\eta(\theta, \theta_0)$ and $\eta \mapsto d(\eta, \eta_0)$, which may be arbitrary.

5.55   Theorem. *Assume that, for arbitrary functions $e_n: \Theta_n \times H_n \mapsto \mathbb{R}$ and $\phi_n: (0, \infty) \mapsto \mathbb{R}$ such that $\delta \mapsto \phi_n(\delta)/\delta^\beta$ is decreasing for some $\beta < 2$, every $(\theta, \eta) \in \Theta_n \times H_n$, and every $\delta > 0$,*

$$P(m_{\theta,\eta} - m_{\theta_0,\eta}) + e_n(\theta, \eta) \leq -d_\eta^2(\theta, \theta_0) + d^2(\eta, \eta_0),$$

$$\mathrm{E}^* \sup_{\substack{d_\eta(\theta,\theta_0)<\delta \\ (\theta,\eta)\in\Theta_n \times H_n}} \left| \mathbb{G}_n(m_{\theta,\eta} - m_{\theta_0,\eta}) - \sqrt{n}\, e_n(\theta, \eta) \right| \leq \phi_n(\delta).$$

Let $\delta_n > 0$ satisfy $\phi_n(\delta_n) \leq \sqrt{n}\, \delta_n^2$ for every n. If $P(\hat{\theta}_n \in \Theta_n, \hat{\eta}_n \in H_n) \to 1$ and $\mathbb{P}_n m_{\hat{\theta}_n,\hat{\eta}_n} \geq \mathbb{P}_n m_{\theta_0,\hat{\eta}_n} - O_P(\delta_n^2)$, then $d_{\hat{\eta}_n}(\hat{\theta}_n, \theta_0) = O_P^*\big(\delta_n + d(\hat{\eta}_n, \eta_0)\big)$.

Proof.   For simplicity assume that $\mathbb{P}_n m_{\hat{\theta}_n,\hat{\eta}_n} \geq \mathbb{P}_n m_{\theta_0,\hat{\eta}_n}$, without a tolerance term. For each $n \in \mathbb{N}$, $j \in \mathbb{Z}$ and $M > 0$, let $S_{n,j,M}$ be the set

$$\big\{(\theta, \eta) \in \Theta_n \times H_n : 2^{j-1}\delta_n < d_\eta(\theta, \theta_0) \leq 2^j \delta_n, d(\eta, \eta_0) \leq 2^{-M} d_\eta(\theta, \theta_0)\big\}.$$

Then the intersection of the events $(\hat{\theta}_n, \hat{\eta}_n) \in \Theta_n \times H_n$, and $d_{\hat{\eta}_n}(\hat{\theta}_n, \theta_0) \geq 2^M\big(\delta_n + d(\hat{\eta}_n, \eta_0)\big)$ is contained in the union of the events $\big\{(\hat{\theta}_n, \hat{\eta}_n) \in S_{n,j,M}\big\}$ over $j \geq M$. By the definition of $\hat{\theta}_n$, the supremum of $\mathbb{P}_n(m_{\theta,\eta} - m_{\theta_0,\eta})$ over the set of parameters $(\theta, \eta) \in S_{n,j,M}$ is nonnegative on the event $\big\{(\hat{\theta}_n, \hat{\eta}_n) \in S_{n,j,M}\big\}$. Conclude that

$$\mathrm{P}^*\Big(d_{\hat{\eta}_n}(\hat{\theta}_n, \theta_0) \geq 2^M\big(\delta_n + d(\hat{\eta}_n, \eta_0)\big), (\hat{\theta}_n, \hat{\eta}_n) \in \Theta_n \times H_n\Big)$$

$$\leq \sum_{j \geq M} \mathrm{P}^*\bigg(\sup_{(\theta,\eta)\in S_{n,j,M}} \mathbb{P}_n(m_{\theta,\eta} - m_{\theta_0,\eta}) \geq 0\bigg).$$

For every j, $(\theta, \eta) \in S_{n,j,M}$, and every sufficiently large M,

$$P(m_{\theta,\eta} - m_{\theta_0,\eta}) + e_n(\theta, \eta) \leq -d_\eta^2(\theta, \theta_0) + d^2(\eta, \eta_0)$$
$$\leq -(1 - 2^{-2M})\, d_\eta^2(\theta, \theta_0) \leq -2^{2j-4}\delta_n^2.$$

From here on the proof is the same as the proof of Theorem 5.52, except that we use that $\phi_n(c\delta) \leq c^\beta \phi_n(\delta)$ for every $c > 1$, by the assumption on ϕ_n. ∎

*5.9 Argmax Theorem

The consistency of a sequence of M-estimators can be understood as the points of maximum $\hat{\theta}_n$ of the criterion functions $\theta \mapsto M_n(\theta)$ converging in probability to a point of maximum of the limit criterion function $\theta \mapsto M(\theta)$. So far we have made no attempt to understand the distributional limit properties of a sequence of M-estimators in a similar way. This is possible, but it is somewhat more complicated and is perhaps best studied after developing the theory of weak convergence of stochastic processes, as in Chapters 18 and 19.

Because the estimators $\hat{\theta}_n$ typically converge to constants, it is necessary to rescale them before studying distributional limit properties. Thus, we start by searching for a sequence of numbers $r_n \mapsto \infty$ such that the sequence $\hat{h}_n = r_n(\hat{\theta}_n - \theta)$ is uniformly tight. The results of the preceding section should be useful. If $\hat{\theta}_n$ maximizes the function $\theta \mapsto M_n(\theta)$, then the rescaled estimators $\hat{h}_n$ are maximizers of the *local criterion functions*

$$h \mapsto M_n\left(\theta + \frac{h}{r_n}\right) - M_n(\theta_0).$$

Suppose that these, if suitably normed, converge to a limit process $h \mapsto M(h)$. Then the general principle is that the sequence $\hat{h}_n$ converges in distribution to the maximizer of this limit process.

For simplicity of notation we shall write the local criterion functions as $h \mapsto M_n(h)$. Let $\{M_n(h): h \in H_n\}$ be arbitrary stochastic processes indexed by subsets H_n of a given metric space. We wish to prove that the argmax-functional is continuous: If $M_n \rightsquigarrow M$ and $H_n \to H$ in a suitable sense, then the (near) maximizers $\hat{h}_n$ of the random maps $h \mapsto M_n(h)$ converge in distribution to the maximizer $\hat{h}$ of the limit process $h \mapsto M(h)$. It is easy to find examples in which this is not true, but given the right definitions it is, under some conditions. Given a set B, set

$$M(B) = \sup_{h \in B} M(h).$$

Then convergence in distribution of the vectors $(M_n(A), M_n(B))$ for given pairs of sets A and B is an appropriate form of convergence of M_n to M. The following theorem gives some flexibility in the choice of the indexing sets. We implicitly either assume that the suprema $M_n(B)$ are measurable or understand the weak convergence in terms of outer probabilities, as in Chapter 18.

The result we are looking for is not likely to be true if the maximizer of the limit process is not well defined. Exactly as in Theorem 5.7, the maximum should be "well separated." Because in the present case the limit is a stochastic process, we require that every sample path $h \mapsto M(h)$ possesses a well-separated maximum (condition (5.57)).

5.56 Theorem (Argmax theorem). *Let M_n and M be stochastic processes indexed by subsets H_n and H of a given metric space such that, for every pair of a closed set F and a set K in a given collection $\mathcal{K}$,*

$$\big(M_n(F \cap K \cap H_n), M_n(K \cap H_n)\big) \rightsquigarrow \big(M(F \cap K \cap H), M(K \cap H)\big).$$

Furthermore, suppose that every sample path of the process $h \mapsto M(h)$ possesses a well-separated point of maximum $\hat{h}$ in that, for every open set G and every $K \in \mathcal{K}$,

$$M(\hat{h}) > M(G^c \cap K \cap H), \qquad \text{if } \hat{h} \in G, \qquad \text{a.s..} \tag{5.57}$$

If $M_n(\hat{h}_n) \geq M_n(H_n) - o_P(1)$ and for every $\varepsilon > 0$ there exists $K \in \mathcal{K}$ such that $\sup_n P(\hat{h}_n \notin K) < \varepsilon$ and $P(\hat{h} \notin K) < \varepsilon$, then $\hat{h}_n \rightsquigarrow \hat{h}$.

Proof. If $\hat{h}_n \in F \cap K$, then $M_n(F \cap K \cap H_n) \geq M_n(B) - o_P(1)$ for any set B. Hence, for every closed set F and every $K \in \mathcal{K}$,

$$P(\hat{h}_n \in F \cap K) \leq P\big(M_n(F \cap K \cap H_n) \geq M_n(K \cap H_n) - o_P(1)\big)$$
$$\leq P\big(M(F \cap K \cap H) \geq M(K \cap H)\big) + o(1),$$

by Slutsky's lemma and the portmanteau lemma. If $\hat{h} \in F^c$, then $M(F \cap K \cap H)$ is strictly smaller than $M(\hat{h})$ by (5.57) and hence on the intersection with the event in the far right side $\hat{h}$ cannot be contained in $K \cap H$. It follows that

$$\limsup P(\hat{h}_n \in F \cap K) \leq P(\hat{h} \in F) + P(\hat{h} \notin K \cap H).$$

By assumption we can choose K such that the left and right sides change by less than ε if we replace K by the whole space. Hence $\hat{h}_n \rightsquigarrow \hat{h}$ by the portmanteau lemma. ∎

The theorem works most smoothly if we can take $\mathcal{K}$ to consist only of the whole space. However, then we are close to assuming some sort of global uniform convergence of M_n to M, and this may not hold or be hard to prove. It is usually more economical in terms of conditions to show that the maximizers $\hat{h}_n$ are contained in certain sets K, with high probability. Then uniform convergence of M_n to M on K is sufficient. The choice of compact sets K corresponds to establishing the uniform tightness of the sequence $\hat{h}_n$ before applying the argmax theorem.

If the sample paths of the processes M_n are bounded on K and $H_n = H$ for every n, then the weak convergence of the processes M_n viewed as elements of the space $\ell^\infty(K)$ implies the convergence condition of the argmax theorem. This follows by the continuous-mapping theorem, because the map

$$z \mapsto \big(z(A \cap K), z(B \cap K)\big)$$

from $\ell^\infty(K)$ to $\mathbb{R}^2$ is continuous, for every pair of sets A and B. The weak convergence in $\ell^\infty(K)$ remains sufficient if the sets H_n depend on n but converge in a suitable way. Write $H_n \to H$ if H is the set of all limits $\lim h_n$ of converging sequences h_n with $h_n \in H_n$ for every n and, moreover, the limit $h = \lim_i h_{n_i}$ of every converging sequence h_{n_i} with $h_{n_i} \in H_{n_i}$ for every i is contained in H.

5.58   Corollary. *Suppose that $M_n \rightsquigarrow M$ in $\ell^\infty(K)$ for every compact subset K of $\mathbb{R}^k$, for a limit process M with continuous sample paths that have unique points of maxima $\hat{h}$. If $H_n \to H$, $M_n(\hat{h}_n) \geq M_n(H_n) - o_P(1)$, and the sequence $\hat{h}_n$ is uniformly tight, then $\hat{h}_n \rightsquigarrow \hat{h}$.*

Proof. The compactness of K and the continuity of the sample paths $h \mapsto M(h)$ imply that the (unique) points of maximum $\hat{h}$ are automatically well separated in the sense of (5.57). Indeed, if this fails for a given open set $G \ni \hat{h}$ and K (and a given ω in the underlying probability space), then there exists a sequence h_m in $G^c \cap K \cap H$ such that $M(h_m) \to M(\hat{h})$. If K is compact, then this sequence can be chosen convergent. The limit h_0 must be in the closed set G^c and hence cannot be $\hat{h}$. By the continuity of M it also has the property that $M(h_0) = \lim M(h_m) = M(\hat{h})$. This contradicts the assumption that $\hat{h}$ is a unique point of maximum.

If we can show that $\big(M_n(F \cap H_n), M_n(K \cap H_n)\big)$ converges to the corresponding limit for every compact sets $F \subset K$, then the theorem is a corollary of Theorem 5.56. If $H_n = H$ for every n, then this convergence is immediate from the weak convergence of M_n to M in $\ell^\infty(K)$, by the continuous-mapping theorem. For H_n changing with n this convergence may fail, and we need to refine the proof of Theorem 5.56. This goes through with minor changes if

$$\limsup_{n \to \infty} P\big(M_n(F \cap H_n) - M_n(\mathring{K} \cap H_n) \geq x\big) \leq P\big(M(F \cap H) - M(\mathring{K} \cap H) \geq x\big),$$

for every x, every compact set F and every large closed ball K. Define functions $g_n : \ell^\infty(K) \mapsto \mathbb{R}$ by

$$g_n(z) = \sup_{h \in F \cap H_n} z(h) - \sup_{h \in \mathring{K} \cap H_n} z(h),$$

and g similarly, but with H replacing H_n. By an argument as in the proof of Theorem 18.11, the desired result follows if $\limsup g_n(z_n) \leq g(z)$ for every sequence $z_n \to z$ in $\ell^\infty(K)$ and continuous function z. (Then $\limsup P(g_n(M_n) \geq x) \leq P(g(M) \geq x)$ for every x, for any weakly converging sequence $M_n \rightsquigarrow M$ with a limit with continuous sample paths.) This in turn follows if for every precompact set $B \subset K$,

$$\sup_{h \in \mathring{B} \cap H} z(h) \leq \varlimsup_{n \to \infty} \sup_{h \in B \cap H_n} z_n(h) \leq \sup_{h \in \overline{B} \cap H} z(h).$$

To prove the upper inequality, select $h_n \in B \cap H_n$ such that

$$\sup_{h \in B \cap H_n} z_n(h) = z_n(h_n) + o(1) = z(h_n) + o(1).$$

Because $\overline{B}$ is compact, every subsequence of h_n has a converging subsequence. Because $H_n \to H$, the limit h must be in $\overline{B} \cap H$. Because $z(h_n) \to z(h)$, the upper bound follows.

To prove the lower inequality, select for given $\varepsilon > 0$ an element $h \in \mathring{B} \cap H$ such that

$$\sup_{h \in \mathring{B} \cap H} z(h) \leq z(h) + \varepsilon.$$

Because $H_n \to H$, there exists $h_n \in H_n$ with $h_n \to h$. This sequence must be in $\mathring{B} \subset B$ eventually, whence $z(h) = \lim z(h_n) = \lim z_n(h_n)$ is bounded above by $\liminf \sup_{h \in B \cap H_n} z_n(h)$. ∎

The argmax theorem can also be used to prove consistency, by applying it to the original criterion functions $\theta \mapsto M_n(\theta)$. Then the limit process $\theta \mapsto M(\theta)$ is degenerate, and has a fixed point of maximum θ_0. Weak convergence becomes convergence in probability, and the theorem now gives conditions for the consistency $\hat{\theta}_n \xrightarrow{P} \theta_0$. Condition (5.57) reduces to the well-separation of θ_0, and the convergence

$$\sup_{\theta \in F \cap K \cap \Theta_n} M_n(\theta) \xrightarrow{P} \sup_{\theta \in F \cap K \cap \Theta} M_n(\theta)$$

is, apart from allowing Θ_n to depend on n, weaker than the uniform convergence of M_n to M.

Notes

In the section on consistency we have given two main results (uniform convergence and Wald's proof) that have proven their value over the years, but there is more to say on this subject. The two approaches can be unified by replacing the uniform convergence by "one-sided uniform convergence," which in the case of i.i.d. observations can be established under the conditions of Wald's theorem by a bracketing approach as in Example 19.8 (but then one-sided). Furthermore, the use of special properties, such as convexity of the ψ or m functions, is often helpful. Examples such as Lemma 5.10, or the treatment of maximum likelihood estimators in exponential families in Chapter 4, appear to indicate that no single approach can be satisfactory.

The study of the asymptotic properties of maximum likelihood estimators and other M-estimators has a long history. Fisher [48], [50] was a strong advocate of the method of maximum likelihood and noted its asymptotic optimality as early as the 1920s. What we have labelled the classical conditions correspond to the rigorous treatment given by Cramér [27] in his authoritative book. Huber initiated the systematic study of M-estimators, with the purpose of developing robust statistical procedures. His paper [78] contains important ideas that are precursors for the application of techniques from the theory of empirical processes by, among others, Pollard, as in [117], [118], and [120]. For one-dimensional parameters these empirical process methods can be avoided by using a maximal inequality based on the L_2-norm (see, e.g., Theorem 2.2.4 in [146]). Surprisingly, then a Lipschitz condition on the Hellinger distance (an integrated quantity) suffices; see for example, [80] or [94]. For higher-dimensional parameters the results are also not the best possible, but I do not know of any simple better ones.

The books by Huber [79] and by Hampel, Ronchetti, Rousseeuw, and Stahel [73] are good sources for applications of M-estimators in robust statistics. These references also discuss the relative efficiency of the different M-estimators, which motivates, for instance, the use of Huber's ψ-function. In this chapter we have derived Huber's estimator as the solution of the problem of minimizing the asymptotic variance under the side condition of a uniformly bounded influence function. Originally Huber derived it as the solution to the problem of minimizing the maximum asymptotic variance $\sup_P \sigma_P^2$ for P ranging over a contamination neighborhood: $P = (1 - \varepsilon)\Phi + \varepsilon Q$ with Q arbitrary. For M-estimators these two approaches turn out to be equivalent.

The one-step method can be traced back to numerical schemes for solving the likelihood equations, including Fisher's method of scoring. One-step estimators were introduced for their asymptotic efficiency by Le Cam in 1956, who later developed them for general locally asymptotically quadratic models, and also introduced the discretization device, (see [93]).

PROBLEMS

1. Let $X_1, \ldots, X_n$ be a sample from a density that is strictly positive and symmetric about some point. Show that the Huber M-estimator for location is consistent for the symmetry point.

2. Find an expression for the asymptotic variance of the Huber estimator for location if the observations are normally distributed.

3. Define $\psi(x) = 1 - p, 0, p$ if $x < 0, 0, > 0$. Show that $\mathrm{E}\psi(X - \theta) = 0$ implies that $\mathrm{P}(X < \theta) \leq p \leq \mathrm{P}(X \leq \theta)$.

4. Let $X_1, \ldots, X_n$ be i.i.d. $N(\mu, \sigma^2)$-distributed. Derive the maximum likelihood estimator for (μ, σ^2) and show that it is asymptotically normal. Calculate the Fisher information matrix for this parameter and its inverse.

5. Let $X_1, \ldots, X_n$ be i.i.d. Poisson$(1/\theta)$-distributed. Derive the maximum likelihood estimator for θ and show that it is asymptotically normal.

6. Let $X_1, \ldots, X_n$ be i.i.d. $N(\theta, \theta)$-distributed. Derive the maximum likelihood estimator for θ and show that it is asymptotically normal.

7. Find a sequence of fixed (nonrandom) functions $M_n: \mathbb{R} \mapsto \mathbb{R}$ that converges pointwise to a limit M_0 and such that each M_n has a unique maximum at a point θ_n, but the sequence θ_n does not converge to θ_0. Can you also find a sequence M_n that converges uniformly?

8. Find a sequence of fixed (nonrandom) functions $M_n: \mathbb{R} \mapsto \mathbb{R}$ that converges pointwise but not uniformly to a limit M_0 such that each M_n has a unique maximum at a point θ_n and the sequence θ_n converges to θ_0.

9. Let $X_1, \ldots, X_n$ be i.i.d. observations from a uniform distribution on $[0, \theta]$. Show that the sequence of maximum likelihood estimators is asymptotically consistent. Show that it is not asymptotically normal.

10. Let $X_1, \ldots, X_n$ be i.i.d. observations from an exponential density $\theta \exp(-\theta x)$. Show that the sequence of maximum likelihood estimators is asymptotically normal.

11. Let $\mathbb{F}_n^{-1}(p)$ be a pth sample quantile of a sample from a cumulative distribution F on $\mathbb{R}$ that is differentiable with positive derivative at the population pth-quantile $F^{-1}(p) = \inf\{x: F(x) \geq p\}$. Show that $\sqrt{n}\big(\mathbb{F}_n^{-1}(p) - F^{-1}(p)\big)$ is asymptotically normal with mean zero and variance $p(1 - p)/f\big(F^{-1}(p)\big)^2$.

12. Derive a minimal condition on the distribution function F that guarantees the consistency of the sample pth quantile.

13. Calculate the asymptotic variance of $\sqrt{n}(\hat{\theta}_n - \theta)$ in Example 5.26.

14. Suppose that we observe a random sample from the distribution of (X, Y) in the following *errors-in-variables* model:

$$X = Z + e$$
$$Y_i = \alpha + \beta Z + f,$$

where (e, f) is bivariate normally distributed with mean 0 and covariance matrix $\sigma^2 I$ and is independent from the unobservable variable Z. In analogy to Example 5.26, construct a system of estimating equations for (α, β) based on a conditional likelihood, and study the limit properties of the corresponding estimators.

15. In Example 5.27, for what point is the least squares estimator $\hat{\theta}_n$ consistent if we drop the condition that $\mathrm{E}(e \mid X) = 0$? Derive an (implicit) solution in terms of the function $\mathrm{E}(e \mid X)$. Is it necessarily θ_0 if $\mathrm{E}e = 0$?

16. In Example 5.27, consider the asymptotic behavior of the least absolute-value estimator $\hat{\theta}$ that minimizes $\sum_{i=1}^{n} |Y_i - \phi_\theta(X_i)|$.

17. Let $X_1, \ldots, X_n$ be i.i.d. with density $f_{\lambda,a}(x) = \lambda e^{-\lambda(x-a)} 1\{x \geq a\}$, where the parameters $\lambda > 0$ and $a \in \mathbb{R}$ are unknown. Calculate the maximum likelihood estimator $(\hat{\lambda}_n, \hat{a}_n)$ of (λ, a) and derive its asymptotic properties.

18. Let X be Poisson-distributed with density $p_\theta(x) = \theta^x e^{-\theta}/x!$. Show by direct calculation that $E_\theta \dot{\ell}_\theta(X) = 0$ and $E_\theta \ddot{\ell}_\theta(X) = -E_\theta \dot{\ell}_\theta^2(X)$. Compare this with the assertions in the introduction. Apparently, differentiation under the integral (sum) is permitted in this case. Is that obvious from results from measure theory or (complex) analysis?

19. Let $X_1, \ldots, X_n$ be a sample from the $N(\theta, 1)$ distribution, where it is known that $\theta \geq 0$. Show that the maximum likelihood estimator is not asymptotically normal under $\theta = 0$. Why does this not contradict the theorems of this chapter?

20. Show that $(\tilde{\theta} - \theta_0) \Psi_n(\tilde{\theta}_n)$ in formula (5.18) converges in probability to zero if $\hat{\theta}_n \xrightarrow{P} \theta_0$, and that there exists an integrable function M and $\delta > 0$ with $|\ddot{\psi}_\theta(x)| \leq M(x)$ for every x and every $\|\theta - \theta_0\| < \delta$.

21. If $\hat{\theta}_n$ maximizes M_n, then it also maximizes M_n^+. Show that this may be used to relax the conditions of Theorem 5.7 to $\sup_\theta |M_n^+ - M^+|(\theta) \to 0$ in probability $\left(\text{if } M(\theta_0) > 0\right)$.

22. Suppose that for every $\varepsilon > 0$ there exists a set Θ_ε with $\liminf P(\hat{\theta}_n \in \Theta_\varepsilon) \geq 1 - \varepsilon$. Then uniform convergence of M_n to M in Theorem 5.7 can be relaxed to uniform convergence on every Θ_ε.

23. Show that Wald's consistency proof yields almost sure convergence of $\hat{\theta}_n$, rather than convergence in probability if the parameter space is compact and $M_n(\hat{\theta}_n) \geq M_n(\theta_0) - o(1)$.

24. Suppose that $(X_1, Y_1), \ldots, (X_n, Y_n)$ are i.i.d. and satisfy the linear regression relationship $Y_i = \theta^T X_i + e_i$ for (unobservable) errors $e_1, \ldots, e_n$ independent of $X_1, \ldots, X_n$. Show that the mean absolute deviation estimator, which minimizes $\sum |Y_i - \theta X_i|$, is asymptotically normal under a mild condition on the error distribution.

25. (i) Verify the conditions of Wald's theorem for m_θ the log likelihood function of the $N(\mu, \sigma^2)$-distribution if the parameter set for $\theta = (\mu, \sigma^2)$ is a compact subset of $\mathbb{R} \times \mathbb{R}^+$.

  (ii) Extend m_θ by continuity to the compactification of $\mathbb{R} \times \mathbb{R}^+$. Show that the conditions of Wald's theorem fail at the points $(\mu, 0)$.

  (iii) Replace m_θ by the log likelihood function of a pair of two independent observations from the $N(\mu, \sigma^2)$-distribution. Show that Wald's theorem now does apply, also with a compactified parameter set.

26. A distribution on $\mathbb{R}^k$ is called *ellipsoidally symmetric* if it has a density of the form $x \mapsto g\big((x - \mu)^T \Sigma^{-1}(x - \mu)\big)$ for a function $g: [0, \infty) \mapsto [0, \infty)$, a vector μ, and a symmetric positive-definite matrix Σ. Study the Z-estimators for location $\hat{\mu}$ that solve an equation of the form

$$\sum_{i=1}^{n} \psi\big((X_i - \mu)^T \hat{\Sigma}_n^{-1}(X_i - \mu)\big),$$

for given estimators $\hat{\Sigma}_n$ and, for instance, Huber's ψ-function. Is the asymptotic distribution of $\hat{\Sigma}_n$ important?

27. Suppose that Θ is a compact metric space and $M : \Theta \to \mathbb{R}$ is continuous. Show that (5.8) is equivalent to the point θ_0 being a point of unique global maximum. Can you relax the continuity of M to some form of "semi-continuity"?

6

Contiguity

"Contiguity" is another name for "asymptotic absolute continuity." Contiguity arguments are a technique to obtain the limit distribution of a sequence of statistics under underlying laws Q_n from a limiting distribution under laws P_n. Typically, the laws P_n describe a null distribution under investigation, and the laws Q_n correspond to an alternative hypothesis.

6.1 Likelihood Ratios

Let P and Q be measures on a measurable space $(\Omega, \mathcal{A})$. Then Q is *absolutely continuous* with respect to P if $P(A) = 0$ implies $Q(A) = 0$ for every measurable set A; this is denoted by $Q \ll P$. Furthermore, P and Q are *orthogonal* if Ω can be partitioned as $\Omega = \Omega_P \cup \Omega_Q$ with $\Omega_P \cap \Omega_Q = \emptyset$ and $P(\Omega_Q) = 0 = Q(\Omega_P)$. Thus P "charges" only Ω_P and Q "lives on" the set Ω_Q, which is disjoint with the "support" of P. Orthogonality is denoted by $P \perp Q$.

In general, two measures P and Q need be neither absolutely continuous nor orthogonal. The relationship between their supports can best be described in terms of densities. Suppose P and Q possess densities p and q with respect to a measure μ, and consider the sets

$$\Omega_P = \{p > 0\}, \qquad \Omega_Q = \{q > 0\}.$$

See Figure 6.1. Because $P(\Omega_P^c) = \int_{p=0} p \, d\mu = 0$, the measure P is supported on the set Ω_P. Similarly, Q is supported on Ω_Q. The intersection $\Omega_P \cap \Omega_Q$ receives positive measure from both P and Q provided its measure under μ is positive. The measure Q can be written as the sum $Q = Q^a + Q^\perp$ of the measures

$$Q^a(A) = Q(A \cap \{p > 0\}); \qquad Q^\perp(A) = Q(A \cap \{p = 0\}). \tag{6.1}$$

As proved in the next lemma, $Q^a \ll P$ and $Q^\perp \perp P$. Furthermore, for every measurable set A

$$Q^a(A) = \int_A \frac{q}{p} \, dP.$$

The decomposition $Q = Q^a + Q^\perp$ is called the *Lebesgue decomposition* of Q with respect to P. The measures Q^a and $Q^\perp$ are called the *absolutely continuous part* and the *orthogonal*

85

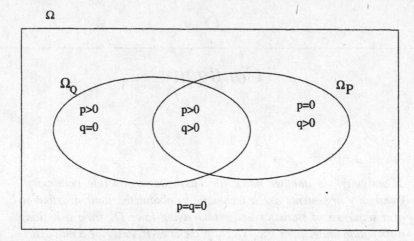

Figure 6.1. Supports of measures.

part (or *singular part*) of Q with respect to P, respectively. In view of the preceding display, the function q/p is a density of Q^a with respect to P. It is denoted dQ/dP (not: dQ^a/dP), so that

$$\frac{dQ}{dP} = \frac{q}{p}, \qquad P-\text{a.s.}$$

As long as we are only interested in the properties of the quotient q/p under P-probability, we may leave the quotient undefined for $p = 0$. The density dQ/dP is only P-almost surely unique by definition. Even though we have used densities to define them, dQ/dP and the Lebesgue decomposition are actually independent of the choice of densities and dominating measure.

In statistics a more common name for a Radon-Nikodym density is *likelihood ratio*. We shall think of it as a random variable $dQ/dP : \Omega \mapsto [0, \infty)$ and shall study its law under P.

6.2   Lemma. *Let P and Q be probability measures with densities p and q with respect to a measure μ. Then for the measures Q^a and $Q^\perp$ defined in (6.1)*

   (i) $Q = Q^a + Q^\perp$, $Q^a \ll P$, $Q^\perp \perp P$.
   (ii) $Q^a(A) = \int_A (q/p)\, dP$ for every measurable set A.
   (iii) $Q \ll P$ if and only if $Q(p = 0) = 0$ if and only if $\int (q/p)\, dP = 1$.

Proof.   The first statement of (i) is obvious from the definitions of Q^a and $Q^\perp$. For the second, we note that $P(A)$ can be zero only if $p(x) = 0$ for μ-almost all $x \in A$. In this case, $\mu\big(A \cap \{p > 0\}\big) = 0$, whence $Q^a(A) = Q\big(A \cap \{p > 0\}\big) = 0$ by the absolute continuity of Q with respect to μ. The third statement of (i) follows from $P(p = 0) = 0$ and $Q^\perp(p > 0) = Q(\emptyset) = 0$.

Statement (ii) follows from

$$Q^a(A) = \int_{A \cap \{p > 0\}} q\, d\mu = \int_{A \cap \{p > 0\}} \frac{q}{p} p\, d\mu = \int_A \frac{q}{p} dP.$$

For (iii) we note first that $Q \ll P$ if and only if $Q^\perp = 0$. By (6.1) the latter happens if and only if $Q(p = 0) = 0$. This yields the first "if and only if." For the second, we note that by (ii) the total mass of Q^a is equal to $Q^a(\Omega) = \int (q/p)dP$. This is 1 if and only if $Q^a = Q$. ∎

It is not true in general that $\int f dQ = \int f(dQ/dP)dP$. For this to be true for every measurable function f, the measure Q must be absolutely continuous with respect to P. On the other hand, for any P and Q and nonnegative f,

$$\int f dQ \geq \int_{p>0} f q \, d\mu = \int_{p>0} f \frac{q}{p} p \, d\mu = \int f \frac{dQ}{dP} dP.$$

This inequality is used freely in the following. The inequality may be strict, because dividing by zero is not permitted.[†]

6.2   Contiguity

If a probability measure Q is absolutely continuous with respect to a probability measure P, then the Q-law of a random vector $X : \Omega \mapsto \mathbb{R}^k$ can be calculated from the P-law of the pair $(X, dQ/dp)$ through the formula

$$E_Q f(X) = E_p f(X) \frac{dQ}{dP}.$$

With $P^{X,V}$ equal to the law of the pair $(X, V) = (X, dQ/dP)$ under P, this relationship can also be expressed as

$$Q(X \in B) = E_P 1_B(X) \frac{dQ}{dP} = \int_{B \times \mathbb{R}} v \, dP^{X,V}(x, v).$$

The validity of these formulas depends essentially on the absolute continuity of Q with respect to P, because a part of Q that is orthogonal with respect to P cannot be recovered from any P-law.

Consider an asymptotic version of the problem. Let $(\Omega_n, \mathcal{A}_n)$ be measurable spaces, each equipped with a pair of probability measures P_n and Q_n. Under what conditions can a Q_n-limit law of random vectors $X_n : \Omega_n \mapsto \mathbb{R}^k$ be obtained from suitable P_n-limit laws? In view of the above it is necessary that Q_n is "asymptotically absolutely continuous" with respect to P_n in a suitable sense. The right concept is contiguity.

6.3   Definition. The sequence Q_n is *contiguous* with respect to the sequence P_n if $P_n(A_n) \to 0$ implies $Q_n(A_n) \to 0$ for every sequence of measurable sets A_n. This is denoted $Q_n \triangleleft P_n$. The sequences P_n and Q_n are *mutually contiguous* if both $P_n \triangleleft Q_n$ and $Q_n \triangleleft P_n$. This is denoted $P_n \triangleleft \triangleright Q_n$.

The name "contiguous" is standard, but perhaps conveys a wrong image. "Contiguity" suggests sequences of probability measures living next to each other, but the correct image is "on top of each other" (in the limit).

[†] The algebraic identity $dQ = (dQ/dP)dP$ is false, because the notation dQ/dP is used as shorthand for dQ^a/dP: If we write dQ/dp, then we are not implicitly assuming that $Q \ll P$.

Before answering the question of interest, we give two characterizations of contiguity in terms of the asymptotic behavior of the likelihood ratios of P_n and Q_n. The likelihood ratios dQ_n/dP_n and dP_n/dQ_n are nonnegative and satisfy

$$E_{P_n}\frac{dQ_n}{dP_n} \le 1 \quad \text{and} \quad E_{Q_n}\frac{dP_n}{dQ_n} \le 1.$$

Thus, the sequences of likelihood ratios dQ_n/dP_n and dP_n/dQ_n are uniformly tight under P_n and Q_n, respectively. By Prohorov's theorem, every subsequence has a further weakly converging subsequence. The next lemma shows that the properties of the limit points determine contiguity. This can be understood in analogy with the nonasymptotic situation. For probability measures P and Q, the following three statements are equivalent by (iii) of Lemma 6.2:

$$Q \ll P, \qquad Q\left(\frac{dP}{dQ} = 0\right) = 0, \qquad E_P\frac{dQ}{dP} = 1.$$

This equivalence persists if the three statements are replaced by their asymptotic counterparts: Sequences P_n and Q_n satisfy $Q_n \triangleleft P_n$, if and only if the weak limit points of dP_n/dQ_n under Q_n give mass 0 to 0, if and only if the weak limit points of dQ_n/dP_n under P_n have mean 1.

6.4   Lemma (Le Cam's first lemma). *Let P_n and Q_n be sequences of probability measures on measurable spaces $(\Omega_n, \mathcal{A}_n)$. Then the following statements are equivalent:*

(i) *$Q_n \triangleleft P_n$.*

(ii) *If $dP_n/dQ_n \overset{Q_n}{\rightsquigarrow} U$ along a subsequence, then $P(U > 0) = 1$.*

(iii) *If $dQ_n/dP_n \overset{P_n}{\rightsquigarrow} V$ along a subsequence, then $EV = 1$.*

(iv) *For any statistics $T_n : \Omega_n \mapsto \mathbb{R}^k$: If $T_n \overset{P_n}{\to} 0$, then $T_n \overset{Q_n}{\to} 0$.*

Proof.   The equivalence of (i) and (iv) follows directly from the definition of contiguity: Given statistics T_n, consider the sets $A_n = \{\|T_n\| > \varepsilon\}$; given sets A_n, consider the statistics $T_n = 1_{A_n}$.

(i) $\Rightarrow$ (ii). For simplicity of notation, we write just $\{n\}$ for the given subsequence along which $dP_n/dQ_n \overset{Q_n}{\rightsquigarrow} U$. For given n, we define the function $g_n(\varepsilon) = Q_n(dP_n/dQ_n < \varepsilon) - P(U < \varepsilon)$. By the portmanteau lemma, $\liminf g_n(\varepsilon) \ge 0$ for every $\varepsilon > 0$. Then, for $\varepsilon_n \downarrow 0$ at a sufficiently slow rate, also $\liminf g_n(\varepsilon_n) \ge 0$. Thus,

$$P(U = 0) = \lim P(U < \varepsilon_n) \le \liminf Q_n\left(\frac{dP_n}{dQ_n} < \varepsilon_n\right).$$

On the other hand,

$$P_n\left(\frac{dP_n}{dQ_n} \le \varepsilon_n \wedge q_n > 0\right) = \int_{dP_n/dQ_n \le \varepsilon_n} \frac{dP_n}{dQ_n}\, dQ_n \le \int \varepsilon_n\, dQ_n \to 0.$$

If Q_n is contiguous with respect to P_n, then the Q_n-probability of the set on the left goes to zero also. But this is the probability on the right in the first display. Combination shows that $P(U = 0) = 0$.

(iii) $\Rightarrow$ (i). If $P_n(A_n) \to 0$, then the sequence $1_{\Omega_n - A_n}$ converges to 1 in P_n-probability. By Prohorov's theorem, every subsequence of $\{n\}$ has a further subsequence along which

$(dQ_n/dP_n, 1_{\Omega_n - A_n}) \rightsquigarrow (V, 1)$ under P_n, for some weak limit V. The function $(v, t) \mapsto vt$ is continuous and nonnegative on the set $[0, \infty) \times \{0, 1\}$. By the portmanteau lemma

$$\liminf Q_n(\Omega_n - A_n) \geq \liminf \int 1_{\Omega_n - A_n} \frac{dQ_n}{dP_n} dP_n \geq \mathrm{E}1 \cdot V.$$

Under (iii) the right side equals $\mathrm{E}V = 1$. Then the left side is 1 as well and the sequence $Q_n(A_n) = 1 - Q_n(\Omega_n - A_n)$ converges to zero.

(ii) $\Rightarrow$ (iii). The probability measures $\mu_n = \frac{1}{2}(P_n + Q_n)$ dominate both P_n and Q_n, for every n. The sum of the densities of P_n and Q_n with respect to μ_n equals 2. Hence, each of the densities takes its values in the compact interval $[0, 2]$. By Prohorov's theorem every subsequence possesses a further subsequence along which

$$\frac{dP_n}{dQ_n} \overset{Q_n}{\rightsquigarrow} U, \qquad \frac{dQ_n}{dP_n} \overset{P_n}{\rightsquigarrow} V, \qquad W_n := \frac{dP_n}{d\mu_n} \overset{R_n}{\rightsquigarrow} W,$$

for certain random variables U, V and W. Every W_n has expectation 1 under μ_n. In view of the boundedness, the weak convergence of the sequence W_n implies convergence of moments, and the limit variable has mean $\mathrm{E}W = 1$ as well. For a given bounded, continuous function f, define a function $g : [0, 2] \mapsto \mathbb{R}$ by $g(w) = f(w/(2-w))(2-w)$ for $0 \leq w < 2$ and $g(2) = 0$. Then g is bounded and continuous. Because $dP_n/dQ_n = W_n/(2 - W_n)$ and $dQ_n/d\mu_n = 2 - W_n$, the portmanteau lemma yields

$$\mathrm{E}_{Q_n} f\left(\frac{dP_n}{dQ_n}\right) = \mathrm{E}_{\mu_n} f\left(\frac{dP_n}{dQ_n}\right) \frac{dQ_n}{d\mu_n} = \mathrm{E}_{\mu_n} g(W_n) \rightarrow \mathrm{E}f\left(\frac{W}{2 - W}\right)(2 - W),$$

where the integrand in the right side is understood to be $g(2) = 0$ if $W = 2$. By assumption, the left side converges to $\mathrm{E}f(U)$. Thus $\mathrm{E}f(U)$ equals the right side of the display for every continuous and bounded function f. Take a sequence of such functions with $1 \geq f_m \downarrow 1_{\{0\}}$, and conclude by the dominated-convergence theorem that

$$\mathrm{P}(U = 0) = \mathrm{E}1_{\{0\}}(U) = \mathrm{E}1_{\{0\}}\left(\frac{W}{2 - W}\right)(2 - W) = 2\mathrm{P}(W = 0).$$

By a similar argument, $\mathrm{E}f(V) = \mathrm{E}f((2 - W)/W)W$ for every continuous and bounded function f, where the integrand on the right is understood to be zero if $W = 0$. Take a sequence $0 \leq f_m(x) \uparrow x$ and conclude by the monotone convergence theorem that

$$\mathrm{E}V = \mathrm{E}\left(\frac{2 - W}{W}\right)W = \mathrm{E}(2 - W)1_{W>0} = 2\mathrm{P}(W > 0) - 1.$$

Combination of the last two displays shows that $\mathrm{P}(U = 0) + \mathrm{E}V = 1$. ∎

6.5 Example (Asymptotic log normality). The following special case plays an important role in the asymptotic theory of smooth parametric models. Let P_n and Q_n be probability measures on arbitrary measurable spaces such that

$$\frac{dP_n}{dQ_n} \overset{Q_n}{\rightsquigarrow} e^{N(\mu, \sigma^2)}.$$

Then $Q_n \triangleleft P_n$. Furthermore, $Q_n \triangleleft \triangleright P_n$ if and only if $\mu = -\frac{1}{2}\sigma^2$.

Because the (log normal) variable on the right is positive, the first assertion is immediate from (ii) of the theorem. The second follows from (iii) with the roles of P_n and Q_n switched, on noting that $\mathrm{E} \exp N(\mu, \sigma^2) = 1$ if and only if $\mu = -\frac{1}{2}\sigma^2$.

A mean equal to minus half times the variance looks peculiar, but we shall see that this situation arises naturally in the study of the asymptotic optimality of statistical procedures.   □

The following theorem solves the problem of obtaining a Q_n-limit law from a P_n-limit law that we posed in the introduction. The result, a version of *Le Cam's third lemma*, is in perfect analogy with the nonasymptotic situation.

6.6   Theorem. *Let P_n and Q_n be sequences of probability measures on measurable spaces $(\Omega_n, \mathcal{A}_n)$, and let $X_n : \Omega_n \mapsto \mathbb{R}^k$ be a sequence of random vectors. Suppose that $Q_n \triangleleft P_n$ and*

$$\left(X_n, \frac{dQ_n}{dP_n} \right) \overset{P_n}{\rightsquigarrow} (X, V).$$

Then $L(B) = \mathrm{E} 1_B(X)\, V$ defines a probability measure, and $X_n \overset{Q_n}{\rightsquigarrow} L$.

Proof. Because $V \geq 0$, it follows with the help of the monotone convergence theorem that L defines a measure. By contiguity, $\mathrm{E}V = 1$ and hence L is a probability measure. It is immediate from the definition of L that $\int f\, dL = \mathrm{E}f(X)\, V$ for every measurable indicator function f. Conclude, in steps, that the same is true for every simple function f, any nonnegative measurable function, and every integrable function.

If f is continuous and nonnegative, then so is the function $(x, v) \mapsto f(x)\, v$ on $\mathbb{R}^k \times [0, \infty)$. Thus

$$\liminf \mathrm{E}_{Q_n} f(X_n) \geq \liminf \int f(X_n) \frac{dQ_n}{dP_n}\, dP_n \geq \mathrm{E}f(X)V,$$

by the portmanteau lemma. Apply the portmanteau lemma in the converse direction to conclude the proof that $X_n \overset{Q_n}{\rightsquigarrow} L$.   ■

6.7   Example (Le Cam's third lemma). The name *Le Cam's third lemma* is often reserved for the following result. If

$$\left(X_n, \log \frac{dQ_n}{dP_n} \right) \overset{P_n}{\rightsquigarrow} N_{k+1} \left(\begin{pmatrix} \mu \\ -\frac{1}{2}\sigma^2 \end{pmatrix}, \begin{pmatrix} \Sigma & \tau \\ \tau^T & \sigma^2 \end{pmatrix} \right),$$

then

$$X_n \overset{Q_n}{\rightsquigarrow} N_k(\mu + \tau, \Sigma).$$

In this situation the asymptotic covariance matrices of the sequence X_n are the same under P_n and Q_n, but the mean vectors differ by the asymptotic covariance τ between X_n and the log likelihood ratios.[†]

The statement is a special case of the preceding theorem. Let (X, W) have the given $(k+1)$-dimensional normal distribution. By the continuous mapping theorem, the sequence $(X_n, dQ_n/dP_n)$ converges in distribution under P_n to (X, e^W). Because W is $N(-\frac{1}{2}\sigma^2, \sigma^2)$-distributed, the sequences P_n and Q_n are mutually contiguous. According to the abstract

---

[†] We set $\log 0 = -\infty$; because the normal distribution does not charge the point $-\infty$ the assumed asymptotic normality of $\log dQ_n/dP_n$ includes the assumption that $P_n(dQ_n/dP_n = 0) \to 0$.

version of Le Cam's third lemma, $X_n \overset{Q_n}{\leadsto} L$ with $L(B) = \mathrm{E} 1_B(X) e^W$. The characteristic function of L is $\int e^{it^T x} \, dL(x) = \mathrm{E} e^{it^T X} e^W$. This is the characteristic function of the given normal distribution at the vector $(t, -i)$. Thus

$$\int e^{it^T x} \, dL(x) = e^{it^T \mu - \frac{1}{2}\sigma^2 - \frac{1}{2}(t^T, -i)\left(\begin{smallmatrix} \Sigma & \tau \\ \tau^T & \sigma^2 \end{smallmatrix}\right)\left(\begin{smallmatrix} t \\ -i \end{smallmatrix}\right)} = e^{it^T(\mu+\tau) - \frac{1}{2}t^T \Sigma t}.$$

The right side is the characteristic function of the $N_k(\mu + \tau, \Sigma)$ distribution. $\qquad\square$

Notes

The concept and theory of contiguity was developed by Le Cam in [92]. In his paper the results that were later to become known as Le Cam's lemmas are listed as a single theorem. The names "first" and "third" appear to originate from [71]. (The second lemma is on product measures and the first lemma is actually only the implication (iii) $\Rightarrow$ (i).)

PROBLEMS

1. Let $P_n = N(0, 1)$ and $Q_n = N(\mu_n, 1)$. Show that the sequences P_n and Q_n are mutually contiguous if and only if the sequence μ_n is bounded.

2. Let P_n and Q_n be the distribution of the mean of a sample of size n from the $N(0, 1)$ and the $N(\theta_n, 1)$ distribution, respectively. Show that $P_n \triangleleft\triangleright Q_n$ if and only if $\theta_n = O(1/\sqrt{n})$.

3. Let P_n and Q_n be the law of a sample of size n from the uniform distribution on $[0, 1]$ or $[0, 1+1/n]$, respectively. Show that $P_n \triangleleft Q_n$. Is it also true that $Q_n \triangleleft P_n$? Use Lemma 6.4 to derive your answers.

4. Suppose that $\|P_n - Q_n\| \to 0$, where $\| \cdot \|$ is the total variation distance $\|P - Q\| = \sup_A |P(A) - Q(A)|$. Show that $P_n \triangleleft\triangleright Q_n$.

5. Given $\varepsilon > 0$ find an example of sequences such that $P_n \triangleleft\triangleright Q_n$, but $\|P_n - Q_n\| \to 1 - \varepsilon$. (The maximum total variation distance between two probability measures is 1.) This exercise shows that it is wrong to think of contiguous sequences as being close. (Try measures that are supported on just two points.)

6. Give a simple example in which $P_n \triangleleft Q_n$, but it is not true that $Q_n \triangleleft P_n$.

7. Show that the constant sequences $\{P\}$ and $\{Q\}$ are contiguous if and only if P and Q are absolutely continuous.

8. If $P \ll Q$, then $Q(A_n) \to 0$ implies $P(A_n) \to 0$ for every sequence of measurable sets. How does this follow from Lemma 6.4?

7

Local Asymptotic Normality

A sequence of statistical models is "locally asymptotically normal" if, asymptotically, their likelihood ratio processes are similar to those for a normal location parameter. Technically, this is if the likelihood ratio processes admit a certain quadratic expansion. An important example in which this arises is repeated sampling from a smooth parametric model. Local asymptotic normality implies convergence of the models to a Gaussian model after a rescaling of the parameter.

7.1 Introduction

Suppose we observe a sample $X_1, \ldots, X_n$ from a distribution P_θ on some measurable space $(\mathcal{X}, \mathcal{A})$ indexed by a parameter θ that ranges over an open subset Θ of $\mathbb{R}^k$. Then the full observation is a single observation from the product P_θ^n of n copies of P_θ, and the statistical model is completely described as the collection of probability measures $\{P_\theta^n : \theta \in \Theta\}$ on the sample space $(\mathcal{X}^n, \mathcal{A}^n)$. In the context of the present chapter we shall speak of a *statistical experiment*, rather than of a statistical model. In this chapter it is shown that many statistical experiments can be approximated by Gaussian experiments after a suitable reparametrization.

The reparametrization is centered around a fixed parameter θ_0, which should be regarded as known. We define a *local parameter* $h = \sqrt{n}(\theta - \theta_0)$, rewrite P_θ^n as $P_{\theta_0 + h/\sqrt{n}}^n$, and thus obtain an experiment with parameter h. In this chapter we show that, for large n, the experiments

$$\left(P_{\theta_0 + h/\sqrt{n}}^n : h \in \mathbb{R}^k\right) \quad \text{and} \quad \left(N\big(h, I_{\theta_0}^{-1}\big) : h \in \mathbb{R}^k\right)$$

are similar in statistical properties, whenever the original experiments $\theta \mapsto P_\theta$ are "smooth" in the parameter. The second experiment consists of observing a single observation from a normal distribution with mean h and known covariance matrix (equal to the inverse of the Fisher information matrix). This is a simple experiment, which is easy to analyze, whence the approximation yields much information about the asymptotic properties of the original experiments. This information is extracted in several chapters to follow and concerns both asymptotic optimality theory and the behavior of statistical procedures such as the maximum likelihood estimator and the likelihood ratio test.

We have taken the local parameter set equal to $\mathbb{R}^k$, which is not correct if the parameter set Θ is a true subset of $\mathbb{R}^k$. If θ_0 is an inner point of the original parameter set, then the vector $\theta = \theta_0 + h/\sqrt{n}$ is a parameter in Θ for a given h, for every sufficiently large n, and the local parameter set converges to the whole of $\mathbb{R}^k$ as $n \to \infty$. Then taking the local parameter set equal to $\mathbb{R}^k$ does not cause errors. To give a meaning to the results of this chapter, the measure $P_{\theta_0 + h/\sqrt{n}}$ may be defined arbitrarily if $\theta_0 + h/\sqrt{n} \notin \Theta$.

7.2 Expanding the Likelihood

The convergence of the local experiments is defined and established later in this chapter. First, we discuss the technical tool: a Taylor expansion of the logarithm of the likelihood. Let p_θ be a density of P_θ with respect to some measure μ. Assume for simplicity that the parameter is one-dimensional and that the log likelihood $\ell_\theta(x) = \log p_\theta(x)$ is twice-differentiable with respect to θ, for every x, with derivatives $\dot{\ell}_\theta(x)$ and $\ddot{\ell}_\theta(x)$. Then, for every fixed x,

$$\log \frac{p_{\theta+h}}{p_\theta}(x) = h\dot{\ell}_\theta(x) + \frac{1}{2}h^2\ddot{\ell}_\theta(x) + o_x(h^2).$$

The subscript x in the remainder term is a reminder of the fact that this term depends on x as well as on h. It follows that

$$\log \prod_{i=1}^n \frac{p_{\theta+h/\sqrt{n}}}{p_\theta}(X_i) = \frac{h}{\sqrt{n}} \sum_{i=1}^n \dot{\ell}_\theta(X_i) + \frac{1}{2}\frac{h^2}{n} \sum_{i=1}^n \ddot{\ell}_\theta(X_i) + \text{Rem}_n.$$

Here the score has mean zero, $P_\theta\dot{\ell}_\theta = 0$, and $-P_\theta\ddot{\ell}_\theta = P_\theta\dot{\ell}_\theta^2 = I_\theta$ equals the Fisher information for θ (see, e.g., section 5.5). Hence the first term can be rewritten as $h\Delta_{n,\theta}$, where $\Delta_{n,\theta} = n^{-1/2} \sum_{i=1}^n \dot{\ell}_\theta(X_i)$ is asymptotically normal with mean zero and variance I_θ, by the central limit theorem. Furthermore, the second term in the expansion is asymptotically equivalent to $-\frac{1}{2}h^2 I_\theta$, by the law of large numbers. The remainder term should behave as $o(1/n)$ times a sum of n terms and hopefully is asymptotically negligible. Consequently, under suitable conditions we have, for every h,

$$\log \prod_{i=1}^n \frac{p_{\theta+h/\sqrt{n}}}{p_\theta}(X_i) = h\Delta_{n,\theta} - \frac{1}{2}I_\theta h^2 + o_{P_\theta}(1).$$

In the next section we see that this is similar in form to the likelihood ratio process of a Gaussian experiment. Because this expansion concerns the likelihood process in a neighborhood of θ, we speak of "local asymptotic normality" of the sequence of models $\{P_\theta^n : \theta \in \Theta\}$.

The preceding derivation can be made rigorous under moment or continuity conditions on the second derivative of the log likelihood. Local asymptotic normality was originally deduced in this manner. Surprisingly, it can also be established under a single condition that only involves a first derivative: differentiability of the root density $\theta \mapsto \sqrt{p_\theta}$ in quadratic mean. This entails the existence of a vector of measurable functions $\dot{\ell}_\theta = (\dot{\ell}_{\theta,1}, \ldots, \dot{\ell}_{\theta,k})^T$ such that

$$\int \left[\sqrt{p_{\theta+h}} - \sqrt{p_\theta} - \frac{1}{2}h^T\dot{\ell}_\theta \sqrt{p_\theta} \right]^2 d\mu = o\left(\|h\|^2\right), \qquad h \to 0. \tag{7.1}$$

If this condition is satisfied, then the model $(P_\theta : \theta \in \Theta)$ is called *differentiable in quadratic mean* at θ.

Usually, $\frac{1}{2} h^T \dot{\ell}_\theta(x) \sqrt{p_\theta(x)}$ is the derivative of the map $h \mapsto \sqrt{p_{\theta+h}(x)}$ at $h = 0$ for (almost) every x. In this case

$$\dot{\ell}_\theta(x) = 2 \frac{1}{\sqrt{p_\theta(x)}} \frac{\partial}{\partial \theta} \sqrt{p_\theta(x)} = \frac{\partial}{\partial \theta} \log p_\theta(x).$$

Condition (7.1) does not require differentiability of the map $\theta \mapsto p_\theta(x)$ for any single x, but rather differentiability in (quadratic) mean. Admittedly, the latter is typically established by pointwise differentiability plus a convergence theorem for integrals. Because the condition is exactly right for its purpose, we establish in the following theorem local asymptotic normality under (7.1). A lemma following the theorem gives easily verifiable conditions in terms of pointwise derivatives.

7.2 Theorem. *Suppose that Θ is an open subset of $\mathbb{R}^k$ and that the model $(P_\theta : \theta \in \Theta)$ is differentiable in quadratic mean at θ. Then $P_\theta \dot{\ell}_\theta = 0$ and the Fisher information matrix $I_\theta = P_\theta \dot{\ell}_\theta \dot{\ell}_\theta^T$ exists. Furthermore, for every converging sequence $h_n \to h$, as $n \to \infty$,*

$$\log \prod_{i=1}^n \frac{p_{\theta+h_n/\sqrt{n}}}{p_\theta}(X_i) = \frac{1}{\sqrt{n}} \sum_{i=1}^n h^T \dot{\ell}_\theta(X_i) - \frac{1}{2} h^T I_\theta h + o_{P_\theta}(1).$$

Proof. Given a converging sequence $h_n \to h$, we use the abbreviations p_n, p, and g for $p_{\theta+h_n/\sqrt{n}}$, p_θ, and $h^T \dot{\ell}_\theta$, respectively. By (7.1) the sequence $\sqrt{n}(\sqrt{p_n} - \sqrt{p})$ converges in quadratic mean (i.e., in $L_2(\mu)$) to $\frac{1}{2} g \sqrt{p}$. This implies that the sequence $\sqrt{p_n}$ converges in quadratic mean to $\sqrt{p}$. By the continuity of the inner product,

$$Pg = \int \frac{1}{2} g \sqrt{p}\, 2\sqrt{p}\, d\mu = \lim \int \sqrt{n}(\sqrt{p_n} - \sqrt{p})(\sqrt{p_n} + \sqrt{p})\, d\mu.$$

The right side equals $\sqrt{n}(1-1) = 0$ for every n, because both probability densities integrate to 1. Thus $Pg = 0$.

The random variable $W_{ni} = 2\left[\sqrt{p_n/p}(X_i) - 1\right]$ is with P-probability 1 well defined. By (7.1)

$$\mathrm{var}\left(\sum_{i=1}^n W_{ni} - \frac{1}{\sqrt{n}} \sum_{i=1}^n g(X_i)\right) \leq \mathrm{E}\left(\sqrt{n} W_{ni} - g(X_i)\right)^2 \to 0,$$

$$\mathrm{E} \sum_{i=1}^n W_{ni} = 2n\left(\int \sqrt{p_n}\sqrt{p}\, d\mu - 1\right) = -n \int [\sqrt{p_n} - \sqrt{p}]^2\, d\mu \to -\frac{1}{4} Pg^2. \tag{7.3}$$

Here $Pg^2 = \int g^2\, dP = h^T I_\theta h$ by the definitions of g and I_θ. If both the means and the variances of a sequence of random variables converge to zero, then the sequence converges to zero in probability. Therefore, combining the preceding pair of displayed equations, we find

$$\sum_{i=1}^n W_{ni} = \frac{1}{\sqrt{n}} \sum_{i=1}^n g(X_i) - \frac{1}{4} Pg^2 + o_P(1). \tag{7.4}$$

Next, we express the log likelihood ratio in $\sum_{i=1}^{n} W_{ni}$ through a Taylor expansion of the logarithm. If we write $\log(1 + x) = x - \frac{1}{2}x^2 + x^2 R(2x)$, then $R(x) \to 0$ as $x \to 0$, and

$$\log \prod_{i=1}^{n} \frac{p_n}{p}(X_i) = 2 \sum_{i=1}^{n} \log\left(1 + \frac{1}{2} W_{ni}\right)$$
$$= \sum_{i=1}^{n} W_{ni} - \frac{1}{4} \sum_{i=1}^{n} W_{ni}^2 + \frac{1}{2} \sum_{i=1}^{n} W_{ni}^2 R(W_{ni}). \tag{7.5}$$

As a consequence of the right side of (7.3), it is possible to write $nW_{ni}^2 = g^2(X_i) + A_{ni}$ for random variables A_{ni} such that $\mathrm{E}|A_{ni}| \to 0$. The averages $\bar{A}_n$ converge in mean and hence in probability to zero. Combination with the law of large numbers yields

$$\sum_{i=1}^{n} W_{ni}^2 = \overline{(g^2)}_n + \bar{A}_n \overset{P}{\to} Pg^2.$$

By the triangle inequality followed by Markov's inequality,

$$nP(|W_{ni}| > \varepsilon\sqrt{2}) \le nP(g^2(X_i) > n\varepsilon^2) + nP(|A_{ni}| > n\varepsilon^2)$$
$$\le \varepsilon^{-2} Pg^2\{g^2 > n\varepsilon^2\} + \varepsilon^{-2}\mathrm{E}|A_{ni}| \to 0.$$

The left side is an upper bound for $P(\max_{1 \le i \le n} |W_{ni}| > \varepsilon\sqrt{2})$. Thus the sequence $\max_{1 \le i \le n} |W_{ni}|$ converges to zero in probability. By the property of the function R, the sequence $\max_{1 \le i \le n} |R(W_{ni})|$ converges in probability to zero as well. The last term on the right in (7.5) is bounded by $\max_{1 \le i \le n} |R(W_{ni})| \sum_{i=1}^{n} W_{ni}^2$. Thus it is $o_P(1)O_P(1)$, and converges in probability to zero. Combine to obtain that

$$\log \prod_{i=1}^{n} \frac{p_n}{p}(X_i) = \sum_{i=1}^{n} W_{ni} - \frac{1}{4} Pg^2 + o_P(1).$$

Together with (7.4) this yields the theorem. ∎

To establish the differentiability in quadratic mean of specific models requires a convergence theorem for integrals. Usually one proceeds by showing differentiability of the map $\theta \mapsto p_\theta(x)$ for almost every x plus μ-equi-integrability (e.g., domination). The following lemma takes care of most examples.

7.6   Lemma. *For every θ in an open subset of $\mathbb{R}^k$ let p_θ be a μ-probability density. Assume that the map $\theta \mapsto s_\theta(x) = \sqrt{p_\theta(x)}$ is continuously differentiable for every x. If the elements of the matrix $I_\theta = \int (\dot{p}_\theta / p_\theta)(\dot{p}_\theta^T / p_\theta)\, p_\theta \, d\mu$ are well defined and continuous in θ, then the map $\theta \mapsto \sqrt{p_\theta}$ is differentiable in quadratic mean (7.1) with $\dot{\ell}_\theta$ given by $\dot{p}_\theta / p_\theta$.*

Proof. By the chain rule, the map $\theta \mapsto p_\theta(x) = s_\theta^2(x)$ is differentiable for every x with gradient $\dot{p}_\theta = 2s_\theta \dot{s}_\theta$. Because s_θ is nonnegative, its gradient $\dot{s}_\theta$ at a point at which $s_\theta = 0$ must be zero. Conclude that we can write $\dot{s}_\theta = \frac{1}{2}(\dot{p}_\theta / p_\theta) \sqrt{p_\theta}$, where the quotient $\dot{p}_\theta / p_\theta$ may be defined arbitrarily if $p_\theta = 0$. By assumption, the map $\theta \mapsto I_\theta = 4 \int \dot{s}_\theta \dot{s}_\theta^T \, d\mu$ is continuous.

Because the map $\theta \mapsto s_\theta(x)$ is continuously differentiable, the difference $s_{\theta+h}(x) - s_\theta(x)$ can be written as the integral $\int_0^1 h^T \dot{s}_{\theta+uh}(x)\, du$ of its derivative. By Jensen's (or Cauchy-Schwarz's) inequality, the square of this integral is bounded by the integral $\int_0^1 (h^T \dot{s}_{\theta+uh}(x))^2$

du of the square. Conclude that

$$\int \left(\frac{s_{\theta+th_t} - s_\theta}{t} \right)^2 d\mu \le \int\int_0^1 \left(h_t^T \dot{s}_{\theta+uth_t} \right)^2 du\, d\mu = \frac{1}{4} \int_0^1 h_t^T I_{\theta+uth_t} h_t\, du,$$

where the last equality follows by Fubini's theorem and the definition of I_θ. For $h_t \to h$ the right side converges to $\frac{1}{4} h^T I_\theta h = \int (h^T \dot{s}_\theta)^2\, d\mu$ by the continuity of the map $\theta \mapsto I_\theta$.

By the differentiability of the map $\theta \mapsto s_\theta(x)$ the integrand in

$$\int \left[\frac{s_{\theta+th_t} - s_\theta}{t} - h^T \dot{s}_\theta \right]^2 d\mu$$

converges pointwise to zero. The result of the preceding paragraph combined with Proposition 2.29 shows that the integral converges to zero. ∎

7.7 Example (Exponential families). The preceding lemma applies to most exponential family models

$$p_\theta(x) = d(\theta) h(x) e^{Q(\theta)^T t(x)}.$$

An exponential family model is smooth in its natural parameter (away from the boundary of the natural parameter space). Thus the maps $\theta \mapsto \sqrt{p_\theta(x)}$ are continuously differentiable if the maps $\theta \mapsto Q(\theta)$ are continuously differentiable and map the parameter set Θ into the interior of the natural parameter space. The score function and information matrix equal

$$\dot{\ell}_\theta(x) = Q_\theta'\big(t(x) - \mathrm{E}_\theta t(X)\big), \qquad I_\theta = Q_\theta' \mathrm{cov}_\theta\, t(X)(Q_\theta')^T.$$

Thus the asymptotic expansion of the local log likelihood is valid for most exponential families. □

7.8 Example (Location models). The preceding lemma also includes all location models $\{f(x - \theta) : \theta \in \mathbb{R}\}$ for a positive, continuously differentiable density f with finite *Fisher information for location*

$$I_f = \int \left(\frac{f'}{f} \right)^2 (x)\, f(x)\, dx.$$

The score function $\dot{\ell}_\theta(x)$ can be taken equal to $-(f'/f)(x - \theta)$. The Fisher information is equal to I_f for every θ and hence certainly continuous in θ.

By a refinement of the lemma, differentiability in quadratic mean can also be established for slightly irregular shapes, such as the Laplace density $f(x) = \frac{1}{2} e^{-|x|}$. For the Laplace density the map $\theta \mapsto \log f(x - \theta)$ fails to be differentiable at the single point $\theta = x$. At other points the derivative exists and equals $\mathrm{sign}(x - \theta)$. It can be shown that the Laplace location model is differentiable in quadratic mean with score function $\dot{\ell}_\theta(x) = \mathrm{sign}(x - \theta)$. This may be proved by writing the difference $\sqrt{f(x-h)} - \sqrt{f(x)}$ as the integral $\int_0^1 \frac{1}{2} h\, \mathrm{sign}(x - uh)\sqrt{f(x - uh)}\, du$ of its derivative, which is possible even though the derivative does not exist everywhere. Next the proof of the preceding lemma applies. □

7.9 Counterexample (Uniform distribution). The family of uniform distributions on $[0, \theta]$ is nowhere differentiable in quadratic mean. The reason is that the support of the uniform distribution depends too much on the parameter. Differentiability in quadratic mean (7.1) does not require that all densities p_θ have the same support. However, restriction of the integral (7.1) to the set $\{p_\theta = 0\}$ yields

$$P_{\theta+h}(p_\theta = 0) = \int_{p_\theta=0} p_{\theta+h}\, d\mu = o(h^2).$$

Thus, under (7.1) the total mass $P_{\theta+h}(p_\theta = 0)$ of $P_{\theta+h}$ that is orthogonal to P_θ must "disappear" as $h \to 0$ at a rate faster than h^2.

This is not true for the uniform distribution, because, for $h \geq 0$,

$$P_{\theta+h}(p_\theta = 0) = \int_{[0,\theta]^c} \frac{1}{\theta + h} 1_{[0,\theta+h]}(x)\, dx = \frac{h}{\theta + h}.$$

The orthogonal part does converge to zero, but only at the rate $O(h)$. $\square$

7.3   Convergence to a Normal Experiment

The true meaning of local asymptotic normality is convergence of the local statistical experiments to a normal experiment. In Chapter 9 the notion of convergence of statistical experiments is introduced in general. In this section we bypass this general theory and establish a direct relationship between the local experiments and a normal limit experiment.

The limit experiment is the experiment that consists of observing a single observation X with the $N(h, I_\theta^{-1})$-distribution. The log likelihood ratio process of this experiment equals

$$\log \frac{dN(h, I_\theta^{-1})}{dN(0, I_\theta^{-1})}(X) = h^T I_\theta X - \frac{1}{2} h^T I_\theta h.$$

The right side is very similar in form to the right side of the expansion of the log likelihood ratio $\log dP^n_{\theta+h/\sqrt{n}}/dP^n_\theta$ given in Theorem 7.2. In view of the similarity, the possibility of a normal approximation is not a complete surprise. The approximation in this section is "local" in nature: We fix θ and think of

$$\left(P^n_{\theta+h/\sqrt{n}} : h \in \mathbb{R}^k\right)$$

as a statistical model with parameter h, for "known" θ. We show that this can be approximated by the statistical model $\left(N(h, I_\theta^{-1}) : h \in \mathbb{R}^k\right)$.

A motivation for studying a local approximation is that, usually, asymptotically, the "true" parameter can be known with unlimited precision. The true statistical difficulty is therefore determined by the nature of the measures P_θ for θ in a small neighbourhood of the true value. In the present situation "small" turns out to be "of size $O(1/\sqrt{n})$."

A relationship between the models that can be statistically interpreted will be described through the possible (limit) distributions of statistics. For each n, let $T_n = T_n(X_1, \ldots, X_n)$ be a statistic in the experiment $(P^n_{\theta+h/\sqrt{n}} : h \in \mathbb{R}^k)$ with values in a fixed Euclidean space. Suppose that the sequence of statistics T_n converges in distribution under every possible (local) parameter:

$$T_n \overset{h}{\rightsquigarrow} L_{\theta,h}, \qquad \text{every } h.$$

Here $\overset{h}{\rightsquigarrow}$ means convergence in distribution under the parameter $\theta + h/\sqrt{n}$, and $L_{\theta,h}$ may be any probability distribution. According to the following theorem, the distributions $\{L_{\theta,h} : h \in \mathbb{R}^k\}$ are necessarily the distributions of a statistic T in the normal experiment $(N(h, I_\theta^{-1}) : h \in \mathbb{R}^k)$. Thus, every weakly converging sequence of statistics is "matched" by a statistic in the limit experiment. (In the present set-up the vector θ is considered known and the vector h is the statistical parameter. Consequently, by "statistics" T_n and T are understood measurable maps that do not depend on h but may depend on θ.)

This principle of matching estimators is a method to give the convergence of models a statistical interpretation. Most measures of quality of a statistic can be expressed in the distribution of the statistic under different parameters. For instance, if a certain hypothesis is rejected for values of a statistic T_n exceeding a number c, then the power function $h \mapsto P_h(T_n > c)$ is relevant; alternatively, if T_n is an estimator of h, then the mean square error $h \mapsto E_h(T_n - h)^2$, or a similar quantity, determines the quality of T_n. Both quality measures depend on the laws of the statistics only. The following theorem asserts that as a function of h the law of a statistic T_n can be well approximated by the law of some statistic T. Then the quality of the approximating T is the same as the "asymptotic quality" of the sequence T_n. Investigation of the possible T should reveal the asymptotic performance of possible sequences T_n. Concrete applications of this principle to testing and estimation are given in later chapters.

A minor technical complication is that it is necessary to allow randomized statistics in the limit experiment. A *randomized statistic* T based on the observation X is defined as a measurable map $T = T(X, U)$ that depends on X but may also depend on an independent variable U with a uniform distribution on $[0, 1]$. Thus, the statistician working in the limit experiment is allowed to base an estimate or test on both the observation and the outcome of an extra experiment that can be run without knowledge of the parameter. In most situations such randomization is not useful, but the following theorem would not be true without it.[†]

7.10 Theorem. *Assume that the experiment $(P_\theta : \theta \in \Theta)$ is differentiable in quadratic mean (7.1) at the point θ with nonsingular Fisher information matrix I_θ. Let T_n be statistics in the experiments $(P_{\theta+h/\sqrt{n}}^n : h \in \mathbb{R}^k)$ such that the sequence T_n converges in distribution under every h. Then there exists a randomized statistic T in the experiment $(N(h, I_\theta^{-1}) : h \in \mathbb{R}^k)$ such that $T_n \overset{h}{\rightsquigarrow} T$ for every h.*

Proof. For later reference, it is useful to use the abbreviations

$$P_{n,h} = P_{\theta+h/\sqrt{n}}^n, \qquad J = I_\theta, \qquad \Delta_n = \frac{1}{\sqrt{n}} \sum \dot{\ell}_\theta(X_i).$$

By assumption, the marginals of the sequence (T_n, Δ_n) converge in distribution under $h = 0$; hence they are uniformly tight by Prohorov's theorem. Because marginal tightness implies joint tightness, Prohorov's theorem can be applied in the other direction to see the existence of a subsequence of $\{n\}$ along which

$$(T_n, \Delta_n) \overset{0}{\rightsquigarrow} (S, \Delta),$$

[†] It is not important that U is uniformly distributed. Any randomization mechanism that is sufficiently rich will do.

jointly, for some random vector (S, Δ). The vector Δ is necessarily a marginal weak limit of the sequence Δ_n and hence it is $N(0, J)$-distributed. Combination with Theorem 7.2 yields

$$\left(T_n, \log \frac{dP_{n,h}}{dP_{n,0}} \right) \overset{0}{\rightsquigarrow} \left(S, h^T \Delta - \frac{1}{2} h^T J h \right).$$

In particular, the sequence $\log dP_{n,h}/dP_{n,0}$ converges to the normal $N(-\frac{1}{2} h^T J h, h^T J h)$-distribution. By Example 6.5, the sequences $P_{n,h}$ and $P_{n,0}$ are contiguous. The limit law L_h of T_n under h can therefore be expressed in the joint law on the right, by the general form of Le Cam's third lemma: For each Borel set B

$$L_h(B) = E1_B(S) e^{h^T \Delta - \frac{1}{2} h^T J h}.$$

We need to find a statistic T in the normal experiment having this law under h (for every h), using only the knowledge that Δ is $N(0, J)$-distributed.

By the lemma below there exists a randomized statistic T such that, with U uniformly distributed and independent of Δ,[†]

$$\left(T(\Delta, U), \Delta \right) \sim (S, \Delta).$$

Because the random vectors on the left and right sides have the same second marginal distribution, this is the same as saying that $T(\delta, U)$ is distributed according to the conditional distribution of S given $\Delta = \delta$, for almost every δ. As shown in the next lemma, this can be achieved by using the quantile transformation.

Let X be an observation in the limit experiment $\left(N(h, J^{-1}) : h \in \mathbb{R}^k \right)$. Then JX is under $h = 0$ normally $N(0, J)$-distributed and hence it is equal in distribution to Δ. Furthermore, by Fubini's theorem,

$$P_h\left(T(JX, U) \in B \right) = \int P\left(T(Jx, U) \in B \right) e^{-\frac{1}{2}(x-h)^T J (x-h)} \sqrt{\frac{\det J}{(2\pi)^k}} \, dx$$

$$= E_0 1_B\left(T(JX, U) \right) e^{h^T JX - \frac{1}{2} h^T J h}.$$

This equals $L_h(B)$, because, by construction, the vector $\left(T(JX, U), JX \right)$ has the same distribution under $h = 0$ as (S, Δ). The randomized statistic $T(JX, U)$ has law L_h under h and hence satisfies the requirements. ∎

7.11 Lemma. *Given a random vector (S, Δ) with values in $\mathbb{R}^d \times \mathbb{R}^k$ and an independent uniformly $[0, 1]$ random variable U (defined on the same probability space), there exists a jointly measurable map T on $\mathbb{R}^k \times [0, 1]$ such that $\left(T(\Delta, U), \Delta \right)$ and (S, Δ) are equal in distribution.*

Proof. For simplicity of notation we only give a construction for $d = 2$. It is possible to produce two independent uniform $[0, 1]$ variables U_1 and U_2 from one given $[0, 1]$ variable U. (For instance, construct U_1 and U_2 from the even and odd numbered digits in the decimal expansion of U.) Therefore it suffices to find a statistic $T = T(\Delta, U_1, U_2)$ such that (T, Δ) and (S, Δ) are equal in law. Because the second marginals are equal, it

---

[†] The symbol $\sim$ means "equal-in-law."

suffices to construct T such that $T(\delta, U_1, U_2)$ is equal in distribution to S given $\Delta = \delta$, for every $\delta \in \mathbb{R}^k$. Let $Q_1(u_1 \mid \delta)$ and $Q_2(u_2 \mid \delta, s_1)$ be the quantile functions of the conditional distributions

$$P^{S_1 \mid \Delta = \delta} \qquad \text{and} \qquad P^{S_2 \mid \Delta = \delta, S_1 = s_1},$$

respectively. These are measurable functions in their two and three arguments, respectively. Furthermore, $Q_1(U_1 \mid \delta)$ has law $P^{S_1 \mid \Delta = \delta}$ and $Q_2(U_2 \mid \delta, s_1)$ has law $P^{S_2 \mid \Delta = \delta, S_1 = s_1}$, for every δ and s_1. Set

$$T(\delta, U_1, U_2) = \Big(Q_1(U_1 \mid \delta), \quad Q_2\big(U_2 \mid \delta, Q_1(U_1 \mid \delta)\big) \Big).$$

Then the first coordinate $Q_1(U_1 \mid \delta)$ of $T(\delta, U_1, U_2)$ possesses the distribution $P^{S_1 \mid \Delta = \delta}$. Given that this first coordinate equals s_1, the second coordinate is distributed as $Q_2(U_2 \mid \delta, s_1)$, which has law $P^{S_2 \mid \Delta = \delta, S_1 = s_1}$ by construction. Thus T satisfies the requirements. ∎

7.4　Maximum Likelihood

Maximum likelihood estimators in smooth parametric models were shown to be asymptotically normal in Chapter 5. The convergence of the local experiments to a normal limit experiment gives an insightful explanation of this fact.

By the representation theorem, Theorem 7.10, every sequence of statistics in the local experiments $(P^n_{\theta + h/\sqrt{n}} : h \in \mathbb{R}^k)$ is matched in the limit by a statistic in the normal experiment. Although this does not follow from this theorem, a sequence of maximum likelihood estimators is typically matched by the maximum likelihood estimator in the limit experiment. Now the maximum likelihood estimator for h in the experiment $\big(N(h, I_\theta^{-1}) : h \in \mathbb{R}^k\big)$ is the observation X itself (the mean of a sample of size one), and this is normally distributed. Thus, we should expect that the maximum likelihood estimators $\hat{h}_n$ for the local parameter h in the experiments $(P^n_{\theta + h/\sqrt{n}} : h \in \mathbb{R}^k)$ converge in distribution to X. In terms of the original parameter θ, the local maximum likelihood estimator $\hat{h}_n$ is the standardized maximum likelihood estimator $\hat{h}_n = \sqrt{n}(\hat{\theta}_n - \theta)$. Furthermore, the local parameter $h = 0$ corresponds to the value θ of the original parameter. Thus, we should expect that under θ the sequence $\sqrt{n}(\hat{\theta}_n - \theta)$ converges in distribution to X under $h = 0$, that is, to the $N(0, I_\theta^{-1})$-distribution.

As a heuristic explanation of the asymptotic normality of maximum likelihood estimators the preceding argument is much more insightful than the proof based on linearization of the score equation. It also explains why, or in what sense, the maximum likelihood estimator is asymptotically optimal: in the same sense as the maximum likelihood estimator of a Gaussian location parameter is optimal.

This heuristic argument cannot be justified under just local asymptotic normality, which is too weak a connection between the sequence of local experiments and the normal limit experiment for this purpose. Clearly, the argument is valid under the conditions of Theorem 5.39, because the latter theorem guarantees the asymptotic normality of the maximum likelihood estimator. This theorem adds a Lipschitz condition on the maps $\theta \mapsto \log p_\theta(x)$, and the "global" condition that $\hat{\theta}_n$ is consistent to differentiability in quadratic mean. In the following theorem, we give a direct argument, and also allow that θ is not an inner point of the parameter set, so that the local parameter spaces may not converge to the full space $\mathbb{R}^k$.

Then the maximum likelihood estimator in the limit experiment is a "projection" of X and the limit distribution of $\sqrt{n}(\hat{\theta}_n - \theta)$ may change accordingly.

Let Θ be an arbitrary subset of $\mathbb{R}^k$ and define H_n as the *local parameter space* $H_n = \sqrt{n}(\Theta - \theta)$. Then $\hat{h}_n$ is the maximizer over H_n of the random function (or "process")

$$h \mapsto \log \frac{dP^n_{\theta + h/\sqrt{n}}}{dP^n_\theta}.$$

If the experiment $(P_\theta : \theta \in \Theta)$ is differentiable in quadratic mean, then this sequence of processes converges (marginally) in distribution to the process

$$h \mapsto \log \frac{dN(h, I_\theta^{-1})}{dN(0, I_\theta^{-1})}(X) = -\frac{1}{2}(X - h)^T I_\theta (X - h) + \frac{1}{2} X^T I_\theta X.$$

If the sequence of sets H_n converges in a suitable sense to a set H, then we should expect, under regularity conditions, that the sequence $\hat{h}_n$ converges to the maximizer $\hat{h}$ of the latter process over H. This maximizer is the projection of the vector X onto the set H relative to the metric $d(x, y) = (x - y)^T I_\theta (x - y)$ (where a "projection" means a closest point); if $H = \mathbb{R}^k$, this projection reduces to X itself.

An appropriate notion of *convergence of sets* is the following. Write $H_n \to H$ if H is the set of all limits $\lim h_n$ of converging sequences h_n with $h_n \in H_n$ for every n and, moreover, the limit $h = \lim_i h_{n_i}$ of every converging sequence h_{n_i} with $h_{n_i} \in H_{n_i}$ for every i is contained in H.[†]

7.12 Theorem. *Suppose that the experiment $(P_\theta : \theta \in \Theta)$ is differentiable in quadratic mean at θ_0 with nonsingular Fisher information matrix I_{θ_0}. Furthermore, suppose that for every θ_1 and θ_2 in a neighborhood of θ_0 and a measurable function $\dot{\ell}$ with $P_{\theta_0} \ell^2 < \infty$,*

$$\left| \log p_{\theta_1}(x) - \log p_{\theta_2}(x) \right| \le \dot{\ell}(x) \, \|\theta_1 - \theta_2\|.$$

If the sequence of maximum likelihood estimators $\hat{\theta}_n$ is consistent and the sets $H_n = \sqrt{n}(\Theta - \theta_0)$ converge to a nonempty, convex set H, then the sequence $I_{\theta_0}^{1/2} \sqrt{n}(\hat{\theta}_n - \theta_0)$ converges under θ_0 in distribution to the projection of a standard normal vector onto the set $I_{\theta_0}^{1/2} H$.

*Proof. Let $\mathbb{G}_n = \sqrt{n}(\mathbb{P}_n - P_{\theta_0})$ be the empirical process. In the proof of Theorem 5.39 it is shown that the map $\theta \mapsto \log p_\theta$ is differentiable at θ_0 in $L_2(P_{\theta_0})$ with derivative $\dot{\ell}_{\theta_0}$ and that the map $\theta \mapsto P_{\theta_0} \log p_\theta$ permits a Taylor expansion of order 2 at θ_0, with "second-derivative matrix" $-I_{\theta_0}$. Therefore, the conditions of Lemma 19.31 are satisfied for $m_\theta = \log p_\theta$, whence, for every M,

$$\sup_{\|h\| \le M} \left| n\mathbb{P}_n \log \frac{p_{\theta_0 + h/\sqrt{n}}}{p_{\theta_0}} - h^T \mathbb{G}_n \dot{\ell}_{\theta_0} + \frac{1}{2} h^T I_{\theta_0} h \right| \xrightarrow{P} 0.$$

By Corollary 5.53 the estimators $\hat{\theta}_n$ are $\sqrt{n}$-consistent under θ_0.

The preceding display is also valid for every sequence M_n that diverges to ∞ sufficiently slowly. Fix such a sequence. By the $\sqrt{n}$-consistency of $\hat{\theta}_n$, the local maximum likelihood

---

[†] See Chapter 16 for examples.

estimators $\hat{h}_n$ are bounded in probability and hence belong to the balls of radius M_n with probability tending to 1. Furthermore, the sequence of intersections $H_n \cap$ ball $(0, M_n)$ converges to H, as the original sets H_n. Thus, we may assume that the $\hat{h}_n$ are the maximum likelihood estimators relative to local parameter sets H_n that are contained in the balls of radius M_n. Fix an arbitrary closed set F. If $\hat{h}_n \in F$, then the log likelihood is maximal on F. Hence $\mathrm{P}(\hat{h}_n \in F)$ is bounded above by

$$\mathrm{P}\left(\sup_{h \in F \cap H_n} \mathbb{P}_n \log \frac{p_{\theta_0 + h/\sqrt{n}}}{p_{\theta_0}} \geq \sup_{h \in H_n} \mathbb{P}_n \log \frac{p_{\theta_0 + h/\sqrt{n}}}{p_{\theta_0}} \right)$$

$$= \mathrm{P}\left(\sup_{h \in F \cap H_n} h^T \mathbb{G}_n \dot{\ell}_{\theta_0} - \frac{1}{2} h^T I_{\theta_0} h \geq \sup_{h \in H_n} h^T \mathbb{G}_n \dot{\ell}_{\theta_0} - \frac{1}{2} h^T I_{\theta_0} h + o_P(1) \right)$$

$$= \mathrm{P}\left(\left\| I_{\theta_0}^{-1/2} \mathbb{G}_n \dot{\ell}_{\theta_0} - I_{\theta_0}^{1/2}(F \cap H_n) \right\| \leq \left\| I_{\theta_0}^{-1/2} \mathbb{G}_n \dot{\ell}_{\theta_0} - I_{\theta_0}^{1/2} H_n \right\| + o_P(1) \right),$$

by completing the square. By Lemma 7.13 (ii) and (iii) ahead, we can replace H_n by H on both sides, at the cost of adding a further $o_P(1)$-term and increasing the probability. Next, by the continuous mapping theorem and the continuity of the map $z \mapsto \|z - A\|$ for every set A, the probability is asymptotically bounded above by, with Z a standard normal vector,

$$\mathrm{P}\left(\left\| Z - I_{\theta_0}^{1/2}(F \cap H) \right\| \leq \left\| Z - I_{\theta_0}^{1/2} H \right\| \right).$$

The projection ΠZ of the vector Z on the set $I_{\theta_0}^{1/2} H$ is unique, because the latter set is convex by assumption and automatically closed. If the distance of Z to $I_{\theta_0}^{1/2}(F \cap H)$ is smaller than its distance to the set $I_{\theta_0}^{1/2} H$, then ΠZ must be in $I_{\theta_0}^{1/2}(F \cap H)$. Consequently, the probability in the last display is bounded by $\mathrm{P}(\Pi Z \in I_{\theta_0}^{1/2} F)$. The theorem follows from the portmanteau lemma. ∎

7.13 Lemma. *If the sequence of subsets H_n of $\mathbb{R}^k$ converges to a nonempty set H and the sequence of random vectors X_n converges in distribution to a random vector X, then*

(i) $\|X_n - H_n\| \rightsquigarrow \|X - H\|$.

(ii) $\|X_n - H_n \cap F\| \geq \|X_n - H \cap F\| + o_P(1)$, *for every closed set F.*

(iii) $\|X_n - H_n \cap G\| \leq \|X_n - H \cap G\| + o_P(1)$, *for every open set G.*

Proof. (i). Because the map $x \mapsto \|x - H\|$ is (Lipschitz) continuous for any set H, we have that $\|X_n - H\| \rightsquigarrow \|X - H\|$ by the continuous-mapping theorem. If we also show that $\|X_n - H_n\| - \|X_n - H\| \overset{P}{\to} 0$, then the proof is complete after an application of Slutsky's lemma. By the uniform tightness of the sequence X_n, it suffices to show that $\|x - H_n\| \to \|x - H\|$ uniformly for x ranging over compact sets, or equivalently that $\|x_n - H_n\| \to \|x - H\|$ for every converging sequence $x_n \to x$.

For every fixed vector x_n, there exists a vector $h_n \in H_n$ with $\|x_n - H_n\| \geq \|x_n - h_n\| - 1/n$. Unless $\|x_n - H_n\|$ is unbounded, we can choose the sequence h_n bounded. Then every subsequence of h_n has a further subsequence along which it converges, to a limit h in H. Conclude that, in any case,

$$\liminf \|x_n - H_n\| \geq \liminf \|x_n - h_n\| \geq \|x - h\| \geq \|x - H\|.$$

Conversely, for every $\varepsilon > 0$ there exists $h \in H$ and a sequence $h_n \to h$ with $h_n \in H_n$ and

$$\|x - H\| \geq \|x - h\| - \varepsilon = \lim \|x_n - h_n\| - \varepsilon \geq \limsup \|x_n - H_n\| - \varepsilon.$$

Combination of the last two displays yields the desired convergence of the sequence $\|x_n - H_n\|$ to $\|x - H\|$.

(ii). The assertion is equivalent to the statement $P\big(\|X_n - H_n \cap F\| - \|X_n - H \cap F\| > -\varepsilon\big) \to 1$ for every $\varepsilon > 0$. In view of the uniform tightness of the sequence X_n, this follows if $\liminf \|x_n - H_n \cap F\| \geq \|x - H \cap F\|$ for every converging sequence $x_n \to x$. We can prove this by the method of the first half of the proof of (i), replacing H_n by $H_n \cap F$.

(iii). Analogously to the situation under (ii), it suffices to prove that $\limsup \|x_n - H_n \cap G\| \leq \|x - H \cap G\|$ for every converging sequence $x_n \to x$. This follows as the second half of the proof of (i). ∎

*7.5 Limit Distributions under Alternatives

Local asymptotic normality is a convenient tool in the study of the behavior of statistics under "contiguous alternatives." Under local asymptotic normality,

$$\log \frac{dP^n_{\theta+h/\sqrt{n}}}{dP^n_\theta} \overset{\theta}{\leadsto} N\left(-\frac{1}{2}h^T I_\theta h, h^T I_\theta h\right).$$

Therefore, in view of Example 6.5 the sequences of distributions $P^n_{\theta+h/\sqrt{n}}$ and P^n_θ are mutually contiguous. This is of great use in many proofs. With the help of Le Cam's third lemma it also allows to obtain limit distributions of statistics under the parameters $\theta + h/\sqrt{n}$, once the limit behavior under θ is known. Such limit distributions are of interest, for instance, in studying the asymptotic efficiency of estimators or tests.

The general scheme is as follows. Many sequences of statistics T_n allow an approximation by an average of the type

$$\sqrt{n}\,(T_n - \mu_\theta) = \frac{1}{\sqrt{n}} \sum_{i=1}^n \psi_\theta(X_i) + o_{P_\theta}(1).$$

According to Theorem 7.2, the sequence of log likelihood ratios can be approximated by an average as well: It is asymptotically equivalent to an affine transformation of $n^{-1/2} \sum \dot\ell_\theta(X_i)$. The sequence of joint averages $n^{-1/2} \sum \big(\psi_\theta(X_i), \dot\ell_\theta(X_i)\big)$ is asymptotically multivariate normal under θ by the central limit theorem (provided ψ_θ has mean zero and finite second moment). With the help of Slutsky's lemma we obtain the joint limit distribution of T_n and the log likelihood ratios under θ:

$$\left(\sqrt{n}\,(T_n - \mu_\theta), \log \frac{dP^n_{\theta+h/\sqrt{n}}}{dP^n_\theta}\right) \overset{\theta}{\leadsto} N\left(\begin{pmatrix} 0 \\ -\frac{1}{2}h^T I_\theta h \end{pmatrix}, \begin{pmatrix} P_\theta \psi_\theta \psi_\theta^T & P_\theta \psi_\theta h^T \dot\ell_\theta \\ P_\theta \psi_\theta^T h^T \dot\ell_\theta & h^T I_\theta h \end{pmatrix}\right).$$

Finally we can apply Le Cam's third lemma, Example 6.7, to obtain the limit distribution of $\sqrt{n}(T_n - \mu_\theta)$ under $\theta + h/\sqrt{n}$. Concrete examples of this scheme are discussed in later chapters.

*7.6 Local Asymptotic Normality

The preceding sections of this chapter are restricted to the case of independent, identically distributed observations. However, the general ideas have a much wider applicability. A wide variety of models satisfy a general form of local asymptotic normality and for that reason allow a unified treatment. These include models with independent, not identically distributed observations, but also models with dependent observations, such as used in time series analysis or certain random fields. Because local asymptotic normality underlies a large part of asymptotic optimality theory and also explains the asymptotic normality of certain estimators, such as maximum likelihood estimators, it is worthwhile to formulate a general concept.

Suppose the observation at "time" n is distributed according to a probability measure $P_{n,\theta}$, for a parameter θ ranging over an open subset Θ of $\mathbb{R}^k$.

7.14 Definition. The sequence of statistical models $(P_{n,\theta} : \theta \in \Theta)$ is *locally asymptotically normal (LAN)* at θ if there exist matrices r_n and I_θ and random vectors $\Delta_{n,\theta}$ such that $\Delta_{n,\theta} \overset{\theta}{\rightsquigarrow} N(0, I_\theta)$ and for every converging sequence $h_n \to h$

$$\log \frac{dP_{n,\theta+r_n^{-1}h_n}}{dP_{n,\theta}} = h^T \Delta_{n,\theta} - \frac{1}{2} h^T I_\theta h + o_{P_{n,\theta}}(1).$$

7.15 *Example.* If the experiment $(P_\theta : \theta \in \Theta)$ is differentiable in quadratic mean, then the sequence of models $(P_\theta^n : \theta \in \Theta)$ is locally asymptotically normal with norming matrices $r_n = \sqrt{n}I$. □

An inspection of the proof of Theorem 7.10 readily reveals that this depends on the local asymptotic normality property only. Thus, the local experiments

$$\left(P_{n,\theta+r_n^{-1}h} : h \in \mathbb{R}^k \right)$$

of a locally asymptotically normal sequence converge to the experiment $\left(N(h, I_\theta^{-1}) : h \in \mathbb{R}^k \right)$, in the sense of this theorem. All results for the case of i.i.d. observations that are based on this approximation extend to general locally asymptotically normal models. To illustrate the wide range of applications we include, without proof, three examples, two of which involve dependent observations.

7.16 *Example (Autoregressive processes).* An autoregressive process $\{X_t : t \in \mathbb{Z}\}$ of order 1 satisfies the relationship $X_t = \theta X_{t-1} + Z_t$ for a sequence of independent, identically distributed variables $\ldots, Z_{-1}, Z_0, Z_1, \ldots$ with mean zero and finite variance. There exists a stationary solution $\ldots, X_{-1}, X_0, X_1, \ldots$ to the autoregressive equation if and only if $|\theta| \neq 1$. To identify the parameter it is usually assumed that $|\theta| < 1$. If the density of the noise variables Z_j has finite Fisher information for location, then the sequence of models corresponding to observing $X_1, \ldots, X_n$ with parameter set $(-1, 1)$ is locally asymptotically normal at θ with norming matrices $r_n = \sqrt{n}I$.

The observations in this model form a stationary Markov chain. The result extends to general ergodic Markov chains with smooth transition densities (see [130]). □

7.17 *Example (Gaussian time series).* This example requires some knowledge of time-series models. Suppose that at time n the observations are a stretch $X_1, \ldots, X_n$ from a stationary, Gaussian time series $\{X_t : t \in \mathbb{Z}\}$ with mean zero. The covariance matrix of n

consecutive variables is given by the (Toeplitz) matrix

$$T_n(f_\theta) = \left(\int_{-\pi}^{\pi} e^{i(t-s)\lambda} f_\theta(\lambda) \, d\lambda \right)_{s,t=1,\dots,n}.$$

The function f_θ is the *spectral density* of the series. It is convenient to let the parameter enter the model through the spectral density, rather than directly through the density of the observations.

Let $P_{n,\theta}$ be the distribution (on $\mathbb{R}^n$) of the vector $(X_1, \dots, X_n)$, a normal distribution with mean zero and covariance matrix $T_n(f_\theta)$. The *periodogram* of the observations is the function

$$I_n(\lambda) = \frac{1}{2\pi n} \left| \sum_{t=1}^{n} X_t e^{it\lambda} \right|^2.$$

Suppose that f_θ is bounded away from zero and infinity, and that there exists a vector-valued function $\dot\ell_\theta : \mathbb{R} \mapsto \mathbb{R}^d$ such that, as $h \to 0$,

$$\int [f_{\theta+h} - f_\theta - h^T \dot\ell_\theta \, f_\theta]^2 \, d\lambda = o(\|h\|^2).$$

Then the sequence of experiments $(P_{n,\theta} : \theta \in \Theta)$ is locally asymptotically normal at θ with

$$r_n = \sqrt{n}, \qquad \Delta_{n,\theta} = \frac{\sqrt{n}}{4\pi} \int (I_n - \mathrm{E}_\theta I_n) \frac{\dot\ell_\theta}{f_\theta} \, d\lambda, \qquad I_\theta = \frac{1}{4\pi} \int \dot\ell_\theta \dot\ell_\theta^T \, d\lambda.$$

The proof is elementary, but involved, because it has to deal with the quadratic forms in the n-variate normal density, which involve vectors whose dimension converges to infinity (see [30]). $\square$

7.18 Example (Almost regular densities). Consider estimating a location parameter θ based on a sample of size n from the density $f(x - \theta)$. If f is smooth, then this model is differentiable in quadratic mean and hence locally asymptotically normal by Example 7.8. If f possesses points of discontinuity, or other strong irregularities, then a locally asymptotically normal approximation is impossible.[†] Examples of densities that are on the boundary between these "extremes" are the triangular density $f(x) = (1 - |x|)^+$ and the gamma density $f(x) = xe^{-x}1\{x > 0\}$. These yield models that are locally asymptotically normal, but with norming rate $\sqrt{n \log n}$ rather than $\sqrt{n}$. The existence of singularities in the density makes the estimation of the parameter θ easier, and hence a faster rescaling rate is necessary. (For the triangular density, the true singularities are the points -1 and 1, the singularity at 0 is statistically unimportant, as in the case of the Laplace density.)

For a more general result, consider densities f that are absolutely continuous except possibly in small neighborhoods $U_1, \dots, U_k$ of finitely many fixed points $c_1, \dots, c_k$. Suppose that $f'/\sqrt{f}$ is square-integrable on the complement of $\cup_j U_j$, that $f(c_j) = 0$ for every j, and that, for fixed constants $a_1, \dots, a_k$ and $b_1, \dots, b_k$, each of the functions

$$x \mapsto f(x) - (a_j 1\{x < c_j\} + b_j 1\{x > c_j\})|x - c_j|, \qquad x \in U_j,$$

[†] See Chapter 9 for some examples.

is twice continuously differentiable. If $\sum(a_j + b_j) > 0$, then the model is locally asymptotically normal at $\theta = 0$ with, for V_n equal to the interval $\left(n^{-1/2}(\log n)^{-1/4}, (\log n)^{-1}\right)$ around zero,[†]

$$r_n = \sqrt{n \log n}, \qquad I_0 = \sum_j \left(a_j + b_j\right),$$

$$\Delta_{n,0} = \frac{1}{\sqrt{n \log n}} \sum_{i=1}^{n} \sum_{j=1}^{k} \left(\frac{1\{X_i - c_j \in V_n\}}{X_i - c_j} - \int_{V_n} \frac{1}{x} f\left(x + c_j\right) dx \right).$$

The sequence $\Delta_{n,0}$ may be thought of as "asymptotically sufficient" for the local parameter h. The definition of $\Delta_{n,0}$ shows that, asymptotically, all the "information" about the parameter is contained in the observations falling into the neighborhoods $V_n + C_j$. Thus, asymptotically, the problem is determined by the points of irregularity.

The remarkable rescaling rate $\sqrt{n \log n}$ can be explained by computing the Hellinger distance between the densities $f(x - \theta)$ and $f(x)$ (see section 14.5). $\square$

Notes

Local asymptotic normality was introduced by Le Cam [92], apparently motivated by the study and construction of asymptotically similar tests. In this paper Le Cam defines two sequences of models $(P_{n,\theta} : \theta \in \Theta)$ and $(Q_{n,\theta} : \theta \in \Theta)$ to be *differentially equivalent* if

$$\sup_{h \in K} \left\| P_{n,\theta+h/\sqrt{n}} - Q_{n,\theta+h/\sqrt{n}} \right\| \to 0,$$

for every bounded set K and every θ. He next shows that a sequence of statistics T_n in a given *asymptotically differentiable* sequence of experiments (roughly LAN) that is asymptotically equivalent to the centering sequence $\Delta_{n,\theta}$ is asymptotically sufficient, in the sense that the original experiments and the experiments consisting of observing the T_n are differentially equivalent. After some interpretation this gives roughly the same message as Theorem 7.10. The latter is a concrete example of an abstract result in [95], with a different (direct) proof.

PROBLEMS

1. Show that the Poisson distribution with mean θ satisfies the conditions of Lemma 7.6. Find the information.
2. Find the Fisher information for location for the normal, logistic, and Laplace distributions.
3. Find the Fisher information for location for the Cauchy distributions.
4. Let f be a density that is symmetric about zero. Show that the Fisher information matrix (if it exists) of the location scale family $f\left((x - \mu)/\sigma\right)/\sigma$ is diagonal.
5. Find an explicit expression for the $o_{P_\theta}(1)$-term in Theorem 7.2 in the case that p_θ is the density of the $N(\theta, 1)$-distribution.
6. Show that the Laplace location family is differentiable in quadratic mean.

[†] See, for example, [80, pp. 133–139] for a proof, and also a discussion of other almost regular situations. For instance, singularities of the form $f(x) \sim f(c_j) + |x - c_j|^{1/2}$ at points c_j with $f(c_j) > 0$.

7. Find the form of the score function for a location-scale family $f\big((x - \mu)/\sigma\big)/\sigma$ with parameter $\theta = (\mu, \sigma)$ and apply Lemma 7.6 to find a sufficient condition for differentiability in quadratic mean.

8. Investigate for which parameters k the location family $f(x - \theta)$ for f the gamma$(k, 1)$ density is differentiable in quadratic mean.

9. Let $P_{n,\theta}$ be the distribution of the vector $(X_1, \ldots, X_n)$ if $\{X_t : t \in \mathbb{Z}\}$ is a stationary Gaussian time series satisfying $X_t = \theta X_{t-1} + Z_t$ for a given number $|\theta| < 1$ and independent standard normal variables Z_t. Show that the model is locally asymptotically normal.

10. Investigate whether the log normal family of distributions with density

$$\frac{1}{\sigma \sqrt{2\pi}(x - \xi)} e^{-\frac{1}{2\sigma^2}(\log(x - \xi) - \mu)^2} \, 1\{x > \xi\}$$

is differentiable in quadratic mean with respect to $\theta = (\xi, \mu, \sigma)$.

8

Efficiency of Estimators

One purpose of asymptotic statistics is to compare the performance of estimators for large sample sizes. This chapter discusses asymptotic lower bounds for estimation in locally asymptotically normal models. These show, among others, in what sense maximum likelihood estimators are asymptotically efficient.

8.1 Asymptotic Concentration

Suppose the problem is to estimate $\psi(\theta)$ based on observations from a model governed by the parameter θ. What is the best asymptotic performance of an estimator sequence T_n for $\psi(\theta)$?

To simplify the situation, we shall in most of this chapter assume that the sequence $\sqrt{n}(T_n - \psi(\theta))$ converges in distribution under every possible value of θ. Next we rephrase the question as: What are the best possible limit distributions? In analogy with the Cramér-Rao theorem a "best" limit distribution is referred to as an *asymptotic lower bound*. Under certain restrictions the normal distribution with mean zero and covariance the inverse Fisher information is an asymptotic lower bound for estimating $\psi(\theta) = \theta$ in a smooth parametric model. This is the main result of this chapter, but it needs to be qualified.

The notion of a "best" limit distribution is understood in terms of concentration. If the limit distribution is a priori assumed to be normal, then this is usually translated into asymptotic unbiasedness and minimum variance. The statement that $\sqrt{n}(T_n - \psi(\theta))$ converges in distribution to a $N(\mu(\theta), \sigma^2(\theta))$-distribution can be roughly understood in the sense that eventually T_n is approximately normally distributed with mean and variance given by

$$\psi(\theta) + \frac{\mu(\theta)}{\sqrt{n}} \quad \text{and} \quad \frac{\sigma^2(\theta)}{n}.$$

Because T_n is meant to estimate $\psi(\theta)$, optimal choices for the asymptotic mean and variance are $\mu(\theta) = 0$ and variance $\sigma^2(\theta)$ as small as possible. These choices ensure not only that the asymptotic mean square error is small but also that the limit distribution $N(\mu(\theta), \sigma^2(\theta))$ is maximally concentrated near zero. For instance, the probability of the interval $(-a, a)$ is maximized by choosing $\mu(\theta) = 0$ and $\sigma^2(\theta)$ minimal.

We do not wish to assume a priori that the estimators are asymptotically normal. That normal limits are best will actually be an interesting conclusion. The concentration of a general limit distribution L_θ cannot be measured by mean and variance alone. Instead, we can employ a variety of concentration measures, such as

$$\int x^2 \, dL_\theta(x); \qquad \int |x| \, dL_\theta(x); \qquad \int 1\{|x| > a\} \, dL_\theta(x); \qquad \int (|x| \wedge a) \, dL_\theta(x).$$

A limit distribution is "good" if quantities of this type are small. More generally, we focus on minimizing $\int \ell \, dL_\theta$ for a given nonnegative function ℓ. Such a function is called a *loss function* and its integral $\int \ell \, dL_\theta$ is the *asymptotic risk* of the estimator. The method of measuring concentration (or rather lack of concentration) by means of loss functions applies to one- and higher-dimensional parameters alike.

The following example shows that a definition of what constitutes asymptotic optimality is not as straightforward as it might seem.

8.1 *Example (Hodges' estimator).* Suppose that T_n is a sequence of estimators for a real parameter θ with standard asymptotic behavior in that, for each θ and certain limit distributions L_θ,

$$\sqrt{n}(T_n - \theta) \overset{\theta}{\rightsquigarrow} L_\theta.$$

As a specific example, let T_n be the mean of a sample of size n from the $N(\theta, 1)$-distribution. Define a second estimator S_n through

$$S_n = \begin{cases} T_n & \text{if } |T_n| \ge n^{-1/4} \\ 0 & \text{if } |T_n| < n^{-1/4} \end{cases}.$$

If the estimator T_n is already close to zero, then it is changed to exactly zero; otherwise it is left unchanged. The truncation point $n^{-1/4}$ has been chosen in such a way that the limit behavior of S_n is the same as that of T_n for every $\theta \ne 0$, but for $\theta = 0$ there appears to be a great improvement. Indeed, for every r_n,

$$r_n S_n \overset{0}{\rightsquigarrow} 0$$
$$\sqrt{n}(S_n - \theta) \overset{\theta}{\rightsquigarrow} L_\theta, \qquad \theta \ne 0.$$

To see this, note first that the probability that T_n falls in the interval $(\theta - Mn^{-1/2}, \theta + Mn^{-1/2})$ converges to $L_\theta(-M, M)$ for most M and hence is arbitrarily close to 1 for M and n sufficiently large. For $\theta \ne 0$, the intervals $(\theta - Mn^{-1/2}, \theta + Mn^{-1/2})$ and $(-n^{-1/4}, n^{-1/4})$ are centered at different places and eventually disjoint. This implies that truncation will rarely occur: $P_\theta(T_n = S_n) \to 1$ if $\theta \ne 0$, whence the second assertion. On the other hand the interval $(-Mn^{-1/2}, Mn^{-1/2})$ is contained in the interval $(-n^{-1/4}, n^{-1/4})$ eventually. Hence under $\theta = 0$ we have truncation with probability tending to 1 and hence $P_0(S_n = 0) \to 1$; this is stronger than the first assertion.

At first sight, S_n is an improvement on T_n. For every $\theta \ne 0$ the estimators behave the same, while for $\theta = 0$ the sequence S_n has an "arbitrarily fast" rate of convergence. However, this reasoning is a bad use of asymptotics.

Consider the concrete situation that T_n is the mean of a sample of size n from the normal $N(\theta, 1)$-distribution. It is well known that $T_n = \overline{X}$ is optimal in many ways for every fixed n and hence it ought to be asymptotically optimal also. Figure 8.1 shows why $S_n = \overline{X}1\{|\overline{X}| \ge n^{-1/4}\}$ is no improvement. It shows the graph of the risk function $\theta \mapsto E_\theta(S_n - \theta)^2$ for three different values of n. These functions are close to 1 on most

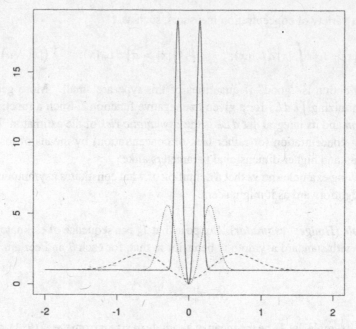

Figure 8.1. Risk function $\theta \mapsto n\mathrm{E}_\theta(S_n - \theta)^2$ of the Hodges estimator based on the means of samples of size 10 (dashed), 100 (dotted), and 1000 (solid) observations from the $N(\theta, 1)$-distribution.

of the domain but possess peaks close to zero. As $n \to \infty$, the locations and widths of the peaks converge to zero but their heights to infinity. The conclusion is that S_n "buys" its better asymptotic behavior at $\theta = 0$ at the expense of erratic behavior close to zero. Because the values of θ at which S_n is bad differ from n to n, the erratic behavior is not visible in the pointwise limit distributions under fixed θ. $\quad\square$

8.2   Relative Efficiency

In order to choose between two estimator sequences, we compare the concentration of their limit distributions. In the case of normal limit distributions and convergence rate $\sqrt{n}$, the quotient of the asymptotic variances is a good numerical measure of their relative efficiency. This number has an attractive interpretation in terms of the numbers of observations needed to attain the same goal with each of two sequences of estimators.

Let $\nu \to \infty$ be a "time" index, and suppose that it is required that, as $\nu \to \infty$, our estimator sequence attains mean zero and variance 1 (or $1/\nu$). Assume that an estimator T_n based on n observations has the property that, as $n \to \infty$,

$$\sqrt{n}\big(T_n - \psi(\theta)\big) \overset{\theta}{\rightsquigarrow} N\big(0, \sigma^2(\theta)\big).$$

Then the requirement is to use at time ν an appropriate number n_ν of observations such that, as $\nu \to \infty$,

$$\sqrt{\nu}\big(T_{n_\nu} - \psi(\theta)\big) \overset{\theta}{\rightsquigarrow} N(0, 1).$$

Given two available estimator sequences, let $n_{\nu,1}$ and $n_{\nu,2}$ be the numbers of observations needed to meet the requirement with each of the estimators. Then, if it exists, the limit

$$\lim_{\nu \to \infty} \frac{n_{\nu,2}}{n_{\nu,1}}$$

is called the *relative efficiency* of the estimators. (In general, it depends on the parameter θ.)

Because $\sqrt{\nu}\big(T_{n_\nu} - \psi(\theta)\big)$ can be written as $\sqrt{\nu/n_\nu}\,\sqrt{n_\nu}\big(T_{n_\nu} - \psi(\theta)\big)$, it follows that necessarily $n_\nu \to \infty$, and also that $n_\nu/\nu \to \sigma^2(\theta)$. Thus, the relative efficiency of two estimator sequences with asymptotic variances $\sigma_i^2(\theta)$ is just

$$\lim_{\nu \to \infty} \frac{n_{\nu,2}/\nu}{n_{\nu,1}/\nu} = \frac{\sigma_2^2(\theta)}{\sigma_1^2(\theta)}.$$

If the value of this quotient is bigger than 1, then the second estimator sequence needs proportionally that many observations more than the first to achieve the same (asymptotic) precision.

8.3 Lower Bound for Experiments

It is certainly impossible to give a nontrivial lower bound on the limit distribution of a standardized estimator $\sqrt{n}\big(T_n - \psi(\theta)\big)$ for a single θ. Hodges' example shows that it is not even enough to consider the behavior under every θ, pointwise for all θ. Different values of the parameters must be taken into account simultaneously when taking the limit as $n \to \infty$. We shall do this by studying the performance of estimators under parameters in a "shrinking" neighborhood of a fixed θ.

We consider parameters $\theta + h/\sqrt{n}$ for θ fixed and h ranging over $\mathbb{R}^k$ and suppose that, for certain limit distributions $L_{\theta,h}$,

$$\sqrt{n}\left(T_n - \psi\left(\theta + \frac{h}{\sqrt{n}}\right)\right) \overset{\theta+h/\sqrt{n}}{\rightsquigarrow} L_{\theta,h}, \quad \text{every } h. \tag{8.2}$$

Then T_n can be considered a good estimator for $\psi(\theta)$ if the limit distributions $L_{\theta,h}$ are maximally concentrated near zero. If they are maximally concentrated for every h and some fixed θ, then T_n can be considered locally optimal at θ. Unless specified otherwise, we assume in the remainder of this chapter that the parameter set Θ is an open subset of $\mathbb{R}^k$, and that ψ maps Θ into $\mathbb{R}^m$. The derivative of $\theta \mapsto \psi(\theta)$ is denoted by $\dot{\psi}_\theta$.

Suppose that the observations are a sample of size n from a distribution P_θ. If P_θ depends smoothly on the parameter, then

$$\left(P_{\theta+h/\sqrt{n}}^n : h \in \mathbb{R}^k\right) \rightsquigarrow \left(N\big(h, I_\theta^{-1}\big) : h \in \mathbb{R}^k\right),$$

as experiments, in the sense of Theorem 7.10. This theorem shows which limit distributions are possible and can be specialized to the estimation problem in the following way.

8.3 Theorem. *Assume that the experiment* $(P_\theta : \theta \in \Theta)$ *is differentiable in quadratic mean (7.1) at the point θ with nonsingular Fisher information matrix I_θ. Let ψ be differentiable at θ. Let T_n be estimators in the experiments* $(P_{\theta+h/\sqrt{n}}^n : h \in \mathbb{R}^k)$ *such that*

(8.2) *holds for every h. Then there exists a randomized statistic T in the experiment* $\left(N(h, I_\theta^{-1}) : h \in \mathbb{R}^k\right)$ *such that* $T - \dot{\psi}_\theta h$ *has distribution* $L_{\theta,h}$ *for every h.*

Proof. Apply Theorem 7.10 to $S_n = \sqrt{n}\left(T_n - \psi(\theta)\right)$. In view of the definition of $L_{\theta,h}$ and the differentiability of ψ, the sequence

$$S_n = \sqrt{n}\left(T_n - \psi\left(\theta + \frac{h}{\sqrt{n}}\right)\right) + \sqrt{n}\left(\psi\left(\theta + \frac{h}{\sqrt{n}}\right) - \psi(\theta)\right)$$

converges in distribution under h to $L_{\theta,h} * \delta_{\dot{\psi}_\theta h}$, where $*\delta_h$ denotes a translation by h. According to Theorem 7.10, there exists a randomized statistic T in the normal experiment such that T has distribution $L_{\theta,h} * \delta_{\dot{\psi}_\theta h}$ for every h. This satisfies the requirements. ∎

This theorem shows that for most estimator sequences T_n there is a randomized estimator T such that the distribution of $\sqrt{n}\left(T_n - \psi(\theta + h/\sqrt{n})\right)$ under $\theta + h/\sqrt{n}$ is, for large n, approximately equal to the distribution of $T - \dot{\psi}_\theta h$ under h. Consequently the standardized distribution of the best possible estimator T_n for $\psi(\theta + h/\sqrt{n})$ is approximately equal to the standardized distribution of the best possible estimator T for $\dot{\psi}_\theta h$ in the limit experiment. If we know the best estimator T for $\dot{\psi}_\theta h$, then we know the "locally best" estimator sequence T_n for $\psi(\theta)$.

In this way, the asymptotic optimality problem is reduced to optimality in the experiment based on one observation X from a $N(h, I_\theta^{-1})$-distribution, in which θ is known and h ranges over $\mathbb{R}^k$. This experiment is simple and easy to analyze. The observation itself is the customary estimator for its expectation h, and the natural estimator for $\dot{\psi}_\theta h$ is $\dot{\psi}_\theta X$. This has several optimality properties: It is minimum variance unbiased, minimax, best equivariant, and Bayes with respect to the noninformative prior. Some of these properties are reviewed in the next section.

Let us agree, at least for the moment, that $\dot{\psi}_\theta X$ is a "best" estimator for $\dot{\psi}_\theta h$. The distribution of $\dot{\psi}_\theta X - \dot{\psi}_\theta h$ is normal with zero mean and covariance $\dot{\psi}_\theta I_\theta^{-1} \dot{\psi}_\theta^T$ for every h. The parameter $h = 0$ in the limit experiment corresponds to the parameter θ in the original problem. We conclude that the "best" limit distribution of $\sqrt{n}\left(T_n - \psi(\theta)\right)$ under θ is the $N(0, \dot{\psi}_\theta I_\theta^{-1} \dot{\psi}_\theta^T)$-distribution.

This is the main result of the chapter. The remaining sections discuss several ways of making this reasoning more rigorous. Because the expression $\dot{\psi}_\theta I_\theta^{-1} \dot{\psi}_\theta^T$ is precisely the Cramér-Rao lower bound for the covariance of unbiased estimators for $\psi(\theta)$, we can think of the results of this chapter as asymptotic Cramér-Rao bounds. This is helpful, even though it does not do justice to the depth of the present results. For instance, the Cramér-Rao bound in no way suggests that normal limiting distributions are best. Also, it is not completely true that an $N(h, I_\theta^{-1})$-distribution is "best" (see section 8.8). We shall see exactly to what extent the optimality statement is false.

8.4   Estimating Normal Means

According to the preceding section, the asymptotic optimality problem reduces to optimality in a normal location (or "Gaussian shift") experiment. This section has nothing to do with asymptotics but reviews some facts about Gaussian models.

Based on a single observation X from a $N(h, \Sigma)$-distribution, it is required to estimate Ah for a given matrix A. The covariance matrix Σ is assumed known and nonsingular. It is well known that AX is minimum variance unbiased. It will be shown that AX is also best-equivariant and minimax for many loss functions.

A randomized estimator T is called *equivariant-in-law* for estimating Ah if the distribution of $T - Ah$ under h does not depend on h. An example is the estimator AX, whose "invariant law" (the law of $AX - Ah$ under h) is the $N(0, A\Sigma A^T)$-distribution. The following proposition gives an interesting characterization of the law of general equivariant-in-law estimators: These are distributed as the sum of AX and an independent variable.

8.4 Proposition. *The null distribution L of any randomized equivariant-in-law estimator of Ah can be decomposed as $L = N(0, A\Sigma A^T) * M$ for some probability measure M. The only randomized equivariant-in-law estimator for which M is degenerate at 0 is AX.*

The measure M can be interpreted as the distribution of a noise factor that is added to the estimator AX. If no noise is best, then it follows that AX is best equivariant-in-law.

A more precise argument can be made in terms of loss functions. In general, convoluting a measure with another measure decreases its concentration. This is immediately clear in terms of variance: The variance of a sum of two independent variables is the sum of the variances, whence convolution increases variance. For normal measures this extends to all "bowl-shaped" symmetric loss functions. The name should convey the form of their graph. Formally, a function is defined to be *bowl-shaped* if the sublevel sets $\{x : \ell(x) \le c\}$ are convex and symmetric about the origin; it is called *subconvex* if, moreover, these sets are closed. A *loss function* is any function with values in $[0, \infty)$. The following lemma quantifies the loss in concentration under convolution (for a proof, see, e.g., [80] or [114].)

8.5 Lemma (Anderson's lemma). *For any bowl-shaped loss function ℓ on $\mathbb{R}^k$, every probability measure M on $\mathbb{R}^k$, and every covariance matrix Σ*

$$\int \ell \, dN(0, \Sigma) \le \int \ell \, d[N(0, \Sigma) * M].$$

Next consider the *minimax criterion*. According to this criterion the "best" estimator, relative to a given loss function, minimizes the maximum risk

$$\sup_h \mathrm{E}_h \ell(T - Ah),$$

over all (randomized) estimators T. For every bowl-shaped loss function ℓ, this leads again to the estimator AX.

8.6 Proposition. *For any bowl-shaped loss function ℓ, the maximum risk of any randomized estimator T of Ah is bounded below by $\mathrm{E}_0\ell(AX)$. Consequently, AX is a minimax estimator for Ah. If Ah is real and $\mathrm{E}_0(AX)^2\ell(AX) < \infty$, then AX is the only minimax estimator for Ah up to changes on sets of probability zero.*

Proofs. For a proof of the uniqueness of the minimax estimator, see [18] or [80]. We prove the other assertions for subconvex loss functions, using a Bayesian argument.

Let H be a random vector with a normal $N(0, \Lambda)$-distribution, and consider the original $N(h, \Sigma)$-distribution as the conditional distribution of X given $H = h$. The randomization variable U in $T(X, U)$ is constructed independently of the pair (X, H). In this notation, the distribution of the variable $T - AH$ is equal to the "average" of the distributions of $T - Ah$ under the different values of h in the original set-up, averaged over h using a $N(0, \Lambda)$-"prior distribution."

By a standard calculation, we find that the "a posteriori" distribution, the distribution of H given X, is the normal distribution with mean $(\Sigma^{-1} + \Lambda^{-1})^{-1}\Sigma^{-1}X$ and covariance matrix $(\Sigma^{-1} + \Lambda^{-1})^{-1}$. Define the random vectors

$$W_\Lambda = T - A(\Sigma^{-1} + \Lambda^{-1})^{-1}\Sigma^{-1}X, \quad G_\Lambda = -A\big(H - (\Sigma^{-1} + \Lambda^{-1})^{-1}\Sigma^{-1}X\big)$$

These vectors are independent, because W_Λ is a function of (X, U) only, and the conditional distribution of G_Λ given X is normal with mean 0 and covariance matrix $A(\Sigma^{-1} + \Lambda^{-1})^{-1}A^T$, independent of X. As $\Lambda = \lambda I$ for a scalar $\lambda \to \infty$, the sequence G_Λ converges in distribution to a $N(0, A\Sigma A^T)$-distributed vector G. The sum of the two vectors yields $T - AH$, for every Λ.

Because a supremum is larger than an average, we obtain, where on the left we take the expectation with respect to the original model,

$$\sup_h \mathrm{E}_h \ell(T - Ah) \geq \mathrm{E}\ell(T - AH) = \mathrm{E}\ell(G_\Lambda + W_\Lambda) \geq \mathrm{E}\ell(G_\Lambda),$$

by Anderson's lemma. This is true for every Λ. The lim inf of the right side as $\Lambda \to \infty$ is at least $\mathrm{E}\ell(G)$, by the portmanteau lemma. This concludes the proof that AX is minimax.

If T is equivariant-in-law with invariant law L, then the distribution of $G_\Lambda + W_\Lambda = T - AH$ is L, for every Λ. It follows that

$$\int e^{it^T x} \, dL(x) = \mathrm{E}e^{it^T G_\Lambda} \mathrm{E}e^{it^T W_\Lambda}, \quad \text{every } t.$$

As $\Lambda \to \infty$, the left side remains fixed; the first factor on the right side converges to the characteristic function of G, which is positive. Conclude that the characteristic functions of W_Λ converge to a continuous function, whence W_Λ converges in distribution to some vector W, by Lévy's continuity theorem. By the independence of G_Λ and W_Λ for every Λ, the sequence (G_Λ, W_Λ) converges in distribution to a pair (G, W) of independent vectors with marginal distributions as before. Next, by the continuous-mapping theorem, the distribution of $G_\Lambda + W_\Lambda$, which is fixed at L, "converges" to the distribution of $G + W$. This proves that L can be written as a convolution, as claimed in Proposition 8.4.

If T is an equivariant-in-law estimator and $\tilde{T}(X) = \mathrm{E}\big(T(X, U) \mid X\big)$, then

$$\mathrm{E}_h(\tilde{T} - AX) = \mathrm{E}_h(T - AX) = \mathrm{E}_h(T - Ah) - \mathrm{E}_h(AX - Ah)$$

is independent of h. By the completeness of the normal location family, we conclude that $\tilde{T} - AX$ is constant, almost surely. If T has the same law as AX, then the constant is zero. Furthermore, T must be equal to its projection $\tilde{T}$ almost surely, because otherwise it would have a bigger second moment than $\tilde{T} = AX$. Thus $T = AX$ almost surely. $\blacksquare$

8.5   Convolution Theorem

An estimator sequence T_n is called *regular* at θ for estimating a parameter $\psi(\theta)$ if, for every h,

$$\sqrt{n}\left(T_n - \psi\left(\theta + \frac{h}{\sqrt{n}}\right)\right) \overset{\theta+h/\sqrt{n}}{\rightsquigarrow} L_\theta.$$

The probability measure L_θ may be arbitrary but should be the same for every h.

A regular estimator sequence attains its limit distribution in a "locally uniform" manner. This type of regularity is common and is often considered desirable: A small change in the parameter should not change the distribution of the estimator too much; a disappearing small change should not change the (limit) distribution at all. However, some estimator sequences of interest, such as shrinkage estimators, are not regular.

In terms of the limit distributions $L_{\theta,h}$ in (8.2), regularity is exactly that all $L_{\theta,h}$ are equal, for the given θ. According to Theorem 8.3, every estimator sequence is matched by an estimator T in the limit experiment $\left(N(h, I_\theta^{-1}) : h \in \mathbb{R}^k\right)$. For a regular estimator sequence this matching estimator has the property

$$T - \dot{\psi}_\theta h \overset{h}{\sim} L_\theta, \quad \text{every } h. \tag{8.7}$$

Thus a regular estimator sequence is matched by an equivariant-in-law estimator for $\dot{\psi}_\theta h$. A more informative name for "regular" is *asymptotically equivariant-in-law*.

It is now easy to determine a best estimator sequence from among the regular estimator sequences (a *best regular* sequence): It is the sequence T_n that corresponds to the best equivariant-in-law estimator T for $\dot{\psi}_\theta h$ in the limit experiment, which is $\dot{\psi}_\theta X$ by Proposition 8.4. The best possible limit distribution of a regular estimator sequence is the law of this estimator, a $N(0, \dot{\psi}_\theta I_\theta^{-1} \dot{\psi}_\theta^T)$-distribution.

The characterization as a convolution of the invariant laws of equivariant-in-law estimators carries over to the asymptotic situation.

8.8   Theorem (Convolution). *Assume that the experiment $(P_\theta : \theta \in \Theta)$ is differentiable in quadratic mean (7.1) at the point θ with nonsingular Fisher information matrix I_θ. Let ψ be differentiable at θ. Let T_n be an at θ regular estimator sequence in the experiments $(P_\theta^n : \theta \in \Theta)$ with limit distribution L_θ. Then there exists a probability measure M_θ such that*

$$L_\theta = N\left(0, \dot{\psi}_\theta I_\theta^{-1} \dot{\psi}_\theta^T\right) * M_\theta.$$

In particular, if L_θ has covariance matrix Σ_θ, then the matrix $\Sigma_\theta - \dot{\psi}_\theta I_\theta^{-1} \dot{\psi}_\theta^T$ is nonnegative-definite.

Proof.   Apply Theorem 8.3 to conclude that L_θ is the distribution of an equivariant-in-law estimator T in the limit experiment, satisfying (8.7). Next apply Proposition 8.4.   ∎

8.6   Almost-Everywhere Convolution Theorem

Hodges' example shows that there is no hope for a nontrivial lower bound for the limit distribution of a standardized estimator sequence $\sqrt{n}\left(T_n - \psi(\theta)\right)$ for *every* θ. It is always

possible to improve on a given estimator sequence for selected parameters. In this section
it is shown that improvement over an $N(0, \dot{\psi}_\theta I_\theta^{-1} \dot{\psi}_\theta{}^T)$-distribution can be made on at most
a Lebesgue null set of parameters. Thus the possibilities for improvement are very much
restricted.

8.9   Theorem. *Assume that the experiment $(P_\theta : \theta \in \Theta)$ is differentiable in quadratic
mean (7.1) at every θ with nonsingular Fisher information matrix I_θ. Let ψ be differentiable
at every θ. Let T_n be an estimator sequence in the experiments $(P_\theta^n : \theta \in \Theta)$ such that
$\sqrt{n}(T_n - \psi(\theta))$ converges to a limit distribution L_θ under every θ. Then there exist
probability distributions M_θ such that for Lebesgue almost every θ*

$$L_\theta = N\big(0, \dot{\psi}_\theta I_\theta^{-1} \dot{\psi}_\theta{}^T\big) * M_\theta.$$

*In particular, if L_θ has covariance matrix Σ_θ, then the matrix $\Sigma_\theta - \dot{\psi}_\theta I_\theta^{-1} \dot{\psi}_\theta{}^T$ is
nonnegative definite for Lebesgue almost every θ.*

The theorem follows from the convolution theorem in the preceding section combined
with the following remarkable lemma. Any estimator sequence with limit distributions is
automatically regular at almost every θ along a subsequence of $\{n\}$.

8.10   Lemma. *Let T_n be estimators in experiments $(P_{n,\theta} : \theta \in \Theta)$ indexed by a measurable
subset Θ of $\mathbb{R}^k$. Assume that the map $\theta \mapsto P_{n,\theta}(A)$ is measurable for every measurable
set A and every n, and that the map $\theta \mapsto \psi(\theta)$ is measurable. Suppose that there exist
distributions L_θ such that for Lebesgue almost every θ*

$$r_n(T_n - \psi(\theta)) \overset{\theta}{\rightsquigarrow} L_\theta.$$

*Then for every $\gamma_n \to 0$ there exists a subsequence of $\{n\}$ such that, for Lebesgue almost
every (θ, h), along the subsequence,*

$$r_n(T_n - \psi(\theta + \gamma_n h)) \overset{\theta + \gamma_n h}{\rightsquigarrow} L_\theta.$$

Proof.   Assume without loss of generality that $\Theta = \mathbb{R}^k$; otherwise, fix some θ_0 and let
$P_{n,\theta} = P_{n,\theta_0}$ for every θ not in Θ. Write $T_{n,\theta} = r_n(T_n - \psi(\theta))$. There exists a countable
collection $\mathcal{F}$ of uniformly bounded, left- or right-continuous functions f such that weak
convergence of a sequence of maps T_n is equivalent to $\mathrm{E}f(T_n) \to \int f \, dL$ for every $f \in \mathcal{F}$.[†]
Suppose that for every f there exists a subsequence of $\{n\}$ along which

$$\mathrm{E}_{\theta + \gamma_n h} f(T_{n,\theta + \gamma_n h}) \to \int f \, dL_\theta, \qquad \lambda^{2k} - \text{a.e. } (\theta, h).$$

Even in case the subsequence depends on f, we can, by a diagonalization scheme, construct a subsequence for which this is valid for every f in the countable set $\mathcal{F}$. Along this
subsequence we have the desired convergence.

[†] For continuous distributions L we can use the indicator functions of cells $(-\infty, c]$ with c ranging over Q^k. For
general L replace every such indicator by an approximating sequence of continuous functions. Alternatively,
see, e.g., Theorem 1.12.2 in [146]. Also see Lemma 2.25.

Setting $g_n(\theta) = E_\theta f(T_{n,\theta})$ and $g(\theta) = \int f \, dL_\theta$, we see that the lemma is proved once we have established the following assertion: Every sequence of bounded, measurable functions g_n that converges almost everywhere to a limit g, has a subsequence along which

$$g_n(\theta + \gamma_n h) \to g(\theta), \qquad \lambda^{2k} - \text{a.e. } (\theta, h).$$

We may assume without loss of generality that the function g is integrable; otherwise we first multiply each g_n and g with a suitable, fixed, positive, continuous function. It should also be verified that, under our conditions, the functions g_n are measurable.

Write p for the standard normal density on $\mathbb{R}^k$ and p_n for the density of the $N(0, I + \gamma_n^2 I)$-distribution. By Scheffé's lemma, the sequence p_n converges to p in L_1. Let Θ and H denote independent standard normal vectors. Then, by the triangle inequality and the dominated-convergence theorem,

$$E\big|g_n(\Theta + \gamma_n H) - g(\Theta + \gamma_n H)\big| = \int \big|g_n(u) - g(u)\big| p_n(u) \, du \to 0.$$

Secondly for any fixed continuous and bounded function g_ε the sequence $E\big|g_\varepsilon(\Theta + \gamma_n H) - g_\varepsilon(\Theta)\big|$ converges to zero as $n \to \infty$ by the dominated convergence theorem. Thus, by the triangle inequality, we obtain

$$E\big|g(\Theta + \gamma_n H) - g(\Theta)\big| \le \int |g - g_\varepsilon|(u)\,(p_n + p)(u)\,du + o(1)$$
$$= 2 \int |g - g_\varepsilon|(u)\, p(u) \, du + o(1).$$

Because any measurable integrable function g can be approximated arbitrarily closely in L_1 by continuous functions, the first term on the far right side can be made arbitrarily small by choice of g_ε. Thus the left side converges to zero.

By combining this with the preceding display, we see that $E\big|g_n(\Theta + \gamma_n H) - g(\Theta)\big| \to 0$. In other words, the sequence of functions $(\theta, h) \mapsto g_n(\theta + \gamma_n h) - g(\theta)$ converges to zero in mean and hence in probability, under the standard normal measure. There exists a subsequence along which it converges to zero almost surely. ∎

*8.7 Local Asymptotic Minimax Theorem

The convolution theorems discussed in the preceding sections are not completely satisfying. The convolution theorem designates a best estimator sequence among the regular estimator sequences, and thus imposes an a priori restriction on the set of permitted estimator sequences. The almost-everywhere convolution theorem imposes no (serious) restriction but yields no information about some parameters, albeit a null set of parameters.

This section gives a third attempt to "prove" that the normal $N(0, \dot\psi_\theta I_\theta^{-1} \dot\psi_\theta^T)$-distribution is the best possible limit. It is based on the minimax criterion and gives a lower bound for the maximum risk over a small neighborhood of a parameter θ. In fact, it bounds the expression

$$\lim_{\delta \to 0} \liminf_{n \to \infty} \sup_{\|\theta' - \theta\| < \delta} E_{\theta'} \ell\Big(\sqrt{n}\big(T_n - \psi(\theta')\big)\Big).$$

This is the asymptotic maximum risk over an arbitrarily small neighborhood of θ. The following theorem concerns an even more refined (and smaller) version of the local maximum risk.

8.11   Theorem.   *Let the experiment* $(P_\theta : \theta \in \Theta)$ *be differentiable in quadratic mean* (7.1) *at* θ *with nonsingular Fisher information matrix* I_θ. *Let* ψ *be differentiable at* θ. *Let* T_n *be any estimator sequence in the experiments* $(P_\theta^n : \theta \in \mathbb{R}^k)$. *Then for any bowl-shaped loss function* ℓ

$$\sup_{I} \liminf_{n \to \infty} \sup_{h \in I} \mathrm{E}_{\theta + h/\sqrt{n}} \ell \left(\sqrt{n} \left(T_n - \psi \left(\theta + \frac{h}{\sqrt{n}} \right) \right) \right) \geq \int \ell \, dN \left(0, \dot{\psi}_\theta I_\theta^{-1} \dot{\psi}_\theta{}^T \right).$$

Here the first supremum is taken over all finite subsets I *of* $\mathbb{R}^k$.

Proof.   We only give the proof under the further assumptions that the sequence $\sqrt{n}(T_n - \psi(\theta))$ is uniformly tight under θ and that ℓ is (lower) semicontinuous.[†] Then Prohorov's theorem shows that every subsequence of $\{n\}$ has a further subsequence along which the vectors

$$\left(\sqrt{n}(T_n - \psi(\theta)), \frac{1}{\sqrt{n}} \sum \dot{\ell}_\theta(X_i) \right)$$

converge in distribution to a limit under θ. By Theorem 7.2 and Le Cam's third lemma, the sequence $\sqrt{n}(T_n - \psi(\theta))$ converges in law also under every $\theta + h/\sqrt{n}$ along the subsequence. By differentiability of ψ, the same is true for the sequence $\sqrt{n}(T_n - \psi(\theta + h/\sqrt{n}))$, whence (8.2) is satisfied. By Theorem 8.3, the distributions $L_{\theta,h}$ are the distributions of $T - \dot{\psi}_\theta h$ under h for a randomized estimator T based on an $N(h, I_\theta^{-1})$-distributed observation. By Proposition 8.6,

$$\sup_{h \in \mathbb{R}^k} E_h \ell(T - \dot{\psi}_\theta h) \geq E_0 \ell(\dot{\psi}_\theta X) = \int \ell \, dN \left(0, \dot{\psi}_\theta I_\theta^{-1} \dot{\psi}_\theta{}^T \right).$$

It suffices to show that the left side of this display is a lower bound for the left side of the theorem.

The complicated construction that defines the asymptotic minimax risk (the lim inf sandwiched between two suprema) requires that we apply the preceding argument to a carefully chosen subsequence. Place the rational vectors in an arbitrary order, and let I_k consist of the first k vectors in this sequence. Then the left side of the theorem is larger than

$$R := \lim_{k \to \infty} \liminf_{n \to \infty} \sup_{h \in I_k} \mathrm{E}_{\theta + h/\sqrt{n}} \ell \left(\sqrt{n} \left(T_n - \psi \left(\theta + \frac{h}{\sqrt{n}} \right) \right) \right).$$

There exists a subsequence $\{n_k\}$ of $\{n\}$ such that this expression is equal to

$$\lim_{k \to \infty} \sup_{h \in I_k} \mathrm{E}_{\theta + h/\sqrt{n_k}} \ell \left(\sqrt{n_k} \left(T_{n_k} - \psi \left(\theta + \frac{h}{\sqrt{n_k}} \right) \right) \right).$$

We apply the preceding argument to this subsequence and find a further subsequence along which T_n satisfies (8.2). For simplicity of notation write this as $\{n'\}$ rather than with a double subscript. Because ℓ is nonnegative and lower semicontinuous, the portmanteau lemma gives, for every h,

$$\liminf_{n' \to \infty} \mathrm{E}_{\theta + h/\sqrt{n'}} \ell \left(\sqrt{n'} \left(T_{n'} - \psi \left(\theta + \frac{h}{\sqrt{n'}} \right) \right) \right) \geq \int \ell \, dL_{\theta,h}.$$

[†] See, for example, [146, Chapter 3.11] for the general result, which can be proved along the same lines, but using a compactification device to induce tightness.

Every rational vector h is contained in I_k for every sufficiently large k. Conclude that

$$R \geq \sup_{h \in \mathbb{Q}^k} \int \ell \, dL_{\theta,h} = \sup_{h \in \mathbb{Q}^k} E_h \ell(T - \dot{\psi}_\theta h).$$

The risk function in the supremum on the right is lower semicontinuous in h, by the continuity of the Gaussian location family and the lower semicontinuity of ℓ. Thus the expression on the right does not change if $\mathbb{Q}^k$ is replaced by $\mathbb{R}^k$. This concludes the proof. ∎

*8.8 Shrinkage Estimators

The theorems of the preceding sections seem to prove in a variety of ways that the best possible limit distribution is the $N(0, \dot{\psi}_\theta I_\theta^{-1} \dot{\psi}_\theta^T)$-distribution. At closer inspection, the situation is more complicated, and to a certain extent optimality remains a matter of taste, asymptotic optimality being no exception. The "optimal" normal limit is the distribution of the estimator $\dot{\psi}_\theta X$ in the normal limit experiment. Because this estimator has several optimality properties, many statisticians consider it best. Nevertheless, one might prefer a Bayes estimator or a shrinkage estimator. With a changed perception of what constitutes "best" in the limit experiment, the meaning of "asymptotically best" changes also. This becomes particularly clear in the example of shrinkage estimators.

8.12 *Example (Shrinkage estimator).* Let $X_1, \dots, X_n$ be a sample from a multivariate normal distribution with mean θ and covariance the identity matrix. The dimension k of the observations is assumed to be at least 3. This is essential! Consider the estimator

$$T_n = \overline{X}_n - (k-2) \frac{\overline{X}_n}{n \|\overline{X}_n\|^2}.$$

Because $\overline{X}_n$ converges in probability to the mean θ, the second term in the definition of T_n is $O_P(n^{-1})$ if $\theta \neq 0$. In that case $\sqrt{n}(T_n - \overline{X}_n)$ converges in probability to zero, whence the estimator sequence T_n is regular at every $\theta \neq 0$. For $\theta = h/\sqrt{n}$, the variable $\sqrt{n}\overline{X}_n$ is distributed as a variable X with an $N(h, I)$-distribution, and for every n the standardized estimator $\sqrt{n}(T_n - h/\sqrt{n})$ is distributed as $T - h$ for

$$T(X) = X - (k-2) \frac{X}{\|X\|^2}.$$

This is the Stein *shrinkage estimator*. Because the distribution of $T - h$ depends on h, the sequence T_n is not regular at $\theta = 0$. The Stein estimator has the remarkable property that, for every h (see, e.g., [99, p. 300]),

$$E_h \|T - h\|^2 < E_h \|X - h\|^2 = k.$$

It follows that, in terms of joint quadratic loss $\ell(x) = \|x\|^2$, the local limit distributions $L_{0,h}$ of the sequence $\sqrt{n}(T_n - h/\sqrt{n})$ under $\theta = h/\sqrt{n}$ are all better than the $N(0, I)$-limit distribution of the best regular estimator sequence $\overline{X}_n$. □

The example of shrinkage estimators shows that, depending on the optimality criterion, a normal $N(0, \dot{\psi}_\theta I_\theta^{-1} \dot{\psi}_\theta^T)$-limit distribution need not be optimal. In this light, is it reasonable to uphold that maximum likelihood estimators are asymptotically optimal? Perhaps not. On the other hand, the possibility of improvement over the $N(0, \dot\psi_\theta I_\theta^{-1} \dot\psi_\theta{}^T)$-limit is restricted in two important ways.

First, improvement can be made only on a null set of parameters by Theorem 8.9. Second, improvement is possible only for special loss functions, and improvement for one loss function necessarily implies worse performance for other loss functions. This follows from the next lemma.

Suppose that we require the estimator sequence to be *locally asymptotically minimax* for a given loss function ℓ in the sense that

$$\sup_I \limsup_{n\to\infty} \sup_{h\in I} E_{\theta+h/\sqrt{n}}\ell\left(\sqrt{n}\left(T_n - \psi\left(\theta + \frac{h}{\sqrt{n}}\right)\right)\right) \le \int \ell\, dN(0, \dot\psi_\theta I_\theta^{-1} \dot\psi_\theta{}^T).$$

This is a reasonable requirement, and few statisticians would challenge it. The following lemma shows that for one-dimensional parameters $\psi(\theta)$ local asymptotic minimaxity for even a single loss function implies regularity. Thus, if it is required that all coordinates of a certain estimator sequence be locally asymptotically minimax for some loss function, then the best regular estimator sequence is optimal without competition.

8.13   Lemma. *Assume that the experiment $(P_\theta : \theta \in \Theta)$ is differentiable in quadratic mean (7.1) at θ with nonsingular Fisher information matrix I_θ. Let ψ be a real-valued map that is differentiable at θ. Then an estimator sequence in the experiments $(P_\theta^n : \theta \in \mathbb{R}^k)$ can be locally asymptotically minimax at θ for a bowl-shaped loss function ℓ such that $0 < \int x^2 \ell(x)\, dN(0, \dot\psi_\theta I_\theta^{-1} \dot\psi_\theta{}^T)(x) < \infty$ only if T_n is best regular at θ.*

Proof. We only give the proof under the further assumption that the sequence $\sqrt{n}(T_n - \psi(\theta))$ is uniformly tight under θ. Then by the same arguments as in the proof of Theorem 8.11, every subsequence of $\{n\}$ has a further subsequence along which the sequence $\sqrt{n}(T_n - \psi(\theta + h/\sqrt{n}))$ converges in distribution under $\theta + h/\sqrt{n}$ to the distribution $L_{\theta,h}$ of $T - \dot\psi_\theta h$ under h, for a randomized estimator T based on an $N(h, I_\theta^{-1})$-distributed observation. Because T_n is locally asymptotically minimax, it follows that

$$\sup_{h\in\mathbb{R}^k} E_h \ell(T - \dot\psi_\theta h) = \sup_{h\in\mathbb{R}^k} \int \ell\, dL_{\theta,h} \le \int \ell\, dN(0, \dot\psi_\theta I_\theta^{-1} \dot\psi_\theta{}^T).$$

Thus T is a minimax estimator for $\dot\psi_\theta h$ in the limit experiment. By Proposition 8.6, $T = \dot\psi_\theta X$, whence $L_{\theta,h}$ is independent of h. ∎

*8.9   Achieving the Bound

If the convolution theorem is taken as the basis for asymptotic optimality, then an estimator sequence is best if it is asymptotically regular with a $N(0, \dot\psi_\theta I_\theta^{-1} \dot\psi_\theta{}^T)$-limit distribution. An estimator sequence has this property if and only if the estimator is asymptotically linear in the score function.

8.14   Lemma. *Assume that the experiment $(P_\theta : \theta \in \Theta)$ is differentiable in quadratic mean (7.1) at θ with nonsingular Fisher information matrix I_θ. Let ψ be differentiable at*

θ. *Let T_n be an estimator sequence in the experiments $(P_\theta^n : \theta \in \mathbb{R}^k)$ such that*

$$\sqrt{n}\big(T_n - \psi(\theta)\big) = \frac{1}{\sqrt{n}} \sum_{i=1}^n \dot{\psi}_\theta I_\theta^{-1} \dot{\ell}_\theta(X_i) + o_{P_\theta}(1).$$

Then T_n is best regular estimator for $\psi(\theta)$ at θ. Conversely, every best regular estimator sequence satisfies this expansion.

Proof. The sequence $\Delta_{n,\theta} = n^{-1/2} \sum \dot{\ell}_\theta(X_i)$ converges in distribution to a vector Δ_θ with a $N(0, I_\theta)$-distribution. By Theorem 7.2 the sequence $\log dP_{\theta+h/\sqrt{n}}^n / dP_\theta^n$ is asymptotically equivalent to $h^T \Delta_{n,\theta} - \frac{1}{2} h^T I_\theta h$. If T_n is asymptotically linear, then $\sqrt{n}\big(T_n - \psi(\theta)\big)$ is asymptotically equivalent to the function $\dot{\psi}_\theta I_\theta^{-1} \Delta_{n,\theta}$. Apply Slutsky's lemma to find that

$$\left(\sqrt{n}\big(T_n - \psi(\theta)\big), \log \frac{dP_{\theta+h/\sqrt{n}}}{dP_\theta^n} \right) \overset{\theta}{\rightsquigarrow} \left(\dot{\psi}_\theta I_\theta^{-1} \Delta_\theta, \, h^T \Delta_\theta - \frac{1}{2} h^T I_\theta h \right)$$

$$\sim N \left(\begin{pmatrix} 0 \\ -\frac{1}{2} h^T I_\theta h \end{pmatrix} \begin{pmatrix} \dot{\psi}_\theta I_\theta^{-1} \dot{\psi}_\theta^T & \dot{\psi}_\theta h \\ \dot{\psi}_\theta h^T & h^T I_\theta h \end{pmatrix} \right).$$

The limit distribution of the sequence $\sqrt{n}\big(T_n - \psi(\theta)\big)$ under $\theta + h/\sqrt{n}$ follows by Le Cam's third lemma, Example 6.7, and is normal with mean $\dot{\psi}_\theta h$ and covariance matrix $\dot{\psi}_\theta I_\theta^{-1} \dot{\psi}_\theta^T$. Combining this with the differentiability of ψ, we obtain that T_n is regular.

Next suppose that S_n and T_n are both best regular estimator sequences. By the same arguments as in the proof of Theorem 8.11 it can be shown that, at least along subsequences, the joint estimators (S_n, T_n) for $\big(\psi(\theta), \psi(\theta)\big)$ satisfy for every h

$$\left(\sqrt{n}\left(S_n - \psi\left(\theta + \frac{h}{\sqrt{n}}\right)\right), \sqrt{n}\left(T_n - \psi\left(\theta + \frac{h}{\sqrt{n}}\right)\right) \right) \overset{\theta+h/\sqrt{n}}{\rightsquigarrow} (S - \dot{\psi}_\theta h, \, T - \dot{\psi}_\theta h),$$

for a randomized estimator (S, T) in the normal-limit experiment. Because S_n and T_n are best regular, the estimators S and T are best equivariant-in-law. Thus $S = T = \dot{\psi}_\theta X$ almost surely by Proposition 8.6, whence $\sqrt{n}(S_n - T_n)$ converges in distribution to $S - T = 0$.

Thus every two best regular estimator sequences are asymptotically equivalent. The second assertion of the lemma follows on applying this to T_n and the estimators

$$S_n = \psi(\theta) + \frac{1}{\sqrt{n}} \dot{\psi}_\theta I_\theta^{-1} \Delta_{n,\theta}.$$

Because the parameter θ is known in the local experiments $(P_{\theta+h/\sqrt{n}}^n : h \in \mathbb{R}^k)$, this indeed defines an estimator sequence within the present context. It is best regular by the first part of the lemma. ∎

Under regularity conditions, for instance those of Theorem 5.39, the maximum likelihood estimator $\hat{\theta}_n$ in a parametric model satisfies

$$\sqrt{n}(\hat{\theta}_n - \theta) = \frac{1}{\sqrt{n}} \sum_{i=1}^n I_\theta^{-1} \dot{\ell}_\theta(X_i) + o_{P_\theta}(1).$$

Then the maximum likelihood estimator is asymptotically optimal for estimating θ in terms of the convolution theorem. By the delta method, the estimator $\psi(\hat{\theta}_n)$ for $\psi(\theta)$ can be seen to be asymptotically linear as in the preceding theorem, so that it is asymptotically regular and optimal as well.

Actually, regular and asymptotically optimal estimators for θ exist in every parametric model $(P_\theta : \theta \in \Theta)$ that is differentiable in quadratic mean with nonsingular Fisher information throughout Θ, provided the parameter θ is identifiable. This can be shown using the discretized one-step method discussed in section 5.7 (see [93]).

*8.10   Large Deviations

Consistency of an estimator sequence T_n entails that the probability of the event $d(T_n, \psi(\theta)) > \varepsilon$ tends to zero under θ, for every $\varepsilon > 0$. This is a very weak requirement. One method to strengthen it is to make ε dependent on n and to require that the probabilities $P_\theta(d(T_n, \psi(\theta)) > \varepsilon_n)$ converge to 0, or are bounded away from 1, for a given sequence $\varepsilon_n \to 0$. The results of the preceding sections address this question and give very precise lower bounds for these probabilities using an "optimal" rate $\varepsilon_n = r_n^{-1}$, typically $n^{-1/2}$.

Another method of strengthening the consistency is to study the speed at which the probabilities $P_\theta(d(T_n, \psi(\theta)) > \varepsilon)$ converge to 0 for a fixed $\varepsilon > 0$. This method appears to be of less importance but is of some interest. Typically, the speed of convergence is exponential, and there is a precise lower bound for the exponential rate in terms of the Kullback-Leibler information.

We consider the situation that T_n is based on a random sample of size n from a distribution P_θ, indexed by a parameter θ ranging over an arbitrary set Θ. We wish to estimate the value of a function $\psi : \Theta \mapsto \mathbb{D}$ that takes its values in a metric space.

8.15   Theorem. *Suppose that the estimator sequence T_n is consistent for $\psi(\theta)$ under every θ. Then, for every $\varepsilon > 0$ and every θ_0,*

$$\limsup_{n \to \infty} -\frac{1}{n} \log P_{\theta_0}\Big(d(T_n, \psi(\theta_0)) > \varepsilon\Big) \leq \inf_{\theta : d(\psi(\theta), \psi(\theta_0)) > \varepsilon} -P_\theta \log \frac{p_{\theta_0}}{p_\theta}.$$

Proof. If the right side is infinite, then there is nothing to prove. The Kullback-Leibler information $-P_\theta \log p_{\theta_0}/p_\theta$ can be finite only if $P_\theta \ll P_{\theta_0}$. Hence, it suffices to prove that $-P_\theta \log p_{\theta_0}/p_\theta$ is an upper bound for the left side for every θ such that $P_\theta \ll P_{\theta_0}$ and $d(\psi(\theta), \psi(\theta_0)) > \varepsilon$. The variable $\Lambda_n = (n^{-1}) \sum_{i=1}^n \log(p_\theta/p_{\theta_0})(X_i)$ is well defined (possibly $-\infty$). For every constant M,

$$P_{\theta_0}\Big(d(T_n, \psi(\theta_0)) > \varepsilon\Big) \geq P_{\theta_0}\Big(d(T_n, \psi(\theta_0)) > \varepsilon, \Lambda_n < M\Big)$$
$$\geq E_\theta 1\Big\{d(T_n, \psi(\theta_0)) > \varepsilon, \Lambda_n < M\Big\}e^{-n\Lambda_n}$$
$$\geq e^{-nM} P_\theta\Big(d(T_n, \psi(\theta_0)) > \varepsilon, \Lambda_n < M\Big).$$

Take logarithms and multiply by $-(1/n)$ to conclude that

$$-\frac{1}{n} \log P_{\theta_0}\Big(d(T_n, \psi(\theta_0)) > \varepsilon\Big) \leq M - \frac{1}{n} \log P_\theta\Big(d(T_n, \psi(\theta_0)) > \varepsilon, \Lambda_n < M\Big).$$

For $M > P_\theta \log p_\theta/p_{\theta_0}$, we have that $P_\theta(\Lambda_n < M) \to 1$ by the law of large numbers. Furthermore, by the consistency of T_n for $\psi(\theta)$, the probability $P_\theta(d(T_n, \psi(\theta_0)) > \varepsilon)$

converges to 1 for every θ such that $d\big(\psi(\theta), \psi(\theta_0)\big) > \varepsilon$. Conclude that the probability in the right side of the preceding display converges to 1, whence the lim sup of the left side is bounded by M. ∎

Notes

Chapter 32 of the famous book by Cramér [27] gives a rigorous proof of what we now know as the Cramér-Rao inequality and next goes on to define the *asymptotic efficiency* of an estimator as the quotient of the inverse Fisher information and the asymptotic variance. Cramér defines an estimator as asymptotically efficient if its efficiency (the quotient mentioned previously) equals one. These definitions lead to the conclusion that the method of maximum likelihood produces asymptotically efficient estimators, as already conjectured by Fisher [48, 50] in the 1920s. That there is a conceptual hole in the definitions was clearly realized in 1951 when Hodges produced his example of a superefficient estimator. Not long after this, in 1953, Le Cam proved that superefficiency can occur only on a Lebesgue null set. Our present result, almost without regularity conditions, is based on later work by Le Cam (see [95].) The asymptotic convolution and minimax theorems were obtained in the present form by Hájek in [69] and [70] after initial work by many authors. Our present proofs follow the approach based on limit experiments, initiated by Le Cam in [95].

PROBLEMS

1. Calculate the asymptotic relative efficiency of the sample mean and the sample median for estimating θ, based on a sample of size n from the normal $N(\theta, 1)$ distribution.

2. As the previous problem, but now for the Laplace distribution (density $p(x) = \frac{1}{2}e^{-|x|}$).

3. Consider estimating the distribution function $P(X \leq x)$ at a fixed point x based on a sample $X_1, \ldots, X_n$ from the distribution of X. The "nonparametric" estimator is $n^{-1}\#(X_i \leq x)$. If it is known that the true underlying distribution is normal $N(\theta, 1)$, another possible estimator is $\Phi(x - \overline{X})$. Calculate the relative efficiency of these estimators.

4. Calculate the relative efficiency of the empirical p-quantile and the estimator $\Phi^{-1}(p)S_n + \overline{X}_n$ for the estimating the p-th quantile of the distribution of a sample from the normal $N(\mu, \sigma^2)$-distribution.

5. Consider estimating the population variance by either the sample variance S^2 (which is unbiased) or else $n^{-1}\sum_{i=1}^{n}(X_i - \overline{X})^2 = (n - 1)/n\, S^2$. Calculate the asymptotic relative efficiency.

6. Calculate the asymptotic relative efficiency of the sample standard deviation and the interquartile range (corrected for unbiasedness) for estimating the standard deviation based on a sample of size n from the normal $N(\mu, \sigma^2)$-distribution.

7. Given a sample of size n from the uniform distribution on $[0, \theta]$, the maximum $X_{(n)}$ of the observations is biased downwards. Because $E_\theta(\theta - X_{(n)}) = E_\theta X_{(1)}$, the bias can be removed by adding the minimum of the observations. Is $X_{(1)} + X_{(n)}$ a good estimator for θ from an asymptotic point of view?

8. Consider the Hodges estimator S_n based on the mean of a sample from the $N(\theta, 1)$-distribution.
   (i) Show that $\sqrt{n}(S_n - \theta_n) \overset{\theta_n}{\leadsto} -\infty$, if $\theta_n \to 0$ in such a way that $n^{1/4}\theta_n \to 0$ and $n^{1/2}\theta_n \to \infty$.
   (ii) Show that S_n is not regular at $\theta = 0$.

(iii) Show that $\sup_{-\delta < \theta < \delta} P_\theta(\sqrt{n}|S_n - \theta| > k_n) \to 1$ for every k_n that converges to infinity sufficiently slowly.

9. Show that a loss function $\ell : \mathbb{R} \mapsto \mathbb{R}$ is bowl-shaped if and only if it has the form $\ell(x) = \ell_0(|x|)$ for a nondecreasing function ℓ_0.

10. Show that a function of the form $\ell(x) = \ell_0(\|x\|)$ for a nondecreasing function ℓ_0 is bowl-shaped.

11. Prove Anderson's lemma for the one-dimensional case, for instance by calculating the derivative of $\int \ell(x + h) \, dN(0, 1)(x)$. Does the proof generalize to higher dimensions?

12. What does Lemma 8.13 imply about the coordinates of the Stein estimator. Are they good estimators of the coordinates of the expectaction vector?

13. All results in this chapter extend in a straightforward manner to general locally asymptotically normal models. Formulate Theorem 8.9 and Lemma 8.14 for such models.

9

Limits of Experiments

A sequence of experiments is defined to converge to a limit experiment if the sequence of likelihood ratio processes converges marginally in distribution to the likelihood ratio process of the limit experiment. A limit experiment serves as an approximation for the converging sequence of experiments. This generalizes the convergence of locally asymptotically normal sequences of experiments considered in Chapter 7. Several examples of nonnormal limit experiments are discussed.

9.1 Introduction

This chapter introduces a notion of convergence of statistical models or "experiments" to a limit experiment. In this notion a sequence of models, rather than just a sequence of estimators or tests, converges to a limit. The limit experiment serves two purposes. First, it provides an absolute standard for what can be achieved asymptotically by a sequence of tests or estimators, in the form of a "lower bound": No sequence of statistical procedures can be asymptotically better than the "best" procedure in the limit experiment. For instance, the best limiting power function is the best power function in the limit experiment; a best sequence of estimators converges to a best estimator in the limit experiment. Statements of this type are true irrespective of the precise meaning of "best." A second purpose of a limit experiment is to explain the asymptotic behaviour of sequences of statistical procedures. For instance, the asymptotic normality or (in)efficiency of maximum likelihood estimators.

Many sequences of experiments converge to normal limit experiments. In particular, the local experiments in a given locally asymptotically normal sequence of experiments, as considered in Chapter 7, converge to a normal location experiment. The asymptotic representation theorem given in the present chapter is therefore a generalization of Theorem 7.10 (for the LAN case) to the general situation. The importance of the general concept is illustrated by several examples of non-Gaussian limit experiments.

In the present context it is customary to speak of "experiment" rather than model, although these terms are interchangeable. Formally an *experiment* is a measurable space $(\mathcal{X}, \mathcal{A})$, the *sample space*, equipped with a collection of probability measures $(P_h : h \in H)$. The set of probability measures serves as a statistical model for the observation, written as X. In this chapter the parameter is denoted by h (and not θ), because the results are typically applied to "local" parameters (such as $h = \sqrt{n}(\theta - \theta_0)$). The experiment is denoted by $(\mathcal{X}, \mathcal{A}, P_h : h \in H)$ and, if there can be no misunderstanding about the sample space, also by $(P_h : h \in H)$.

Given a fixed parameter $h_0 \in H$, the *likelihood ratio process* with base h_0 is formed as

$$\left(\frac{dP_h}{dP_{h_0}}(X)\right)_{h \in H} \equiv \left(\frac{p_h}{p_{h_0}}(X)\right)_{h \in H}.$$

Each likelihood ratio process is a (typically infinite-dimensional) vector of random variables $dP_h/dP_{h_0}(X)$. According to the results of section 6.1, the right side of the display is P_{h_0}-almost surely the same for any given densities p_h and p_{h_0} with respect to any measure μ. Because we are interested only in the laws under P_{h_0} of finite subvectors of the likelihood processes, the nonuniqueness is best left unresolved.

9.1 Definition. A sequence $\mathcal{E}_n = (\mathcal{X}_n, \mathcal{A}_n, P_{n,h} : h \in H)$ of experiments *converges to a limit experiment* $\mathcal{E} = (\mathcal{X}, \mathcal{A}, P_h : h \in H)$ if, for every finite subset $I \subset H$ and every $h_0 \in H$,

$$\left(\frac{dP_{n,h}}{dP_{n,h_0}}(X_n)\right)_{h \in I} \overset{h_0}{\rightsquigarrow} \left(\frac{dP_h}{dP_{h_0}}(X)\right)_{h \in I}.$$

The objects in this display are random vectors of length $|I|$. The requirement is that each of these vectors converges in law, under the assumption that h_0 is the true parameter, in the ordinary sense of convergence in distribution in $\mathbb{R}^I$. This type of convergence is sometimes called *marginal weak convergence*: The finite-dimensional marginal distributions of the likelihood processes converge in distribution to the corresponding marginals in the limit experiment.

Because a weak limit of a sequence of random vectors is unique, the marginal distributions of the likelihood ratio process of a limit experiment are unique. The limit experiment itself is not unique; even its sample space is not uniquely determined. This causes no problems. Two experiments of which the likelihood ratio processes are equal in marginal distributions are called *equivalent* or of the same type. Many examples of equivalent experiments arise through sufficiency.

9.2 Example (Equivalence by sufficiency). Let $S : \mathcal{X} \mapsto \mathcal{Y}$ be a statistic in the statistical experiment $(\mathcal{X}, \mathcal{A}, P_h : h \in H)$ with values in the measurable space $(\mathcal{Y}, \mathcal{B})$. The experiment of image laws $(\mathcal{Y}, \mathcal{B}, P_h \circ S^{-1} : h \in H)$ corresponds to observing S. If S is a sufficient statistic, then this experiment is equivalent to the original experiment $(\mathcal{X}, \mathcal{A}, P_h : h \in H)$. This may be proved using the Neyman factorization criterion of sufficiency. This shows that there exist measurable functions g_h and f such that $p_h(x) = g_h(S(x)) f(x)$, so that the likelihood ratio $p_h/p_{h_0}(X)$ is the function $g_h/g_{h_0}(S)$ of S. The likelihood ratios of the measures $P_h \circ S^{-1}$ take the same form.

Consequently, if $(P_h : h \in H)$ is a limit experiment, then so is $(P_h \circ S^{-1} : h \in H)$. A very simple example that we encounter frequently is as follows: For a given invertible matrix J the experiments $(N(Jh, J) : h \in \mathbb{R}^d)$ and $(N(h, J^{-1}) : h \in \mathbb{R}^d)$ are equivalent. $\square$

9.2 Asymptotic Representation Theorem

In this section it is shown that a limit experiment is always statistically easier than a given sequence. Suppose that a sequence of statistical problems involves experiments

$\mathcal{E}_n = (P_{n,h} : h \in H)$ and statistics T_n. For instance, the statistics are test statistics for testing certain hypotheses concerning the parameter h, or estimators of some function of h. Most of the quality measures of the procedures based on the statistics T_n can be expressed in their laws under the different parameters. For simplicity we assume that the sequence of statistics T_n converges under a given parameter h in distribution to a limit L_h, for every parameter h. Then the asymptotic quality of the sequence T_n may be judged from the set of limit laws $\{L_h : h \in H\}$. According to the following theorem the only possible sets of limit laws are the laws of randomized statistics in the limit experiment: Every weakly converging sequence of statistics converges to a statistic in the limit experiment. One consequence is that asymptotically no sequence of statistical procedures can be better than the best procedure in the limit experiment. This is true for every meaning of "good" that is expressible in terms of laws. In this way the limit experiment obtains the character of an asymptotic lower bound.

We assume that the limit experiment $\mathcal{E} = (P_h : h \in H)$ is *dominated*: This requires the existence of a σ-finite measure μ such that $P_h \ll \mu$ for every h. Recall that a *randomized statistic* T in the experiment $(\mathcal{X}, \mathcal{A}, P_h : h \in H)$ with values in $\mathbb{R}^k$ is a measurable map $T : \mathcal{X} \times [0, 1] \mapsto \mathbb{R}^k$ for the product σ-field $\mathcal{A} \times$ Borel sets on the space $\mathcal{X} \times [0, 1]$. Its law under h is to be computed under the product measure $P_h \times$ uniform$[0, 1]$.

9.3   Theorem. *Let $\mathcal{E}_n = (\mathcal{X}_n, \mathcal{A}_n, P_{n,h} : h \in H)$ be a sequence of experiments that converges to a dominated experiment $\mathcal{E} = (\mathcal{X}, \mathcal{A}, P_h : h \in H)$. Let T_n be a sequence of statistics in $\mathcal{E}_n$ that converges in distribution for every h. Then there exists a randomized statistic T in $\mathcal{E}$ such that $T_n \overset{h}{\rightsquigarrow} T$ for every h.*

Proof.   The proof of the theorem starting from the definition of convergence of experiments is long and can best be broken up into parts of independent interest. This goes beyond the scope of this book.

The proof for the case of local asymptotic normal sequences of experiments is given in Chapter 7. (It is shown in Theorem 9.4 that such a sequence of experiments converges to a Gaussian location experiment.) Many other examples can be treated by the same method of proof.[†]   ∎

9.3   Asymptotic Normality

As in much of statistics, normal limits are of prime importance. In Chapter 7 a sequence of statistical models $(P_{n,\theta} : \theta \in \Theta)$ indexed by an open subset $\Theta \subset \mathbb{R}^d$ is defined to be locally asymptotically normal at θ if the log likelihood ratios $\log dP_{n,\theta+r_n^{-1}h_n}/dP_{n,\theta}$ allow a certain quadratic expansion. This is shown to be valid in the case that $P_{n,\theta}$ is the distribution of a sample of size n from a smooth parametric model. Such experiments converge to simple normal limit experiments if they are reparametrized in terms of the "local parameter" h. This follows from the following theorem.

9.4   Theorem. *Let $\mathcal{E}_n = (P_{n,h} : h \in H)$ be a sequence of experiments indexed by a subset H of $\mathbb{R}^d$ (with $0 \in H$) such that*

$$\log \frac{dP_{n,h}}{dP_{n,0}} = h^T \Delta_n - \frac{1}{2} h^T J h + o_{P_{n,0}}(1),$$

[†] For a proof of the general theorem see, for instance, [141].

for a sequence of statistics Δ_n that converges weakly under $h = 0$ to a $N(0, J)$-distribution. Then the sequence $\mathcal{E}_n$ converges to the experiment $\big(N(Jh, J) : h \in H\big)$.

Proof. The log likelihood ratio process with base h_0 for the normal experiment has coordinates

$$\log \frac{dN(Jh, J)}{dN(Jh_0, J)}(X) = (h - h_0)^T X - \frac{1}{2} h^T J h + \frac{1}{2} h_0^T J h_0.$$

If J is nonsingular, then this follows by simple algebra, because the left side is the quotient of two normal densities. The case that J is singular perhaps requires some thought.

By the assumption combined with Slutsky's lemma, the sequence $\log p_{n,h}/p_{n,0}$ is under $h = 0$ asymptotically normal with mean $-\frac{1}{2} h^T J h$ and variance $h^T J h$. This implies contiguity of the sequences of measures $P_{n,h}$ and $P_{n,0}$ for every h, by Example 6.5. Therefore, the probability of the set on which one of $p_{n,0}$, $p_{n,h}$, or p_{n,h_0} is zero converges to zero. Outside this set we can write

$$\log \frac{p_{n,h}}{p_{n,h_0}} = \log \frac{p_{n,h}}{p_{n,0}} - \log \frac{p_{n,h_0}}{p_{n,0}}.$$

Because this is true with probability tending to 1, the difference between the left and the right sides converges to zero in probability. Apply the (local) asymptotic normality assumption twice to obtain that

$$\log \frac{p_{n,h}}{p_{n,h_0}} = (h - h_0)^T \Delta_n - \frac{1}{2} h^T J h + \frac{1}{2} h_0^T J h_0 + o_{P_{n,h_0}}(1).$$

On comparing this to the expression for the normal likelihood ratio process, we see that it suffices to show that the sequence Δ_n converges under h_0 in law to X: In that case the vector $(p_{n,h}/p_{n,h_0})_{h \in I}$ converges in distribution to $\big(dN(Jh, J)/dN(0, J)(X)\big)_{h \in I}$, by Slutsky's lemma and the continuous-mapping theorem.

By assumption, the sequence $(\Delta_n, h_0^T \Delta_n)$ converges in distribution under $h = 0$ to a vector $(\Delta, h_0^T \Delta)$, where Δ is $N(0, J)$-distributed. By local asymptotic normality and Slutsky's lemma, the sequence of vectors $(\Delta_n, \log p_{n,h_0}/p_{n,0})$ converges to the vector $(\Delta, h_0^T \Delta - \frac{1}{2} h_0^T J h_0)$. In other words

$$\left(\Delta_n, \log \frac{p_{n,h_0}}{p_{n,0}} \right) \overset{0}{\rightsquigarrow} N\left(\begin{pmatrix} 0 \\ -\frac{1}{2} h_0^T J h_0 \end{pmatrix}, \begin{pmatrix} J & Jh_0 \\ h_0^T J & h_0^T J h_0 \end{pmatrix} \right).$$

By the Gaussian form of Le Cam's third lemma, Example 6.5, the sequence Δ_n converges in distribution under h_0 to a $N(Jh_0, J)$-distribution. This is equal to the distribution of X under h_0. ∎

9.5 Corollary. *Let Θ be an open subset of $\mathbb{R}^d$, and let the sequence of statistical models $(P_{n,\theta} : \theta \in \Theta)$ be locally asymptotically normal at θ with norming matrices r_n and a nonsingular matrix I_θ. Then the sequence of experiments $(P_{n,\theta + r_n^{-1}h} : h \in \mathbb{R}^d)$ converges to the experiment $\big(N(h, I_\theta^{-1}) : h \in \mathbb{R}^d\big)$.*

9.4 Uniform Distribution

The model consisting of the uniform distributions on $[0, \theta]$ is not differentiable in quadratic mean (see Example 7.9.) In this case an asymptotically normal approximation is impossible. Instead, we have convergence to an exponential experiment.

9.6 Theorem. *Let P_θ^n be the distribution of a random sample of size n from a uniform distribution on $[0, \theta]$. Then the sequence of experiments $(P_{\theta - h/n}^n : h \in \mathbb{R})$ converges for each fixed $\theta > 0$ to the experiment consisting of observing one observation from the shifted exponential density $z \mapsto e^{-(z-h)/\theta} 1\{z > h\}/\theta$.[†]*

Proof. If Z is distributed according to the given exponential density, then

$$\frac{dP_h^Z}{dP_{h_0}^Z}(Z) = \frac{e^{-(Z-h)/\theta} 1\{Z > h\}/\theta}{e^{-(Z-h_0)/\theta} 1\{Z > h_0\}/\theta} = e^{(h-h_0)/\theta} 1\{Z > h\},$$

almost surely under h_0, because the indicator $1\{z > h_0\}$ in the denominator equals 1 almost surely if h_0 is the true parameter.

The joint density of a random sample $X_1, \ldots, X_n$ from the uniform $[0, \theta]$ distribution can be written in the form $(1/\theta)^n 1\{X_{(n)} \leq \theta\}$. The likelihood ratios take the form

$$\frac{dP_{\theta - h/n}^n}{dP_{\theta - h_0/n}^n}(X_1, \ldots, X_n) = \frac{(\theta - h/n)^{-n} 1\{X_{(n)} \leq \theta - h/n\}}{(\theta - h_0/n)^{-n} 1\{X_{(n)} \leq \theta - h_0/n\}}.$$

Under the parameter $\theta - h_0/n$, the maximum of the observations is certainly bounded above by $\theta - h_0/n$ and the indicator in the denominator equals 1. Thus, with probability 1 under $\theta - h_0/n$, the likelihood ratio in the preceding display can be written

$$\left(e^{(h-h_0)/\theta} + o(1)\right) 1\{-n(X_{(n)} - \theta) \geq h\}.$$

By direct calculation, $-n(X_{(n)} - \theta) \overset{h_0}{\leadsto} Z$. By the continuous-mapping theorem and Slutsky's lemma, the sequence of likelihood processes converges under $\theta - h_0/n$ marginally in distribution to the likelihood process of the exponential experiment. ∎

Along the same lines it may be proved that in the case of uniform distributions with both endpoints unknown a limit experiment based on observation of two independent exponential variables pertains. These types of experiments are completely determined by the discontinuities of the underlying densities at their left and right endpoints. It can be shown more generally that exponential limit experiments are obtained for any densities that have jumps at one or both of their endpoints and are smooth in between. For densities with discontinuities in the middle, or weaker singularities, other limit experiments pertain.

The convergence to a limit experiment combined with the asymptotic representation theorem, Theorem 9.3, allows one to obtain asymptotic lower bounds for sequences of estimators, much as in the locally asymptotically normal case in Chapter 8. We give only one concrete statement.

[†] Define P_θ arbitrarily for $\theta < 0$.

9.7 Corollary. *Let T_n be estimators based on a sample $X_1, \ldots, X_n$ from the uniform distribution on $[0, \theta]$ such that the sequence $n(T_n - \theta)$ converges under θ in distribution to a limit L_θ, for every θ. Then for Lebesgue almost-every θ we have $\int |x| \, dL_\theta(x) \geq$ $\mathrm{E}|Z - \mathrm{med}\, Z|$ and $\int x^2 \, dL_\theta(x) \geq \mathrm{E}(Z - \mathrm{E}Z)^2$ for the random variable Z exponentially distributed with mean θ.*

Proof (Sketch). By Lemma 8.10, the estimator sequence T_n is automatically almost regular in the sense that $n(T_n - \theta + h/n)$ converges under $\theta - h/n$ in distribution to L_θ for Lebesgue almost every θ and h, at least along a subsequence. Thus, it is matched in the limit experiment by an equivariant-in-law estimator for almost every θ. More precisely, for almost every θ there exists a randomized statistic T_θ such that the law of $T_\theta(Z + h, U) - h$ does not depend on h (if Z is exponentially distributed with mean θ). By classical statistical decision theory the given lower bounds are the (constant) risks of the best equivariant-in-law estimators in the exponential limit experiment in terms of absolute error and mean-square error loss functions, respectively. ∎

In view of this lemma, the maximum likelihood estimator $X_{(n)}$ is asymptotically inefficient. This is not surprising given its bias downwards, but it is encouraging for the present approach that the small bias, which is of the order $1/n$, is visible in the "first-order" asymptotics. The bias can be corrected by a multiplicative factor, which, unfortunately, must depend on the loss function. The sequences of estimators

$$\frac{n + \log 2}{n} X_{(n)} \quad \text{and} \quad \frac{n + 1}{n} X_{(n)}$$

are asymptotically efficient in terms of absolute value and quadratic loss, respectively.

9.5 Pareto Distribution

The Pareto distributions are a two-parameter family of distributions on the real line with parameters $\alpha > 0$ and $\mu > 0$ and density

$$x \mapsto \frac{\alpha \mu^\alpha}{x^{\alpha+1}} 1\{x > \mu\}.$$

This density is smooth in α, but it resembles a uniform distribution as discussed in the preceding section in its dependence on μ. The limit experiment consists of a combination of a normal experiment and an exponential experiment.

The likelihood ratios for a sample of size n from the Pareto distributions with parameters $(\alpha + g/\sqrt{n}, \mu + h/n)$ and $(\alpha + g_0/\sqrt{n}, \mu + h_0/n)$, respectively, is equal to

$$\left(\frac{\alpha + g/\sqrt{n}}{\alpha + g_0/\sqrt{n}} \right)^n \frac{(\mu + h/n)^{n\alpha + \sqrt{n}g}}{(\mu + h_0/n)^{n\alpha + \sqrt{n}g_0}} \left(\prod_{i=1}^n X_i \right)^{(g_0 - g)/\sqrt{n}} 1\left\{ X_{(1)} > \mu + \frac{h}{n} \right\}$$

$$= \exp\left((g - g_0)\Delta_n - \frac{1}{2} \frac{g^2 + g_0^2}{\alpha^2} + o(1) \right) \left(e^{(h - h_0)\alpha/\mu} + o(1) \right) 1\{Z_n > h\}.$$

Here, under the parameters $(\alpha + g_0/\sqrt{n}, \mu + h_0/n)$, the sequence

$$\Delta_n = -\frac{1}{\sqrt{n}} \sum_{i=1}^{n} \left(\log \frac{X_i}{\mu} - \frac{1}{\alpha} \right)$$

converges weakly to a normal distribution with mean g_0/α^2 and variance $1/\alpha^2$; and the sequence $Z_n = n(X_{(1)} - \mu)$ converges in distribution to the (shifted) exponential distribution with mean $\mu/\alpha + h_0$ and variance $(\mu/\alpha)^2$. The two sequences are asymptotically independent. Thus the likelihood is a product of a locally asymptotically normal and a "locally asymptotically exponential" factor. The local limit experiment consists of observing a pair (Δ, Z) of independent variables Δ and Z with a $N(g, \alpha^2)$-distribution and an $\exp(\alpha/\mu) + h$-distribution, respectively.

The maximum likelihood estimators for the parameters α and μ are given by

$$\hat{\alpha}_n = \frac{n}{\sum_{i=1}^{n} \log(X_i / X_{(1)})}, \quad \text{and} \quad \hat{\mu}_n = X_{(1)}.$$

The sequence $\sqrt{n}(\hat{\alpha}_n - \alpha)$ converges in distribution under the parameters $(\alpha + g/\sqrt{n}, \mu + h/n)$ to the variable $\Delta - g$. Because the distribution of Z does not depend on g, and Δ follows a normal location model, the variable Δ can be considered an optimal estimator for g based on the observation (Δ, Z). This optimality is carried over into the asymptotic optimality of the maximum likelihood estimator $\hat{\alpha}_n$. A precise formulation could be given in terms of a convolution or a minimax theorem.

On the other hand, the maximum likelihood estimator for μ is asymptotically inefficient. Because the sequence $n(\hat{\mu}_n - \mu - h/n)$ converges in distribution to $Z - h$, the estimators $\hat{\mu}_n$ are asymptotically biased upwards.

9.6 Asymptotic Mixed Normality

The likelihood ratios of some models allow an approximation by a two-term Taylor expansion without the linear term being asymptotically normal and the quadratic term being deterministic. Then a generalization of local asymptotic normality is possible. In the most important example of this situation, the linear term is asymptotically distributed as a mixture of normal distributions.

A sequence of experiments $(P_{n,\theta} : \theta \in \Theta)$ indexed by an open subset Θ of $\mathbb{R}^d$ is called *locally asymptotically mixed normal* at θ if there exist matrices $\gamma_{n,\theta} \to 0$ such that

$$\log \frac{dP_{n,\theta+\gamma_{n,\theta}h_n}}{dP_{n,\theta}} = h^T \Delta_{n,\theta} - \frac{1}{2} h^T J_{n,\theta} h + o_{P_{n,\theta}}(1),$$

for every converging sequence $h_n \to h$, and random vectors $\Delta_{n,\theta}$ and random matrices $J_{n,\theta}$ such that $(\Delta_{n,\theta}, J_{n,\theta}) \overset{\theta}{\rightsquigarrow} (\Delta_\theta, J_\theta)$ for a random vector such that the conditional distribution of Δ_θ given that $J_\theta = J$ is normal $N(0, J)$.

Locally asymptotically mixed normal is often abbreviated to LAMN. Locally asymptotically normal, or LAN, is the special case in which the matrix J_θ is deterministic. Sequences of experiments whose likelihood ratios allow a quadratic approximation as in the preceding display (but without the specific limit distribution of $(\Delta_{n,\theta}, J_{n,\theta})$) and that are such that $P_{n,\theta+\gamma_{n,\theta}h} \lhd \rhd P_{n,\theta}$ are called *locally asymptotically quadratic*, or LAQ. We note that LAQ or LAMN requires much more than the mere existence of two derivatives of the likelihood: There is no reason why, in general, the remainder would be negligible.

9.8   Theorem. *Assume that the sequence of experiments* $(P_{n,\theta} : \theta \in \Theta)$ *is locally asymptotically mixed normal at* θ. *Then the sequence of experiments* $(P_{n,\theta+\gamma_{n,\theta}h} : h \in \mathbb{R}^d)$ *converges to the experiment consisting of observing a pair* (Δ, J) *such that* J *is marginally distributed as* J_θ *for every* h *and the conditional distribution of* Δ *given* J *is normal* $N(Jh, J)$.

Proof.   Write $P_{\theta,h}$ for the distribution of (Δ, J) under h. Because the marginal distribution of J does not depend on h and the conditional distribution of Δ given J is Gaussian

$$\frac{dP_{\theta,h}}{dP_{\theta,h_0}}(\Delta, J) = \frac{dN(Jh, J)}{dN(Jh_0, J)}(\Delta) = e^{(h-h_0)^T \Delta - \frac{1}{2}h^T Jh + \frac{1}{2}h_0^T Jh_0}.$$

By Slutsky's lemma and the assumptions, the sequence $dP_{n,\theta+\gamma_{n,\theta}h}/dP_{n,\theta}$ converges under θ in distribution to $\exp(h^T \Delta_\theta - \frac{1}{2}h^T J_\theta h)$. Because the latter variable has mean one, it follows that the sequences of distributions $P_{n,\theta+\gamma_{n,\theta}h}$ and $P_{n,\theta}$ are mutually contiguous. In particular, the probability under θ that $dP_{n,\theta+\gamma_{n,\theta}h}$ is zero converges to zero for every h, so that

$$\log \frac{dP_{n,\theta+\gamma_{n,\theta}h}}{dP_{n,\theta+\gamma_{n,\theta}h_0}} = \log \frac{dP_{n,\theta+\gamma_{n,\theta}h}}{dP_{n,\theta}} - \log \frac{dP_{n,\theta+\gamma_{n,\theta}h_0}}{dP_{n,\theta}} + o_{P_{n,\theta}}(1)$$

$$= (h - h_0)^T \Delta_{n,\theta} - \frac{1}{2}h^T J_{n,\theta}h + \frac{1}{2}h_0^T J_{n,\theta}h_0 + o_{P_{n,\theta}}(1).$$

Conclude that it suffices to show that the sequence $(\Delta_{n,\theta}, J_{n,\theta})$ converges under $\theta + \gamma_{n,\theta}h_0$ to the distribution of (Δ, J) under h_0.

Using the general form of Le Cam's third lemma we obtain that the limit distribution of the sequence $(\Delta_{n,\theta}, J_{n,\theta})$ under $\theta + \gamma_{n,\theta}h$ takes the form

$$L_h(B) = E1_B(\Delta_\theta, J_\theta)e^{h^T \Delta_\theta - \frac{1}{2}h^T J_\theta h}.$$

On noting that the distribution of (Δ, J) under $h = 0$ is the same as the distribution of $(\Delta_\theta, J_\theta)$, we see that this is equal to $E_0 1_B(\Delta, J) dP_{\theta,h}/dP_{\theta,0}(\Delta, J) = P_h((\Delta, J) \in B)$.   ∎

It is possible to develop a theory of asymptotic "lower bounds" for LAMN models, much as is done for LAN models in Chapter 8. Because conditionally on the ancillary statistic J, the limit experiment is a Gaussian shift experiment, the lower bounds take the form of mixtures of the lower bounds for the LAN case. We give only one example, leaving the details to the reader.

9.9   Corollary. *Let* T_n *be an estimator sequence in a LAMN sequence of experiments* $(P_{n,\theta} : \theta \in \Theta)$ *such that* $\gamma_{n,\theta}^{-1}\left(T_n - \psi(\theta + \gamma_{n,\theta}h)\right)$ *converges weakly under every* $\theta + \gamma_{n,\theta}h$ *to a limit distribution* L_θ, *for every* h. *Then there exist probability distributions* M_j *(or rather a Markov kernel) such that* $L_\theta = EN(0, \dot{\psi}_\theta J_\theta^{-1} \dot{\psi}_\theta^T) * M_{J_\theta}$. *In particular,* $\mathrm{cov}_\theta\, L_\theta \geq E\dot{\psi}_\theta J_\theta^{-1} \dot{\psi}_\theta^T$.

We include two examples to give some idea of the application of local asymptotic mixed normality. In both examples the sequence of models is LAMN rather than LAN due to an explosive growth of information, occurring at certain supercritical parameter values. The second derivative of the log likelihood, the information, remains random. In both examples there is also (approximate) Gaussianity present in every single observation. This appears to be typical, unlike the situation with LAN, in which the normality results from sums over (approximately) independent observations. In explosive models of this type the likelihood is dominated by a few observations, and normality cannot be brought in through (martingale) central limit theorems.

9.10   Example (Branching processes).   In a Galton-Watson branching process the "nth generation" is formed by replacing each element of the $(n-1)$-th generation by a random number of elements, independently from the rest of the population and from the preceding generations. This random number is distributed according to a fixed distribution, called the *offspring distribution*. Thus, conditionally on the size X_{n-1} of the $(n-1)$th generation the size X_n of the nth generation is distributed as the sum of X_{n-1} i.i.d. copies of an offspring variable Z. Suppose that $X_0 = 1$, that we observe $(X_1, \ldots, X_n)$, and that the offspring distribution is known to belong to an exponential family of the form

$$P_\theta(Z = z) = a_z \, \theta^z c(\theta), \quad z = 0, 1, 2, \ldots,$$

for given numbers $a_0, a_1, \ldots$. The natural parameter space is the set of all θ such that $c(\theta)^{-1} = \sum_z a_z \theta^z$ is finite (an interval). We shall concentrate on parameters in the interior of the natural parameter space such that $\mu(\theta) := E_\theta Z > 1$. Set $\sigma^2(\theta) = \mathrm{var}_\theta Z$.

The sequence $X_1, X_2, \ldots$ is a Markov chain with transition density

$$p_\theta(y \mid x) = P_\theta(X_n = y \mid X_{n-1} = x) = \overbrace{a * \cdots * a}^{x \text{ times}} \, \theta^y c(\theta)^x.$$

To obtain a two-term Taylor expansion of the log likelihood ratios, let $\ell_\theta(y \mid x)$ be the log transition density, and calculate that

$$\dot{\ell}_\theta(y \mid x) = \frac{y - x\mu(\theta)}{\theta}, \qquad \ddot{\ell}_\theta(y \mid x) = -\frac{y - x\mu(\theta)}{\theta^2} - \frac{x\dot{\mu}(\theta)}{\theta}.$$

(The fact that the score function of the model $\theta \mapsto P_\theta(Z = z)$ has derivative zero yields the identity $\mu(\theta) = -\theta(c/\dot{c})(\theta)$, as is usual for exponential families.) Thus, the Fisher information in the observation $(X_1, \ldots, X_n)$ equals (note that $E_\theta(X_j \mid X_{j-1}) = X_{j-1}\mu(\theta)$)

$$-E_\theta \sum_{j=1}^n \ddot{\ell}_\theta(X_j \mid X_{j-1}) = E_\theta \sum_{j=1}^n X_{j-1} \frac{\dot{\mu}(\theta)}{\theta}$$

$$= \frac{\dot{\mu}(\theta)}{\theta} \sum_{j=1}^n \mu(\theta)^{j-1} = \frac{\dot{\mu}(\theta)}{\theta} \frac{\mu(\theta)^n - 1}{\mu(\theta) - 1}.$$

For $\mu(\theta) > 1$, this converges to infinity at a much faster rate than "usually." Because the total information in $(X_1, \ldots, X_n)$ is of the same order as the information in the last observation X_n, the model is "explosive" in terms of growth of information. The calculation suggests the rescaling rate $\gamma_{n,\theta} = \mu(\theta)^{-n/2}$, which is roughly the inverse root of the information.

A Taylor expansion of the log likelihood ratio yields the existence of a point θ_n between θ and $\theta + \gamma_{n,\theta} h$ such that

$$\log \prod_{j=1}^{n} \frac{p_{\theta + \gamma_{n,\theta} h}}{p_\theta}(X_j \mid X_{j-1})$$

$$= \frac{h}{\mu(\theta)^{n/2}} \sum_{j=1}^{n} \dot{\ell}_\theta(X_j \mid X_{j-1}) + \frac{1}{2} \frac{h^2}{\mu(\theta)^n} \sum_{j=1}^{n} \ddot{\ell}_{\theta_n}(X_j \mid X_{j-1}).$$

This motivates the definitions

$$\Delta_{n,\theta} = \frac{1}{\mu(\theta)^{n/2}} \sum_{j=1}^{n} \frac{X_j - \mu(\theta) X_{j-1}}{\theta}$$

$$J_{n,\theta} = \frac{1}{\mu(\theta)^n} \sum_{j=1}^{n} \left[\frac{X_j - \mu(\theta) X_{j-1}}{\theta^2} + \frac{X_{j-1} \dot{\mu}(\theta)}{\theta} \right].$$

Because $E_\theta(X_n \mid X_{n-1}, \ldots, X_1) = X_{n-1} \mu(\theta)$, the sequence of random variables $\mu(\theta)^{-n} X_n$ is a martingale under θ. Some algebra shows that its second moments are bounded as $n \to \infty$. Thus, by a martingale convergence theorem (e.g., Theorem 10.5.4 of [42]), there exists a random variable V such that $\mu(\theta)^{-n} X_n \to V$ almost surely. By the Toeplitz lemma (Problem 9.6) and again some algebra, we obtain that, almost surely under θ,

$$\frac{1}{\mu(\theta)^n} \sum_{j=1}^{n} X_j \to \frac{\mu(\theta)}{\mu(\theta) - 1} V, \qquad \frac{1}{\mu(\theta)^n} \sum_{j=1}^{n} X_{j-1} \to \frac{1}{\mu(\theta) - 1} V.$$

It follows that the point θ_n in the expansion of the log likelihood can be replaced by θ at the cost of adding a term that converges to zero in probability under θ. Furthermore,

$$J_{n,\theta} \mapsto \frac{\dot{\mu}(\theta)}{\theta(\mu(\theta) - 1)} V, \qquad P_\theta\text{-almost surely.}$$

It remains to derive the limit distribution of the sequence $\Delta_{n,\theta}$. If we write $X_j = \sum_{i=1}^{X_{j-1}} Z_{j,i}$ for independent copies $Z_{j,i}$ of the offspring variable Z, then

$$\Delta_{n,\theta} = \frac{1}{\theta \mu(\theta)^{n/2}} \sum_{j=1}^{n} \sum_{i=1}^{X_{j-1}} \left(Z_{j,i} - \mu(\theta) \right) = \frac{1}{\theta \mu(\theta)^{n/2}} \sum_{i=1}^{v_n} \left(Z_i - \mu(\theta) \right),$$

for independent copies Z_i of Z and $v_n = \sum_{i=1}^{n} X_{j-1}$. Even though $Z_1, Z_2, \ldots$ and the total number v_n of variables in the sum are dependent, a central limit theorem applies to the right side: conditionally on the event $\{V > 0\}$ (on which $v_n \to \infty$), the sequence $v_n^{-1/2} \sum_{i=1}^{v_n} (Z_i - \mu(\theta))$ converges in distribution to $\sigma(\theta)$ times a standard normal variable G. Furthermore, if we define G independent of V, conditionally on $\{V > 0\}$,[†]

$$(\Delta_{n,\theta}, J_{n,\theta}) \rightsquigarrow \left(\frac{\sigma(\theta)}{\theta} \sqrt{\frac{V}{\mu(\theta) - 1}} G, \frac{\dot{\mu}(\theta)}{\theta(\mu(\theta) - 1)} V \right). \tag{9.11}$$

[†] See the appendix of [81] or, e.g., Theorem 3.5.1 and its proof in [146].

It is well known that the event $\{V = 0\}$ coincides with the event $\{\lim X_n = 0\}$ of extinction of the population. (This occurs with positive probability if and only if $a_0 > 0$.) Thus, on the set $\{V = 0\}$ the series $\sum_{j=1}^{\infty} X_j$ converges almost surely, whence $\Delta_{n,\theta} \to 0$. Interpreting zero as the product of a standard normal variable and zero, we see that again (9.11) is valid. Thus the sequence $(\Delta_{n,\theta}, J_{n,\theta})$ converges also unconditionally to this limit. Finally, note that $\sigma^2(\theta)/\theta = \dot{\mu}(\theta)$, so that the limit distribution has the right form.

The maximum likelihood estimator for $\mu(\theta)$ can be shown to be asymptotically efficient, (see, e.g., [29] or [81]). $\quad \square$

9.12 Example (Gaussian AR). The canonical example of an LAMN sequence of experiments is obtained from an explosive autoregressive process of order one with Gaussian innovations. (The Gaussianity is essential.) Let $|\theta| > 1$ and $\varepsilon_1, \varepsilon_2, \ldots$ be an i.i.d. sequence of standard normal variables independent of a fixed variable X_0. We observe the vector $(X_0, X_1, \ldots, X_n)$ generated by the recursive formula $X_t = \theta X_{t-1} + \varepsilon_t$.

The observations form a Markov chain with transition density $p(\cdot \mid x_{t-1})$ equal to the $N(\theta x_{t-1}, 1)$-density. Therefore, the log likelihood ratio process takes the form

$$\log \frac{p_{n,\theta+\gamma_{n,\theta}h}}{p_{n,\theta}}(X_0, \ldots, X_n) = h \gamma_{n,\theta} \sum_{t=1}^{n} (X_t - \theta X_{t-1}) X_{t-1} - \frac{1}{2} h^2 \gamma_{n,\theta}^2 \sum_{t=1}^{n} X_{t-1}^2.$$

This has already the appropriate quadratic structure. To establish LAMN, it suffices to find the right rescaling rate and to establish the joint convergence of the linear and the quadratic term. The rescaling rate may be chosen proportional to the Fisher information and is taken $\gamma_{n,\theta} = \theta^{-n}$.

By repeated application of the defining autoregressive relationship, we see that

$$\theta^{-t} X_t = X_0 + \sum_{j=1}^{t} \theta^{-j} \varepsilon_j \to V := X_0 + \sum_{j=1}^{\infty} \theta^{-j} \varepsilon_j,$$

almost surely as well as in second mean. Given the variable X_0, the limit is normally distributed with mean X_0 and variance $(\theta^2 - 1)^{-1}$. An application of the Toeplitz lemma (Problem 9.6) yields

$$\frac{1}{\theta^{2n}} \sum_{t=1}^{n} X_{t-1}^2 \to \frac{V^2}{\theta^2 - 1}.$$

The linear term in the quadratic representation of the log likelihood can (under θ) be rewritten as $\theta^{-n} \sum_{t=1}^{n} \varepsilon_t X_{t-1}$, and satisfies, by the Cauchy-Schwarz inequality and the Toeplitz lemma,

$$\mathrm{E} \left| \frac{1}{\theta^n} \sum_{t=1}^{n} \varepsilon_t X_{t-1} - \frac{1}{\theta^n} \sum_{t=1}^{n} \varepsilon_t \theta^{t-1} V \right| \leq \frac{1}{|\theta|^n} \sum_{t=1}^{n} |\theta|^{t-1} \left(\mathrm{E}(\theta^{-t+1} X_{t-1} - V)^2 \right)^{1/2} \to 0.$$

It follows that the sequence of vectors $(\Delta_{n,\theta}, J_{n,\theta})$ has the same limit distribution as the sequence of vectors $\left(\theta^{-n} \sum_{t=1}^{n} \varepsilon_t \theta^{t-1} V, V^2/(\theta^2 - 1) \right)$. For every n the vector $\left(\theta^{-n} \sum_{t=1}^{n} \varepsilon_t \right.$

θ^{t-1}, V) possesses, conditionally on X_0, a bivariate-normal distribution. As $n \to \infty$ these distributions converge to a bivariate-normal distribution with mean $(0, X_0)$ and covariance matrix $I/(\theta^2 - 1)$. Conclude that the sequence $(\Delta_{n,\theta}, J_{n,\theta})$ converges in distribution as required by the LAMN criterion.   □

9.7   Heuristics

The asymptotic representation theorem, Theorem 9.3, shows that every sequence of statistics in a converging sequence of experiments is matched by a statistic in the limit experiment. It is remarkable that this is true under the present definition of convergence of experiments, which involves only marginal convergence and is very weak.

Under appropriate stronger forms of convergence more can be said about the nature of the matching procedure in the limit experiment. For instance, a sequence of maximum likelihood estimators converges to the maximum likelihood estimator in the limit experiment, or a sequence of likelihood ratio statistics converges to the likelihood ratio statistic in the limit experiment. We do not introduce such stronger convergence concepts in this section but only note the potential of this argument as a heuristic principle. See section 5.9 for rigorous results.

For the maximum likelihood estimator the heuristic argument takes the following form. If $\hat{h}_n$ maximizes the likelihood $h \mapsto dP_{n,h}$, then it also maximizes the likelihood ratio process $h \mapsto dP_{n,h}/dP_{n,h_0}$. The latter sequence of processes converges (marginally) in distribution to the likelihood ratio process $h \mapsto dP_h/dP_{h_0}$ of the limit experiment. It is reasonable to expect that the maximizer $\hat{h}_n$ converges in distribution to the maximizer of the process $h \mapsto dP_h/dP_{h_0}$, which is the maximum likelihood estimator for h in the limit experiment. (Assume that this exists and is unique.) If the converging experiments are the local experiments corresponding to a given sequence of experiments with a parameter θ, then the argument suggests that the sequence of local maximum likelihood estimators $\hat{h}_n = r_n(\hat{\theta}_n - \theta)$ converges, under θ, in distribution to the maximum likelihood estimator in the local limit experiment, under $h = 0$.

Besides yielding the limit distribution of the maximum likelihood estimator, the argument also shows to what extent the estimator is asymptotically efficient. It is efficient, or inefficient, in the same sense as the maximum likelihood estimator is efficient or inefficient in the limit experiment. That maximum likelihood estimators are often asymptotically efficient is a consequence of the fact that often the limit experiment is Gaussian and the maximum likelihood estimator of a Gaussian location parameter is optimal in a certain sense. If the limit experiment is not Gaussian, there is no a priori reason to expect that the maximum likelihood estimators are asymptotically efficient.

A variety of examples shows that the conclusions of the preceding heuristic arguments are often but not universally valid. The reason for failures is that the convergence of experiments is not well suited to allow claims about maximum likelihood estimators. Such claims require stronger forms of convergence than marginal convergence only.

For the case of experiments consisting of a random sample from a smooth parametric model, the argument is made precise in section 7.4. Next to the convergence of experiments, it is required only that the maximum likelihood estimator is consistent and that the log density is locally Lipschitz in the parameter. The preceding heuristic argument also extends to the other examples of convergence to limit experiments considered in this chapter. For instance, the maximum likelihood estimator based on a sample from the uniform distribution on $[0, \theta]$

is asymptotically inefficient, because it corresponds to the estimator Z for h (the maximum likelihood estimator) in the exponential limit experiment. The latter is biased upwards and inefficient for every of the usual loss functions.

Notes

This chapter presents a few examples from a large body of theory. The notion of a limit experiment was introduced by Le Cam in [95]. He defined convergence of experiments through convergence of all finite subexperiments relative to his *deficiency distance*, rather than through convergence of the likelihood ratio processes. This deficiency distance introduces a "strong topology" next to the "weak topology" corresponding to convergence of experiments. For experiments with a finite parameter set, the two topologies coincide. There are many general results that can help to prove the convergence of experiments and to find the limits (also in the examples discussed in this chapter). See [82], [89], [96], [97], [115], [138], [142] and [144] for more information and more examples. For nonlocal approximations in the strong topology see, for example, [96] or [110].

PROBLEMS

1. Let $X_1, \ldots, X_n$ be an i.i.d. sample from the normal $N(h/\sqrt{n}, 1)$ distribution, in which $h \in \mathbb{R}$. The corresponding sequence of experiments converges to a normal experiment by the general results. Can you see this directly?

2. If the nth experiment corresponds to the observation of a sample of size n from the uniform $[0, 1-h/n]$, then the limit experiment corresponds to observation of a shifted exponential variable Z. The sequences $-n(X_{(n)} - 1)$ and $\sqrt{n}(2\overline{X}_n - 1)$ both converge in distribution under every h. According to the representation theorem their sets of limit distributions are the distributions of randomized statistics based on Z. Find these randomized statistics explicitly. Any implications regarding the quality of $X_{(n)}$ and $\overline{X}_n$ as estimators?

3. Let the nth experiment consist of one observation from the binomial distribution with parameters n and success probability h/n with $0 < h < 1$ unknown. Show that this sequence of experiments converges to the experiment consisting of observing a Poisson variable with mean h.

4. Let the nth experiment consists of observing an i.i.d. sample of size n from the uniform $[-1 - h/n, 1 + h/n]$ distribution. Find the limit experiment.

5. Prove the asymptotic representation theorem for the case in which the nth experiment corresponds to an i.i.d. sample from the uniform $[0, \theta - h/n]$ distribution with $h > 0$ by mimicking the proof of this theorem for the locally asymptotically normal case.

6. (Toeplitz lemma.) If a_n is a sequence of nonnegative numbers with $\sum a_n = \infty$ and $x_n \to x$ an arbitrary converging sequence of numbers, then the sequence $\sum_{j=1}^n a_j x_j / \sum_{j=1}^n a_j$ converges to x as well. Show this.

7. Derive a limit experiment in the case of Galton-Watson branching with $\mu(\theta) < 1$.

8. Derive a limit experiment in the case of a Gaussian AR(1) process with $\theta = 1$.

9. Derive a limit experiment for sampling from a $U[\sigma, \tau]$ distribution with both endpoints unknown.

10. In the case of sampling from the $U[0, \theta]$ distribution show that the maximum likelihood estimator for θ converges to the maximum likelihood estimator in the limit experiment. Why is the latter not a good estimator?

11. Formulate and prove a local asymptotic minimax theorem for estimating θ from a sample from a $U[0, \theta]$ distribution, using $\ell(x) = x^2$ as loss function.

10

Bayes Procedures

In this chapter Bayes estimators are studied from a frequentist perspective. Both posterior measures and Bayes point estimators in smooth parametric models are shown to be asymptotically normal.

10.1 Introduction

In Bayesian terminology the distribution $P_{n,\theta}$ of an observation $\vec{X}_n$ under a parameter θ is viewed as the conditional law of $\vec{X}_n$ given that a random variable $\overline{\Theta}_n$ is equal to θ. The distribution Π of the "random parameter" $\overline{\Theta}_n$ is called the *prior distribution*, and the conditional distribution of $\overline{\Theta}_n$ given $\vec{X}_n$ is the *posterior distribution*. If $\overline{\Theta}_n$ possesses a density π and $P_{n,\theta}$ admits a density $p_{n,\theta}$ (relative to given dominating measures), then the density of the posterior distribution is given by Bayes' formula

$$p_{\overline{\Theta}_n \mid \vec{X}_n = x}(\theta) = \frac{p_{n,\theta}(x)\,\pi(\theta)}{\int p_{n,\theta}(x)\,d\Pi(\theta)}.$$

This expression may define a probability density even if π is not a probability density itself. A prior distribution with infinite mass is called *improper*.

The calculation of the posterior measure can be considered the ultimate aim of a Bayesian analysis. Alternatively, one may wish to obtain a "point estimator" for the parameter θ, using the posterior distribution. The posterior mean $\mathrm{E}(\overline{\Theta}_n \mid \vec{X}_n) = \int \theta\, p_{\overline{\Theta}_n \mid \vec{x}_n}(\theta)\, d\theta$ is often used for this purpose, but other location estimators are also reasonable.

A choice of point estimator may be motivated by a loss function. The *Bayes risk* of an estimator T_n relative to the loss function ℓ and prior measure Π is defined as

$$\int \mathrm{E}_\theta \ell(T_n - \theta)\, d\Pi(\theta) = \mathrm{E}\ell(T_n - \overline{\Theta}_n).$$

Here the expectation $\mathrm{E}_\theta \ell(T_n - \theta)$ is the risk function of T_n in the usual set-up and is identical to the conditional risk $\mathrm{E}\big(\ell(T_n - \overline{\Theta}_n) \mid \overline{\Theta}_n = \theta\big)$ in the Bayesian notation. The corresponding *Bayes estimator* is the estimator T_n that minimizes the Bayes risk. Because the Bayes risk can be written in the form $\mathrm{E}\mathrm{E}\big(\ell(T_n - \overline{\Theta}_n) \mid \vec{X}_n\big)$, the value $T_n = T_n(x)$ minimizes, for every fixed x, the "posterior risk"

$$\mathrm{E}\big(\ell(T_n - \overline{\Theta}_n) \mid \vec{X}_n = x\big) = \frac{\int \ell(T_n - \theta)\, p_{n,\theta}(x)\, d\Pi(\theta)}{\int p_{n,\theta}(x)\, d\Pi(\theta)}.$$

138

Minimizing this expression may again be a well-defined problem even for prior densities of infinite total mass. For the loss function $\ell(y) = \|y\|^2$, the solution T_n is the posterior mean $E(\overline{\Theta}_n \mid \vec{X}_n)$, for absolute loss $\ell(y) = \|y\|$, the solution is the posterior median.

Other Bayesian point estimators are the posterior mode, which reduces to the maximum likelihood estimator in the case of a uniform prior density; or a maximum probability estimator, such as the center of the smallest ball that contains at least posterior mass $1/2$ (the "posterior shorth" in dimension one).

If the underlying experiments converge, in a suitable sense, to a Gaussian location experiment, then all these possibilities are typically asymptotically equivalent. Consider the case that the observation consists of a random sample of size n from a density p_θ that depends smoothly on a Euclidean parameter θ. Thus the density $p_{n,\theta}$ has a product form, and, for a given prior Lebesgue density π, the posterior density takes the form

$$P_{\overline{\Theta}_n \mid X_1,\dots,X_n}(\theta) = \frac{\prod_{i=1}^n p_\theta(X_i)\pi(\theta)}{\int \prod_{i=1}^n p_\theta(X_i)\pi(\theta)\,d\theta}.$$

Typically, the distribution corresponding to this measure converges to the measure that is degenerate at the true parameter value θ_0, as $n \to \infty$. In this sense Bayes estimators are usually consistent. A further discussion is given in sections 10.2 and 10.4. To obtain a more interesting limit, we rescale the parameter in the usual way and study the sequence of posterior distributions of $\sqrt{n}(\overline{\Theta}_n - \theta_0)$, whose densities are given by

$$P_{\sqrt{n}(\overline{\Theta}_n - \theta_0) \mid X_1,\dots,X_n}(h) = \frac{\prod_{i=1}^n p_{\theta_0+h/\sqrt{n}}(X_i)\,\pi(\theta_0 + h/\sqrt{n})}{\int \prod_{i=1}^n p_{\theta_0+h/\sqrt{n}}(X_i)\,\pi(\theta_0 + h/\sqrt{n})\,dh}.$$

If the prior density π is continuous, then $\pi(\theta_0 + h/\sqrt{n})$, for large n, behaves like the constant $\pi(\theta_0)$, and π cancels from the expression for the posterior density. For densities p_θ that are sufficiently smooth in the parameter, the sequence of models $(P_{\theta_0+h/\sqrt{n}} : h \in \mathbb{R}^k)$ is locally asymptotically normal, as discussed in Chapter 7. This means that the likelihood ratio processes $h \mapsto \prod_{i=1}^n p_{\theta_0+h/\sqrt{n}}/p_{\theta_0}(X_i)$ behave asymptotically as the likelihood ratio process of the normal experiment $\left(N(h, I_{\theta_0}^{-1}) : h \in \mathbb{R}^k\right)$. Then we may expect the preceding display to be asymptotically equivalent in distribution to

$$\frac{dN\left(h, I_{\theta_0}^{-1}\right)(X)}{\int dN\left(h, I_{\theta_0}^{-1}\right)(X)dh} = dN\left(X, I_{\theta_0}^{-1}\right)(h),$$

where $dN(\mu, \Sigma)$ denotes the density of the normal distribution. The expression in the preceding display is exactly the posterior density for the experiment $\left(N(h, I_{\theta_0}^{-1}) : h \in \mathbb{R}^k\right)$, relative to the (improper) Lebesgue prior distribution. The expression on the right shows that this is a normal distribution with mean X and covariance matrix $I_{\theta_0}^{-1}$.

This heuristic argument leads us to expect that the posterior distribution of $\sqrt{n}(\overline{\Theta}_n - \theta_0)$ "converges" under the true parameter θ_0 to the posterior distribution of the Gaussian limit experiment relative to the Lebesgue prior. The latter is equal to the $N(X, I_{\theta_0}^{-1})$-distribution, for X possessing the $N(0, I_{\theta_0}^{-1})$-distribution. The notion of convergence in this statement is a complicated one, because a posterior distribution is a conditional, and hence stochastic, probability measure, but there is no need to make the heuristics precise at this point. On the other hand, the convergence should certainly include that "nice" Euclidean-valued functionals applied to the posterior laws converge in distribution in the usual sense.

Consequently, a sequence of Bayes point estimators, which can be viewed as location functionals applied to the posterior distributions, should converge to the corresponding Bayes point estimator in the limit experiment. Most location estimators (all reasonable ones) map symmetric distributions, such as the normal distribution, into their center of symmetry. Then, the Bayes point estimator in the limit experiment is X, and we should expect Bayes point estimators to converge in distribution to the random vector X, that is, to a $N(0, I_{\theta_0}^{-1})$-distribution under θ_0. In particular, they are asymptotically efficient and asymptotically equivalent to maximum likelihood estimators (under regularity conditions).

A remarkable fact about this conclusion is that the limit distribution of a sequence of Bayes estimators does not depend on the prior measure. Apparently, for an increasing number of observations one's prior beliefs are erased (or corrected) by the observations. To make this true an essential assumption is that the prior distribution possesses a density that is smooth and positive in a neighborhood of the true value of the parameter. Without this property the conclusion fails. For instance, in the case in which one rigorously sticks to a fixed discrete distribution that does not charge θ_0, the sequence of posterior distributions of $\overline{\Theta}_n$ cannot even be consistent.

In the next sections we make the preceding heuristic argument precise. For technical reasons we separately consider the distributional approximation of the posterior distributions by a Gaussian one and the weak convergence of Bayes point estimators.

Even though the heuristic extends to convergence to other than Gaussian location experiments, we limit ourselves in this chapter to the locally asymptotically normal case. More precisely, we even assume that the observations are a random sample $X_1, \ldots, X_n$ from a distribution P_θ that admits a density p_θ with respect to a measure μ on a measurable space $(\mathcal{X}, \mathcal{A})$. The parameter θ is assumed to belong to a measurable subset Θ of $\mathbb{R}^k$ that contains the true parameter θ_0 as an interior point, and we assume that the maps $(\theta, x) \mapsto p_\theta(x)$ are jointly measurable.

All theorems in this chapter are frequentist in character in that we study the posterior laws under the assumption that the observations are a random sample from P_{θ_0} for some fixed, nonrandom θ_0. The alternative, which we do not consider, would be to make probability statements relative to the joint distribution of $(X_1, \ldots, X_n, \overline{\Theta}_n)$, given a fixed prior marginal measure for $\overline{\Theta}_n$ and with P_θ^n being the conditional law of $(X_1, \ldots, X_n)$ given $\overline{\Theta}_n$.

10.2 Bernstein–von Mises Theorem

The heuristic argument in the preceding section indicates that posterior distributions in differentiable parametric models converge to the Gaussian posterior distribution $N(X, I_{\theta_0}^{-1})$. The Bernstein–von Mises theorem makes this approximation rigorous and actually yields the approximation in a stronger sense than discussed so far. In Chapter 7 it is seen that the observation X in the limit experiment is the asymptotic analogue of the "locally sufficient" statistics

$$\Delta_{n,\theta_0} = \frac{1}{\sqrt{n}} \sum_{i=1}^n I_{\theta_0}^{-1} \dot{\ell}_{\theta_0}(X_i),$$

where $\dot{\ell}_\theta$ is the score function of the model. The Bernstein–von Mises theorem asserts that the total variation distance between the posterior distribution of $\sqrt{n}(\overline{\Theta}_n - \theta_0)$ and the random distribution $N(\Delta_{n,\theta_0}, I_{\theta_0}^{-1})$ converges to zero. Because $\Delta_{n,\theta_0} \rightsquigarrow X$, this has as a consequence that the posterior distribution of $\sqrt{n}(\bar{\Theta}_n - \theta_0)$ converges, in any reasonable sense, in distribution to $N(X, I_{\theta_0}^{-1})$.

The conditions of the following version of the Bernstein–von Mises theorem are remarkably weak. Besides differentiability in quadratic mean of the model, it is assumed that there exists a sequence of uniformly consistent tests for testing $H_0 : \theta = \theta_0$ against $H_1 : \|\theta - \theta_0\| \geq \varepsilon$, for every $\varepsilon > 0$. In other words, it must be possible to separate the true value θ_0 from the complements of balls centered at θ_0. Because the theorem implies that the posterior distributions eventually concentrate on balls of radii $M_n/\sqrt{n}$ around θ_0, for every $M_n \to \infty$, this separation hypothesis appears to be very reasonable. Even more so, since, as is noted in Lemmas 10.4 and 10.6, under continuity and identifiability of the model, separation by tests of $H_0 : \theta = \theta_0$ from $H_1 : \|\theta - \theta_0\| \geq \varepsilon$ for a single (large) $\varepsilon > 0$ already implies separation for every $\varepsilon > 0$. Furthermore, if Θ is compact and the model continuous and identifiable, then even the separation condition is superfluous (because it is automatically satisfied).[†]

10.1 *Theorem (Bernstein-von Mises). Let the experiment $(P_\theta : \theta \in \Theta)$ be differentiable in quadratic mean at θ_0 with nonsingular Fisher information matrix I_{θ_0}, and suppose that for every $\varepsilon > 0$ there exists a sequence of tests ϕ_n such that*

$$P_{\theta_0}^n \phi_n \to 0, \qquad \sup_{\|\theta - \theta_0\| \geq \varepsilon} P_\theta^n (1 - \phi_n) \to 0. \tag{10.2}$$

Furthermore, let the prior measure be absolutely continuous in a neighborhood of θ_0 with a continuous positive density at θ_0. Then the corresponding posterior distributions satisfy, with $\|\cdot\|$ the total variation norm,

$$\left\| P_{\sqrt{n}(\bar{\Theta}_n - \theta_0)|X_1,\ldots,X_n} - N\big(\Delta_{n,\theta_0}, I_{\theta_0}^{-1}\big) \right\| \overset{P_{\theta_0}^n}{\to} 0.$$

Proof. Throughout the proof we rescale the parameter θ to the local parameter $h = \sqrt{n}(\theta - \theta_0)$. Let Π_n be the corresponding prior distribution on h (hence $\Pi_n(B) = \Pi(\theta_0 + B/\sqrt{n})$), and for a given set C let Π_n^C be the probability measure obtained by restricting Π_n to C and next renormalizing. Write $P_{n,h}$ for the distribution of $\vec{X}_n = (X_1, \ldots, X_n)$ under the original parameter $\theta_0 + h/\sqrt{n}$, and let $P_{n,C} = \int P_{n,h} d\Pi_n^C(h)$. Finally, let $\overline{H}_n = \sqrt{n}(\bar{\Theta}_n - \theta_0)$, and denote the posterior distributions relative to Π_n and Π_n^C by $P_{\overline{H}_n|\vec{X}_n}$ and $P_{\overline{H}_n|\vec{X}_n}^C$, respectively.

The proof consists of two steps. First, it is shown that the difference between the posterior measures relative to the priors Π_n and $\Pi_n^{C_n}$, for C_n the ball with radius M_n, is asymptotically negligible, for any $M_n \to \infty$. Next it is shown that the difference between $N(\Delta_{n,\theta_0}, I_{\theta_0}^{-1})$ and the posterior measures relative to the priors $\Pi_n^{C_n}$ converges to zero in probability, for some $M_n \to \infty$.

For U, a ball of fixed radius around zero, we have $P_{n,U} \lhd \rhd P_{n,0}$, because $P_{n,h_n} \lhd \rhd P_{n,0}$ for every bounded sequence h_n, by Theorem 7.2. Thus, when showing convergence to zero in probability, we may always exchange $P_{n,0}$ and $P_{n,U}$.

[†] Recall that a test is a measurable function of the observations taking values in the interval $[0, 1]$; in the present context this means a measurable function $\phi_n : \mathcal{X}^n \mapsto [0, 1]$.

Let C_n be the ball of radius M_n. By writing out the conditional densities we see that, for any measurable set B,

$$P_{\overline{H}_n \mid \vec{X}_n}(B) - P_{\overline{H}_n \mid \vec{X}_n}^{C_n}(B) = P_{\overline{H}_n \mid \vec{X}_n}(C_n^c \cap B) - P_{\overline{H}_n \mid \vec{X}_n}(C_n^c)\, P_{\overline{H}_n \mid \vec{X}_n}^{C_n}(B).$$

Taking the supremum over B yields the bound

$$\left\| P_{\overline{H}_n \mid \vec{X}_n} - P_{\overline{H}_n \mid \vec{X}_n}^{C_n} \right\| \le 2 P_{\overline{H}_n \mid \vec{X}_n}(C_n^c).$$

The right side will be shown to converge to zero in mean under $P_{n,U}$ for U a ball of fixed radius around zero. First, by assumption and because $P_{n,U} \lhd P_{n,0}$,

$$P_{n,U} P_{\overline{H}_n \mid \vec{X}_n}(C_n^c) = P_{n,U} P_{\overline{H}_n \mid \vec{X}_n}(C_n^c)(1 - \phi_n) + o(1).$$

Manipulating again the expressions for the posterior densities, we can rewrite the first term on the right as

$$\frac{\Pi_n(C_n^c)}{\Pi_n(U)} P_{n, C_n^c} P_{\overline{H}_n \mid \vec{X}_n}(U)(1 - \phi_n) \le \frac{1}{\Pi_n(U)} \int_{C_n^c} P_{n,h}(1 - \phi_n)\, d\Pi_n(h).$$

For the tests given in the statement of the theorem, the integrand on the right converges to zero pointwise, but this is not enough. By Lemma 10.3, there automatically exist tests ϕ_n for which the convergence is exponentially fast. For the tests given by the lemma the preceding display is bounded above by

$$\frac{1}{\Pi_n(U)} \int_{\|h\| \ge M_n} e^{-c(\|h\|^2 \wedge n)}\, d\Pi_n(h).$$

Here $\Pi_n(U) = \Pi(\theta_0 + U/\sqrt{n})$ is bounded below by a term of the order $1/\sqrt{n}^k$, by the positivity and continuity of the density π at θ_0. Splitting the integral into the domains $M_n \le \|h\| \le D\sqrt{n}$ and $\|h\| \ge D\sqrt{n}$ for $D \le 1$ sufficiently small that $\pi(\theta)$ is uniformly bounded on $\|\theta - \theta_0\| \le D$, we see that the expression is bounded above by a multiple of

$$\int_{\|h\| \ge M_n} e^{-c\|h\|^2} dh + \sqrt{n}^k e^{-cD^2 n}.$$

This converges to zero as $n, M_n \to \infty$.

In the second part of the proof, let C be the ball of fixed radius M around zero, and let $N^C(\mu, \Sigma)$ be the normal distribution restricted and renormalized to C. The total variation distance between two arbitrary probability measures P and Q can be expressed in the form $\|P - Q\| = 2 \int (1 - p/q)^+ dQ$. It follows that

$$\frac{1}{2} \left\| N^C\left(\Delta_{n,\theta_0}, I_{\theta_0}^{-1}\right) - P_{\overline{H}_n \mid \vec{X}_n}^C \right\|$$

$$= \int \left(1 - \frac{dN^C\left(\Delta_{n,\theta_0}, I_{\theta_0}^{-1}\right)(h)}{1_C(h) p_{n,h}(\vec{X}_n) \pi_n(h) / \int_C p_{n,g}(\vec{X}_n) \pi_n(g)\, dg} \right)^+ dP_{\overline{H}_n \mid \vec{X}_n}^C(h)$$

$$\le \iint \left(1 - \frac{p_{n,g}(\vec{X}_n) \pi_n(g)\, dN^C\left(\Delta_{n,\theta_0}, I_{\theta_0}^{-1}\right)(h)}{p_{n,h}(\vec{X}_n) \pi_n(h)\, dN^C\left(\Delta_{n,\theta_0}, I_{\theta_0}^{-1}\right)(g)} \right)^+ dN^C\left(\Delta_{n,\theta_0}, I_{\theta_0}^{-1}\right)(g)\, dP_{\overline{H}_n \mid \vec{X}_n}^C(h),$$

because $(1 - \mathrm{E}Y)^+ \le \mathrm{E}(1 - Y)^+$. This can be further bounded by replacing the third occurrence of $N^C(\Delta_{n,\theta_0}, I_{\theta_0}^{-1})$ by a multiple of the uniform measure λ_C on C. By the dominated-convergence theorem, the double integral on the right side converges to zero in mean under $P_{n,C}$ if the integrand converges to zero in probability under the measure

$$P_{n,C}(dx)\, P^C_{\vec{H}_n \mid \vec{X}_n = x}(dh)\, \lambda_C(dg) = \Pi^C_n(dh)\, P_{n,h}(dx)\, \lambda_C(dg).$$

(Note that $P_{n,C}$ is the marginal distribution of $\vec{X}_n$ under the Bayesian model with prior Π^C_n.) Here Π^C_n is bounded up to a constant by λ_C for every sufficiently large n. Because $P_{n,h} \lhd \rhd P_{n,0}$ for every h, the sequence of measures on the right is contiguous with respect to the measures $\lambda_C(dh)\, P_{n,0}(dx)\, \lambda_C(dg)$. The integrand converges to zero in probability under the latter measure by Theorem 7.2 and the continuity of π at θ_0.

This is true for every ball C of fixed radius M and hence also for some $M_n \to \infty$. $\blacksquare$

10.3 Lemma. *Under the conditions of Theorem 10.1, there exists for every $M_n \to \infty$ a sequence of tests ϕ_n and a constant $c > 0$ such that, for every sufficiently large n and every $\|\theta - \theta_0\| \ge M_n/\sqrt{n}$,*

$$P^n_{\theta_0}\phi_n \to 0, \qquad P^n_\theta(1 - \phi_n) \le e^{-cn(\|\theta - \theta_0\|^2 \wedge 1)}.$$

Proof. We shall construct two sequences of tests, which "work" for the ranges $M_n/\sqrt{n} \le \|\theta - \theta_0\| \le \varepsilon$ and $\|\theta - \theta_0\| > \varepsilon$, respectively, and a given $\varepsilon > 0$. Then the ϕ_n of the lemma can be defined as the maximum of the two sequences.

First consider the range $M_n/\sqrt{n} \le \|\theta - \theta_0\| \le \varepsilon$. Let $\dot{\ell}^L_{\theta_0}$ be the score function truncated (coordinatewise) to the interval $[-L, L]$. By the dominated convergence theorem, $P_{\theta_0}\dot{\ell}^L_{\theta_0}\dot{\ell}^T_{\theta_0} \to I_{\theta_0}$ as $L \to \infty$. Hence, there exists $L > 0$ such that the matrix $P_{\theta_0}\dot{\ell}^L_{\theta_0}\dot{\ell}^T_{\theta_0}$ is nonsingular. Fix such an L and define

$$\omega_n = 1\left\{ \left\| (\mathbb{P}_n - P_{\theta_0})\dot{\ell}^L_{\theta_0} \right\| \ge \sqrt{M_n/n} \right\}.$$

By the central limit theorem, $P^n_{\theta_0}\omega_n \to 0$, so that ω_n satisfies the first requirement. By the triangle inequality,

$$\left\| (\mathbb{P}_n - P_\theta)\dot{\ell}^L_{\theta_0} \right\| \ge \left\| (P_{\theta_0} - P_\theta)\dot{\ell}^L_{\theta_0} \right\| - \left\| (\mathbb{P}_n - P_{\theta_0})\dot{\ell}^L_{\theta_0} \right\|.$$

Because, by the differentiability of the model, $P_\theta\dot{\ell}^L_{\theta_0} - P_{\theta_0}\dot{\ell}^L_{\theta_0} = \left(P_{\theta_0}\dot{\ell}^L_{\theta_0}\dot{\ell}^T_{\theta_0} + o(1) \right)(\theta - \theta_0)$, the first term on the right is bounded below by $c\|\theta - \theta_0\|$ for some $c > 0$, for every θ that is sufficiently close to θ_0, say for $\|\theta - \theta_0\| < \varepsilon$. If $\omega_n = 0$, then the second term (without the minus sign) is bounded above by $\sqrt{M_n/n}$. Consequently, for every $c\|\theta - \theta_0\| \ge 2\sqrt{M_n/n}$, and hence for every $\|\theta - \theta_0\| \ge M_n/\sqrt{n}$ and every sufficiently large n,

$$P^n_\theta(1 - \omega_n) \le P_\theta\left(\left\| (\mathbb{P}_n - P_\theta)\dot{\ell}^L_{\theta_0} \right\| \ge \tfrac{1}{2}c\|\theta - \theta_0\| \right) \le e^{-Cn\|\theta - \theta_0\|^2},$$

by Hoeffding's inequality (e.g., Appendix B in [117]), for a sufficiently small constant C.

Next, consider the range $\|\theta - \theta_0\| > \varepsilon$ for an arbitrary fixed $\varepsilon > 0$. By assumption there exist tests ϕ_n such that

$$P^n_{\theta_0}\phi_n \to 0, \qquad \sup_{\|\theta - \theta_0\| > \varepsilon} P^n_\theta(1 - \phi_n) \to 0.$$

It suffices to show that these tests can be replaced, if necessary, by tests for which the convergence to zero is exponentially fast. Fix k large enough such that $P_{\theta_0}^k \phi_k$ and $P_\theta^k (1 - \phi_k)$ are smaller than $1/4$ for every $\|\theta - \theta_0\| > \varepsilon$. Let $n = mk + r$ for $0 \leq r < k$, and define $Y_{n,1}, \ldots, Y_{n,m}$ as ϕ_k applied in turn to $X_1, \ldots, X_k$, to $X_{k+1}, \ldots, X_{2k}$, and so forth. Let $\overline{Y}_{n,m}$ be their average and then define $\omega_n = 1\{\overline{Y}_{n,m} \geq 1/2\}$. Because $E_\theta Y_{n,j} \geq 3/4$ for every $\|\theta - \theta_0\| > \varepsilon$ and every j, Hoeffding's inequality implies that

$$P_\theta^n (1 - \omega_n) = P_\theta(\overline{Y}_{n,m} < 1/2) \leq e^{-2m(\frac{1}{2} - \frac{3}{4})^2} \leq e^{-m/8}.$$

Because m is proportional to n, this gives the desired exponential decay. Because $E_{\theta_0} Y_{n,j} \leq 1/4$, the expectations $P_{\theta_0}^n \omega_n$ are similarly bounded. ∎

The Bernstein–von Mises theorem is sometimes written with a different "centering sequence." By Theorem 8.14 any sequence of standardized asymptotically efficient estimators $\sqrt{n}(\hat{\theta}_n - \theta)$ is asymptotically equivalent in probability to $\Delta_{n,\theta}$. Because the total variation distance

$$\left\| N\left(\Delta_{n,\theta}, I_\theta^{-1}\right) - N\left(\sqrt{n}(\hat{\theta}_n - \theta), I_\theta^{-1}\right) \right\|$$

is bounded by a multiple of $\left\| \Delta_{n,\theta} - \sqrt{n}(\hat{\theta}_n - \theta) \right\|$, any such sequence $\sqrt{n}(\hat{\theta}_n - \theta)$ may replace $\Delta_{n,\theta}$ in the Bernstein–von Mises theorem. By the invariance of the total variation norm under location and scale changes, the resulting statement can be written

$$\left\| P_{\overline{\Theta}_n \mid X_1, \ldots, X_n} - N\left(\hat{\theta}_n, \frac{1}{n} I_\theta^{-1}\right) \right\| \xrightarrow{P_\theta^n} 0.$$

Under regularity conditions this is true for the maximum likelihood estimators $\hat{\theta}_n$. Combining this with Theorem 5.39 we then have, informally,

$$P_{\overline{\Theta}_n \mid \hat{\theta}_n} \approx N\left(\hat{\theta}_n, \frac{1}{n} I_{\hat{\theta}_n}^{-1}\right) \quad \text{and} \quad P_{\hat{\theta}_n \mid \overline{\Theta}_n} \approx N\left(\overline{\Theta}_n, \frac{1}{n} I_{\overline{\Theta}_n}^{-1}\right),$$

since conditioning $\hat{\theta}_n$ on $\overline{\Theta}_n = \theta$ gives the usual "frequentist" distribution of $\hat{\theta}_n$ under θ. This gives a remarkable symmetry.

Le Cam's version of the Bernstein–von Mises theorem requires the existence of tests that are uniformly consistent for testing $H_0 : \theta = \theta_0$ versus $H_1 : \|\theta - \theta_0\| \geq \varepsilon$, for every $\varepsilon > 0$. Such tests certainly exist if there exist estimators T_n that are uniformly consistent, in that, for every $\varepsilon > 0$,

$$\sup_\theta P_\theta\left(\|T_n - \theta\| \geq \varepsilon\right) \to 0.$$

In that case, we can define $\phi_n = 1\{\|T_n - \theta_0\| \geq \varepsilon/2\}$. Thus the condition of the Bernstein–von Mises theorem that certain tests exist can be replaced by the condition that uniformly consistent estimators exist. This is often the case. For instance, the next lemma shows that this is the case for a Euclidean sample space $\mathcal{X}$ provided, for F_θ the distribution functions corresponding to the P_θ,

$$\inf_{\|\theta - \theta'\| > \varepsilon} \|F_\theta - F_{\theta'}\|_\infty > 0.$$

For compact parameter sets, this is implied by identifiability and continuity of the maps $\theta \mapsto F_\theta$. We generalize and formalize this in a second lemma, which shows that uniformity on compact subsets is always achievable if the model $(P_\theta : \theta \in \Theta)$ is differentiable in quadratic mean at every θ and the parameter θ is identifiable.

A class of measurable functions $\mathcal{F}$ is a *uniform Glivenko-Cantelli* class (in probability) if, for every $\varepsilon > 0$,

$$\sup_P P_P\big(\|\mathbb{P}_n - P\|_{\mathcal{F}} > \varepsilon\big) \to 0.$$

Here the supremum is taken over all probability measures P on the sample space, and $\|Q\|_{\mathcal{F}} = \sup_{f \in \mathcal{F}} |Qf|$. An example is the collection of indicators of all cells $(-\infty, t]$ in a Euclidean sample space.

10.4 Lemma. *Suppose that there exists a uniform Glivenko–Cantelli class $\mathcal{F}$ such that, for every $\varepsilon > 0$,*

$$\inf_{d(\theta, \theta') > \varepsilon} \|P_\theta - P_{\theta'}\|_{\mathcal{F}} > 0. \tag{10.5}$$

Then there exists a sequence of estimators that is uniformly consistent on Θ for estimating θ.

10.6 Lemma. *Suppose that Θ is σ-compact, $P_\theta \neq P_{\theta'}$ for every pair $\theta \neq \theta'$, and the maps $\theta \mapsto P_\theta$ are continuous for the total variation norm. Then there exists a sequence of estimators that is uniformly consistent on every compact subset of Θ.*

Proof. For the proof of the first lemma, define $\hat{\theta}_n$ to be a point of (near) minimum of the map $\theta \mapsto \|\mathbb{P}_n - P_\theta\|_{\mathcal{F}}$. Then, by the triangle inequality and the definition of $\hat{\theta}_n$, $\|P_{\hat{\theta}_n} - P_\theta\|_{\mathcal{F}} \leq 2\|\mathbb{P}_n - P_\theta\|_{\mathcal{F}} + 1/n$, if the near minimum is chosen within distance $1/n$ of the true infimum. Fix $\varepsilon > 0$, and let δ be the positive number given in condition (10.5). Then

$$P_\theta\big(d(\hat{\theta}_n, \theta) > \varepsilon\big) \leq P_\theta\big(\|P_{\hat{\theta}_n} - P_\theta\|_{\mathcal{F}} \geq \delta\big) \leq P_\theta\Big(2\|\mathbb{P}_n - P_\theta\|_{\mathcal{F}} \geq \delta - \frac{1}{n}\Big).$$

By assumption, the right side converges to zero uniformly in θ.

For the proof of the second lemma, first assume that Θ is compact. Then there exists a uniform Glivenko-Cantelli class that satisfies the condition of the first lemma. To see this, first find a sequence $A_1, A_2, \ldots$ of measurable sets that separates the points P_θ. Thus, for every pair $\theta, \theta' \in \Theta$, if $P_\theta(A_i) = P_{\theta'}(A_i)$ for every i, then $\theta = \theta'$. A separating collection exists by the identifiability of the parameter, and it can be taken to be countable by the continuity of the maps $\theta \mapsto P_\theta$. (For a Euclidean sample space, we can use the cells $(-\infty, t]$ for t ranging over the vectors with rational coordinates. More generally, see the lemma below.) Let $\mathcal{F}$ be the collection of functions $x \mapsto i^{-1}1_{A_i}(x)$. Then the map $h : \Theta \mapsto \ell^\infty(\mathcal{F})$ given by $\theta \mapsto (P_\theta f)_{f \in \mathcal{F}}$ is continuous and one-to-one. By the compactness of Θ, the inverse $h^{-1} : h(\Theta) \mapsto \Theta$ is automatically uniformly continuous. Thus, for every $\varepsilon > 0$ there exists $\delta > 0$ such that

$$\|h(\theta) - h(\theta')\|_{\mathcal{F}} \leq \delta \quad \text{implies} \quad d(\theta, \theta') \leq \varepsilon.$$

This means that (10.5) is satisfied. The class $\mathcal{F}$ is also a uniform Glivenko-Cantelli class, because by Chebyshev's inequality,

$$P_P\big(\|\mathbb{P}_n - P\|_{\mathcal{F}} > \varepsilon\big) \le \sum_f P_P\big(|\mathbb{P}_n f - Pf| > \varepsilon\big) \le \sum_i \frac{1}{n\varepsilon^2 i^2}.$$

This concludes the proof of the second lemma for compact Θ.

To remove the compactness condition, write Θ as the union of an increasing sequence of compact sets $K_1 \subset K_2 \subset \cdots$. For every m there exists a sequence of estimators $T_{n,m}$ that is uniformly consistent on K_m, by the preceding argument. Thus, for every fixed m,

$$a_{n,m} := \sup_{\theta \in K_m} P_\theta\bigg(d(T_{n,m}, \theta) \ge \frac{1}{m}\bigg) \to 0, \qquad n \to \infty.$$

Then there exists a sequence $m_n \to \infty$ such that $a_{n,m_n} \to 0$ as $n \to \infty$. It is not hard to see that $\hat\theta_n = T_{n,m_n}$ satisfies the requirements. ∎

As a consequence of the second lemma, if there exists a sequence of tests ϕ_n such that (10.2) holds for some $\varepsilon > 0$, then it holds for every $\varepsilon > 0$. In that case we can replace the given sequence ϕ_n by the minimum of ϕ_n and the tests $1\big\{\|T_n - \theta_0\| \ge \varepsilon/2\big\}$ for a sequence of estimators T_n that is uniformly consistent on a sufficiently large subset of Θ.

10.7   Lemma. *Let the set of probability measures $\mathcal{P}$ on a measurable space $(\mathcal{X}, \mathcal{A})$ be separable for the total variation norm. Then there exists a countable subset $\mathcal{A}_0 \subset \mathcal{A}$ such that $P_1 = P_2$ on $\mathcal{A}_0$ implies $P_1 = P_2$ for every $P_1, P_2 \in \mathcal{P}$.*

Proof.   The set $\mathcal{P}$ can be identified with a subset of $L_1(\mu)$ for a suitable probability measure μ. For instance, μ can be taken a convex linear combination of a countable dense set. Let $\mathcal{P}_0$ be a countable dense subset, and let $\mathcal{A}_0$ be the set of all finite intersections of the sets $p^{-1}(B)$ for p ranging over a choice of densities of the set $\mathcal{P}_0 \subset L_1(\mu)$ and B ranging over a countable generator of the Borel sets in $\mathbb{R}$.

Then every density $p \in \mathcal{P}_0$ is $\sigma(\mathcal{A}_0)$-measurable by construction. A density of a measure $P \in P - P_0$ can be approximated in $L_1(\mu)$ by a sequence from P_0 and hence can be chosen $\sigma(\mathcal{A}_0)$-measurable, without loss of generality.

Because $\mathcal{A}_0$ is intersection-stable (a "π-system"), two probability measures that agree on $\mathcal{A}_0$ automatically agree on the σ-field $\sigma(\mathcal{A}_0)$ generated by $\mathcal{A}_0$. Then they also give the same expectation to every $\sigma(\mathcal{A}_0)$-measurable function $f: \mathcal{X} \mapsto [0, 1]$. If the measures have $\sigma(\mathcal{A}_0)$-measurable densities, then they must agree on $\mathcal{A}$, because $P(A) = E_\mu 1_A p = E_\mu E_\mu\big(1_A \mid \sigma(\mathcal{A}_0)\big)p$ if p is $\sigma(\mathcal{A}_0)$-measurable. ∎

10.3   Point Estimators

The Bernstein–von Mises theorem shows that the posterior laws converge in distribution to a Gaussian posterior law in total variation distance. As a consequence, any location functional that is suitably continuous relative to the total variation norm applied to the sequence of posterior laws converges to the same location functional applied to the limiting Gaussian posterior distribution. For most choices this means to X, or a $N(0, I_{\theta_0}^{-1})$-distribution.

In this section we consider more general Bayes point estimators that are defined as the minimizers of the posterior risk functions relative to some loss function. For a given loss function $\ell : \mathbb{R}^k \mapsto [0, \infty)$, let T_n, for fixed $X_1, \ldots, X_n$, minimize the posterior risk

$$t \mapsto \frac{\int \ell\big(\sqrt{n}(t - \theta)\big) \prod_{i=1}^n p_\theta(X_i) \, d\Pi(\theta)}{\int \prod_{i=1}^n p_\theta(X_i) \, d\Pi(\theta)}.$$

It is not immediately clear that the minimizing values T_n can be selected as a measurable function of the observations. This is an implicit assumption, or otherwise the statements are to be understood relative to outer probabilities. We also make it an implicit assumption that the integrals in the preceding display exist, for almost every sequence of observations.

To derive the limit distribution of $\sqrt{n}(T_n - \theta_0)$, we apply general results on M-estimators, in particular the argmax continuous-mapping theorem, Theorem 5.56.

We restrict ourselves to loss functions with the property, for every $M > 0$,

$$\sup_{\|h\| \leq M} \ell(h) \leq \inf_{\|h\| \geq 2M} \ell(h),$$

with strict inequality for at least one M.[†] This is true, for instance, for loss functions of the form $\ell(h) = \ell_0\big(\|h\|\big)$ for a nondecreasing function $\ell_0 : [0, \infty) \mapsto [0, \infty)$ that is not constant on $(0, \infty)$. Furthermore, we suppose that ℓ grows at most polynomially: For some constant $p \geq 0$,

$$\ell(h) \leq 1 + \|h\|^p.$$

10.8   Theorem.   *Let the conditions of Theorem 10.1 hold, and let ℓ satisfy the conditions as listed, for a p such that $\int \|\theta\|^p \, d\Pi(\theta) < \infty$. Then the sequence $\sqrt{n}(T_n - \theta_0)$ converges under θ_0 in distribution to the minimizer of $t \mapsto \int \ell(t - h) \, dN(X, I_{\theta_0}^{-1})(h)$, for X possessing the $N(0, I_{\theta_0}^{-1})$-distribution, provided that any two minimizers of this process coincide almost surely. In particular, for every nonzero, subconvex loss function it converges to X.*

*Proof.   We adopt the notation as listed in the first paragraph of the proof of Theorem 10.1. The last assertion of the theorem is a consequence of Anderson's lemma, Lemma 8.5.

The standardized estimator $\sqrt{n}(T_n - \theta_0)$ minimizes the function

$$t \mapsto Z_n(t) = \frac{\int \ell(t - h) \, p_{n,h}(\vec{X}_n) \, d\Pi_n(h)}{\int p_{n,h}(\vec{X}_n) \, d\Pi_n(h)} = P_{\overline{H}_n \mid \vec{X}_n} \ell_t,$$

where ℓ_t is the function $h \mapsto \ell(t - h)$. The proof consists of three parts. First it is shown that integrals over the sets $\|h\| \geq M_n$ can be neglected for every $M_n \to \infty$. Next, it is proved that the sequence $\sqrt{n}(T_n - \theta_0)$ is uniformly tight. Finally, it is shown that the stochastic processes $t \mapsto Z_n(t)$ converge in distribution in the space $\ell^\infty(K)$, for every compact K, to the process

$$t \mapsto Z(t) = \int \ell(t - h) \, dN\big(X, I_{\theta_0}^{-1}\big)(h).$$

[†] The 2 is for convenience, any other number would do.

The sample paths of this limit process are continuous in t, in view of the subexponential growth of ℓ and the smoothness of the normal density. Hence the theorem follows from the argmax theorem, Corollary 5.58.

Let C_n be the ball of radius M_n for a given, arbitrary sequence $M_n \to \infty$. We first show that, for every measurable function f that grows subpolynomially of order p,

$$P_{\bar{H}_n \mid \vec{X}_n}(f 1_{C_n^c}) \overset{P_{n,0}}{\to} 0. \tag{10.9}$$

To see this, we utilize the tests ϕ_n for testing $H_0 : \theta = \theta_0$ that exist by assumption. In view of Lemma 10.3, these may be assumed without loss of generality to satisfy the stronger property as given in the statement of this lemma. Furthermore, they can be constructed to be nonrandomized (i.e., to have range $\{0, 1\}$). Then it is immediate that $(P_{\bar{H}_n \mid \vec{X}_n} f)\phi_n$ converges to zero in $P_{n,0}$-probability for every measurable function f. Next, by writing out the posterior densities, we see that, for U a fixed ball around the origin,

$$P_{n,U} P_{\bar{H}_n \mid \vec{X}_n}(f 1_{C_n^c})(1 - \phi_n) = \frac{1}{\Pi_n(U)} \int_{C_n^c} f(h) P_{n,h} \left[P_{\bar{H}_n \mid \vec{X}_n}(U)(1 - \phi_n) \right] d\Pi_n(h)$$

$$\leq \frac{1}{\Pi_n(U)} \int_{C_n^c} \left(1 + \|h\|^p \right) e^{-c(\|h\|^2 \wedge n)} d\Pi_n(h).$$

Here $\Pi_n(U)$ is bounded below by a term of the order $1/\sqrt{n}^k$, by the positivity and continuity at θ_0 of the prior density π. Split the integral over the domains $M_n \leq \|h\| \leq D\sqrt{n}$ and $\|h\| \geq D\sqrt{n}$, and use the fact that $\int \|\theta\|^p d\Pi(\theta) < \infty$ to bound the right side of the display by terms of the order $e^{-AM_n^2}$ and $\sqrt{n}^{k+p} e^{-Bn}$, for some $A, B > 0$. These converge to zero, whence (10.9) has been proved.

Define $\bar{\ell}(M)$ as the supremum of $\ell(h)$ over the ball of radius M, and $\underline{\ell}(M)$ as the infimum over the complement of this ball. By assumption, there exists $\delta > 0$ such that $\eta := \underline{\ell}(2\delta) - \bar{\ell}(\delta) > 0$. Let U be the ball of radius δ around 0. For every $\|t\| \geq 3M_n$ and sufficiently large M_n, we have $\ell(t - h) - \ell(-h) \geq \eta$ if $h \in U$, and $\ell(t - h) - \ell(-h) \geq \underline{\ell}(2M_n) - \bar{\ell}(M_n) \geq 0$ if $h \in U^c \cap C_n$, by assumption. Therefore,

$$Z_n(t) - Z_n(0) = P_{\bar{H}_n \mid \vec{X}_n} \left[\left(\ell(t - h) - \ell(-h) \right) \left(1_U + 1_{U^c \cap C_n} + 1_{C_n^c} \right) \right]$$

$$\geq \eta P_{\bar{H}_n \mid \vec{X}_n}(U) - P_{\bar{H}_n \mid \vec{X}_n} \left(\ell(-h) 1_{C_n^c} \right).$$

Here the posterior probability $P_{\bar{H}_n \mid \vec{X}_n}(U)$ of U converges in distribution to $N(X, I_{\theta_0}^{-1})(U)$, by the Bernstein–von Mises theorem. This limit is positive almost surely. The second term in the preceding display converges to zero in probability by (10.9). Conclude that the infimum of $Z_n(t) - Z_n(0)$ over the set of t with $\|t\| \geq 3M_n$ is bounded below by variables that converge in distribution to a strictly positive variable. Thus this infimum is positive with probability tending to one. This implies that the probability that $t \mapsto Z_n(t)$ has a minimizer in the set $\|t\| \geq 3M_n$ converges to zero. Because this is true for any $M_n \to \infty$, it follows that the sequence $\sqrt{n}(T_n - \theta_0)$ is uniformly tight.

Let C be the ball of fixed radius M around 0, and fix some compact set $K \subset \mathbb{R}^k$. Define stochastic processes

$$Z_{n,M}(t) = P_{\bar{H}_n \mid \vec{X}_n}(\ell_t 1_C), \qquad W_{n,M} = N\left(\Delta_{n,\theta_0}, I_{\theta_0}^{-1} \right)(\ell_t 1_C),$$

$$W_M = N\left(X, I_{\theta_0}^{-1} \right)(\ell_t 1_C).$$

The function $h \mapsto \ell(t-h)1_C(h)$ is bounded, uniformly if t ranges over the compact K. Hence, by the Bernstein–von Mises theorem, $Z_{n,M} - W_{n,M} \overset{P}{\to} 0$ in $\ell^\infty(K)$ as $n \to \infty$, for every fixed M. Second, by the continuous-mapping theorem, $W_{n,M} \rightsquigarrow W_M$ in $\ell^\infty(K)$, as $n \to \infty$, for fixed M. Next $W_M \overset{P}{\to} Z$ in $\ell^\infty(K)$ as $M \to \infty$, or equivalently $C \uparrow \mathbb{R}^k$. Conclude that there exists a sequence $M_n \to \infty$ such that the processes $Z_{n,M_n} \rightsquigarrow Z$ in $\ell^\infty(K)$. Because, by (10.9), $Z_n(t) - Z_{n,M_n}(t) \overset{P}{\to} 0$, we finally conclude that $Z_n \rightsquigarrow Z$ in $\ell^\infty(K)$. ∎

*10.4   Consistency

A sequence of posterior measures $P_{\overline{\Theta}_n \mid X_1, \ldots, X_n}$ is called *consistent* under θ if under P_θ^∞-probability it converges in distribution to the measure δ_θ that is degenerate at θ, in probability; it is *strongly consistent* if this happens for almost every sequence $X_1, X_2, \ldots$.

Given that, usually, ordinarily consistent point estimators of θ exist, consistency of posterior measures is a modest requirement. If we could know θ with almost complete accuracy as $n \to \infty$, then we would use a Bayes estimator only if this would also yield the true value with similar accuracy. Fortunately, posterior measures are usually consistent. The following famous theorem by Doob shows that under hardly any conditions we already have consistency under almost every parameter.

Recall that Θ is assumed to be Euclidean and the maps $\theta \mapsto P_\theta(A)$ to be measurable for every measurable set A.

10.10   Theorem (Doob's consistency theorem). *Suppose that the sample space $(\mathcal{X}, \mathcal{A})$ is a subset of Euclidean space with its Borel σ-field. Suppose that $P_\theta \neq P_{\theta'}$ whenever $\theta \neq \theta'$. Then for every prior probability measure Π on Θ the sequence of posterior measures is consistent for Π-almost every θ.*

Proof. On an arbitrary probability space construct random vectors $\overline{\Theta}$ and $X_1, X_2, \ldots$ such that $\overline{\Theta}$ is marginally distributed according to Π and such that given $\overline{\Theta} = \theta$ the vectors $X_1, X_2, \ldots$ are i.i.d. according to P_θ. Then the posterior distribution based on the first n observations is $P_{\overline{\Theta} \mid X_1, \ldots, X_n}$. Let Q be the distribution of $(X_1, X_2, \ldots, \overline{\Theta})$ on $\mathcal{X}^\infty \times \Theta$.

The main part of the proof consists of showing that there exists a measurable function $h : \mathcal{X}^\infty \mapsto \Theta$ with

$$h(x_1, x_2, \ldots) = \theta, \qquad Q\text{-a.s..} \tag{10.11}$$

Suppose that this is true. Then, for any bounded, measurable function $f : \Theta \mapsto \mathbb{R}$, by Doob's martingale convergence theorem,

$$\mathrm{E}\big(f(\overline{\Theta}) \mid X_1, \ldots, X_n\big) \to \mathrm{E}\big(f(\overline{\Theta}) \mid X_1, X_2, \ldots\big)$$
$$= f\big(h(X_1, X_2, \ldots)\big), \qquad Q\text{-a.s..}$$

By Lemma 2.25 there exists a countable collection $\mathcal{F}$ of bounded, continuous functions f that are determining for convergence in distribution. Because the countable union of the associated null sets on which the convergence of the preceding display fails is a null set, we have that

$$P_{\overline{\Theta} \mid X_1, \ldots, X_n} \rightsquigarrow \delta_{h(X_1, X_2, \ldots)}, \qquad Q\text{-a.s..}$$

This statement refers to the marginal distribution of $(X_1, X_2, \ldots)$ under Q. We wish to translate it into a statement concerning the P_θ^∞-measures. Let $C \subset \mathcal{X}^\infty \times \Theta$ be the intersection of the sets on which the weak convergence holds and on which (10.11) is valid. By Fubini's theorem

$$1 = Q(C) = \iint 1_C(x, \theta) \, dP_\theta^\infty(x) \, d\Pi(\theta) = \int P_\theta^\infty(C_\theta) \, d\Pi(\theta),$$

where $C_\theta = \{x : (x, \theta) \in C\}$ is the horizontal section of C at height θ. It follows that $P_\theta^\infty(C_\theta) = 1$ for Π-almost every θ. For every θ such that $P_\theta^\infty(C_\theta) = 1$, we have that $(x, \theta) \in C$ for P_θ^∞-almost every sequence $x_1, x_2, \ldots$ and hence

$$P_{\bar{\Theta}|X_1 = x_1, \ldots, X_n = x_n} \rightsquigarrow \delta_{h(x_1, x_2, \ldots)} = \delta_\theta.$$

This is the assertion of the theorem.

In order to establish (10.11), call a measurable function $f : \Theta \mapsto \mathbb{R}$ *accessible* if there exists a sequence of measurable functions $h_n : \mathcal{X}^n \mapsto \mathbb{R}$ such that

$$\iint |h_n(x) - f(\theta)| \wedge 1 \, dQ(x, \theta) \to 0.$$

(Here we abuse notation in viewing h_n also as a measurable function on $\mathcal{X}^\infty \times \Theta$.) Then there also exists a (sub)sequence with $h_n(x) \to f(\theta)$ almost surely under Q, whence every accessible function f is almost everywhere equal to an $\mathcal{A}^\infty \times \{\emptyset, \Theta\}$-measurable function. This is a measurable function of $x = (x_1, x_2, \ldots)$ alone. If we can show that the functions $f(\theta) = \theta_i$ are accessible, then (10.11) follows. We shall in fact show that every Borel measurable function is accessible.

By the strong law of large numbers, $h_n(x) = \sum_{i=1}^n 1_A(x_i) \to P_\theta(A)$ almost surely under P_θ^∞, for every θ and measurable set A. Consequently, by the dominated convergence theorem,

$$\iint |h_n(x) - P_\theta(A)| \, dQ(x, \theta) \to 0.$$

Thus each of the functions $\theta \mapsto P_\theta(A)$ is accessible.

Because $(\mathcal{X}, \mathcal{A})$ is Euclidean by assumption, there exists a countable measure-determining subcollection $\mathcal{A}_0 \subset \mathcal{A}$. The functions $\theta \mapsto P_\theta(A)$ are measurable by assumption and separate the points of Θ as A ranges over $\mathcal{A}_0$, in view of the choice of $\mathcal{A}_0$ and the identifiability of the parameter θ. This implies that these functions generate the Borel σ-field on Θ, in view of Lemma 10.12.

The proof is complete once it is shown that every function that is measurable in the σ-field generated by the accessible functions (which is the Borel σ-field) is accessible. From the definition it follows easily that the set of accessible functions is a vector space, contains the constant functions, is closed under monotone limits, and is a lattice. The desired result therefore follows by a monotone class argument, as in Lemma 10.13. ∎

The merit of the preceding theorem is that it imposes hardly any conditions, but its drawback is that it gives the consistency only up to null sets of possible parameters (depending on the prior). In certain ways these null sets can be quite large, and examples have been constructed where Bayes estimators behave badly. To guarantee consistency under every parameter it is necessary to impose some further conditions. Because in this chapter we are mainly concerned with asymptotic normality of Bayes estimators (which implies consistency with a rate), we omit a discussion.

10.12   Lemma. *Let $\mathcal{F}$ be a countable collection of measurable functions $f : \Theta \subset \mathbb{R}^k \mapsto \mathbb{R}$ that separates the points of Θ. Then the Borel σ-field and the σ-field generated by $\mathcal{F}$ on Θ coincide.*

Proof. By assumption, the map $h : \Theta \mapsto \mathbb{R}^{\mathcal{F}}$ defined by $h(\theta)f = f(\theta)$ is measurable and one-to-one. Because $\mathcal{F}$ is countable, the Borel σ-field on $\mathbb{R}^{\mathcal{F}}$ (for the product topology) is equal to the σ-field generated by the coordinate projections. Hence the σ-fields generated by h and $\mathcal{F}$ (viewed as Borel measurable maps in $\mathbb{R}^{\mathcal{F}}$ and $\mathbb{R}$, respectively) on Θ are identical. Now h^{-1}, defined on the range of h, is automatically Borel measurable, by Proposition 8.3.5 in [24], and hence Θ and $h(\Theta)$ are Borel isomorphic. ∎

10.13   Lemma. *Let $\mathcal{F}$ be a linear subspace of $\mathcal{L}_1(\Pi)$ with the properties*
 (i) *if $f, g \in \mathcal{F}$, then $f \wedge g \in \mathcal{F}$;*
 (ii) *if $0 \le f_1 \le f_2 \le \cdots \in \mathcal{F}$ and $f_n \uparrow f \in \mathcal{L}_1(\Pi)$, then $f \in \mathcal{F}$;*
 (iii) *$1 \in \mathcal{F}$.*
Then $\mathcal{F}$ contains every $\sigma(\mathcal{F})$-measurable function in $\mathcal{L}_1(\Pi)$.

Proof. Because any $\sigma(\mathcal{F})$-measurable nonnegative function is the monotone limit of a sequence of simple functions, it suffices to prove that $1_A \in \mathcal{F}$ for every $A \in \sigma(\mathcal{F})$. Define $\mathcal{A}_0 = \{A : 1_A \in \mathcal{F}\}$. Then $\mathcal{A}_0$ is an intersection-stable Dynkin system and hence a σ-field. Furthermore, for every $f \in \mathcal{F}$ and $\alpha \in \mathbb{R}$, the functions $n(f - \alpha)^+ \wedge 1$ are contained in $\mathcal{F}$ and increase pointwise to $1_{\{f > \alpha\}}$. It follows that $\{f > \alpha\} \in \mathcal{A}_0$. Hence $\sigma(\mathcal{F}) \subset \mathcal{A}_0$. ∎

Notes

The Bernstein–von Mises theorem has that name, because, as Le Cam and Yang [97] write, it was first discovered by Laplace. The theorem that is presented in this chapter is considerably more elegant than the results by these early authors, and also much better than the result in Le Cam [91], who revived the theorem in order to prove results on superefficiency. We adapted it from Le Cam [96] and Le Cam and Yang [97].

Ibragimov and Hasminskii [80] discuss the convergence of Bayes point estimators in greater generality, and also cover non-Gaussian limit experiments, but their discussion of the i.i.d. case as discussed in the present chapter is limited to bounded parameter sets and requires stronger assumptions. Our treatment uses some elements of their proof, but is heavily based on Le Cam's Bernstein–von Mises theorem. Inspection of the proof shows that the conditions on the loss function can be relaxed significantly, for instance allowing exponential growth.

Doob's theorem originates in [39]. The potential null sets of inconsistency that it leaves open really exist in some situations particularly if the parameter set is infinite dimensional, and have attracted much attention. See [34], which is accompanied by evaluations of the phenomenon by many authors, including Bayesians.

PROBLEMS

1. Verify the conditions of the Bernstein–von Mises theorem for the experiment where P_θ is the Poisson measure of mean θ.

2. Let P_θ be the k-dimensional normal distribution with mean θ and covariance matrix the identify. Find the a posteriori law for the prior $\Pi = N(\tau, \Lambda)$ and some nonsingular matrix Λ. Can you see directly that the Bernstein–von Mises theorem is true in this case?

3. Let P_θ be the Bernoulli distribution with mean θ. Find the posterior distribution relative to the beta-prior measure, which has density

$$\theta \mapsto \frac{\Gamma(\alpha)\Gamma(\beta)}{\Gamma(\alpha + \beta)} \theta^{\alpha-1}(1-\theta)^{\beta-1} 1_{(0,1)}(\theta).$$

4. Suppose that, in the case of a one-dimensional parameter, we use the loss function $\ell(h) = 1_{(-1,2)}(h)$. Find the limit distribution of the corresponding Bayes point estimator, assuming that the conditions of the Bernstein–von Mises theorem hold.

11
Projections

A projection of a random variable is defined as a closest element in a given set of functions. We can use projections to derive the asymptotic distribution of a sequence of variables by comparing these to projections of a simple form. Conditional expectations are special projections. The Hájek projection is a sum of independent variables; it is the leading term in the Hoeffding decomposition.

11.1 Projections

A common method to derive the limit distribution of a sequence of statistics T_n is to show that it is asymptotically equivalent to a sequence S_n of which the limit behavior is known. The basis of this method is Slutsky's lemma, which shows that the sequence $T_n = T_n - S_n + S_n$ converges in distribution to S if both $T_n - S_n \overset{P}{\to} 0$ and $S_n \rightsquigarrow S$.

How do we find a suitable sequence S_n? First, the variables S_n must be of a simple form, because the limit properties of the sequence S_n must be known. Second, S_n must be close enough. One solution is to search for the closest S_n of a certain predetermined form. In this chapter, "closest" is taken as closest in square expectation.

Let T and $\{S : S \in \mathcal{S}\}$ be random variables (defined on the same probability space) with finite second-moments. A random variable $\hat{S}$ is called a *projection* of T onto $\mathcal{S}$ (or L_2-projection) if $\hat{S} \in \mathcal{S}$ and minimizes

$$S \mapsto \mathrm{E}(T - S)^2, \qquad S \in \mathcal{S}.$$

Often $\mathcal{S}$ is a linear space in the sense that $\alpha_1 S_1 + \alpha_2 S_2$ is in $\mathcal{S}$ for every $\alpha_1, \alpha_2 \in R$, whenever $S_1, S_2 \in \mathcal{S}$. In this case $\hat{S}$ is the projection of T if and only if $T - \hat{S}$ is *orthogonal* to $\mathcal{S}$ for the inner product $\langle S_1, S_2 \rangle = \mathrm{E}S_1 S_2$. This is the content of the following theorem.

11.1 Theorem. *Let $\mathcal{S}$ be a linear space of random variables with finite second moments. Then $\hat{S}$ is the projection of T onto $\mathcal{S}$ if and only if $\hat{S} \in \mathcal{S}$ and*

$$\mathrm{E}(T - \hat{S})S = 0, \qquad every \ S \in \mathcal{S}.$$

Every two projections of T onto $\mathcal{S}$ are almost surely equal. If the linear space $\mathcal{S}$ contains the constant variables, then $\mathrm{E}T = \mathrm{E}\hat{S}$ and $\mathrm{cov}(T - \hat{S}, S) = 0$ for every $S \in \mathcal{S}$.

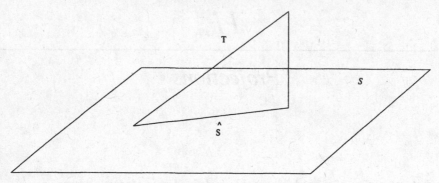

Figure 11.1. A variable T and its projection $\hat{S}$ on a linear space.

Proof. For any S and $\hat{S}$ in $\mathcal{S}$,

$$\mathrm{E}(T - S)^2 = \mathrm{E}(T - \hat{S})^2 + 2\mathrm{E}(T - \hat{S})(\hat{S} - S) + \mathrm{E}(\hat{S} - S)^2.$$

If $\hat{S}$ satisfies the orthogonality condition, then the middle term is zero, and we conclude that $\mathrm{E}(T - S)^2 \geq \mathrm{E}(T - \hat{S})^2$, with strict inequality unless $\mathrm{E}(\hat{S} - S)^2 = 0$. Thus, the orthogonality condition implies that $\hat{S}$ is a projection, and also that it is unique.

Conversely, for any number α,

$$\mathrm{E}(T - \hat{S} - \alpha S)^2 - \mathrm{E}(T - \hat{S})^2 = -2\alpha\mathrm{E}(T - \hat{S})S + \alpha^2\mathrm{E}S^2.$$

If $\hat{S}$ is a projection, then this expression is nonnegative for every α. But the parabola $\alpha \mapsto \alpha^2\mathrm{E}S^2 - 2\alpha\mathrm{E}(T - \hat{S})S$ is nonnegative if and only if the orthogonality condition $\mathrm{E}(T - \hat{S})S = 0$ is satisfied.

If the constants are in $\mathcal{S}$, then the orthogonality condition implies $\mathrm{E}(T - \hat{S})c = 0$, whence the last assertions of the theorem follow. ■

The theorem does not assert that projections always exist. This is not true: The infimum $\inf_S \mathrm{E}(T - S)^2$ need not be achieved. A sufficient condition for existence is that $\mathcal{S}$ is closed for the second-moment norm, but existence is usually more easily established directly.

The orthogonality of $T - \hat{S}$ and $\hat{S}$ yields the Pythagorean rule $\mathrm{E}T^2 = \mathrm{E}(T - \hat{S})^2 + \mathrm{E}\hat{S}^2$. (See Figure 11.1.) If the constants are contained in $\mathcal{S}$, then this is also true for variances instead of second moments.

Now suppose a sequence of statistics T_n and linear spaces $\mathcal{S}_n$ is given. For each n, let $\hat{S}_n$ be the projection of T_n on $\mathcal{S}_n$. Then the limiting behavior of the sequence T_n follows from that of $\hat{S}_n$, and vice versa, provided the quotient $\mathrm{var}T_n/\mathrm{var}\hat{S}_n$ converges to 1.

11.2   Theorem. *Let $\mathcal{S}_n$ be linear spaces of random variables with finite second moments that contain the constants. Let T_n be random variables with projections $\hat{S}_n$ onto $\mathcal{S}_n$. If $\mathrm{var}T_n/\mathrm{var}\hat{S}_n \to 1$ then*

$$\frac{T_n - \mathrm{E}T_n}{\mathrm{sd}\,T_n} - \frac{\hat{S}_n - \mathrm{E}\hat{S}_n}{\mathrm{sd}\,\hat{S}_n} \overset{\mathrm{P}}{\to} 0.$$

Proof. We shall prove convergence in second mean, which is stronger. The expectation of the difference is zero. Its variance is equal to

$$2 - 2\frac{\text{cov}(T_n, \hat{S}_n)}{\text{sd } T_n \text{ sd } \hat{S}_n}.$$

By the orthogonality of $T_n - \hat{S}_n$ and $\hat{S}_n$, it follows that $ET_n\hat{S}_n = E\hat{S}_n^2$. Because the constants are in $\mathcal{S}_n$, this implies that $\text{cov}(T_n, \hat{S}_n) = \text{var}\hat{S}_n$, and the theorem follows. ∎

The condition $\text{var}T_n/\text{var}\hat{S}_n \to 1$ in the theorem implies that the projections $\hat{S}_n$ are asymptotically of the same size as the original T_n. This explains that "nothing is lost" in the limit, and that the difference between T_n and its projection converges to zero. In the preceding theorem it is essential that the $\hat{S}_n$ are the projections of the variables T_n, because the condition $\text{var}T_n/\text{var}S_n \to 1$ for general sequences S_n and T_n does not imply anything.

11.2 Conditional Expectation

The expectation EX of a random variable X minimizes the quadratic form $a \mapsto \text{E}(X - a)^2$ over the real numbers a. This may be expressed as follows: EX is the best prediction of X, given a quadratic loss function, and in the absence of additional information.

The *conditional expectation* $\text{E}(X \mid Y)$ of a random variable X given a random vector Y is defined as the best "prediction" of X given knowledge of Y. Formally, $\text{E}(X \mid Y)$ is a measurable function $g_0(Y)$ of Y that minimizes

$$\text{E}\big(X - g(Y)\big)^2$$

over all measurable functions g. In the terminology of the preceding section, $\text{E}(X \mid Y)$ is the projection of X onto the linear space of all measurable functions of Y. It follows that the conditional expectation is the unique measurable function $\text{E}(X \mid Y)$ of Y that satisfies the orthogonality relation

$$\text{E}\big(X - \text{E}(X \mid Y)\big)g(Y) = 0, \qquad \text{every } g.$$

If $\text{E}(X \mid Y) = g_0(Y)$, then it is customary to write $\text{E}(X \mid Y = y)$ for $g_0(y)$. This is interpreted as the expected value of X given that $Y = y$ is observed. By Theorem 11.1 the projection is unique only up to changes on sets of probability zero. This means that the function $g_0(y)$ is unique up to sets B of values y such that $\text{P}(Y \in B) = 0$. (These could be very big sets.)

The following examples give some properties and also describe the relationship with conditional densities.

11.3 Example. The orthogonality relationship with $g \equiv 1$ yields the formula $\text{E}X = \text{EE}(X \mid Y)$. Thus, "the expectation of a conditional expectation is the expectation." □

11.4 Example. If $X = f(Y)$ for a measurable function f, then $\text{E}(X \mid Y) = X$. This follows immediately from the definition, in which the minimum can be reduced to zero. The interpretation is that X is perfectly predictable given knowledge of Y. □

11.5   *Example.* Suppose that (X, Y) has a joint probability density $f(x, y)$ with respect to a σ-finite product measure $\mu \times \nu$, and let $f(x \mid y) = f(x, y)/f_Y(y)$ be the conditional density of X given $Y = y$. Then

$$\mathrm{E}(X \mid Y) = \int x f(x \mid Y) \, d\mu(x).$$

(This is well defined only if $f_Y(Y) > 0$.) Thus the conditional expectation as defined above concurs with our intuition.

The formula can be established by writing

$$\mathrm{E}\big(X - g(Y)\big)^2 = \int \left[\int \big(x - g(y)\big)^2 f(x \mid y) \, d\mu(x) \right] f_Y(y) \, d\nu(y).$$

To minimize this expression over g, it suffices to minimize the inner integral (between square brackets) by choosing the value of $g(y)$ for every y separately. For each y, the integral $\int (x - a)^2 f(x \mid y) \, d\mu(x)$ is minimized for a equal to the mean of the density $x \mapsto f(x \mid y)$. $\square$

11.6   *Example.* If X and Y are independent, then $\mathrm{E}(X \mid Y) = \mathrm{E}X$. Thus, the extra knowledge of an unrelated variable Y does not change the expectation of X.

The relationship follows from the fact that independent random variables are uncorrelated: Because $\mathrm{E}(X - \mathrm{E}X)g(Y) = 0$ for all g, the orthogonality relationship holds for $g_0(Y) = \mathrm{E}X$. $\square$

11.7   *Example.* If f is measurable, then $\mathrm{E}\big(f(Y)X \mid Y\big) = f(Y)\mathrm{E}(X \mid Y)$ for any X and Y. The interpretation is that, given Y, the factor $f(Y)$ behaves like a constant and can be "taken out" of the conditional expectation.

Formally, the rule can be established by checking the orthogonality relationship. For every measurable function g,

$$\mathrm{E}\big(f(Y)X - f(Y)\mathrm{E}(X \mid Y)\big) g(Y) = \mathrm{E}\big(X - \mathrm{E}(X \mid Y)\big) f(Y)g(Y) = 0,$$

because $X - \mathrm{E}(X \mid Y)$ is orthogonal to all measurable functions of Y, including those of the form $f(Y)g(Y)$. Because $f(Y)\mathrm{E}(X \mid Y)$ is a measurable function of Y, it must be equal to $\mathrm{E}\big(f(Y)X \mid Y\big)$. $\square$

11.8   *Example.* If X and Y are independent, then $\mathrm{E}\big(f(X, Y) \mid Y = y\big) = \mathrm{E}f(X, y)$ for every measurable f. This rule may be remembered as follows: The known value y is substituted for Y; next, because Y carries no information concerning X, the unconditional expectation is taken with respect to X.

The rule follows from the equality

$$\mathrm{E}\big(f(X, Y) - g(Y)\big)^2 = \iint \big(f(x, y) - g(y)\big)^2 \, dP_X(x) \, dP_Y(y).$$

Once again, this is minimized over g by choosing for each y separately the value $g(y)$ to minimize the inner integral. $\square$

11.9   *Example.* For any random vectors X, Y and Z,

$$\mathrm{E}\big(\mathrm{E}(X \mid Y, Z) \mid Y\big) = \mathrm{E}(X \mid Y).$$

This expresses that a projection can be carried out in steps: The projection onto a smaller set can be obtained by projecting the projection onto a bigger set a second time.

Formally, the relationship can be proved by verifying the orthogonality relationship $E(E(X \mid Y, Z) - E(X \mid Y))g(Y) = 0$ for all measurable functions g. By Example 11.7, the left side of this equation is equivalent to $EE(Xg(Y) \mid Y, Z) - EE(g(Y)X \mid Y) = 0$, which is true because conditional expectations retain expectations. $\square$

11.3 Projection onto Sums

Let $X_1, \ldots, X_n$ be independent random vectors, and let S be the set of all variables of the form

$$\sum_{i=1}^n g_i(X_i),$$

for arbitrary measurable functions g_i with $Eg_i^2(X_i) < \infty$. This class is of interest, because the convergence in distribution of the sums can be derived from the central limit theorem. The projection of a variable onto this class is known as its *Hájek projection*.

11.10 Lemma. *Let $X_1, \ldots, X_n$ be independent random vectors. Then the projection of an arbitrary random variable T with finite second moment onto the class S is given by*

$$\hat{S} = \sum_{i=1}^n E(T \mid X_i) - (n-1)ET.$$

Proof. The random variable on the right side is certainly an element of S. Therefore, the assertion can be verified by checking the orthogonality relation. Because the variables X_i are independent, the conditional expectation $E(E(T \mid X_i) \mid X_j)$ is equal to the expectation $EE(T \mid X_i) = ET$ for every $i \neq j$. Consequently, $E(\hat{S} \mid X_j) = E(T \mid X_j)$ for every j, whence

$$E(T - \hat{S})g_j(X_j) = EE(T - \hat{S} \mid X_j)g_j(X_j) = E0g_j(X_j) = 0.$$

This shows that $T - \hat{S}$ is orthogonal to S. ∎

Consider the special case that $X_1, \ldots, X_n$ are not only independent but also identically distributed, and that $T = T(X_1, \ldots, X_n)$ is a permutation-symmetric, measurable function of the X_i. Then

$$E(T \mid X_i = x) = ET(x, X_2, \ldots, X_n).$$

Because this does not depend on i, the projection $\hat{S}$ is also the projection of T onto the smaller set of variables of the form $\sum_{i=1}^n g(X_i)$, where g is an arbitrary measurable function.

*11.4 Hoeffding Decomposition

The Hájek projection gives a best approximation by a sum of functions of one X_i at a time. The approximation can be improved by using sums of functions of two, or more, variables. This leads to the *Hoeffding decomposition*.

Because a projection onto a sum of orthogonal spaces is the sum of the projections onto the individual spaces, it is convenient to decompose the proposed projection space into a sum of orthogonal spaces. Given independent variables $X_1, \ldots, X_n$ and a subset $A \subset \{1, \ldots, n\}$, let H_A denote the set of all square-integrable random variables of the type

$$g_A(X_i : i \in A),$$

for measurable functions g_A of $|A|$ arguments such that

$$\mathrm{E}\big(g_A(X_i : i \in A) \mid X_j : j \in B\big) = 0, \qquad \text{every } B : |B| < |A|.$$

(Define $\mathrm{E}(T \mid \emptyset) = \mathrm{E}T$.) By the independence of $X_1, \ldots, X_n$ the condition in the last display is automatically valid for any $B \subset \{1, 2, \ldots, n\}$ that does not contain A. Consequently, the spaces H_A, when A ranges over all subsets of $\{1, \ldots, n\}$, are pairwise orthogonal. Stated in its present form, the condition reflects the intention to build approximations of increasing complexity by projecting a given variable in turn onto the spaces

$$[1], \qquad \left[\sum_i g_{\{i\}}(X_i)\right], \qquad \left[\sum\sum_{i<j} g_{\{i,j\}}(X_i, X_j)\right], \qquad \ldots,$$

where $g_{\{i\}} \in H_{\{i\}}$, $g_{\{i,j\}} \in H_{\{i,j\}}$, and so forth. Each new space is chosen orthogonal to the preceding spaces.

Let $P_A T$ denote the projection of T onto H_A. Then, by the orthogonality of the H_A, the projection onto the sum of the first r spaces is the sum $\sum_{|A| \le r} P_A T$ of the projections onto the individual spaces. The projection onto the sum of the first two spaces is the Hájek projection. More generally, the projections of zero, first, and second order can be seen to be

$$P_\emptyset T = \mathrm{E}T,$$
$$P_{\{i\}}T = \mathrm{E}(T \mid X_i) - \mathrm{E}T,$$
$$P_{\{i,j\}}T = \mathrm{E}(T \mid X_i, X_j) - \mathrm{E}(T \mid X_i) - \mathrm{E}(T \mid X_j) + \mathrm{E}T.$$

Now the general formula given by the following lemma should not be surprising.

11.11 Lemma. *Let $X_1, \ldots, X_n$ be independent random variables, and let T be an arbitrary random variable with $\mathrm{E}T^2 < \infty$. Then the projection of T onto H_A is given by*

$$P_A T = \sum_{B \subset A} (-1)^{|A|-|B|} \mathrm{E}(T \mid X_i : i \in B).$$

If $T \perp H_B$ for every subset $B \subset A$ of a given set A, then $\mathrm{E}(T \mid X_i : i \in A) = 0$. Consequently, the sum of the spaces H_B with $B \subset A$ contains all square-integrable functions of $(X_i : i \in A)$.

Proof. Abbreviate $\mathrm{E}(T \mid X_i : i \in A)$ to $\mathrm{E}(T \mid A)$ and $g_A(X_i : i \in A)$ to g_A. By the independence of $X_1, \ldots, X_n$ it follows that $\mathrm{E}\big(\mathrm{E}(T \mid A) \mid B\big) = \mathrm{E}(T \mid A \cap B)$ for every subsets A

and B of $\{1, \ldots, n\}$. Thus, for $P_A T$ as defined in the lemma and a set C strictly contained in A,

$$E(P_A T \mid C) = \sum_{B \subset A} (-1)^{|A|-|B|} E(T \mid B \cap C)$$

$$= \sum_{D \subset C} \sum_{j=0}^{|A|-|C|} (-1)^{|A|-|D|-j} \binom{|A| - |C|}{j} E(T \mid D).$$

By the binomial formula, the inner sum is zero for every D. Thus the left side is zero. In view of the form of $P_A T$, it was not a loss of generality to assume that $C \subset A$. Hence $P_A T$ is contained in H_A.

Next we verify the orthogonality relationship. For any measurable function g_A,

$$E(T - P_A T) g_A = E(T - E(T \mid A)) g_A - \sum_{\substack{B \subset A \\ B \neq A}} (-1)^{|A|-|B|} E E(T \mid B) E(g_A \mid B).$$

This is zero for any $g_A \in H_A$. This concludes the proof that $P_A T$ is as given.

We prove the second assertion of the lemma by induction on $r = |A|$. If $T \perp H_\emptyset$, then $E(T \mid \emptyset) = ET = 0$. Thus the assertion is true for $r = 0$. Suppose that it is true for $0, \ldots, r - 1$, and consider a set A of r elements. If $T \perp H_B$ for every $B \subset A$, then certainly $T \perp H_C$ for every $C \subset B$. Consequently, the induction hypothesis shows that $E(T \mid B) = 0$ for every $B \subset A$ of $r - 1$ or fewer elements. The formula for $P_A T$ now shows that $P_A T = E(T \mid A)$. By assumption the left side is zero. This concludes the induction argument.

The final assertion of the lemma follows if the variable $T_A := T - \sum_{B \subset A} P_B T$ is zero for every T that depends on $(X_i : i \in A)$ only. But in this case T_A depends on $(X_i : i \in A)$ only and hence equals $E(T_A \mid A)$, which is zero, because $T_A \perp H_B$ for every $B \subset A$. ∎

If $T = T(X_1, \ldots, X_n)$ is permutation-symmetric and $X_1, \ldots, X_n$ are independent and identically distributed, then the Hoeffding decomposition of T can be simplified to

$$T = \sum_{r=0}^{n} \sum_{|A|=r} g_r(X_i : i \in A),$$

for

$$g_r(x_1, \ldots, x_r) = \sum_{B \subset \{1, \ldots, r\}} (-1)^{r-|B|} ET(x_i \in B, X_i \notin B).$$

The inner sum in the representation of T is for each r a U-statistic of order r (as discussed in the Chapter 12), with degenerate kernel. All terms in the sum are orthogonal, whence the variance of T can be found as $\operatorname{var} T = \sum_{r=1}^{n} \binom{n}{r} E g_r^2(X_1, \ldots, X_r)$.

Notes

Orthogonal projections in Hilbert spaces (complete inner product spaces) are a classical subject in functional analysis. We have limited our discussion to the Hilbert space $L_2(\Omega, \mathcal{U}, P)$ of all square-integrable random variables on a probability space. Another popular method to introduce conditional expectation is based on the Radon-Nikodym theorem. Then $E(X \mid Y)$ is naturally defined for every integrable X. Hájek stated his projection lemma in [68] when proving the asymptotic normality of rank statistics under alternatives. Hoeffding [75] had already used it implicitly when proving the asymptotic normality of U-statistics. The "Hoeffding" decomposition appears to have received its name (for instance in [151]) in honor of Hoeffding's 1948 paper, but we have not been able to find it there. It is not always easy to compute a projection or its variance, and, if applied to a sequence of statistics, a projection may take the form $\sum g_n(X_i)$ for a function g_n depending on n even though a simpler approximation of the form $\sum g(X_i)$ with g fixed is possible.

PROBLEMS

1. Show that "projecting decreases second moment": If $\hat{S}$ is the projection of T onto a linear space, then $E\hat{S}^2 \leq ET^2$. If S contains the constants, then also $\text{var}\hat{S} \leq \text{var}T$.

2. Another idea of projection is based on minimizing variance instead of second moment. Show that $\text{var}(T - S)$ is minimized over a linear space S by $\hat{S}$ if and only if $\text{cov}(T - \hat{S}, S) = 0$ for every $S \in S$.

3. If $X \geq Y$ almost surely, then $E(X \mid Z) \geq E(Y \mid Z)$.

4. For an arbitrary random variable $X \geq 0$ (not necessarily square-integrable), define a conditional expectation $E(X \mid Y)$ by $\lim_{M \to \infty} E(X \wedge M \mid Y)$.
   (i) Show that this is well defined (the limit exists almost surely).
   (ii) Show that this coincides with the earlier definition if $EX^2 < \infty$.
   (iii) If $EX < \infty$ show that $E\big(X - E(X \mid Y)\big)g(Y) = 0$ for every bounded, measurable function g.
   (iv) Show that $E(X \mid Y)$ is the almost surely unique measurable function of Y that satisfies the orthogonality relationship of (iii).
   How would you define $E(X \mid Y)$ for a random variable with $E|X| < \infty$?

5. Show that a projection $\hat{S}$ of a variable T onto a convex set S is almost surely unique.

6. Find the conditional expectation $E(X \mid Y)$ if (X, Y) possesses a bivariate normal distribution.

7. Find the conditional expectation $E(X_1 \mid X_{(n)})$ if $X_1, \ldots, X_n$ are a random sample of standard uniform variables.

8. Find the conditional expectation $E(X_1 \mid \overline{X}_n)$ if $X_1, \ldots, X_n$ are i.i.d.

9. Show that for any random variables S and T (i) $\text{sd}(S+T) \leq \text{sd}\,S + \text{sd}\,T$, and (ii) $|\,\text{sd}\,S - \text{sd}\,T\,| \leq \text{sd}(S - T)$.

10. If S_n and T_n are arbitrary sequences of random variables such that $\text{var}(S_n - T_n)/\text{var}T_n \to 0$, then

$$\frac{S_n - ES_n}{\text{sd}\,S_n} - \frac{T_n - ET_n}{\text{sd}\,T_n} \xrightarrow{P} 0.$$

Moreover, $\text{var}S_n/\text{var}T_n \to 1$. Show this.

11. Show that $P_A h(X_j : X_j \in B) = 0$ for every set B that does not contain A.

12

U-Statistics

One-sample U-statistics can be regarded as generalizations of means. They are sums of dependent variables, but we show them to be asymptotically normal by the projection method. Certain interesting test statistics, such as the Wilcoxon statistics and Kendall's τ-statistic, are one-sample U-statistics. The Wilcoxon statistic for testing a difference in location between two samples is an example of a two-sample U-statistic. The Cramér–von Mises statistic is an example of a degenerate U-statistic.

12.1 One-Sample U-Statistics

Let $X_1, \ldots, X_n$ be a random sample from an unknown distribution. Given a known function h, consider estimation of the "parameter"

$$\theta = \mathrm{E}h(X_1, \ldots, X_r).$$

In order to simplify the formulas, it is assumed throughout this section that the function h is permutation symmetric in its r arguments. (A given h could always be replaced by a symmetric one.) The statistic $h(X_1, \ldots, X_r)$ is an unbiased estimator for θ, but it is unnatural, as it uses only the first r observations. A *U-statistic with kernel h* remedies this; it is defined as

$$U = \frac{1}{\binom{n}{r}} \sum_\beta h(X_{\beta_1}, \ldots, X_{\beta_r}),$$

where the sum is taken over the set of all unordered subsets β of r different integers chosen from $\{1, \ldots, n\}$. Because the observations are i.i.d., U is an unbiased estimator for θ also. Moreover, U is permutation symmetric in $X_1, \ldots X_n$, and has smaller variance than $h(X_1, \ldots, X_r)$. In fact, if $X_{(1)}, \ldots, X_{(n)}$ denote the values $X_1, \ldots, X_n$ stripped from their order (the order statistics in the case of real-valued variables), then

$$U = \mathrm{E}\big(h(X_1, \ldots, X_r) \mid X_{(1)}, \ldots, X_{(n)}\big).$$

Because a conditional expectation is a projection, and projecting decreases second moments, the variance of the U-statistic U is smaller than the variance of the naive estimator $h(X_1, \ldots, X_r)$.

161

In this section it is shown that the sequence $\sqrt{n}(U - \theta)$ is asymptotically normal under the condition that $\mathrm{E}h^2(X_1, \ldots, X_r) < \infty$.

12.1 Example. A U-statistic of degree $r = 1$ is a mean $n^{-1}\sum_{i=1}^{n}h(X_i)$. The asserted asymptotic normality is then just the central limit theorem. □

12.2 Example. For the kernel $h(x_1, x_2) = \frac{1}{2}(x_1 - x_2)^2$ of degree 2, the parameter $\theta = \mathrm{E}h(X_1, X_2) = \mathrm{var}\,X_1$ is the variance of the observations. The corresponding U-statistic can be calculated to be

$$U = \frac{1}{\binom{n}{2}}\sum\sum_{i<j}\frac{1}{2}(X_i - X_j)^2 = \frac{1}{n-1}\sum_{i=1}^{n}(X_i - \overline{X})^2.$$

Thus, the sample variance is a U-statistic of order 2. □

The asymptotic normality of a sequence of U-statistics, if $n \to \infty$ with the kernel remaining fixed, can be established by the projection method. The projection of $U - \theta$ onto the set of all statistics of the form $\sum_{i=1}^{n}g_i(X_i)$ is given by

$$\hat{U} = \sum_{i=1}^{n}\mathrm{E}(U - \theta \mid X_i) = \frac{r}{n}\sum_{i=1}^{n}h_1(X_i),$$

where the function h_1 is given by

$$h_1(x) = \mathrm{E}h(x, X_2, \ldots, X_r) - \theta.$$

The first equality in the formula for $\hat{U}$ is the Hájek projection principle. The second equality is established in the proof below.

The sequence of projections $\hat{U}$ is asymptotically normal by the central limit theorem, provided $\mathrm{E}h_1^2(X_1) < \infty$. The difference between $U - \theta$ and its projection is asymptotically negligible.

12.3 Theorem. *If* $\mathrm{E}h^2(X_1, \ldots, X_r) < \infty$, *then* $\sqrt{n}(U - \theta - \hat{U}) \xrightarrow{\mathrm{P}} 0$. *Consequently, the sequence* $\sqrt{n}(U - \theta)$ *is asymptotically normal with mean* 0 *and variance* $r^2\zeta_1$, *where, with* $X_1, \ldots, X_r, X_1', \ldots, X_r'$ *denoting i.i.d. variables,*

$$\zeta_1 = \mathrm{cov}\big(h(X_1, X_2, \ldots, X_r), h(X_1, X_2', \ldots, X_r')\big).$$

Proof. We first verify the formula for the projection $\hat{U}$. It suffices to show that $\mathrm{E}(U - \theta \mid X_i) = h_1(X_i)$. By the independence of the observations and permutation symmetry of h,

$$\mathrm{E}\big(h(X_{\beta_1}, \ldots, X_{\beta_r}) - \theta \mid X_i = x\big) = \begin{cases} h_1(x) & \text{if } i \in \beta \\ 0 & \text{if } i \notin \beta. \end{cases}$$

To calculate $\mathrm{E}(U - \theta \mid X_i)$, we take the average over all β. Then the first case occurs for $\binom{n-1}{r-1}$ of the vectors β in the definition of U. The factor r/n in the formula for the projection $\hat{U}$ arises as $r/n = \binom{n-1}{r-1}/\binom{n}{r}$.

The projection $\hat{U}$ has mean zero, and variance equal to

$$\text{var}\,\hat{U} = \frac{r^2}{n}\text{E}h_1^2(X_1)$$

$$= \frac{r^2}{n}\int \text{E}\big(h(x, X_2, \ldots, X_r) - \theta\big)\text{E}h(x, X_2', \ldots, X_r')\,dP_{X_1}(x) = \frac{r^2}{n}\zeta_1.$$

Because this is finite, the sequence $\sqrt{n}\,\hat{U}$ converges weakly to the $N(0, r^2\zeta_1)$-distribution by the central limit theorem. By Theorem 11.2 and Slutsky's lemma, the sequence $\sqrt{n}(U - \theta - \hat{U})$ converges in probability to zero, provided $\text{var}\,U/\text{var}\,\hat{U} \to 1$.

In view of the permutation symmetry of the kernel h, an expréssion of the type $\text{cov}\big(h(X_{\beta_1}, \ldots, X_{\beta_r}), h(X_{\beta_1'}, \ldots, X_{\beta_r'})\big)$ depends only on the number of variables X_i that are common to $X_{\beta_1}, \ldots, X_{\beta_r}$ and $X_{\beta_1'}, \ldots, X_{\beta_r'}$. Let ζ_c be this covariance if c variables are in common. Then

$$\text{var}\,U = \binom{n}{r}^{-2}\sum_{\beta}\sum_{\beta'}\text{cov}\big(h(X_{\beta_1}, \ldots, X_{\beta_r}), h(X_{\beta_1'}, \ldots, X_{\beta_r'})\big)$$

$$= \binom{n}{r}^{-2}\sum_{c=0}^{r}\binom{n}{r}\binom{r}{c}\binom{n-r}{r-c}\zeta_c.$$

The last step follows, because a pair (β, β') with c indexes in common can be chosen by first choosing the r indexes in β, next the c common indexes from β, and finally the remaining $r - c$ indexes in β' from $\{1, \ldots, n\} - \beta$. The expression can be simplified to

$$\text{var}\,U = \sum_{c=1}^{r}\frac{r!^2}{c!(r-c)!^2}\frac{(n-r)(n-r-1)\cdots(n-2r+c+1)}{n(n-1)\cdots(n-r+1)}\zeta_c.$$

In this sum the first term is $O(1/n)$, the second term is $O(1/n^2)$, and so forth. Because n times the first term converges to $r^2\zeta_1$, the desired limit result $\text{var}\,U/\text{var}\,\hat{U} \to 1$ follows. ∎

12.4 Example (Signed rank statistic). The parameter $\theta = P(X_1 + X_2 > 0)$ corresponds to the kernel $h(x_1, x_2) = 1\{x_1 + x_2 > 0\}$. The corresponding U-statistic is

$$U = \frac{1}{\binom{n}{2}}\sum\sum_{i<j}1\{X_i + X_j > 0\}.$$

This statistic is the average number of pairs (X_i, X_j) with positive sum $X_i + X_j > 0$, and can be used as a test statistic for investigating whether the distribution of the observations is located at zero. If many pairs (X_i, X_j) yield a positive sum (relative to the total number of pairs), then we have an indication that the distribution is centered to the right of zero.

The sequence $\sqrt{n}(U - \theta)$ is asymptotically normal with mean zero and variance $4\zeta_1$. If F denotes the cumulative distribution function of the observations, then the projection of $U - \theta$ can be written

$$\hat{U} = -\frac{2}{n}\sum_{i=1}^{n}\big(F(-X_i) - \text{E}F(-X_i)\big).$$

This formula is useful in subsequent discussion and is also convenient to express the asymptotic variance in F.

The statistic is particularly useful for testing the null hypothesis that the underlying distribution function is continuous and symmetric about zero: $F(x) = 1 - F(-x)$ for every x. Under this hypothesis the parameter θ equals $\theta = 1/2$, and the asymptotic variance reduces to $4 \operatorname{var} F(X_1) = 1/3$, because $F(X_1)$ is uniformly distributed. Thus, under the null hypothesis of continuity and symmetry, the limit distribution of the sequence $\sqrt{n}(U - 1/2)$ is normal $N(0, 1/3)$, independent of the underlying distribution. The last property means that the sequence U_n is *asymptotically distribution free* under the null hypothesis of symmetry and makes it easy to set critical values. The test that rejects H_0 if $\sqrt{3n}(U - 1/2) \geq z_\alpha$ is asymptotically of level α for every F in the null hypothesis.

This test is asymptotically equivalent to the *signed rank test* of Wilcoxon. Let $R_1^+, \ldots, R_n^+$ denote the *ranks* of the absolute values $|X_1|, \ldots, |X_n|$ of the observations: $R_i^+ = k$ means that $|X_i|$ is the kth smallest in the sample of absolute values. More precisely, $R_i^+ = \sum_{j=1}^n 1\{|X_j| \leq |X_i|\}$. Suppose that there are no pairs of tied observations $X_i = X_j$. Then the signed rank statistic is defined as $W^+ = \sum_{i=1}^n R_i^+ 1\{X_i > 0\}$. Some algebra shows that

$$W^+ = \binom{n}{2} U + \sum_{i=1}^n 1\{X_i > 0\}.$$

The second term on the right is of much lower order than the first and hence it follows that $n^{-3/2}(W^+ - EW^+) \rightsquigarrow N(0, 1/12)$. $\quad\square$

12.5   *Example (Kendall's τ)*. The U-statistic theorem requires that the observations $X_1, \ldots, X_n$ are independent, but they need not be real-valued. In this example the observations are a sample of bivariate vectors, for convenience (somewhat abusing notation) written as $(X_1, Y_1), \ldots, (X_n, Y_n)$. *Kendall's τ-statistic* is

$$\tau = \frac{4}{n(n-1)} \sum \sum_{i<j} 1\{(Y_j - Y_i)(X_j - X_i) > 0\} - 1.$$

This statistic is a measure of dependence between X and Y and counts the number of concordant pairs (X_i, Y_i) and (X_j, Y_j) in the observations. Two pairs are *concordant* if the indicator in the definition of τ is equal to 1. Large values indicate positive dependence (or concordance), whereas small values indicate negative dependence. Under independence of X and Y and continuity of their distributions, the distribution of τ is centered about zero, and in the extreme cases that all or none of the pairs are concordant τ is identically 1 or -1, respectively.

The statistic $\tau + 1$ is a U-statistic of order 2 for the kernel

$$h\left(\binom{x_1}{y_1}, \binom{x_2}{y_2}\right) = 21\{(y_2 - y_1)(x_2 - x_1) > 0\}.$$

Hence the sequence $\sqrt{n}(\tau + 1 - 2P((Y_2 - Y_1)(X_2 - X_1) > 0))$ is asymptotically normal with mean zero and variance $4\zeta_1$. With the notation $F^l(x, y) = P(X < x, Y < y)$ and $F^r(x, y) = P(X > x, Y > y)$, the projection of $U - \theta$ takes the form

$$\hat{U} = \frac{4}{n} \sum_{i=1}^n \left(F^l(X_i, Y_i) + F^r(X_i, Y_i) - EF^l(X_i, Y_i) - EF^r(X_i, Y_i)\right).$$

If X and Y are independent and have continuous marginal distribution functions, then $E\tau = 0$ and the asymptotic variance $4\zeta_1$ can be calculated to be $4/9$, independent of the

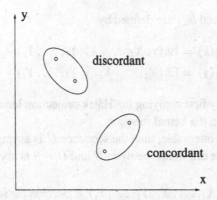

Figure 12.1. Concordant and discordant pairs of points.

marginal distributions. Then $\sqrt{n}\tau \rightsquigarrow N(0, 4/9)$ which leads to the test for "independence": Reject independence if $\sqrt{9n/4}\,|\tau| > z_{\alpha/2}$. $\square$

12.2 Two-Sample U-statistics

Suppose the observations consist of two independent samples $X_1, \ldots, X_m$ and $Y_1, \ldots, Y_n$, i.i.d. within each sample, from possibly different distributions. Let $h(x_1, \ldots, x_r, y_1, \ldots, y_s)$ be a known function that is permutation symmetric in $x_1, \ldots, x_r$ and $y_1, \ldots, y_s$ separately. A *two-sample U-statistic* with kernel h has the form

$$U = \frac{1}{\binom{m}{r}\binom{n}{s}} \sum_{\alpha} \sum_{\beta} h(X_{\alpha_1}, \ldots, X_{\alpha_r}, Y_{\beta_1}, \ldots, Y_{\beta_s}),$$

where α and β range over the collections of all subsets of r different elements from $\{1, 2, \ldots, m\}$ and of s different elements from $\{1, 2, \ldots, n\}$, respectively. Clearly, U is an unbiased estimator of the parameter

$$\theta = Eh(X_1, \ldots, X_r, Y_1, \ldots, Y_s).$$

The sequence $U_{m,n}$ can be shown to be asymptotically normal by the same arguments as for one-sample U-statistics. Here we let both $m \to \infty$ and $n \to \infty$, in such a way that the number of X_i and Y_j are of the same order. Specifically, if $N = m + n$ is the total number of observations we assume that, as $m, n \to \infty$,

$$\frac{m}{N} \to \lambda, \qquad \frac{n}{N} \to 1 - \lambda, \qquad 0 < \lambda < 1.$$

To give an exact meaning to $m, n \to \infty$, we may think of $m = m_\nu$ and $n = n_\nu$ indexed by a third index $\nu \in \mathbb{N}$. Next, we let $m_\nu \to \infty$ and $n_\nu \to \infty$ as $\nu \to \infty$ in such a way that $m_\nu/N_\nu \to \lambda$.

The projection of $U - \theta$ onto the set of all functions of the form $\sum_{i=1}^m k_i(X_i) + \sum_{j=1}^n l_j(Y_j)$ is given by

$$\hat{U} = \frac{r}{m} \sum_{i=1}^m h_{1,0}(X_i) + \frac{s}{n} \sum_{j=1}^n h_{0,1}(Y_j),$$

where the functions $h_{1,0}$ and $\hat{h}_{0,1}$ are defined by

$$h_{1,0}(x) = \mathrm{E}h(x, X_2, \ldots, X_r, Y_1, \ldots, Y_s) - \theta,$$
$$h_{0,1}(y) = \mathrm{E}h(X_1, \ldots, X_r, y, Y_2, \ldots, Y_s) - \theta.$$

This follows, as before, by first applying the Hájek projection lemma, and next expressing $\mathrm{E}(U \mid X_i)$ and $\mathrm{E}(U \mid Y_j)$ in the kernel function.

If the kernel is square-integrable, then the sequence $\hat{U}$ is asymptotically normal by the central limit theorem. The difference between $\hat{U}$ and $U - \theta$ is asymptotically negligible.

12.6   Theorem. *If $\mathrm{E}h^2(X_1, \ldots, X_r, Y_1, \ldots, Y_s) < \infty$, then the sequence $\sqrt{N}(U - \theta - \hat{U})$ converges in probability to zero. Consequently, the sequence $\sqrt{N}(U - \theta)$ converges in distribution to the normal law with mean zero and variance $r^2\zeta_{1,0}/\lambda + s^2\zeta_{0,1}/(1 - \lambda)$, where, with the X_i being i.i.d. variables independent of the i.i.d. variables Y_j,*

$$\zeta_{c,d} = \mathrm{cov}\big(h(X_1, \ldots, X_r, Y_1, \ldots, Y_s),$$
$$h(X_1, \ldots, X_c, X'_{c+1}, \ldots, X'_r, Y_1, \ldots, Y_d, Y'_{d+1}, \ldots, Y'_s)\big).$$

Proof.   The argument is similar to the one given previously for one-sample U-statistics. The variances of U and its projection are given by

$$\mathrm{var}\,\hat{U} = \frac{r^2}{m}\zeta_{1,0} + \frac{s^2}{n}\zeta_{0,1}$$

$$\mathrm{var}\,U = \frac{1}{\binom{m}{r}^2\binom{n}{s}^2} \sum_{c=0}^{r}\sum_{d=0}^{s} \binom{m}{r}\binom{r}{c}\binom{m-r}{r-c}\binom{n}{s}\binom{s}{d}\binom{n-s}{s-d}\zeta_{c,d}.$$

It can be checked from this that both the sequence $N\mathrm{var}\,\hat{U}$ and the sequence $N\mathrm{var}\,U$ converge to the number $r^2\zeta_{1,0}/\lambda + s^2\zeta_{0,1}/(1 - \lambda)$.   ∎

12.7   Example (Mann-Whitney statistic).   The kernel for the parameter $\theta = \mathrm{P}(X \leq Y)$ is $h(x, y) = 1\{X \leq Y\}$, which is of order 1 in both x and y. The corresponding U-statistic is

$$U = \frac{1}{mn}\sum_{i=1}^{m}\sum_{j=1}^{n}1\{X_i \leq Y_j\}.$$

The statistic mnU is known as the *Mann-Whitney statistic* and is used to test for a difference in location between the two samples. A large value indicates that the Y_j are "stochastically larger" than the X_i.

If the X_i and Y_j have cumulative distribution functions F and G, respectively, then the projection of $U - \theta$ can be written

$$\hat{U} = -\frac{1}{m}\sum_{i=1}^{m}\big(G_-(X_i) - \mathrm{E}G_-(X_i)\big) + \frac{1}{n}\sum_{j=1}^{n}\big(F(Y_i) - \mathrm{E}F(Y_i)\big).$$

It is easy to obtain the limit distribution of the projections $\hat{U}$ (and hence of U) from this formula. In particular, under the null hypothesis that the pooled sample $X_1, \ldots, X_m, Y_1, \ldots, Y_n$ is i.i.d. with continuous distribution function $F = G$, the sequence $\sqrt{12mn/N}(U - 1/2)$

converges to a standard normal distribution. (The parameter equals $\theta = 1/2$ and $\zeta_{0,1} = \zeta_{1,0} = 1/12$.)

If no observations in the pooled sample are tied, then $mnU + \frac{1}{2}n(n+1)$ is equal to the sum of the ranks of the Y_j in the pooled sample (see Chapter 13). Hence the latter statistic, the *Wilcoxon two-sample statistic*, is asymptotically normal as well. $\square$

*12.3 Degenerate U-Statistics

A sequence of U-statistics (or, better, their kernel function) is called *degenerate* if the asymptotic variance $r^2\zeta_1$ (found in Theorem 12.3) is zero. The formula for the variance of a U-statistic (in the proof of Theorem 12.3) shows that var U is of the order n^{-c} if $\zeta_1 = \cdots = \zeta_{c-1} = 0 < \zeta_c$. In this case, the sequence $n^{c/2}(U_n - \theta)$ is asymptotically tight. In this section we derive its limit distribution.

Consider the Hoeffding decomposition as discussed in section 11.4. For a U-statistic U_n with kernel h of order r, based on observations $X_1, \ldots, X_n$, this can be simplified to

$$U_n = \sum_{c=0}^{r} \sum_{|A|=c} \frac{1}{\binom{n}{r}} \sum_{\beta} P_A h(X_{\beta_1}, \ldots, X_{\beta_r}) = \sum_{c=0}^{r} \binom{r}{c} U_{n,c} \quad \text{(say)}.$$

Here, for each $0 \le c \le r$, the variable $U_{n,c}$ is a U-statistic of order c with kernel

$$h_c(X_1, \ldots, X_c) = P_{\{1,\ldots,c\}} h(X_1, \ldots, X_r).$$

To see this, fix a set A with c elements. Because the space H_A is orthogonal to all functions $g(X_j : j \in B)$ (i.e., the space $\sum_{C \subset B} H_C$) for every set B that does not contain A, the projection $P_A h(X_{\beta_1}, \ldots, X_{\beta_r})$ is zero unless $A \subset \beta = \{\beta_1, \ldots, \beta_r\}$. For the remaining β the projection $P_A h(X_{\beta_1}, \ldots, X_{\beta_r})$ does not depend on β (i.e., on the $r - c$ elements of $\beta - A$) and is a fixed function h_c of $(X_j : j \in A)$. This follows by symmetry, or explicitly from the formula for the projections in section 11.4. The function h_c is indeed the function as given previously. There are $\binom{n-c}{r-c}$ vectors β that contain the set A. The claim that $U_{n,c}$ is a U-statistic with kernel h_c now follows by simple algebra, using the fact that $\binom{n-c}{r-c}/\binom{r}{c}\binom{n}{r}$ $= 1/\binom{n}{c}$.

By the defining properties of the space $H_{\{1,\ldots,c\}}$, it follows that the kernel h_c is degenerate for $c \ge 2$. In fact, it is *strongly degenerate* in the sense that the conditional expectation of $h_c(X_1, \ldots, X_c)$ given any strict subset of the variables $X_1, \ldots, X_c$ is zero. In other words, the integral $\int h(x, X_2, \ldots, X_c) \, dP(x)$ with respect to any single argument vanishes. By the same reasoning, $U_{n,c}$ is uncorrelated with every measurable function that depends on strictly fewer than c elements of $X_1, \ldots, X_n$.

We shall show that the sequence $n^{c/2} U_{n,c}$ converges in distribution to a limit with variance $c! \, Eh_c^2(X_1, \ldots, X_c)$ for every $c \ge 1$. Then it follows that the sequence $n^{c/2}(U_n - \theta)$ converges in distribution for c equal to the smallest value such that $h_c \not\equiv 0$. For $c \ge 2$ the limit distribution is not normal but is known as *Gaussian chaos*.

Because the idea is simple, but the statement of the theorem (apparently) necessarily complicated, first consider a special case: $c = 3$ and a "product kernel" of the form

$$h_3(x_1, x_2, x_3) = f_1(x_1) f_2(x_2) f_3(x_3).$$

A U-statistic corresponding to a product kernel can be rewritten as a polynomial in sums of the observations. For ease of notation, let $\mathbb{P}_n f = n^{-1} \sum_{i=1}^n f(X_i)$ (the empirical measure), and let $\mathbb{G}_n f = \sqrt{n}(\mathbb{P}_n - P)f$ (the empirical process), for P the distribution of the observations $X_1, \ldots, X_n$. If the kernel h_3 is strongly degenerate, then each function f_i has mean zero and hence $\mathbb{G}_n f_i = \sqrt{n}\mathbb{P}_n f_i$ for every i. Then, with (i_1, i_2, i_3) ranging over all triplets of three different integers from $\{1, \ldots, n\}$ (taking position into account),

$$\frac{3!}{n^{3/2}} \binom{n}{3} U_{n,3} = \frac{1}{n^{3/2}} \sum_{(i_1, i_2, i_3)} f_1(x_{i_1}) f_2(x_{i_2}) f_3(x_{i_3})$$

$$= \mathbb{G}_n f_1 \mathbb{G}_n f_2 \mathbb{G}_n f_3 - \mathbb{P}_n (f_1 f_2) \mathbb{G}_n f_3$$

$$- \mathbb{P}_n (f_1 f_3) \mathbb{G}_n f_2 - \mathbb{P}_n (f_2 f_3) \mathbb{G}_n f_1 + 2 \frac{\mathbb{P}_n (f_1 f_2 f_3)}{\sqrt{n}}.$$

By the law of large numbers, $\mathbb{P}_n f \to P f$ almost surely for every f, while, by the central limit theorem, the marginal distributions of the stochastic processes $f \mapsto \mathbb{G}_n f$ converge weakly to multivariate Gaussian laws. If $\{\mathbb{G}f : f \in L_2(\mathcal{X}, \mathcal{A}, P)\}$ denotes a Gaussian process with mean zero and covariance function $E\mathbb{G}f\mathbb{G}_g = Pfg - PfPg$ (a *P-Brownian bridge process*), then $\mathbb{G}_n \rightsquigarrow \mathbb{G}$. Consequently,

$$n^{3/2} U_{n,3} \rightsquigarrow \mathbb{G}f_1 \mathbb{G}f_2 \mathbb{G}f_3 - P(f_1 f_2) \mathbb{G}f_3 - P(f_1 f_3) \mathbb{G}f_2 - P(f_2 f_3) \mathbb{G}f_1.$$

The limit is a polynomial of order 3 in the Gaussian vector $(\mathbb{G}f_1, \mathbb{G}f_2, \mathbb{G}f_3)$.

There is no similarly simple formula for the limit of a general sequence of degenerate U-statistics. However, any kernel can be written as an infinite linear combination of product kernels. Because a U-statistic is linear in its kernel, the limit of a general sequence of degenerate U-statistics is a linear combination of limits of the previous type.

To carry through this program, it is convenient to employ a decomposition of a given kernel in terms of an orthonormal basis of product kernels. This is always possible. We assume that $L_2(\mathcal{X}, \mathcal{A}, P)$ is separable, so that it has a countable basis.

12.8 *Example (General kernel).* If $1 = f_0, f_1, f_2, \ldots$ is an orthonormal basis of $L_2(\mathcal{X}, \mathcal{A}, P)$, then the functions $f_{k_1} \times \cdots \times f_{k_c}$ with $(k_1, \ldots, k_c)$ ranging over the nonnegative integers form an orthonormal basis of $L_2(\mathcal{X}^c, \mathcal{A}^c, P^c)$. Any square-integrable kernel can be written in the form $h_c(x_1, \ldots, x_c) = \sum a(k_1, \ldots, k_c) f_{k_1} \times \cdots \times f_{k_c}$ for $a(k_1, \ldots, k_c) = \langle h_c, f_{k_1} \times \cdots \times f_{k_c} \rangle$ the inner products of h_c with the basis functions. □

12.9 *Example (Second-order kernel).* In the case that $c = 2$, there is a choice that is specially adapted to our purposes. Because the kernel h_2 is symmetric and square-integrable by assumption, the integral operator $K : L_2(\mathcal{X}, \mathcal{A}, P) \mapsto L_2(\mathcal{X}, \mathcal{A}, P)$ defined by $Kf(x) = \int h_2(x, y) f(y) dP(y)$ is self-adjoint and Hilbert-Schmidt. Therefore, it has at most countably many eigenvalues $\lambda_0, \lambda_1, \ldots$, satisfying $\sum \lambda_k^2 < \infty$, and there exists an orthonormal basis of eigenfunctions $f_0, f_1, \ldots$ (See, for instance, Theorem VI.16 in [124].) The kernel h_2 can be expressed relatively to this basis as

$$h_2(x, y) = \sum_{k=0}^{\infty} \lambda_k f_k(x) f_k(y).$$

For a degenerate kernel h_2 the function 1 is an eigenfunction for the eigenvalue 0, and we can take $f_0 = 1$ without loss of generality.

The gain over the decomposition in the general case is that only product functions of the type $f \times f$ are needed. □

The (nonnormalized) *Hermite polynomial* H_j is a polynomial of degree j with leading coefficient x^j such that $\int H_i(x) H_j(x) \phi(x)\, dx = 0$ whenever $i \neq j$. The Hermite polynomials of lowest degrees are $H_0 = 1$, $H_1(x) = x$, $H_2(x) = x^2 - 1$ and $H_3(x) = x^3 - 3x$.

12.10 Theorem. *Let $h_c : \mathcal{X}^c \mapsto \mathbb{R}$ be a permutation-symmetric, measurable function of c arguments such that $\mathrm{E}h_c^2(X_1, \ldots, X_c) < \infty$ and $\mathrm{E}h_c(x_1, \ldots, x_{c-1}, X_c) \equiv 0$. Let $1 = f_0, f_1, f_2, \ldots$ be an orthonormal basis of $L_2(\mathcal{X}, \mathcal{A}, P)$. Then the sequence of U-statistics $U_{n,c}$ with kernel h_c based on n observations from P satisfies*

$$n^{c/2} U_{n,c} \rightsquigarrow \sum_{k=(k_1,\ldots,k_c)\in\mathbb{N}^c} \langle h_c, f_{k_1} \times \cdots \times f_{k_c}\rangle \prod_{i=1}^{d(k)} H_{a_i(k)}\big(\mathbb{G}\psi_i(k)\big).$$

Here $\mathbb{G}$ is a P-Brownian bridge process, the functions $\psi_1(k), \ldots, \psi_{d(k)}(k)$ are the different elements in $f_{k_1}, \ldots, f_{k_c}$, and $a_i(k)$ is number of times $\psi_i(k)$ occurs among $f_{k_1}, \ldots, f_{k_c}$. The variance of the limit variable is equal to $c!\, \mathrm{E}h_c^2(X_1, \ldots, X_c)$.

Proof. The function h_c can be represented in $L_2(\mathcal{X}^c, \mathcal{A}^c, P^c)$ as the series $\sum_k \langle h_c, f_{k_1} \times \cdots, f_{k_c}\rangle f_{k_1} \times \cdots \times f_{k_c}$. By the degeneracy of h_c the sum can be restricted to $k = (k_1, \ldots, k_c)$ with every $k_j \geq 1$. If $U_{n,c}h$ denotes the U-statistic with kernel $h(x_1, \ldots, x_c)$, then, for a pair of degenerate kernels h and g,

$$\mathrm{cov}(U_{n,c}h, U_{n,c}g) = \frac{c!}{n(n-1)\cdots(n-c+1)} P^c hg.$$

This means that the map $h \mapsto n^{c/2}\sqrt{c!}\, U_{n,c}h$ is close to being an isometry between $L_2(P^c)$ and $L_2(P^n)$. Consequently, the series $\sum_k \langle h_c, f_{k_1} \times \cdots \times f_{k_c}\rangle U_{n,c}f_{k_1} \times \cdots \times f_{k_c}$ converges in $L_2(P^n)$ and equals $U_{n,c}h_c = U_{n,c}$. Furthermore, if it can be shown that the finite-dimensional distributions of the sequence of processes $\{U_{n,c}f_{k_1} \times \cdots \times f_{k_c} : k \in \mathbb{N}^c\}$ converge weakly to the corresponding finite-dimensional distributions of the process $\{\prod_{i=1}^{d(k)} H_{a_i(k)}\big(\mathbb{G}\psi_i(k)\big) : k \in \mathbb{N}^c\}$, then the partial sums of the series converge, and the proof can be concluded by approximation arguments.

There exists a polynomial $\hat{P}_{n,c}$ of degree c, with random coefficients, such that

$$\frac{c!}{n^{c/2}} \binom{n}{c} U_{n,c} f_{k_1} \times \cdots \times f_{k_c} = \hat{P}_{n,c}\big(\mathbb{G}_n f_{k_1}, \ldots, \mathbb{G}_n f_{k_c}\big).$$

(See the example for $c = 3$ and problem 12.13). The only term of degree c in this polynomial is equal to $\mathbb{G}_n f_{k_1} \mathbb{G}_n f_{k_2} \cdots \mathbb{G}_n f_{k_c}$. The coefficients of the polynomials $\hat{P}_{n,c}$ converge in probability to constants. Conclude that the sequence $n^{c/2} c!\, U_{n,c} f_{k_1} \times \cdots \times f_{k_c}$ converges in distribution to $P_c\big(\mathbb{G}f_{k_1}, \ldots, \mathbb{G}f_{k_c}\big)$ for a polynomial P_c of degree c with leading term, and only term of degree c, equal to $\mathbb{G}f_{k_1}\mathbb{G}f_{k_2} \cdots \mathbb{G}f_{k_c}$. This convergence is simultaneous in sets of finitely many k.

It suffices to establish the representation of this limit in terms of Hermite polynomials. This could be achieved directly by algebraic and combinatorial arguments, but then the occurrence of the Hermite polynomials would remain somewhat mysterious. Alternatively, the representation can be derived from the definition of the Hermite polynomials and co-variance calculations. By the degeneracy of the kernel $f_{k_1} \times \cdots \times f_{k_c}$, the U-statistic $U_{n,c} f_{k_1} \times \cdots \times f_{k_c}$ is orthogonal to all measurable functions of $c - 1$ or fewer elements of $X_1, \ldots, X_n$, and their linear combinations. This includes the functions $\prod_i (\mathbb{G}_n g_i)^{a_i}$ for arbitrary functions g_i and nonnegative integers a_i with $\sum a_i < c$. Taking limits, we conclude that $P_c(\mathbb{G} f_{k_1}, \ldots, \mathbb{G} f_{k_c})$ must be orthogonal to every polynomial in $\mathbb{G} f_{k_1}, \ldots, \mathbb{G} f_{k_c}$ of degree less than $c - 1$. By the orthonormality of the basis f_i, the variables $\mathbb{G} f_i$ are independent standard normal variables. Because the Hermite polynomials form a basis for the polynomials in one variable, their (tensor) products form a basis for the polynomials of more than one argument. The polynomial P_c can be written as a linear combination of elements from this basis. By the orthogonality, the coefficients of base elements of degree $< c$ vanish. From the base elements of degree c only the product as in the theorem can occur, as follows from consideration of the leading term of P_c. ∎

12.11 Example. For $c = 2$ and a basis $1 = f_0, f_1, \ldots$ of eigenfunctions of the kernel h_2, we obtain a limit of the form $\sum_k \langle h_2, f_k \times f_k \rangle H_2(\mathbb{G} f_k)$. By the orthonormality of the basis this variable is distributed as $\sum_k \lambda_k (Z_k^2 - 1)$ for $Z_1, Z_2, \ldots$ a sequence of independent standard normal variables. □

12.12 Example (Sample variance). The kernel $h(x_1, x_2) = \frac{1}{2}(x_1 - x_2)^2$ yields the sample variance S_n^2. Because this has asymptotic variance $\mu_4 - \mu_2^2$ (see Example 3.2), the kernel is degenerate if and only if $\mu_4 = \mu_2^2$. This can happen only if $(X_1 - \alpha_1)^2$ is constant, for $\alpha_1 = \mathrm{E} X_i$. If we center the observations, so that $\alpha_1 = 0$, then this means that X_1 only takes the values $-\sigma$ and $\sigma = \sqrt{\mu_2}$, each with probability $1/2$. This is a very degenerate situation, and it is easy to find the limit distribution directly, but perhaps it is instructive to apply the general theorem. The kernels h_c take the forms (See section 11.4),

$$
\begin{aligned}
h_0 &= \mathrm{E}\tfrac{1}{2}(X_1 - X_2)^2 = \sigma^2, \\
h_1(x_1) &= \mathrm{E}\tfrac{1}{2}(x_1 - X_2)^2 - \sigma^2, \\
h_2(x_1, x_2) &= \tfrac{1}{2}(x_1 - x_2)^2 - \mathrm{E}\tfrac{1}{2}(x_1 - X_2)^2 - \mathrm{E}\tfrac{1}{2}(X_1 - x_2)^2 + \sigma^2.
\end{aligned}
$$

The kernel is degenerate if $h_1 = 0$ almost surely, and then the second-order kernel is $h_2(x_1, x_2) = \frac{1}{2}(x_1 - x_2)^2 - \sigma^2$. Because the underlying distribution has only two support points, the eigenfunctions f of the corresponding integral operator can be identified with vectors $(f(-\sigma), f(\sigma))$ in $\mathbb{R}^2$. Some linear algebra shows that they are $(1, 1)$ and $(-1, 1)$, corresponding to the eigenvalues 0 and $-\sigma^2$, respectively. Correspondingly, under degeneracy the kernel allows the decomposition

$$
h_2(x_1, x_2) = \tfrac{1}{2}(x_1^2 + x_2^2) - \sigma^2 - x_1 x_2 = -\sigma^2 \left(\frac{x_1}{\sigma}\right)\left(\frac{x_2}{\sigma}\right).
$$

We can conclude that the sequence $n(S_n^2 - \mu_2)$ converges in distribution to $-\sigma^2(Z_1^2 - 1)$. □

12.13 Example (Cramér–von Mises). Let $\mathbb{F}_n(x) = n^{-1} \sum_{i=1}^n 1\{X_i \le x\}$ be the empirical distribution function of a random sample $X_1, \ldots, X_n$ of real-valued random variables. The *Cramér–Von Mises statistic* for testing the (null) hypothesis that the underlying cumulative distribution is a given function F is given by

$$n \int (\mathbb{F}_n - F)^2 \, dF = \frac{1}{n} \sum_{i=1}^{n} \sum_{j=1}^{n} \int \big(1_{X_i \leq x} - F(x)\big)\big(1_{X_j \leq x} - F(x)\big) \, dF(x).$$

The double sum restricted to the off-diagonal terms is a U-statistic, with, under H_0, a degenerate kernel. Thus, this statistic converges to a nondegenerate limit distribution. The diagonal terms contribute the constant $\int F(1 - F) \, dF$ to the limit distribution, by the law of large numbers. If F is uniform, then the kernel of the U-statistic is

$$h(x, y) = \tfrac{1}{2}x^2 + \tfrac{1}{2}y^2 - x \vee y + \tfrac{1}{3}.$$

To find the eigenvalues of the corresponding integral operator K, we differentiate the identity $Kf = \lambda f$ twice, to find the equation $\lambda f'' + f = \int f(s) \, ds$. Because the kernel is degenerate, the constants are eigenfunctions for the eigenvalue 0. The eigenfunctions corresponding to nonzero eigenvalues are orthogonal to this eigenspace, whence $\int f(s) \, ds = 0$. The equation $\lambda f'' + f = 0$ has solutions $\cos ax$ and $\sin ax$ for $a^2 = \lambda^{-1}$. Reinserting these in the original equation or utilizing the relation $\int f(s) \, ds = 0$, we find that the nonzero eigenvalues are equal to $j^{-2}\pi^{-2}$ for $j \in \mathbb{N}$, with eigenfunctions $\sqrt{2}\cos \pi j x$. Thus, the Cramér–Von Mises statistic converges in distribution to $1/6 + \sum_{j=1}^{\infty} j^{-2}\pi^{-2}(Z_j^2 - 1)$. For another derivation of the limit distribution, see Chapter 19. $\square$

Notes

The main part of this chapter has its roots in the paper by Hoeffding [76]. Because the asymptotic variance is smaller than the true variance of a U-statistic, Hoeffding recommends to apply a standard normal approximation to $(U - \mathrm{E}U)/ \operatorname{sd} U$. Degenerate U-statistics were considered, among others, in [131] within the context of more general linear combinations of symmetric kernels. Arcones and Giné [2] have studied the weak convergence of "U-processes", stochastic processes indexed by classes of kernels, in spaces of bounded functions as discussed in Chapter 18.

PROBLEMS

1. Derive the asymptotic distribution of *Gini's mean difference*, which is defined as $\binom{n}{2}^{-1} \sum \sum_{i<j} |X_i - X_j|$.

2. Derive the projection of the sample variance.

3. Find a kernel for the parameter $\theta = \mathrm{E}(X - \mathrm{E}X)^3$.

4. Find a kernel for the parameter $\theta = \operatorname{cov}(X, Y)$. Show that the corresponding U-statistic is the sample covariance $\sum_{i=1}^{n}(X_i - \overline{X})(Y_i - \overline{Y})/(n-1)$.

5. Find the limit distribution of $U = \binom{n}{2}^{-1} \sum \sum_{i<j}(Y_j - Y_i)(X_j - X_i)$.

6. Let U_{n1} and U_{n2} be U-statistics with kernels h_1 and h_2, respectively. Derive the joint asymptotic distribution of (U_{n1}, U_{n2}).

7. Suppose $\mathrm{E}X_1^2 < \infty$. Derive the asymptotic distribution of the sequence $n^{-1} \sum \sum_{i \neq j} X_i X_j$. Can you give a two line proof without using the U-statistic theorem? What happens if $\mathrm{E}X_1 = 0$?

8. (**Mann's test against trend.**) To test the null hypothesis that a sample $X_1, \ldots, X_n$ is i.i.d. against the alternative hypothesis that the distributions of the X_i are stochastically increasing in i, Mann

suggested to reject the null hypothesis if the number of pairs (X_i, X_j) with $i < j$ and $X_i < X_j$ is large. How can we choose the critical value for large n?

9. Show that the U-statistic U with kernel $1\{x_1 + x_2 > 0\}$, the signed rank statistic W^+, and the positive-sign statistic $S = \sum_{i=1}^n 1\{X_i > 0\}$ are related by $W^+ = \binom{n}{2} U + S$ in the case that there are no tied observations.

10. A V-*statistic* of order 2 is of the form $n^{-2} \sum_{i=1}^n \sum_{j=1}^n h(X_i, X_j)$ where $h(x, y)$ is symmetric in x and y. Assume that $E h^2(X_1, X_1) < \infty$ and $E h^2(X_1, X_2) < \infty$. Obtain the asymptotic distribution of a V-statistic from the corresponding result for a U-statistic.

11. Define a V-statistic of general order r and give conditions for its asymptotic normality.

12. Derive the asymptotic distribution of $n(S_n^2 - \mu_2)$ in the case that $\mu_4 = \mu_2^2$ by using the delta-method (see Example 12.12). Does it make a difference whether we divide by n or $n - 1$?

13. For any $(n \times c)$ matrix a_{ij} we have

$$\sum_i a_{i_1, 1} \cdots a_{i_c, c} = \sum_B \prod_{B \in \mathcal{B}} (-1)^{|B|-1} (|B| - 1)! \sum_{i=1}^n \prod_{j \in B} a_{ij}.$$

Here the sum on the left ranges over all ordered subsets $(i_1, \ldots, i_c)$ of different integers from $\{1, \ldots, n\}$ and the first sum on the right ranges over all partitions $\mathcal{B}$ of $\{1, \ldots, c\}$ into nonempty sets (see Example [131]).

14. Given a sequence of i.i.d. random variables $X_1, X_2, \ldots$, let $\mathcal{A}_n$ be the σ-field generated by all functions of $(X_1, X_2, \ldots)$ that are symmetric in their first n arguments. Prove that a sequence U_n of U-statistics with a fixed kernel h of order r is a reverse martingale (for $n \geq r$) with respect to the filtration $\mathcal{A}_r \supset \mathcal{A}_{r+1} \supset \cdots$.

15. (**Strong law.**) If $E\big|h(X_1, \cdots, X_r)\big| < \infty$, then the sequence U_n of U-statistics with kernel h converges almost surely to $E h(X_1, \cdots, X_r)$. (For $r > 1$ the condition is not necessary, but a simple necessary and sufficient condition appears to be unknown.) Prove this. (Use the preceding problem, the martingale convergence theorem, and the Hewitt-Savage 0-1 law.)

13

Rank, Sign, and Permutation Statistics

*Statistics that depend on the observations only through their ranks can be
used to test hypotheses on departures from the null hypothesis that the observations are identically distributed. Such rank statistics are attractive,
because they are distribution-free under the null hypothesis and need not
be less efficient, asymptotically. In the case of a sample from a symmetric
distribution, statistics based on the ranks of the absolute values and the
signs of the observations have a similar property. Rank statistics are a
special example of permutation statistics.*

13.1 Rank Statistics

The *order statistics* $X_{N(1)} \leq X_{N(2)} \leq \cdots \leq X_{N(N)}$ of a set of real-valued observations
$X_1, \ldots, X_N$ ith order statistic are the values of the observations positioned in increasing
order. The *rank R_{Ni}* of X_i among $X_1, \ldots, X_N$ is its position number in the order statistics.
More precisely, if $X_1, \ldots, X_N$ are all different, then R_{Ni} is defined by the equation

$$X_i = X_{N(R_{Ni})}.$$

If X_i is tied with some other observations, this definition is invalid. Then the rank R_{Ni} is
defined as the average of all indices j such that $X_i = X_{N(j)}$ (sometimes called the *midrank*),
or alternatively as $\sum_{j=1}^{N} 1\{X_j \leq X_i\}$ (which is something like an *uprank*).

In this section it is assumed that the random variables $X_1, \ldots, X_N$ have continuous
distribution functions, so that ties in the observations occur with probability zero. We shall
neglect the latter null set. The ranks and order statistics are written with double subscripts,
because N varies and we shall consider order statistics of samples of different sizes. The
vectors of order statistics and ranks are abbreviated to $X_{N()}$ and R_N, respectively.

A *rank statistic* is any function of the ranks. A linear rank statistic is a rank statistic of
the special form $\sum_{i=1}^{N} a_N(i, R_{Ni})$ for a given $(N \times N)$ matrix $\big(a_N(i, j)\big)$. In this chapter
we are be concerned with the subclass of *simple linear rank statistics*, which take the form

$$\sum_{i=1}^{N} c_{Ni}\, a_{N, R_{Ni}}.$$

Here $(c_{N1}, \ldots, c_{NN})$ and $(a_{N1}, \ldots, a_{NN})$ are given vectors in $\mathbb{R}^N$ and are called the *coefficients* and *scores*, respectively. The class of simple linear rank statistics is sufficiently large

to contain interesting statistics for testing a variety of hypotheses. In particular, we shall see that it contains all "locally most powerful" rank statistics, which in another chapter are shown to be asymptotically efficient within the class of all tests.

Some elementary properties of ranks and order statistics are gathered in the following lemma.

13.1   Lemma. *Let $X_1, \ldots, X_N$ be a random sample from a continuous distribution function F with density f. Then*

(i) *the vectors $X_{N()}$ and R_N are independent;*

(ii) *the vector $X_{N()}$ has density $N! \prod_{i=1}^{N} f(x_i)$ on the set $x_1 < \cdots < x_N$;*

(iii) *the variable $X_{N(i)}$ has density $N \binom{N-1}{i-1} F(x)^{i-1} \left(1 - F(x)\right)^{N-i} f(x)$; for F the uniform distribution on $[0, 1]$, it has mean $i/(N + 1)$ and variance $i(N - i + 1)/ \left((N + 1)^2(N + 2)\right)$;*

(iv) *the vector R_N is uniformly distributed on the set of all $N!$ permutations of $1, 2, \ldots, N$;*

(v) *for any statistic T and permutation $r = (r_1, \ldots, r_N)$ of $1, 2, \ldots, N$,*

$$\mathrm{E}\big(T(X_1, \ldots, X_N) \mid R_N = r\big) = \mathrm{E} T\big(X_{N(r_1)}, \ldots, X_{N(r_N)}\big);$$

(vi) *for any simple linear rank statistic $T = \sum_{i=1}^{N} c_{Ni} a_{N, R_{Ni}}$,*

$$\mathrm{E}T = N \bar{c}_N \bar{a}_N; \qquad \mathrm{var}\, T = \frac{1}{N-1} \sum_{i=1}^{N} (c_{Ni} - \bar{c}_N)^2 \sum_{i=1}^{N} (a_{Ni} - \bar{a}_N)^2.$$

Proof. Statements (i) through (iv) are well-known and elementary. For the proof of (v), it is helpful to write $T(X_1, \ldots, X_N)$ as a function of the ranks and the order statistics. Next, we apply (i). For the proof of statement (vi), we use that the distributions of the variables R_{Ni} and the vectors (R_{Ni}, R_{Nj}) for $i \neq j$ are uniform on the sets $I = \{1, \ldots, N\}$ and $\big\{(i, j) \in I^2 : i \neq j\big\}$, respectively. Furthermore, a double sum of the form $\sum_{i \neq j}(b_i - \bar{b})(b_j - \bar{b})$ is equal to $-\sum_i (b_i - \bar{b})^2$. ∎

It follows that rank statistics are *distribution-free* over the set of all models in which the observations are independent and identically distributed. On the one hand, this makes them statistically useless in situations in which the observations are, indeed, a random sample from some distribution. On the other hand, it makes them of great interest to detect certain differences in distribution between the observations, such as in the two-sample problem. If the null hypothesis is taken to assert that the observations are identically distributed, then the critical values for a rank test can be chosen in such a way that the probability of an error of the first kind is equal to a given level α, for any probability distribution in the null hypothesis. Somewhat surprisingly, this gain is not necessarily counteracted by a loss in asymptotic efficiency, as we see in Chapter 14.

13.2   Example (Two-sample location problem). Suppose that the total set of observations consists of two independent random samples, inconsistently with the preceding notation written as $X_1, \ldots, X_m$ and $Y_1, \ldots, Y_n$. Set $N = m + n$ and let R_N be the rank vector of the *pooled sample* $X_1, \ldots, X_m, Y_1, \ldots, Y_n$.

We are interested in testing the null hypothesis that the two samples are identically distributed (according to a continuous distribution) against the alternative that the distribution of the second sample is stochastically larger than the distribution of the first sample. Even without a more precise description of the alternative hypothesis, we can discuss a collection of useful rank statistics. If the Y_j are a sample from a stochastically larger distribution, then the ranks of the Y_j in the pooled sample should be relatively large. Thus, any measure of the size of the ranks $R_{N,m+1}, \ldots, R_{NN}$ can be used as a test statistic. It will be distribution-free under the null hypothesis.

The most popular choice in this problem is the *Wilcoxon statistic*

$$W = \sum_{i=m+1}^{N} R_{Ni}.$$

This is a simple linear rank statistic with coefficients $c = (0, \ldots, 0, 1, \ldots, 1)$, and scores $a = (1, \ldots, N)$. The null hypothesis is rejected for large values of the Wilcoxon statistic. (The Wilcoxon statistic is equivalent to the *Mann-Whitney statistic* $U = \sum_{i,j} 1\{X_i \leq Y_j\}$ in that $W = U + \frac{1}{2}n(n+1)$.)

There are many other reasonable choices of rank statistics, some of which are of special interest and have names. For instance, the *van der Waerden statistic* is defined as

$$\sum_{i=m+1}^{N} \Phi^{-1}(R_{Ni}).$$

Here Φ^{-1} is the standard normal quantile function. We shall see ahead that this statistic is particularly attractive if it is believed that the underlying distribution of the observations is approximately normal. A general method to generate useful rank statistics is discussed below. $\square$

A critical value for a test based on a (distribution-free) rank statistic can be found by simply tabulating its null distribution. For a large number of observations this is a bit tedious. In most cases it is also unnecessary, because there exist accurate asymptotic approximations. The remainder of this section is concerned with proving asymptotic normality of simple linear rank statistics under the null hypothesis. Apart from being useful for finding critical values, the theorem is used subsequently to study the asymptotic efficiency of rank tests.

Consider a rank statistic of the form $T_N = \sum_{i=1}^{N} c_{Ni} a_{N, R_{Ni}}$. For a sequence of this type to be asymptotically normal, some restrictions on the coefficients c and scores a are necessary. In most cases of interest, the scores are "generated" through a given function $\phi : [0, 1] \mapsto \mathbb{R}$ in one of two ways. Either

$$a_{Ni} = \mathrm{E}\phi(U_{N(i)}), \tag{13.3}$$

where $U_{N(1)}, \ldots, U_{N(N)}$ are the order statistics of a sample of size N from the uniform distribution on $[0, 1]$; or

$$a_{Ni} = \phi\left(\frac{i}{N+1}\right). \tag{13.4}$$

For well-behaved functions ϕ, these definitions are closely related and almost identical, because $i/(N+1) = \mathrm{E}U_{N(i)}$. Scores of the first type correspond to the locally most powerful rank tests that are discussed ahead; scores of the second type are attractive in view of their simplicity.

13.5   Theorem. *Let R_N be the rank vector of an i.i.d. sample $X_1, \ldots, X_N$ from the continuous distribution function F. Let the scores a_N be generated according to (13.3) for a measurable function ϕ that is not constant almost everywhere, and satisfies $\int_0^1 \phi^2(u)\, du < \infty$. Define the variables*

$$T_N = \sum_{i=1}^N c_{Ni} a_{N,R_{Ni}}, \qquad \tilde{T}_N = N\bar{c}_N \bar{a}_N + \sum_{i=1}^N (c_{Ni} - \bar{c}_N)\phi\big(F(X_i)\big).$$

Then the sequences T_N and $\tilde{T}_N$ are asymptotically equivalent in the sense that $\mathrm{E}T_N = \mathrm{E}\tilde{T}_N$ and $\mathrm{var}\,(T_N - \tilde{T}_N)/\mathrm{var}\,T_N \to 0$. The same is true if the scores are generated according to (13.4) for a function ϕ that is continuous almost everywhere, is nonconstant, and satisfies $N^{-1}\sum_{i=1}^N \phi^2\big(i/(N+1)\big) \to \int_0^1 \phi^2(u)\, du < \infty$.

Proof.   Set $U_i = F(X_i)$, and view the rank vector R_N as the ranks of the first N elements of the infinite sequence $U_1, U_2, \ldots$. In view of statement (v) of the Lemma 13.1 the definition (13.3) is equivalent to

$$a_{N,R_{Ni}} = \mathrm{E}\big(\phi(U_i)\mid R_N\big).$$

This immediately yields that the projection of $\tilde{T}_N$ onto the set of all square-integrable functions of R_N is equal to $T_N = \mathrm{E}(\tilde{T}_N \mid R_N)$. It is straightforward to compute that

$$\frac{\mathrm{var}\,T_N}{\mathrm{var}\,\tilde{T}_N} = \frac{1/(N-1)\sum(c_{Ni} - \bar{c}_N)^2 \sum(a_{Ni} - \bar{a}_N)^2}{\sum(c_{Ni} - \bar{c}_N)^2 \mathrm{var}\,\phi(U_1)} = \frac{N}{N-1}\frac{\mathrm{var}\,a_{N,R_{N1}}}{\mathrm{var}\,\phi(U_1)}.$$

If it can be shown that the right side converges to 1, then the sequences T_N and $\tilde{T}_N$ are asymptotically equivalent by the projection theorem, Theorem 11.2, and the proof for the scores (13.3) is complete.

Using a martingale convergence theorem, we shall show the stronger statement

$$\mathrm{E}\big(a_{N,R_{N1}} - \phi(U_1)\big)^2 \to 0. \tag{13.6}$$

Because each rank vector R_{j-1} is a function of the next rank vector R_j (for one observation more), it follows that $a_{N,R_{N1}} = \mathrm{E}\big(\phi(U_1)\mid R_1, \ldots, R_N\big)$ almost surely. Because ϕ is square-integrable, a martingale convergence theorem (e.g., Theorem 10.5.4 in [42]) yields that the sequence $a_{N,R_{N1}}$ converges in second mean and almost surely to $\mathrm{E}\big(\phi(U_1)\mid R_1, R_2, \ldots\big)$. If $\phi(U_1)$ is measurable with respect to the σ-field generated by $R_1, R_2, \ldots$, then the conditional expectation reduces to $\phi(U_1)$ and (13.6) follows.

The projection of U_1 onto the set of measurable functions of R_{N1} equals the conditional expectation $\mathrm{E}(U_1 \mid R_{N1}) = R_{N1}/(N+1)$. By a straightforward calculation, the sequence $\mathrm{var}\big(R_{N1}/(N+1)\big)$ converges to $1/12 = \mathrm{var}\,U_1$. By the projection Theorem 11.2 it follows that $R_{N1}/(N+1) \to U_1$ in quadratic mean. Because R_{N1} is measurable in the σ-field generated by $R_1, R_2, \ldots$, for every N, so must be its limit U_1. This concludes the proof that $\phi(U_1)$ is measurable with respect to the σ-field generated by $R_1, R_2, \ldots$ and hence the proof of the theorem for the scores 13.3.

Next, consider the case that the scores are generated by (13.4). To avoid confusion, write these scores as $b_{Ni} = \phi(1/(N+1))$, and let a_{Ni} be defined by (13.3) as before. We shall prove that the sequences of rank statistics S_N and T_N defined from the scores a_N and b_N, respectively, are asymptotically equivalent.

Because $R_{N1}/(N+1)$ converges in probability to U_1 and ϕ is continuous almost everywhere, it follows that $\phi(R_{N1}/(N+1)) \to \phi(U_1)$. The assumption on ϕ is exactly that $E\phi^2(R_{N1}/(N+1))$ converges to $E\phi^2(U_1)$. By Proposition 2.29, we conclude that $\phi(R_{N1}/(N+1)) \to \phi(U_1)$ in second mean. Combining this with (13.6), we obtain that

$$\frac{1}{N}\sum_{i=1}^{N}(a_{Ni} - b_{Ni})^2 = E\left(a_{N,R_{N1}} - \phi\left(\frac{R_{N1}}{N+1}\right)\right)^2 \to 0.$$

By the formula for the variance of a linear rank statistic, we obtain that

$$\frac{\text{var}(S_N - T_N)}{\text{var}\, T_N} = \frac{\sum_{i=1}^{N}(a_{Ni} - b_{Ni} - (\bar{a}_N - \bar{b}_N))^2}{\sum_{i=1}^{N}(a_{Ni} - \bar{a}_N)^2} \to 0,$$

because $\text{var}\, a_{N,R_{N1}} \to \text{var}\, \phi(U_1) > 0$. This implies that $\text{var}\, S_N/\text{var}\, T_N \to 1$. The proof is complete. ∎

Under the conditions of the preceding theorem, the sequence of rank statistics $\sum c_{Ni} a_{N,R_{Ni}}$ is asymptotically equivalent to a sum of independent variables. This sum is asymptotically normal under the Lindeberg-Feller condition, given in Proposition 2.27. In the present case, because the variables $\phi(F(X_i))$ are independent and identically distributed, this is implied by

$$\frac{\max_{1 \le i \le N}(c_{Ni} - \bar{c}_N)^2}{\sum_{i=1}^{N}(c_{Ni} - \bar{c}_N)^2} \to 0. \tag{13.7}$$

This is satisfied by the most important choices of vectors of coefficients.

13.8 Corollary. *If the vector of coefficients c_N satisfies (13.7), and the scores are generated according to (13.3) for a measurable, nonconstant, square-integrable function ϕ, then the sequence of standardized rank statistics $(T_N - ET_N)/\text{sd}\, T_N$ converges weakly to an $N(0, 1)$-distribution. The same is true if the scores are generated by (13.4) for a function ϕ that is continuous almost everywhere, is nonconstant, and satisfies $N^{-1}\sum_{i=1}^{N}\phi^2(i/(N+1)) \to \int_0^1 \phi^2(u)\, du$.*

13.9 Example (Monotone score generating functions). Any nondecreasing, nonconstant function ϕ satisfies the conditions imposed on score-generating functions of the type (13.4) in the preceding theorem and corollary. The same is true for every ϕ that is of bounded variation, because any such ϕ is a difference of two monotone functions.

To see this, we recall from the preceding proof that it is always true that $R_{N1}/(N+1) \to U_1$, almost surely. Furthermore,

$$E\phi^2\left(\frac{R_{N1}}{N+1}\right) = \frac{1}{N}\sum_{i=1}^{N}\phi^2\left(\frac{i}{N+1}\right) \le \frac{N+1}{N}\sum_{i=1}^{N}\int_{i/(N+1)}^{(i+1)/(N+1)}\phi^2(u)\, du.$$

The right side converges to $\int \phi^2(u)\, du$. Because ϕ is continuous almost everywhere, it follows by Proposition 2.29 that $\phi(R_{N1}/(N+1)) \to \phi(U_1)$ in quadratic mean. □

13.10   Example (Two-sample problem). In a two-sample problem, in which the first m observations constitute the first sample and the remaining observations $n = N - m$ the second, the coefficients are usually chosen to be

$$c_{Ni} = \begin{cases} 0 & i = 1, \ldots, m \\ 1 & i = m + 1, \ldots, m + n. \end{cases}$$

In this case $\bar{c}_N = n/N$ and $\sum_{i=1}^{N}(c_{Ni} - \bar{c}_N)^2 = mn/N$. The Lindeberg condition is satisfied provided both $m \to \infty$ and $n \to \infty$. □

13.11   Example (Wilcoxon test). The function $\phi(u) = u$ generates the scores $a_{Ni} = i/(N+1)$. Combined with "two-sample coefficients," it yields a multiple of the Wilcoxon statistic. According to the preceding theorem, the sequence of Wilcoxon statistics $W_N = \sum_{i=m+1}^{N} R_{Ni}/(N+1)$ is asymptotically equivalent to

$$\tilde{W}_N = -\frac{n}{N}\sum_{i=1}^{m} F(X_i) + \frac{m}{N}\sum_{j=1}^{n} F(Y_j) + N\frac{n}{N}\frac{1}{2}.$$

The expectations and variances of these statistics are given by $EW_N = E\tilde{W}_N = n/2$, var $W_N = mn/(12(N+1))$, and var $\tilde{W}_N = mn/(12N)$. □

13.12   Example (Median test). The *median test* is a two-sample rank test with scores of the form $a_{Ni} = \phi(i/(N+1))$ generated by the function $\phi(u) = 1_{(0,1/2]}(u)$. The corresponding test statistic is

$$\sum_{i=m+1}^{N} 1\left\{R_{Ni} \leq \frac{N+1}{2}\right\}.$$

This counts the number of Y_j less than the median of the pooled sample. Large values of this test statistic indicate that the distribution of the second sample is stochastically smaller than the distribution of the first sample. □

The examples of rank statistics discussed so far have a direct intuitive meaning as statistics measuring a difference in location. It is not always obvious to find a rank statistic appropriate for testing certain hypotheses. Which rank statistics measure a difference in scale, for instance?

A general method of generating rank statistics for a specific situation is as follows. Suppose that it is required to test the null hypothesis that $X_1, \ldots, X_N$ are i.i.d. versus the alternative that $X_1, \ldots, X_N$ are independent with X_i having a distribution with density $f_{c_{Ni}\theta}$, for a given one-dimensional parametric model $\theta \mapsto f_\theta$. According to the Neyman-Pearson lemma, the most powerful rank test for testing $H_0 : \theta = 0$ against the simple alternative $H_1 : \theta = \theta$ rejects the null hypothesis for large values of the quotient

$$\frac{P_\theta(R_N = r)}{P_0(R_N = r)} = N! P_\theta(R_N = r).$$

Equivalently, the null hypothesis is rejected for large values of $P_\theta(R_N = r)$. This test depends on the alternative θ, but this dependence disappears if we restrict ourselves to alternatives θ that are sufficiently close to 0. Indeed, under regularity conditions,

$$P_\theta(R_N = r) - P_0(R_N = r)$$

$$= \int \cdots \int_{R_N = r} \left(\prod_{i=1}^{N} f_{c_{Ni}\theta}(x_i) - \prod_{i=1}^{N} f_0(x_i) \right) dx_1 \cdots dx_N$$

$$= \theta \int \cdots \int_{R_N = r} \sum_{i=1}^{N} c_{Ni} \frac{\dot{f}_0}{f_0}(x_i) \prod_{i=1}^{N} f_0(x_i) \, dx_1 \cdots dx_N + o(\theta)$$

$$= \theta \frac{1}{N!} \sum_{i=1}^{N} c_{Ni} E_0\left(\frac{\dot{f}_0}{f_0}(X_i) \mid R_N = r \right) + o(\theta).$$

Conclude that, for small $\theta > 0$, large values of $P_\theta(R_N = r)$ correspond to large values of the simple linear rank statistic $T_N = \sum_{i=1}^{N} c_{Ni} a_{N,R_{Ni}}$, for the vector a_N of scores given by

$$a_{Ni} = E_0 \frac{\dot{f}_0}{f_0}(X_{N_{(i)}}) = E \frac{\dot{f}_0}{f_0}\left(F_0^{-1}(U_{N_{(i)}}) \right).$$

These scores are of the form (13.3), with score-generating function $\phi = (\dot{f}_0/f_0) \circ F_0^{-1}$. Thus the corresponding rank statistics are asymptotically equivalent to the statistics $\sum_{i=1}^{N} c_{Ni} \dot{f}_0/f_0(X_i)$.

Rank statistics with scores generated as in the preceding display yield *locally most powerful* rank tests. They are most powerful within the class of all rank tests, uniformly in a sufficiently small neighbourhood $(0, \varepsilon)$ of 0. (For a precise statement, see problem 13.1). Such a local optimality property may seem weak, but it is actually of considerable importance, particularly if the number of observations is large. In the latter situation, any reasonable test can discriminate well between the null hypothesis and "distant" alternatives. A good test proves itself by having high power in discriminating "close" alternatives.

13.13 Example (Two-sample scale). To generate a test statistic for the two-sample scale problem, let $f_\theta(x) = e^{-\theta} f(e^{-\theta} x)$ for a fixed density f. If X_i has density $f_{c_{Ni}\theta}$ and the vector c is chosen equal to the usual vector of two-sample coefficients, then the first m observations have density $f_0 = f$; the last $n = N - m$ observations have density f_θ. The alternative hypothesis that the second sample has larger scale corresponds to $\theta > 0$. The scores for the locally most powerful rank test are given by

$$a_{Ni} = -E\left(1 + F^{-1}(U_{N_{(i)}}) \frac{f'}{f}\left(F^{-1}(U_{N_{(i)}}) \right) \right).$$

For instance, for f equal to the standard normal density this leads to the rank statistic $\sum_{i=m+1}^{N} a_{N,R_{Ni}}$ with scores

$$a_{Ni} = E\Phi^{-1}(U_{N_{(i)}})^2 - 1.$$

The same test is found for f equal to a normal density with a different mean or variance. This follows by direct calculation, or alternatively from the fact that rank statistics are location and scale invariant. The latter implies that the probabilities $P_{\mu,\sigma,\theta}(R_N = r)$ of the rank vector R_N of a sample of independent variables $X_1, \ldots, X_N$ with X_i distributed according to $e^{-\theta} f(e^{-\theta}(x - \mu)/\sigma)/\sigma$ do not depend on (μ, σ). Thus the procedure to generate locally most powerful scores yields the same result for any (μ, σ). $\square$

13.14 *Example (Two-sample location).* In order to find locally most powerful tests for location, we choose $f_\theta(x) = f(x - \theta)$ for a fixed density f and the coefficients c equal to the two-sample coefficients. Then the first m observations have density $f(x)$ and the last $n = N - m$ observations have density $f(x - \theta)$. The scores for a locally most powerful rank test are

$$a_{Ni} = -\mathrm{E}\left(\frac{f'}{f}\left(F^{-1}\big(U_{N_{(i)}}\big) \right) \right).$$

For the standard normal density, this leads to a variation of the van der Waerden statistic. The Wilcoxon statistic corresponds to the logistic density. □

13.15 *Example (Log rank test).* The *cumulative hazard function* corresponding to a continuous distribution function F is the function $\Lambda = -\log(1 - F)$. This is an important modeling tool in survival analysis. Suppose that we wish to test the null hypothesis that two samples with cumulative hazard functions Λ_X and Λ_Y are identically distributed against the alternative that they are not. The hypothesis of *proportional hazards* postulates that $\Lambda_Y = \theta \Lambda_X$ for a constant θ, meaning that the second sample is a factor θ more "at risk" at any time. If we wish to have large power against alternatives that satisfy this postulate, then it makes sense to use the locally most powerful scores corresponding to a family defined by $\Lambda_\theta = \theta \Lambda_1$. The corresponding family of cumulative distribution functions F_θ satisfies $1 - F_\theta = (1 - F_1)^\theta$ and is known as the family of *Lehmann alternatives*. The locally most powerful scores for this family correspond to the generating function

$$\phi(u) = \frac{\partial}{\partial \theta} \log \frac{\partial}{\partial x}(1 - F_\theta)(x)_{|\theta=1, x=F_1^{-1}(u)} = 1 - \log(1 - u).$$

It is fortunate that the score-generating function does not depend on the baseline hazard function Λ_1. The resulting test is known as the *log rank test*. The test is related to the *Savage test*, which uses the scores

$$a_{N,i} = \sum_{j=N-i+1}^{N} \frac{1}{j} \approx -\log\left(1 - \frac{i}{N+1} \right).$$

The log rank test is a very popular test in survival analysis. Then usually it needs to be extended to the situation that the observations are censored. □

13.16 *Example (More-sample problem).* Suppose the problem is to test the hypothesis that k independent random samples $X_1, \ldots, X_{N_1}, X_{N_1+1}, \ldots, X_{N_2}, \ldots, X_{N_{k-1}+1}, \ldots, X_{N_k}$ are identical in distribution. Let $N = N_k$ be the total number of observations, and let R_N be the rank vector of the pooled sample $X_1, \ldots, X_N$. Given scores a_N inference can be based on the rank statistics

$$T_{N1} = \sum_{i=1}^{N_1} a_{N, R_{Ni}}, \quad T_{N2} = \sum_{i=N_1+1}^{N_2} a_{N, R_{Ni}}, \ldots, T_{Nk} = \sum_{i=N_{k-1}+1}^{N_k} a_{N, R_{Ni}}.$$

The testing procedure can consist of several two-sample tests, comparing pairs of (pooled) subsamples, or on an overall statistic. One possibility for an overall statistic is the chi-square statistic. For $n_j = N_j - N_{j-1}$ equal to the number of observations in the jth sample, define

$$C_N^2 = \sum_{j=1}^{k} \frac{(T_{N_j} - n_j \bar{a}_N)^2}{n_j \operatorname{var} \phi(U_1)}.$$

If the scores are generated by (13.3) or (13.4) and all sample sizes n_j tend to infinity, then every sequence T_{N_j} is asymptotically normal under the null hypothesis, under the conditions of Theorem 13.5. In fact, because the approximations $\tilde{T}_{N_j}$ are jointly asymptotically normal by the multivariate central limit theorem, the vector $T_N = (T_{N1}, \ldots, T_{Nk})$ is asymptotically normal as well. By elementary calculations, if $n_i/N \to \lambda_i$,

$$\frac{T_N - ET_N}{\sqrt{N} \operatorname{sd} \phi(U_1)} \rightsquigarrow N_k \left(0, \begin{pmatrix} \lambda_1(1-\lambda_1) & -\lambda_1\lambda_2 & \cdots & -\lambda_1\lambda_k \\ -\lambda_2\lambda_1 & \lambda_2(1-\lambda_2) & \cdots & -\lambda_2\lambda_k \\ \vdots & \vdots & & \vdots \\ -\lambda_k\lambda_1 & -\lambda_k\lambda_2 & \cdots & \lambda_k(1-\lambda_k) \end{pmatrix} \right).$$

This limit distribution is similar to the limit distribution of a sequence of multinomial vectors. Analogously to the situation in the case of Pearson's chi-square tests for a multinomial distribution (see Chapter 17), the sequence C_N^2 converges in distribution to a chi-square distribution with $k - 1$ degrees of freedom.

There are many reasonable choices of scores. The most popular choice is based on $\phi(u) = u$ and leads to the *Kruskal-Wallis* test. Its test statistic is usually written in the form

$$\frac{12}{N(N-1)} \sum_{j=1}^{k} n_j \left(\bar{R}_{j\cdot} - \frac{N+1}{2} \right)^2, \qquad \bar{R}_{j\cdot} = \frac{\sum_{i=N_{j-1}+1}^{N_j} R_{Ni}}{n_j}.$$

This test statistic measures the distance of the average scores of the k samples to the average score $(N + 1)/2$ of the pooled sample.

An alternative is to use locally asymptotically powerful scores for a family of distributions of interest. Also, choosing the same score generating function for all subsamples is convenient, but not necessary, provided the chi-square statistic is modified. $\square$

13.2 Signed Rank Statistics

The *sign* of a number x, denoted $\operatorname{sign}(x)$, is defined to be -1, 0, or 1 if $x < 0$, $x = 0$ or $x > 0$, respectively. The *absolute rank* R_{Ni}^+ of an observation X_i in a sample $X_1, \ldots, X_N$ is defined as the rank of $|X_i|$ in the sample of absolute values $|X_1|, \ldots, |X_N|$. A simple linear *signed rank statistic* has the form

$$\sum_{i=1}^{N} a_{N, R_{Ni}^+} \operatorname{sign}(X_i).$$

The ordinary ranks of a sample can always be derived from the combined set of absolute ranks and signs. Thus, the vectors of absolute ranks and signs are together statistically more informative than the ordinary ranks. The difference is dramatic if testing the location of a symmetric density of a given form, in which case the class of signed rank statistics contains asymptotically efficient test statistics in great generality.

The main attraction of signed rank statistics is their simplicity, particularly their being distribution-free over the set of all symmetric distributions. Write $|X|$, R_N^+, and $\text{sign}_N(X)$ for the vectors of absolute values, absolute ranks, and signs.

13.17 Lemma. *Let $X_1, \ldots, X_N$ be a random sample from a continuous distribution that is symmetric about zero. Then*
  (i) *the vectors $\big(|X|, R_N^+\big)$ and $\text{sign}_N(X)$ are independent;*
  (ii) *the vector R_N^+ is uniformly distributed over $\{1, \ldots, N\}$;*
  (iii) *the vector $\text{sign}_N(X)$ is uniformly distributed over $\{-1, 1\}^N$;*
  (iv) *for any signed rank statistic, $\text{var} \sum_{i=1}^N a_{N, R_{Ni}^+} \text{sign}(X_i) = \sum_{i=1}^N a_{Ni}^2$.*

Consequently, for testing the null hypothesis that a sample is i.i.d. from a continuous, symmetric distribution, the critical level of a signed rank statistic can be set without further knowledge of the "shape" of the underlying distribution.

The null hypothesis of symmetry arises naturally in the two-sample problem with paired observations. Suppose that, given independent observations $(X_1, Y_1), \ldots, (X_N, Y_N)$, it is desired to test the hypothesis that the distribution of $X_i - Y_i$ is "centered at zero." If the observations (X_i, Y_i) are exchangeable, that is, the pairs (X_i, Y_i) and (Y_i, X_i) are equal in distribution, then $X_i - Y_i$ is symmetrically distributed about zero. This is the case, for instance, if, given a third variable (usually called "factor"), the observations X_i and Y_i are conditionally independent and identically distributed. For the vector of absolute ranks to be uniformly distributed on the set of all permutations it is necessary to assume in addition that the differences are identically distributed.

For the signs alone to be distribution-free, it suffices, of course, that the pairs are independent and that $P(X_i < Y_i) = P(X_i > Y_i) = \frac{1}{2}$ for every i. Consequently, tests based on only the signs have a wider applicability than the more general signed rank tests. However, depending on the model they may be less efficient.

13.18 Theorem. *Let $X_1, \ldots, X_N$ be a random sample from a continuous distribution that is symmetric about zero. Let the scores a_N be generated according to (13.3) for a measurable function ϕ such that $\int_0^1 \phi^2(u)\, du < \infty$. For F^+ the distribution function of $|X_1|$, define*

$$T_N = \sum_{i=1}^N a_{N, R_{Ni}^+} \text{sign}(X_i), \qquad \tilde{T}_N = \sum_{i=1}^N \phi\big(F^+(|X_i|)\big) \text{sign}(X_i).$$

Then the sequences T_N and $\tilde{T}_N$ are asymptotically equivalent in the sense that $N^{-1}\text{var}\,(T_N - \tilde{T}_N) \to 0$. Consequently, the sequence $N^{-1/2} T_N$ is asymptotically normal with mean zero and variance $\int_0^1 \phi^2(u)\, du$. The same is true if the scores are generated according to (13.4) for a function ϕ that is continuous almost everywhere and satisfies $N^{-1}\sum_{i=1}^N \phi^2\big(i/(N+1)\big) \to \int_0^1 \phi^2(u)\, du < \infty$.

Proof. Because the vectors $\text{sign}_N(X)$ and $\big(|X|, R_N^+\big)$ are independent and $\text{E}\,\text{sign}_N(X) = 0$, the means of both T_N and $\tilde{T}_N$ are zero. Furthermore, by the independence and the orthogonality of the signs,

$$\text{E}(\tilde{T}_N - T_N)^2 = N\text{E}\big(a_{N, R_{N1}^+} - \phi\big(F^+(|X_1|)\big)\big)^2.$$

The expectation on the right side is exactly the expression in (13.6), evaluated for the special choice $U_1 = F^+(|X_1|)$. This can be shown to converge to zero as in the proof of Theorem 13.5. ∎

13.19 *Example (Wilcoxon signed rank statistic).* The *Wilcoxon signed rank statistic* $W_N = \sum_{i=1}^{N} R_{Ni}^+ \operatorname{sign}(X_i)$ is obtained from the score-generating function $\phi(u) = u$. Large values of this statistic indicate that large absolute values $|X_i|$ tend to go together with positive X_i. Thus large values of the Wilcoxon statistic suggest that the location of the X_i is larger than zero. Under the null hypothesis that $X_1, \ldots, X_N$ are i.i.d. and symmetrically distributed about zero, the sequence $N^{-3/2} W_N$ is asymptotically normal $N(0, 1/3)$. The variance of W_N is equal to $N(2N+1)(N+1)/6$.

The signed rank statistic is asymptotically equivalent to the U-statistic with kernel $h(x_1, x_2) = 1\{x_1 + x_2 > 0\}$. (See problem 12.9.) This connection yields the limit distribution also under nonsymmetric distributions. □

Signed rank statistics that are locally most powerful can be obtained in a similar fashion as locally most powerful rank statistics were obtained in the previous section. Let f be a symmetric density, and let $X_1, \ldots, X_N$ be a random sample from the density $f(\cdot - \theta)$. Then, under regularity conditions,

$$P_\theta\big(\operatorname{sign}_N(X) = s, R_N^+ = r\big) - P_0\big(\operatorname{sign}_N(X) = s, R_N^+ = r\big)$$

$$= -\theta\, E_0 \sum_{i=1}^{N} \operatorname{sign}(X_i) \frac{f'}{f}(|X_i|)\big\{\operatorname{sign}_N(x) = s, R_N^+ = r\big\} + o(\theta)$$

$$= -\theta\, \frac{1}{2^N N!} \sum_{i=1}^{N} s_i E_0\left(\frac{f'}{f}(|X_i|) \mid R_{Ni}^+ = r_i\right) + o(\theta).$$

In the second equality it is used that $f'/f(x)$ is equal to $\operatorname{sign}(x) f'/f(|x|)$ by the skew symmetry of f'/f. It follows that *locally most powerful signed rank statistics* for testing f against $f(\cdot - \theta)$ are obtained from the scores

$$a_{Ni} = -E\frac{f'}{f}\big((F^+)^{-1}(U_{N(i)})\big).$$

These scores are of the form (13.3) with score-generating function $\phi = -(f'/f) \circ (F^+)^{-1}$, whence locally most powerful rank statistics are asymptotically linear by Theorem 13.18. By the symmetry of F, we have $(F^+)^{-1}(u) = F^{-1}((u+1)/2)$.

13.20 *Example.* The Laplace density has score function $f'/f(x) = \operatorname{sign}(x) = 1$, for $x \geq 0$. This leads to the locally most powerful scores $a_{Ni} \equiv 1$. The corresponding test statistic is the *sign statistic* $T_N = \sum_{i=1}^{N} \operatorname{sign}(X_i)$. Is it surprising that this simple statistic possesses an optimality property? It is shown to be asymptotically optimal for testing $H_0 : \theta = 0$ in Chapter 15. □

13.21 *Example.* The locally most powerful score for the normal distribution are $a_{Ni} = E\Phi^{-1}((U_{N(i)} + 1)/2)$. These are appropriately known as the normal (absolute) scores. □

13.3   Rank Statistics for Independence

Let $(X_1, Y_1), \ldots, (X_N, Y_N)$ be independent, identically distributed bivariate vectors, with continuous marginal distributions. The problem is to determine whether, within each pair, X_i and Y_i are independent.

Let R_N and S_N be the rank vectors of the samples $X_1, \ldots, X_N$ and $Y_1, \ldots, Y_N$, respectively. If X_i and Y_i are positively dependent, then we expect the vectors R_N and S_N to be roughly parallel. Therefore, rank statistics of the form

$$\sum_{i=1}^{N} a_{N,R_{Ni}} b_{N,S_{Ni}},$$

with a_N and b_N increasing vectors, are reasonable choices for testing independence.

Under the null hypothesis of independence of X_i and Y_i, the vectors R_N and S_N are independent and both uniformly distributed on the permutations of $\{1, \ldots, N\}$. Let R_N^o be the vector of ranks of $X_1, \ldots, X_N$ if first the pairs $(X_1, Y_1), \ldots, (X_N, Y_N)$ have been put in increasing order of $Y_1 < Y_2 < \cdots < Y_N$. The coordinates of R_N^o are called the *antiranks*. Under the null hypothesis, the antiranks are also uniformly distributed on the permutations of $\{1, \ldots, N\}$. By the definition of the antiranks,

$$\sum_{i=1}^{N} a_{N,R_{Ni}} b_{N,S_{Ni}} = \sum_{i=1}^{N} a_{N,R_{Ni}^o} b_{Ni}.$$

The right side is a simple linear rank statistic and can be shown to be asymptotically normal by Theorem 13.5.

13.22   *Example (Spearman rank correlation).* The simplest choice of scores corresponds to the generating function $\phi(u) = u$. This leads to the *rank correlation coefficient* ρ_N, which is the ordinary sample correlation coefficient of the rank vectors R_N and S_N. Indeed, because the rank vectors are permutations of the numbers $1, 2, \ldots, N$, their sample mean and variance are fixed, at $(N+1)/2$ and $N(N+1)/12$, respectively, and hence

$$\rho_N = \frac{\sum_{i=1}^{N}(R_{Ni} - \overline{R}_N)(S_{Ni} - \overline{S}_N)}{\left(\sum_{i=1}^{N}(R_{Ni} - \overline{R}_N)^2 \sum_{i=1}^{N}(S_{Ni} - \overline{S}_N)^2\right)^{1/2}}$$

$$= \frac{12}{N(N-1)(N+1)} \sum_{i=1}^{N} R_{Ni} S_{Ni} - 3\frac{N+1}{N-1}.$$

Thus the tests based on the rank correlation coefficient ρ_N are equivalent to tests based on the signed rank statistic $\sum R_{Ni} S_{Ni}$.

It is straightforward to derive from Theorem 13.5 that the sequence $\sqrt{N}\rho_N$ is asymptotically standard normal under the null hypothesis of independence. $\square$

*13.4   Rank Statistics under Alternatives

Let R_N be the rank vector of the independent random variables $X_1, \ldots, X_N$ with continuous distribution functions $F_1, \ldots, F_N$. Theorem 13.5 gives the asymptotic distribution of simple, linear rank statistics under very mild conditions on the score-generating function,

but under the strong assumption that the distribution functions F_i are all equal. This is sufficient for setting critical values of rank tests for the null hypothesis of identical distributions,
but for studying their asymptotic efficiency we also need the asymptotic behavior under
alternatives. For instance, in the two-sample problem we are interested in the asymptotic
distributions under alternatives of the form $F, \ldots, F, G, \ldots, G$, where F and G are the
distributions of the two samples.

For alternatives that converge to the null hypothesis "sufficiently fast," the best approach
is to use Le Cam's third lemma. In particular, if the log likelihood ratios of the alternatives
$F_n, \ldots, F_n, G_n, \ldots, G_n$ with respect to the null distributions $F, \ldots, F, F, \ldots, F$ allow
an asymptotic approximation by a sum of the type $\sum \ell_i(X_i)$, then the joint asymptotic
distribution of the rank statistics and the log likelihood ratios under the null hypothesis
can be obtained from the multivariate central limit theorem and Slutsky's lemma, because
Theorem 13.5 yields a similar approximation for the rank statistics. Next, we can apply Le
Cam's third lemma, as in Example 6.7, to find the limit distribution of the rank statistics
under the alternatives. This approach is relatively easy, and is sufficiently general for most
of the questions of interest. See sections 7.5 and 14.1.1 for examples.

More general alternatives must be handled directly and appear to require stronger conditions on the score-generating function. One possibility is to write the rank statistic as a
functional of the empirical distribution function $\mathbb{F}_N$, and the weighted empirical distribution
$\mathbb{F}_N^c(x) = N^{-1}\sum_{i=1}^{N}c_{Ni}1\{X_i \le x\}$ of the observations. Because $R_{Ni} = N\mathbb{F}_N(X_i)$, we have

$$\frac{1}{N}\sum_{i=1}^{N}c_{Ni}a_{N,R_{Ni}} = \int a_{N,N\mathbb{F}_N(x)}\,d\mathbb{F}_N^c(x).$$

Next, we can apply a von Mises analysis, using the convergence of the empirical distribution
functions to Brownian bridges. This method is explained in general in Chapter 20.

In this section we illustrate another method, based on Hájek's projection lemma. To
avoid technical complications, we restrict ourselves to smooth score-generating functions.
Let $\overline{F}_N$ be the average of $F_1, \ldots, F_N$ and let $\overline{F}_N^c$ be the weighted sum $N^{-1}\sum_{i=1}^{N}c_{Ni}F_i$, and
define

$$T_N = \sum_{i=1}^{N}c_{Ni}\phi\left(\frac{R_{Ni}}{N+1}\right),$$

$$\hat{T}_N = \sum_{i=1}^{N}\left[c_{Ni}\phi\left(\overline{F}_N(X_i)\right) + \int_{X_i}^{\infty}\phi'\left(\overline{F}_N(x)\right)d\overline{F}_N^c(x)\right].$$

We shall show that the variables $\hat{T}_N$ are the Hájek projections of approximations to the
variables T_N, up to centering at mean zero. The Hájek projections of the variables T_N
themselves give a better approximation but are more complicated.

13.23   Lemma. *If $\phi : [0, 1] \mapsto \mathbb{R}$ is twice continuously differentiable, then there exists a
universal constant K such that*

$$\mathrm{var}\,(T_N - \hat{T}_N) \le K\frac{1}{N}\sum_{i=1}^{N}(c_{Ni} - \overline{c}_N)^2\left(\|\phi'\|_\infty^2 + \|\phi''\|_\infty^2\right).$$

Proof.   Because the inequality is for every fixed N, we delete the index N in the proof.
Furthermore, because the assertion concerns a variance and both T_N and $\hat{T}_N$ change by a

constant if the c_{Ni} are replaced by $c_{Ni} - \bar{c}_N$, it is not a loss of generality to assume that $\bar{c}_N = 0$. (Evaluate the integral defining $\hat{T}_N$ to see this.)

The rank of X_i can be written as $R_i = 1 + \sum_{k \neq i} 1\{X_k \leq X_i\}$. This representation and a little algebra show that

$$\left| \mathrm{E}\left(\frac{R_i}{N+1} \,\bigg|\, X_i \right) - \overline{F}(X_i) \right| = \frac{1}{N+1} \left| 1 - \overline{F}(X_i) - F_i(X_i) \right| \leq \frac{1}{N}.$$

Furthermore, applying the Marcinkiewitz-Zygmund inequality (e.g., [23, p. 356]) conditionally on X_i, we obtain that

$$\mathrm{E}\left(\frac{R_i}{N+1} - \overline{F}(X_i) \right)^4$$

$$= \frac{1}{(N+1)^4} \mathrm{E}\left(\sum_{k \neq i} (1\{X_k \leq X_i\} - F_k(X_i)) + 1 - \overline{F}(X_i) - F_i(X_i) \right)^4$$

$$\lesssim \frac{1}{N^2} \mathrm{E}\mathrm{E}\left(\frac{1}{N} \sum_{k \neq i} (1\{X_k \leq X_i\} - F_k(X_i))^4 \,\bigg|\, X_i \right) + \frac{1}{N^4} \lesssim \frac{1}{N^2}.$$

Next, developing ϕ in a two-term Taylor expansion around $\overline{F}(X_i)$, for each term in the sum that defines T, we see that there exist random variables K_i that are bounded by $\|\phi''\|_\infty$ such that

$$T = \sum_{i=1}^{N} c_i \phi(\overline{F}(X_i)) + \sum_{i=1}^{N} c_i \left(\frac{R_{Ni}}{N+1} - \overline{F}(X_i) \right) \phi'(\overline{F}(X_i))$$

$$+ \sum_{i=1}^{N} c_i \left(\frac{R_{Ni}}{N+1} - \overline{F}(X_i) \right)^2 K_i \quad =: T_0 + T_1 + T_2.$$

Using the Cauchy-Schwarz inequality and the fourth-moment bound obtained previously, we see that the quadratic term T_2 is bounded above in second mean as in the lemma. The leading term T_0 is a sum of functions of the single variables X_i, and is the first part of $\hat{T}$. We shall show that the linear term T_1 is asymptotically equivalent to its Hájek projection, which, moreover, is asymptotically equivalent to the second part of $\hat{T}$, up to a constant. The Hájek projection of T_1 is equal to, up to a constant,

$$\sum_i c_i \sum_j \mathrm{E}\left[\frac{R_i}{N+1} \phi'(\overline{F}(X_i)) \,\bigg|\, X_j \right] - \sum_i c_i \overline{F}(X_i) \phi'(\overline{F}(X_i))$$

$$= \sum_i c_i \left[\sum_{j \neq i} \mathrm{E}\left[\frac{R_i}{N+1} \phi'(\overline{F}(X_i)) \,\bigg|\, X_j \right] \right]$$

$$+ \sum_i c_i \left(\mathrm{E}\left(\frac{R_i}{N+1} \,\bigg|\, X_i \right) - \overline{F}(X_i) \right) \phi'(\overline{F}(X_i)).$$

The second term is bounded in second mean as in the lemma; the first term is equal to

$$\frac{1}{N+1} \sum_i c_i \sum_{j \neq i} \mathrm{E}\left(1\{X_j \leq X_i\} \phi'(\overline{F}(X_i)) \,|\, X_j \right) + \text{constant}.$$

If we replace $(N + 1)$ by N, write out the conditional expectation, add the diagonal terms, and remove the constant, then we obtain the second term in the definition of $\hat{T}$. The difference between these two expressions is bounded above in second mean as in the lemma.

To conclude the proof it suffices to show that the difference between T_1 and its Hájek projection is negligible. We employ the Hoeffding decomposition. Because each of the variables $R_i \phi'\big(\overline{F}(X_i)\big)$ is contained in the space $\sum_{|A| \le 2} H_A$, the difference between T_1 and its Hájek projection is equal to the projection of T_1 onto the space $\sum_{|A| = 2} H_A$. This projection has second moment

$$\frac{1}{(N + 1)^2} \sum_{|A| = 2} \mathrm{E}\left(P_A \sum_i c_i \sum_k 1\{X_k \le X_i\} \phi'\big(\overline{F}(X_i)\big) \right)^2.$$

The projection of the variable $1\{X_k \le X_i\}\phi'\big(\overline{F}(X_i)\big)$, which is contained in the space $H_{\{k,i\}}$, onto the space $H_{\{a,b\}}$ is zero unless $\{a, b\} \subset \{k, i\}$. Thus, the expression in the preceding display is equal to

$$\frac{1}{(N + 1)^2} \sum_{a < b} \mathrm{E}\Big(c_b 1\{X_a \le X_b\} \phi'\big(\overline{F}(X_b)\big) + c_a 1\{X_b \le X_a\} \phi'\big(\bar{F}(X_a)\big) \Big)^2.$$

This is bounded by the upper bound of the lemma, as desired. The proof is complete. ∎

As a consequence of the lemma, the sequences $(T_N - \mathrm{E}T_N)/\mathrm{sd}\, T_N$ and $(\hat{T}_N - \mathrm{E}\hat{T}_N)/\mathrm{sd}\, \hat{T}_N$ have the same limiting distribution (if any) if

$$\frac{\sum_{i=1}^{N} (c_{Ni} - \bar{c}_N)^2}{N \operatorname{var} \hat{T}_N} \to 0.$$

This condition is certainly satisfied if the observations are identically distributed. Then the rank vector is uniformly distributed on the permutations, and the explicit expression for $\operatorname{var} T_N$ given by Lemma 13.1 shows that the left side (with $\operatorname{var} T_N$ instead of $\operatorname{var} \hat{T}_N$) is of the order $O(1/N)$. Because this leaves much too spare, the condition remains satisfied under small departures from identical distributions, but the general situation requires a calculation.

Under the conditions of the lemma we have the approximation

$$\mathrm{E}T_N \approx \bar{c}_N \sum_{i=1}^{N} \phi\left(\frac{i}{N+1} \right) + \sum_{i=1}^{N} (c_{Ni} - \bar{c}_N) \,\mathrm{E}\phi\big(\overline{F}_N(X_i)\big).$$

The square of the difference is bounded by the upper bound of the lemma.

The preceding lemma is restricted to smooth score-generating functions. One possibility to extend the result to more general scores is to show that the difference between the rank statistics of interest and suitable approximations by rank statistics with smooth scores is small. The following lemma is useful for this purpose, although it is suboptimal if the observations are identically distributed. (For a proof, see Theorem 3.1, in [68].)

13.24 **Lemma (Variance inequality).** *For nondecreasing coefficients* $a_{N1} \le \cdots \le a_{NN}$ *and arbitrary scores* $c_{N1}, \ldots, c_{NN}$,

$$\operatorname{var} \sum_{i=1}^{N} c_{Ni} a_{N, R_{N,i}} \le 21 \max_{1 \le i \le N} (c_{Ni} - \bar{c}_N)^2 \sum_{i=1}^{N} (a_{Ni} - \bar{a}_N)^2.$$

13.5 Permutation Tests

Rank tests are examples of *permutation tests*. General permutation tests also possess a distribution-free level but still use the values of the observations next to their ranks. In this section we illustrate this for the two-sample problem.

Suppose that the null hypothesis H_0 that two independent random samples $X_1, \ldots, X_m$ and $Y_1, \ldots, Y_n$ are identically distributed is rejected for large values of a test statistic $T_N(X_1, \ldots, X_m, Y_1, \ldots, Y_n)$. Write $Z_{(1)}, \ldots, Z_{(N)}$ for the values of the pooled sample stripped of its original order. ($N = m + n$.) Under the null hypothesis each permutation $Z_{\pi_1}, \ldots, Z_{\pi_N}$ of the N values is equally likely to lead back to the original observations. More precisely, the conditional null distribution of $X_1, \ldots, X_m, Y_1, \ldots, Y_n$ given $Z_{(1)}, \ldots, Z_{(N)}$ is uniform on the $N!$ permutations of the latter sample. Thus, it would be reasonable to reject H_0 if the observed value $T_N(x_1, \ldots, x_m, y_1, \ldots, y_n)$ is among the $100\alpha\%$ largest values $T_N(z_{\pi_1}, \ldots, z_{\pi_N})$ as π ranges over all permutations. Then we obtain a test of level α, conditionally given the observed values and hence also unconditionally.

Does this procedure work? Does the test have the desired power? The answer is affirmative for statistics T_N that are sums, in the sense that, asymptotically, the permutation test is equivalent to the test based on the normal approximation to T_N. If the latter test performs well, then so does the permutation test.

We consider statistics of the form, for a given measurable function f,

$$T_N(X_1, \ldots, X_m, Y_1, \ldots, Y_n) = \frac{1}{m}\sum_{i=1}^{m} f(X_i) - \frac{1}{n}\sum_{j=1}^{n} f(Y_j).$$

These statistics include, for instance, the score statistics for testing that the two samples have distributions p_0 and p_θ, respectively, for which we take f equal to the score function $\dot{p}_0/p_0$ of the model. Because a permutation test is conditional on the observed values, and T_N is fixed once $\sum_j f(Y_j)$ and $\sum_i f(Z_i)$ are fixed, it would be equivalent to consider statistics of the form $\sum_j f(Y_j)$.

Let $(\pi_{N1}, \ldots, \pi_{NN})$ be uniformly distributed on the $N!$ permutations of the numbers $1, 2, \ldots, N$, and be independent of $X_1, \ldots, X_m, Y_1, \ldots, Y_n$.

13.25   Theorem. *Let both* $\mathrm{E}f^2(X_1)$ *and* $\mathrm{E}f^2(Y_1)$ *be finite, and suppose that* $m, n \to \infty$ *such that* $m/N \to \lambda \in (0, 1)$. *Then, given almost every sequence* $X_1, X_2, \ldots, Y_1, Y_2, \ldots$, *the sequence* $\sqrt{N}T_N(Z_{\pi_{N1}}, \ldots, Z_{\pi_{NN}})$ *is asymptotically normal with mean zero. Under the null hypothesis the asymptotic variance is equal to* $\mathrm{var}\, f(X_1)/(\lambda(1 - \lambda))$.

Proof. Conditionally on the values of the pooled sample, the statistic $NT_N(Z_{\pi_{N1}}, \ldots, Z_{\pi_{NN}})$ is distributed as the simple linear rank statistic $\sum_{i=1}^{N} c_{Ni}a_{N,R_{Ni}}$ with coefficients and scores

$$c_{Ni} = f(Z_i), \qquad a_{Ni} = \begin{cases} \dfrac{N}{m}, & i \le m \\[2mm] -\dfrac{N}{n}, & i > m \end{cases}$$

Here $R_{N1}, \ldots, R_{NN}$ are the antiranks of $\pi_{N1}, \ldots, \pi_{NN}$ defined by the equation $\sum c_{N,\pi_{Ni}}a_{Ni} = \sum c_{Ni}a_{N,R_{Ni}}$ (for any numbers c_{Ni} and a_{Ni}).

The coefficients satisfy relation (13.7) for almost every sequence $X_1, X_2, \ldots, Y_1, Y_2, \ldots$, because, by the law of large numbers,

$$\overline{c_N^k} \overset{as}{\to} \lambda E f^k(X_1) + (1-\lambda) E f^k(Y_1), \qquad k = 1, 2,$$

$$\frac{1}{N} \max_{1 \le i \le N} c_{Ni}^2 \overset{as}{\to} 0.$$

The scores are generated as $a_{Ni} = \phi_N\big(i/(N+1)\big)$ for the functions

$$\phi_N(u) = \begin{cases} \dfrac{N}{m}, & u \le \dfrac{m}{N+1}, \\[2mm] -\dfrac{N}{n}, & u > \dfrac{m}{N+1}. \end{cases}$$

These functions depend on N, unlike the situation of Theorem 13.5, but they converge to the fixed function $\phi = \lambda^{-1} 1_{[0,\lambda)} - (1-\lambda)^{-1} 1_{(\lambda,1]}$. By a minor extension of Theorem 13.5, the sequence $\sum c_{Ni} a_{N,R_{Ni}}$ is asymptotically equivalent to $\sum (c_{Ni} - \bar{c}_N) \phi(U_i)$, for a uniform sample $U_1, \ldots, U_N$. The (asymptotic) variance of the latter variable is easy to compute. ∎

By the central limit theorem, under the null hypothesis,

$$\sqrt{N} T_N(X_1, \ldots, X_m, Y_1, \ldots, Y_n) \rightsquigarrow N(0, \sigma^2), \qquad \sigma^2 = \frac{\operatorname{var} f(X_1)}{\lambda(1-\lambda)}.$$

The limit is the same as the conditional limit distribution of the sequence $\sqrt{N} T_N(Z_{\pi_{N1}}, \ldots, Z_{\pi_{NN}})$ under the null hypothesis. Thus, we have a choice of two sequences of tests, both of asymptotic level α, rejecting H_0 if:
- $\sqrt{N} T_N(X_1, \ldots, X_m, Y_1, \ldots, Y_n) \ge z_\alpha \sigma$; or
- $\sqrt{N} T_N(X_1, \ldots, X_m, Y_1, \ldots, Y_n) \ge c_N(X_1, \ldots, X_m, Y_1, \ldots, Y_n)$, where
  $c_N(X_1, \ldots, X_m, Y_1, \ldots, Y_n)$ is the upper α-quantile of the conditional
  distribution of $\sqrt{N} T_N (Z_{\pi_{N1}}, \ldots, Z_{\pi_{NN}})$ given $Z_{(1)}, \ldots, Z_{(N)}$.

The second test is just the permutation test discussed previously. By the preceding theorem the "random critical values" $c_N(X_1, \ldots, X_m, Y_1, \ldots, Y_n)$ converge in probability to $z_\alpha \sigma$ under H_0. Therefore the two tests are asymptotically equivalent under the null hypothesis. Furthermore, this equivalence remains under "contiguous alternatives" (for which again $c_N(X_1, \ldots, X_m, Y_1, \ldots, Y_n) \overset{P}{\to} z_\alpha \sigma$; see Chapter 6), and hence the local asymptotic power functions as discussed in Chapter 14 are the same for the two sequences of tests.

The preceding theorem also shows that the sequence of "critical values" $c_N(X_1, \ldots, X_m, Y_1, \ldots, Y_n)$ remains bounded in probability under every alternative. Because $\sqrt{N} T_N (X_1, \ldots, X_m, Y_1, \ldots, Y_n) \rightsquigarrow \infty$ if $E f(X_1) > E f(Y_1)$, the power at any alternative with this property converges to 1. Thus, permutation tests are an attractive alternative to both rank and classical tests. Their main drawback is computational complexity. The dependence of the null distribution on the observed values means that it cannot be tabulated and must be computed for every new data set.

*13.6   Rank Central Limit Theorem

The rank central limit theorem Theorem 13.5, is slightly special in that the scores a_{Ni} are assumed to be of one of the forms (13.3) or (13.4). In this section we record what is commonly viewed as the rank central limit theorem. For a proof see [67]. For given coefficients and scores, let

$$C_n^2 = \sum_{i=1}^n (c_{Ni} - \bar{c}_N)^2, \qquad A_n^2 = \sum_{i=1}^n (a_{Ni} - \bar{a}_N)^2.$$

13.26   Theorem (Rank central limit theorem). *Let* $T_N = \sum c_{Ni} a_{N,R_{Ni}}$ *be the simple linear rank statistic with coefficients and scores such that* $\max_{1 \le i \le N} |a_{Ni} - \bar{a}_N|/A_N \to 0$ *and* $\max_{1 \le i \le N} |c_{Ni} - \bar{c}_N|/C_N \to 0$, *and let the rank vector* R_N *be uniformly distributed on the set of all* $N!$ *permutations of* $\{1, 2, \ldots, N\}$. *Then the sequence* $(T_N - ET_N)/\operatorname{sd} T_N$ *converges in distribution to a standard normal distribution if and only if, for every* $\varepsilon > 0$,

$$\sum_{(i,j): \sqrt{N}|a_{Ni}-\bar{a}_N||c_{Ni}-\bar{c}_N| > \varepsilon A_N C_N} \frac{|a_{Ni} - \bar{a}_N|^2 |c_{Ni} - \bar{c}_N|^2}{A_N^2 C_N^2} \to 0.$$

Notes

The classical reference on rank statistics is the book by Hájek and Šidák [71], which still makes wonderful reading and gives extensive references. Its treatment of rank statistics for nonidentically distributed observations is limited to contiguous alternatives, as in the first sections of this chapter. The papers [43] and [68] remedied this, shortly after the publication of the book. Section 13.4 reports only a few of the results from these papers, which, as does the book, use the projection method. An alternative approach to obtaining the limit distribution of rank statistics, initiated by Chernoff and Savage in the late 1950s and refined many times, is to write them as functions of empirical measures and next apply the von Mises method. We discuss examples of this approach in Chapter 20. See [134] for a more comprehensive treatment and further references.

PROBLEMS

1. This problem asks one to give a precise meaning to the notion of a *locally most powerful test*. Let T_N be a rank statistic based on the "locally most powerful scores." Let $\alpha = P_0(T_N > c_\alpha)$ for a given number c_α. (Then α is a *natural level* of the test statistic, a level that is attained without randomization.) Then there exists $\varepsilon > 0$ such that the test that rejects the null hypothesis if $T_N > c_\alpha$ is most powerful within the class of all rank tests at level α uniformly in the alternatives $\theta \in (0, \varepsilon)$.
   (i) Prove the statement.
   (ii) Can the statement be extended to arbitrary levels?

2. Find the asymptotic distribution of the median test statistic under the null hypothesis that the two samples are identically distributed and continuous.

3. Show that $\sqrt{n}$ times Spearman's rank correlation coefficient is asymptotically standard normal.

4. Find the scores for a locally most powerful two-sample rank test for location for the Laplace family of densities.

5. Find the scores for a locally most powerful two-sample rank test for location for the Cauchy family of densities.

6. For which density is the Wilcoxon signed rank statistic locally most powerful?

7. Show that Spearman's rank correlation coefficient is a linear combination of Kendall's τ and the U-statistic with (asymmetric) kernel $h(x, y, z) = \text{sign}(x_1 - y_1) \, \text{sign}(x_2 - z_2)$. This decomposition yields another method to prove the asymptotic normality.

8. The symmetrized *Siegel-Tukey test* is a two-sample test with score vector of the form $a_N = (1, 3, 5, \ldots, 5, 3, 1)$. For which type of alternative hypothesis would you use this test?

9. For any a_{Ni} given by (13.3), show that $\bar{a}_N = \int_0^1 \phi(u) \, du$.

14

Relative Efficiency of Tests

The quality of sequences of tests can be judged from their power at alternatives that become closer and closer to the null hypothesis. This motivates the study of local asymptotic power functions. The relative efficiency of two sequences of tests is the quotient of the numbers of observations needed with the two tests to obtain the same level and power. We discuss several types of asymptotic relative efficiencies.

14.1 Asymptotic Power Functions

Consider the problem of testing a null hypothesis $H_0 : \theta \in \Theta_0$ versus the alternative $H_1 : \theta \in \Theta_1$. The power function of a test that rejects the null hypothesis if a test statistic falls into a *critical region* K_n is the function $\theta \mapsto \pi_n(\theta) = P_\theta(T_n \in K_n)$, which gives the probability of rejecting the null hypothesis. The test is of *level* α if its *size* $\sup\{\pi_n(\theta) : \theta \in \Theta_0\}$ does not exceed α. A sequence of tests is called *asymptotically of level* α if

$$\limsup_{n \to \infty} \sup_{\theta \in \Theta_0} \pi_n(\theta) \leq \alpha.$$

(An alternative definition is to drop the supremum and require only that $\limsup \pi_n(\theta) \leq \alpha$ for every $\theta \in \Theta_0$.) A test with power function π_n is better than a test with power function $\underline{\pi}_n$ if both

$$\pi_n(\theta) \leq \underline{\pi}_n(\theta), \qquad \theta \in \Theta_0,$$
$$\text{and} \quad \pi_n(\theta) \geq \underline{\pi}_n(\theta), \qquad \theta \in \Theta_1.$$

The aim of this chapter is to compare tests asymptotically. We consider sequences of tests with power functions π_n and $\underline{\pi}_n$ and wish to decide which of the sequences is best as $n \to \infty$. Typically, the tests corresponding to a sequence $\pi_1, \pi_2, \ldots$ are of the same type. For instance, they are all based on a certain U-statistic or rank statistic, and only the number of observations changes with n. Otherwise the comparison would have little relevance.

A first idea is to consider limiting power functions of the form

$$\pi(\theta) = \lim_{n \to \infty} \pi_n(\theta).$$

If this limit exists for all θ, and the same is true for the competing tests $\underline{\pi}_n$, then the sequence π_n is better than the sequence $\underline{\pi}_n$ if the limiting power function π is better than the limiting power function $\underline{\pi}$. It turns out that this approach is too naive. The limiting power functions typically exist, but they are trivial and identical for all reasonable sequences of tests.

14.1 Example (Sign test). Suppose the observations $X_1, \ldots, X_n$ are a random sample from a distribution with unique median θ. The null hypothesis $H_0 : \theta = 0$ can be tested against the alternative $H_1 : \theta > 0$ by means of the *sign statistic* $S_n = n^{-1} \sum_{i=1}^{n} 1\{X_i > 0\}$. If $F(x - \theta)$ is the distribution function of the observations, then the expectation and variance of S_n are equal to $\mu(\theta) = 1 - F(-\theta)$ and $\sigma^2(\theta)/n = \big(1 - F(-\theta)\big) F(-\theta)/n$, respectively. By the normal approximation to the binomial distribution, the sequence $\sqrt{n}\big(S_n - \mu(\theta)\big)$ is asymptotically normal $N\big(0, \sigma^2(\theta)\big)$. Under the null hypothesis the mean and variance are equal to $\mu(0) = 1/2$ and $\sigma^2(0) = 1/4$, respectively, so that $\sqrt{n}(S_n - 1/2) \overset{0}{\rightsquigarrow} N(0, 1/4)$. The test that rejects the null hypothesis if $\sqrt{n}(S_n - 1/2)$ exceeds the critical value $\frac{1}{2} z_\alpha$ has power function

$$\pi_n(\theta) = P_\theta \Big(\sqrt{n}\big(S_n - \mu(\theta)\big) > \tfrac{1}{2} z_\alpha - \sqrt{n}\big(\mu(\theta) - \mu(0)\big) \Big)$$

$$= 1 - \Phi \left(\frac{\tfrac{1}{2} z_\alpha - \sqrt{n}\big(F(0) - F(-\theta)\big)}{\sigma(\theta)} \right) + o(1).$$

Because $F(0) - F(-\theta) > 0$ for every $\theta > 0$, it follows that for $\alpha = \alpha_n \to 0$ sufficiently slowly

$$\pi_n(\theta) \to \begin{cases} 0 & \text{if } \theta = 0, \\ 1 & \text{if } \theta > 0. \end{cases}$$

The limit power function corresponds to the perfect test with all error probabilities equal to zero. $\square$

The example exhibits a sequence of tests whose (pointwise) limiting power function is the perfect power function. This type of behavior is typical for all reasonable tests. The point is that, with arbitrarily many observations, it should be possible to tell the null and alternative hypotheses apart with complete accuracy. The power at every fixed alternative should therefore converge to 1.

14.2 Definition. A sequence of tests with power functions $\theta \mapsto \pi_n(\theta)$ is asymptotically *consistent* at level α at (or against) the alternative θ if it is asymptotically of level α and $\pi_n(\theta) \to 1$. If a family of sequences of tests contains for every level $\alpha \in (0, 1)$ a sequence that is consistent against every alternative, then the corresponding tests are simply called consistent.

Consistency is an optimality criterion for tests, but because most sequences of tests are consistent, it is too weak to be really useful. To make an informative comparison between sequences of (consistent) tests, we shall study the performance of the tests in problems that become harder as more observations become available. One way of making a testing problem harder is to choose null and alternative hypotheses closer to each other. In this section we fix the null hypothesis and consider the power at sequences of alternatives that converge to the null hypothesis.

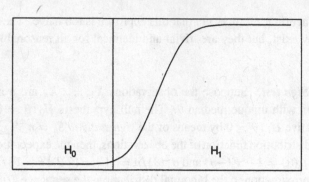

Figure 14.1. Asymptotic power function.

14.3 *Example (Sign test, continued).* Consider the power of the sign test at sequences of alternatives $\theta_n \downarrow 0$. Suppose that the null hypothesis $H_0 : \theta = 0$ is rejected if $\sqrt{n}(S_n - \frac{1}{2}) \geq \frac{1}{2}z_\alpha$. Extension of the argument of the preceding example yields

$$\pi_n(\theta_n) = 1 - \Phi\left(\frac{\frac{1}{2}z_\alpha - \sqrt{n}\big(F(0) - F(-\theta_n)\big)}{\sigma(\theta_n)}\right) + o(1).$$

Since $\sigma(0) = \frac{1}{2}$, the levels $\pi_n(0)$ of the tests converge to $\Phi(z_\alpha) = \alpha$. The asymptotic power at θ_n depends on the rate at which $\theta_n \to 0$. If θ_n converges to zero fast enough to ensure that $\sqrt{n}\big(F(0) - F(-\theta_n)\big) \to 0$, then the power $\pi_n(\theta_n)$ converges to α: the sign test is not able to discriminate these alternatives from the null hypothesis. If θ_n converges to zero at a slow rate, then $\sqrt{n}\big(F(0) - F(-\theta_n)\big) \to \infty$, and the asymptotic power is equal to 1: these alternatives are too easy. The intermediate rates, which yield a nontrivial asymptotic power, appear to be of most interest. Suppose that the underlying distribution function F is differentiable at zero with positive derivative $f(0) > 0$. Then

$$\sqrt{n}\big(F(0) - F(-\theta_n)\big) = \sqrt{n}\,\theta_n f(0) + \sqrt{n}\,o(\theta_n).$$

This is bounded away from zero and infinity if θ_n converges to zero at rate $\theta_n = O(n^{-1/2})$. For such rates the power $\pi_n(\theta_n)$ is asymptotically strictly between α and 1. In particular, for every h,

$$\pi_n\left(\frac{h}{\sqrt{n}}\right) \to 1 - \Phi\big(z_\alpha - 2hf(0)\big).$$

The form of the limit power function is shown in Figure 14.1. $\square$

In the preceding example only alternatives θ_n that converge to the null hypothesis at rate $O(1/\sqrt{n})$ lead to a nontrivial asymptotic power. This is typical for parameters that depend "smoothly" on the underlying distribution. In this situation a reasonable method for asymptotic comparison of two sequences of tests for $H_0 : \theta = 0$ versus $H_0 : \theta > 0$ is to consider *local limiting power* functions, defined as

$$\pi(h) = \lim_{n \to \infty} \pi_n\left(\frac{h}{\sqrt{n}}\right), \qquad h \geq 0.$$

These limits typically exist and can be derived by the same method as in the preceding example. A general scheme is as follows.

Let θ be a real parameter and let the tests reject the null hypothesis $H_0 : \theta = 0$ for large values of a test statistic T_n. Assume that the sequence T_n is asymptotically normal in the sense that, for all sequences of the form $\theta_n = h/\sqrt{n}$,

$$\frac{\sqrt{n}\big(T_n - \mu(\theta_n)\big)}{\sigma(\theta_n)} \overset{\theta_n}{\rightsquigarrow} N(0, 1). \tag{14.4}$$

Often $\mu(\theta)$ and $\sigma^2(\theta)$ can be taken to be the mean and the variance of T_n, but this is not necessary. Because the convergence (14.4) is under a law indexed by θ_n that changes with n, the convergence is not implied by

$$\frac{\sqrt{n}\big(T_n - \mu(\theta)\big)}{\sigma(\theta)} \overset{\theta}{\rightsquigarrow} N(0, 1), \qquad \text{every } \theta. \tag{14.5}$$

On the other hand, this latter convergence uniformly in the parameter θ is more than is needed in (14.4). The convergence (14.4) is sometimes referred to as "locally uniform" asymptotic normality. "Contiguity arguments" can reduce the derivation of asymptotic normality under $\theta_n = h/\sqrt{n}$ to derivation under $\theta = 0$. (See section 14.1.1).

Assumption (14.4) includes that the sequence $\sqrt{n}\big(T_n - \mu(0)\big)$ converges in distribution to a normal $N\big(0, \sigma^2(0)\big)$-distribution under $\theta = 0$. Thus, the tests that reject the null hypothesis $H_0 : \theta = 0$ if $\sqrt{n}\big(T_n - \mu(0)\big)$ exceeds $\sigma(0)z_\alpha$ are asymptotically of level α. The power functions of these tests can be written

$$\pi_n(\theta_n) = P_{\theta_n}\Big(\sqrt{n}\big(T_n - \mu(\theta_n)\big) > \sigma(0)z_\alpha - \sqrt{n}\big(\mu(\theta_n) - \mu(0)\big)\Big).$$

For $\theta_n = h/\sqrt{n}$, the sequence $\sqrt{n}\big(\mu(\theta_n) - \mu(0)\big)$ converges to $h\mu'(0)$ if μ is differentiable at zero. If $\sigma(\theta_n) \to \sigma(0)$, then under (14.4)

$$\pi_n\!\left(\frac{h}{\sqrt{n}}\right) \to 1 - \Phi\!\left(z_\alpha - h\frac{\mu'(0)}{\sigma(0)}\right). \tag{14.6}$$

For easy reference we formulate this result as a theorem.

14.7   Theorem. *Let μ and σ be functions of θ such that (14.4) holds for every sequence $\theta_n = h/\sqrt{n}$. Suppose that μ is differentiable and that σ is continuous at $\theta = 0$. Then the power functions π_n of the tests that reject $H_0 : \theta = 0$ for large values of T_n and are asymptotically of level α satisfy (14.6) for every h.*

The limiting power function depends on the sequence of test statistics only through the quantity $\mu'(0)/\sigma(0)$. This is called the *slope* of the sequence of tests. Two sequences of tests can be asymptotically compared by just comparing the sizes of their slopes. The bigger the slope, the better the test for $H_0 : \theta = 0$ versus $H_1 : \theta > 0$. The size of the slope depends on the rate $\mu'(0)$ of change of the asymptotic mean of the test statistics relative to their asymptotic dispersion $\sigma(0)$. A good quantitative measure of comparison is the square of the quotient of two slopes. This quantity is called the *asymptotic relative efficiency* and is discussed in section 14.3.

If θ is the only unknown parameter in the problem, then the available tests can be ranked in asymptotic quality simply by the value of their slopes. In many problems there are also nuisance parameters (for instance the shape of a density), and the slope is a function of the nuisance parameter rather than a number. This complicates the comparison considerably. For every value of the nuisance parameter a different test may be best, and additional criteria are needed to choose a particular test.

14.8   Example (Sign test). According to Example 14.3, the sign test has slope $2f(0)$. This can also be obtained from the preceding theorem, in which we can choose $\mu(\theta) = 1 - F(-\theta)$ and $\sigma^2(\theta) = \big(1 - F(-\theta)\big) F(-\theta)$.   □

14.9   Example (t-test). Let $X_1, \ldots, X_n$ be a random sample from a distribution with mean θ and finite variance. The t-test rejects the null hypothesis for large values of Σ. The sample variance S^2 converges in probability to the variance σ^2 of a single observation. The central limit theorem and Slutsky's lemma give

$$\sqrt{n}\left(\frac{\overline{X}}{S} - \frac{h/\sqrt{n}}{\sigma}\right) = \frac{\sqrt{n}(\overline{X} - h/\sqrt{n})}{S} + h\left(\frac{1}{S} - \frac{1}{\sigma}\right) \overset{h/\sqrt{n}}{\rightsquigarrow} N(0, 1).$$

Thus Theorem 14.7 applies with $\mu(\theta) = \theta/\sigma$ and $\sigma(\theta) = 1$. The slope of the t-test equals $1/\sigma$.[†]   □

14.10   Example (Sign versus t-test). Let $X_1, \ldots, X_n$ be a random sample from a density $f(x - \theta)$, where f is symmetric about zero. We shall compare the performance of the sign test and the t-test for testing the hypothesis $H_0 : \theta = 0$ that the observations are symmetrically distributed about zero. Assume that the distribution with density f has a unique median and a finite second moment.

It suffices to compare the slopes of the two tests. By the preceding examples these are $2f(0)$ and $\left(\int x^2 f(x)\, dx\right)^{-1/2}$, respectively. Clearly the outcome of the comparison depends on the shape f. It is interesting that the two slopes depend on the underlying shape in an almost orthogonal manner. The slope of the sign test depends only on the height of f at zero; the slope of the t-test depends mainly on the tails of f. For the standard normal distribution the slopes are $\sqrt{2/\pi}$ and 1. The superiority of the t-test in this case is not surprising, because the t-test is uniformly most powerful for every n. For the Laplace distribution, the ordering is reversed: The slopes are 1 and $\frac{1}{2}\sqrt{2}$. The superiority of the sign test has much to do with the "unsmooth" character of the Laplace density at its mode.

The relative efficiency of the sign test versus the t-test is equal to

$$4f^2(0) \int x^2 f(x)\, dx.$$

Table 14.1 summarizes these numbers for a selection of shapes. For the uniform distribution, the relative efficiency of the sign test with respect to the t-test equals $1/3$. It can be shown that this is the minimal possible value over all densities with mode zero (problem 14.7). On the other hand, it is possible to construct distributions for which this relative efficiency is arbitrarily large, by shifting mass into the tails of the distribution. The sign test is "robust" against heavy tails, the t-test is not.   □

The simplicity of comparing slopes is attractive on the one hand, but indicates the potential weakness of asymptotics on the other. For instance, the slope of the sign test was seen to be $f(0)$, but it is clear that this value alone cannot always give an accurate indication

---

[†] Although (14.4) holds with this choice of μ and σ, it is not true that the sequence $\sqrt{n}(\overline{X}/S - \theta/\sigma)$ is asymptotically standard normal for every fixed θ. Thus (14.5) is false for this choice of μ and σ. For fixed θ the contribution of $S - \sigma$ to the limit distribution cannot be neglected, but for our present purpose it can.

Table 14.1. *Relative efficiencies of the sign test versus the t-test for some distributions.*

Distribution	Efficiency (sign/t-test)
Logistic	$\pi^2/12$
Normal	$2/\pi$
Laplace	2
Uniform	$1/3$

of the quality of the sign test. Consider a density that is basically a normal density, but a tiny proportion of $10^{-10}\%$ of its total mass is located under an extremely thin but enormously high peak at zero. The large value $f(0)$ would strongly favor the sign test. However, at moderate sample sizes the observations would not differ significantly from a sample from a normal distribution, so that the t-test is preferable. In this situation the asymptotics are only valid for unrealistically large sample sizes.

Even though asymptotic approximations should always be interpreted with care, in the present situation there is actually little to worry about. Even for $n = 20$, the comparison of slopes of the sign test and the t-test gives the right message for the standard distributions listed in Table 14.1.

14.11 *Example (Mann-Whitney).* Suppose we observe two independent random samples $X_1, \ldots, X_m$ and $Y_1, \ldots, Y_n$ from distributions $F(x)$ and $G(y - \theta)$, respectively. The base distributions F and G are fixed, and it is desired to test the null hypothesis $H_0 : \theta = 0$ versus the alternative $H_1 : \theta > 0$. Set $N = m + n$ and assume that $m/N \to \lambda \in (0, 1)$. Furthermore, assume that G has a bounded density g.

The Mann-Whitney test rejects the null hypothesis for large numbers of $U = (mn)^{-1} \sum_i \sum_j 1\{X_i \le Y_j\}$. By the two-sample U-statistic theorem

$$\sqrt{N}\big(U - P_\theta(X \le Y)\big) = -\frac{\sqrt{N}}{m} \sum_{i=1}^{m} \big(G(X_i - \theta) - EG(X_i - \theta)\big)$$
$$+ \frac{\sqrt{N}}{n} \sum_{j=1}^{n} \big(F(Y_i) - EF(Y_i)\big) + o_{P_\theta}(1).$$

This readily yields the asymptotic normality (14.5) for every fixed θ, with

$$\mu(\theta) = 1 - \int G(x - \theta)\, dF(x), \qquad \sigma^2(\theta) = \frac{1}{\lambda}\, \text{var}\, G(X - \theta) + \frac{1}{1 - \lambda} \text{var}\, F(Y).$$

To obtain the local asymptotic power function, this must be extended to sequences $\theta_N = h/\sqrt{N}$. It can be checked that the U-statistic theorem remains valid and that the Lindeberg central limit theorem applies to the right side of the preceding display with θ_N replacing θ. Thus, we find that (14.4) holds with the same functions μ and σ. (Alternatively we can use contiguity and Le Cam's third lemma.) Hence, the slope of the Mann-Whitney test equals $\mu'(0)/\sigma(0) = \int g\, dF/\sigma(0)$. $\square$

14.12 *Example (Two-sample t-test).* In the set-up of the preceding example suppose that the base distributions F and G have equal means and finite variances. Then $\theta = E(Y - X)$

Table 14.2. *Relative efficiencies of the Mann-Whitney*
test versus the two-sample t-test if $f = g$ equals
a number of distributions.

Distribution	Efficiency (Mann-Whitney/two-sample t-test)
Logistic	$\pi^2/9$
Normal	$3/\pi$
Laplace	$3/2$
Uniform	1
t_3	1.24
t_5	1.90
$c(1 - x^2) \vee 0$	$108/125$

and the t-test rejects the null hypothesis $H_0 : \theta = 0$ for large values of the statistic $(\bar{Y} - \bar{X})/S$, where $S^2/N = S_X^2/m + S_Y^2/n$ is the unbiased estimator of $\operatorname{var}(\bar{Y} - \bar{X})$. The sequence S^2 converges in probability to $\sigma^2 = \operatorname{var} X/\lambda + \operatorname{var} Y/(1 - \lambda)$. By Slutsky's lemma and the central limit theorem

$$\sqrt{N}\left(\frac{\bar{Y} - \bar{X}}{S} - \frac{h/\sqrt{N}}{\sigma}\right) \overset{h/\sqrt{N}}{\rightsquigarrow} N(0, 1).$$

Thus (14.4) is satisfied and Theorem 14.7 applies with $\mu(\theta) = \theta/\sigma$ and $\sigma(\theta) = 1$. The slope of the t-test equals $\mu'(0)/\sigma(0) = 1/\sigma$. $\square$

14.13   Example (*t-Test versus Mann-Whitney test*). Suppose we observe two independent random samples $X_1, \ldots, X_m$ and $Y_1, \ldots, Y_n$ from distributions $F(x)$ and $G(x - \theta)$, respectively. The base distributions F and G are fixed and are assumed to have equal means and bounded densities. It is desired to test the null hypothesis $H_0 : \theta = 0$ versus the alternative $H_1 : \theta > 0$. Set $N = m + n$ and assume that $m/N \to \lambda \in (0, 1)$.

The slopes of the Mann-Whitney test and the t-test depend on the nuisance parameters F and G. According to the preceding examples the relative efficiency of the two sequences of tests equals

$$\frac{\big((1 - \lambda) \operatorname{var} X + \lambda \operatorname{var} Y\big)\big(\int g \, dF\big)^2}{(1 - \lambda) \operatorname{var}_0 G(X) + \lambda \operatorname{var}_0 F(Y)}.$$

In the important case that $F = G$, this expression simplifies. Then the variables $G(X)$ and $F(Y)$ are uniformly distributed on $[0, 1]$. Hence they have variance $1/12$ and the relative efficiency reduces to $12 \operatorname{var} X \big(\int f^2(y) \, dy\big)^2$. Table 14.2 gives the relative efficiency if $F = G$ are both equal to a number of standard distributions. The Mann-Whitney test is inferior to the t-test if $F = G$ equals the normal distribution, but better for the logistic, Laplace, and t-distribution. Even for the normal distribution the Mann-Whitney test does remarkably well, with a relative efficiency of $3/\pi \approx 95\%$. The density that is proportional to $(1 - x^2) \vee 0$ (and any member of its scale family) is least favorable for the Mann-Whitney test. This density yields the lowest possible relative efficiency, which is still equal to $108/125 \approx 86\%$ (problem 14.8). On the other hand, the relative efficiency of the Mann-Whitney test is large for heavy-tailed distributions; the supremum value is infinite. Together with the fact that the Mann-Whitney test is distribution-free under the null hypothesis, this makes the Mann-Whitney test a strong competitor to the t-test, even in situations in which the underlying distribution is thought to be approximately normal. $\square$

*14.1.1 *Using Le Cam's Third Lemma*

In the preceding examples the asymptotic normality of sequences of test statistics was established by direct methods. For more complicated test statistics the validity of (14.4) is easier checked by means of Le Cam's third lemma. This is illustrated by the following example.

14.14 Example (Median test). In the two-sample set-up of Example 14.11, suppose that $F = G$ is a continuous distribution function with finite Fisher information for location I_g. The median test rejects the null hypothesis $H_0 : \theta = 0$ for large values of the rank statistic $T_N = N^{-1} \sum_{i=m+1}^{N} 1\{R_{Ni} \leq (N+1)/2\}$. By the rank central limit theorem, Theorem 13.5, under the null hypothesis,

$$\sqrt{N}\left(T_N - \frac{n}{2N}\right) = -\frac{n}{N\sqrt{N}} \sum_{i=1}^{m} 1\{F(X_i) \leq 1/2\}$$

$$+ \frac{m}{N\sqrt{N}} \sum_{j=1}^{n} 1\{F(Y_j) \leq 1/2\} + o_P(1).$$

Under the null hypothesis the sequence of variables on the right side is asymptotically normal with mean zero and variance $\sigma^2(0) = \lambda(1 - \lambda)/4$. By Theorem 7.2, for every $\theta_N = h/\sqrt{N}$,

$$\log \frac{\prod_i f(X_i) \prod_j g(Y_j - \theta_N)}{\prod_i f(X_i) \prod_j g(Y_j)} = -\frac{h\sqrt{1-\lambda}}{\sqrt{n}} \sum_{j=1}^{n} \frac{g'}{g}(Y_i) - \frac{1}{2}h^2(1-\lambda)I_g + o_P(1).$$

By the multivariate central limit theorem, the linear approximations on the right sides of the two preceding displays are jointly asymptotically normal. By Slutsky's lemma the same is true for the left sides. Consequently, by Le Cam's third lemma the sequence $\sqrt{N}(T_N - n/(2N))$ converges under the alternatives $\theta_N = h/\sqrt{N}$ in distribution to a normal distribution with variance $\sigma^2(0)$ and mean the asymptotic covariance $\tau(h)$ of the linear approximations. This is given by

$$\tau(h) = -h\lambda(1-\lambda) \int_{F(y)\leq 1/2} \frac{f'}{f}(y) \, dF(y).$$

Conclude that (14.4) is valid with $\mu(\theta) = \tau(\theta)$ and $\sigma(\theta) = \sigma(0)$. (Use the test statistics $T_N - n/(2N)$ rather than T_N.) The slope of the median test is given by $-2\sqrt{\lambda(1-\lambda)} \int_0^{1/2} (f'/f)(F^{-1}(u)) \, du$. $\square$

14.2 Consistency

After noting that the power at fixed alternatives typically tends to 1, we focused attention on the performance of tests at alternatives converging to the null hypothesis. The comparison of local power functions is only of interest if the sequences of tests are consistent at fixed alternatives. Fortunately, establishing consistency is rarely a problem. The following lemmas describe two basic methods.

14.15   Lemma. *Let T_n be a sequence of statistics such that $T_n \xrightarrow{P_\theta} \mu(\theta)$ for every θ. Then the family of tests that reject the null hypothesis $H_0 : \theta = 0$ for large values of T_n is consistent against every θ such that $\mu(\theta) > \mu(0)$.*

14.16   Lemma. *Let μ and σ be functions of θ such that (14.4) holds for every sequence $\theta_n = h/\sqrt{n}$. Suppose that μ is differentiable and that σ is continuous at zero, with $\mu'(0) > 0$ and $\sigma(0) > 0$. Suppose that the tests that reject the null hypothesis for large values of T_n possess nondecreasing power functions $\theta \mapsto \pi_n(\theta)$. Then this family of tests is consistent against every alternative $\theta > 0$. Moreover, if $\pi_n(0) \to \alpha$, then $\pi_n(\theta_n) \to \alpha$ or $\pi_n(\theta_n) \to 1$ when $\sqrt{n}\,\theta_n \to 0$ or $\sqrt{n}\,\theta_n \to \infty$, respectively.*

Proofs.   For the first lemma, suppose that the tests reject the null hypothesis if T_n exceeds the critical value c_n. By assumption, the probability under $\theta = 0$ that T_n is outside the interval $\big(\mu(0) - \varepsilon, \mu(0) + \varepsilon\big)$ converges to zero as $n \to \infty$, for every fixed $\varepsilon > 0$. If the asymptotic level $\lim P_0(T_n > c_n)$ is positive, then it follows that $c_n < \mu(0) + \varepsilon$ eventually. On the other hand, under θ the probability that T_n is in $\big(\mu(\theta) - \varepsilon, \mu(\theta) + \varepsilon\big)$ converges to 1. For sufficiently small ε and $\mu(\theta) > \mu(0)$, this interval is to the right of $\mu(0) + \varepsilon$. Thus for sufficiently large n, the power $P_\theta(T_n > c_n)$ can be bounded below by $P_\theta\big(T_n \in \big(\mu(\theta) - \varepsilon, \mu(\theta) + \varepsilon\big)\big) \to 1$.

For the proof of the second lemma, first note that by Theorem 14.7 the sequence of local power functions $\pi_n(h/\sqrt{n})$ converges to $\pi(h) = 1 - \Phi\big(z_\alpha - h\mu'(0)/\sigma(0)\big)$, for every h, if the asymptotic level is α. If $\sqrt{n}\,\theta_n \to 0$, then eventually $\theta_n < h/\sqrt{n}$ for every given $h > 0$. By the monotonicity of the power functions, $\pi_n(\theta_n) \le \pi_n(h/\sqrt{n})$ for sufficiently large n. Thus $\limsup \pi_n(\theta_n) \le \pi(h)$ for every $h > 0$. For $h \downarrow 0$ the right side converges to $\pi(0) = \alpha$. Combination with the inequality $\pi_n(\theta_n) \ge \pi_n(0) \to \alpha$ gives $\pi_n(\theta_n) \to \alpha$. The case that $\sqrt{n}\,\theta_n \to \infty$ can be handled similarly. Finally, the power $\pi_n(\theta)$ at fixed alternatives is bounded below by $\pi_n(\theta_n)$ eventually, for every sequence $\theta_n \downarrow 0$. Thus $\pi_n(\theta) \to 1$, and the sequence of tests is consistent at θ. ∎

The following examples show that the t-test and Mann-Whitney test are both consistent against large sets of alternatives, albeit not exactly the same sets. They are both tests to compare the locations of two samples, but the pertaining definitions of "location" are not the same. The t-test can be considered a test to detect a difference in mean; the Mann-Whitney test is designed to find a difference of $P(X \le Y)$ from its value $1/2$ under the null hypothesis. This evaluation is justified by the following examples and is further underscored by the consideration of asymptotic efficiency in nonparametric models. It is shown in Section 25.6 that the tests are asymptotically efficient for testing the parameters $EY - EX$ or $P(X \le Y)$ if the underlying distributions F and G are completely unknown.

14.17   Example (t-test). The two-sample t-statistic $(\bar{Y} - \bar{X})/S$ converges in probability to $E(Y - X)/\sigma$, where $\sigma^2 = \lim \operatorname{var}(\bar{Y} - \bar{X})$. If the null hypothesis postulates that $EY = EX$, then the test that rejects the null hypothesis for large values of the t-statistic is consistent against every alternative for which $EY > EX$. □

14.18   *Example (Mann-Whitney test).* The Mann-Whitney statistic U converges in probability to $P(X \leq Y)$, by the two-sample U-statistic theorem. The probability $P(X \leq Y)$ is equal to $1/2$ if the two samples are equal in distribution and possess a continuous distribution function. If the null hypothesis postulates that $P(X \leq Y) = 1/2$, then the test that rejects for large values of U is consistent against any alternative for which $P(X \leq Y) > 1/2$. $\square$

14.3   Asymptotic Relative Efficiency

Sequences of tests can be ranked in quality by comparing their asymptotic power functions. For the test statistics we have considered so far, this comparison only involves the "slopes" of the tests. The concept of relative efficiency yields a method to quantify the interpretation of the slopes.

Consider a sequence of testing problems consisting of testing a null hypothesis $H_0 : \theta = 0$ versus the alternative $H_1 : \theta = \theta_\nu$. We use the parameter ν to describe the asymptotics; thus $\nu \to \infty$. We require a priori that our tests attain asymptotically level α and power $\gamma \in (\alpha, 1)$. Usually we can meet this requirement by choosing an appropriate number of observations at "time" ν. A larger number of observations allows smaller level and higher power. If π_n is the power function of a test if n observations are available, then we define n_ν to be the minimal number of observations such that both

$$\pi_{n_\nu}(0) \leq \alpha, \quad \text{and} \quad \pi_{n_\nu}(\theta_\nu) \geq \gamma.$$

If two sequences of tests are available, then we prefer the sequence for which the numbers n_ν are smallest. Suppose that $n_{\nu,1}$ and $n_{\nu,2}$ observations are needed for two given sequences of tests. Then, if it exists, the limit

$$\lim_{\nu \to \infty} \frac{n_{\nu,2}}{n_{\nu,1}}$$

is called the (asymptotic) *relative efficiency* or *Pitman efficiency* of the first with respect to the second sequence of tests. A relative efficiency larger than 1 indicates that fewer observations are needed with the first sequence of tests, which may then be considered the better one.

In principle, the relative efficiency may depend on α, γ and the sequence of alternatives θ_ν. The concept is mostly of interest if the relative efficiency is the same for all possible choices of these parameters. This is often the case. In particular, in the situations considered previously, the relative efficiency turns out to be the square of the quotient of the slopes.

14.19   *Theorem.* *Consider statistical models* $(P_{n,\theta} : \theta \geq 0)$ *such that* $\| P_{n,\theta} - P_{n,0} \| \to 0$ *as* $\theta \to 0$, *for every* n. *Let* $T_{n,1}$ *and* $T_{n,2}$ *be sequences of statistics that satisfy (14.4) for every sequence* $\theta_n \downarrow 0$ *and functions* μ_i *and* σ_i *such that* μ_i *is differentiable at zero and* σ_i *is continuous at zero, with* $\mu'_i(0) > 0$ *and* $\sigma_i(0) > 0$. *Then the relative efficiency of the tests that reject the null hypothesis* $H_0 : \theta = 0$ *for large values of* $T_{n,i}$ *is equal to*

$$\left(\frac{\mu'_1(0)/\sigma_1(0)}{\mu'_2(0)/\sigma_2(0)} \right)^2 ,$$

for every sequence of alternatives $\theta_\nu \downarrow 0$, *independently of* $\alpha > 0$ *and* $\gamma \in (\alpha, 1)$. *If the power functions of the tests based on* $T_{n,i}$ *are nondecreasing for every* n, *then the assumption* *of asymptotic normality of $T_{n,i}$ can be relaxed to asymptotic normality under every sequence $\theta_n = O(1/\sqrt{n})$ only.*

Proof. Fix α and γ as in the introduction and, given alternatives $\theta_\nu \downarrow 0$, let $n_{\nu,i}$ observations be used with each of the two tests. The assumption that $\|P_{n,\theta_\nu} - P_{n,0}\| \to 0$ as $\nu \to \infty$ for each fixed n forces $n_{\nu,i} \to \infty$. Indeed, the sum of the probabilities of the first and second kind of the test with critical region K_n equals

$$\int_{K_n} dP_{n,0} + \int_{K_n^c} dP_{n,\theta_\nu} = 1 + \int_{K_n} (p_{n,0} - p_{n,\theta_\nu}) \, d\mu_n.$$

This sum is minimized for the critical region $K_n = \{p_{n,0} - p_{n,\theta_\nu} < 0\}$, and then equals $1 - \frac{1}{2}\|P_{n,\theta_\nu} - P_{n,0}\|$. By assumption, this converges to 1 as $\nu \to \infty$ uniformly in every finite set of n. Thus, for every bounded sequence $n = n_\nu$ and any sequence of tests, the sum of the error probabilities is asymptotically bounded below by 1 and cannot be bounded above by $\alpha + 1 - \gamma < 1$, as required.

Now that we have ascertained that $n_{\nu,i} \to \infty$ as $\nu \to \infty$, we can use the asymptotic normality of the test statistics $T_{n,i}$. The convergence to a continuous distribution implies that the asymptotic level and power attained for the minimal numbers of observations (minimal for obtaining at most level α and at least power γ) is exactly α and γ. In order to obtain asymptotic level α the tests must reject H_0 if $\sqrt{n_\nu}\big(T_{n_\nu,i} - \mu_i(0)\big) > \sigma_i(0)z_\alpha + o(1)$. The powers of these tests are equal to

$$\pi_{n_{\nu,i}}(\theta_\nu) = 1 - \Phi\bigg(z_\alpha + o(1) - \sqrt{n_{\nu,i}}\, \theta_\nu \frac{\mu_i'(0)}{\sigma_i(0)}\big(1 + o(1)\big) \bigg) + o(1).$$

This sequence of powers tends to $\gamma < 1$ if and only if the argument of Φ tends to z_γ. Thus the relative efficiency of the two sequences of tests equals

$$\lim_{\nu \to \infty} \frac{n_{\nu,2}}{n_{\nu,1}} = \lim_{\nu \to \infty} \frac{n_{\nu,2}\theta_\nu^2}{n_{\nu,1}\theta_\nu^2} = \frac{(z_\alpha - z_\gamma)^2}{\big(\mu_2'(0)/\sigma_2(0)\big)^2} \bigg/ \frac{(z_\alpha - z_\gamma)^2}{\big(\mu_1'(0)/\sigma_1(0)\big)^2}.$$

This proves the first assertion of the theorem.

If the power functions of the tests are monotone and the test statistics are asymptotically normal for every sequence $\theta_n = O(1/\sqrt{n})$, then $\pi_{n,i}(\theta_n) \to \alpha$ or 1 if $\sqrt{n}\,\theta_n \to 0$ or ∞, respectively (see Lemma 14.16). In that case the sequences of tests can only meet the (α, γ) requirement for testing alternatives θ_ν such that $\sqrt{n_{\nu,i}}\,\theta_\nu = O(1)$. For such sequences the preceding argument is valid and gives the asserted relative efficiency. ∎

*14.4 Other Relative Efficiencies

The asymptotic relative efficiency defined in the preceding section is known as the *Pitman relative efficiency*. In this section we discuss some other types of relative efficiencies. Define $n_i(\alpha, \gamma, \theta)$ as the minimal numbers of observations needed, with $i \in \{1, 2\}$ for two given sequences of tests, to test a null hypothesis $H_0: \theta = 0$ versus the alternative $H_1: \theta = \theta$ at level α and with power at least γ. Then the Pitman efficiency against a sequence of alternatives $\theta_\nu \to 0$ is defined as (if the limits exists)

$$\lim_{\nu \to \infty} \frac{n_2(\alpha, \gamma, \theta_\nu)}{n_1(\alpha, \gamma, \theta_\nu)}.$$

The device to let the alternatives θ_ν tend to the null hypothesis was introduced to make the testing problems harder and harder, so that the required numbers of observations tend to infinity, and the comparison becomes an asymptotic one. There are other possibilities that can serve the same end. The testing problem is harder as α is smaller, as γ is larger, and (typically) as θ is closer to the null hypothesis. Thus, we could also let α tend to zero, or γ tend to one, keeping the other parameters fixed, or even let two or all three of the parameters vary. For each possible method we could define the relative efficiency of two sequences of tests as the limit of the quotient of the minimal numbers of observations that are needed. Most of these possibilities have been studied in the literature. Next to the Pitman efficiency the most popular efficiency measure appears to be the *Bahadur efficiency*, which is defined as

$$\lim_{\nu \to \infty} \frac{n_2(\alpha_\nu, \gamma, \theta)}{n_1(\alpha_\nu, \gamma, \theta)}.$$

Here α_ν tends to zero, but γ and θ are fixed. Typically, the Bahadur efficiency depends on θ, but not on γ, and not on the particular sequence $\alpha_\nu \downarrow 0$ that is used.

Whereas the calculation of Pitman efficiencies is most often based on distributional limit theorems, Bahadur efficiencies are derived from large deviations results. The reason is that the probabilities of first or second kind for testing a fixed null hypothesis against a fixed alternative usually tend to zero at an exponential speed. Large deviations theorems quantify this speed. Suppose that the null hypothesis $H_0 : \theta = 0$ is rejected for large values of a test statistic T_n, and that

$$-\frac{2}{n} \log P_0(T_n \geq t) \to e(t), \qquad \text{every } t, \qquad (14.20)$$

$$T_n \overset{P_\theta}{\to} \mu(\theta). \qquad (14.21)$$

The first result is a *large deviation* type result, and the second a "law of large numbers." The *observed significance level* of the test is defined as $P_0(T_n \geq t)_{|t=T_n}$. Under the null hypothesis, this random variable is uniformly distributed if T_n possesses a continuous distribution function. For a fixed alternative θ, it typically converges to zero at an exponential rate. For instance, under the preceding conditions, if e is continuous at $\mu(\theta)$, then (because e is necessarily monotone) it is immediate that

$$-\frac{2}{n} \log P_0(T_n \geq t)_{|t=T_n} \overset{P_\theta}{\to} e\big(\mu(\theta)\big).$$

The quantity $e\big(\mu(\theta)\big)$ is called the *Bahadur slope* of the test (or rather the limit in probability of the left side if it exists). The quotient of the slopes of two sequences of test statistics gives the *Bahadur relative efficiency*.

14.22   Theorem. *Let $T_{n,1}$ and $T_{n,2}$ be sequences of statistics in statistical models $(P_{n,0},$ $P_{n,\theta})$ that satisfy (14.20) and (14.21) for functions e_i and numbers $\mu_i(\theta)$ such that e_i is continuous at $\mu_i(\theta)$. Then the Bahadur relative efficiency of the sequences of tests that reject for large values of $T_{n,i}$ is equal to $e_1\big(\mu_1(\theta)\big)/e_2\big(\mu_2(\theta)\big)$, for every $\alpha_\nu \downarrow 0$ and every $1 > \gamma > \sup_n P_{n,\theta}(p_{n,0} = 0)$.*

Proof.   For simplicity of notation, we drop the index $i \in \{1, 2\}$ and write n_ν for the minimal numbers of observations needed to obtain level α_ν and power γ with the test statistics T_n.

The sample sizes n_ν necessarily converge to ∞ as $\nu \to \infty$. If not, then there would exist a fixed value n and a (sub)sequence of tests with levels tending to 0 and powers at least γ. However, for any fixed n, and any sequence of measurable sets K_m with $P_{n,0}(K_m) \to 0$ as $m \to \infty$, the probabilities $P_{n,\theta}(K_m) = P_{n,\theta}(K_m \cap p_{n,0} = 0) + o(1)$ are eventually strictly smaller than γ, by assumption.

The most powerful level α_ν-test that rejects for large values of T_n has critical region $\{T_n \geq c_n\}$ or $\{T_n > c_n\}$ for $c_n = \inf\{c : P_0(T_n \geq c) \leq \alpha_\nu\}$, where we use $\geq$ if $P_0(T_n \geq c_n) \leq \alpha_\nu$ and $>$ otherwise. Equivalently, with the notation $L_n = P_0(T_n \geq t)_{|t=T_n}$, this is the test with critical region $\{L_n \leq \alpha_\nu\}$. By the definition of n_ν we conclude that

$$P_{n,\theta}\left(-\frac{2}{n}\log L_n \geq -\frac{2}{n}\log\alpha_\nu\right) \begin{cases} \geq \gamma & \text{for } n = n_\nu, \\ < \gamma & \text{for } n = n_\nu - 1. \end{cases}$$

By (14.20) and (14.21), the random variable inside the probability converges in probability to the number $e(\mu(\theta))$ as $n \to \infty$. Thus, the probability converges to 0 or 1 if $-(2/n)\log\alpha_\nu$ is asymptotically strictly bigger or smaller than $e(\mu(\theta))$, respectively. Conclude that

$$\limsup_{\nu\to\infty} -\frac{2}{n_\nu}\log\alpha_\nu \leq e(\mu(\theta))$$

$$\liminf_{\nu\to\infty} -\frac{2}{n_\nu - 1}\log\alpha_\nu \geq e(\mu(\theta)).$$

Combined, this yields the asymptotic equivalence $n_\nu \sim -2\log\alpha_\nu/e(\mu(\theta))$. Applying this for both $n_{\nu,1}$ and $n_{\nu,2}$ and taking the quotient, we obtain the theorem. ∎

Bahadur and Pitman efficiencies do not always yield the same ordering of sequences of tests. In numerical comparisons, the Pitman efficiencies appear to be more relevant for moderate sample sizes. This is explained by their method of calculation. By the preceding theorem, Bahadur efficiencies follow from a large deviations result under the null hypothesis and a law of large numbers under the alternative. A law of large numbers is of less accuracy than a distributional limit result. Furthermore, large deviation results, while mathematically interesting, often yield poor approximations for the probabilities of interest. For instance, condition (14.20) shows that $P_0(T_n \geq t) = \exp\left(-\frac{1}{2}ne(t)\right)\exp o(n)$. Nothing guarantees that the term $\exp o(n)$ is close to 1.

On the other hand, often the Bahadur efficiencies as a function of θ are more informative than Pitman efficiencies. The Pitman slopes are obtained under the condition that the sequence $\sqrt{n}(T_n - \mu(0))$ is asymptotically normal with mean zero and variance $\sigma^2(0)$. Suppose, for the present argument, that T_n is normally distributed for every finite n, with the parameters $\mu(0)$ and $\sigma^2(0)/n$. Then, because $1 - \Phi(t) \sim \phi(t)/t$ as $t \to \infty$,

$$-\frac{2}{n}\log P_0(T_n \geq \mu(0) + t) = -\frac{2}{n}\log\left(1 - \Phi\left(\frac{t\sqrt{n}}{\sigma(0)}\right)\right) \to \frac{t^2}{\sigma^2(0)}, \qquad \text{every } t.$$

The Bahadur slope would be equal to $(\mu(\theta) - \mu(0))^2/\sigma^2(0)$. For $\theta \to 0$, this is approximately equal to θ^2 times the square of the Pitman slope $\mu'(0)^2/\sigma^2(0)$. Consequently, the limit of the Bahadur efficiencies as $\theta \to 0$ would yield the Pitman efficiency.

Now, the preceding argument is completely false if T_n is only approximately normally distributed: Departures from normality that are negligible in the sense of weak convergence need not be so for large-deviation probabilities. The difference between the "approximate

Bahadur slopes" just obtained and the true slopes is often substantial. However, the argument tends to be "more correct" as t approaches $\mu(0)$, and the conclusion that limiting Bahadur efficiencies are equal to Pitman efficiencies is often correct.[†]

The main tool needed to evaluate Bahadur efficiencies is the large-deviation result (14.20). For averages T_n, this follows from the Cramér-Chernoff theorem, which can be thought of as the analogue of the central limit theorem for large deviations. It is a refinement of the weak law of large numbers that yields exponential convergence of probabilities of deviations from the mean.

The *cumulant generating function* of a random variable Y is the function $u \mapsto K(u) = \log Ee^{uY}$. If we allow the value ∞, then this is well-defined for every $u \in \mathbb{R}$. The set of u such that $K(u)$ is finite is an interval that may or may not contain its boundary points and may be just the point $\{0\}$.

14.23 Proposition (Cramér-Chernoff theorem). *Let $Y_1, Y_2, \ldots$ be i.i.d. random variables with cumulant generating function K. Then, for every t,*

$$\frac{1}{n} \log \mathrm{P}(\bar{Y} \geq t) \to \inf_{u \geq 0}(K(u) - tu).$$

Proof. The cumulant generating function of the variables $Y_i - t$ is equal to $u \mapsto K(u) - ut$. Therefore, we can restrict ourselves to the case $t = 0$. The proof consists of separate upper and lower bounds on the probabilities $\mathrm{P}(\bar{Y} \geq 0)$.

The upper bound is easy and is valid for every n. By Markov's inequality, for every $u \geq 0$,

$$\mathrm{P}(\bar{Y} \geq 0) = \mathrm{P}(e^{un\bar{Y}_n} \geq 1) \leq Ee^{un\bar{Y}_n} = e^{nK(u)}.$$

Take logarithms, divide by n, and take the infimum over $u \geq 0$ to find one half of the proposition.

For the proof of the lower bound, first consider the cases that Y_i is nonnegative or nonpositive. If $\mathrm{P}(Y_i < 0) = 0$, then the function $u \mapsto K(u)$ is monotonely increasing on $\mathbb{R}$ and its infimum on $u \geq 0$ is equal to 0 (attained at $u = 0$); this is equal to $n^{-1} \log \mathrm{P}(\bar{Y} \geq 0)$ for every n. Second, if $\mathrm{P}(Y_i > 0) = 0$, then the function $u \mapsto K(u)$ is monotonely decreasing on $\mathbb{R}$ with $K(\infty) = \log \mathrm{P}(Y_1 = 0)$; this is equal to $n^{-1} \log \mathrm{P}(\bar{Y} \geq 0)$ for every n. Thus, the theorem is valid in both cases, and we may exclude them from now on.

First, assume that $K(u)$ is finite for every $u \in \mathbb{R}$. Then the function $u \mapsto K(u)$ is analytic on $\mathbb{R}$, and, by differentiating under the expectation, we see that $K'(0) = EY_1$. Because Y_i takes both negative and positive values, $K(u) \to \infty$ as $u \to \pm\infty$. Thus, the infimum of the function $u \mapsto K(u)$ over $u \in \mathbb{R}$ is attained at a point u_0 such that $K'(u_0) = 0$.

The case that $u_0 < 0$ is trivial, but requires an argument. By the convexity of the function $u \mapsto K(u)$, K is nondecreasing on $[u_0, \infty)$. If $u_0 < 0$, then it attains its minimum value over $u \geq 0$ at $u = 0$, which is $K(0) = 0$. Furthermore, in this case $EY_1 = K'(0) > K'(u_0) = 0$ (strict inequality under our restrictions, for instance because $K''(0) = \mathrm{var}\, Y_1 > 0$) and hence $\mathrm{P}(\bar{Y} \geq 0) \to 1$ by the law of large numbers. Thus, the limit of the left side of the proposition (with $t = 0$) is 0 as well.

[†] In [85] a precise argument is given.

For $u_0 \geq 0$, let $Z_1, Z_2, \ldots$ be i.i.d. random variables with the distribution given by

$$dP_Z(z) = e^{-K(u_0)} e^{u_0 z} \, dP_Y(z).$$

Then Z_1 has cumulant generating function $u \mapsto K(u_0 + u) - K(u_0)$, and, as before, its mean can be found by differentiating this function at $u = 0$: $EZ_1 = K'(u_0) = 0$. For every $\varepsilon > 0$,

$$P(\overline{Y} \geq 0) = E1\{\overline{Z}_n \geq 0\}e^{-u_0 n \overline{Z}_n} e^{nK(u_0)}$$
$$\geq P(0 \leq \overline{Z}_n \leq \varepsilon) e^{-u_0 n \varepsilon} e^{nK(u_0)}.$$

Because $\overline{Z}_n$ has mean 0, the sequence $P(0 \leq \overline{Z}_n \leq \varepsilon)$ is bounded away from 0, by the central limit theorem. Conclude that n^{-1} times the limit inferior of the logarithm of the left side is bounded below by $-u_0 \varepsilon + K(u_0)$. This is true for every $\varepsilon > 0$ and hence also for $\varepsilon = 0$.

Finally, we remove the restriction that $K(u)$ is finite for every u, by a truncation argument. For a fixed, large M, let $Y_1^M, Y_2^M, \ldots$ be distributed as the variables $Y_1, Y_2, \ldots$ given that $|Y_i| \leq M$ for every i, that is, they are i.i.d. according to the conditional distribution of Y_1 given $|Y_1| \leq M$. Then, with $u \mapsto K_M(u) = \log E e^{uY_1} 1\{|Y_1| \leq M\}$,

$$\liminf \frac{1}{n} \log P(\overline{Y} \geq 0) \geq \frac{1}{n} \log \left(P(\overline{Y}_n^M \geq 0) P(|Y_i^M| \leq M)^n \right)$$
$$\geq \inf_{u \geq 0} K_M(u),$$

by the preceding argument applied to the truncated variables. Let s be the limit of the right side as $M \to \infty$, and let A_M be the set $\{u \geq 0 : K_M(u) \leq s\}$. Then the sets A_M are nonempty and compact for sufficiently large M (as soon as $K_M(u) \to \infty$ as $u \to \pm\infty$), with $A_1 \supset A_2 \supset \cdots$, whence $\cap A_M$ is nonempty as well. Because K_M converges pointwise to K as $M \to \infty$, any point $u_1 \in \cap A_M$ satisfies $K(u_1) = \lim K_M(u_1) \leq s$. Conclude that s is bigger than the right side of the proposition (with $t = 0$). ■

14.24   Example (Sign statistic). The cumulant generating function of a variable Y that is -1 and 1, each with probability $\frac{1}{2}$, is equal to $K(u) = \log \cosh u$. Its derivative is $K'(u) = \tanh u$ and hence the infimum of $K(u) - tu$ over $u \in \mathbb{R}$ is attained for $u = \operatorname{arctanh} t$. By the Cramér-Chernoff theorem, for $0 < t < 1$,

$$-\frac{2}{n} \log P(\overline{Y} \geq t) \to e(t) := -2 \log \cosh \operatorname{arctanh} t + 2t \operatorname{arctanh} t.$$

We can apply this result to find the Bahadur slope of the sign statistic $T_n = n^{-1} \sum_{i=1}^{n} \operatorname{sign}(X_i)$. If the null distribution of the random variables $X_1, \ldots, X_n$ is continuous and symmetric about zero, then (14.20) is valid with $e(t)$ as in the preceding display and with $\mu(\theta) = E_\theta \operatorname{sign}(X_1)$. Figure 14.2 shows the slopes of the sign statistic and the sample mean for testing the location of the Laplace distribution. The local optimality of the sign statistic is reflected in the Bahadur slopes, but for detecting large differences of location the mean is better than the sign statistic. However, it should be noted that the power of the sign test in this range is so close to 1 that improvement may be irrelevant; for example, the power is 0.999 at level 0.007 for $n = 25$ at $\theta = 2$. □

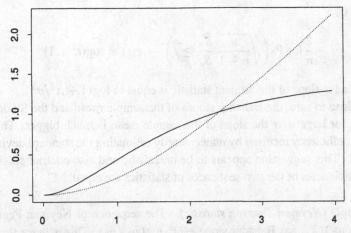

Figure 14.2. Bahadur slopes of the sign statistic (*solid line*) and the sample mean (*dotted line*) for testing that a random sample from the Laplace distribution has mean zero versus the alternative that the mean is θ, as a function of θ.

14.25 *Example (Student statistic).* Suppose that $X_1, \ldots, X_n$ are a random sample from a normal distribution with mean μ and variance σ^2. We shall consider σ known and compare the slopes of the sample mean and the Student statistic $\overline{X}_n/S_n$ for testing $H_0 : \mu = 0$.

The cumulant generating function of the normal distribution is equal to $K(u) = u\mu + \frac{1}{2}u^2\sigma^2$. By the Cramér-Chernoff theorem, for $t > 0$,

$$-\frac{2}{n}\log P_0(\overline{X}_n \geq t) \to e(t) := \frac{t^2}{\sigma^2}.$$

Thus, the Bahadur slope of the sample mean is equal to μ^2/σ^2, for every $\mu > 0$.

Under the null hypothesis, the statistic $\sqrt{n}\,\overline{X}_n/S_n$ possesses the t-distribution with $(n-1)$ degrees of freedom. Thus, for a random sample $Z_0, Z_1, \ldots$ of standard normal variables, for every $t > 0$,

$$P_0\!\left(\sqrt{\frac{n}{n-1}}\,\frac{\overline{X}_n}{S_n} \geq t\right) = \frac{1}{2}P\!\left(\frac{t_{n-1}^2}{n-1} \geq t^2\right) = \frac{1}{2}P\!\left(Z_0^2 - t^2\sum_{i=1}^{n-1} Z_i^2 \geq 0\right).$$

This probability is not of the same form as in the Cramér-Chernoff theorem, but it concerns almost an average, and we can obtain the large deviation probabilities from the cumulant generating function in an analogous way. The cumulant generating function of a square of a standard normal variable is equal to $u \mapsto -\frac{1}{2}\log(1 - 2u)$, and hence the cumulant generating function of the variable $Z_0^2 - t^2\sum_{i=1}^{n-1} Z_i^2$ is equal to

$$K_n(u) = -\tfrac{1}{2}\log(1 - 2u) - \tfrac{1}{2}(n-1)\log(1 + 2t^2 u).$$

This function is nicely differentiable and, by straightforward calculus, its minimum value can be found to be

$$\inf_u K_n(u) = -\frac{1}{2}\log\!\left(\frac{t^2 + 1}{t^2 n}\right) - \frac{1}{2}(n-1)\log\!\left(\frac{(n-1)(t^2+1)}{n}\right).$$

The minimum is achieved on $[0, \infty)$ for $t^2 \geq (n-1)^{-1}$. This expression divided by n is the analogue of $\inf_u K(u)$ in the Cramér-Chernoff theorem. By an extension of this theorem, for every $t > 0$,

$$-\frac{2}{n} \log P_0\left(\sqrt{\frac{n}{n-1}} \frac{\bar{X}_n}{S_n} \geq t\right) \to e(t) = \log(t^2 + 1).$$

Thus, the Bahadur slope of the Student statistic is equal to $\log(1 + \mu^2/\sigma^2)$.

For μ/σ close to zero, the Bahadur slopes of the sample mean and the Student statistic are close, but for large μ/σ the slope of the sample mean is much bigger. This suggests that the loss in efficiency incurred by unnecessarily estimating the standard deviation σ can be substantial. This suggestion appears to be unrealistic and also contradicts the fact that the Pitman efficiencies of the two sequences of statistics are equal. □

14.26   Example (Neyman-Pearson statistic).   The sequence of Neyman-Pearson statistics $\prod_{i=1}^n (p_\theta/p_{\theta_0})(X_i)$ has Bahadur slope $-2P_\theta \log(p_{\theta_0}/p_\theta)$. This is twice the Kullback-Leibler divergence of the measures P_{θ_0} and P_θ and shows an important connection between large deviations and the Kullback-Leibler divergence.

In regular cases this result is a consequence of the Cramér-Chernoff theorem. The variable $Y = \log p_\theta/p_{\theta_0}$ has cumulant generating function $K(u) = \log \int p_\theta^u p_{\theta_0}^{1-u} \, d\mu$ under P_{θ_0}. The function $K(u)$ is finite for $0 \leq u \leq 1$, and, at least by formal calculus, $K'(1) = P_\theta \log(p_\theta/p_{\theta_0}) = \mu(\theta)$, where $\mu(\theta)$ is the asymptotic mean of the sequence $n^{-1} \sum \log(p_\theta/p_{\theta_0})(X_i)$. Thus the infimum of the function $u \mapsto K(u) - u\mu(\theta)$ is attained at $u = 1$ and the Bahadur slope is given by

$$e(\mu(\theta)) = -2(K(1) - \mu(\theta)) = 2P_\theta \log \frac{p_\theta}{p_{\theta_0}}.$$

In section 16.6 we obtain this result by a direct, and rigorous, argument. □

For statistics that are not means, the Cramér-Chernoff theorem is not applicable, and we need other methods to compute the Bahadur efficiencies. An important approach applies to functions of means and is based on more general versions of Cramér's theorem. A first generalization asserts that, for certain sets B, not necessarily of the form $[t, \infty)$,

$$\frac{1}{n} \log P(\bar{Y} \in B) \to -\inf_{y \in B} I(y), \qquad I(y) = \sup_u (uy - K(u)).$$

For a given statistic of the form $\phi(\bar{Y})$, the large deviation probabilities of interest $P(\phi(\bar{Y}) \geq t)$ can be written in the form $P(\bar{Y} \in B_t)$ for the inverse images $B_t = \phi^{-1}[t, \infty)$. If B_t is an eligible set in the preceding display, then the desired large deviations result follows, although we shall still have to evaluate the repeated "inf sup" on the right side. Now, according to Cramér's theorem, the display is valid for every set such that the right side does not change if B is replaced by its interior or its closure. In particular, if ϕ is continuous, then B_t is closed and its interior $\mathring{B}_t$ contains the set $\phi^{-1}(t, \infty)$. Then we obtain a large deviations result if the difference set $\phi^{-1}\{t\}$ is "small" in that it does not play a role when evaluating the right side of the display.

Transforming a univariate mean $\bar{Y}$ into a statistic $\phi(\bar{Y})$ can be of interest (for example, to study the two-sided test statistics $|\bar{Y}|$), but the real promise of this approach is in its applications to multivariate and infinite-dimensional means. Cramér's theorem has been generalized to these situations. General large deviation theorems can best be formulated as separate upper and lower bounds. A sequence of random maps $X_n : \Omega \mapsto \mathbb{D}$ from a probability space $(\Omega, \mathcal{U}, P)$ into a topological space $\mathbb{D}$ is said to satisfy the *large deviation principle with rate function I* if, for every closed set F and for every open set G,

$$\limsup_{n \to \infty} \frac{1}{n} \log P^* (X_n \in F) \leq - \inf_{y \in F} I(y),$$

$$\liminf_{n \to \infty} \frac{1}{n} \log P_* (X_n \in G) \geq - \inf_{y \in G} I(y).$$

The rate function $I : \mathbb{D} \mapsto [0, \infty]$ is assumed to be lower semicontinuous and is called a *good rate function* if the sublevel sets $\{y : I(y) \leq M\}$ are compact, for every $M \in \mathbb{R}$. The inner and outer probabilities that X_n belongs to a general set B are sandwiched between the probabilities that it belongs to the interior $\overset{\circ}{B}$ and the closure $\overline{B}$. Thus, we obtain a large deviation result with equality for every set B such that $\inf \{I(y) : y \in \overline{B}\} = \inf \{I(y) : y \in \overset{\circ}{B}\}$. An implication for the slopes of test statistics of the form $\phi(X_n)$ is as follows.

14.27 Lemma. *Suppose that $\phi : \mathbb{D} \mapsto \mathbb{R}$ is continuous at every y such that $I(y) < \infty$ and suppose that $\inf \{I(y) : \phi(y) > t\} = \inf \{I(y) : \phi(y) \geq t\}$. If the sequence X_n satisfies the large-deviation principle with the rate function I under P_0, then $T_n = \phi(X_n)$ satisfies (14.20) with $e(t) = 2 \inf \{I(y) : \phi(y) \geq t\}$. Furthermore, if I is a good rate function, then e is continuous at t.*

Proof. Define sets $A_t = \phi^{-1}(t, \infty)$ and $B_t = \phi^{-1}[t, \infty)$, and let $\mathbb{D}_0$ be the set where I is finite. By the continuity of ϕ, $\overline{B}_t \cap \mathbb{D}_0 = B_t \cap \mathbb{D}_0$ and $\overset{\circ}{B} \cap \mathbb{D}_0 \supset A_t \cap \mathbb{D}_0$. (If $y \notin \overset{\circ}{B}_t$, then there is a net $y_n \in B_t^c$ with $y_n \to y$; if also $y \in \mathbb{D}_0$, then $\phi(y) = \lim \phi(y_n) \leq t$ and hence $y \notin A_t$.) Consequently, the infimum of I over $\overset{\circ}{B}_t$ is at least the infimum over A_t, which is the infimum over B_t by assumption, and also the infimum over $\overline{B}_t$. Condition (14.20) follows upon applying the large deviation principle to $\overset{\circ}{B}_t$ and $\overline{B}_t$.

The function e is nondecreasing. The condition on the pair (I, ϕ) is exactly that e is right-continuous, because $e(t+) = \inf \{I(y) : \phi(y) > t\}$. To prove the left-continuity of e, let $t_m \uparrow t$. Then $e(t_m) \uparrow a$ for some $a \leq e(t)$. If $a = \infty$, then $e(t) = \infty$ and e is left-continuous. If $a < \infty$, then there exists a sequence y_m with $\phi(y_m) \geq t_m$ and $2I(y_m) \leq a + 1/m$. By the goodness of I, this sequence has a converging subnet $y_{m'} \to y$. Then $2I(y) \leq \liminf 2I(y_{m'}) \leq a$ by the lower semicontinuity of I, and $\phi(y) \geq t$ by the continuity of ϕ. Thus $e(t) \leq 2I(y) \leq a$. ∎

Empirical distributions can be viewed as means (of Dirac measures), and are therefore potential candidates for a large-deviation theorem. Cramér's theorem for empirical distributions is known as *Sanov's theorem*. Let $\mathbb{L}_1(\mathcal{X}, \mathcal{A})$ be the set of all probability measures on the measurable space $(\mathcal{X}, \mathcal{A})$, which we assume to be a complete, separable metric space with its Borel σ-field. The *τ-topology* on $\mathbb{L}_1(\mathcal{X}, \mathcal{A})$ is defined as the weak topology generated by the collection of all maps $P \mapsto Pf$ for f ranging over the set of all bounded, measurable functions on $f : \mathcal{X} \mapsto \mathbb{R}$.[†]

14.28 Theorem (Sanov's theorem). *Let $\mathbb{P}_n$ be the empirical measure of a random sample of size n from a fixed measure P. Then the sequence $\mathbb{P}_n$ viewed as maps into $\mathbb{L}_1(\mathcal{X}, \mathcal{A})$*

[†] For a proof of the following theorem, see [31], [32]. or [65].

satisfies the large deviation principle relative to the τ-topology, with the good rate function
$I(Q) = -Q \log p/q$.

For $\mathcal{X}$ equal to the real line, $L_1(\mathcal{X}, \mathcal{A})$ can be identified with the set of cumulative distribution functions. The τ-topology is stronger than the topology obtained from the uniform norm on the distribution functions. This follows from the fact that if both $F_n(x) \to F(x)$ and $F_n\{x\} \to F\{x\}$ for every $x \in \mathbb{R}$, then $\|F_n - F\|_\infty \to 0$. (see problem 19.9). Thus any function ϕ that is continuous with respect to the uniform norm is also continuous with respect to the τ-topology, and we obtain a large collection of functions to which we can apply the preceding lemma. Trimmed means are just one example.

14.29   *Example (Trimmed means).* Let $\mathbb{F}_n$ be the empirical distribution function of a random sample of size n from the distribution function F, and let $\mathbb{F}_n^{-1}$ be the corresponding quantile function. The function $\phi(\mathbb{F}_n) = (1-2\alpha)^{-1} \int_\alpha^{1-\alpha} \mathbb{F}_n^{-1}(s) \, ds$ yields a version of the α-trimmed mean (see Chapter 22). We assume that $0 < \alpha < \frac{1}{2}$ and (partly for simplicity) that the null distribution F_0 is continuous.

If we show that the conditions of Lemma 14.27 are fulfilled, then we can conclude, by Sanov's theorem,

$$-\frac{2}{n} \log P_{F_0}\big(\phi(\mathbb{F}_n) \ge t\big) \to e(t) := 2 \inf_{G \,:\, \phi(G) \ge t} -G \log \frac{f_0}{g}.$$

Because $\mathbb{F}_n \xrightarrow{P} F$ uniformly by the Glivenko-Cantelli theorem, Theorem 19.1, and ϕz is continuous, $\phi(\mathbb{F}_n) \xrightarrow{P} \phi(F)$, and the Bahadur slope of the α-trimmed mean at an alternative F is equal to $e\big(\phi(F)\big)$.

Finally, we show that ϕ is continuous with respect to the uniform topology and that the function $t \mapsto \inf\{-G \log(f_0/g)) : \phi(G) \ge t\}$ is right-continuous at t if F_0 is continuous at t. The map ϕ is even continuous with respect to the weak topology on the set of distribution functions: If a sequence of measures G_m converges weakly to a measure G, then the corresponding quantile functions G_m^{-1} converge weakly to the quantile function G^{-1} (see Lemma 21.2) and hence $\phi(G_m) \to \phi(G)$ by the dominated convergence theorem.

The function $t \mapsto \inf\{-G \log(f_0/g) : \phi(G) \ge t\}$ is right-continuous at t if for every G with $\phi(G) = t$ there exists a sequence G_m with $\phi(G_m) > t$ and $G_m \log(f_0/g_m) \to G \log(f_0/g)$. If $G \log(f_0/g) = -\infty$, then this is easy, for we can choose any fixed G_m that is singular with respect to F_0 and has a trimmed mean bigger than t. Thus, we may assume that $G |\log(f_0/g)| < \infty$, that $G \ll F_0$ and hence that G is continuous. Then there exists a point c such that $\alpha < G(c) < 1 - \alpha$. Define

$$\frac{dG_m}{dG}(x) = \begin{cases} 1 - \frac{1}{m} & \text{if } x \le c, \\ 1 + \varepsilon_m & \text{if } x > c. \end{cases}$$

Then G_m is a probability distribution for suitably chosen $\varepsilon_m > 0$, and, by the dominated convergence $G_m \log(f_0/g_m) \to G \log(f_0/g)$ as $m \to \infty$. Because $G_m(x) \le G(x)$ for all x, with strict inequality (at least) for all $x \le c$ such that $G(x) > 0$, we have that $G_m^{-1}(s) \ge G^{-1}(s)$ for all s, with strict inequality for all $s \in (0, G(c)]$. Hence the trimmed mean $\phi(G_m)$ is strictly bigger than the trimmed mean $\phi(G)$, for every m. $\square$

*14.5 Rescaling Rates

The asymptotic power functions considered earlier in this chapter are the limits of "local power functions" of the form $h \mapsto \pi_n(h/\sqrt{n})$. The rescaling rate $\sqrt{n}$ is typical for testing smooth parameters of the model. In this section we have a closer look at the rescaling rate and discuss some nonregular situations.

Suppose that in a given sequence of models $(\mathcal{X}_n, \mathcal{A}_n, P_{n,\theta} : \theta \in \Theta)$ it is desired to test the null hypothesis $H_0 : \theta = \theta_0$ versus the alternatives $H_1 : \theta = \theta_n$. For probability measures P and Q define the *total variation distance* $\|P - Q\|$ as the L_1-distance $\int |p - q| \, d\mu$ between two densities of P and Q.

14.30   Lemma. *The power function π_n of any test in $(\mathcal{X}_n, \mathcal{A}_n, P_{n,\theta} : \theta \in \Theta)$ satisfies*

$$\pi_n(\theta) - \pi_n(\theta_0) \leq \tfrac{1}{2} \|P_{n,\theta} - P_{n,\theta_0}\|.$$

For any θ and θ_0 there exists a test whose power function attains equality.

Proof. If π_n is the power function of the test ϕ_n, then the difference on the left side can be written as $\int \phi_n(p_{n,\theta} - p_{n,\theta_0}) \, d\mu_n$. This expression is maximized for the test function $\phi_n = 1\{p_{n,\theta} > p_{n,\theta_0}\}$. Next, for any pair of probability densities p and q we have $\int_{q>p} (q - p) \, d\mu = \tfrac{1}{2} \int |p - q| \, d\mu$, since $\int (p - q) \, d\mu = 0$.   ∎

This lemma implies that for any sequence of alternatives θ_n:

(i) If $\|P_{n,\theta_n} - P_{n,\theta_0}\| \to 2$, then there exists a sequence of tests with power $\pi_n(\theta_n)$ tending to 1 and size $\pi_n(\theta_0)$ tending to 0 (a *perfect* sequence of tests).

(ii) If $\|P_{n,\theta_n} - P_{n,\theta_0}\| \to 0$, then the power of any sequence of tests is asymptotically less than the level (every sequence of tests is worthless).

(iii) If $\|P_{n,\theta_n} - P_{n,\theta_0}\|$ is bounded away from 0 and 2, then there exists no perfect sequence of tests, but not every test is worthless.

The rescaling rate $h/\sqrt{n}$ used earlier sections corresponds to the third possibility. These examples concern models with independent observations. Because the total variation distance between product measures cannot be easily expressed in the distances for the individual factors, we translate the results into the Hellinger distance and next study the implications for product experiments.

The *Hellinger distance* $H(P, Q)$ between two probability measures is the L_2-distance between the square roots of the corresponding densities. Thus, its square $H^2(P, Q)$ is equal to $\int (\sqrt{p} - \sqrt{q})^2 \, d\mu$. The distance is convenient if considering product measures. First, the Hellinger distance can be expressed in the *Hellinger affinity* $A(P, Q) = \int \sqrt{p}\sqrt{q} \, d\mu$, through the formula

$$H^2(P, Q) = 2 - 2A(P, Q).$$

Next, by Fubini's theorem, the affinity of two product measures is the product of the affinities. Thus we arrive at the formula

$$H^2(P^n, Q^n) = 2 - 2\big(1 - \tfrac{1}{2} H^2(P, Q)\big)^n.$$

14.31   Lemma. *Given a statistical model* $(P_\theta : \theta \geq \theta_0)$ *set* $P_{n,\theta} = P_\theta^n$. *Then the possibilities (i), (ii), and (iii) arise when* $n H^2(P_{\theta_n}, P_{\theta_0})$ *converges to* ∞, *converges to* 0, *or is bounded away from* 0 *and* ∞, *respectively. In particular, if* $H^2(P_\theta, P_{\theta_0}) = O(|\theta - \theta_0|^\alpha)$ *as* $\theta \to \theta_0$, *then the possibilities (i), (ii), and (iii) are valid when* $n^{1/\alpha}|\theta_n - \theta_0|$ *converges to* ∞, *converges to* 0, *or is bounded away from* 0 *and* ∞, *respectively.*

Proof.   The possibilities (i), (ii), and (iii) can equivalently be described by replacing the total variation distance $\| P_{\theta_n}^n - P_{\theta_0}^n \|$ by the squared Hellinger distance $H^2(P_{\theta_n}^n, P_{\theta_0}^n)$. This follows from the inequalities, for any probability measures P and Q,

$$H^2(P, Q) \leq \| P - Q \| \leq \left(2 - A^2(P, Q) \right) \wedge 2H(P, Q).$$

The inequality on the left is immediate from the inequality $|\sqrt{p} - \sqrt{q}|^2 \leq |p - q|$, valid for any nonnegative numbers p and q. For the inequality on the right, first note that $pq = (p \vee q)(p \wedge q) \leq (p + q)(p \wedge q)$, whence $A^2(P, Q) \leq 2 \int (p \wedge q) \, d\mu$, by the Cauchy-Schwarz inequality. Now $\int (p \wedge q) \, d\mu$ is equal to $1 - \frac{1}{2} \| P - Q \|$, as can be seen by splitting the domains of both integrals in the sets $p < q$ and $p \geq q$. This shows that $\| P - Q \| \leq 2 - A^2(P, Q)$. That $\| P - Q \| \leq 2H(P, Q)$ is a direct consequence of the Cauchy-Schwarz inequality.

We now express the Hellinger distance of the product measures in the Hellinger distance of P_{θ_n} and P_{θ_0} and manipulate the nth power function to conclude the proof. ∎

14.32   Example (Smooth models). If the model $(\mathcal{X}, \mathcal{A}, P_\theta : \theta \in \Theta)$ is differentiable in quadratic mean at θ_0, then $H^2(P_\theta, P_{\theta_0}) = O(|\theta - \theta_0|^2)$. The intermediate rate of convergence (case (iii)) is $\sqrt{n}$. □

14.33   Example (Uniform law). If P_θ is the uniform measure on $[0, \theta]$; then $H^2(P_\theta, P_{\theta_0}) = O(|\theta - \theta_0|)$. The intermediate rate of convergence is n. In this case we would study asymptotic power functions defined as the limits of the local power functions of the form $h \mapsto \pi_n(\theta_0 + h/n)$. For instance, the level α tests that reject the null hypothesis $H_0 : \theta = \theta_0$ for large values of the maximum $X_{(n)}$ of the observations have power functions

$$\pi_n\left(\theta_0 + \frac{h}{n} \right) = P_{\theta_0 + h/n}\left(X_{(n)} \geq \theta_0 (1 - \alpha)^{1/n} \right) \to 1 - (1 - \alpha)e^{-h/\theta_0}.$$

Relative to this rescaling rate, the level α tests that reject the null hypothesis for large values of the mean $\bar{X}_n$ have asymptotic power function α (no power). □

14.34   Example (Triangular law). Let P_θ be the probability distribution with density $x \mapsto \left(1 - |x - \theta| \right)^+$ on the real line. Some clever integrations show that $H^2(P_\theta, P_0) = \frac{1}{2}\theta^2 \log(1/\theta) + O(\theta^2)$ as $\theta \to 0$. (It appears easiest to compute the affinity first.) This leads to the intermediate rate of convergence $\sqrt{n \log n}$. □

The preceding lemmas concern testing a given simple null hypothesis against a simple alternative hypothesis. In many cases the rate obtained from considering simple hypotheses does not depend on the hypotheses and is also globally attainable at every parameter in the parameter space. If not, then the global problems have to be taken into account from the beginning. One possibility is discussed within the context of density estimation in section 24.3.

Lemma 14.31 gives rescaling rates for problems with independent observations. In models with dependent observations quite different rates may pertain.

14.35 *Example (Branching).* Consider the Galton-Watson branching process, discussed in Example 9.10. If the offspring distribution has mean $\mu(\theta)$ larger than 1, then the parameter is estimable at the exponential rate $\mu(\theta)^n$. This is also the right rescaling rate for defining asymptotic power functions. □

Notes

Apparently, E.J.G. Pitman introduced the efficiencies that are named for him in an unpublished set of lecture notes in 1949. A published proof of a slightly more general result can be found in [109].

Cramér [26] was interested in preciser approximations to probabilities of large deviations than are presented in this chapter and obtained the theorem under the condition that the moment-generating function is finite on $\mathbb{R}$. Chernoff [20] proved the theorem as presented here, by a different argument. Chernoff used it to study the minimum weighted sums of error probabilities of tests that reject for large values of a mean and showed that, for any $0 < \pi < 1$,

$$\frac{1}{n} \log \inf_t \big(\pi P_0(\bar{Y} > t) + (1 - \pi) P_1(\bar{Y} \le t)\big)$$
$$\to \inf_{E_0 Y_1 < t < E_1 Y_1} \inf_u \big(K_0(u) - ut\big) \vee \inf_u \big(K_1(u) - ut\big).$$

Furthermore, for $\bar{Y}$ the likelihood ratio statistic for testing P_0 versus P_1, the right side of this display can be expressed in the *Hellinger integral* of the experiment (P_0, P_1) as

$$\inf_{0 < u < 1} \log \int dP_0^u dP_1^{1-u}.$$

Thus, this expression is a lower bound for the $\liminf_{n\to\infty} n^{-1} \log(\alpha_n + \beta_n)$ for α_n and β_n the error probabilities of any test of P_0 versus P_1. That the Bahadur slope of Neyman-Pearson tests is twice the Kullback-Leibler divergence (Example 14.26) is essentially known as *Stein's lemma* and is apparently among those results by Stein that he never cared to publish.

A first version of Sanov's theorem was proved by Sanov in 1957. Subsequently, many authors contributed to strengthening the result, the version presented here being given in [65]. Large-deviation theorems are subject of current research by probabilists, particularly with extensions to more complicated objects than sums of independent variables. See [31] and [32]. For further information and references concerning applications in statistics, we refer to [4] and [61], as well as to Chapters 8, 16, and 17.

For applications and extensions of the results on rescaling rates, see [37].

PROBLEMS

1. Show that the power function of the Wilcoxon two sample test is monotone under shift of location.
2. Let $X_1, \ldots, X_n$ be a random sample from the $N(\mu, \sigma^2)$-distribution, where σ^2 is known. A test for $H_0 : \mu = 0$ against $H_1 : \mu > 0$ can be based on either $\bar{X}/\sigma$ or $\bar{X}/S$. Show that the asymptotic

relative efficiency of the two sequences of tests is 1. Does it make a difference whether normal or t-critical values are used?

3. Let $X_1, \ldots, X_n$ be a random sample from a density $f(x - \theta)$ where f is symmetric about zero. Calculate the relative efficiency of the t-test and the test that rejects for large values of $\sum \sum_{i < j} 1\{X_i + X_j > 0\}$ for f equal to the logistic, normal, Laplace, and uniform shapes.

4. Calculate the relative efficiency of the van der Waerden test with respect to the t-test in the two-sample problem.

5. Calculate the relative efficiency of the tests based on Kendall's τ and the sample correlation coefficient to test independence for bivariate normal pairs of observations.

6. Suppose $\phi : \mathcal{F} \mapsto \mathbb{R}$ and $\psi : \mathcal{F} \mapsto \mathbb{R}^k$ are arbitrary maps on an arbitrary set $\mathcal{F}$ and we wish to find the minimum value of ϕ over the set $\{f \in \mathcal{F} : \psi(f) = 0\}$. If the map $f \mapsto \phi(f) + a^T \psi(f)$ attains its minimum over $\mathcal{F}$ at f_a, for each fixed a in an arbitrary set A, and there exists $a_0 \in A$ such that $\psi(f_{a_0}) = 0$, then the desired minimum value is $\phi(f_{a_0})$. This is a rather trivial use of Lagrange multipliers, but it is helpful to solve the next problems. $(\phi(f_{a_0}) = \phi(f_{a_0}) + a_0^T \psi(f_{a_0})$ is the minimum of $\phi(f) + a_0^T \psi(f)$ over $\mathcal{F}$ and hence smaller than the minimum of $\phi(f) + a_0^T \psi(f)$ over $\{f \in \mathcal{F} : \psi(f) = 0\}$.)

7. Show that $4 f(0)^2 \int y^2 f(y) \, dy \geq 1/3$ for every probability density f that has its mode at 0. (The minimum is equal to the minimum of $4 \int y^2 f(y) \, dy$ over all probability densities f that are bounded by 1.)

8. Show that $12 \left(\int f^2(y) \, dy \right)^2 \int y^2 f(y) \, dy \geq 108/125$ for every probability density f with mean zero. (The minimum is equal to 12 times the minimum of the square of $\phi(f) = \int f^2(y) \, dy$ over all probability densities with mean 0 and variance 1.)

9. Study the asymptotic power function of the sign test if the observations are a sample from a distribution that has a positive mass at its median. Is it good or bad to have a nonsmooth distribution?

10. Calculate the Hellinger and total variation distance between two uniform $U[0, \theta]$ measures.

11. Calculate the Hellinger and total variation distance between two normal $N(\mu, \sigma^2)$ measures.

12. Let $X_1, \ldots, X_n$ be a sample from the uniform distribution on $[-\theta, \theta]$.
   (i) Calculate the asymptotic power functions of the tests that reject $H_0 : \theta = \theta_0$ for large values of $X_{(n)}$, $X_{(n)} \vee (-X_{(1)})$ and $X_{(n)} - X_{(1)}$.
   (ii) Calculate the asymptotic relative efficiencies of these tests.

13. If two sequences of test statistics satisfied (14.4) for every $\theta_n \downarrow 0$, but with norming rate n^α instead of $\sqrt{n}$, how would Theorem 14.19 have to be modified to find the Pitman relative efficiency?

15

Efficiency of Tests

It is shown that, given converging experiments, every limiting power function is the power function of a test in the limit experiment. Thus, uniformly most powerful tests in the limit experiment give absolute upper bounds for the power of a sequence of tests. In normal experiments such uniformly most powerful tests exist for linear hypotheses of codimension one. The one-sample location problem and the two-sample problem are discussed in detail, and appropriately designed (signed) rank tests are shown to be asymptotically optimal.

15.1 Asymptotic Representation Theorem

A randomized *test* (or *test function*) ϕ in an experiment $(\mathcal{X}, \mathcal{A}, P_h : h \in H)$ is a measurable map $\phi : \mathcal{X} \mapsto [0, 1]$ on the sample space. The interpretation is that if x is observed, then a null hypothesis is rejected with probability $\phi(x)$. The *power function* of a test ϕ is the function

$$h \mapsto \pi(h) = E_h \phi(X).$$

This gives the probabilities that the null hypothesis is rejected. A test is of level α for testing a null hypothesis H_0 if its size $\sup \{\pi(h) : h \in H_0\}$ does not exceed α. The quality of a test can be judged from its power function, and classical testing theory is aimed at finding, among the tests of level α, a test with high power at every alternative.

The asymptotic quality of a sequence of tests may be judged from the limit of the sequence of local power functions. If the tests are defined in experiments that converge to a limit experiment, then a pointwise limit of power functions is necessarily a power function in the limit experiment. This follows from the following theorem, which specializes the asymptotic representation theorem, Theorem 9.3, to the testing problem. Applied to the special case of the local experiments $\mathcal{E}_n = (P_{\theta+h/\sqrt{n}}^n : h \in \mathbb{R}^k)$ of a differentiable parametric model as considered in Chapter 7, which converge to the Gaussian experiment $(N(h, I_\theta^{-1}), h \in \mathbb{R}^k)$, the theorem is the parallel for testing of Theorem 7.10.

15.1 Theorem. *Let the sequence of experiments $\mathcal{E}_n = (P_{n,h} : h \in H)$ converge to a dominated experiment $\mathcal{E} = (P_h : h \in H)$. Suppose that a sequence of power functions π_n of tests in $\mathcal{E}_n$ converges pointwise: $\pi_n(h) \to \pi(h)$, for every h and some arbitrary function π. Then π is a power function in the limit experiment: There exists a test ϕ in $\mathcal{E}$ with $\pi(h) = E_h \phi(X)$ for every h.*

215

Proof. We give the proof for the special case of experiments that satisfy the following assumption: Every sequence of statistics T_n that is tight under every given parameter h possesses a subsequence (not depending on h) that converges in distribution to a limit under every h. See problem 15.2 for a method to extend the proof to the general situation.

The additional condition is valid in the case of local asymptotic normality. With the notation of the proof of Theorem 7.10, we argue first that the sequence (T_n, Δ_n) is uniformly tight under $h = 0$ and hence possesses a weakly convergent subsequence by Prohorov's theorem. Next, by the expansion of the likelihood and Slutsky's lemma, the sequence $(T_n, \log dP_{n,h}/dP_{n,0})$ converges under $h = 0$ along the same sequence, for every h. Finally, we conclude by Le Cam's third lemma that the sequence T_n converges under h, along the subsequence.

Let ϕ_n be tests with power functions π_n. Because each ϕ_n takes its values in the compact interval $[0, 1]$, the sequence of random variables ϕ_n is certainly uniformly tight. By assumption, there exists a subsequence of $\{n\}$ along which ϕ_n converges in distribution under every h. Thus, the assumption of the asymptotic representation theorem, Theorem 9.3 or Theorem 7.10, is satisfied along some subsequence of the statistics ϕ_n. By this theorem, there exists a randomized statistic $T = T(X, U)$ in the limit experiment such that $\phi_n \overset{h}{\leadsto} T$ along the subsequence, for every h. The randomized statistic may be assumed to take its values in $[0, 1]$. Because the ϕ_n are uniformly bounded, $E_h\phi_n \to E_hT$. Combination with the assumption yields $\pi(h) = E_hT$ for every h. The randomized statistic T is not a test function (it is a "doubly randomized" test). However, the test $\phi(x) = E\big(T(X, U) \mid X = x\big)$ satisfies the requirements. ∎

The theorem suggests that the best possible limiting power function is the power function of the best test in the limit experiment. In classical testing theory an "absolutely best" test is defined as a *uniformly most powerful test* of the required level. Depending on the experiment, such a test may or may not exist. If it does not exist, then the classical solution is to find a uniformly most powerful test in a restricted class, such as the class of all unbiased or invariant tests; to use the maximin criterion; or to use a conditional test. In combination with the preceding theorem, each of these approaches leads to a criterion for asymptotic quality. We do not pursue this in detail but note that, in general, we would avoid any sequence of tests that is matched in the limit experiment by a test that is considered suboptimal.

In the remainder of this chapter we consider the implications for locally asymptotically normal models in more detail. We start with reviewing testing in normal location models.

15.2 Testing Normal Means

Suppose that the observation X is $N_k(h, \Sigma)$-distributed, for a known covariance matrix Σ and unknown mean vector h. First consider testing the null hypothesis $H_0 : c^Th = 0$ versus the alternative $H_1 : c^Th > 0$, for a known vector c. The "natural" test, which rejects H_0 for large values of c^TX, is uniformly most powerful. In other words, if π is a power function such that $\pi(h) \le \alpha$ for every h with $c^Th = 0$, then for every h with $c^Th > 0$,

$$\pi(h) \le P_h(c^TX > z_\alpha \sqrt{c^T\Sigma c}) = 1 - \Phi\left(z_\alpha - \frac{c^Th}{\sqrt{c^T\Sigma c}}\right).$$

15.2 Proposition. *Suppose that X be $N_k(h, \Sigma)$-distributed for a known nonnegative-definite matrix Σ, and let c be a fixed vector with $c^T \Sigma c > 0$. Then the test that rejects H_0 if $c^T X > z_\alpha \sqrt{c^T \Sigma c}$ is uniformly most powerful at level α for testing $H_0 : c^T h = 0$ versus $H_1 : c^T h > 0$, based on X.*

Proof. Fix h_1 with $c^T h_1 > 0$. Define $h_0 = h_1 - (c^T h_1 / c^T \Sigma c) \Sigma c$. Then $c^T h_0 = 0$. By the Neyman-Pearson lemma, the most powerful test for testing the simple hypotheses $H_0 : h = h_0$ and $H_1 : h = h_1$ rejects H_0 for large values of

$$\log \frac{dN(h_1, \Sigma)}{dN(h_0, \Sigma)}(X) = \frac{c^T h_1}{c^T \Sigma c} c^T X - \frac{1}{2} \frac{(c^T h_1)^2}{c^T \Sigma c}.$$

This is equivalent to the test that rejects for large values of $c^T X$. More precisely, the most powerful level α test for $H_0 : h = h_0$ versus $H_1 : h = h_1$ is the test given by the proposition. Because this test does not depend on h_0 or h_1, it is uniformly most powerful for testing $H_0 : c^T h = 0$ versus $H_1 : c^T h > 0$. ∎

The natural test for the two-sided problem $H_0 : c^T h = 0$ versus $H_1 : c^T h \neq 0$ rejects the null hypothesis for large values of $|c^T X|$. This test is not uniformly most powerful, because its power is dominated by the uniformly most powerful tests for the two one-sided alternatives whose union is H_1. However, the test with critical region $\{x : |c^T x| \geq z_{\alpha/2} \sqrt{c^T \Sigma c}\}$ is uniformly most powerful among the unbiased level α tests (see problem 15.1).

A second problem of interest is to test a simple null hypothesis $H_0 : h = 0$ versus the alternative $H_1 : h \neq 0$. If the parameter set is one-dimensional, then this reduces to the problem in the preceding paragraph. However, if θ is of dimension $k > 1$, then there exists no uniformly most powerful test, not even among the unbiased tests. A variety of tests are reasonable, and whether a test is "good" depends on the alternatives at which we desire high power. For instance, the test that is most sensitive to detect the alternatives such that $c^T h > 0$ (for a given c) is the test given in the preceding theorem. Probably in most situations no particular "direction" is of special importance, and we would use a test that distributes the power over all directions. It is known that any test with as critical region the complement of a closed, convex set C is admissible (see, e.g., [138, p. 137]). In particular, complements of closed, convex, and symmetric sets are admissible critical regions and cannot easily be ruled out a priori. The shape of C determines the power function, the directions in which C extends little receiving large power (although the power also depends on Σ).

The most popular test rejects the null hypothesis for large values of $X^T \Sigma^{-1} X$. This test arises as the limit version of the Wald test, the score test, and the likelihood ratio test. One advantage is a simple choice of critical values, because $X^T \Sigma^{-1} X$ is chi square–distributed with k degrees of freedom. The power function of this test is, with Z a standard normal vector,

$$\pi(h) = P_h\big(X^T \Sigma^{-1} X > \chi^2_{k,\alpha}\big) = P\big(\|Z + \Sigma^{-1/2} h\|^2 > \chi^2_{k,\alpha}\big).$$

By the rotational symmetry of the standard normal distribution, this depends only on the *noncentrality parameter* $\|\Sigma^{-1/2} h\|$. The power is relatively large in the directions h for which $\|\Sigma^{-1/2} h\|$ is large. In particular, it increases most steeply in the direction of the eigenvector corresponding to the smallest eigenvalue of Σ. Note that the test does not distribute the power evenly, but dependent on Σ. Two optimality properties of this test are given in problems 15.3 and 15.4, but these do not really seem convincing.

Due to the lack of an acceptable optimal test in the limit problem, a satisfactory asymptotic optimality theory of testing simple hypotheses on multidimensional parameters is impossible.

15.3   Local Asymptotic Normality

A normal limit experiment arises, among others, in the situation of repeated sampling from a differentiable parametric model. If the model $(P_\theta : \theta \in \Theta)$ is differentiable in quadratic mean, then the local experiments converge to a Gaussian limit:

$$\left(P^n_{\theta_0 + h/\sqrt{n}} : h \in \mathbb{R}^k\right) \to \left(N\left(h, I^{-1}_{\theta_0}\right) : h \in \mathbb{R}^k\right).$$

A sequence of power functions $\theta \mapsto \pi_n(\theta)$ in the original experiments induces the sequence of power functions $h \mapsto \pi_n(\theta_0 + h/\sqrt{n})$ in the local experiments. Suppose that $\pi_n(\theta_0 + h/\sqrt{n}) \to \pi(h)$ for every h and some function π. Then, by the asymptotic representation theorem, the limit π is a power function in the Gaussian limit experiment.

Suppose for the moment that θ is real and that the sequence π_n is of asymptotic level α for testing $H_0 : \theta \le \theta_0$ versus $H_1 : \theta > \theta_0$. Then $\pi(0) = \lim \pi_n(\theta_0) \le \alpha$ and hence π corresponds to a level α test for $H_0 : h = 0$ versus $H_1 : h > 0$ in the limit experiment. It must be bounded above by the power function of the uniformly most powerful level α test in the limit experiment, which is given by Proposition 15.2. Conclude that

$$\lim_{n \to \infty} \pi_n\left(\theta_0 + \frac{h}{\sqrt{n}}\right) \le 1 - \Phi(z_\alpha - h\sqrt{I_{\theta_0}}), \quad \text{every } h > 0.$$

(Apply the proposition with $c = 1$ and $\Sigma = I^{-1}_{\theta_0}$.) We have derived an absolute upper bound on the local asymptotic power of level α tests.

In Chapter 14 a sequence of power functions such that $\pi_n(\theta_0 + h/\sqrt{n}) \to 1 - \Phi(z_\alpha - hs)$ for every h is said to have slope s. It follows from the present upper bound that the square root $\sqrt{I_{\theta_0}}$ of the Fisher information is the largest possible slope. The quantity

$$\frac{I_{\theta_0}}{s^2}$$

is the relative efficiency of the best test and the test with slope s. It can be interpreted as the number of observations needed with the given sequence of tests with slope s divided by the number of observations needed with the best test to obtain the same power.

With a bit of work, the assumption that $\pi_n(\theta_0 + h/\sqrt{n})$ converges to a limit for every h can be removed. Also, the preceding derivation does not use the special structure of i.i.d. observations but only uses the convergence to a Gaussian experiment. We shall rederive the result within the context of local asymptotic normality and also indicate how to construct optimal tests.

Suppose that at "time" n the observation is distributed according to a distribution $P_{n,\theta}$ with parameter ranging over an open subset Θ of $\mathbb{R}^k$. The sequence of experiments $(P_{n,\theta} : \theta \in \Theta)$ is locally asymptotically normal at θ_0 if

$$\log \frac{dP_{n,\theta_0 + r_n^{-1} h}}{dP_{n,\theta_0}} = h^T \Delta_{n,\theta_0} - \frac{1}{2} h^T I_{\theta_0} h + o_{P_{n,\theta_0}}(1), \tag{15.3}$$

for a sequence of statistics Δ_{n,θ_0} that converges in distribution under θ_0 to a normal $N_k(0, I_{\theta_0})$-distribution.

15.4 Theorem. *Let $\Theta \subset \mathbb{R}^k$ be open and let $\psi : \Theta \mapsto \mathbb{R}$ be differentiable at θ_0, with nonzero gradient $\dot{\psi}_{\theta_0}$ and such that $\psi(\theta_0) = 0$. Let the sequence of experiments $(P_{n,\theta} : \theta \in \Theta)$ be locally asymptotically normal at θ_0 with nonsingular Fisher information, for constants $r_n \to \infty$. Then the power functions $\theta \mapsto \pi_n(\theta)$ of any sequence of level α tests for testing $H_0 : \psi(\theta) \leq 0$ versus $H_1 : \psi(\theta) > 0$ satisfy, for every h such that $\dot{\psi}_{\theta_0} h > 0$,*

$$\limsup_{n \to \infty} \pi_n\left(\theta_0 + \frac{h}{r_n}\right) \leq 1 - \Phi\left(z_\alpha - \frac{\dot{\psi}_{\theta_0} h}{\sqrt{\dot{\psi}_{\theta_0} I_{\theta_0}^{-1} \dot{\psi}_{\theta_0}^T}}\right).$$

15.5 Addendum. *Let T_n be statistics such that*

$$T_n = \frac{\dot{\psi}_{\theta_0} I_{\theta_0}^{-1} \Delta_{n,\theta_0}}{\sqrt{\dot{\psi}_{\theta_0} I_{\theta_0}^{-1} \dot{\psi}_{\theta_0}^T}} + o_{P_{n,\theta_0}}(1).$$

Then the sequence of tests that reject for values of T_n exceeding z_α is asymptotically optimal in the sense that the sequence $P_{\theta_0 + r_n^{-1} h}(T_n \geq z_\alpha)$ converges to the right side of the preceding display, for every h.

Proofs. The sequence of localized experiments $(P_{n,\theta_0 + r_n^{-1} h} : h \in \mathbb{R}^k)$ converges by Theorem 7.10, or Theorem 9.4, to the Gaussian location experiment $(N_k(h, I_{\theta_0}^{-1}) : h \in \mathbb{R}^k)$.

Fix some h_1 such that $\dot{\psi}_{\theta_0} h_1 > 0$, and a subsequence of $\{n\}$ along which the lim sup $\pi(\theta_0 + h_1 / r_n)$ is taken. There exists a further subsequence along which $\pi_n(\theta_0 + r_n^{-1} h)$ converges to a limit $\pi(h)$ for every $h \in \mathbb{R}^k$ (see the proof of Theorem 15.1). The function $h \mapsto \pi(h)$ is a power function in the Gaussian limit experiment. For $\dot{\psi}_{\theta_0} h < 0$, we have $\psi(\theta_0 + r_n^{-1} h) = r_n^{-1}(\dot{\psi}_{\theta_0} h + o(1)) < 0$ eventually, whence $\pi(h) \leq \limsup \pi_n(\theta_0 + r_n^{-1} h) \leq \alpha$. By continuity, the inequality $\pi(h) \leq \alpha$ extends to all h such that $\dot{\psi}_{\theta_0} h \leq 0$. Thus, π is of level α for testing $H_0 : \dot{\psi}_{\theta_0} h \leq 0$ versus $H_1 : \dot{\psi}_{\theta_0} h > 0$. Its power function is bounded above by the power function of the uniformly most powerful test, which is given by Proposition 15.2. This concludes the proof of the theorem.

The asymptotic optimality of the sequence T_n follows by contiguity arguments. We start by noting that the sequence $(\Delta_{n,\theta_0}, \Delta_{n,\theta_0})$ converges under θ_0 in distribution to a (degenerate) normal vector (Δ, Δ). By Slutsky's lemma and local asymptotic normality,

$$\left(\Delta_{n,\theta_0}, \log \frac{dP_{n,\theta_0 + r_n^{-1} h}}{dP_{n,\theta_0}}\right) \stackrel{\theta_0}{\rightsquigarrow} \left(\Delta, h^T \Delta - \tfrac{1}{2} h^T I_{\theta_0} h\right)$$

$$\sim N\left(\begin{pmatrix} 0 \\ -\tfrac{1}{2} h^T I_{\theta_0} h \end{pmatrix}, \begin{pmatrix} I_{\theta_0} & I_{\theta_0} h \\ h^T I_{\theta_0} & h^T I_{\theta_0} h \end{pmatrix}\right).$$

By Le Cam's third lemma, the sequence Δ_{n,θ_0} converges in distribution under $\theta_0 + r_n^{-1} h$ to a $N(I_{\theta_0} h, I_{\theta_0})$-distribution. Thus, the sequence T_n converges under $\theta_0 + r_n^{-1} h$ in distribution to a normal distribution with mean $\dot{\psi}_{\theta_0} h / (\dot{\psi}_{\theta_0} I_{\theta_0}^{-1} \dot{\psi}_{\theta_0}^T)^{1/2}$ and variance 1. ∎

The point θ_0 in the preceding theorem is on the boundary of the null and the alternative hypotheses. If the dimension k is larger than 1, then this boundary is typically $(k-1)$-dimensional, and there are many possible values for θ_0. The upper bound is valid at every possible choice.

If $k = 1$, the boundary point θ_0 is typically unique and hence known, and we could use $T_n = I_{\theta_0}^{-1/2} \Delta_{n,\theta_0}$ to construct an optimal sequence of tests for the problem $H_0 : \theta = \theta_0$. These are known as *score tests*.

Another possibility is to base a test on an estimator sequence. Not surprisingly, efficient estimators yield efficient tests.

15.6   Example (Wald tests). Let $X_1, \ldots, X_n$ be a random sample in an experiment $(P_\theta : \theta \in \Theta)$ that is differentiable in quadratic mean with nonsingular Fisher information. Then the sequence of local experiments $(P_{\theta+h/\sqrt{n}}^n : h \in \mathbb{R}^k)$ is locally asymptotically normal with $r_n = \sqrt{n}$, I_θ the Fisher information matrix, and

$$\Delta_{n,\theta} = \frac{1}{\sqrt{n}} \sum_{i=1}^{n} \dot{\ell}_\theta(X_i).$$

A sequence of estimators $\hat{\theta}_n$ is asymptotically efficient for estimating θ if (see Chapter 8)

$$\sqrt{n}(\hat{\theta}_n - \theta) = I_\theta^{-1} \Delta_{n,\theta} + o_{P_\theta}(1).$$

Under regularity conditions, the maximum likelihood estimator qualifies. Suppose that $\theta \mapsto I_\theta$ is continuous, and that ψ is continuously differentiable with nonzero gradient. Then the sequence of tests that reject $H_0 : \psi(\theta) \leq 0$ if

$$\sqrt{n}\psi(\hat{\theta}_n) \geq z_\alpha \sqrt{\dot{\psi}_{\hat{\theta}_n} I_{\hat{\theta}_n}^{-1} \dot{\psi}_{\hat{\theta}_n}^T}$$

is asymptotically optimal at every point θ_0 on the boundary of H_0. Furthermore, this sequence of tests is consistent at every θ with $\psi(\theta) > 0$.

These assertions follow from the preceding theorem, upon using the delta method and Slutsky's lemma. The resulting tests are called *Wald tests* if $\hat{\theta}_n$ is the maximum likelihood estimator.   □

15.4   One-Sample Location

Let $X_1, \ldots, X_n$ be a sample from a density $f(x - \theta)$, where f is symmetric about zero and has finite Fisher information for location I_f. It is required to test $H_0 : \theta = 0$ versus $H_1 : \theta > 0$. The density f may be known or (partially) unknown. For instance, it may be known to belong to the normal scale family.

For fixed f, the sequence of experiments $(\prod_{i=1}^n f(x_i - \theta) : \theta \in \mathbb{R})$ is locally asymptotically normal at $\theta = 0$ with $\Delta_{n,0} = -n^{-1/2} \sum_{i=1}^n (f'/f)(X_i)$, norming rate $\sqrt{n}$, and Fisher information I_f. By the results of the preceding section, the best asymptotic level α power function (for known f) is

$$1 - \Phi(z_\alpha - h\sqrt{I_f}).$$

This function is an upper bound for $\limsup \pi_n(h/\sqrt{n})$, for every $h > 0$, for every sequence of level α power functions. Suppose that T_n are statistics with

$$T_n = -\frac{1}{\sqrt{n}} \frac{1}{\sqrt{I_f}} \sum_{i=1}^{n} \frac{f'}{f}(X_i) + o_{P_0}(1). \tag{15.7}$$

Then, according to the second assertion of Theorem 15.4, the sequence of tests that reject the null hypothesis if $T_n \geq z_\alpha$ attains the bound and hence is asymptotically optimal. We shall discuss several ways of constructing test statistics with this property.

If the shape of the distribution is completely known, then the test statistics T_n can simply be taken equal to the right side of (15.7), without the remainder term, and we obtain the *score test*. It is more realistic to assume that the underlying distribution is only known up to scale. If the underlying density takes the form $f(x) = f_0(x/\sigma)/\sigma$ for a known density f_0 that is symmetric about zero, but for an unknown scale parameter σ, then

$$\frac{f'}{f}(x) = \frac{1}{\sigma} \frac{f_0'}{f_0}\left(\frac{x}{\sigma}\right), \qquad I_f = \frac{1}{\sigma^2} I_{f_0}, \qquad \frac{1}{\sqrt{I_f}} \frac{f'}{f}(x) = \frac{1}{\sqrt{I_{f_0}}} \frac{f_0'}{f_0}\left(\frac{x}{\sigma}\right).$$

15.8 Example (t-test). The standard normal density f_0 possesses score function f_0'/f_0 $(x) = -x$ and Fisher information $I_{f_0} = 1$. Consequently, if the underlying distribution is normal, then the optimal test statistics should satisfy $T_n = \sqrt{n}\overline{X}_n/\sigma + o_{P_0}(n^{-1/2})$. The t-statistics $\overline{X}_n/S_n$ fulfill this requirement. This is not surprising, because in the case of normally distributed observations the t-test is uniformly most powerful for every finite n and hence is certainly asymptotically optimal. $\square$

The t-statistic in the preceding example simply replaces the unknown standard deviation σ by an estimate. This approach can be followed for most scale families. Under some regularity conditions, the statistics

$$T_n = -\frac{1}{\sqrt{n}} \frac{1}{\sqrt{I_{f_0}}} \sum_{i=1}^{n} \frac{f_0'}{f_0}\left(\frac{X_i}{\hat{\sigma}_n}\right)$$

should yield asymptotically optimal tests, given a consistent sequence of scale estimators $\hat{\sigma}_n$.

Rather than using score-type tests, we could use a test based on an efficient estimator for the unknown symmetry point and efficient estimators for possible nuisance parameters, such as the scale – for instance, the maximum likelihood estimators. This method is indicated in general in Example 15.8 and leads to the Wald test.

Perhaps the most attractive approach is to use signed rank statistics. We summarize some definitions and conclusions from Chapter 13. Let $R_{n1}^+, \ldots, R_{nn}^+$ be the ranks of the absolute values $|X_1|, \ldots, |X_n|$ in the ordered sample of absolute values. A *linear signed rank statistic* takes the form

$$T_n = \frac{1}{\sqrt{n}} \sum_{i=1}^{n} a_{n,R_{ni}^+} \operatorname{sign}(X_i),$$

for given numbers $a_{n1}, \ldots, a_{nn}$, which are called the *scores* of the statistic. Particular examples are the *Wilcoxon signed rank statistic*, which has scores $a_{ni} = i$, and the *sign statistic*, which corresponds to scores $a_{ni} = 1$. In general, the scores can be chosen to weigh the influence of the different observations. A convenient method of generating scores is through a fixed function $\phi : [0, 1] \mapsto \mathbb{R}$, by

$$a_{ni} = \mathrm{E}\phi(U_{n(i)}).$$

(Here $U_{n(1)}, \ldots, U_{n(n)}$ are the order statistics of a random sample of size n from the uniform distribution on $[0, 1]$.) Under the condition that $\int_0^1 \phi^2(u)\, du < \infty$, Theorem 13.18 shows that, under the null hypothesis, and with $F^+(x) = 2F(x) - 1$ denoting the distribution function of $|X_1|$,

$$T_n = \frac{1}{\sqrt{n}} \sum_{i=1}^n \phi\Big(F^+\big(|X_i|\big)\Big) \operatorname{sign}(X_i) + o_{P_0}(1).$$

Because the score-generating function ϕ can be chosen freely, this allows the construction of an asymptotically optimal rank statistic for any given shape f. The choice

$$\phi(u) = -\frac{1}{\sqrt{I_f}} \frac{f'}{f}\big((F^+)^{-1}(u)\big). \tag{15.9}$$

yields the *locally most powerful scores*, as discussed in Chapter 13. Because $f'/f\big(|x|\big) \operatorname{sign}(x) = f'/f(x)$ by the symmetry of f, it follows that the signed rank statistics T_n satisfy (15.7). Thus, the locally most powerful scores yield asymptotically optimal signed rank tests. This surprising result, that the class of signed rank statistics contains asymptotically efficient tests for every given (symmetric) shape of the underlying distribution, is sometimes expressed by saying that the signs and absolute ranks are "asymptotically sufficient" for testing the location of a symmetry point.

15.10 Corollary. *Let T_n be the simple linear signed rank statistic with scores $a_{ni} = \mathrm{E}\phi(U_{n(i)})$ generated by the function ϕ defined in (15.9). Then T_n satisfies (15.7) and hence the sequence of tests that reject $H_0 : \theta = 0$ if $T_n \geq z_\alpha$ is asymptotically optimal at $\theta = 0$.*

Signed rank statistics were originally constructed because of their attractive property of being distribution free under the null hypothesis. Apparently, this can be achieved without losing (asymptotic) power. Thus, rank tests are strong competitors of classical parametric tests. Note also that signed rank statistics automatically adapt to the unknown scale: Even though the definition of the optimal scores appears to depend on f, they are actually identical for every member of a scale family $f(x) = f_0(x/\sigma)/\sigma$ (since $(F^+)^{-1}(u) = \sigma(F_0^+)^{-1}(u)$). Thus, no auxiliary estimate for σ is necessary for their definition.

15.11 Example (Laplace). The sign statistic $T_n = n^{-1/2} \sum_{i=1}^n \operatorname{sign}(X_i)$ satisfies (15.7) for f equal to the Laplace density. Thus the sign test is asymptotically optimal for testing location in the Laplace scale family. □

15.12 Example (Normal). The standard normal density has score function for location $f_0'/f_0(x) = -x$ and Fisher information $I_{f_0} = 1$. The optimal signed rank statistic for the normal scale family has score-generating function

$$\phi(u) = \mathrm{E}(\Phi^+)^{-1}(U_{n(i)}) = \mathrm{E}\Phi^{-1}\left(\frac{U_{n(i)} + 1}{2}\right) \approx \Phi^{-1}\left(\frac{i}{2n+2} + \frac{1}{2}\right).$$

We conclude that the corresponding sequence of rank tests has the same asymptotic slope as the t-test if the underlying distribution is normal. (For other distributions the two sequences of tests have different asymptotic behavior.) ☐

Even the assumption that the underlying distribution of the observations is known up to scale is often unrealistic. Because rank statistics are distribution-free under the null hypothesis, the level of a rank test is independent of the underlying distribution, which is the best possible protection of the level against misspecification of the model. On the other hand, the power of a rank test is not necessarily robust against deviations from the postulated model. This might lead to the use of the best test for the wrong model. The dependence of the power on the underlying distribution may be relaxed as well, by a procedure known as *adaptation*. This entails estimating the underlying density from the data and next using an optimal test for the estimated density. A remarkable fact is that this approach can be completely successful: There exist test statistics that are asymptotically optimal for any shape f. In fact, without prior knowledge of f (other than that it is symmetric with finite and positive Fisher information for location), estimators $\hat{\theta}_n$ and I_n can be constructed such that, for every θ and f,

$$\sqrt{n}(\hat{\theta}_n - \theta) = -\frac{1}{\sqrt{n}} \frac{1}{I_f} \sum_{i=1}^{n} \frac{f'}{f}(X_i - \theta) + o_{P_\theta}(1); \qquad I_n \overset{P_\theta}{\leadsto} I_f.$$

We give such a construction in section 25.8.1. Then the test statistics $T_n = \sqrt{n}\hat{\theta}_n I_n^{1/2}$ satisfy (15.7) and hence are asymptotically (locally) optimal at $\theta = 0$ for every given shape f. Moreover, for every $\theta > 0$, and every f,

$$P_\theta(T_n > z_\alpha) = P_\theta\big(\sqrt{n}(\hat{\theta}_n - \theta) > z_\alpha I_n^{-1/2} - \sqrt{n}\theta\big) \to 1.$$

Hence, the sequence of tests based on T_n is also consistent at every (θ, f) in the alternative hypothesis $H_1 : \theta > 0$.

15.5 Two-Sample Problems

Suppose we observe two independent random samples $X_1, \ldots, X_m$ and $Y_1, \ldots, Y_n$ from densities p_μ and q_ν, respectively. The problem is to test the null hypothesis $H_0 : \nu \leq \mu$ versus the alternative $H_1 : \nu > \mu$. There may be other unknown parameters in the model besides μ and ν, but we shall initially ignore such "nuisance parameters" and parametrize the model by $(\mu, \nu) \in \mathbb{R}^2$. Null and alternative hypotheses are shown graphically in Figure 15.1. We let $N = m + n$ be the total number of observations and assume that $m/N \to \lambda$ as $m, n \to \infty$.

15.13 *Example (Testing shift).* If $p_\mu(x) = f(x - \mu)$ and $q_\nu(y) = g(y - \nu)$ for two densities f and g that have the same "location," then we obtain the two-sample location problem. The alternative hypothesis asserts that the second sample is "stochastically larger." ☐

The alternatives of greatest interest for the study of the asymptotic performance of tests are sequences (μ_N, ν_N) that converge to the boundary between null and alternative hypotheses. In the study of relative efficiency, in Chapter 14, we restricted ourselves to

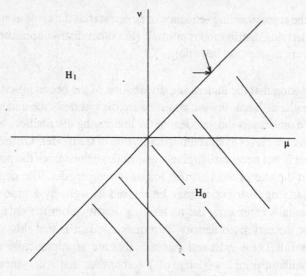

Figure 15.1. Null and alternative hypothesis.

vertical perturbations $(\theta, \theta + h/\sqrt{N})$. Here we shall use the sequences $(\theta + g/\sqrt{N}, \theta + h/\sqrt{N})$, which approach the boundary in the direction of a general vector (g, h).

If both p_μ and q_ν define differentiable models, then the sequence of experiments $\left(P_\mu^m \otimes P_\nu^n : (\mu, \nu) \in \mathbb{R}^2\right)$ is locally asymptotically normal with norming rate $\sqrt{N}$. If the score functions are denoted by $\dot\kappa_\mu$ and $\dot\ell_\nu$, and the Fisher informations by I_μ and J_ν, respectively, then the parameters of local asymptotic normality are

$$\Delta_{n,(\mu,\nu)} = \begin{pmatrix} \dfrac{\sqrt{\lambda}}{\sqrt{m}} \displaystyle\sum_{i=1}^m \dot\kappa_\mu(X_i) \\ \dfrac{\sqrt{1-\lambda}}{\sqrt{n}} \displaystyle\sum_{j=1}^n \dot\ell_\nu(Y_j) \end{pmatrix}, \qquad I_{(\mu,\nu)} = \begin{pmatrix} \lambda I_\mu & 0 \\ 0 & (1-\lambda)J_\nu \end{pmatrix}.$$

The corresponding limit experiment consists of observing two independent normally distributed variables with means g and h and variances $\lambda^{-1}I_\mu^{-1}$ and $(1-\lambda)^{-1}J_\nu^{-1}$, respectively.

15.14   Corollary. *Suppose that the models* $(P_\mu : \mu \in \mathbb{R})$ *and* $(Q_\nu : \nu \in \mathbb{R})$ *are differentiable in quadratic mean, and let* $m, n \to \infty$ *such that* $m/N \to \lambda \in (0, 1)$. *Then the power functions of any sequence of level* α *tests for* $H_0 : \nu = \mu$ *satisfies, for every* μ *and for every* $h > g$,

$$\limsup_{n,m\to\infty} \pi_{m,n}\left(\mu + \frac{g}{\sqrt{N}}, \mu + \frac{h}{\sqrt{N}}\right) \le 1 - \Phi\left(z_\alpha - (h-g)\sqrt{\frac{\lambda(1-\lambda)I_\mu J_\mu}{\lambda I_\mu + (1-\lambda)J_\mu}}\right).$$

Proof.   This is a special case of Theorem 15.4, with $\psi(\mu, \nu) = \nu - \mu$ and Fisher information matrix $\mathrm{diag}\left(\lambda I_\mu, (1 - \lambda)J_\mu\right)$. It is slightly different in that the null hypothesis $H_0 : \psi(\theta) = 0$ takes the form of an equality, which gives a weaker requirement on the sequence T_n. The proof goes through because of the linearity of ψ. ∎

It follows that the optimal slope of a sequence of tests is equal to

$$s_{\text{opt}}(\mu) = \sqrt{\frac{\lambda(1-\lambda)I_\mu J_\mu}{\lambda I_\mu + (1-\lambda)J_\mu}}.$$

The square of the quotient of the actual slope of a sequence of tests and this number is a good absolute measure of the asymptotic quality of the sequence of tests.

According to the second assertion of Theorem 15.4, an optimal sequence of tests can be based on any sequence of statistics such that

$$T_N = s_{\text{opt}}(\mu) \left(\frac{1}{\sqrt{1-\lambda}J_\mu} \frac{1}{\sqrt{n}} \sum_{j=1}^{n} \dot{\ell}_\mu(Y_j) - \frac{1}{\sqrt{\lambda}I_\mu} \frac{1}{\sqrt{m}} \sum_{i=1}^{m} \dot{\kappa}_\mu(X_i) \right) + o_P(1).$$

(The multiplicative factor $s_{\text{opt}}(\mu)$ ensures that the sequence T_N is asymptotically normally distributed with variance 1.) Test statistics with this property can be constructed using a variety of methods. For instance, in many cases we can use asymptotically efficient estimators for the parameters μ and ν, combined with estimators for possible nuisance parameters, along the lines of Example 15.6.

If $p_\mu = q_\mu = f_\mu$ are equal and are densities on the real line, then rank statistics are attractive. Let $R_{N1}, \ldots, R_{NN}$ be the ranks of the pooled sample $X_1, \ldots, X_m, Y_1, \ldots, Y_n$. Consider the two-sample rank statistics

$$T_N = \frac{1}{\sqrt{N}} \sum_{i=m+1}^{N} a_{N,R_{Ni}}, \qquad a_{Ni} = \mathrm{E}\phi(U_{N(i)}),$$

for the score generating function

$$\phi(u) = \frac{1}{\sqrt{\lambda(1-\lambda)}\sqrt{I_\mu}} \dot{\ell}_\mu\big(F_\mu^{-1}(u)\big).$$

Up to a constant these are the locally most powerful scores introduced in Chapter 13. By Theorem 13.5, because $\bar{a}_N = \int_0^1 \phi(u)\,du = 0$,

$$T_N = -\frac{1}{\sqrt{I_\mu}} \left(\sqrt{1-\lambda}\frac{1}{\sqrt{m}} \sum_{i=1}^{m} \dot{\ell}_\mu(X_i) - \sqrt{\lambda}\frac{1}{\sqrt{n}} \sum_{j=1}^{n} \dot{\ell}_\mu(Y_j) \right) + o_{P_\mu}(1).$$

Thus, the locally most powerful rank statistics yield asymptotically optimal tests. In general, the optimal rank test depends on μ, and other parameters in the model, which must be estimated from the data, but in the most interesting cases this is not necessary.

15.15 *Example (Wilcoxon statistic).* For f_μ equal to the logistic density with mean μ, the scores $a_{N,i}$ are proportional to i. Thus, the Wilcoxon (or Mann-Whitney) two-sample statistic is asymptotically uniformly most powerful for testing a difference in location between two samples from logistic densities with different means. □

15.16 *Example (Log rank test).* The log rank test is asymptotically optimal for testing proportional hazard alternatives, given any baseline distribution. □

Notes

Absolute bounds on asymptotic power functions as developed in this chapter are less known than the absolute bounds on estimator sequences given in Chapter 8. Testing problems were nevertheless an important subject in Wald [149], who is credited by Le Cam for having first conceived of the method of approximating experiments by Gaussian experiments, albeit in a somewhat different way than later developed by Le Cam. From the point of view of statistical decision theory, there is no difference between testing and estimating, and hence the asymptotic bounds for tests in this chapter fit in the general theory developed in [99]. Wald appears to use the Gaussian approximation to transfer the optimality of the likelihood ratio and the Wald test (that is now named for him) in the Gaussian experiment to the sequence of experiments. In our discussion we use the Gaussian approximation to show that, in the multidimensional case, "asymptotic optimality" can only be defined in a somewhat arbitrary manner, because optimality in the Gaussian experiment is not easy to define. That is a difference of taste.

PROBLEMS

1. Consider the two-sided testing problem $H_0 : c^T h = 0$ versus $H_1 : c^T h \neq 0$ based on an $N_k(h, \Sigma)$-distributed observation X. A test for testing H_0 versus H_1 is called *unbiased* if $\sup_{h \in H_0} \pi(h) \leq \inf_{h \in H_1} \pi(h)$. The test that rejects H_0 for large values of $|c^T X|$ is uniformly most powerful among the unbiased tests. More precisely, for every power function π of a test based on X the conditions

$$\pi(h) \leq \alpha \quad \text{if } h^T c = 0 \quad \text{and} \quad \pi(h) \geq \alpha \quad \text{if } h^T c \neq 0,$$

   imply that, for every $c^T h \neq 0$,

$$\pi(h) \leq P\big(|c^T X| > z_{\alpha/2} \sqrt{c^T \Sigma c}\big) = 1 - \Phi\left(z_{\alpha/2} - \frac{c^T h}{\sqrt{c^T \Sigma c}}\right) + 1 - \Phi\left(z_{\alpha/2} + \frac{c^T h}{\sqrt{c^T \Sigma c}}\right).$$

   Formulate an asymptotic upper bound theorem for two-sided testing problems in the spirit of Theorem 15.4.

2. (i) Show that the set of power functions $h \mapsto \pi_n(h)$ in a dominated experiment $(P_h : h \in H)$ is compact for the topology of pointwise convergence (on H).

   (ii) Give a full proof of Theorem 15.1 along the following lines. First apply the proof as given for every finite subset $I \subset H$. This yields power functions π_I in the limit experiment that coincide with π on I.

3. Consider testing $H_0 : h = 0$ versus $H_1 : h \neq 0$ based on an observation X with an $N(h, \Sigma)$-distribution. Show that the testing problem is invariant under the transformations $x \mapsto \Sigma^{1/2} O \Sigma^{-1/2} x$ for O ranging over the orthonormal group. Find the best invariant test.

4. Consider testing $H_0 : h = 0$ versus $H_1 : h \neq 0$ based on an observation X with an $N(h, \Sigma)$-distribution. Find the test that maximizes the minimum power over $\{h : \|\Sigma^{-1/2} h\| = c\}$. (By the Hunt-Stein theorem the best invariant test is maximin, so one can apply the preceding problem. Alternatively, one can give a direct derivation along the following lines. Let π be the distribution of $\Sigma^{1/2} U$ if U is uniformly distributed on the set $\{h : \|h\| = c\}$. Derive the Neyman-Pearson test for testing $H_0 : N(0, \Sigma)$ versus $H_1 : \int N(h, \Sigma) \, d\pi(h)$. Show that its power is constant on $\{h : \|\Sigma^{-1/2} h\| = c\}$. The minimum power of any test on this set is always smaller than the average power over this set, which is the power at $\int N(h, \Sigma) \, d\pi(h)$.)

16

Likelihood Ratio Tests

The critical values of the likelihood ratio test are usually based on an asymptotic approximation. We derive the asymptotic distribution of the likelihood ratio statistic and investigate its asymptotic quality through its asymptotic power function and its Bahadur efficiency.

16.1 Introduction

Suppose that we observe a sample $X_1, \ldots, X_n$ from a density p_θ, and wish to test the null hypothesis $H_0 : \theta \in \Theta_0$ versus the alternative $H_1 : \theta \in \Theta_1$. If both the null and the alternative hypotheses consist of single points, then a most powerful test can be based on the log likelihood ratio, by the Neyman-Pearson theory. If the two points are θ_0 and θ_1, respectively, then the optimal test statistic is given by

$$\log \frac{\prod_{i=1}^n p_{\theta_1}(X_i)}{\prod_{i=1}^n p_{\theta_0}(X_i)}.$$

For certain special models and hypotheses, the most powerful test turns out not to depend on θ_1, and the test is uniformly most powerful for a composite hypothesis Θ_1. Sometimes the null hypothesis can be extended as well, and the testing problem has a fully satisfactory solution. Unfortunately, in many situations there is no single best test, not even in an asymptotic sense (see Chapter 15). A variety of ideas lead to reasonable tests. A sensible extension of the idea behind the Neyman-Pearson theory is to base a test on the log likelihood ratio

$$\tilde{\Lambda}_n = \log \frac{\sup_{\theta \in \Theta_1} \prod_{i=1}^n p_\theta(X_i)}{\sup_{\theta \in \Theta_0} \prod_{i=1}^n p_\theta(X_i)}.$$

The single points are replaced by maxima over the hypotheses. As before, the null hypothesis is rejected for large values of the statistic.

Because the distributional properties of $\tilde{\Lambda}_n$ can be somewhat complicated, one usually replaces the supremum in the numerator by a supremum over the whole parameter set $\Theta = \Theta_0 \cup \Theta_1$. This changes the test statistic only if $\tilde{\Lambda}_n \leq 0$, which is inessential, because in most cases the critical value will be positive. We study the asymptotic properties of the (log) *likelihood ratio statistic*

$$\Lambda_n = 2 \log \frac{\sup_{\theta \in \Theta} \prod_{i=1}^{n} p_\theta (X_i)}{\sup_{\theta \in \Theta_0} \prod_{i=1}^{n} p_\theta (X_i)} = 2(\tilde{\Lambda}_n \vee 0).$$

The most important conclusion of this chapter is that, under the null hypothesis, the sequence Λ_n is asymptotically chi squared-distributed. The main conditions are that the model is differentiable in θ and that the null hypothesis Θ_0 and the full parameter set Θ are (locally) equal to linear spaces. The number of degrees of freedom is equal to the difference of the (local) dimensions of Θ and Θ_0. Then the test that rejects the null hypothesis if Λ_n exceeds the upper α-quantile of the chi-square distribution is asymptotically of level α. Throughout the chapter we assume that $\Theta \subset \mathbb{R}^k$.

The "local linearity" of the hypotheses is essential for the chi-square approximation, which fails already in a number of simple examples. An open set is certainly locally linear at every of its points, and so is a relatively open subset of an affine subspace. On the other hand, a half line or space, which arises, for instance, if testing a one-sided hypothesis $H_0 : \theta \leq 0$, or a ball $H_0 : \|\theta\| \leq 1$, is not locally linear at its boundary points. In that case the asymptotic null distribution of the likelihood ratio statistic is not chi-square, but the distribution of a certain functional of a Gaussian vector.

Besides for testing, the likelihood ratio statistic is often used for constructing confidence regions for a parameter $\psi(\theta)$. These are defined, as usual, as the values τ for which a null hypothesis $H_0 : \psi(\theta) = \tau$ is not rejected. Asymptotic confidence sets obtained by using the chi-square approximation are thought to be of better coverage accuracy than those obtained by other asymptotic methods.

The likelihood ratio test has the desirable property of automatically achieving reduction of the data by sufficiency: The test statistic depends on a minimal sufficient statistic only. This is immediate from its definition as a quotient and the characterization of sufficiency by the factorization theorem. Another property of the test is also immediate: The likelihood ratio statistic is invariant under transformations of the parameter space that leave the null and alternative hypotheses invariant. This requirement is often imposed on test statistics but is not necessarily desirable.

16.1 Example (Multinomial vector). A vector $N = (N_1, \ldots, N_k)$ that possesses the multinomial distribution with parameters n and $p = (p_1, \ldots, p_k)$ can be viewed as the sum of n independent multinomial vectors with parameters 1 and p. By the sufficiency reduction, the likelihood ratio statistic based on N is the same as the statistic based on the single observations. Thus our asymptotic results apply to the likelihood ratio statistic based on N, if $n \to \infty$.

If the success probabilities are completely unknown, then their maximum likelihood estimator is N/n. Thus, the log likelihood ratio statistic for testing a null hypothesis $H_0 : p \in \mathcal{P}_0$ against the alternative $H_1 : p \notin \mathcal{P}_0$ is given by

$$2 \log \frac{\binom{n}{N_1 \cdots N_k} (N_1/n)^{N_1} \cdots (N_k/n)^{N_k}}{\sup_{p \in \mathcal{P}_0} \binom{n}{N_1 \cdots N_k} p_1^{N_1} \cdots p_k^{N_k}} = 2 \inf_{p \in \mathcal{P}_0} \sum_{i=1}^{k} N_i \log \left(\frac{N_i}{n p_i} \right).$$

The full parameter set can be identified with an open subset of $\mathbb{R}^{k-1}$, if p with zero coordinates are excluded. The null hypothesis may take many forms. For a simple null hypothesis the statistic is asymptotically chi-square distributed with $k - 1$ degrees of freedom. This follows from the general results in this chapter.[†]

Multinomial variables arise, among others, in testing goodness-of-fit. Suppose we wish to test that the true distribution of a sample of size n belongs to a certain parametric model $\{P_\theta : \theta \in \Theta\}$. Given a partition of the sample space into sets $\mathcal{X}_1, \ldots, \mathcal{X}_k$, define $N_1, \ldots, N_k$ as the numbers of observations falling into each of the sets of the partition. Then the vector $N = (N_1, \ldots, N_k)$ possesses a multinomial distribution, and the original problem can be translated in testing the null hypothesis that the success probabilities p have the form $\left(P_\theta(\mathcal{X}_1), \ldots, P_\theta(\mathcal{X}_k)\right)$ for some θ. $\square$

16.2   Example (Exponential families). Suppose that the observations are sampled from a density p_θ in the k-dimensional exponential family

$$p_\theta(x) = c(\theta)h(x)e^{\theta^T t(x)}.$$

Let $\Theta \subset \mathbb{R}^k$ be the natural parameter space, and consider testing a null hypothesis $\Theta_0 \subset \overset{\circ}{\Theta}$ versus its complement $\Theta - \Theta_0$. The log likelihood ratio statistic is given by

$$\Lambda_n = 2n \sup_{\theta \in \Theta} \inf_{\theta \in \Theta_0} \left[(\theta - \theta_0)^T \bar{t}_n + \log c(\theta) - \log c(\theta_0) \right].$$

This is closely related to the Kullback-Leibler divergence of the measures P_{θ_0} and P_θ, which is equal to

$$K(\theta, \theta_0) = P_\theta \log \frac{p_\theta}{p_{\theta_0}} = (\theta - \theta_0)^T P_\theta t + \log c(\theta) - \log c(\theta_0).$$

If the maximum likelihood estimator $\hat{\theta}$ exists and is contained in the interior of Θ, which is the case with probability tending to 1 if the true parameter is contained in $\overset{\circ}{\Theta}$, then $\hat{\theta}$ is the moment estimator that solves the equation $P_\theta t = \bar{t}_n$. Comparing the two preceding displays, we see that the likelihood ratio statistic can be written as $\Lambda_n = 2nK(\hat{\theta}, \Theta_0)$, where $K(\theta, \Theta_0)$ is the infimum of $K(\theta, \theta_0)$ over $\theta_0 \in \Theta_0$. This pretty formula can be used to study the asymptotic properties of the likelihood ratio statistic directly. Alternatively, the general results obtained in this chapter are applicable to exponential families. $\square$

*16.2   Taylor Expansion

Write $\hat{\theta}_{n,0}$ and $\hat{\theta}_n$ for the maximum likelihood estimators for θ if the parameter set is taken equal to Θ_0 or Θ, respectively, and set $\ell_\theta = \log p_\theta$. In this section assume that the true value of the parameter ϑ is an inner point of Θ. The likelihood ratio statistic can be rewritten as

$$\Lambda_n = -2 \sum_{i=1}^{n} \left(\ell_{\hat{\theta}_{n,0}}(X_i) - \ell_{\hat{\theta}_n}(X_i) \right).$$

To find the limit behavior of this sequence of random variables, we might replace $\sum \ell_\theta(X_i)$ by its Taylor expansion around the maximum likelihood estimator $\theta = \hat{\theta}_n$. If $\theta \mapsto \ell_\theta(x)$

[†] It is also proved in Chapter 17 by relating the likelihood ratio statistic to the chi-square statistic.

is twice continuously differentiable for every x, then there exists a vector $\tilde{\theta}_n$ between $\hat{\theta}_{n,0}$ and $\hat{\theta}_n$ such that the preceding display is equal to

$$-2(\hat{\theta}_{n,0} - \hat{\theta}_n) \sum_{i=1}^n \dot{\ell}_{\tilde{\theta}_n}(X_i) - (\hat{\theta}_{n,0} - \hat{\theta}_n)^T \sum \ddot{\ell}_{\tilde{\theta}_n}(X_i)(\hat{\theta}_{n,0} - \hat{\theta}_n).$$

Because $\hat{\theta}_n$ is the maximum likelihood estimator in the unrestrained model, the linear term in this expansion vanishes as soon as $\hat{\theta}_n$ is an inner point of Θ. If the averages $-n^{-1} \sum \ddot{\ell}_{\tilde{\theta}}(X_i)$ converge in probability to the Fisher information matrix I_ϑ and the sequence $\sqrt{n}(\hat{\theta}_{n,0} - \hat{\theta}_n)$ is bounded in probability, then we obtain the approximation

$$\Lambda_n = \sqrt{n}(\hat{\theta}_n - \hat{\theta}_{n,0})^T I_\vartheta \sqrt{n}(\hat{\theta}_n - \hat{\theta}_{n,0}) + o_{P_\vartheta}(1). \tag{16.3}$$

In view of the results of Chapter 5, the latter conditions are reasonable if $\vartheta \in \Theta_0$, for then both $\hat{\theta}_n$ and $\hat{\theta}_{n,0}$ can be expected to be $\sqrt{n}$-consistent. The preceding approximation, if it can be justified, sheds some light on the quality of the likelihood ratio test. It shows that, asymptotically, the likelihood ratio test measures a certain distance between the maximum likelihood estimators under the null and the full hypotheses. Such a procedure is intuitively reasonable, even though many other distance measures could be used as well. The use of the likelihood ratio statistic entails a choice as to how to weigh the different "directions" in which the estimators may differ, and thus a choice of weights for "distributing power" over different deviations. This is further studied in section 16.4.

If the null hypothesis is a single point $\Theta_0 = \{\theta_0\}$, then $\hat{\theta}_{n,0} = \theta_0$, and the quadratic form in the preceding display reduces under $H_0 : \theta = \theta_0$ (i.e., $\vartheta = \theta_0$) to $\hat{h}_n I_\vartheta \hat{h}_n$ for $\hat{h}_n = \sqrt{n}(\hat{\theta}_n - \vartheta)^T$. In view of the results of Chapter 5, the sequence $\hat{h}_n$ can be expected to converge in distribution to a variable $\hat{h}$ with a normal $N(0, I_\vartheta^{-1})$-distribution. Then the sequence Λ_n converges under the null hypothesis in distribution to the quadratic form $\hat{h}^T I_\vartheta \hat{h}$. This is the squared length of the standard normal vector $I_\vartheta^{1/2}\hat{h}$, and possesses a chi-square distribution with k degrees of freedom. Thus the chi-square approximation announced in the introduction follows.

The situation is more complicated if the null hypothesis is composite. If the sequence $\sqrt{n}(\hat{\theta}_{n,0} - \vartheta, \hat{\theta}_n - \vartheta)$ converges jointly to a variable $(\hat{h}_0, \hat{h})$, then the sequence Λ_n is asymptotically distributed as $(\hat{h} - \hat{h}_0)^T I_\vartheta (\hat{h} - \hat{h}_0)$. A null hypothesis Θ_0 that is (a segment of) a lower dimensional affine linear subspace is itself a "regular" parametric model. If it contains ϑ as a relative inner point, then the maximum likelihood estimator $\hat{\theta}_{n,0}$ may be expected to be asymptotically normal within this affine subspace, and the pair $\sqrt{n}(\hat{\theta}_{n,0} - \vartheta, \hat{\theta}_n - \vartheta)$ may be expected to be jointly asymptotically normal. Then the likelihood ratio statistic is asymptotically distributed as a quadratic form in normal variables. Closer inspection shows that this quadratic form possesses a chi-square distribution with $k - l$ degrees of freedom, where k and l are the dimensions of the full and null hypotheses. In comparison with the case of a simple null hypothesis, l degrees of freedom are "lost."

Because we shall rigorously derive the limit distribution by a different approach in the next section, we make this argument precise only in the particular case that the null hypothesis Θ_0 consists of all points $(\theta_1, \ldots, \theta_l, 0, \ldots, 0)$, if θ ranges over an open subset Θ of $\mathbb{R}^k$. Then the score function for θ under the null hypothesis consists of the first l coordinates of the score function $\dot{\ell}_\vartheta$ for the whole model, and the information matrix under the null hypothesis is equal to the $(l \times l)$ principal submatrix of I_ϑ. Write these as $\dot{\ell}_{\vartheta,\leq l}$ and $I_{\vartheta,\leq l,\leq l}$, respectively, and use a similar partitioning notation for other vectors and matrices.

Under regularity conditions we have the linear approximations (see Theorem 5.39)

$$\sqrt{n}(\hat{\theta}_{n,0,\leq l} - \vartheta_{\leq l}) = \frac{1}{\sqrt{n}} \sum_{i=1}^{n} I_{\vartheta,\leq l,\leq l}^{-1} \dot{\ell}_{\vartheta,\leq l}(X_i) + o_{P_\vartheta}(1),$$

$$\sqrt{n}(\hat{\theta}_n - \vartheta) = \frac{1}{\sqrt{n}} \sum_{i=1}^{n} I_\vartheta^{-1} \dot{\ell}_\vartheta(X_i) + o_{P_\vartheta}(1).$$

Given these approximations, the multivariate central limit theorem and Slutsky's lemma yield the joint asymptotic normality of the maximum likelihood estimators. From the form of the asymptotic covariance matrix we see, after some matrix manipulation,

$$\sqrt{n}(\hat{\theta}_{n,\leq l} - \hat{\theta}_{n,0,\leq l}) = -I_{\vartheta,\leq l,\leq l}^{-1} I_{\vartheta,\leq l,>l} \sqrt{n}\hat{\theta}_{n,>l} + o_P(1).$$

(Alternatively, this approximation follows from a Taylor expansion of $0 = \sum_{i=1}^{n} \dot{\ell}_{\hat{\theta}_{n,\leq l}}$ around $\hat{\theta}_{n,0,\leq l}$.) Substituting this in (16.3) and again carrying out some matrix manipulations, we find that the likelihood ratio statistic is asymptotically equivalent to (see problem 16.5)

$$\sqrt{n}\hat{\theta}_{n,>l}^T \left((I_\vartheta^{-1})_{>l,>l} \right)^{-1} \sqrt{n}\hat{\theta}_{n,>l}. \tag{16.4}$$

The matrix $(I_\vartheta^{-1})_{>l,>l}$ is the asymptotic covariance matrix of the sequence $\sqrt{n}\hat{\theta}_{n,>l}$, whence we obtain an asymptotic chi-square distribution with $k - l$ degrees of freedom, by the same argument as before.

We close this section by relating the likelihood ratio statistic to two other test statistics.

Under the simple null hypothesis $\Theta_0 = \{\theta_0\}$, the likelihood ratio statistic is asymptotically equivalent to both the *maximum likelihood statistic* (or *Wald statistic*) and the *score statistic*. These are given by

$$n(\hat{\theta}_n - \theta_0)^T I_{\theta_0}(\hat{\theta}_n - \theta_0) \quad \text{and} \quad \frac{1}{n}\left[\sum_{i=1}^{n} \dot{\ell}_{\theta_0}(X_i) \right]^T I_{\theta_0}^{-1} \left[\sum_{i=1}^{n} \dot{\ell}_{\theta_0}(X_i) \right].$$

The Wald statistic is a natural statistic, but it is often criticized for necessarily yielding ellipsoidal confidence sets, even if the data are not symmetric. The score statistic has the advantage that calculation of the supremum of the likelihood is unnecessary, but it appears to perform less well for smaller values of n.

In the case of a composite hypothesis, a Wald statistic is given in (16.4) and a score statistic can be obtained by substituting $n\hat{\theta}_{n,>l} \approx (I_\vartheta^{-1})_{>l,>l} \sum \dot{\ell}_{\hat{\theta}_{n,0,>l}}(X_i)$ in (16.4). (This approximation is obtainable from linearizing $\sum (\dot{\ell}_{\hat{\theta}_n} - \dot{\ell}_{\hat{\theta}_{n,0}})$.) In both cases we also replace the unknown parameter ϑ by an estimator.

16.3   Using Local Asymptotic Normality

An insightful derivation of the asymptotic distribution of the likelihood ratio statistic is based on convergence of experiments. This approach is possible for general experiments, but this section is restricted to the case of local asymptotic normality. The approach applies also in the case that the (local) parameter spaces are not linear.

Introducing the local parameter spaces $H_n = \sqrt{n}(\Theta - \vartheta)$ and $H_{n,0} = \sqrt{n}(\Theta_0 - \vartheta)$, we can write the likelihood ratio statistic in the form

$$\Lambda_n = 2 \sup_{h \in H_n} \log \frac{\prod_{i=1}^{n} p_{\vartheta + h/\sqrt{n}}(X_i)}{\prod_{i=1}^{n} p_\vartheta(X_i)} - 2 \sup_{h \in H_{n,0}} \log \frac{\prod_{i=1}^{n} p_{\vartheta + h/\sqrt{n}}(X_i)}{\prod_{i=1}^{n} p_\vartheta(X_i)}.$$

In Chapter 7 it is seen that, for large n, the rescaled likelihood ratio process in this display is similar to the likelihood ratio process of the normal experiment $(N(h, I_\vartheta^{-1}) : h \in \mathbb{R}^k)$. This suggests that, if the sets H_n and $H_{n,0}$ converge in a suitable sense to sets H and H_0, the sequence Λ_n converges in distribution to the random variable Λ obtained by substituting the normal likelihood ratios, given by

$$\Lambda = 2 \sup_{h \in H} \log \frac{dN(h, I_\vartheta^{-1})}{dN(0, I_\vartheta^{-1})}(X) - 2 \sup_{h \in H_0} \log \frac{dN(h, I_\vartheta^{-1})}{dN(0, I_\vartheta^{-1})}(X).$$

This is exactly the likelihood ratio statistic for testing the null hypothesis $H_0 : h \in H_0$ versus the alternative $H_1 : h \in H - H_0$ based on the observation X in the normal experiment. Because the latter experiment is simple, this heuristic is useful not only to derive the asymptotic distribution of the sequence Λ_n, but also to understand the asymptotic quality of the corresponding sequence of tests.

The likelihood ratio statistic for the normal experiment is

$$\begin{aligned} \Lambda &= \inf_{h \in H_0} (X - h)^T I_\vartheta (X - h) - \inf_{h \in H} (X - h)^T I_\vartheta (X - h) \\ &= \left\| I_\vartheta^{1/2} X - I_\vartheta^{1/2} H_0 \right\|^2 - \left\| I_\vartheta^{1/2} X - I_\vartheta^{1/2} H \right\|^2. \end{aligned} \tag{16.5}$$

The distribution of the sequence Λ_n under ϑ corresponds to the distribution of Λ under $h = 0$. Under $h = 0$ the vector $I_\vartheta^{1/2} X$ possesses a standard normal distribution. The following lemma shows that the squared distance of a standard normal variable to a linear subspace is chi square–distributed and hence explains the chi-square limit when H_0 is a linear space.

16.6   Lemma. *Let X be a k-dimensional random vector with a standard normal distribution and let H_0 be an l-dimensional linear subspace of $\mathbb{R}^k$. Then $\|X - H_0\|^2$ is chi square–distributed with $k - l$ degrees of freedom.*

Proof.   Take an orthonormal base of $\mathbb{R}^k$ such that the first l elements span H_0. By Pythagoras' theorem, the squared distance of a vector z to the space H_0 equals the sum of squares $\sum_{i > l} z_i^2$ of its last $k - l$ coordinates with respect to this basis. A change of base corresponds to an orthogonal transformation of the coordinates. Because the standard normal distribution is invariant under orthogonal transformations, the coordinates of X with respect to any orthonormal base are independent standard normal variables. Thus $\|X - H_0\|^2 = \sum_{i > l} X_i^2$ is chi square–distributed.   ∎

If ϑ is an inner point of Θ, then the set H is the full space $\mathbb{R}^k$ and the second term on the right of (16.5) is zero. Thus, if the local null parameter spaces $\sqrt{n}(\Theta_0 - \vartheta)$ converge to a linear subspace of dimension l, then the asymptotic null distribution of the likelihood ratio statistic is chi-square with $k - l$ degrees of freedom.

The following theorem makes the preceding informal derivation rigorous under the same mild conditions employed to obtain the asymptotic normality of the maximum likelihood estimator in Chapter 5. It uses the following notion of *convergence of sets*. Write $H_n \to H$ if H is the set of all limits $\lim h_n$ of converging sequences h_n with $h_n \in H_n$ for every n and, moreover, the limit $h = \lim_i h_{n_i}$ of every converging sequence h_{n_i} with $h_{n_i} \in H_{n_i}$ for every i is contained in H.

16.7   Theorem. *Let the model* $(P_\theta : \theta \in \Theta)$ *be differentiable in quadratic mean at* ϑ *with nonsingular Fisher information matrix, and suppose that for every* θ_1 *and* θ_2 *in a neighborhood of* ϑ *and for a measurable function* $\dot{\ell}$ *such that* $P_\vartheta \dot{\ell}^2 < \infty$,

$$\left| \log p_{\theta_1}(x) - \log p_{\theta_2}(x) \right| \leq \dot{\ell}(x) \, \|\theta_1 - \theta_2\|.$$

If the maximum likelihood estimators $\hat{\theta}_{n,0}$ *and* $\hat{\theta}_n$ *are consistent under* ϑ *and the sets* $H_{n,0}$ *and* H_n *converge to sets* H_0 *and* H, *then the sequence of likelihood ratio statistics* Λ_n *converges under* $\vartheta + h/\sqrt{n}$ *in distribution to* Λ *given in (16.5), for* X *normally distributed with mean* h *and covariance matrix* I_ϑ^{-1}.

*Proof.   Let $\mathbb{G}_n = \sqrt{n}(\mathbb{P}_n - P_\vartheta)$ be the empirical process, and define stochastic processes $\mathbb{Z}_n$ by

$$\mathbb{Z}_n(h) = n\mathbb{P}_n \log \frac{p_{\vartheta + h/\sqrt{n}}}{p_\vartheta} - h^T \mathbb{G}_n \dot{\ell}_\vartheta + \frac{1}{2} h^T I_\vartheta h.$$

The differentiability of the model implies that $\mathbb{Z}_n(h) \overset{P}{\to} 0$ for every h. In the proof of Theorem 7.12 this is strengthened to the uniform convergence

$$\sup_{\|h\| \leq M} |\mathbb{Z}_n(h)| \overset{P}{\to} 0, \qquad \text{every } M.$$

Furthermore, it follows from this proof that both $\hat{\theta}_{n,0}$ and $\hat{\theta}_n$ are $\sqrt{n}$-consistent under ϑ. (These statements can also be proved by elementary arguments, but under stronger regularity conditions.)

The preceding display is also valid for every sequence M_n that increases to ∞ sufficiently slowly. Fix such a sequence. By the $\sqrt{n}$-consistency, the estimators $\hat{\theta}_{n,0}$ and $\hat{\theta}_n$ are contained in the ball of radius $M_n/\sqrt{n}$ around ϑ with probability tending to 1. Thus, the limit distribution of Λ_n does not change if we replace the sets H_n and $H_{n,0}$ in its definition by the sets $H_n \cap \text{ball}(0, M_n)$ and $H_{n,0} \cap \text{ball}(0, M_n)$. These "truncated" sequences of sets still converge to H and H_0, respectively. Now, by the uniform convergence to zero of the processes $\mathbb{Z}_n(h)$ on H_n and $H_{n,0}$, and simple algebra,

$$\begin{aligned}
\Lambda_n &= 2 \sup_{h \in H_n} n\mathbb{P}_n \log \frac{p_{\vartheta + h/\sqrt{n}}}{p_\vartheta} - 2 \sup_{h \in H_{n,0}} n\mathbb{P}_n \log \frac{p_{\vartheta + h/\sqrt{n}}}{p_\vartheta} \\
&= 2 \sup_{h \in H_n} \left(h^T \mathbb{G}_n \dot{\ell}_\vartheta - \tfrac{1}{2} h^T I_\vartheta h \right) - 2 \sup_{h \in H_{n,0}} \left(h^T \mathbb{G}_n \dot{\ell}_\vartheta - \tfrac{1}{2} h^T I_\vartheta h \right) + o_P(1) \\
&= \left\| I_\vartheta^{-1/2} \mathbb{G}_n \dot{\ell}_\vartheta - I_\vartheta^{1/2} H_0 \right\|^2 - \left\| I_\vartheta^{-1/2} \mathbb{G}_n \dot{\ell}_\vartheta - I_\vartheta^{1/2} H \right\|^2 + o_P(1)
\end{aligned}$$

by Lemma 7.13 (ii) and (iii). The theorem follows by the continuous-mapping theorem. ∎

16.8   Example (Generalized linear models).   In a generalized linear model a typical observation (X, Y), consisting of a "covariate vector" X and a "response" Y, possesses a density of the form

$$p_\beta(x, y) = e^{yk(\beta^T x)\phi - bok(\beta^T x)\phi} c_\phi(y) p_X(x).$$

(It may be more natural to model the covariates as (observed) constants, but to fit the model into our i.i.d. setup, we consider them to be a random sample from a density p_X.) Thus, given

X, the variable Y follows an exponential family density $e^{y\theta\phi - b(\theta)\phi}c_\phi(y)$ with parameters $\theta = k(\beta^T X)$ and ϕ. Using the identities for exponential families based on Lemma 4.5, we obtain

$$\mathrm{E}_\beta(Y \mid X) = b' \circ k(\beta^T X), \qquad \mathrm{var}_{\beta,\phi}(Y \mid X) = \frac{b'' \circ k(\beta^T X)}{\phi}.$$

The function $(b' \circ k)^{-1}$ is called the *link function* of the model and is assumed known. To make the parameter β identifiable, we assume that the matrix $\mathrm{E} X X^T$ exists and is nonsingular.

To judge the goodness-of-fit of a generalized linear model to a given set of data $(X_1, Y_1), \ldots, (X_n, Y_n)$, it is customary to calculate, for fixed ϕ, the log likelihood ratio statistic for testing the model as described previously within the model in which each Y_i, given X_i, still follows the given exponential family density, but in which the parameters θ (and hence the conditional means $\mathrm{E}(Y_i \mid X_i)$) are allowed to be arbitrary values θ_i, unrelated across the n observations (X_i, Y_i). This statistic, with the parameter ϕ set to 1, is known as the *deviance*, and takes the form, with $\hat{\beta}_n$ the maximum likelihood estimator for β,[†]

$$D(\vec{Y}_n, \hat{\mu}) = -2 \log \frac{\sup_\beta \prod_{i=1}^n e^{Y_i k(\beta^T X_i) - b \circ k(\beta^T X_i)}}{\sup_{\theta_1, \ldots, \theta_n} \prod_{i=1}^n e^{Y_i \theta_i - b(\theta_i)}}$$

$$= -2 \sum_{i=1}^n \left[Y_i \left(k(\hat{\beta}_n^T X_i) - (b')^{-1}(Y_i) \right) - b \circ k(\hat{\beta}_n^T X_i) + b \circ (b')^{-1}(Y_i) \right].$$

In our present setup, the codimension of the null hypothesis within the "full model" is equal to $n - k$, if β is k-dimensional, and hence the preceding theory does not apply to the deviance. (This could be different if there are multiple responses for every given covariate and the asymptotics are relative to the number of responses.) On the other hand, the preceding theory allows an "analysis of deviance" to test nested sequences of regression models corresponding to inclusion or exclusion of a given covariate (i.e., column of the regression matrix). For instance, if $D_i(\vec{Y}_n, \hat{\mu}_{(i)})$ is the deviance of the model in which the $i + 1, i + 2, \ldots, k$th coordinates of β are a priori set to zero, then the difference $D_{i-1}(\vec{Y}_n, \hat{\mu}_{(i-1)}) - D_i(\vec{Y}_n, \hat{\mu}_{(i)})$ is the log likelihood ratio statistic for testing that the ith coordinate of β is zero within the model in which all higher coordinates are zero. According to the theory of this chapter, ϕ times this statistic is asymptotically chi square–distributed with one degree of freedom under the smaller of the two models.

To see this formally, it suffices to verify the conditions of the preceding theorem. Using the identities for exponential families based on Lemma 4.5, the score function and Fisher information matrix can be computed to be

$$\dot{\ell}_\beta(x, y) = (y - b' \circ k(\beta^T x)) k'(\beta^T x) x,$$
$$I_\beta = \mathrm{E} b'' \circ k(\beta^T X) k'(\beta^T X)^2 X X^T.$$

Depending on the function k, these are very well-behaved functions of β, because b is a strictly convex, analytic function on the interior of the natural parameter space of the family, as is seen in section 4.2. Under reasonable conditions the function $\sup_{\beta \in U} \| \dot{\ell}_\beta \|$ is

---

[†] The arguments $\vec{Y}_n$ and $\hat{\mu}$ of D are the vectors of estimated (conditional) means of Y given the full model and the generalized linear model, respectively. Thus $\hat{\mu}_i = b' \circ k(\hat{\beta}_n^T X_i)$.

square-integrable, for every small neighborhood U, and the Fisher information is continuous. Thus, the local conditions on the model are easily satisfied.

Proving the consistency of the maximum likelihood estimator may be more involved, depending on the link function. If the parameter β is restricted to a compact set, then most approaches to proving consistency apply without further work, including Wald's method, Theorem 5.7, and the classical approach of section 5.7. The last is particularly attractive in the case of *canonical link functions*, which correspond to setting k equal to the identity. Then the second-derivative matrix $\ddot{\ell}_\beta$ is equal to $-b''(\beta^T x)xx^T$, whence the likelihood is a strictly concave function of β whenever the observed covariate vectors are of full rank. Consequently, the point of maximum of the likelihood function is unique and hence consistent under the conditions of Theorem 5.14.[†] □

16.9   Example (Location scale). Suppose we observe a sample from the density $f\big((x - \mu)/\sigma\big)/\sigma$ for a given probability density f, and a location-scale parameter $\theta = (\mu, \sigma)$ ranging over the set $\Theta = \mathbb{R} \times \mathbb{R}^+$. We consider two testing problems.

(i). Testing $H_0 : \mu = 0$ versus $H_1 : \mu \neq 0$ corresponds to setting $\Theta_0 = \{0\} \times \mathbb{R}^+$. For a given point $\vartheta = (0, \sigma)$ from the null hypothesis the set $\sqrt{n}(\Theta_0 - \vartheta)$ equals $\{0\} \times (-\sqrt{n}\sigma, \infty)$ and converges to the linear space $\{0\} \times \mathbb{R}$. Under regularity conditions on f, the sequence of likelihood ratio statistics is asymptotically chi square–distributed with 1 degree of freedom.

(ii). Testing $H_0 : \mu \leq 0$ versus $H_1 : \mu > 0$ corresponds to setting $\Theta_0 = (-\infty, 0] \times \mathbb{R}^+$. For a given point $\vartheta = (0, \sigma)$ on the boundary of the null hypothesis, the sets $\sqrt{n}(\Theta_0 - \vartheta)$ converge to $H_0 = (-\infty, 0] \times \mathbb{R}$. In this case, the limit distribution of the likelihood ratio statistics is not chi-square but equals the distribution of the square distance of a standard normal vector to the set $I_\vartheta^{1/2} H_0 = \{h : \langle h, I_\vartheta^{-1/2} e_1 \rangle \leq 0\}$. The latter is a half-space with boundary line through the origin. Because a standard normal vector is rotationally symmetric, the distribution of its distance to a half-space of this type does not depend on the orientation of the half-space. Thus the limit distribution is equal to the distribution of the squared distance of a standard normal vector to the half-space $\{h : h_2 \leq 0\}$: the distribution of $(Z \vee 0)^2$ for a standard normal variable Z. Because $P\big((Z \vee 0)^2 > c\big) = \frac{1}{2}P(Z^2 > c)$ for every $c > 0$, we must choose the critical value of the test equal to the upper 2α-quantile of the chi-square distribution with 1 degree of freedom. Then the asymptotic level of the test is α for every ϑ on the boundary of the null hypothesis (provided $\alpha < 1/2$).

For a point ϑ in the interior of the null hypothesis $H_0 : \mu \leq 0$ the sets $\sqrt{n}(\Theta_0 - \vartheta)$ converge to $\mathbb{R} \times \mathbb{R}$ and the sequence of likelihood ratio statistics converges in distribution to the squared distance to the whole space, which is zero. This means that the probability of an error of the first kind converges to zero for every ϑ in the interior of the null hypothesis. □

16.10   Example (Testing a ball). Suppose we wish to test the null hypothesis $H_0 : \|\theta\| \leq 1$ that the parameter belongs to the unit ball versus the alternative $H_1 : \|\theta\| > 1$ that this is not case.

If the true parameter ϑ belongs to the interior of the null hypothesis, then the sets $\sqrt{n}(\Theta_0 - \vartheta)$ converge to the whole space, whence the sequence of likelihood ratio statistics converges in distribution to zero.

[†] For a detailed study of sufficient conditions for consistency see [45].

For ϑ on the boundary of the unit ball, the sets $\sqrt{n}(\Theta_0 - \vartheta)$ grow to the half-space $H_0 = \{h : \langle h, \vartheta \rangle \le 0\}$. The sequence of likelihood ratio statistics converges in distribution to the distribution of the square distance of a standard normal vector to the half-space $I_\vartheta^{1/2} H_0 = \{h : \langle h, I_\vartheta^{-1/2} \vartheta \rangle \le 0\}$. By the same argument as in the preceding example, this is the distribution of $(Z \vee 0)^2$ for a standard normal variable Z. Once again we find an asymptotic level-α test by using a 2α-quantile. $\square$

16.11 *Example (Testing a range).* Suppose that the null hypothesis is equal to the image $\Theta_0 = g(T)$ of an open subset T of a Euclidean space of dimension $l \le k$. If g is a homeomorphism, continuously differentiable, and of full rank, then the sets $\sqrt{n}(\Theta_0 - g(\tau))$ converge to the range of the derivative of g at τ, which is a subspace of dimension l.

Indeed, for any $\eta \in \mathbb{R}^l$ the vectors $\tau + \eta/\sqrt{n}$ are contained in T for sufficiently large n, and the sequence $\sqrt{n}(g(\tau + \eta/\sqrt{n}) - g(\tau))$ converges to $g_\tau' \eta$. Furthermore, if a subsequence of $\sqrt{n}(g(t_n) - g(\tau))$ converges to a point h for a given sequence t_n in T, then the corresponding subsequence of $\sqrt{n}(t_n - \tau)$ converges to $\eta = (g^{-1})_{g(\tau)}' h$ by the differentiability of the inverse mapping g^{-1} and hence $\sqrt{n}(g(t_n) - g(\tau)) \to g_\tau' \eta$. (We can use the rank theorem to give a precise definition of the differentiability of the map g^{-1} on the manifold $g(T)$.) $\square$

16.4 Asymptotic Power Functions

Because the sequence of likelihood ratio statistics converges to the likelihood ratio statistic in the Gaussian limit experiment, the likelihood ratio test is "asymptotically efficient" in the same way as the likelihood ratio statistic in the limit experiment is "efficient." If the local limit parameter set H_0 is a half-space or a hyperplane, then the latter test is uniformly most powerful, and hence the likelihood ratio tests are asymptotically optimal (see Proposition 15.2). This is the case, in particular, for testing a simple null hypothesis in a one-dimensional parametric model. On the other hand, if the hypotheses are higher-dimensional, then there is often no single best test, not even under reasonable restrictions on the class of admitted tests. For different (one-dimensional) deviations of the null hypothesis, different tests are optimal (see the discussion in Chapter 15). The likelihood ratio test is an omnibus test that gives reasonable power in all directions. In this section we study its local asymptotic power function more closely.

We assume that the parameter ϑ is an inner point of the parameter set and denote the true parameter by $\vartheta + h/\sqrt{n}$. Under the conditions of Theorem 16.7, the sequence of likelihood ratio statistics is asymptotically distributed as

$$\Lambda = \left\| Z + I_\vartheta^{1/2} h - I_\vartheta^{1/2} H_0 \right\|^2$$

for a standard normal vector Z. Suppose that the limiting local parameter set H_0 is a linear subspace of dimension l, and that the null hypothesis is rejected for values of Λ_n exceeding the critical value $\chi_{k-l,\alpha}^2$. Then the local power functions of the resulting tests satisfy

$$\pi_n \left(\vartheta + \frac{h}{\sqrt{n}} \right) = \mathrm{P}_{\vartheta + h/\sqrt{n}} (\Lambda_n > \chi_{k-l,\alpha}^2) \to \mathrm{P}_h (\Lambda > \chi_{k-l,\alpha}^2) =: \pi(h).$$

The variable Λ is the squared distance of the vector Z to the affine subspace $-I_\vartheta^{1/2} h + I_\vartheta^{1/2} H_0$. By the rotational invariance of the normal distribution, the distribution of Λ does not depend on the orientation of the affine subspace, but only on its codimension and its distance

$\delta = \|I_\vartheta^{1/2}h - I_\vartheta^{1/2}H_0\|$ to the origin. This distribution is known as the *noncentral chi-square distribution* with noncentrality parameter δ. Thus

$$\pi(h) = P\left(\chi^2_{k-l}\left(\|I_\vartheta^{1/2}h - I_\vartheta^{1/2}H_0\|\right) > \chi^2_{k-l,\alpha}\right).$$

The noncentral chi-square distributions are stochastically increasing in the noncentrality parameter. It follows that the likelihood ratio test has good (local) power at h that yield a large value of the noncentrality parameter.

The shape of the asymptotic power function is easiest to understand in the case of a simple null hypothesis. Then $H_0 = \{0\}$, and the noncentrality parameter reduces to the square root of $h^T I_\vartheta h$. For $h = \mu h_e$ equal to a multiple of an eigenvector h_e (of unit norm) of I_ϑ with eigenvalue λ_e, the noncentrality parameter equals $\sqrt{\lambda_e}\mu$. The asymptotic power function in the direction of h_e equals

$$\pi(\mu h_e) = P\left(\chi^2_k(\sqrt{\lambda_e}\mu) > \chi^2_{k,\alpha}\right).$$

The test performs best for departures from the null hypothesis in the direction of the eigenvector corresponding to the largest eigenvalue. Even though the likelihood ratio test gives power in all directions, it does not treat the directions equally. This may be worrisome if the eigenvalues are very inhomogeneous.

Further insight is gained by comparing the likelihood ratio test to tests that are designed to be optimal in given directions. Let X be an observation in the limit experiment, having a $N(h, I_\vartheta^{-1})$-distribution. The test that rejects the null hypothesis $H_0 = \{0\}$ if $|\sqrt{\lambda_e}\, h_e^T X| > z_{\alpha/2}$ has level α and power function

$$\pi_{h_e}(\mu h_e) = P\left(\chi^2_1(\sqrt{\lambda_e}\mu) > \chi^2_{1,\alpha}\right).$$

For large k this is a considerably better power function than the power function of the likelihood ratio test (Figure 16.1), but the forms of the power functions are similar. In particular, the optimal power functions show a similar dependence on the eigenvalues of

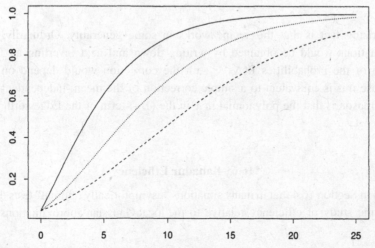

Figure 16.1. The functions $\mu^2 \to P\left(\chi^2_k(\mu) > \chi^2_{k,\alpha}\right)$ for $k = 1$ (*solid*), $k = 5$ (*dotted*) and $k = 15$ (*dashed*), respectively, for $\alpha = 0.05$.

the covariance matrix. In this sense, the apparently unequal distribution of power over the different directions is not unfair in that it reflects the intrinsic difficulty of detecting changes in different directions. This is not to say that we should never change the (automatic) emphasis given by the likelihood ratio test.

16.5   Bartlett Correction

The chi-square approximation to the distribution of the likelihood ratio statistic is relatively accurate but can be much improved by a correction. This was first noted in the example of testing for inequality of the variances in the one-way layout by Bartlett and has since been generalized. Although every approximation can be improved, the *Bartlett correction* appears to enjoy a particular popularity.

The correction takes the form of a correction of the (asymptotic) mean of the likelihood ratio statistic. In regular cases the distribution of the likelihood ratio statistic is asymptotically chi-square with, say, r degrees of freedom, whence its mean ought to be approximately equal to r. Bartlett's correction is intended to make the mean exactly equal to r, by replacing the likelihood ratio statistic Λ_n by

$$\frac{r\Lambda_n}{E_{\theta_0}\Lambda_n}.$$

The distribution of this statistic is next approximated by a chi-square distribution with r degrees of freedom. Unfortunately, the mean $E_{\theta_0}\Lambda_n$ may be hard to calculate, and may depend on an unknown null parameter θ_0. Therefore, one first obtains an expression for the mean of the form

$$E_{\theta_0}\Lambda_n = 1 + \frac{b(\theta_0)}{n} + \cdots.$$

Next, with $\hat{b}_n$ an appropriate estimator for the parameter $b(\theta_0)$, the corrected statistic takes the form

$$\frac{r\Lambda_n}{1 + \hat{b}_n/n}.$$

The surprising fact is that this recipe works in some generality. Ordinarily, improved approx-imations would be obtained by writing down and next inverting an Edgeworth expansion of the probabilities $P(\Lambda_n \leq x)$; the correction would depend on x. In the present case this is equivalent to a simple correction of the mean, independent of x. The technical reason is that the polynomial in x in the $(1/n)$-term of the Edgeworth expansion is of degree 1.[†]

*16.6   Bahadur Efficiency

The claim in Section 16.4 that in many situations "asymptotically optimal" tests do not exist refers to the study of efficiency relative to the local Gaussian approximations described

---

[†] For a further discussion, see [5], [9], and [83], and the references cited there.

in Chapter 7. The purpose of this section is to show that, under regularity conditions, the likelihood ratio test is asymptotically optimal in a different setting, the one of Bahadur efficiency.

For simplicity we restrict ourselves to the testing of finite hypotheses. Given finite sets $\mathcal{P}_0$ and $\mathcal{P}_1$ of probability measures on a measurable space $(\mathcal{X}, \mathcal{A})$ and a random sample $X_1, \ldots, X_n$, we study the log likelihood ratio statistic

$$\tilde{\Lambda}_n = \log \frac{\sup_{Q \in \mathcal{P}_1} \prod_{i=1}^n q(X_i)}{\sup_{P \in \mathcal{P}_0} \prod_{i=1}^n p(X_i)}.$$

More general hypotheses can be treated, under regularity conditions, by finite approximation (see e.g., Section 10 of [4]).

The *observed level* of a test that rejects for large values of a statistic T_n is defined as

$$L_n = \sup_{P \in \mathcal{P}_0} P_P(T_n \geq t)_{|t=T_n}.$$

The test that rejects the null hypothesis if $L_n \leq \alpha$ has level α. The power of this test is maximal if L_n is "minimal" under the alternative (in a stochastic sense). The *Bahadur slope* under the alternative Q is defined as the limit in probability under Q (if it exists) of the sequence $(-2/n) \log L_n$. If this is "large," then L_n is small and hence we prefer sequences of test statistics that have a large slope. The same conclusion is reached in section 14.4 by considering the asymptotic relative Bahadur efficiencies. It is indicated there that the Neyman-Pearson tests for testing the simple null and alternative hypotheses P and Q have Bahadur slope $-2Q \log(p/q)$. Because these are the most powerful tests, this is the maximal slope for testing P versus Q. (We give a precise proof in the following theorem.) Consequently, the slope for a general null hypothesis cannot be bigger than $\inf_{P \in \mathcal{P}_0} -2Q \log(p/q)$. The sequence of likelihood ratio statistics attains equality, even if the alternative hypothesis is composite.

16.12 Theorem. *The Bahadur slope of any sequence of test statistics for testing an arbitrary null hypothesis $H_0 : P \in \mathcal{P}_0$ versus a simple alternative $H_1 : P = Q$ is bounded above by $\inf_{P \in \mathcal{P}_0} -2Q \log(p/q)$, for any probability measure Q. If $\mathcal{P}_0$ and $\mathcal{P}_1$ are finite sets of probability measures, then the sequence of likelihood ratio statistics for testing $H_0 : P \in \mathcal{P}_0$ versus $H_1 : P \in \mathcal{P}_1$ attains equality for every $Q \in \mathcal{P}_1$.*

Proof. Because the observed level is a supremum over $\mathcal{P}_0$, it suffices to prove the upper bound of the theorem for a simple null hypothesis $\mathcal{P}_0 = \{P\}$. If $-2Q \log(p/q) = \infty$, then there is nothing to prove. Thus, we can assume without loss of generality that Q is absolutely continuous with respect to P. Write Λ_n for $\log \prod_{i=1}^n (q/p)(X_i)$. Then, for any constants $B > A > Q \log(q/p)$,

$$P_Q(L_n < e^{-nB}, \Lambda_n < nA) = E_P 1\{L_n < e^{-nB}, \Lambda_n < nA\} e^{\Lambda_n}$$
$$\leq e^{nA} P_P(L_n < e^{-nB}).$$

Because L_n is superuniformly distributed under the null hypothesis, the last expression is bounded above by $\exp -n(B - A)$. Thus, the sum of the probabilities on the left side over $n \in \mathbb{N}$ is finite, whence $-(2/n) \log L_n \leq 2B$ or $\Lambda_n \geq nA$ for all sufficiently large n, almost surely under Q, by the Borel-Cantelli lemma. Because the sequence $n^{-1} \Lambda_n$

converges almost surely under Q to $Q \log(q/p) < A$, by the strong law of large numbers, the second possibility can occur only finitely many times. It follows that $-(2/n) \log L_n \leq 2B$ eventually, almost surely under Q. This having been established for any $B > Q \log(q/p)$, the proof of the first assertion is complete.

To prove that the likelihood ratio statistic attains equality, it suffices to prove that its slope is bigger than the upper bound. Write $\tilde{\Lambda}_n$ for the log likelihood ratio statistic, and write $\sup_P$ and $\sup_Q$ for suprema over the null and alternative hypotheses. Because $(1/n)\tilde{\Lambda}_n$ is bounded above by $\sup_Q \mathbb{P}_n \log(q/p)$, we have, by Markov's inequality,

$$\mathrm{P}_P\left(\frac{1}{n}\tilde{\Lambda}_n \geq t\right) \leq \sum_Q \mathrm{P}_P\left(\mathbb{P}_n \log \frac{q}{p} \geq t\right) \leq |\mathcal{P}_1| \max_Q e^{-nt} \mathrm{E}_P e^{n\mathbb{P}_n \log(q/p)}.$$

The expectation on the right side is the nth power of the integral $\int (q/p)\, dP = Q(p > 0) \leq 1$. Take logarithms left and right and multiply with $-(2/n)$ to find that

$$-\frac{2}{n} \log \mathrm{P}_P\left(\frac{1}{n}\tilde{\Lambda}_n \geq t\right) \geq 2t - \frac{2 \log |\mathcal{P}_1|}{n}.$$

Because this is valid uniformly in t and P, we can take the infimum over P on the left side; next evaluate the left and right sides at $t = (1/n)\tilde{\Lambda}_n$. By the law of large numbers, $\mathbb{P}_n \log(q/p) \to Q \log(q/p)$ almost surely under Q, and this remains valid if we first add the infimum over the (finite) set $\mathcal{P}_0$ on both sides. Thus, the limit inferior of the sequence $(1/n)\tilde{\Lambda}_n \geq \inf_P \mathbb{P}_n \log(q/p)$ is bounded below by $\inf_P Q \log(q/p)$ almost surely under Q, where we interpret $Q \log(q/p)$ as ∞ if $Q(p = 0) > 0$. Insert this lower bound in the preceding display to conclude that the Bahadur slope of the likelihood ratio statistics is bounded below by $2 \inf_P Q \log(q/p)$. ∎

Notes

The classical references on the asymptotic null distribution of likelihood ratio statistic are papers by Chernoff [21] and Wilks [150]. Our main theorem appears to be better than Chernoff's, who uses the "classical regularity conditions" and a different notion of approximation of sets, but is not essentially different. Wilks' treatment would not be acceptable to present-day referees but maybe is not so different either. He appears to be saying that we can replace the original likelihood by the likelihood for having observed only the maximum likelihood estimator (the error is asymptotically negligible), next refers to work by Doob to infer that this is a Gaussian likelihood, and continues to compute the likelihood ratio statistic for a Gaussian likelihood, which is easy, as we have seen. The approach using a Taylor expansion and the asymptotic distributions of both likelihood estimators is one way to make the argument rigorous, but it seems to hide the original intuition.

Bahadur [3] presented the efficiency of the likelihood ratio statistic at the fifth Berkeley symposium. Kallenberg [84] shows that the likelihood ratio statistic remains asymptotically optimal in the setting in which both the desired level and the alternative tend to zero, at least in exponential families. As the proof of Theorem 16.12 shows, the composite nature of the alternative hypothesis "disappears" elegantly by taking $(1/n)$ log of the error probabilities – too elegantly to attach much value to this type of optimality?

PROBLEMS

1. Let $(X_1, Y_1), \ldots, (X_n, Y_n)$ be a sample from the bivariate normal distribution with mean vector (μ, ν) and covariance matrix the diagonal matrix with entries σ^2 and τ^2. Calculate (or characterize) the likelihood ratio statistic for testing $H_0 : \mu = \nu$ versus $H_1 : \mu \neq \nu$.

2. Let N be a kr-dimensional multinomial variable written as a $(k \times r)$ matrix (N_{ij}). Calculate the likelihood ratio statistic for testing the null hypothesis of independence $H_0 : p_{ij} = p_{i \cdot} p_{\cdot j}$ for every i and j. Here the dot denotes summation over all columns and rows, respectively. What is the limit distribution under the null hypothesis?

3. Calculate the likelihood ratio statistic for testing $H_0 : \mu = \nu$ based on independent samples of size n from multivariate normal distributions $N_r(\mu, \Sigma)$ and $N_r(\nu, \Sigma)$. The matrix Σ is unknown. What is the limit distribution under the null hypothesis?

4. Calculate the likelihood ratio statistic for testing $H_0 : \mu_1 = \cdots = \mu_k$ based on k independent samples of size n from $N(\mu_j, \sigma^2)$-distributions. What is the asymptotic distribution under the null hypothesis?

5. Show that $(I_{\vartheta}^{-1})_{>l, >l}$ is the inverse of the matrix $I_{\vartheta, >l, >l} - I_{\vartheta, >l, \leq l} I_{\vartheta, \leq l, \leq l}^{-1} I_{\vartheta, \leq l, >l}$.

6. Study the asymptotic distribution of the sequence $\tilde{\Lambda}_n$ if the true parameter is contained in both the null and alternative hypotheses.

7. Study the asymptotic distribution of the likelihood ratio statistics for testing the hypothesis $H_0 : \sigma = -\tau$ based on a sample of size n from the uniform distribution on $[\sigma, \tau]$. Does the asymptotic distribution correspond to a likelihood ratio statistic in a limit experiment?

17

Chi-Square Tests

The chi-square statistic for testing hypotheses concerning multinomial distributions derives its name from the asymptotic approximation to its distribution. Two important applications are the testing of independence in a two-way classification and the testing of goodness-of-fit. In the second application the multinomial distribution is created artificially by grouping the data, and the asymptotic chi-square approximation may be lost if the original data are used to estimate nuisance parameters.

17.1 Quadratic Forms in Normal Vectors

The *chi-square distribution* with k degrees of freedom is (by definition) the distribution of $\sum_{i=1}^{k} Z_i^2$ for i.i.d. $N(0, 1)$-distributed variables $Z_1, \ldots, Z_k$. The sum of squares is the squared norm $\|Z\|^2$ of the standard normal vector $Z = (Z_1, \ldots, Z_k)$. The following lemma gives a characterization of the distribution of the norm of a general zero-mean normal vector.

17.1 Lemma. *If the vector X is $N_k(0, \Sigma)$-distributed, then $\|X\|^2$ is distributed as $\sum_{i=1}^{k} \lambda_i Z_i^2$ for i.i.d. $N(0, 1)$-distributed variables $Z_1, \ldots, Z_k$ and $\lambda_1, \ldots, \lambda_k$ the eigenvalues of Σ.*

Proof. There exists an orthogonal matrix O such that $O\Sigma O^T = \operatorname{diag}(\lambda_i)$. Then the vector OX is $N_k(0, \operatorname{diag}(\lambda_i))$-distributed, which is the same as the distribution of the vector $(\sqrt{\lambda_1} Z_1, \ldots, \sqrt{\lambda_k} Z_k)$. Now $\|X\|^2 = \|OX\|^2$ has the same distribution as $\sum (\sqrt{\lambda_i} Z_i)^2$. ∎

The distribution of a quadratic form of the type $\sum_{i=1}^{k} \lambda_i Z_i^2$ is complicated in general. However, in the case that every λ_i is either 0 or 1, it reduces to a chi-square distribution. If this is not naturally the case in an application, then a statistic is often transformed to achieve this desirable situation. The definition of the Pearson statistic illustrates this.

17.2 Pearson Statistic

Suppose that we observe a vector $X_n = (X_{n,1}, \ldots, X_{n,k})$ with the multinomial distribution corresponding to n trials and k classes having probabilities $p = (p_1, \ldots, p_k)$. The *Pearson statistic* for testing the null hypothesis $H_0 : p = a$ is given by

$$C_n(a) = \sum_{i=1}^{k} \frac{(X_{n,i} - na_i)^2}{na_i}.$$

We shall show that the sequence $C_n(a)$ converges in distribution to a chi-square distribution if the null hypothesis is true. The practical relevance is that we can use the chi-square table to find critical values for the test. The proof shows why Pearson divided the squares by na_i and did not propose the simpler statistic $\|X_n - na\|^2$.

17.2   Theorem. *If the vectors X_n are multinomially distributed with parameters n and $a = (a_1, \ldots, a_k) > 0$, then the sequence $C_n(a)$ converges under a in distribution to the χ^2_{k-1}-distribution.*

Proof. The vector X_n can be thought of as the sum of n independent multinomial vectors $Y_1, \ldots, Y_n$ with parameters 1 and $a = (a_1, \ldots, a_k)$. Then

$$\mathrm{E}Y_i = a, \qquad \mathrm{Cov}\, Y_i = \begin{pmatrix} a_1(1-a_1) & -a_1a_2 & \cdots & -a_1a_k \\ -a_2a_1 & a_2(1-a_2) & \cdots & -a_2a_k \\ \vdots & \vdots & & \vdots \\ -a_ka_1 & -a_ka_2 & \cdots & a_k(1-a_k) \end{pmatrix}.$$

By the multivariate central limit theorem, the sequence $n^{-1/2}(X_n - na)$ converges in distribution to the $N_k(0, \mathrm{Cov}\, Y_1)$-distribution. Consequently, with $\sqrt{a}$ the vector with coordinates $\sqrt{a_i}$,

$$\left(\frac{X_{n,1} - na_1}{\sqrt{na_1}}, \ldots, \frac{X_{n,k} - na_k}{\sqrt{na_k}} \right) \rightsquigarrow N(0, I - \sqrt{a}\sqrt{a}^T).$$

Because $\sum a_i = 1$, the matrix $I - \sqrt{a}\sqrt{a}^T$ has eigenvalue 0, of multiplicity 1 (with eigenspace spanned by $\sqrt{a}$), and eigenvalue 1, of multiplicity $(k-1)$ (with eigenspace equal to the orthocomplement of $\sqrt{a}$). An application of the continuous-mapping theorem and next Lemma 17.1 conclude the proof.   ∎

The number of degrees of freedom in the chi-squared approximation for Pearson's statistic is the number of cells of the multinomial vector that have positive probability. However, the quality of the approximation also depends on the size of the cell probabilities a_j. For instance, if 1001 cells have null probabilities $10^{-23}, \ldots, 10^{-23}, 1 - 10^{-20}$, then it is clear that for moderate values of n all cells except one are empty, and a huge value of n is necessary to make a χ^2_{1000}-approximation work. As a rule of thumb, it is often advised to choose the partitioning sets such that each number na_j is at least 5. This criterion depends on the (possibly unknown) null distribution and is not the same as saying that the number of observations in each cell must satisfy an absolute lower bound, which could be very unlikely if the null hypothesis is false. The rule of thumb means to protect the level.

The Pearson statistic is oddly asymmetric in the observed and the true frequencies (which is motivated by the form of the asymptotic covariance matrix). One method to symmetrize the statistic leads to the *Hellinger statistic*

$$H_n^2(a) = 4\sum_{i=1}^{k} \frac{(X_{n,i} - na_i)^2}{(\sqrt{X_{n,i}} + \sqrt{na_i})^2} = 4\sum_{i=1}^{n}(\sqrt{X_{n,i}} - \sqrt{na_i})^2.$$

Up to a multiplicative constant this is the Hellinger distance between the discrete probability distributions on $\{1, \ldots, k\}$ with probability vectors a and X_n/n, respectively. Because $X_n/n - a \overset{\text{P}}{\to} 0$, the Hellinger statistic is asymptotically equivalent to the Pearson statistic.

17.3 Estimated Parameters

Chi-square tests are used quite often, but usually to test more complicated hypotheses. If the null hypothesis of interest is composite, then the parameter a is unknown and cannot be used in the definition of a test statistic. A natural extension is to replace the parameter by an estimate $\hat{a}_n$ and use the statistic

$$C_n(\hat{a}_n) = \sum_{i=1}^{k} \frac{(X_{n,i} - n\hat{a}_{n,i})^2}{n\hat{a}_{n,i}}.$$

The estimator $\hat{a}_n$ is constructed to be a good estimator if the null hypothesis is true. The asymptotic distribution of this modified Pearson statistic is not necessarily chi-square but depends on the estimators $\hat{a}_n$ being used. Most often the estimators are asymptotically normal, and the statistics

$$\frac{X_{n,i} - n\hat{a}_{n,i}}{\sqrt{n\hat{a}_{n,i}}} = \frac{X_{n,i} - na_{n,i}}{\sqrt{n\hat{a}_{n,i}}} - \frac{\sqrt{n}(\hat{a}_{n,i} - a_{n,i})}{\sqrt{\hat{a}_{n,i}}}$$

are asymptotically normal as well. Then the modified chi-square statistic is asymptotically distributed as a quadratic form in a multivariate-normal vector. In general, the eigenvalues determining this form are not restricted to 0 or 1, and their values may depend on the unknown parameter. Then the critical value cannot be taken from a table of the chi-square distribution. There are two popular possibilities to avoid this problem.

First, the Pearson statistic is a certain quadratic form in the observations that is motivated by the asymptotic covariance matrix of a multinomial vector. If the parameter a is estimated, the asymptotic covariance matrix changes in form, and it is natural to change the quadratic form in such a way that the resulting statistic is again chi-square distributed. This idea leads to the Rao-Robson-Nikulin modification of the Pearson statistic, of which we discuss an example in section 17.5.

Second, we can retain the form of the Pearson statistic but use special estimators $\hat{a}$. In particular, the maximum likelihood estimator based on the multinomial vector X_n, or the *minimum–chi square estimator* $\bar{a}_n$ defined by, with $\mathcal{P}_0$ being the null hypothesis,

$$\sum_{i=1}^{k} \frac{(X_{n,i} - n\bar{a}_{n,i})^2}{n\bar{a}_{n,i}} = \inf_{p \in \mathcal{P}_0} \sum_{i=1}^{k} \frac{(X_{n,i} - np_i)^2}{np_i}.$$

The right side of this display is the "minimum–chi square distance" of the observed frequencies to the null hypothesis and is an intuitively reasonable test statistic. The null hypothesis is rejected if the distance of the observed frequency vector X_n/n to the set $\mathcal{P}_0$ is large. A disadvantage is greater computational complexity.

These two modifications, using the minimum–chi square estimator or the maximum likelihood estimator based on X_n, may seem natural but are artificial in some applications. For instance, in goodness-of-fit testing, the multinomial vector is formed by grouping the "raw data," and it is more natural to base the estimators on the raw data rather than on the grouped data. On the other hand, using the maximum likelihood or minimum–chi square estimator based on X_n has the advantage of a remarkably simple limit theory: If the null hypothesis is "locally linear," then the modified Pearson statistic is again asymptotically chi-square distributed, but with the number of degrees of freedom reduced by the (local) dimension of the estimated parameter.

This interesting asymptotic result is most easily explained in terms of the minimum–chi square statistic, as the loss of degrees of freedom corresponds to a projection (i.e., a minimum distance) of the limiting normal vector. We shall first show that the two types of modifications are asymptotically equivalent and are asymptotically equivalent to the likelihood ratio statistic as well. The likelihood ratio statistic for testing the null hypothesis $H_0 : p \in \mathcal{P}_0$ is given by (see Example 16.1)

$$L_n(\hat{a}_n) = \inf_{p \in \mathcal{P}_0} L_n(p), \qquad L_n(p) = 2\sum_{i=1}^{k} X_{n,i} \log \frac{X_{n,i}}{np_i}.$$

17.3   Lemma. *Let $\mathcal{P}_0$ be a closed subset of the unit simplex, and let $\hat{a}_n$ be the maximum likelihood estimator of a under the null hypothesis $H_0 : a \in \mathcal{P}_0$ (based on X_n). Then*

$$\inf_{p \in \mathcal{P}_0} \sum_{i=1}^{k} \frac{(X_{n,i} - np_i)^2}{np_i} = C_n(\hat{a}_n) + o_P(1) = L_n(\hat{a}_n) + o_P(1).$$

Proof.   Let $\bar{a}_n$ be the minimum–chi square estimator of a under the null hypothesis. Both sequences of estimators $\bar{a}_n$ and $\hat{a}_n$ are $\sqrt{n}$-consistent. For the maximum likelihood estimator this follows from Corollary 5.53. The minimum–chi square estimator satisfies by its definition

$$\sum_{i=1}^{k} \frac{(X_{n,i} - n\bar{a}_{n,i})^2}{n\bar{a}_{n,i}} \le \sum_{i=1}^{k} \frac{(X_{n,i} - na_i)^2}{na_i} = O_P(1).$$

This implies that each term in the sum on the left is $O_P(1)$, whence $n|\bar{a}_{n,i} - a_i|^2 = O_P(\bar{a}_{n,i}) + O_P(|X_{n,i} - na_i|^2/n)$ and hence the $\sqrt{n}$-consistency.

Next, the two-term Taylor expansion $\log(1 + x) = x - \frac{1}{2}x^2 + o(x^2)$ combined with Lemma 2.12 yields, for any $\sqrt{n}$-consistent estimator sequence $\hat{p}_n$,

$$\sum_{i=1}^{k} X_{n,i} \log \frac{X_{n,i}}{n\hat{p}_{n,i}} = -\sum_{i=1}^{k} X_{n,i} \left(\frac{n\hat{p}_{n,i}}{X_{n,i}} - 1 \right) + \frac{1}{2}\sum_{i=1}^{k} X_{n,i} \left(\frac{n\hat{p}_{n,i}}{X_{n,i}} - 1 \right)^2 + o_P(1)$$

$$= 0 + \frac{1}{2}\sum_{i=1}^{k} \frac{(X_{n,i} - n\hat{p}_{n,i})^2}{X_{n,i}} + o_P(1).$$

In the last expression we can also replace $X_{n,i}$ in the denominator by $n\hat{p}_{n,i}$, so that we find the relation $L_n(\hat{p}_n) = C_n(\hat{p}_n)$ between the likelihood ratio and the Pearson statistic, for every $\sqrt{n}$-consistent estimator sequence $\hat{p}_n$. By the definitions of $\overline{a}_n$ and $\hat{a}_n$, we conclude that, up to $o_P(1)$-terms, $C_n(\overline{a}_n) \leq C_n(\hat{a}_n) = L_n(\hat{a}_n) \leq L_n(\overline{a}_n) = C_n(\overline{a}_n)$. The lemma follows. ∎

The asymptotic behavior of likelihood ratio statistics is discussed in general in Chapter 16. In view of the preceding lemma, we can now refer to this chapter to obtain the asymptotic distribution of the chi-square statistics. Alternatively, a direct study of the minimum–chi square statistic gives additional insight (and a more elementary proof).

As in Chapter 16, say that a sequence of sets H_n converges to a set H if H is the set of all limits $\lim h_n$ of converging sequences h_n with $h_n \in H_n$ for every n and, moreover, the limit $h = \lim_i h_{n_i}$ of every converging subsequence h_{n_i} with $h_{n_i} \in H_{n_i}$ for every i is contained in H.

17.4   Theorem. *Let $\mathcal{P}_0$ be a subset of the unit simplex such that the sequence of sets $\sqrt{n}(\mathcal{P}_0 - a)$ converges to a set H (in $\mathbb{R}^k$), and suppose that $a > 0$. Then, under a,*

$$\inf_{p \in \mathcal{P}_0} \sum_{i=1}^{k} \frac{(X_{n,i} - np_i)^2}{np_i} \rightsquigarrow \inf_{h \in H} \left\| X - \frac{1}{\sqrt{a}} H \right\|^2,$$

for a vector X with the $N(0, I - \sqrt{a}\sqrt{a}^T)$-distribution. Here $(1/\sqrt{a})H$ is the set of vectors $(h_1/\sqrt{a_1}, \ldots, h_k/\sqrt{a_k})$ as h ranges over H.

17.5   Corollary. *Let $\mathcal{P}_0$ be a subset of the unit simplex such that the sequence of sets $\sqrt{n}(\mathcal{P}_0 - a)$ converges to a linear subspace of dimension l (of $\mathbb{R}^k$), and let $a > 0$. Then both the sequence of minimum–chi square statistics and the sequence of modified Pearson statistics $C_n(\hat{a}_n)$ converge in distribution to the chi-square distribution with $k - 1 - l$ degrees of freedom.*

Proof. Because the minimum–chi square estimator $\overline{a}_n$ (relative to $\overline{\mathcal{P}}_0$) is $\sqrt{n}$-consistent, the asymptotic distribution of the minimum–chi square statistic is not changed if we replace $n\overline{a}_{n,i}$ in its denominator by the true value na_i. Next, we decompose,

$$\frac{X_{n,i} - np_i}{\sqrt{na_i}} = \frac{X_{n,i} - na_i}{\sqrt{na_i}} - \frac{\sqrt{n}(p_i - a_i)}{\sqrt{a_i}}.$$

The first vector on the right converges in distribution to X. The (modified) minimum–chi square statistics are the distances of these vectors to the sets $H_n = \sqrt{n}(\mathcal{P}_0 - a)/\sqrt{a}$, which converge to the set $H/\sqrt{a}$. The theorem now follows from Lemma 7.13.

The vector X is distributed as $Z - \Pi_{\sqrt{a}} Z$ for $\Pi_{\sqrt{a}}$ the projection onto the linear space spanned by the vector $\sqrt{a}$ and Z a k-dimensional standard normal vector. Because every element of H is the limit of a multiple of differences of probability vectors, $1^T h = 0$ for every $h \in H$. Therefore, the space $(1/\sqrt{a})H$ is orthogonal to the vector $\sqrt{a}$, and $\Pi \Pi_{\sqrt{a}} = 0$ for Π the projection onto the space $(1/\sqrt{a})H$. The distance of X to the space $(1/\sqrt{a})H$ is equal to the norm of $X - \Pi X$, which is distributed as the norm of $Z - \Pi_{\sqrt{a}} Z - \Pi Z$. The latter projection is multivariate normally distributed with mean zero and covariance matrix the projection matrix $I - \Pi_{\sqrt{a}} - \Pi$ with $k - l - 1$ eigenvalues 1. The corollary follows from Lemma 17.1 or 16.6. ∎

17.6   *Example (Parametric model).* If the null hypothesis is a parametric family $\mathcal{P}_0 = \{p_\theta : \theta \in \Theta\}$ indexed by a subset Θ of $\mathbb{R}^l$ with $l \le k$ and the maps $\theta \mapsto p_\theta$ from Θ into the unit simplex are differentiable and of full rank, then $\sqrt{n}(\mathcal{P}_0 - p_\theta) \to \dot{p}_\theta(\mathbb{R}^l)$ for every $\theta \in \overset{\circ}{\Theta}$ (see Example 16.11). Then the chi-square statistics $C_n(\hat{p}_\theta)$ are asymptotically χ^2_{k-l-1}-distributed.

This situation is common in testing the goodness-of-fit of parametric families, as discussed in section 17.5 and Example 16.1. $\square$

17.4   Testing Independence

Suppose that each element of a population can be classified according to two characteristics, having k and r levels, respectively. The full information concerning the classification can be given by a $(k \times r)$ *table* of the form given in Table 17.1.

Often the full information is not available, but we do know the classification $X_{n,ij}$ for a random sample of size n from the population. The matrix $X_{n,ij}$, which can also be written in the form of a $(k \times r)$ table, is multinomially distributed with parameters n and probabilities $p_{ij} = N_{ij}/N$. The null *hypothesis of independence* asserts that the two categories are independent: $H_0 : p_{ij} = a_i b_j$ for (unknown) probability vectors a_i and b_j.

The maximum likelihood estimators for the parameters a and b (under the null hypothesis) are $\hat{a}_i = X_{n,i.}/n$ and $\hat{b}_j = X_{n,.j}/n$. With these estimators the modified Pearson statistic takes the form

$$C_n(\hat{a}_n \otimes \hat{b}_n) = \sum_{i=1}^{k} \sum_{j=1}^{r} \frac{(X_{n,ij} - n\hat{a}_i \hat{b}_j)^2}{n\hat{a}_i \hat{b}_j}.$$

The null hypothesis is a $(k + r - 2)$-dimensional submanifold of the unit simplex in $\mathbb{R}^{kr}$. In a shrinking neighborhood of a parameter in its interior this manifold looks like its tangent space, a linear space of dimension $k + r - 2$. Thus, the sequence $C_n(\hat{a}_n \otimes \hat{b}_n)$ is asymptotically chi square–distributed with $kr - 1 - (k + r - 2) = (k - 1)(r - 1)$ degrees of freedom.

Table 17.1. *Classification of a population of N elements according to two categories, N_{ij} elements having value i on the first category and value j on the second. The borders give the sums over each row and column, respectively.*

N_{11}	N_{12}	$\cdots$	N_{1r}	$N_{1.}$
N_{21}	N_{22}	$\cdots$	N_{1r}	$N_{2.}$
$\vdots$	$\vdots$		$\vdots$	$\vdots$
N_{k1}	N_{k2}	$\cdots$	N_{1r}	$N_{k.}$
$N_{.1}$	$N_{.2}$	$\cdots$	$N_{.r}$	N

17.7   Corollary. *If the $(k \times r)$ matrices X_n are multinomially distributed with parameters n and $p_{ij} = a_i b_j > 0$, then the sequence $C_n(\hat{a}_n \otimes \hat{b}_n)$ converges in distribution to the $\chi^2_{(k-1)(r-1)}$-distribution.*

Proof.   The map $(a_1, \ldots, a_{k-1}, b_1, \ldots, b_{r-1}) \mapsto (a \times b)$ from $\mathbb{R}^{k+r-2}$ into $\mathbb{R}^{kr}$ is continuously differentiable and of full rank. The true values $(a_1, \ldots, a_{k-1}, b_1 \ldots, b_{r-1})$ are interior to the domain of this map. Thus the sequence of sets $\sqrt{n}(\mathcal{P}_0 - a \times b)$ converges to a $(k + r - 2)$-dimensional linear subspace of $\mathbb{R}^{kr}$. ∎

*17.5   Goodness-of-Fit Tests

Chi-square tests are often applied to test goodness-of-fit. Given a random sample $X_1, \ldots, X_n$ from a distribution P, we wish to test the null hypothesis $H_0 : P \in \mathcal{P}_0$ that P belongs to a given class $\mathcal{P}_0$ of probability measures. There are many possible test statistics for this problem, and a particular statistic might be selected to attain high power against certain alternatives. Testing goodness-of-fit typically focuses on no particular alternative. Then chi-square statistics are intuitively reasonable.

The data can be reduced to a multinomial vector by "grouping." We choose a partition $\mathcal{X} = \cup_j \mathcal{X}_j$ of the sample space into finitely many sets and base the test only on the observed numbers of observations falling into each of the sets $\mathcal{X}_j$. For ease of notation, we express these numbers into the empirical measure of the data. For a given set A we denote by $\mathbb{P}_n(A) = n^{-1}(1 \leq i \leq n : X_i \in A)$ the fraction of observations that fall into A. Then the vector $n\big(\mathbb{P}_n(\mathcal{X}_1), \ldots, \mathbb{P}_n(\mathcal{X}_k)\big)$ possesses a multinomial distribution, and the corresponding modified chi-square statistic is given by

$$\sum_{i=1}^{k} \frac{n\big(\mathbb{P}_n(\mathcal{X}_j) - \hat{P}(\mathcal{X}_j)\big)^2}{\hat{P}(\mathcal{X}_j)}.$$

Here $\hat{P}(\mathcal{X}_j)$ is an estimate of $P(\mathcal{X}_j)$ under the null hypothesis and can take a variety of forms.

Theorem 17.4 applies but is restricted to the case that the estimates $\hat{P}(\mathcal{X}_j)$ are based on the frequencies $n\big(\mathbb{P}_n(\mathcal{X}_1), \ldots, \mathbb{P}_n(\mathcal{X}_k)\big)$ only. In the present situation it is more natural to base the estimates on the original observations $X_1, \ldots, X_n$. Usually, this results in a non–chi square limit distribution. For instance, Table 17.2 shows the "errors" in the level of a chi-square test for testing normality, if the unknown mean and variance are estimated by the sample mean and the sample variance but the critical value is chosen from the chi-square distribution. The size of the errors depends on the numbers of cells, the errors being small if there are many cells and few estimated parameters.

17.8   Example (Parametric model).   Consider testing the null hypothesis that the true distribution belongs to a regular parametric model $\{P_\theta : \theta \in \Theta\}$. It appears natural to estimate the unknown parameter θ by an estimator $\hat{\theta}_n$ that is asymptotically efficient under the null hypothesis and is based on the original sample $X_1, \ldots, X_n$, for instance the maximum likelihood estimator. If $\mathbb{G}_n = \sqrt{n}(\mathbb{P}_n - P_\theta)$ denotes the empirical process, then efficiency entails the approximation $\sqrt{n}(\hat{\theta}_n - \theta) = I_\theta^{-1} \mathbb{G}_n \dot{\ell}_\theta + o_P(1)$. Applying the delta method to

Table 17.2. *True levels of the chi-square test for normality using $\chi^2_{k-3,\alpha}$-quantiles as critical values but estimating unknown mean and variance by sample mean and sample variance. Chi square statistic based on partitions of $[-10, 10]$ into $k = 5, 10,$ or 20 equiprobable cells under the standard normal law.*

	$\alpha = 0.20$	$\alpha = 0.10$	$\alpha = 0.05$	$\alpha = 0.01$
$k = 5$	0.30	0.15	0.08	0.02
$k = 10$	0.22	0.11	0.06	0.01
$k = 20$	0.21	0.10	0.05	0.01

Note: Values based on 2000 simulations of standard normal samples of size 100.

the variables $\sqrt{n}\left(P_{\hat{\theta}}(\mathcal{X}_j) - P_\theta(\mathcal{X}_j)\right)$ and using Slutsky's lemma, we find

$$\frac{\sqrt{n}\left(\mathbb{P}_n(\mathcal{X}_j) - P_{\hat{\theta}}(\mathcal{X}_j)\right)}{\sqrt{P_\theta(\mathcal{X}_j)}} = \frac{\mathbb{G}_n 1_{\mathcal{X}_j} - (P_\theta 1_{\mathcal{X}_j}\dot{\ell}_\theta)^T I_\theta^{-1}\mathbb{G}_n\dot{\ell}_\theta}{\sqrt{P_\theta(\mathcal{X}_j)}} + o_P(1).$$

(The map $\theta \mapsto P_\theta(A)$ has derivative $P_\theta 1_A\dot{\ell}_\theta$.) The sequence of vectors $(\mathbb{G}_n 1_{\mathcal{X}_j}, \mathbb{G}_n\dot{\ell}_\theta)$ converges in distribution to a multivariate-normal distribution. Some matrix manipulations show that the vectors in the preceding display are asymptotically distributed as a Gaussian vector X with mean zero and covariance matrix

$$I - \sqrt{a_\theta}\sqrt{a_\theta}^T - C_\theta^T I_\theta^{-1} C_\theta, \qquad (a_\theta)_j = P_\theta(\mathcal{X}_j), \qquad (C_\theta)_{ij} = \frac{P_\theta 1_{\mathcal{X}_j}\dot{\ell}_{\theta,i}}{\sqrt{(a_\theta)_j}}.$$

In general, the covariance matrix of X is not a projection matrix, and the variable $\|X\|^2$ does not possess a chi-square distribution.

Because $P_\theta\dot{\ell}_\theta = 0$, we have that $C_\theta\sqrt{a_\theta} = 0$ and hence the covariance matrix of X can be rewritten as the product $(I - \sqrt{a_\theta}\sqrt{a_\theta}^T)(I - C_\theta^T I_\theta^{-1} C_\theta)$. Here the first matrix is the projection onto the orthocomplement of the vector $\sqrt{a_\theta}$ and the second matrix is a positive-definite transformation that leaves $\sqrt{a_\theta}$ invariant, thus acting only on the orthocomplement $\sqrt{a_\theta}^\perp$. This geometric picture shows that $\mathrm{Cov}_\theta\, X$ has the same system of eigenvectors as the matrix $I - C_\theta^T I_\theta^{-1} C_\theta$, and also the same eigenvalues, except for the eigenvalue corresponding to the eigenvector $\sqrt{a_\theta}$, which is 0 for $\mathrm{Cov}_\theta\, X$ and 1 for $I - C_\theta^T I_\theta^{-1} C_\theta$. Because both matrices $C_\theta^T I_\theta^{-1} C_\theta$ and $I - C_\theta^T I_\theta^{-1} C_\theta$ are nonnegative-definite, the eigenvalues are contained in $[0, 1]$. One eigenvalue (corresponding to eigenvector $\sqrt{a_\theta}$) is 0, $\dim N(C_\theta) - 1$ eigenvalues (corresponding to eigenspace $N(C_\theta) \cap \sqrt{a_\theta}^\perp$) are 1, but the other eigenvalues may be contained in $(0, 1)$ and then typically depend on θ. By Lemma 17.1, the variable $\|X\|^2$ is distributed as

$$\sum_{i=1}^{\dim N(C_\theta)-1} Z_i^2 + \sum_{i=\dim N(C_\theta)}^{k-1} \lambda_i(\theta) Z_i^2.$$

This means that it is stochastically "between" the chi-square distributions with $\dim N(C_\theta) - 1$ and $k - 1$ degrees of freedom.

The inconvenience that this distribution is not standard and depends on θ can be remedied by not using efficient estimators $\hat{\theta}_n$ or, alternatively, by not using the Pearson statistic.

The square root of the matrix $I - C_\theta^T I_\theta^{-1} C_\theta$ is the positive-definite matrix with the same eigenvectors, but with the square roots of the eigenvalues. Thus, it also leaves the vector $\sqrt{a_\theta}$ invariant and acts only on the orthocomplement $\sqrt{a_\theta}^\perp$. It follows that this square root commutes with the matrix $I - \sqrt{a_\theta}\sqrt{a_\theta}^T$ and hence

$$\left(I - C_\theta^T I_\theta^{-1} C_\theta\right)^{-1/2} \frac{\sqrt{n}\left(\mathbb{P}_n(\mathcal{X}_j) - P_\theta(\mathcal{X}_j)\right)}{\sqrt{P_\theta(\mathcal{X}_j)}} \rightsquigarrow N_k\left(0, I - \sqrt{a_\theta}\sqrt{a_\theta}^T\right).$$

(We assume that the matrix $I - C_\theta^T I_\theta^{-1} C_\theta$ is nonsingular, which is typically the case; see problem 17.6). By the continuous-mapping theorem, the squared norm of the left side is asymptotically chi square–distributed with $k - 1$ degrees of freedom. This squared norm is the *Rao-Robson-Nikulin statistic*. $\square$

It is tempting to choose the partitioning sets $\mathcal{X}_j$ dependent on the observed data $X_1, \ldots, X_n$, for instance to ensure that all cells have positive probability under the null hypothesis. This is permissible under some conditions: The choice of a "random partition" typically does not change the distributional properties of the chi-square statistic. Consider partitioning sets $\hat{\mathcal{X}}_j = \mathcal{X}_j(X_1, \ldots, X_n)$ that possibly depend on the data, and a further modified Pearson statistic of the type

$$\sum_{i=1}^k \frac{n\left(\mathbb{P}_n(\hat{\mathcal{X}}_j) - \hat{P}(\hat{\mathcal{X}}_j)\right)^2}{\hat{P}(\hat{\mathcal{X}}_j)}.$$

If the random partitions settle down to a fixed partition eventually, then this statistic is asymptotically equivalent to the statistic for which the partition had been set equal to the limit partition in advance. We discuss this for the case that the null hypothesis is a model $\{P_\theta : \theta \in \Theta\}$ indexed by a subset Θ of a normed space. We use the language of Donsker classes as discussed in Chapter 19.

17.9   Theorem. *Suppose that the sets $\hat{\mathcal{X}}_j$ belong to a P_{θ_0}-Donsker class $\mathcal{C}$ of sets and that $P_{\theta_0}(\hat{\mathcal{X}}_j \bigtriangleup \mathcal{X}_j) \overset{P}{\to} 0$ under P_{θ_0}, for given nonrandom sets $\mathcal{X}_j$ such that $P_{\theta_0}(\mathcal{X}_j) > 0$. Furthermore, assume that $\sqrt{n}\|\hat{\theta} - \theta_0\| = O_P(1)$, and suppose that the map $\theta \mapsto P_\theta$ from Θ into $\ell^\infty(\mathcal{C})$ is differentiable at θ_0 with derivative $\dot{P}_{\theta_0}$ such that $\dot{P}_{\theta_0}(\hat{\mathcal{X}}_j) - \dot{P}_{\theta_0}(\mathcal{X}_j) \overset{P}{\to} 0$ for every j. Then*

$$\sum_{i=1}^k \frac{n\left(\mathbb{P}_n(\hat{\mathcal{X}}_j) - P_{\hat{\theta}}(\hat{\mathcal{X}}_j)\right)^2}{P_{\hat{\theta}}(\hat{\mathcal{X}}_j)} = \sum_{i=1}^k \frac{n\left(\mathbb{P}_n(\mathcal{X}_j) - P_{\hat{\theta}}(\mathcal{X}_j)\right)^2}{P_{\hat{\theta}}(\mathcal{X}_j)} + o_P(1).$$

Proof.   Let $\mathbb{G}_n = \sqrt{n}(\mathbb{P}_n - P_{\theta_0})$ be the empirical process and define $\mathbb{H}_n = \sqrt{n}(P_{\hat{\theta}} - P_{\theta_0})$. Then $\sqrt{n}\left(\mathbb{P}_n(\hat{\mathcal{X}}_j) - P_{\hat{\theta}}(\hat{\mathcal{X}}_j)\right) = (\mathbb{G}_n - \mathbb{H}_n)(\hat{\mathcal{X}}_j)$, and similarly with $\mathcal{X}_j$ replacing $\hat{\mathcal{X}}_j$. The condition that the sets $\mathcal{X}_j$ belong to a Donsker class combined with the continuity condition $P_{\theta_0}(\hat{\mathcal{X}}_j \bigtriangleup \mathcal{X}_j) \overset{P}{\to} 0$, imply that $\mathbb{G}_n(\hat{\mathcal{X}}_j) - \mathbb{G}_n(\mathcal{X}_j) \overset{P}{\to} 0$ (see Lemma 19.24). The differentiability of the map $\theta \mapsto P_\theta$ implies that

$$\sup_C \left| P_{\hat{\theta}}(C) - P_{\theta_0}(C) - \dot{P}_{\theta_0}(C)(\hat{\theta} - \theta_0) \right| = o_P\left(\|\hat{\theta} - \theta_0\|\right).$$

Together with the continuity $\dot{P}_{\theta_0}(\hat{\mathcal{X}}_j) - \dot{P}_{\theta_0}(\mathcal{X}_j) \overset{P}{\to} 0$ and the $\sqrt{n}$-consistency of $\hat{\theta}$, this shows that $\mathbb{H}_n(\hat{\mathcal{X}}_j) - \mathbb{H}_n(\mathcal{X}_j) \overset{P}{\to} 0$. In particular, because $P_{\theta_0}(\hat{\mathcal{X}}_j) \overset{P}{\to} P_{\theta_0}(\mathcal{X}_j)$, both $P_{\hat{\theta}}(\hat{\mathcal{X}}_j)$ and $P_{\hat{\theta}}(\mathcal{X}_j)$ converge in probability to $P_{\theta_0}(\mathcal{X}_j) > 0$. The theorem follows. ∎

The conditions on the random partitions that are imposed in the preceding theorem are mild. An interesing choice is a partition in sets $\mathcal{X}_j(\hat{\theta})$ such that $P_\theta\big(\mathcal{X}_j(\theta)\big) = a_j$ is independent of θ. The corresponding modified Pearson statistic is known as the *Watson-Roy statistic* and takes the form

$$\sum_{i=1}^{k} \frac{n\Big(\mathbb{P}_n\big(\mathcal{X}_j(\hat{\theta})\big) - a_j\Big)^2}{a_j}.$$

Here the null probabilities have been reduced to fixed values again, but the cell frequencies are "doubly random." If the model is smooth and the parameter and the sets $\mathcal{X}_j(\theta)$ are not too wild, then this statistic has the same null limit distribution as the modified Pearson statistic with a fixed partition.

17.10   *Example (Location-scale).* Consider testing a null hypothesis that the true underlying measure of the observations belongs to a location-scale family $\big\{F_0\big((\cdot - \mu)/\sigma\big) : \mu \in \mathbb{R}, \sigma > 0\big\}$, given a fixed distribution F_0 on $\mathbb{R}$. It is reasonable to choose a partition in sets $\hat{\mathcal{X}}_j = \hat{\mu} + \hat{\sigma}(c_{j-1}, c_j]$, for a fixed partition $-\infty = c_0 < c_1 < \cdots < c_k = \infty$ and estimators $\hat{\mu}$ and $\hat{\sigma}$ of the location and scale parameter. The partition could, for instance, be chosen equal to $c_j = F_0^{-1}(j/k)$, although, in general, the partition should depend on the type of deviation from the null hypothesis that one wants to detect.

If we use the same location and scale estimators to "estimate" the null probabilities $F_0\big((\hat{\mathcal{X}}_j - \mu)/\sigma\big)$ of the random cells $\hat{\mathcal{X}}_j = \hat{\mu} + \hat{\sigma}(c_{j-1}, c_j]$, then the estimators cancel, and we find the fixed null probabilities $F_0(c_j) - F_0(c_{j-1})$. □

*17.6   Asymptotic Efficiency

The asymptotic null distributions of various versions of the Pearson statistic enable us to set critical values but by themselves do not give information on the asymptotic power of the tests. Are these tests, which appear to be mostly motivated by their asymptotic null distribution, sufficiently powerful?

The asymptotic power can be measured in various ways. Probably the most important method is to consider local limiting power functions, as in Chapter 14. For the likelihood ratio test these are obtained in Chapter 16. Because, in the local experiments, chi-square statistics are asymptotically equivalent to the likelihood ratio statistics (see Theorem 17.4), the results obtained there also apply to the present problem, and we shall not repeat the discussion.

A second method to evaluate the asymptotic power is by Bahadur efficiencies. For this nonlocal criterion, chi-square tests and likelihood ratio tests are not equivalent, the second being better and, in fact, optimal (see Theorem 16.12).

We shall compute the slopes of the Pearson and likelihood ratio tests for testing the simple hypothesis $H_0 : p = a$. A multinomial vector X_n with parameters n and $p = (p_1, \ldots, p_k)$ can be thought of as n times the empirical measure $\mathbb{P}_n$ of a random sample of size n from the distribution P on the set $\{1, \ldots, k\}$ defined by $P\{i\} = p_i$. Thus we can view both the

Pearson and the likelihood ratio statistics as functions of an empirical measure and next can apply Sanov's theorem to compute the desired limits of large deviations probabilities. Define maps C and K by

$$C(p,a) = \sum_{i=1}^{k} \frac{(p_i - a_i)^2}{a_i},$$

$$K(p,a) = -P \log \frac{a}{p} = \sum_{i=1}^{k} p_i \log \frac{p_i}{a_i}.$$

Then the Pearson and likelihood ratio statistics are equivalent to $C(\mathbb{P}_n, a)$ and $K(\mathbb{P}_n, a)$, respectively.

Under the assumption that $a > 0$, both maps are continuous in p on the k-dimensional unit simplex. Furthermore, for t in the interior of the ranges of C and K, the sets $B_t = \{p : C(p,a) \geq t\}$ and $\tilde{B}_t = \{p : K(p,a) \geq t\}$ are equal to the closures of their interiors. Two applications of Sanov's theorem yield

$$\frac{1}{n} \log P_a\big(C(\mathbb{P}_n, a) \geq t\big) \to -\inf_{p \in B_t} K(p,a),$$

$$\frac{1}{n} \log P_a\big(K(\mathbb{P}_n, a) \geq t\big) \to -\inf_{p \in \tilde{B}_t} K(p,a) = -t.$$

We take the function $e(t)$ of (14.20) equal to minus two times the right sides. Because $\mathbb{P}_n\{i\} \to p_i$ by the law of large numbers, whence $C(\mathbb{P}_n, a) \overset{P}{\to} C(P, a)$ and $K(\mathbb{P}_n, a) \overset{P}{\to} K(P, a)$, the Bahadur slopes of the Pearson and likelihood ratio tests at the alternative $H_1 : p = q$ are given by

$$2 \inf_{p : C(p,a) \geq C(q,a)} K(p,a)$$

and

$$2K(q,a).$$

It is clear from these expressions that the likelihood ratio test has a bigger slope. This is in agreement with the fact that the likelihood ratio test is asymptotically Bahadur optimal in any smooth parametric model. Figure 17.1 shows the difference of the slopes in one particular case. The difference is small in a neighborhood of the null hypothesis a, in agreement with the fact that the Pitman efficiency is equal to 1, but can be substantial for alternatives away from a.

Notes

Pearson introduced his statistic in 1900 in [112] The modification with estimated parameters, using the multinomial frequencies, was considered by Fisher [49], who corrected the mistaken belief that estimating the parameters does not change the limit distribution. Chernoff and Lehmann [22] showed that using maximum likelihood estimators based on the original data for the parameter in a goodness-of-fit statistic destroys the asymptotic chi-square distribution. They note that the errors in the level are small in the case of testing a Poisson distribution and somewhat larger when testing normality.

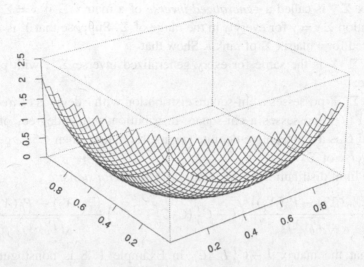

Figure 17.1. The difference of the Bahadur slopes of the likelihood ratio and Pearson tests for testing $H_0 : p = (1/3, 1/3, 1/3)$ based on a multinomial vector with parameters n and $p = (p_1, p_2, p_3)$, as a function of (p_1, p_2).

The choice of the partition in chi-square goodness-of-fit tests is an important issue that we have not discussed. Several authors have studied the optimal number of cells in the partition. This number depends, of course, on the alternative for which one desires large power. The conclusions of these studies are not easily summarized. For alternatives p such that the likelihood ratio p/p_{θ_0} with respect to the null distribution is "wild," the number of cells k should tend to infinity with n. Then the chi-square approximation of the null distribution needs to be modified. Normal approximations are used, because a chi-square distribution with a large number of degrees of freedom is approximately a normal distribution. See [40], [60], and [86] for results and further references.

PROBLEMS

1. Let $N = (N_{ij})$ be a multinomial matrix with success probabilities p_{ij}. Design a test statistic for the null hypothesis of symmetry $H_0 : p_{ij} = p_{ji}$ and derive its asymptotic null distribution.

2. Derive the limit distribution of the chi-square goodness-of-fit statistic for testing normality if using the sample mean and sample variance as estimators for the unknown mean and variance. Use two or three cells to keep the calculations simple. Show that the limit distribution is not chi-square.

3. Suppose that X_m and Y_n are independent multinomial vectors with parameters $(m, a_1, \ldots, a_k)$ and $(n, b_1, \ldots, b_k)$, respectively. Under the null hypothesis $H_0 : a = b$, a natural estimator of the unknown probability vector $a = b$ is $\hat{c} = (m+n)^{-1}(X_m + Y_n)$, and a natural test statistic is given by

$$\sum_{i=1}^{k} \frac{(X_{m,i} - m\hat{c}_i)^2}{m\hat{c}_i} + \sum_{i=1}^{k} \frac{(Y_{n,i} - n\hat{c}_i)^2}{n\hat{c}_i}.$$

Show that $\hat{c}$ is the maximum likelihood estimator and show that the sequence of test statistics is asymptotically chi square–distributed if $m, n \to \infty$.

4. A matrix Σ^- is called a *generalized inverse* of a matrix Σ if $x = \Sigma^- y$ solves the equation $\Sigma x = y$ for every y in the range of Σ. Suppose that Y is $N_k(0, \Sigma)$-distributed for a matrix Σ of rank r. Show that

   (i) $Y^T \Sigma^- Y$ is the same for every generalized inverse Σ^-, with probability one;

   (ii) $Y^T \Sigma^- Y$ possesses a chi-square distribution with r degrees of freedom;

   (iii) if $Y^T C Y$ possesses a chi-square distribution with r degrees of freedom and C is a nonnegative-definite symmetric matrix, then C is a generalized inverse of Σ.

5. Find the limit distribution of the *Dzhaparidze-Nikulin statistic*

$$n\frac{\left(\mathbb{P}_n(\mathcal{X}_j) - P_{\hat{\theta}}(\mathcal{X}_j)\right)}{\sqrt{P_{\hat{\theta}}(\mathcal{X}_j)}} \left(I - C_{\hat{\theta}}^T (C_{\hat{\theta}} C_{\hat{\theta}}^T)^{-1} C_{\hat{\theta}}\right) \frac{\left(\mathbb{P}_n(\mathcal{X}_j) - P_{\hat{\theta}}(\mathcal{X}_j)\right)}{\sqrt{P_{\hat{\theta}}(\mathcal{X}_j)}}.$$

6. Show that the matrix $I - C_\theta^T I_\theta^{-1} C_\theta$ in Example 17.8 is nonsingular unless the empirical estimator $\left(\mathbb{P}_n(\mathcal{X}_1), \ldots, \mathbb{P}_n(\mathcal{X}_k)\right)$ is asymptotically efficient. (The estimator $\left(P_{\hat{\theta}}(\mathcal{X}_1), \ldots, P_{\hat{\theta}}(\mathcal{X}_k)\right)$ is asymptotically efficient and has asymptotic covariance matrix diag $(\sqrt{a_\theta}) C_\theta^T I_\theta^{-1} C_\theta$ diag $(\sqrt{a_\theta})$; the empirical estimator has asymptotic covariance matrix diag $(\sqrt{a_\theta})(I - \sqrt{a_\theta}\sqrt{a_\theta}^T)$ diag $(\sqrt{a_\theta})$.)

18

Stochastic Convergence in Metric Spaces

This chapter extends the concepts of convergence in distribution, in probability, and almost surely from Euclidean spaces to more abstract metric spaces. We are particularly interested in developing the theory for random functions, or stochastic processes, viewed as elements of the metric space of all bounded functions.

18.1 Metric and Normed Spaces

In this section we recall some basic topological concepts and introduce a number of examples of metric spaces.

A *metric space* is a set $\mathbb{D}$ equipped with a metric. A *metric* or *distance function* is a map $d : \mathbb{D} \times \mathbb{D} \mapsto [0, \infty)$ with the properties

- (i) $d(x, y) = d(y, x)$;
- (ii) $d(x, z) \le d(x, y) + d(y, z)$ (triangle inequality);
- (iii) $d(x, y) = 0$ if and only if $x = y$.

A *semimetric* satisfies (i) and (ii), but not necessarily (iii). An *open ball* is a set of the form $\{y : d(x, y) < r\}$. A subset of a metric space is *open* if and only if it is the union of open balls; it is *closed* if and only if its complement is open. A sequence x_n *converges* to x if and only if $d(x_n, x) \to 0$; this is denoted by $x_n \to x$. The *closure* $\overline{A}$ of a set $A \subset \mathbb{D}$ consists of all points that are the limit of a sequence in A; it is the smallest closed set containing A. The *interior* $\mathring{A}$ is the collection of all points x such that $x \in G \subset A$ for some open set G; it is the largest open set contained in A. A function $f : \mathbb{D} \mapsto \mathbb{E}$ between two metric spaces is *continuous* at a point x if and only if $f(x_n) \to f(x)$ for every sequence $x_n \to x$; it is continuous at every x if and only if the inverse image $f^{-1}(G)$ of every open set $G \subset \mathbb{E}$ is open in $\mathbb{D}$. A subset of a metric space is *dense* if and only if its closure is the whole space. A metric space is *separable* if and only if it has a countable dense subset. A subset K of a metric space is *compact* if and only if it is closed and every sequence in K has a converging subsequence. A subset K is *totally bounded* if and only if for every $\varepsilon > 0$ it can be covered by finitely many balls of radius ε. A semimetric space is *complete* if every *Cauchy sequence*, a sequence such that $d(x_n, x_m) \to 0$ as $n, m \to \infty$, has a limit. A subset of a complete semimetric space is compact if and only if it is totally bounded and closed.

A *normed space* $\mathbb{D}$ is a vector space equipped with a norm. A *norm* is a map $\| \cdot \| : \mathbb{D} \mapsto [0, \infty)$ such that, for every x, y in $\mathbb{D}$, and $\alpha \in \mathbb{R}$,

(i) $\|x + y\| \le \|x\| + \|y\|$ (triangle inequality);

(ii) $\|\alpha x\| = |\alpha| \|x\|$;

(iii) $\|x\| = 0$ if and only if $x = 0$.

A *seminorm* satisfies (i) and (ii), but not necessarily (iii). Given a norm, a metric can be defined by $d(x, y) = \|x - y\|$.

18.1   Definition. The *Borel σ-field* on a metric space $\mathbb{D}$ is the smallest σ-field that contains the open sets (and then also the closed sets). A function defined relative to (one or two) metric spaces is called *Borel-measurable* if it is measurable relative to the Borel σ-field(s). A Borel-measurable map $X : \Omega \mapsto \mathbb{D}$ defined on a probability space $(\Omega, \mathcal{U}, P)$ is referred to as a *random element* with values in $\mathbb{D}$.

For Euclidean spaces, Borel measurability is just the usual measurability. Borel measurability is probably the natural concept to use with metric spaces. It combines well with the topological structure, particularly if the metric space is separable. For instance, continuous maps are Borel-measurable.

18.2   Lemma. *A continuous map between metric spaces is Borel-measurable.*

Proof.   A map $g : \mathbb{D} \mapsto \mathbb{E}$ is continuous if and only if the inverse image $g^{-1}(G)$ of every open set $G \subset \mathbb{E}$ is open in $\mathbb{D}$. In particular, for every open G the set $g^{-1}(G)$ is a Borel set in $\mathbb{D}$. By definition, the open sets in $\mathbb{E}$ generate the Borel σ-field. Thus, the inverse image of a generator of the Borel sets in $\mathbb{E}$ is contained in the Borel σ-field in $\mathbb{D}$. Because the inverse image $g^{-1}(\mathcal{G})$ of a generator $\mathcal{G}$ of a σ-field $\mathcal{B}$ generates the σ-field $g^{-1}(\mathcal{B})$, it follows that the inverse image of every Borel set is a Borel set. ∎

18.3   Example (Euclidean spaces). The Euclidean space $\mathbb{R}^k$ is a normed space with respect to the Euclidean norm (whose square is $\|x\|^2 = \sum_{i=1}^{k} x_i^2$), but also with respect to many other norms, for instance $\|x\| = \max_i |x_i|$, all of which are equivalent. By the Heine-Borel theorem a subset of $\mathbb{R}^k$ is compact if and only if it is closed and bounded. A Euclidean space is separable, with, for instance, the vectors with rational coordinates as a countable dense subset.

The Borel σ-field is the usual σ-field, generated by the intervals of the type $(-\infty, x]$. □

18.4   Example (Extended real line). The extended real line $\overline{\mathbb{R}} = [-\infty, \infty]$ is the set consisting of all real numbers and the additional elements $-\infty$ and ∞. It is a metric space with respect to

$$d(x, y) = |\Phi(x) - \Phi(y)|.$$

Here Φ can be any fixed, bounded, strictly increasing continuous function. For instance, the normal distribution function (with $\Phi(-\infty) = 0$ and $\Phi(\infty) = 1$). Convergence of a sequence $x_n \to x$ with respect to this metric has the usual meaning, also if the limit x is $-\infty$ or ∞ (normally we would say that x_n "diverges"). Consequently, every sequence has a converging subsequence and hence the extended real line is compact. □

18.5   Example (Uniform norm). Given an arbitrary set T, let $\ell^\infty(T)$ be the collection of all bounded functions $z : T \mapsto \mathbb{R}$. Define sums $z_1 + z_2$ and products with scalars az pointwise. For instance, $z_1 + z_2$ is the element of $\ell^\infty(T)$ such that $(z_1 + z_2)(t) = z_1(t) + z_2(t)$ for every t. The *uniform norm* is defined as

$$\|z\|_T = \sup_{t \in T} |z(t)|.$$

With this notation the space $\ell^\infty(T)$ consists exactly of all functions $z : T \mapsto \mathbb{R}$ such that $\|z\|_T < \infty$. The space $\ell^\infty(T)$ is separable if and only if T is finite. $\square$

18.6   Example (Skorohod space). Let $T = [a, b]$ be an interval in the extended real line. We denote by $C[a, b]$ the set of all continuous functions $z : [a, b] \mapsto \mathbb{R}$ and by $D[a, b]$ the set of all functions $z : [a, b] \mapsto \mathbb{R}$ that are right continuous and whose limits from the left exist everywhere in $[a, b]$. (The functions in $D[a, b]$ are called *cadlag: continue à droite, limites à gauche.*) It can be shown that $C[a, b] \subset D[a, b] \subset \ell^\infty[a, b]$. We always equip the spaces $C[a, b]$ and $D[a, b]$ with the uniform norm $\|z\|_T$, which they "inherit" from $\ell^\infty[a, b]$.

The space $D[a, b]$ is referred to here as the *Skorohod space*, although Skorohod did not consider the uniform norm but equipped the space with the "Skorohod metric" (which we do not use or discuss).

The space $C[a, b]$ is separable, but the space $D[a, b]$ is not (relative to the uniform norm). $\square$

18.7   Example (Uniformly continuous functions). Let T be a totally bounded semimetric space with semimetric ρ. We denote by $UC(T, \rho)$ the collection of all uniformly continuous functions $z : T \mapsto \mathbb{R}$. Because a uniformly continuous function on a totally bounded set is necessarily bounded, the space $UC(T, \rho)$ is a subspace of $\ell^\infty(T)$. We equip $UC(T, \rho)$ with the uniform norm.

Because a compact semimetric space is totally bounded, and a continuous function on a compact space is automatically uniformly continuous, the spaces $C(T, \rho)$ for a compact semimetric space T, for instance $C[a, b]$, are special cases of the spaces $UC(T, \rho)$. Actually, every space $UC(T, \rho)$ can be identified with a space $c(\overline{T}, \rho)$, because the *completion* $\overline{T}$ of a totally bounded semimetric T space is compact, and every uniformly continuous function on T has a unique continuous extension to the completion.

The space $UC(T, \rho)$ is separable. Furthermore, the Borel σ-field is equal to the σ-field generated by all coordinate projections (see Problem 18.3). The *coordinate projections* are the maps $z \mapsto z(t)$ with t ranging over T. These are continuous and hence always Borel-measurable. $\square$

18.8   Example (Product spaces). Given a pair of metric spaces $\mathbb{D}$ and $\mathbb{E}$ with metrics d and e, the *Cartesian product* $\mathbb{D} \times \mathbb{E}$ is a metric space with respect to the metric

$$f\big((x_1, y_1), (x_2, y_2)\big) = d(x_1, x_2) \vee e(y_1, y_2).$$

For this metric, convergence of a sequence $(x_n, y_n) \to (x, y)$ is equivalent to both $x_n \to x$ and $y_n \to y$.

For a product metric space, there exist two natural σ-fields: The product of the Borel σ-fields and the Borel σ-field of the product metric. In general, these are not the same, the second one being bigger. A sufficient condition for them to be equal is that the metric spaces $\mathbb{D}$ and $\mathbb{E}$ are separable (e.g., Chapter 1.4 in [146])).

The possible inequality of the two σ-fields causes an inconvenient problem. If $X : \Omega \mapsto \mathbb{D}$ and $Y : \Omega \mapsto \mathbb{E}$ are Borel-measurable maps, defined on some measurable space $(\Omega, \mathcal{U})$, then $(X, Y) : \Omega \mapsto \mathbb{D} \times \mathbb{E}$ is always measurable for the product of the Borel σ-fields. This is an easy fact from measure theory. However, if the two σ-fields are different, then the map (X, Y) need not be Borel-measurable. If they have separable range, then they are. $\square$

18.2 Basic Properties

In Chapter 2 convergence in distribution of random vectors is defined by reference to their distribution functions. Distribution functions do not extend in a natural way to random elements with values in metric spaces. Instead, we define convergence in distribution using one of the characterizations given by the portmanteau lemma.

A sequence of random elements X_n with values in a metric space $\mathbb{D}$ is said to *converge in distribution* to a random element X if $\mathrm{E}f(X_n) \to \mathrm{E}f(X)$ for every bounded, continuous function $f : \mathbb{D} \mapsto \mathbb{R}$. In some applications the "random elements" of interest turn out not to be Borel-measurable. To accomodate this situation, we extend the preceding definition to a sequence of *arbitrary maps* $X_n : \Omega_n \mapsto \mathbb{D}$, defined on probability spaces $(\Omega_n, \mathcal{U}_n, P_n)$. Because $\mathrm{E}f(X_n)$ need no longer make sense, we replace expectations by outer expectations. For an arbitrary map $X : \Omega \mapsto \mathbb{D}$, define

$$\mathrm{E}^* f(X) = \inf\{\mathrm{E}U : U : \Omega \mapsto \mathbb{R}, \text{measurable}, U \geq f(X), \mathrm{E}U \text{ exists}\}.$$

Then we say that a sequence of arbitrary maps $X_n : \Omega_n \mapsto \mathbb{D}$ *converges in distribution* to a random element X if $\mathrm{E}^* f(X_n) \to \mathrm{E}f(X)$ for every bounded, continuous function $f : \mathbb{D} \mapsto \mathbb{R}$. Here we insist that the limit X be Borel-measurable.

In the following, we do not stress the measurability issues. However, throughout we do write stars, if necessary, as a reminder that there are measurability issues that need to be taken care of. Although Ω_n may depend on n, we do not let this show up in the notation for E^* and P^*.

Next consider convergence in probability and almost surely. An arbitrary sequence of maps $X_n : \Omega_n \mapsto \mathbb{D}$ *converges in probability* to X if $\mathrm{P}^*\big(d(X_n, X) > \varepsilon\big) \to 0$ for all $\varepsilon > 0$. This is denoted by $X_n \overset{\mathrm{P}}{\to} X$. The sequence X_n *converges almost surely* to X if there exists a sequence of (measurable) random variables Δ_n such that $d(X_n, X) \leq \Delta_n$ and $\Delta_n \overset{\mathrm{as}}{\to} 0$. This is denoted by $X_n \overset{\mathrm{as}^*}{\to} X$.

These definitions also do not require the X_n to be Borel-measurable. In the definition of convergence in probability we solved this by adding a star, for *outer probability*. On the other hand, the definition of almost-sure convergence is unpleasantly complicated. This cannot be avoided easily, because, even for Borel-measurable maps X_n and X, the distance $d(X_n, X)$ need not be a random variable.

The portmanteau lemma, the continuous-mapping theorem and the relations among the three modes of stochastic convergence extend without essential changes to the present definitions. Even the proofs, as given in Chapter 2, do not need essential modifications. However, we seize the opportunity to formulate and prove a refinement of the continuous-mapping theorem. The continuous-mapping theorem furnishes a more intuitive interpretation of weak convergence in terms of weak convergence of random vectors: $X_n \rightsquigarrow X$ in the metric space $\mathbb{D}$ if and only if $g(X_n) \rightsquigarrow g(X)$ for every continuous map $g : \mathbb{D} \mapsto \mathbb{R}^k$.

18.9   Lemma (Portmanteau). *For arbitrary maps $X_n : \Omega_n \mapsto \mathbb{D}$ and every random element X with values in $\mathbb{D}$, the following statements are equivalent.*

    (i) $\mathrm{E}^* f(X_n) \to \mathrm{E} f(X)$ for all bounded, continuous functions f.

    (ii) $\mathrm{E}^* f(X_n) \to \mathrm{E} f(X)$ for all bounded, Lipschitz functions f.

    (iii) $\liminf \mathrm{P}_*(X_n \in G) \geq \mathrm{P}(X \in G)$ for every open set G.

    (iv) $\limsup \mathrm{P}^*(X_n \in F) \leq \mathrm{P}(X \in F)$ for every closed set F.

    (v) $\mathrm{P}^*(X_n \in B) \to \mathrm{P}(X \in B)$ for all Borel sets B with $\mathrm{P}(X \in \delta B) = 0$.

18.10   Theorem. *For arbitrary maps $X_n, Y_n : \Omega_n \mapsto \mathbb{D}$ and every random element X with values in $\mathbb{D}$:*

    (i) $X_n \overset{as*}{\to} X$ implies $X_n \overset{\mathrm{P}}{\to} X$.

    (ii) $X_n \overset{\mathrm{P}}{\to} X$ implies $X_n \rightsquigarrow X$.

    (iii) $X_n \overset{\mathrm{P}}{\to} c$ for a constant c if and only if $X_n \rightsquigarrow c$.

    (iv) if $X_n \rightsquigarrow X$ and $d(X_n, Y_n) \overset{\mathrm{P}}{\to} 0$, then $Y_n \rightsquigarrow X$.

    (v) if $X_n \rightsquigarrow X$ and $Y_n \overset{\mathrm{P}}{\to} c$ for a constant c, then $(X_n, Y_n) \rightsquigarrow (X, c)$.

    (vi) if $X_n \overset{\mathrm{P}}{\to} X$ and $Y_n \overset{\mathrm{P}}{\to} Y$, then $(X_n, Y_n) \overset{\mathrm{P}}{\to} (X, Y)$.

18.11   Theorem (Continuous mapping). *Let $\mathbb{D}_n \subset \mathbb{D}$ be arbitrary subsets and $g_n : \mathbb{D}_n \mapsto \mathbb{E}$ be arbitrary maps $(n \geq 0)$ such that for every sequence $x_n \in \mathbb{D}_n$: if $x_{n'} \to x$ along a subsequence and $x \in \mathbb{D}_0$, then $g_{n'}(x_{n'}) \to g_0(x)$. Then, for arbitrary maps $X_n : \Omega_n \mapsto \mathbb{D}_n$ and every random element X with values in $\mathbb{D}_0$ such that $g_0(X)$ is a random element in $\mathbb{E}$:*

    (i) If $X_n \rightsquigarrow X$, then $g_n(X_n) \rightsquigarrow g_0(X)$.

    (ii) If $X_n \overset{\mathrm{P}}{\to} X$, then $g_n(X_n) \overset{\mathrm{P}}{\to} g_0(X)$.

    (iii) If $X_n \overset{as*}{\to} X$, then $g_n(X_n) \overset{as*}{\to} g_0(X)$.

Proof.   The proofs for $\mathbb{D}_n = \mathbb{D}$ and $g_n = g$ fixed, where g is continuous at every point of $\mathbb{D}_0$, are the same as in the case of Euclidean spaces. We prove the refinement only for (i). The other refinements are not needed in the following.

For every closed set F, we have the inclusion

$$\bigcap_{k=1}^{\infty} \overline{\bigcup_{m=k}^{\infty} \left\{ x \in \mathbb{D}_m : g_m(x) \in F \right\}} \subset g_0^{-1}(F) \cup (\mathbb{D} - \mathbb{D}_0).$$

Indeed, suppose that x is in the set on the left side. Then for every k there is an $m_k \geq k$ and an element $x_{m_k} \in g_{m_k}^{-1}(F)$ with $d(x_{m_k}, x) < 1/k$. Thus, there exist a sequence $m_k \to \infty$ and elements $x_{m_k} \in \mathbb{D}_{m_k}$ with $x_{m_k} \to x$. Then either $g_{m_k}(x_{m_k}) \to g_0(x)$ or $x \notin \mathbb{D}_0$. Because the set F is closed, this implies that $g_0(x) \in F$ or $x \notin \mathbb{D}_0$.

Now, for every fixed k, by the portmanteau lemma,

$$\limsup_{n \to \infty} \mathrm{P}^*\big(g_n(X_n) \in F\big) \leq \limsup_{n \to \infty} \mathrm{P}^*\left(X_n \in \overline{\bigcup_{m=k}^{\infty}\{x \in \mathbb{D}_m : g_m(x) \in F\}}\right)$$

$$\leq \mathrm{P}\left(X \in \overline{\bigcup_{m=k}^{\infty} g_m^{-1}(F)}\right).$$

As $k \to \infty$, the last probability converges to $\mathrm{P}\big(X \in \bigcap_{k=1}^{\infty} \overline{\bigcup_{m=k}^{\infty} g_m^{-1}(F)}\big)$, which is smaller than or equal to $\mathrm{P}\big(g_0(X) \in F\big)$, by the preceding paragraph. Thus, $g_n(X_n) \rightsquigarrow g_0(X)$ by the portmanteau lemma in the other direction. ∎

The extension of Prohorov's theorem requires more care.[†] In a Euclidean space, a set is compact if and only if it is closed and bounded. In general metric spaces, a compact set is closed and bounded, but a closed, bounded set is not necessarily compact. It is the compactness that we employ in the definition of tightness. A Borel-measurable random element X into a metric space is *tight* if for every $\varepsilon > 0$ there exists a compact set K such that $\mathrm{P}(X \notin K) < \varepsilon$. A sequence of arbitrary maps $X_n : \Omega_n \mapsto \mathbb{D}$ is called *asymptotically tight* if for every $\varepsilon > 0$ there exists a compact set K such that

$$\limsup_{n \to \infty} \mathrm{P}^*(X_n \notin K^\delta) < \varepsilon, \qquad \text{every } \delta > 0.$$

Here K^δ is the δ-enlargement $\{y : d(y, K) < \delta\}$ of the set K. It can be shown that, for Borel-measurable maps in $\mathbb{R}^k$, this is identical to "uniformly tight," as defined in Chapter 2. In order to obtain a theory that applies to a sufficient number of applications, again we do not wish to assume that the X_n are Borel-measurable. However, Prohorov's theorem is true only under, at least, "measurability in the limit." An arbitrary sequence of maps X_n is called *asymptotically measurable* if

$$E^* f(X_n) - E_* f(X_n) \to 0, \qquad \text{every } f \in C_b(D).$$

Here E_* denotes the inner expectation, which is defined in analogy with the outer expectation, and $C_b(\mathbb{D})$ is the collection of all bounded, continuous functions $f : \mathbb{D} \mapsto \mathbb{R}$. A Borel-measurable sequence of random elements X_n is certainly asymptotically measurable, because then both the outer and the inner expectations in the preceding display are equal to the expectation, and the difference is identically zero.

18.12   Theorem (Prohorov's theorem). *Let $X_n : \Omega_n \to \mathbb{D}$ be arbitrary maps into a metric space.*
  (i) *If $X_n \rightsquigarrow X$ for some tight random element X, then $\{X_n : n \in \mathbb{N}\}$ is asymptotically tight and asymptotically measurable.*
  (ii) *If X_n is asymptotically tight and asymptotically measurable, then there is a subsequence and a tight random element X such that $X_{n_j} \rightsquigarrow X$ as $j \to \infty$.*

18.3   Bounded Stochastic Processes

A *stochastic process* $X = \{X_t : t \in T\}$ is a collection of random variables $X_t : \Omega \mapsto \mathbb{R}$, indexed by an arbitrary set T and defined on the same probability space $(\Omega, \mathcal{U}, \mathrm{P})$. For a fixed ω, the map $t \mapsto X_t(\omega)$ is called a *sample path*, and it is helpful to think of X as a random function, whose realizations are the sample paths, rather than as a collection of random variables. If every sample path is a bounded function, then X can be viewed as a

---

[†] The following Prohorov's theorem is not used in this book. For a proof see, for instance, [146].

map $X : \Omega \mapsto \ell^{\infty}(T)$. If $T = [a, b]$ and the sample paths are continuous or cadlag, then X is also a map with values in $C[a, b]$ or $D[a, b]$.

Because $C[a, b] \subset D[a, b] \subset \ell^{\infty}[a, b]$, we can consider the weak convergence of a sequence of maps with values in $C[a, b]$ relative to $C[a, b]$, but also relative to $D[a, b]$, or $\ell^{\infty}[a, b]$. The following lemma shows that this does not make a difference, as long as we use the uniform norm for all three spaces.

18.13   Lemma. *Let $\mathbb{D}_0 \subset \mathbb{D}$ be arbitrary metric spaces equipped with the same metric. If X and every X_n take their values in $\mathbb{D}_0$, then $X_n \leadsto X$ as maps in $\mathbb{D}_0$ if and only if $X_n \leadsto X$ as maps in $\mathbb{D}$.*

Proof.   Because a set G_0 in $\mathbb{D}_0$ is open if and only if it is of the form $G \cap \mathbb{D}_0$ for an open set G in $\mathbb{D}$, this is an easy corollary of (iii) of the portmanteau lemma. ∎

Thus, we may concentrate on weak convergence in the space $\ell^{\infty}(T)$, and automatically obtain characterizations of weak convergence in $C[a, b]$ or $D[a, b]$. The next theorem gives a characterization by *finite approximation*. It is required that, for any $\varepsilon > 0$, the index set T can be partitioned into finitely many sets $T_1, \dots, T_k$ such that (asymptotically) the variation of the sample paths $t \mapsto X_{n,t}$ is less than ε on every one of the sets T_i, with large probability. Then the behavior of the process can be described, within a small error margin, by the behavior of the *marginal vectors* $(X_{n,t_1}, \dots, X_{n,t_k})$ for arbitrary fixed points $t_i \in T_i$. If these marginals converge, then the processes converge.

18.14   Theorem. *A sequence of arbitrary maps $X_n : \Omega_n \mapsto \ell^{\infty}(T)$ converges weakly to a tight random element if and only if both of the following conditions hold:*
   (i) *The sequence $(X_{n,t_1}, \dots, X_{n,t_k})$ converges in distribution in $\mathbb{R}^k$ for every finite set of points $t_1, \dots, t_k$ in T;*
   (ii) *for every $\varepsilon, \eta > 0$ there exists a partition of T into finitely many sets $T_1, \dots, T_k$ such that*

$$\limsup_{n \to \infty} \mathrm{P}^* \left(\sup_i \sup_{s,t \in T_i} |X_{n,s} - X_{n,t}| \geq \varepsilon \right) \leq \eta.$$

Proof.   We only give the proof of the more constructive part, the sufficiency of (i) and (ii). For each natural number m, partition T into sets $T_1^m, \dots, T_{k_m}^m$, as in (ii) corresponding to $\varepsilon = \eta = 2^{-m}$. Because the probabilities in (ii) decrease if the partition is refined, we can assume without loss of generality that the partitions are successive refinements as m increases. For fixed m define a semimetric ρ_m on T by $\rho_m(s, t) = 0$ if s and t belong to the same partioning set T_j^m, and by $\rho_m(s, t) = 1$ otherwise. Every ρ_m-ball of radius $0 < \varepsilon < 1$ coincides with a partitioning set. In particular, T is totally bounded for ρ_m, and the ρ_m-diameter of a set T_j^m is zero. By the nesting of the partitions, $\rho_1 \leq \rho_2 \leq \cdots$. Define $\rho(s, t) = \sum_{m=1}^{\infty} 2^{-m} \rho_m(s, t)$. Then ρ is a semimetric such that the ρ-diameter of T_j^m is smaller than $\sum_{k>m} 2^{-k} = 2^{-m}$, and hence T is totally bounded for ρ. Let T_0 be the countable ρ-dense subset constructed by choosing an arbitrary point t_j^m from every T_j^m.

By assumption (i) and Kolmogorov's consistency theorem (e.g., [133, p. 244] or [42, p. 347]), we can construct a stochastic process $\{X_t : t \in T_0\}$ on some probability space such that $(X_{n,t_1}, \dots, X_{n,t_k}) \leadsto (X_{t_1}, \dots, X_{t_k})$ for every finite set of points $t_1, \dots, t_k$ in T_0. By the portmanteau lemma and assumption (ii), for every finite set $S \subset T_0$,

$$P\left(\sup_j \sup_{\substack{s,t \in T_j^m \\ s,t \in S}} |X_s - X_t| > 2^{-m}\right) \leq 2^{-m}.$$

By the monotone convergence theorem this remains true if S is replaced by T_0. If $\rho(s, t) < 2^{-m}$, then $\rho_m(s, t) < 1$ and hence s and t belong to the same partitioning set T_j^m. Consequently, the event in the preceding display with $S = T_0$ contains the event in the following display, and

$$P\left(\sup_{\substack{\rho(s,t)<2^{-m} \\ s,t \in T_0}} |X_s - X_t| > 2^{-m}\right) \leq 2^{-m}.$$

This sums to a finite number over $m \in \mathbb{N}$. Hence, by the Borel-Cantelli lemma, for almost all ω, $|X_s(\omega) - X_t(\omega)| \leq 2^{-m}$ for all $\rho(s, t) < 2^{-m}$ and all sufficiently large m. This implies that almost all sample paths of $\{X_t : t \in T_0\}$ are contained in $UC(T_0, \rho)$. Extend the process by continuity to a process $\{X_t : t \in T\}$ with almost all sample paths in $UC(T, \rho)$. This process is Borel measurable and hence tight, because the latter space is complete and separable.

Define $\pi_m : T \mapsto T$ as the map that maps every partioning set T_j^m onto the point $t_j^m \in T_j^m$. Then, by the uniform continuity of X, and the fact that the ρ-diameter of T_j^m is smaller than 2^{-m}, $X \circ \pi_m \rightsquigarrow X$ in $\ell^\infty(T)$ as $m \to \infty$ (even almost surely). The processes $\{X_n \circ \pi_m(t) : t \in T\}$ are essentially k_m-dimensional vectors. By (i), $X_n \circ \pi_m \rightsquigarrow X \circ \pi_m$ in $\ell^\infty(T)$ as $n \to \infty$, for every fixed m. Consequently, for every Lipschitz function $f : \ell^\infty(T) \mapsto [0, 1]$, $E^* f(X_n \circ \pi_m) \to Ef(X)$ as $n \to \infty$, followed by $m \to \infty$. Conclude that, for every $\varepsilon > 0$,

$$\left|E^* f(X_n) - Ef(X)\right| \leq \left|E^* f(X_n) - E^* f(X_n \circ \pi_m)\right| + o(1)$$
$$\leq \|f\|_{\text{lip}}\varepsilon + P^*\left(\|X_n - X_n \circ \pi_m\|_T > \varepsilon\right) + o(1).$$

For $\varepsilon = 2^{-m}$ this is bounded by $\|f\|_{\text{lip}} 2^{-m} + 2^{-m} + o(1)$, by the construction of the partitions. The proof is complete. ■

In the course of the proof of the preceding theorem a semimetric ρ is constructed such that the weak limit X has uniformly ρ-continuous sample paths, and such that (T, ρ) is totally bounded. This is surprising: even though we are discussing stochastic processes with values in the very large space $\ell^\infty(T)$, the limit is concentrated on a much smaller space of continuous functions. Actually, this is a consequence of imposing the condition (ii), which can be shown to be equivalent to asymptotic tightness. It can be shown, more generally, that every tight random element X in $\ell^\infty(T)$ necessarily concentrates on $UC(T, \rho)$ for some semimetric ρ (depending on X) that makes T totally bounded.

In view of this connection between the partitioning condition (ii), continuity, and tightness, we shall sometimes refer to this condition as the condition of *asymptotic tightness* or *asymptotic equicontinuity*.

We record the existence of the semimetric for later reference and note that, for a Gaussian limit process, this can always be taken equal to the "intrinsic" standard deviation semimetric.

18.15 Lemma. *Under the conditions (i) and (ii) of the preceding theorem there exists a semimetric ρ on T for which T is totally bounded, and such that the weak limit of the sequence X_n can be constructed to have almost all sample paths in UC (T, ρ). Furthermore, if the weak limit X is zero-mean Gaussian, then this semimetric can be taken equal to $\rho(s, t) = \mathrm{sd}(X_s - X_t)$.*

Proof. A semimetric ρ is constructed explicitly in the proof of the preceding theorem. It suffices to prove the statement concerning Gaussian limits X.

Let ρ be the semimetric obtained in the proof of the theorem and let ρ_2 be the standard deviation semimetric. Because every uniformly ρ-continuous function has a unique continuous extension to the ρ-completion of T, which is compact, it is no loss of generality to assume that T is ρ-compact. Furthermore, assume that *every* sample path of X is ρ-continuous.

An arbitrary sequence t_n in T has a ρ-converging subsequence $t_{n'} \to t$. By the ρ-continuity of the sample paths, $X_{t_{n'}} \to X_t$ almost surely. Because every X_t is Gaussian, this implies convergence of means and variances, whence $\rho_2(t_{n'}, t)^2 = \mathrm{E}(X_{t_{n'}} - X_t)^2 \to 0$ by Proposition 2.29. Thus $t_{n'} \to t$ also for ρ_2 and hence T is ρ_2–compact.

Suppose that a sample path $t \mapsto X_t(\omega)$ is not ρ_2-continuous. Then there exists an $\varepsilon > 0$ and a $t \in T$ such that $\rho_2(t_n, t) \to 0$, but $|X_{t_n}(\omega) - X_t(\omega)| \geq \varepsilon$ for every n. By the ρ-compactness and continuity, there exists a subsequence such that $\rho(t_{n'}, s) \to 0$ and $X_{t_{n'}}(\omega) \to X_s(\omega)$ for some s. By the argument of the preceding paragraph, $\rho_2(t_{n'}, s) \to 0$, so that $\rho_2(s, t) = 0$ and $|X_s(\omega) - X_t(\omega)| \geq \varepsilon$. Conclude that the path $t \mapsto X_t(\omega)$ can only fail to be ρ_2-continuous for ω for which there exist $s, t \in T$ with $\rho_2(s, t) = 0$, but $X_s(\omega) \neq X_t(\omega)$. Let N be the set of ω for which there do exist such s, t. Take a countable, ρ-dense subset A of $\{(s, t) \in T \times T : \rho_2(s, t) = 0\}$. Because $t \mapsto X_t(\omega)$ is ρ-continuous, N is also the set of all ω such that there exist $(s, t) \in A$ with $X_s(\omega) \neq X_t(\omega)$. From the definition of ρ_2, it is clear that for every fixed (s, t), the set of ω such that $X_s(\omega) \neq X_t(\omega)$ is a null set. Conclude that N is a null set. Hence, almost all paths of X are ρ_2-continuous. ∎

Notes

The theory in this chapter was developed in increasing generality over the course of many years. Work by Donsker around 1950 on the approximation of the empirical process and the partial sum process by the Brownian bridge and Brownian motion processes was an important motivation. The first type of approximation is discussed in Chapter 19. For further details and references concerning the material in this chapter, see, for example, [76] or [146].

PROBLEMS

1. (i) Show that a compact set is totally bounded.

   (ii) Show that a compact set is separable.

2. Show that a function $f : \mathbb{D} \mapsto \mathbb{E}$ is continuous at every $x \in \mathbb{D}$ if and only if $f^{-1}(G)$ is open in $\mathbb{D}$ for every open $G \in \mathbb{E}$.

3. **(Projection σ-field.)** Show that the σ-field generated by the coordinate projections $z \mapsto z(t)$ on $C[a, b]$ is equal to the Borel σ-field generated by the uniform norm. (First, show that the space

$C[a, b]$ is separable. Next show that every open set in a separable metric space is a *countable* union of open balls. Next, it suffices to prove that every open ball is measurable for the projection σ-field.)

4. Show that $D[a, b]$ is not separable for the uniform norm.

5. Show that every function in $D[a, b]$ is bounded.

6. Let h be an arbitrary element of $D[-\infty, \infty]$ and let $\varepsilon > 0$. Show that there exists a grid $u_0 = -\infty < u_1 < \cdots u_m = \infty$ such that h varies at most ε on every interval $[u_i, u_{i+1})$. Here "varies at most ε" means that $|h(u) - h(v)|$ is less than ε for every u, v in the interval. (Make sure that all points at which h jumps more than ε are grid points.)

7. Suppose that H_n and H_0 are subsets of a semimetric space H such that $H_n \to H_0$ in the sense that
    (i) Every $h \in H_0$ is the limit of a sequence $h_n \in H_n$;
    (ii) If a subsequence h_{n_j} converges to a limit h, then $h \in H_0$.

    Suppose that Λ_n are stochastic processes indexed by H that converge in distribution in the space $\ell^\infty(H)$ to a stochastic process Λ that has uniformly continuous sample paths. Show that $\sup_{h \in H_n} \Lambda_n(h) \rightsquigarrow \sup_{h \in H_0} \Lambda(h)$.

19

Empirical Processes

The empirical distribution of a random sample is the uniform discrete measure on the observations. In this chapter, we study the convergence of this measure and in particular the convergence of the corresponding distribution function. This leads to laws of large numbers and central limit theorems that are uniform in classes of functions. We also discuss a number of applications of these results.

19.1 Empirical Distribution Functions

Let $X_1, \ldots, X_n$ be a random sample from a distribution function F on the real line. The *empirical distribution function* is defined as

$$\mathbb{F}_n(t) = \frac{1}{n} \sum_{i=1}^{n} 1\{X_i \le t\}.$$

It is the natural estimator for the underlying distribution F if this is completely unknown. Because $n\mathbb{F}_n(t)$ is binomially distributed with mean $nF(t)$, this estimator is unbiased. By the law of large numbers it is also consistent,

$$\mathbb{F}_n(t) \overset{\text{as}}{\to} F(t), \qquad \text{every } t.$$

By the central limit theorem it is asymptotically normal,

$$\sqrt{n}\big(\mathbb{F}_n(t) - F(t)\big) \rightsquigarrow N\Big(0, F(t)\big(1 - F(t)\big)\Big).$$

In this chapter we improve on these results by considering $t \mapsto \mathbb{F}_n(t)$ as a random function, rather than as a real-valued estimator for each t separately. This is of interest on its own account but also provides a useful starting tool for the asymptotic analysis of other statistics, such as quantiles, rank statistics, or trimmed means.

The *Glivenko-Cantelli theorem* extends the law of large numbers and gives uniform convergence. The uniform distance

$$\|\mathbb{F}_n - F\|_\infty = \sup_t \big|\mathbb{F}_n(t) - F(t)\big|$$

is known as the *Kolmogorov-Smirnov statistic*.

19.1   Theorem (Glivenko-Cantelli). *If $X_1, X_2, \ldots$ are i.i.d. random variables with distribution function F, then $\|\mathbb{F}_n - F\|_\infty \overset{as}{\to} 0$.*

Proof.   By the strong law of large numbers, both $\mathbb{F}_n(t) \overset{as}{\to} F(t)$ and $\mathbb{F}_n(t-) \overset{as}{\to} F(t-)$ for every t. Given a fixed $\varepsilon > 0$, there exists a partition $-\infty = t_0 < t_1 < \cdots < t_k = \infty$ such that $F(t_i-) - F(t_{i-1}) < \varepsilon$ for every i. (Points at which F jumps more than ε are points of the partition.) Now, for $t_{i-1} \le t < t_i$,

$$\mathbb{F}_n(t) - F(t) \le \mathbb{F}_n(t_i-) - F(t_i-) + \varepsilon,$$
$$\mathbb{F}_n(t) - F(t) \ge \mathbb{F}_n(t_{i-1}) - F(t_{i-1}) - \varepsilon.$$

The convergence of $\mathbb{F}_n(t)$ and $\mathbb{F}_n(t-)$ for every fixed t is certainly uniform for t in the finite set $\{t_1, \ldots, t_{k-1}\}$. Conclude that $\limsup \|\mathbb{F}_n - F\|_\infty \le \varepsilon$, almost surely. This is true for every $\varepsilon > 0$ and hence the limit superior is zero. ∎

The extension of the central limit theorem to a "uniform" or "functional" central limit theorem is more involved. A first step is to prove the joint weak convergence of finitely many coordinates. By the multivariate central limit theorem, for every $t_1, \ldots, t_k$,

$$\sqrt{n}\big(\mathbb{F}_n(t_i) - F(t_i), \ldots, \mathbb{F}_n(t_k) - F(t_k)\big) \rightsquigarrow \big(\mathbb{G}_F(t_1), \ldots, \mathbb{G}_F(t_k)\big),$$

where the vector on the right has a multivariate-normal distribution, with mean zero and covariances

$$E\mathbb{G}_F(t_i)\mathbb{G}_F(t_j) = F(t_i \wedge t_j) - F(t_i)F(t_j). \tag{19.2}$$

This suggests that the sequence of *empirical processes* $\sqrt{n}(\mathbb{F}_n - F)$, viewed as random functions, converges in distribution to a Gaussian process $\mathbb{G}_F$ with zero mean and covariance functions as in the preceding display. According to an extension of Donsker's theorem, this is true in the sense of weak convergence of these processes in the Skorohod space $D[-\infty, \infty]$ equipped with the uniform norm. The limit process $\mathbb{G}_F$ is known as an F-*Brownian bridge* process, and as a *standard (or uniform) Brownian bridge* if F is the uniform distribution λ on $[0, 1]$. From the form of the covariance function it is clear that the F-Brownian bridge is obtainable as $\mathbb{G}_\lambda \circ F$ from a standard bridge $\mathbb{G}_\lambda$. The name "bridge" results from the fact that the sample paths of the process are zero (one says "tied down") at the endpoints $-\infty$ and ∞. This is a consequence of the fact that the difference of two distribution functions is zero at these points.

19.3   Theorem (Donsker). *If $X_1, X_2, \ldots$ are i.i.d. random variables with distribution function F, then the sequence of empirical processes $\sqrt{n}(\mathbb{F}_n - F)$ converges in distribution in the space $D[-\infty, \infty]$ to a tight random element $\mathbb{G}_F$, whose marginal distributions are zero-mean normal with covariance function (19.2).*

Proof.   The proof of this theorem is long. Because there is little to be gained by considering the special case of cells in the real line, we deduce the theorem from a more general result in the next section. ∎

Figure 19.1 shows some realizations of the uniform empirical process. The roughness of the sample path for $n = 5000$ is remarkable, and typical. It is carried over onto the limit

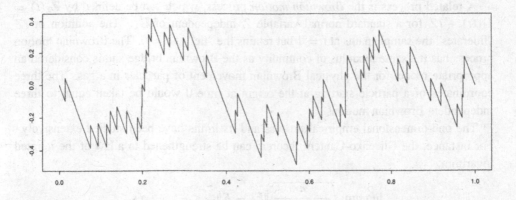

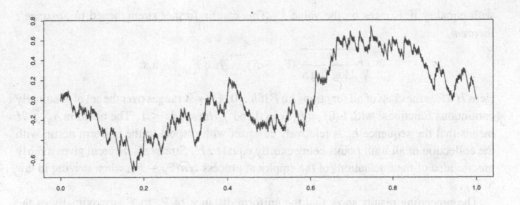

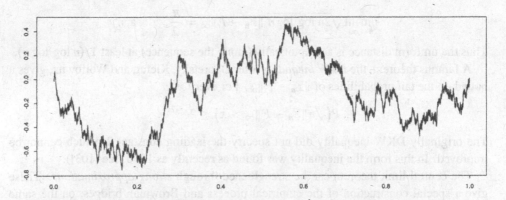

Figure 19.1.   Three realizations of the uniform empirical process, of 50 (top), 500 (middle), and 5000 (bottom) observations, respectively.

process, for it can be shown that, for every t,

$$0 < \liminf_{h \to 0} \frac{|\mathbb{G}_\lambda(t+h) - \mathbb{G}_\lambda(t)|}{\sqrt{|h \log \log h|}} \le \limsup_{h \to 0} \frac{|\mathbb{G}_\lambda(t+h) - \mathbb{G}_\lambda(t)|}{\sqrt{|h \log \log h|}} < \infty, \qquad \text{a.s.}$$

Thus, the increments of the sample paths of a standard Brownian bridge are close to being of the order $\sqrt{|h|}$. This means that the sample paths are continuous, but nowhere differentiable.

A related process is the *Brownian motion* process, which can be defined by $\mathbb{Z}_\lambda(t) = \mathbb{G}_\lambda(t) + tZ$ for a standard normal variable Z independent of $\mathbb{G}_\lambda$. The addition of tZ "liberates" the sample paths at $t = 1$ but retains the "tie" at $t = 0$. The Brownian motion process has the same modulus of continuity as the Brownian bridge and is considered an appropriate model for the physical Brownian movement of particles in a gas. The three coordinates of a particle starting at the origin at time 0 would be taken equal to three independent Brownian motions.

The one-dimensional empirical process and its limits have been studied extensively.[†] For instance, the Glivenko-Cantelli theorem can be strengthened to a law of the iterated logarithm,

$$\limsup_{n \to \infty} \sqrt{\frac{n}{2 \log \log n}} \, \|\mathbb{F}_n - F\|_\infty \le \tfrac{1}{2}, \qquad \text{a.s.},$$

with equality if F takes on the value $\tfrac{1}{2}$. This can be further strengthened to *Strassen's theorem*

$$\sqrt{\frac{n}{2 \log \log n}} \, (\mathbb{F}_n - F) \overset{\leadsto}{\underset{\leadsto}{}} \mathcal{H} \circ F, \qquad \text{a.s.}$$

Here $\mathcal{H} \circ F$ is the class of all functions $h \circ F$ if $h : [0, 1] \mapsto \mathbb{R}$ ranges over the set of absolutely continuous functions[‡] with $h(0) = h(1) = 0$ and $\int_0^1 h'(s)^2 \, ds \le 1$. The notation $h_n \overset{\leadsto}{\underset{\leadsto}{}} \mathcal{H}$ means that the sequence h_n is relatively compact with respect to the uniform norm, with the collection of all limit points being exactly equal to $\mathcal{H}$. Strassen's theorem gives a fairly precise idea of the fluctuations of the empirical process $\sqrt{n}(\mathbb{F}_n - F)$, when striving in law to $\mathbb{G}_F$.

The preceding results show that the uniform distance of $\mathbb{F}_n$ to F is maximally of the order $\sqrt{\log \log n / n}$ as $n \to \infty$. It is also known that

$$\liminf_{n \to \infty} \sqrt{2n \log \log n} \, \|\mathbb{F}_n - F\|_\infty = \frac{\pi}{2}, \qquad \text{a.s.}$$

Thus the uniform distance is asymptotically (along the sequence) at least $1/(n \log \log n)$.

A famous theorem, the *DKW inequality* after Dvoretsky, Kiefer, and Wolfowitz, gives a bound on the tail probabilities of $\|\mathbb{F}_n - F\|_\infty$. For every x

$$P\big(\sqrt{n}\|\mathbb{F}_n - F\|_\infty > x\big) \le 2e^{-2x^2}.$$

The originally DKW inequality did not specify the leading constant 2, which cannot be improved. In this form the inequality was found as recently as 1990 (see [103]).

The central limit theorem can be strengthened through *strong approximations*. These give a special construction of the empirical process and Brownian bridges, on the same probability space, that are close not only in a distributional sense but also in a pointwise sense. One such result asserts that there exists a probability space carrying i.i.d. random variables $X_1, X_2, \ldots$ with law F and a sequence of Brownian bridges $\mathbb{G}_{F,n}$ such that

$$\limsup_{n \to \infty} \frac{\sqrt{n}}{(\log n)^2} \, \big\|\sqrt{n}(\mathbb{F}_n - F) - \mathbb{G}_{F,n}\big\|_\infty < \infty, \qquad \text{a.s.}$$

[†] See [134] for the following and many other results on the univariate empirical process.

[‡] A function is *absolutely continuous* if it is the primitive function $\int_0^t g(s) \, ds$ of an integrable function g. Then it is almost-everywhere differentiable with derivative g.

Because, by construction, every $\mathbb{G}_{F,n}$ is equal in law to $\mathbb{G}_F$, this implies that $\sqrt{n}(\mathbb{F}_n - F) \rightsquigarrow \mathbb{G}_F$ as a process (Donsker's theorem), but it implies a lot more. Apparently, the distance between the sequence and its limit is of the order $O\big((\log n)^2/\sqrt{n}\big)$. After the method of proof and the country of origin, results of this type are also known as *Hungarian embeddings*. Another construction yields the estimate, for fixed constants a, b, and c and every $x > 0$,

$$\mathrm{P}\bigg(\big\| \sqrt{n}(\mathbb{F}_n - F) - \mathbb{G}_{F,n} \big\|_\infty > \frac{a \log n + x}{\sqrt{n}} \bigg) \leq b e^{-cx}.$$

19.2 Empirical Distributions

Let $X_1, \ldots, X_n$ be a random sample from a probability distribution P on a measurable space $(\mathcal{X}, \mathcal{A})$. The *empirical distribution* is the discrete uniform measure on the observations. We denote it by $\mathbb{P}_n = n^{-1}\sum_{i=1}^n \delta_{X_i}$, where δ_x is the probability distribution that is degenerate at x. Given a measurable function $f : \mathcal{X} \mapsto \mathbb{R}$, we write $\mathbb{P}_n f$ for the expectation of f under the empirical measure, and Pf for the expectation under P. Thus

$$\mathbb{P}_n f = \frac{1}{n} \sum_{i=1}^n f(X_i), \qquad Pf = \int f \, dP.$$

Actually, this chapter is concerned with these maps rather than with $\mathbb{P}_n$ as a measure.

By the law of large numbers, the sequence $\mathbb{P}_n f$ converges almost surely to Pf, for every f such that Pf is defined. The abstract Glivenko-Cantelli theorems make this result uniform in f ranging over a class of functions. A class $\mathcal{F}$ of measurable functions $f : \mathcal{X} \mapsto \mathbb{R}$ is called *P-Glivenko-Cantelli* if

$$\|\mathbb{P}_n f - Pf\|_{\mathcal{F}} = \sup_{f \in \mathcal{F}} |\mathbb{P}_n f - Pf| \overset{\text{as*}}{\to} 0.$$

The *empirical process* evaluated at f is defined as $\mathbb{G}_n f = \sqrt{n}(\mathbb{P}_n f - Pf)$. By the multivariate central limit theorem, given any finite set of measurable functions f_i with $Pf_i^2 < \infty$,

$$(\mathbb{G}_n f_1, \ldots, \mathbb{G}_n f_k) \rightsquigarrow (\mathbb{G}_P f_1, \ldots, \mathbb{G}_P f_k),$$

where the vector on the right possesses a multivariate-normal distribution with mean zero and covariances

$$\mathrm{E}\mathbb{G}_P f \, \mathbb{G}_P g = Pfg - Pf Pg.$$

The abstract Donsker theorems make this result "uniform" in classes of functions. A class $\mathcal{F}$ of measurable functions $f : \mathcal{X} \mapsto \mathbb{R}$ is called *P-Donsker* if the sequence of processes $\{\mathbb{G}_n f : f \in \mathcal{F}\}$ converges in distribution to a tight limit process in the space $\ell^\infty(\mathcal{F})$. Then the limit process is a Gaussian process $\mathbb{G}_P$ with zero mean and covariance function as given in the preceding display and is known as a *P-Brownian bridge*. Of course, the Donsker property includes the requirement that the sample paths $f \mapsto \mathbb{G}_n f$ are uniformly bounded for every n and every realization of $X_1, \ldots, X_n$. This is the case, for instance, if the class $\mathcal{F}$

has a finite and integrable *envelope function* F: a function such that $|f(x)| \leq F(x) < \infty$, for every x and f. It is not required that the function $x \mapsto F(x)$ be uniformly bounded.

For convenience of terminology we define a class $\mathcal{F}$ of vector-valued functions $f : x \mapsto \mathbb{R}^k$ to be Glivenko-Cantelli or Donsker if each of the classes of coordinates $f_i : x \mapsto \mathbb{R}$ with $f = (f_i, \ldots, f_k)$ ranging over $\mathcal{F}$ ($i = 1, 2, \ldots, k$) is Glivenko-Cantelli or Donsker. It can be shown that this is equivalent to the union of the k coordinate classes being Glivenko-Cantelli or Donsker.

Whether a class of functions is Glivenko-Cantelli or Donsker depends on the "size" of the class. A finite class of integrable functions is always Glivenko-Cantelli, and a finite class of square-integrable functions is always Donsker. On the other hand, the class of all square-integrable functions is Glivenko-Cantelli, or Donsker, only in trivial cases. A relatively simple way to measure the size of a class $\mathcal{F}$ is in terms of entropy. We shall mainly consider the bracketing entropy relative to the $L_r(P)$-norm

$$\|f\|_{P,r} = (P|f|^r)^{1/r}.$$

Given two functions l and u, the *bracket* $[l, u]$ is the set of all functions f with $l \leq f \leq u$. An ε-*bracket* in $L_r(P)$ is a bracket $[l, u]$ with $P(u - l)^r < \varepsilon^r$. The *bracketing number* $N_{[]}(\varepsilon, \mathcal{F}, L_r(P))$ is the minimum number of ε-brackets needed to cover $\mathcal{F}$. (The bracketing functions l and u must have finite $L_r(P)$-norms but need not belong to $\mathcal{F}$.) The *entropy with bracketing* is the logarithm of the bracketing number.

A simple condition for a class to be P-Glivenko-Cantelli is that the bracketing numbers in $L_1(P)$ are finite for every $\varepsilon > 0$. The proof is a straightforward generalization of the proof of the classical Glivenko-Cantelli theorem, Theorem 19.1, and is omitted.

19.4 Theorem (Glivenko-Cantelli). *Every class $\mathcal{F}$ of measurable functions such that $N_{[]}(\varepsilon, \mathcal{F}, L_1(P)) < \infty$ for every $\varepsilon > 0$ is P-Glivenko-Cantelli.*

For most classes of interest, the bracketing numbers $N_{[]}(\varepsilon, \mathcal{F}, L_r(P))$ grow to infinity as $\varepsilon \downarrow 0$. A sufficient condition for a class to be Donsker is that they do not grow too fast. The speed can be measured in terms of the *bracketing integral*

$$J_{[]}(\delta, \mathcal{F}, L_2(P)) = \int_0^\delta \sqrt{\log N_{[]}(\varepsilon, \mathcal{F}, L_2(P))} \, d\varepsilon.$$

If this integral is finite-valued, then the class $\mathcal{F}$ is P-Donsker. The integrand in the integral is a decreasing function of ε. Hence, the convergence of the integral depends only on the size of the bracketing numbers for $\varepsilon \downarrow 0$. Because $\int_0^1 \varepsilon^{-r} \, d\varepsilon$ converges for $r < 1$ and diverges for $r \geq 1$, the integral condition roughly requires that the entropies grow of slower order than $(1/\varepsilon)^2$.

19.5 Theorem (Donsker). *Every class $\mathcal{F}$ of measurable functions with $J_{[]}(1, \mathcal{F}, L_2(P)) < \infty$ is P-Donsker.*

Proof. Let $\mathcal{G}$ be the collection of all differences $f - g$ if f and g range over $\mathcal{F}$. With a given set of ε-brackets $[l_i, u_i]$ over $\mathcal{F}$ we can construct 2ε-brackets over $\mathcal{G}$ by taking differences $[l_i - u_j, u_i - l_j]$ of upper and lower bounds. Therefore, the bracketing numbers $N_{[]}(\varepsilon, \mathcal{G}, L_2(P))$ are bounded by the squares of the bracketing numbers

$N_{[]}\big(\varepsilon/2, \mathcal{F}, L_2(P)\big)$. Taking a logarithm turns the square into a multiplicative factor 2, and hence the entropy integrals of $\mathcal{F}$ and $\mathcal{G}$ are proportional.

For a given, small $\delta > 0$ choose a minimal number of brackets of size δ that cover $\mathcal{F}$, and use them to form a partition of $\mathcal{F} = \cup_i \mathcal{F}_i$ in sets of diameters smaller than δ. The subset of $\mathcal{G}$ consisting of differences $f - g$ of functions f and g belonging to the same partitioning set consists of functions of $L_2(P)$-norm smaller than δ. Hence, by Lemma 19.34 ahead, there exists a finite number $a(\delta)$ such that

$$\mathrm{E}^* \sup_i \sup_{f,g \in \mathcal{F}_i} \big|\mathbb{G}_n(f - g)\big| \lesssim J_{[]}\big(\delta, \mathcal{F}, L_2(P)\big) + \sqrt{n}\, P F 1\{F > a(\delta)\sqrt{n}\}.$$

Here the envelope function F can be taken equal to the supremum of the absolute values of the upper and lower bounds of finitely many brackets that cover $\mathcal{F}$, for instance a minimal set of brackets of size 1. This F is square-integrable.

The second term on the right is bounded by $a(\delta)^{-1} P F^2 1\{F > a(\delta)\sqrt{n}\}$ and hence converges to zero as $n \to \infty$ for every fixed δ. The integral converges to zero as $\delta \to 0$. The theorem follows from Theorem 18.14, in view of Markov's inequality. ∎

19.6 Example (Distribution function). If $\mathcal{F}$ is equal to the collection of all indicator functions of the form $f_t = 1_{(-\infty, t]}$, with t ranging over $\mathbb{R}$, then the empirical process $\mathbb{G}_n f_t$ is the classical empirical process $\sqrt{n}\big(\mathbb{F}_n(t) - F(t)\big)$. The preceding theorems reduce to the classical theorems by Glivenko-Cantelli and Donsker. We can see this by bounding the bracketing numbers of the set of indicator functions f_t.

Consider brackets of the form $[1_{(-\infty, t_{i-1}]}, 1_{(-\infty, t_i]}]$ for a grid of points $-\infty = t_0 < t_1 < \cdots < t_k = \infty$ with the property $F(t_i-) - F(t_{i-1}) < \varepsilon$ for each i. These brackets have $L_1(F)$-size ε. Their total number k can be chosen smaller than $2/\varepsilon$. Because $F f^2 \le F f$ for every $0 \le f \le 1$, the $L_2(F)$-size of the brackets is bounded by $\sqrt{\varepsilon}$. Thus $N_{[]}\big(\sqrt{\varepsilon}, \mathcal{F}, L_2(F)\big) \le (2/\varepsilon)$, whence the bracketing numbers are of the polynomial order $(1/\varepsilon)^2$. This means that this class of functions is very small, because a function of the type $\log(1/\varepsilon)$ satisfies the entropy condition of Theorem 19.5 easily. □

19.7 Example (Parametric class). Let $\mathcal{F} = \{f_\theta : \theta \in \Theta\}$ be a collection of measurable functions indexed by a bounded subset $\Theta \subset \mathbb{R}^d$. Suppose that there exists a measurable function m such that

$$\big|f_{\theta_1}(x) - f_{\theta_2}(x)\big| \le m(x)\|\theta_1 - \theta_2\|, \quad \text{every } \theta_1, \theta_2.$$

If $P|m|^r < \infty$, then there exists a constant K, depending on Θ and d only, such that the bracketing numbers satisfy

$$N_{[]}\big(\varepsilon\|m\|_{P,r}, \mathcal{F}, L_r(P)\big) \le K\left(\frac{\operatorname{diam}\Theta}{\varepsilon}\right)^d, \quad \text{every } 0 < \varepsilon < \operatorname{diam}\Theta.$$

Thus the entropy is of smaller order than $\log(1/\varepsilon)$. Hence the bracketing entropy integral certainly converges, and the class of functions $\mathcal{F}$ is Donsker.

To establish the upper bound we use brackets of the type $[f_\theta - \varepsilon m, f_\theta + \varepsilon m]$ for θ ranging over a suitably chosen subset of Θ. These brackets have $L_r(P)$-size $2\varepsilon\|m\|_{P,r}$. If θ ranges over a grid of meshwidth ε over Θ, then the brackets cover $\mathcal{F}$, because by the Lipschitz condition, $f_{\theta_1} - \varepsilon m \le f_{\theta_2} \le f_{\theta_1} + \varepsilon m$ if $\|\theta_1 - \theta_2\| \le \varepsilon$. Thus, we need as many brackets as we need balls of radius $\varepsilon/2$ to cover Θ.

The size of Θ in every fixed dimension is at most diam Θ. We can cover Θ with fewer than $(\text{diam}\,\Theta/\varepsilon)^d$ cubes of size ε. The circumscribed balls have radius a multiple of ε and also cover Θ. If we replace the centers of these balls by their projections into Θ, then the balls of twice the radius still cover Θ. $\quad\square$

19.8   *Example (Pointwise Compact Class)*. The parametric class in Example 19.7 is certainly Glivenko-Cantelli, but for this a much weaker continuity condition also suffices. Let $\mathcal{F} = \{f_\theta : \theta \in \Theta\}$ be a collection of measurable functions with integrable envelope function F indexed by a compact metric space Θ such that the map $\theta \mapsto f_\theta(x)$ is continuous for every x. Then the L_1-bracketing numbers of $\mathcal{F}$ are finite and hence $\mathcal{F}$ is Glivenko-Cantelli.

We can construct the brackets in the obvious way in the form $[f_B, f^B]$, where B is an open ball and f_B and f^B are the infimum and supremum of f_θ for $\theta \in B$, respectively. Given a sequence of balls B_m with common center a given θ and radii decreasing to 0, we have $f^{B_m} - f_{B_m} \downarrow f_\theta - f_\theta = 0$ by the continuity, pointwise in x and hence also in L_1 by the dominated-convergence theorem and the integrability of the envelope. Thus, given $\varepsilon > 0$, for every θ there exists an open ball B around θ such that the bracket $[f_B, f^B]$ has size at most ε. By the compactness of Θ, the collection of balls constructed in this way has a finite subcover. The corresponding brackets cover $\mathcal{F}$.

This construction shows that the bracketing numbers are finite, but it gives no control on their sizes. $\quad\square$

19.9   *Example (Smooth functions)*. Let $\mathbb{R}^d = \cup_j I_j$ be a partition in cubes of volume 1 and let $\mathcal{F}$ be the class of all functions $f : \mathbb{R}^d \to \mathbb{R}$ whose partial derivatives up to order α exist and are uniformly bounded by constants M_j on each of the cubes I_j. (The condition includes bounds on the "zero-th derivative," which is f itself.) Then the bracketing numbers of $\mathcal{F}$ satisfy, for every $V \geq d/\alpha$ and every probability measure P,

$$\log N_{[\,]}\big(\varepsilon, \mathcal{F}, L_r(P)\big) \leq K \left(\frac{1}{\varepsilon}\right)^V \left(\sum_{j=1}^{\infty} \big(M_j^r P(I_j)\big)^{\frac{V}{V+r}}\right)^{\frac{V+r}{r}}.$$

The constant K depends on α, V, r, and d only. If the series on the right converges for $r = 2$ and some $d/\alpha \leq V < 2$, then the bracketing entropy integral of the class $\mathcal{F}$ converges and hence the class is P-Donsker.[†] This requires sufficient smoothness $\alpha > d/2$ and sufficiently small tail probabilities $P(I_j)$ relative to the uniform bounds M_j. If the functions f have compact support (equivalently $M_j = 0$ for all large j), then smoothness of order $\alpha > d/2$ suffices. $\quad\square$ •

19.10   *Example (Sobolev classes)*. Let $\mathcal{F}$ be the set of all functions $f : [0, 1] \mapsto \mathbb{R}$ such that $\|f\|_\infty \leq 1$ and the $(k-1)$-th derivative is absolutely continuous with $\int (f^{(k)})^2(x)\,dx \leq 1$ for some fixed $k \in \mathbb{N}$. Then there exists a constant K such that, for every $\varepsilon > 0$,[‡]

$$\log N_{[\,]}\big(\varepsilon, \mathcal{F}, \|\cdot\|_\infty\big) \leq K \left(\frac{1}{\varepsilon}\right)^{1/k}.$$

Thus, the class $\mathcal{F}$ is Donsker for every $k \geq 1$ and every P. $\quad\square$

[†] The upper bound and this sufficient condition can be slightly improved. For this and a proof of the upper bound, see e.g., [146, Corollary 2.74].

[‡] See [16].

19.11 *Example (Bounded variation).* Let $\mathcal{F}$ be the collection of all monotone functions $f : \mathbb{R} \mapsto [-1, 1]$, or, bigger, the set of all functions that are of variation bounded by 1. These are the differences of pairs of monotonely increasing functions that together increase at most 1. Then there exists a constant K such that, for every $r \geq 1$ and probability measure P,[†]

$$\log N_{[\,]}\big(\varepsilon, \mathcal{F}, L_2(P)\big) \leq K\left(\frac{1}{\varepsilon}\right).$$

Thus, this class of functions is P-Donsker for every P. $\square$

19.12 *Example (Weighted distribution function).* Let $w : (0, 1) \mapsto \mathbb{R}^+$ be a fixed, continuous function. The *weighted empirical process* of a sample of real-valued observations is the process

$$t \mapsto \mathbb{G}_n^w(t) = \sqrt{n}(\mathbb{F}_n - F)(t)w\big(F(t)\big)$$

(defined to be zero if $F(t) = 0$ or $F(t) = 1$). For a bounded function w, the map $z \mapsto z \cdot w \circ F$ is continuous from $\ell^\infty[-\infty, \infty]$ into $\ell^\infty[-\infty, \infty]$ and hence the weak convergence of the weighted empirical process follows from the convergence of the ordinary empirical process and the continuous-mapping theorem. Of more interest are weight functions that are unbounded at 0 or 1, which can be used to rescale the empirical process at its two extremes $-\infty$ and ∞. Because the difference $(\mathbb{F}_n - F)(t)$ converges to 0 as $t \to \pm\infty$, the sample paths of the process $t \mapsto \mathbb{G}_n^w(t)$ may be bounded even for unbounded w, and the rescaling increases our knowledge of the behavior at the two extremes.

A simple condition for the weak convergence of the weighted empirical process in $\ell^\infty(-\infty, \infty)$ is that the weight function w is monotone around 0 and 1 and satisfies $\int_0^1 w^2(s)\, ds < \infty$. The square-integrability is almost necessary, because the convergence is known to fail for $w(t) = 1/\sqrt{t(1 - t)}$. The *Chibisov-O'Reilly theorem* gives necessary and sufficient conditions but is more complicated.

We shall give the proof for the case that w is unbounded at only one endpoint and decreases from $w(0) = \infty$ to $w(1) = 0$. Furthermore, we assume that F is the uniform measure on $[0, 1]$. (The general case can be treated in the same way, or by the quantile transformation.) Then the function $v(s) = w^2(s)$ with domain $[0, 1]$ has an inverse $v^{-1}(t) = w^{-1}(\sqrt{t})$ with domain $[0, \infty]$. A picture of the graphs shows that $\int_0^\infty w^{-1}(\sqrt{t})\, dt = \int_0^1 w^2(t)\, dt$, which is finite by assumption. Thus, given an $\varepsilon > 0$, we can choose partitions $0 = s_0 < s_1 < \cdots < s_k = 1$ and $0 = t_0 < t_1 < \cdots < t_l = \infty$ such that, for every i,

$$\int_{s_{i-1}}^{s_i} w^2(s)\, ds < \varepsilon^2, \qquad \int_{t_{i-1}}^{t_i} w^{-1}(\sqrt{t})\, dt < \varepsilon^2.$$

This corresponds to slicing the area under w^2 both horizontally and vertically in pieces of size ε^2. Let the partition $0 = u_0 < u_1 < \cdots < u_m = 1$ be the partition consisting of all points s_i and all points $w^{-1}(\sqrt{t}_j)$. Then, for every i,

$$\big(w^2(u_{i-1}) - w^2(u_i)\big)u_{i-1} \leq \int_{w^2(u_i)}^{w^2(u_{i-1})} w^{-1}(\sqrt{t})\, dt < \varepsilon^2.$$

[†] See, e.g., [146, Theorem 2.75].

It follows that the brackets

$$\left[w^2(u_i) 1_{[0, u_{i-1}]}, \; w^2(u_{i-1}) 1_{[0, u_{i-1}]} + w^2 1_{[(u_{i-1}, u_i]} \right]$$

have $L_1(\lambda)$-size $2\varepsilon^2$. Their square roots are brackets for the functions of interest $x \mapsto w(t)$ $1_{[0,t]}(x)$, and have $L_2(\lambda)$-size $\sqrt{2}\varepsilon$, because $P|\sqrt{u} - \sqrt{l}|^2 \leq P|u - l|$. Because the number m of points in the partitions can be chosen of the order $(1/\varepsilon)^2$ for small ε, the bracketing integral of the class of functions $x \mapsto w(t) 1_{[0,t]}(x)$ converges easily. $\quad\square$

The conditions given by the preceding theorems are not necessary, but the theorems cover many examples. Simple necessary and sufficient conditions are not known and may not exist. An alternative set of relatively simple conditions is based on "uniform covering numbers." The *covering number* $N(\varepsilon, \mathcal{F}, L_2(Q))$ is the minimal number of $L_2(Q)$-balls of radius ε needed to cover the set $\mathcal{F}$. The *entropy* is the logarithm of the covering number. The following theorems show that the bracketing numbers in the preceding Glivenko-Cantelli and Donsker theorems can be replaced by the *uniform covering numbers*

$$\sup_Q N(\varepsilon \|F\|_{Q,r}, \mathcal{F}, L_r(Q)).$$

Here the supremum is taken over all probability measures Q for which the class $\mathcal{F}$ is not identically zero (and hence $\|F\|_{Q,r}^r = QF^r > 0$). The uniform covering numbers are relative to a given envelope function F. This is fortunate, because the covering numbers under different measures Q typically are more stable if standardized by the norm $\|F\|_{Q,r}$ of the envelope function. In comparison, in the case of bracketing numbers we consider a single distribution P, and standardization by an envelope does not make much of a difference. The *uniform entropy integral* is defined as

$$J(\delta, \mathcal{F}, L_2) = \int_0^\delta \sqrt{\log \sup_Q N(\varepsilon \|F\|_{Q,2}, \mathcal{F}, L_2(Q))} \, d\varepsilon.$$

19.13   Theorem (Glivenko-Cantelli). *Let $\mathcal{F}$ be a suitably measurable class of measurable functions with $\sup_Q N(\varepsilon \|F\|_{Q,1}, \mathcal{F}, L_1(Q)) < \infty$ for every $\varepsilon > 0$. If $P^*F < \infty$, then $\mathcal{F}$ is P-Glivenko-Cantelli.*

19.14   Theorem (Donsker). *Let $\mathcal{F}$ be a suitably measurable class of measurable functions with $J(1, \mathcal{F}, L_2) < \infty$. If $P^*F^2 < \infty$, then $\mathcal{F}$ is P-Donsker.*

The condition that the class $\mathcal{F}$ be "suitably measurable" is satisfied in most examples but cannot be omitted. We do not give a general definition here but note that it suffices that there exists a countable collection $\mathcal{G}$ of functions such that each f is the pointwise limit of a sequence g_m in $\mathcal{G}$.[†]

An important class of examples for which good estimates on the uniform covering numbers are known are the so-called *Vapnik-Červonenkis classes*, or *VC classes*, which are defined through combinatorial properties and include many well-known examples.

[†] See, for example, [117], [120], or [146] for proofs of the preceding theorems and other unproven results in this section.

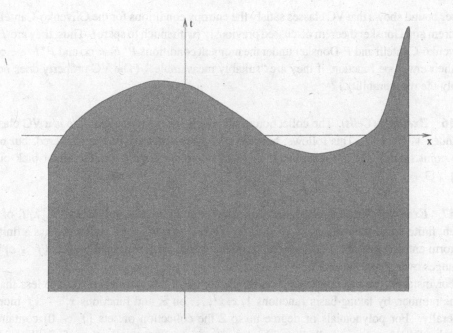

Figure 19.2. The subgraph of a function.

Say that a collection C of subsets of the sample space $\mathcal{X}$ *picks out* a certain subset A of the finite set $\{x_1, \ldots, x_n\} \subset \mathcal{X}$ if it can be written as $A = \{x_1, \ldots, x_n\} \cap C$ for some $C \in C$. The collection C is said to *shatter* $\{x_1, \ldots, x_n\}$ if C picks out each of its 2^n subsets. The *VC index* $V(C)$ of C is the smallest n for which no set of size n is shattered by C. A collection C of measurable sets is called a *VC class* if its index $V(C)$ is finite.

More generally, we can define VC classes of functions. A collection $\mathcal{F}$ is a *VC class* of functions if the collection of all *subgraphs* $\{(x, t) : f(x) < t\}$, if f ranges over $\mathcal{F}$, forms a VC class of sets in $\mathcal{X} \times \mathbb{R}$ (Figure 19.2). It is not difficult to see that a collection of sets C is a VC class of sets if and only if the collection of corresponding indicator functions 1_C is a VC class of functions. Thus, it suffices to consider VC classes of functions.

By definition, a VC class of sets picks out strictly less than 2^n subsets from any set of $n \geq V(C)$ elements. The surprising fact, known as Sauer's lemma, is that such a class can necessarily pick out only a polynomial number $O(n^{V(C)-1})$ of subsets, well below the $2^n - 1$ that the definition appears to allow. Now, the number of subsets picked out by a collection C is closely related to the covering numbers of the class of indicator functions $\{1_C : C \in C\}$ in $L_1(Q)$ for discrete, empirical type measures Q. By a clever argument, Sauer's lemma can be used to bound the uniform covering (or entropy) numbers for this class.

19.15 Lemma. *There exists a universal constant K such that for any VC class $\mathcal{F}$ of functions, any $r \geq 1$ and $0 < \varepsilon < 1$,*

$$\sup_Q N\big(\varepsilon \|F\|_{Q,r}, \mathcal{F}, L_r(Q)\big) \leq K V(\mathcal{F})(16e)^{V(\mathcal{F})} \left(\frac{1}{\varepsilon}\right)^{r(V(\mathcal{F})-1)}.$$

Consequently, VC classes are examples of *polynomial classes* in the sense that their covering numbers are bounded by a polynomial in $1/\varepsilon$. They are relatively small. The upper bound shows that VC classes satisfy the entropy conditions for the Glivenko-Cantelli theorem and Donsker theorem discussed previously (with much to spare). Thus, they are P-Glivenko-Cantelli and P-Donsker under the moment conditions $P^*F < \infty$ and $P^*F^2 < \infty$ on their envelope function, if they are "suitably measurable." (The VC property does not imply the measurability.)

19.16   *Example (Cells).* The collection of all cells $(-\infty, t]$ in the real line is a VC class of index $V(\mathcal{C}) = 2$. This follows, because every one-point set $\{x_1\}$ is shattered, but no two-point set $\{x_1, x_2\}$ is shattered: If $x_1 < x_2$, then the cells $(-\infty, t]$ cannot pick out $\{x_2\}$. $\square$

19.17   *Example (Vector spaces).* Let $\mathcal{F}$ be the set of all linear combinations $\sum \lambda_i f_i$ of a given, finite set of functions $f_1, \ldots, f_k$ on $\mathcal{X}$. Then $\mathcal{F}$ is a VC class and hence has a finite uniform entropy integral. Furthermore, the same is true for the class of all sets $\{f > c\}$ if f ranges over f and c over $\mathbb{R}$.

For instance, we can construct $\mathcal{F}$ to be the set of all polynomials of degree less than some number, by taking basis functions $1, x, x^2, \ldots$ on $\mathbb{R}$ and functions $x_1^{i_1} \cdots x_d^{i_d}$ more generally. For polynomials of degree up to 2 the collection of sets $\{f > 0\}$ contains already all half-spaces and all ellipsoids. Thus, for instance, the collection of all ellipsoids is Glivenko-Cantelli and Donsker for any P.

To prove that $\mathcal{F}$ is a VC class, consider any collection of $n = k + 2$ points $(x_1, t_1), \ldots, (x_n, t_n)$ in $\mathcal{X} \times \mathbb{R}$. We shall show this set is not shattered by $\mathcal{F}$, whence $V(\mathcal{F}) \leq n$.

By assumption, the vectors $\left(f(x_1) - t_1, \ldots, f(x_n) - t_n\right)^T$ are contained in a $(k + 1)$-dimensional subspace of $\mathbb{R}^n$. Any vector a that is orthogonal to this subspace satisfies

$$\sum_{i \,:\, a_i > 0} a_i \big(f(x_i) - t_i\big) = \sum_{i \,:\, a_i < 0} (-a_i)\big(f(x_i) - t_i\big).$$

(Define a sum over the empty set to be zero.) There exists a vector a with at least one strictly positive coordinate. Then the set $\{(x_i, t_i) : a_i > 0\}$ is nonempty and is not picked out by the subgraphs of $\mathcal{F}$. If it were, then it would be of the form $\{(x_i, t_i) : t_i < f(t_i)\}$ for some f, but then the left side of the display would be strictly positive and the right side nonpositive. $\square$

A number of operations allow to build new VC classes or Donsker classes out of known VC classes or Donsker classes.

19.18   *Example (Stability properties).* The class of all complements C^c, all intersections $C \cap D$, all unions $C \cup D$, and all Cartesian products $C \times D$ of sets C and D that range over VC classes $\mathcal{C}$ and $\mathcal{D}$ is VC.

The class of all suprema $f \vee g$ and infima $f \wedge g$ of functions f and g that range over VC classes $\mathcal{F}$ and $\mathcal{G}$ is VC.

The proof that the collection of all intersections is VC is easy upon using Sauer's lemma, according to which a VC class can pick out only a polynomial number of subsets. From n given points $\mathcal{C}$ can pick out at most $O(n^{V(\mathcal{C})})$ subsets. From each of these subsets $\mathcal{D}$ can pick out at most $O(n^{V(\mathcal{D})})$ further subsets. A subset picked out by $C \cap D$ is equal to the subset picked out by C intersected with D. Thus we get all subsets by following the two-step procedure and hence $C \cap D$ can pick out at most $O(n^{V(C)+V(D)})$ subsets. For large n this is well below 2^n, whence $C \cap D$ cannot pick out all subsets.

That the set of all complements is VC is an immediate consequence of the definition. Next the result for the unions follows by combination, because $C \cup D = (C^c \cap D^c)^c$.

The results for functions are consequences of the results for sets, because the subgraphs of suprema and infima are the intersections and unions of the subgraphs, respectively. $\square$

19.19   Example (Uniform entropy). If $\mathcal{F}$ and $\mathcal{G}$ possess a finite uniform entropy integral, relative to envelope functions F and G, then so does the class $\mathcal{F}\mathcal{G}$ of all functions $x \mapsto f(x)g(x)$, relative to the envelope function FG.

More generally, suppose that $\phi : \mathbb{R}^2 \mapsto \mathbb{R}$ is a function such that, for given functions L_f and L_g and every x,

$$\left| \phi\big(f_1(x), g_1(x)\big) - \phi\big(f_2(x), g_2(x)\big) \right| \le L_f(x)|f_1 - f_2|(x) + L_g(x)|g_1 - g_2|(x).$$

Then the class of all functions $\phi(f, g) - \phi(f_0, g_0)$ has a finite uniform entropy integral relative to the envelope function $L_f F + L_g G$, whenever $\mathcal{F}$ and $\mathcal{G}$ have finite uniform entropy integrals relative to the envelopes F and G. $\square$

19.20   Example (Lipschitz transformations). For any fixed Lipschitz function $\phi : \mathbb{R}^2 \mapsto \mathbb{R}$, the class of all functions of the form $\phi(f, g)$ is Donsker, if f and g range over Donsker classes $\mathcal{F}$ and $\mathcal{G}$ with integrable envelope functions.

For example, the class of all sums $f + g$, all minima $f \wedge g$, and all maxima $f \vee g$ are Donsker. If the classes $\mathcal{F}$ and $\mathcal{G}$ are uniformly bounded, then also the products fg form a Donsker class, and if the functions f are uniformly bounded away from zero, then the functions $1/f$ form a Donsker class. $\square$

19.3   Goodness-of-Fit Statistics

An important application of the empirical distribution is the testing of goodness-of-fit. Because the empirical distribution $\mathbb{P}_n$ is always a reasonable estimator for the underlying distribution P of the observations, any measure of the discrepancy between $\mathbb{P}_n$ and P can be used as a test statistic for testing the hypothesis that the true underlying distribution is P.

Some popular global measures of discrepancy for real-valued observations are

$$\sqrt{n}\|\mathbb{F}_n - F\|_\infty, \qquad \textit{(Kolmogorov-Smirnov)},$$

$$n \int (\mathbb{F}_n - F)^2 \, dF, \qquad \textit{(Cramér–von Mises)}.$$

These statistics, as well as many others, are continuous functions of the empirical process. The continuous-mapping theorem and Theorem 19.3 immediately imply the following result.

19.21   Corollary. *If $X_1, X_2, \ldots$ are i.i.d. random variables with distribution function F, then the sequences of Kolmogorov-Smirnov statistics and Cramér–von Mises statistics converge in distribution to $\|\mathbb{G}_F\|_\infty$ and $\int \mathbb{G}_F^2 \, dF$, respectively. The distributions of these limits are the same for every continuous distribution function F.*

Proof. The maps $z \mapsto \|z\|_\infty$ and $z \mapsto \int z^2(t)\,dF(t)$ from $D[-\infty, \infty]$ into $\mathbb{R}$ are continuous with respect to the supremum norm. Consequently, the first assertion follows from the continuous-mapping theorem. The second assertion follows by the change of variables $F(t) \mapsto u$ in the representation $\mathbb{G}_F = \mathbb{G}_\lambda \circ F$ of the Brownian bridge. Alternatively, use the quantile transformation to see that the Kolmogorov-Smirnov and Cramér-von Mises statistics are distribution-free for every fixed n. ∎

It is probably practically more relevant to test the goodness-of-fit of compositive null hypotheses, for instance the hypothesis that the underlying distribution P of a random sample is normal, that is, it belongs to the normal location-scale family. To test the null hypothesis that P belongs to a certain family $\{P_\theta : \theta \in \Theta\}$, it is natural to use a measure of the discrepancy between $\mathbb{P}_n$ and $P_{\hat\theta}$, for a reasonable estimator $\hat\theta$ of θ. For instance, a modified Kolmogorov-Smirnov statistic for testing normality is

$$\sup_t \sqrt{n} \left| \mathbb{F}_n(t) - \Phi\left(\frac{t - \overline{X}}{S}\right) \right|.$$

For many goodness-of-fit statistics of this type, the limit distribution follows from the limit distribution of $\sqrt{n}(\mathbb{P}_n - P_{\hat\theta})$. This is not a Brownian bridge but also contains a "drift," due to $\hat\theta$. Informally, if $\theta \mapsto P_\theta$ has a derivative $\dot{P}_\theta$ in an appropriate sense, then

$$\sqrt{n}(\mathbb{P}_n - P_{\hat\theta}) = \sqrt{n}(\mathbb{P}_n - P_\theta) - \sqrt{n}(P_{\hat\theta} - P_\theta),$$
$$\approx \sqrt{n}(\mathbb{P}_n - P_\theta) - \sqrt{n}(\hat\theta - \theta)^T \dot{P}_\theta. \qquad (19.22)$$

By the continuous-mapping theorem, the limit distribution of the last approximation can be derived from the limit distribution of the sequence $\sqrt{n}(\mathbb{P}_n - P_\theta, \hat\theta - \theta)$. The first component converges in distribution to a Brownian bridge. Its joint behavior with $\sqrt{n}(\hat\theta - \theta)$ can most easily be obtained if the latter sequence is asymptotically linear. Assume that

$$\sqrt{n}(\hat\theta_n - \theta) = \frac{1}{\sqrt{n}} \sum_{i=1}^n \psi_\theta(X_i) + o_{P_\theta}(1),$$

for "influence functions" ψ_θ with $P_\theta \psi_\theta = 0$ and $P_\theta \|\psi_\theta\|^2 < \infty$.

19.23   Theorem. *Let $X_1, \ldots, X_n$ be a random sample from a distribution P_θ indexed by $\theta \in \mathbb{R}^k$. Let $\mathcal{F}$ be a P_θ-Donsker class of measurable functions and let $\hat\theta_n$ be estimators that are asymptotically linear with influence function ψ_θ. Assume that the map $\theta \mapsto P_\theta$ from $\mathbb{R}^k$ to $\ell^\infty(\mathcal{F})$ is Fréchet differentiable at θ.[†] Then the sequence $\sqrt{n}(\mathbb{P}_n - P_{\hat\theta})$ converges under θ in distribution in $\ell^\infty(F)$ to the process $f \mapsto \mathbb{G}_{P_\theta} f - \mathbb{G}_{P_\theta} \psi_\theta^T \dot{P}_\theta f$.*

Proof. In view of the differentiability of the map $\theta \mapsto P_\theta$ and Lemma 2.12,

$$\left\| P_{\hat\theta_n} - P_\theta - (\hat\theta - \theta)^T \dot{P}_\theta \right\|_{\mathcal{F}} = o_P(\|\hat\theta_n - \theta\|).$$

This justifies the approximation (19.22). The class $\mathcal{G}$ obtained by adding the k components of ψ_θ to $\mathcal{F}$ is Donsker. (The union of two Donsker classes is Donsker, in general. In

---

[†] This means that there exists a map $\dot{P}_\theta : \mathcal{F} \mapsto \mathbb{R}^k$ such that $\|P_{\theta+h} = P_\theta - h^T \dot{P}_\theta\|_{\mathcal{F}} = o(\|h\|)$ as $h \to 0$; see Chapter 20.

the present case, the result also follows directly from Theorem 18.14.) The variables $\left(\sqrt{n}(\mathbb{P}_n - P_\theta), n^{-1/2}\sum \psi_\theta(X_i)\right)$ are obtained from the empirical process seen as an element of $\ell^\infty(\mathcal{G})$ by a continuous map. Finally, apply Slutsky's lemma. ∎

The preceding theorem implies, for instance, that the sequences of modified Kolmogorov-Smirnov statistic $\sqrt{n}\|\mathbb{F}_n - F_{\hat\theta}\|_\infty$ converge in distribution to the supremum of a certain Gaussian process. The distribution of the limit may depend on the model $\theta \mapsto F_\theta$, the estimators $\hat\theta_n$, and even on the parameter value θ. Typically, this distribution is not known in closed form but has to be approximated numerically or by simulation. On the other hand, the limit distribution of the true Kolmogorov-Smirnov statistic under a continuous distribution can be derived from properties of the Brownian bridge, and is given by[†]

$$P\left(\|\mathbb{G}_\lambda\|_\infty > x\right) = 2\sum_{j=1}^\infty (-1)^{j+1} e^{-2j^2 x^2}.$$

With the Donsker theorem in hand, the route via the Brownian bridge is probably the most convenient. In the 1940s Smirnov obtained the right side as the limit of an explicit expression for the distribution function of the Kolmogorov-Smirnov statistic.

19.4 Random Functions

The language of Glivenko-Cantelli classes, Donsker classes, and entropy appears to be convenient to state the "regularity conditions" needed in the asymptotic analysis of many statistical procedures. For instance, in the analysis of Z- and M-estimators, the theory of empirical processes is a powerful tool to control remainder terms. In this section we consider the key element in this application: controlling random sequences of the form $\sum_{i=1}^n f_{n,\hat\theta_n}(X_i)$ for functions $f_{n,\theta}$ that change with n and depend on an estimated parameter.

If a class $\mathcal{F}$ of functions is P-Glivenko-Cantelli, then the difference $|\mathbb{P}_n f - Pf|$ converges to zero uniformly in f varying over $\mathcal{F}$, almost surely. Then it is immediate that also $|\mathbb{P}_n \hat f_n - P\hat f_n| \overset{as}{\to} 0$ for every sequence of random functions $\hat f_n$ that are contained in $\mathcal{F}$. If $\hat f_n$ converges almost surely to a function f_0 and the sequence is dominated (or uniformly integrable), so that $P\hat f_n \overset{as}{\to} Pf_0$, then it follows that $\mathbb{P}_n \hat f_n \overset{as}{\to} Pf_0$.

Here by "random functions" we mean measurable functions $x \mapsto \hat f_n(x; \omega)$ that, for every fixed x, are real maps defined on the same probability space as the observations $X_1(\omega), \ldots, X_n(\omega)$. In many examples the function $\hat f_n(x) = \hat f_n(x; X_1, \ldots, X_n)$ is a function of the observations, for every fixed x. The notations $\mathbb{P}_n \hat f_n$ and $P\hat f_n$ are abbreviations for the expectations of the functions $x \mapsto \hat f_n(x; \omega)$ with ω fixed.

A similar principle applies to Donsker classes of functions. For a Donsker class $\mathcal{F}$, the empirical process $\mathbb{G}_n f$ converges in distribution to a P-Brownian bridge process $\mathbb{G}_P f$ "uniformly in $f \in \mathcal{F}$." In view of Lemma 18.15, the limiting process has uniformly continuous sample paths with respect to the variance semimetric. The uniform convergence combined with the continuity yields the weak convergence $\mathbb{G}_n \hat f_n \rightsquigarrow \mathbb{G}_P f_0$ for every sequence $\hat f_n$ of random functions that are contained in $\mathcal{F}$ and that converges in the variance semimetric to a function f_0.

[†] See, for instance, [42, Chapter 12], or [134].

19.24   Lemma. *Suppose that $\mathcal{F}$ is a P-Donsker class of measurable functions and $\hat{f}_n$ is a sequence of random functions that take their values in $\mathcal{F}$ such that $\int \left(\hat{f}_n(x) - f_0(x) \right)^2 dP(x)$ converges in probability to 0 for some $f_0 \in L_2(P)$. Then $\mathbb{G}_n(\hat{f}_n - f_0) \overset{P}{\to} 0$ and hence $\mathbb{G}_n \hat{f}_n \rightsquigarrow \mathbb{G}_P f_0$.*

Proof.   Assume without of loss of generality that f_0 is contained in $\mathcal{F}$. Define a function $g : \ell^\infty(\mathcal{F}) \times \mathcal{F} \mapsto \mathbb{R}$ by $g(z, f) = z(f) - z(f_0)$. The set $\mathcal{F}$ is a semimetric space relative to the $L_2(P)$-metric. The function g is continuous with respect to the product semimetric at every point (z, f) such that $f \mapsto z(f)$ is continuous. Indeed, if $(z_n, f_n) \to (z, f)$ in the space $\ell^\infty(\mathcal{F}) \times \mathcal{F}$, then $z_n \to z$ uniformly and hence $z_n(f_n) = z(f_n) + o(1) \to z(f)$ if z is continuous at f.

By assumption, $\hat{f}_n \overset{P}{\to} f_0$ as maps in the metric space $\mathcal{F}$. Because $\mathcal{F}$ is Donsker, $\mathbb{G}_n \rightsquigarrow \mathbb{G}_P$ in the space $\ell^\infty(\mathcal{F})$, and it follows that $(\mathbb{G}_n, \hat{f}_n) \rightsquigarrow (\mathbb{G}_P, f_0)$ in the space $\ell^\infty(\mathcal{F}) \times \mathcal{F}$. By Lemma 18.15, almost all sample paths of $\mathbb{G}_P$ are continuous on $\mathcal{F}$. Thus the function g is continuous at almost every point $(\mathbb{G}_P, f_0)$. By the continuous-mapping theorem, $\mathbb{G}_n(\hat{f}_n - f_0) = g(\mathbb{G}_n, \hat{f}_n) \rightsquigarrow g(\mathbb{G}_P, f_0) = 0$. The lemma follows, because convergence in distribution and convergence in probability are the same for a degenerate limit. ∎

The preceding lemma can also be proved by reference to an almost sure representation for the converging sequence $\mathbb{G}_n \rightsquigarrow \mathbb{G}_P$. Such a representation, a generalization of Theorem 2.19 exists. However, the correct handling of measurability issues makes its application involved.

19.25   Example (Mean absolute deviation). The *mean absolute deviation* of a random sample $X_1, \ldots, X_n$ is the scale estimator

$$M_n = \frac{1}{n} \sum_{i=1}^n |X_i - \overline{X}_n|.$$

The absolute value bars make the derivation of its asymptotic distribution surprisingly difficult. (Try and do it by elementary means.) Denote the distribution function of the observations by F, and assume for simplicity of notation that they have mean Fx equal to zero. We shall write $\mathbb{F}_n |x - \theta|$ for the stochastic process $\theta \mapsto n^{-1} \sum_{i=1}^n |X_i - \theta|$, and use the notations $\mathbb{G}_n |x - \theta|$ and $F|x - \theta|$ in a similar way.

If $Fx^2 < \infty$, then the set of functions $x \mapsto |x - \theta|$ with θ ranging over a compact, such as $[-1, 1]$, is F-Donsker by Example 19.7. Because, by the triangle inequality, $F\big(|x - \overline{X}_n| - |x|\big)^2 \leq |\overline{X}_n|^2 \overset{P}{\to} 0$, the preceding lemma shows that $\mathbb{G}_n |x - \overline{X}_n| - \mathbb{G}_n |x| \overset{P}{\to} 0$. This can be rewritten as

$$\sqrt{n}\big(M_n - F|x|\big) = \sqrt{n}\big(F|x - \overline{X}_n| - F|x|\big) + \mathbb{G}_n|x| + o_P(1).$$

If the map $\theta \mapsto F|x - \theta|$ is differentiable at 0, then, with the derivative written in the form $2F(0) - 1$, the first term on the right is asymptotically equivalent to $\big(2F(0) - 1\big)\mathbb{G}_n x$, by the delta method. Thus, the mean absolute deviation is asymptotically normal with mean zero and asymptotic variance equal to the variance of $\big(2F(0) - 1\big)X_1 + |X_1|$.

If the mean and median of the observations are equal (i.e., $F(0) = \frac{1}{2}$), then the first term is 0 and hence the centering of the absolute values at the sample mean has the same effect as centering at the true mean. In this case not knowing the true mean does not hurt the scale estimator. In comparison, for the sample variance this is true for any F. □

Perhaps the most important application of the preceding lemma is to the theory of Z-estimators. In Theorem 5.21 we imposed a pointwise Lipschitz condition on the maps $\theta \mapsto \psi_\theta$ to ensure the convergence 5.22:

$$\mathbb{G}_n\big(\hat\psi_{\hat\theta_n} - \psi_{\theta_0}\big) \xrightarrow{P} 0.$$

In view of Example 19.7, this is now seen to be a consequence of the preceding lemma. The display is valid if the class of functions $\big\{\psi_\theta : \|\theta - \theta_0\| < \delta\big\}$ is Donsker for some $\delta > 0$ and $\psi_\theta \to \psi_{\theta_0}$ in quadratic mean. Imposing a Lipschitz condition is just one method to ensure these conditions, and hence Theorem 5.21 can be extended considerably. In particular, in its generalized form the theorem covers the sample median, corresponding to the choice $\psi_\theta(x) = \text{sign}(x - \theta)$. The sign functions can be bracketed just as the indicator functions of cells considered in Example 19.6 and thus form a Donsker class.

For the treatment of semiparametric models (see Chapter 25), it is useful to extend the results on Z-estimators to the case of infinite-dimensional parameters. A differentiability or Lipschitz condition on the maps $\theta \mapsto \psi_\theta$ would preclude most applications of interest. However, if we use the language of Donsker classes, the extension is straightforward and useful.

If the parameter θ ranges over a subset of an infinite-dimensional normed space, then we use an infinite number of estimating equations, which we label by some set H and assume to be sums. Thus the estimator $\hat\theta_n$ (nearly) solves an equation $\mathbb{P}_n \psi_{\theta,h} = 0$ for every $h \in H$. We assume that, for every fixed x and θ, the map $h \mapsto \psi_{\theta,h}(x)$, which we denote by $\psi_\theta(x)$, is uniformly bounded, and the same for the map $h \mapsto P\psi_{\theta,h}$, which we denote by $P\psi_\theta$.

19.26 Theorem. *For each θ in a subset Θ of a normed space and every h in an arbitrary set H, let $x \mapsto \psi_{\theta,h}(x)$ be a measurable function such that the class $\big\{\psi_{\theta,h} : \|\theta - \theta_0\| < \delta, h \in H\big\}$ is P-Donsker for some $\delta > 0$, with finite envelope function. Assume that, as a map into $\ell^\infty(H)$, the map $\theta \mapsto P\psi_\theta$ is Fréchet-differentiable at a zero θ_0, with a derivative $V : \text{lin } \Theta \mapsto \ell^\infty(H)$ that has a continuous inverse on its range. Furthermore, assume that $\big\|P(\psi_{\theta,h} - \psi_{\theta_0,h})^2\big\|_H \to 0$ as $\theta \to \theta_0$. If $\|\mathbb{P}_n \psi_{\hat\theta_n}\|_H = o_P(n^{-1/2})$ and $\hat\theta_n \xrightarrow{P} \theta_0$, then*

$$V\sqrt{n}(\hat\theta_n - \theta_0) = -\mathbb{G}_n \psi_{\theta_0} + o_P(1).$$

Proof. This follows the same lines as the proof of Theorem 5.21. The only novel aspect is that a uniform version of Lemma 19.24 is needed to ensure that $\mathbb{G}_n(\psi_{\hat\theta_n} - \psi_{\theta_0})$ converges to zero in probability in $\ell^\infty(H)$. This is proved along the same lines.

Assume without loss of generality that $\hat\theta_n$ takes its values in $\Theta_\delta = \big\{\theta \in \Theta : \|\theta - \theta_0\| < \delta\big\}$ and define a map $g : \ell^\infty(\Theta_\delta \times H) \times \Theta_\delta \mapsto \ell^\infty(H)$ by $g(z, \theta)h = z(\theta, h) - z(\theta_0, h)$. This map is continuous at every point (z, θ_0) such that $\big\|z(\theta, h) - z(\theta_0, h)\big\|_H \to 0$ as $\theta \to \theta_0$. The sequence $(\mathbb{G}_n \psi_\theta, \hat\theta_n)$ converges in distribution in the space $\ell^\infty(\Theta_\delta \times H) \times \Theta_\delta$ to a pair $(\mathbb{G}\psi_\theta, \theta_0)$. As $\theta \to \theta_0$, we have that $\sup_h P(\psi_{\theta,h} - \psi_{\theta_0,h})^2 \to 0$ by assumption, and thus $\|\mathbb{G}\psi_\theta - \mathbb{G}\psi_{\theta_0}\|_H \to 0$ almost surely, by the uniform continuity of the sample paths of the Brownian bridge. Thus, we can apply the continuous-mapping theorem and conclude that $g(\mathbb{G}_n \psi_\theta, \hat\theta_n) \rightsquigarrow g(\mathbb{G}\psi_\theta, \theta_0) = 0$, which is the desired result. ∎

19.5 Changing Classes

The Glivenko-Cantelli and Donsker theorems concern the empirical process for different n, but each time with the same indexing class $\mathcal{F}$. This is sufficient for a large number of applications, but in other cases it may be necessary to allow the class $\mathcal{F}$ to change with n. For instance, the range of the random function $\hat{f}_n$ in Lemma 19.24 might be different for every n. We encounter one such a situation in the treatment of M-estimators and the likelihood ratio statistic in Chapters 5 and 16, in which the random functions of interest $\sqrt{n}(m_{\hat{\theta}_n} - m_{\theta_0}) - \sqrt{n}(\tilde{\theta}_n - \theta_0)\dot{m}_{\theta_0}$ are obtained by rescaling a given class of functions. It turns out that the convergence of random variables such as $\mathbb{G}_n \hat{f}_n$ does not require the ranges $\mathcal{F}_n$ of the functions $\hat{f}_n$ to be constant but depends only on the sizes of the ranges to stabilize. The nature of the functions inside the classes could change completely from n to n (apart from a Lindeberg condition).

Directly or indirectly, all the results in this chapter are based on the maximal inequalities obtained in section 19.6. The most general results can be obtained by applying these inequalities, which are valid for every fixed n, directly. The conditions for convergence of quantities such as $\mathbb{G}_n \hat{f}_n$ are then framed in terms of (random) entropy numbers. In this section we give an intermediate treatment, starting with an extension of the Donsker theorems, Theorems 19.5 and 19.14, to the weak convergence of the empirical process indexed by classes that change with n.

Let $\mathcal{F}_n$ be a sequence of classes of measurable functions $f_{n,t} : \mathcal{X} \mapsto \mathbb{R}$ indexed by a parameter t, which belongs to a common index set T. Then we can consider the weak convergence of the stochastic processes $t \mapsto \mathbb{G}_n f_{n,t}$ as elements of $\ell^\infty(T)$, assuming that the sample paths are bounded. By Theorem 18.14 weak convergence is equivalent to marginal convergence and asymptotic tightness. The marginal convergence to a Gaussian process follows under the conditions of the Lindeberg theorem, Proposition 2.27. Sufficient conditions for tightness can be given in terms of the entropies of the classes $\mathcal{F}_n$.

We shall assume that there exists a semimetric ρ that makes T into a totally bounded space and that relates to the L_2-metric in that

$$\sup_{\rho(s,t)<\delta_n} P(f_{n,s} - f_{n,t})^2 \to 0, \quad \text{every } \delta_n \downarrow 0. \tag{19.27}$$

Furthermore, we suppose that the classes $\mathcal{F}_n$ possess envelope functions F_n that satisfy the Lindeberg condition

$$PF_n^2 = O(1),$$
$$PF_n^2\{F_n > \varepsilon\sqrt{n}\} \to 0, \quad \text{every } \varepsilon > 0.$$

Then the central limit theorem holds under an entropy condition. As before, we can use either bracketing or uniform entropy.

19.28   Theorem. *Let $\mathcal{F}_n = \{f_{n,t} : t \in T\}$ be a class of measurable functions indexed by a totally bounded semimetric space (T, ρ) satisfying (19.27) and with envelope function that satisfies the Lindeberg condition. If $J_{[]}(\delta_n, \mathcal{F}_n, L_2(P)) \to 0$ for every $\delta_n \downarrow 0$, or alternatively, every $\mathcal{F}_n$ is suitably measurable and $J(\delta_n, \mathcal{F}_n, L_2) \to 0$ for every $\delta_n \downarrow 0$, then the sequence $\{\mathbb{G}_n f_{n,t} : t \in T\}$ converges in distribution to a tight Gaussian process, provided the sequence of covariance functions $Pf_{n,s}f_{n,t} - Pf_{n,s}Pf_{n,t}$ converges pointwise on $T \times T$.*

Proof. Under bracketing the proof of the following theorem is similar to the proof of Theorem 19.5. We omit the proof under uniform entropy.

For every given $\delta > 0$ we can use the semimetric ρ and condition (19.27) to partition T into finitely many sets $T_1, \ldots, T_k$ such that, for every sufficiently large n,

$$\sup_i \sup_{s,t \in T_i} P(f_{n,s} - f_{n,t})^2 < \delta^2.$$

(This is the only role for the totally bounded semimetric ρ; alternatively, we could assume the existence of partitions as in this display directly.) Next we apply Lemma 19.34 to obtain the bound

$$E \sup_i \sup_{s,t \in T_i} |\mathbb{G}_n(f_{n,s} - f_{n,t})| \lesssim J_{[]}(\delta, \mathcal{F}_n, L_2(P)) + \frac{PF_n^2 1\{F_n > a_n(\delta)\sqrt{n}\}}{a_n(\delta)}.$$

Here $a_n(\delta)$ is the number given in Lemma 19.34 evaluated for the class of functions $\mathcal{F}_n - \mathcal{F}_n$ and F_n is its envelope, but the corresponding number and envelope of the class $\mathcal{F}_n$ differ only by constants. Because $J_{[]}(\delta_n, \mathcal{F}_n, L_2(P)) \to 0$ for every $\delta_n \downarrow 0$, we must have that $J_{[]}(\delta, \mathcal{F}_n, L_2(P)) = O(1)$ for every $\delta > 0$ and hence $a_n(\delta)$ is bounded away from zero. Then the second term in the preceding display converges to zero for every fixed $\delta > 0$, by the Lindeberg condition. The first term can be made arbitrarily small as $n \to \infty$ by choosing δ small, by assumption. $\blacksquare$

19.29 **Example (Local empirical measure).** Consider the functions $f_{n,t} = r_n 1_{(a, a+t\delta_n]}$ for t ranging over a compact in $\mathbb{R}$, say $[0, 1]$, a fixed number a, and sequences $\delta_n \downarrow 0$ and $r_n \to \infty$. This leads to a multiple of the *local empirical measure* $\mathbb{P}_n f_{n,t} = (1/n)\#(X_i \in (a, a+t\delta_n])$, which counts the fraction of observations falling into the shrinking intervals $(a, a+t\delta_n]$.

Assume that the distribution of the observations is continuous with density p. Then

$$Pf_{n,t}^2 = r_n^2 P(a, a+t\delta_n] = r_n^2 p(a) t\delta_n + o(r_n^2 \delta_n).$$

Thus, we obtain an interesting limit only if $r_n^2 \delta_n \sim 1$. From now on, set $r_n^2 \delta_n = 1$. Then the variance of every $\mathbb{G}_n f_{n,t}$ converges to a nonzero limit. Because the envelope function is $F_n = f_{n,1}$, the Lindeberg condition reduces to $r_n^2 P(a, a+\delta_n] 1_{r_n > \varepsilon\sqrt{n}} \to 0$, which is true provided $n\delta_n \to \infty$. This requires that we do not localize too much. If the intervals become too small, then catching an observation becomes a rare event and the problem is not within the domain of normal convergence.

The bracketing numbers of the cells $1_{(a, a+t\delta_n]}$ with $t \in [0, 1]$ are of the order $O(\delta_n/\varepsilon^2)$. Multiplication with r_n changes this in $O(1/\varepsilon^2)$. Thus Theorem 19.28 applies easily, and we conclude that the sequence of processes $t \mapsto \mathbb{G}_n f_{n,t}$ converges in distribution to a Gaussian process for every $\delta_n \downarrow 0$ such that $n\delta_n \to \infty$.

The limit process is not a Brownian bridge, but a Brownian motion process, as follows by computing the limit covariance of $(\mathbb{G}_n f_{n,s}, \mathbb{G}_n f_{n,t})$. Asymptotically the local empirical process "does not know" that it is tied down at its extremes. In fact, it is an interesting exercise to check that two different local empirical processes (fixed at two different numbers a and b) converge jointly to two independent Brownian motions. $\square$

In the treatment of M-estimators and the likelihood ratio statistic in Chapters 5 and 16, we encountered random functions resulting from rescaling a given class of functions. Given functions $x \mapsto m_\theta(x)$ indexed by a Euclidean parameter θ, we needed conditions that ensure that, for a given sequence $r_n \to \infty$ and any random sequence $\tilde{h}_n = O_P^*(1)$,

$$\mathbb{G}_n\big(r_n(m_{\theta_0 + \tilde{h}_n/r_n} - m_{\theta_0}) - \tilde{h}_n^T \dot{m}_{\theta_0}\big) \overset{P}{\to} 0. \tag{19.30}$$

We shall prove this under a Lipschitz condition, but it should be clear from the following proof and the preceding theorem that there are other possibilities.

19.31 **Lemma.** *For each θ in an open subset of Euclidean space let $x \mapsto m_\theta(x)$ be a measurable function such that the map $\theta \mapsto m_\theta(x)$ is differentiable at θ_0 for almost every x (or in probability) with derivative $\dot{m}_{\theta_0}(x)$ and such that, for every θ_1 and θ_2 in a neighborhood of θ_0, and for a measurable function $\dot{m}$ such that $P\dot{m}^2 < \infty$,*

$$\big\| m_{\theta_1}(x) - m_{\theta_2}(x) \big\| \le \dot{m}(x) \, \|\theta_1 - \theta_2\|.$$

Then (19.30) is valid for every random sequence $\tilde{h}_n$ that is bounded in probability.

Proof. The random variables $\mathbb{G}_n\big(r_n(m_{\theta_0+h/r_n} - m_{\theta_0}) - h^T \dot{m}_{\theta_0}\big)$ have mean zero and their variance converges to 0, by the differentiability of the maps $\theta \mapsto m_\theta$ and the Lipschitz condition, which allows application of the dominated-convergence theorem. In other words, this sequence seen as stochastic processes indexed by h converges marginally in distribution to zero. Because the sequence $\tilde{h}_n$ is bounded in probability, it suffices to strengthen this to uniform convergence in $\|h\| \le 1$. This follows if the sequence of processes converges weakly in the space $\ell^\infty\big(h : \|h\| \le 1\big)$, because taking a supremum is a continuous operation and, by the marginal convergence, the weak limit is then necessarily zero. By Theorem 18.14, we can confine ourselves to proving asymptotic tightness (i.e., condition (ii) of this theorem). Because the linear processes $h \mapsto h^T \mathbb{G}_n \dot{m}_{\theta_0}$ are trivially tight, we may concentrate on the processes $h \mapsto \mathbb{G}_n\big(r_n(m_{\theta_0+h/r_n} - m_{\theta_0})\big)$, the empirical process indexed by the classes of functions $r_n \mathcal{M}_{1/r_n}$, for $\mathcal{M}_\delta = \{m_\theta - m_{\theta_0} : \|\theta - \theta_0\| \le \delta\}$.

By Example 19.7, the bracketing numbers of the classes of functions $\mathcal{M}_\delta$ satisfy

$$N_{[\,]}\big(\varepsilon\delta\|\dot{m}\|_{P,2}, \mathcal{M}_\delta, L_2(P)\big) \le C\left(\frac{1}{\varepsilon}\right)^d, \qquad 0 < \varepsilon < \delta.$$

The constant C is independent of ε and δ. The function $M_\delta = \delta\dot{m}$ is an envelope function of $\mathcal{M}_\delta$. The left side also gives the bracketing numbers of the rescaled classes $\mathcal{M}_\delta/\delta$ relative to the envelope functions $M_\delta/\delta = \dot{m}$. Thus, we compute

$$J_{[\,]}\big(\delta_n, \mathcal{M}_\delta/\delta, L_2(P)\big) \lesssim \int_0^{\delta_n} \sqrt{d \operatorname{Log}\left(\frac{1}{\varepsilon}\right) + \operatorname{Log} C} \, d\varepsilon.$$

The right side converges to zero as $\delta_n \downarrow 0$ uniformly in δ. The envelope functions $M_\delta/\delta = \dot{m}$ also satisfy the Lindeberg condition. The lemma follows from Theorem 19.28. ∎

19.6 Maximal Inequalities

The main aim of this section is to derive the maximal inequality that is used in the proofs of Theorems 19.5 and 19.28. We use the notation $\lesssim$ for "smaller than up to a universal constant" and denote the function $1 \vee \log x$ by $\operatorname{Log} x$.

A *maximal inequality* bounds the tail probabilities or moments of a supremum of random variables. A maximal inequality for an infinite supremum can be obtained by combining two devices: a *chaining argument* and maximal inequalities for finite maxima. The chaining argument bounds every element in the supremum by a (telescoping) sum of small deviations. In order that a sum of small terms is small, each of the terms must be exponentially small. So we start with an *exponential inequality*. Next we apply this to obtain bounds on finite suprema, and finally we derive the desired maximal inequality.

19.32 **Lemma (Bernstein's inequality).** *For any bounded, measurable function f[†]*

$$P_P\big(|\mathbb{G}_n f| > x\big) \le 2\exp\left(-\frac{1}{4}\frac{x^2}{Pf^2 + x\|f\|_\infty/\sqrt{n}}\right), \qquad every\ x > 0.$$

Proof. The leading term 2 results from separate bounds on the right and left tail probabilities. It suffices to bound the right tail probabilities by the exponential, because the left tail inequality follows from the right tail inequality applied to $-f$. By Markov's inequality, for every $\lambda > 0$,

$$P(\mathbb{G}_n f > x) \le e^{-\lambda x}\,\mathbb{E}e^{\lambda \mathbb{G}_n f} = e^{-\lambda x}\left(1 + \sum_{k=1}^\infty \frac{1}{k!}\left(\frac{\lambda}{\sqrt{n}}\right)^k P(f - Pf)^k\right)^n,$$

by Fubini's theorem and next developing the exponential function in its power series. The term for $k = 1$ vanishes because $P(f - Pf) = 0$, so that a factor $1/n$ can be moved outside the sum. We apply this inequality with the choice

$$\lambda = \frac{1}{2}\frac{x}{Pf^2 + x\|f\|_\infty/\sqrt{n}} \le \frac{1}{2}\left(\frac{x}{Pf^2} \wedge \frac{\sqrt{n}}{\|f\|_\infty}\right) =: \lambda_1 \wedge \lambda_2.$$

Next, with λ_1 and λ_2 defined as in the preceding display, we insert the bound $\lambda^k \le \lambda_1 \lambda_2^{k-2}\lambda$ and use the inequality $\big|P(f - Pf)^k\big| \le Pf^2\big(2\|f\|_\infty\big)^{k-2}$, and we obtain

$$P(\mathbb{G}_n f > x) \le e^{-\lambda x}\left(1 + \frac{1}{n}\sum_{k=2}^\infty \frac{1}{k!}\frac{1}{2}\lambda x\right)^n.$$

Because $\sum (1/k!) \le e - 2 \le 1$ and $(1 + a)^n \le e^{an}$, the right side of this inequality is bounded by $\exp(-\lambda x/2)$, which is the exponential in the lemma. ∎

19.33 **Lemma.** *For any finite class $\mathcal{F}$ of bounded, measurable, square-integrable functions, with $|\mathcal{F}|$ elements,*

$$E_P\|\mathbb{G}_n\|_{\mathcal{F}} \lesssim \max_f \frac{\|f\|_\infty}{\sqrt{n}}\log\big(1 + |\mathcal{F}|\big) + \max_f \|f\|_{P,2}\sqrt{\log\big(1 + |\mathcal{F}|\big)}.$$

Proof. Define $a = 24\|f\|_\infty/\sqrt{n}$ and $b = 24Pf^2$. For $x \ge b/a$ and $x \le b/a$ the exponent in Bernstein's inequality is bounded above by $-3x/a$ and $-3x^2/b$, respectively.

[†] The constant $1/4$ can be replaced by $1/2$ (which is the best possible constant) by a more precise argument.

For the truncated variables $A_f = \mathbb{G}_n f 1\{|\mathbb{G}_n f| > b/a\}$ and $B_f = \mathbb{G}_n f 1\{|\mathbb{G}_n f| \leq b/a\}$, Bernstein's inequality yields the bounds, for *all* $x > 0$,

$$P(|A_f| > x) \leq 2\exp\left(\frac{-3x}{a}\right), \qquad P(|B_f| > x) \leq 2\exp\left(\frac{-3x^2}{b}\right).$$

Combining the first inequality with Fubini's theorem, we obtain, with $\psi_p(x) = \exp x^p - 1$,

$$E\psi_1\left(\frac{|A_f|}{a}\right) = E\int_0^{|A_f|/a} e^x \, dx = \int_0^\infty P(|A_f| > xa) \, e^x \, dx \leq 1.$$

By a similar argument we find that $E\psi_2(|B_f|/\sqrt{b}) \leq 1$. Because the function ψ_1 is convex and nonnegative, we next obtain, by Jensen's inequality,

$$\psi_1\left(E\max_f \frac{|A_f|}{a}\right) \leq E\psi_1\left(\frac{\max_f |A_f|}{a}\right) \leq E\sum_f \psi_1\left(\frac{|A_f|}{a}\right) \leq |\mathcal{F}|.$$

Because $\psi_1^{-1}(u) = \log(1 + u)$ is increasing, we can apply it across the display, and find a bound on $E\max|A_f|$ that yields the first term on the right side of the lemma. An analogous inequality is valid for $\max_f |B_f|/\sqrt{b}$, but with ψ_2 instead of ψ_1. An application of the triangle inequality concludes the proof. ∎

19.34   Lemma. *For any class $\mathcal{F}$ of measurable functions $f : \mathcal{X} \mapsto \mathbb{R}$ such that $Pf^2 < \delta^2$ for every f, we have, with $a(\delta) = \delta/\sqrt{\operatorname{Log} N_{[]}(\delta, \mathcal{F}, L_2(P))}$, and F an envelope function,*

$$E_P^* \|\mathbb{G}_n\|_{\mathcal{F}} \lesssim J_{[]}(\delta, \mathcal{F}, L_2(P)) + \sqrt{n} P^* F\{F > \sqrt{n} a(\delta)\}.$$

Proof.   Because $|\mathbb{G}_n f| \leq \sqrt{n}(\mathbb{P}_n + P)g$ for every pair of functions $|f| \leq g$, we obtain, for F an envelope function of $\mathcal{F}$,

$$E^* \left\|\mathbb{G}_n f\{F > \sqrt{n} a(\delta)\}\right\|_{\mathcal{F}} \leq 2\sqrt{n} PF\{F > \sqrt{n} a(\delta)\}.$$

The right side is twice the second term in the bound of the lemma. It suffices to bound $E^* \|\mathbb{G}_n f\{F \leq \sqrt{n} a(\delta)\}\|_{\mathcal{F}}$ by a multiple of the first term. The bracketing numbers of the class of functions $f\{F \leq a(\delta)\sqrt{n}\}$ if f ranges over $\mathcal{F}$ are smaller than the bracketing numbers of the class $\mathcal{F}$. Thus, to simplify the notation, we can assume that every $f \in \mathcal{F}$ is bounded by $\sqrt{n} a(\delta)$.

Fix an integer q_0 such that $4\delta \leq 2^{-q_0} \leq 8\delta$. There exists a nested sequence of partitions $\mathcal{F} = \cup_{i=1}^{N_q} \mathcal{F}_{qi}$ of $\mathcal{F}$, indexed by the integers $q \geq q_0$, into N_q disjoint subsets and measurable functions $\Delta_{qi} \leq 2F$ such that

$$\sum_{q \geq q_0} 2^{-q} \sqrt{\operatorname{Log} N_q} \lesssim \int_0^\delta \sqrt{\operatorname{Log} N_{[]}(\varepsilon, \mathcal{F}, L_2(P))} \, d\varepsilon,$$

$$\sup_{f,g \in \mathcal{F}_{qi}} |f - g| \leq \Delta_{qi}, \qquad P\Delta_{qi}^2 < 2^{-2q}.$$

To see this, first cover $\mathcal{F}$ with minimal numbers of $L_2(P)$-brackets of size 2^{-q} and replace these by as many disjoint sets, each of them equal to a bracket minus "previous" brackets. This gives partitions that satisfy the conditions with Δ_{qi} equal to the difference of the upper and lower brackets. If this sequence of partitions does not yet consist of successive refinements, then replace the partition at stage q by the set of all intersections of the form $\cap_{p=q_0}^q \mathcal{F}_{p,i_p}$. This gives partitions into $\overline{N}_q = N_{q_0} \cdots N_q$ sets. Using the inequality $\left(\log \prod N_p\right)^{1/2} \le \sum (\log N_p)^{1/2}$ and rearranging sums, we see that the first of the two displayed conditions is still satisfied.

Choose for each $q \ge q_0$ a fixed element f_{qi} from each partitioning set $\mathcal{F}_{qi}$, and set

$$\pi_q f = f_{qi}, \qquad \Delta_q f = \Delta_{qi}, \quad \text{if } f \in \mathcal{F}_{qi}.$$

Then $\pi_q f$ and $\Delta_q f$ run through a set of N_q functions if f runs through $\mathcal{F}$. Define for each fixed n and $q \ge q_0$ numbers and indicator functions

$$a_q = 2^{-q}/\sqrt{\text{Log } N_{q+1}},$$
$$A_{q-1} f = 1\{\Delta_{q_0} f \le \sqrt{n} a_{q_0}, \ldots, \Delta_{q-1} f \le \sqrt{n} a_{q-1}\},$$
$$B_q f = 1\{\Delta_{q_0} f \le \sqrt{n} a_{q_0}, \ldots, \Delta_{q-1} f \le \sqrt{n} a_{q-1}, \Delta_q f > \sqrt{n} a_q\}.$$

Then $A_q f$ and $B_q f$ are constant in f on each of the partitioning sets $\mathcal{F}_{qi}$ at level q, because the partitions are nested. Our construction of partitions and choice of q_0 also ensure that $2a(\delta) \le a_{q_0}$, whence $A_{q_0} f = 1$. Now decompose, pointwise in x (which is suppressed in the notation),

$$f - \pi_{q_0} f = \sum_{q_0+1}^{\infty} (f - \pi_q f) B_q f + \sum_{q_0+1}^{\infty} (\pi_q f - \pi_{q-1} f) A_{q-1} f.$$

The idea here is to write the left side as the sum of $f - \pi_{q_1} f$ and the telescopic sum $\sum_{q_0+1}^{q_1} (\pi_q f - \pi_{q-1} f)$ for the largest $q_1 = q_1(f, x)$ such that each of the bounds $\Delta_q f$ on the "links" $\pi_q f - \pi_{q-1} f$ in the "chain" is uniformly bounded by $\sqrt{n} a_q$ (with q_1 possibly infinite). We note that either all $B_q f$ are 1 or there is a unique $q_1 > q_0$ with $B_{q_1} f = 1$. In the first case $A_q f = 1$ for every q; in the second case $A_q f = 1$ for $q < q_1$ and $A_q f = 0$ for $q \ge q_1$.

Next we apply the empirical process $\mathbb{G}_n$ to both series on the right separately, take absolute values, and next take suprema over $f \in \mathcal{F}$. We shall bound the means of the resulting two variables.

First, because the partitions are nested, $\Delta_q f B_q f \le \Delta_{q-1} f B_q f \le \sqrt{n} a_{q-1}$ trivially $P(\Delta_q f)^2 B_q f \le 2^{-2q}$. Because $|\mathbb{G}_n f| \le \mathbb{G}_n g + 2\sqrt{n} P g$ for every pair of functions $|f| \le g$, we obtain, by the triangle inequality and next Lemma 19.33,

$$\text{E}^* \left\| \sum_{q_0+1}^{\infty} \mathbb{G}_n (f - \pi_q f) B_q f \right\|_{\mathcal{F}} \le \sum_{q_0+1}^{\infty} \text{E}^* \|\mathbb{G}_n \Delta_q f B_q f\|_{\mathcal{F}} + \sum_{q_0+1}^{\infty} 2\sqrt{n} \|P \Delta_q f B_q f\|_{\mathcal{F}}$$

$$\lesssim \sum_{q_0+1}^{\infty} \left[a_{q-1} \text{Log } N_q + 2^{-q} \sqrt{\text{Log } N_q} + \frac{4}{a_q} 2^{-2q} \right].$$

In view of the definition of a_q, the series on the right can be bounded by a multiple of the series $\sum_{q_0+1}^{\infty} 2^{-q} \sqrt{\text{Log } N_q}$.

Second, there are at most N_q functions $\pi_q f - \pi_{q-1} f$ and at most N_{q-1} indicator functions $A_{q-1} f$. Because the partitions are nested, the function $|\pi_q f - \pi_{q-1} f| A_{q-1} f$ is bounded by $\Delta_{q-1} f A_{q-1} f \leq \sqrt{n} \, a_{q-1}$. The $L_2(P)$-norm of $|\pi_q f - \pi_{q-1} f|$ is bounded by 2^{-q+1}. Apply Lemma 19.33 to find

$$\mathrm{E}^* \left\| \sum_{q_0+1}^{\infty} \mathbb{G}_n (\pi_q f - \pi_{q-1} f) A_{q-1} f \right\|_{\mathcal{F}} \lesssim \sum_{q_0+1}^{\infty} [a_{q-1} \operatorname{Log} N_q + 2^{-q} \sqrt{\operatorname{Log} N_q} \,].$$

Again this is bounded above by a multiple of the series $\sum_{q_0+1}^{\infty} 2^{-q} \sqrt{\operatorname{Log} N_q}$.

To conclude the proof it suffices to consider the terms $\pi_{q_0} f$. Because $|\pi_{q_0} f| \leq F \leq a(\delta) \sqrt{n} \leq \sqrt{n} a_{q_0}$ and $P(\pi_{q_0} f)^2 \leq \delta^2$ by assumption, another application of Lemma 19.33 yields

$$\mathrm{E}^* \| \mathbb{G}_n \pi_{q_0} f \|_{\mathcal{F}} \lesssim a_{q_0} \operatorname{Log} N_{q_0} + \delta \sqrt{\operatorname{Log} N_{q_0}}.$$

By the choice of q_0, this is bounded by a multiple of the first few terms of the series $\sum_{q_0+1}^{\infty} 2^{-q} \sqrt{\operatorname{Log} N_q}$. ∎

19.35 Corollary. *For any class $\mathcal{F}$ of measurable functions with envelope function F,*

$$\mathrm{E}_P^* \| \mathbb{G}_n \|_{\mathcal{F}} \lesssim J_{[]}\big(\| F \|_{P,2}, \mathcal{F}, L_2(P) \big).$$

Proof. Because $\mathcal{F}$ is contained in the single bracket $[-F, F]$, we have $N_{[]}\big(\delta, \mathcal{F}, L_2(P)\big) = 1$ for $\delta = 2\| F \|_{P,2}$. Then the constant $a(\delta)$ as defined in the preceding lemma reduces to a multiple of $\| F \|_{P,2}$, and $\sqrt{n} P^* F \{ F > \sqrt{n} a(\delta) \}$ is bounded above by a multiple of $\| F \|_{P,2}$, by Markov's inequality. ∎

The second term in the maximal inequality Lemma 19.34 results from a crude majorization in the first step of its proof. This bound can be improved by taking special properties of the class of functions $\mathcal{F}$ into account, or by using different norms to measure the brackets. The following lemmas, which are used in Chapter 25, exemplify this.[†] The first uses the $L_2(P)$-norm but is limited to uniformly bounded classes; the second uses a stronger norm, which we call the "Bernstein norm" as it relates to a strengthening of Bernstein's inequality. Actually, this is not a true norm, but it can be used in the same way to measure the size of brackets. It is defined by

$$\| f \|_{P,B}^2 = 2P\big(e^{|f|} - 1 - |f| \big).$$

19.36 Lemma. *For any class $\mathcal{F}$ of measurable functions $f : \mathcal{X} \mapsto \mathbb{R}$ such that $Pf^2 < \delta^2$ and $\| f \|_\infty \leq M$ for every f,*

$$\mathrm{E}_P^* \| \mathbb{G}_n \|_{\mathcal{F}} \lesssim J_{[]}\big(\delta, \mathcal{F}, L_2(P) \big) \left(1 + \frac{J_{[]}\big(\delta, \mathcal{F}, L_2(P)\big)}{\delta^2 \sqrt{n}} M \right).$$

[†] For a proof of the following lemmas and further results, see Lemmas 3.4.2 and 3.4.3 and Chapter 2.14, in [146] Also see [14], [15], and [51].

19.37 **Lemma.** *For any class $\mathcal{F}$ of measurable functions $f : \mathcal{X} \mapsto \mathbb{R}$ such that $\|f\|_{P,B} < \delta$ for every f,*

$$E_P^* \|\mathbb{G}_n\|_\mathcal{F} \lesssim J_{[]}(\delta, \mathcal{F}, \|\cdot\|_{P,B}) \left(1 + \frac{J_{[]}(\delta, \mathcal{F}, \|\cdot\|_{P,B})}{\delta^2 \sqrt{n}}\right).$$

Instead of brackets, we may also use uniform covering numbers to obtain maximal inequalities. As is the case for the Glivenko-Cantelli and Donsker theorem, the inequality given by Corollary 19.35 has a complete uniform entropy counterpart. This appears to be untrue for the inequality given by Lemma 19.34, for it appears difficult to use the information that a class $\mathcal{F}$ is contained in a small $L_2(P)$-ball directly in a uniform entropy maximal inequality.[†]

19.38 **Lemma.** *For any suitably measurable class $\mathcal{F}$ of measurable functions $f : \mathcal{X} \mapsto \mathbb{R}$, we have, with $\theta_n^2 = \sup_{f \in \mathcal{F}} \mathbb{P}_n f^2 / \mathbb{P}_n F^2$,*

$$E_P^* \|\mathbb{G}_n\|_\mathcal{F} \lesssim E\big(J(\theta_n, \mathcal{F}, L_2) \|F\|_{\mathbb{P}_n, 2}\big) \lesssim J(1, \mathcal{F}, L_2) \|F\|_{P,2}.$$

Notes

The law of large numbers for the empirical distribution function was derived by Glivenko [59] and Cantelli [19] in the 1930s. The Kolmogorov-Smirnov and Cramér–von Mises statistics were introduced and studied in the same period. The limit distributions of these statistics were obtained by direct methods. That these were the same as the distribution of corresponding functions of the Brownian bridge was noted and proved by Doob before Donsker [38] formalized the theory of weak convergence in the space of continuous functions in 1952. Donsker's main examples were the empirical process on the real line, and the partial sum process. Abstract empirical processes were studied more recently. The bracketing central limit presented here was obtained by Ossiander [111] and the uniform entropy central limit theorem by Pollard [116] and Kolčinskii [88]. In both cases these were generalizations of earlier results by Dudley, who also was influential in developing a theory of weak convergence that can deal with the measurability problems, which were partly ignored by Donsker. The maximal inequality Lemma 19.34 was proved in [119]. The first Vapnik-Červonenkis classes were considered in [147].

For further results on the classical empirical process, including an introduction to strong approximations, see [134] . For the abstract empirical process, see [57], [117], [120] and [146]. For connections with limit theorems for random elements with values in Banach spaces, see [98].

PROBLEMS

1. Derive a formula for the covariance function of the Gaussian process that appears in the limit of the modified Kolmogorov-Smirnov statistic for estimating normality.

[†] For a proof of the following lemma, see, for example, [120], or Theorem 2.14.1 in [146].

2. Find the covariance function of the Brownian motion process.

3. If $\mathbb{Z}$ is a standard Brownian motion, then $\mathbb{Z}(t) - t\mathbb{Z}(1)$ is a Brownian bridge.

4. Suppose that $X_1, \ldots, X_m$ and $Y_1, \ldots, Y_n$ are independent samples from distribution functions F and G, respectively. The Kolmogorov-Smirnov statistic for testing the null hypothesis $H_0 : F = G$ is the supremum distance $K_{m,n} = \|\mathbb{F}_m - \mathbb{G}_n\|_\infty$ between the empirical distribution functions of the two samples.

   (i) Find the limit distribution of $K_{m,n}$ under the null hypothesis.

  (ii) Show that the Kolmogorov-Smirnov test is asymptotically consistent against every alternative $F \neq G$.

 (iii) Find the asymptotic power function as a function of (g, h) for alternatives $(F_{g/\sqrt{m}}, G_{h/\sqrt{n}})$ belonging to smooth parametric models $\theta \mapsto F_\theta$ and $\theta \mapsto G_\theta$.

5. Consider the class of all functions $f : [0, 1] \mapsto [0, 1]$ such that $|f(x) - f(y)| \leq |x - y|$. Construct a set of ε-brackets for this class of functions of cardinality bounded by $\exp(C/\varepsilon)$.

6. Determine the VC index of

   (i) The collection of all cells $(a, b]$ in the real line;

  (ii) The collection of all cells $(-\infty, t]$ in the plane;

 (iii) The collection of all translates $\{\psi(\cdot - \theta) : \theta \in \mathbb{R}\}$ of a monotone function $\psi : \mathbb{R} \mapsto \mathbb{R}$.

7. Suppose that the class of functions $\mathcal{F}$ is VC. Show that the following classes are VC as well:

   (i) The collection of sets $\{f > 0\}$ as f ranges over $\mathcal{F}$;

  (ii) The collection of functions $x \mapsto f(x) + g(x)$ as f ranges over $\mathcal{F}$ and g is fixed;

 (iii) The collection of functions $x \mapsto f(x)g(x)$ as f ranges over $\mathcal{F}$ and g is fixed.

8. Show that a collection of sets is a VC class of sets if and only if the corresponding class of indicator functions is a VC class of functions.

9. Let F_n and F be distribution functions on the real line. Show that:

   (i) If $F_n(x) \to F(x)$ for every x and F is continuous, then $\|F_n - F\|_\infty \to 0$.

  (ii) If $F_n(x) \to F(x)$ and $F_n\{x\} \to F\{x\}$ for every x, then $\|F_n - F\|_\infty \to 0$.

10. Find the asymptotic distribution of the mean absolute deviation from the median.

20

Functional Delta Method

The delta method was introduced in Chapter 3 as an easy way to turn the weak convergence of a sequence of random vectors $r_n(T_n - \theta)$ into the weak convergence of transformations of the type $r_n\big(\phi(T_n) - \phi(\theta)\big)$. It is useful to apply a similar technique in combination with the more powerful convergence of stochastic processes. In this chapter we consider the delta method at two levels. The first section is of a heuristic character and limited to the case that T_n is the empirical distribution. The second section establishes the delta method rigorously and in general, completely parallel to the delta method for $\mathbb{R}^k$, for Hadamard differentiable maps between normed spaces.

20.1 von Mises Calculus

Let $\mathbb{P}_n$ be the empirical distribution of a random sample $X_1, \ldots, X_n$ from a distribution P. Many statistics can be written in the form $\phi(\mathbb{P}_n)$, where ϕ is a function that maps every distribution of interest into some space, which for simplicity is taken equal to the real line. Because the observations can be regained from $\mathbb{P}_n$ completely (unless there are ties), any statistic can be expressed in the empirical distribution. The special structure assumed here is that the statistic can be written as a fixed function ϕ of $\mathbb{P}_n$, independent of n, a strong assumption.

Because $\mathbb{P}_n$ converges to P as n tends to infinity, we may hope to find the asymptotic behavior of $\phi(\mathbb{P}_n) - \phi(P)$ through a differential analysis of ϕ in a neighborhood of P. A first-order analysis would have the form

$$\phi(\mathbb{P}_n) - \phi(P) = \phi'_P(\mathbb{P}_n - P) + \cdots,$$

where ϕ'_P is a "derivative" and the remainder is hopefully negligible. The simplest approach towards defining a derivative is to consider the function $t \mapsto \phi(P + tH)$ for a fixed perturbation H and as a function of the real-valued argument t. If ϕ takes its values in $\mathbb{R}$, then this function is just a function from the reals to the reals. Assume that the ordinary derivatives of the map $t \mapsto \phi(P + tH)$ at $t = 0$ exist for $k = 1, 2, \ldots, m$. Denoting them by $\phi_P^{(k)}(H)$, we obtain, by Taylor's theorem,

$$\phi(P + tH) - \phi(P) = t\phi'_P(H) + \cdots + \frac{1}{m!}t^m \phi_P^{(m)}(H) + o(t^m).$$

291

Substituting $t = 1/\sqrt{n}$ and $H = \mathbb{G}_n$, for $\mathbb{G}_n = \sqrt{n}(\mathbb{P}_n - P)$ the empirical process of the observations, we obtain the *von Mises expansion*

$$\phi(\mathbb{P}_n) - \phi(P) = \frac{1}{\sqrt{n}}\phi_P'(\mathbb{G}_n) + \cdots + \frac{1}{m!}\frac{1}{n^{m/2}}\phi_P^{(m)}(\mathbb{G}_n) + \cdots.$$

Actually, because the empirical process $\mathbb{G}_n$ is dependent on n, it is not a legal choice for H under the assumed type of differentiability: There is no guarantee that the remainder is small. However, we make this our working hypothesis. This is reasonable, because the remainder has one factor $1/\sqrt{n}$ more, and the empirical process $\mathbb{G}_n$ shares at least one property with a fixed H: It is "bounded." Then the asymptotic distribution of $\phi(\mathbb{P}_n) - \phi(P)$ should be determined by the first nonzero term in the expansion, which is usually the first-order term $\phi_P'(\mathbb{G}_n)$. A method to make our wishful thinking rigorous is discussed in the next section. Even in cases in which it is hard to make the differentation operation rigorous, the von Mises expansion still has heuristic value. It may suggest the type of limiting behavior of $\phi(\mathbb{P}_n) - \phi(P)$, which can next be further investigated by ad-hoc methods.

We discuss this in more detail for the case that $m = 1$. A first derivative typically gives a *linear* approximation to the original function. If, indeed, the map $H \mapsto \phi_P'(H)$ is linear, then, writing $\mathbb{P}_n$ as the linear combination $\mathbb{P}_n = n^{-1}\sum \delta_{X_i}$ of the Dirac measures at the observations, we obtain

$$\phi(\mathbb{P}_n) - \phi(P) \approx \frac{1}{\sqrt{n}}\phi_P'(\mathbb{G}_n) = \frac{1}{n}\sum_{i=1}^{n}\phi_P'(\delta_{X_i} - P). \tag{20.1}$$

Thus, the difference $\phi(\mathbb{P}_n) - \phi(P)$ behaves as an average of the independent random variables $\phi_P'(\delta_{X_i} - P)$. If these variables have zero means and finite second moments, then a normal limit distribution of $\sqrt{n}(\phi(\mathbb{P}_n) - \phi(P))$ may be expected. Here the zero mean ought to be automatic, because we may expect that

$$\int \phi_P'(\delta_x - P)\, dP(x) = \phi_P'\left(\int (\delta_x - P)\, dP(x)\right) = \phi_P'(0) = 0.$$

The interchange of order of integration and application of ϕ_P' is motivated by linearity (and continuity) of this derivative operator.

The function $x \mapsto \phi_P'(\delta_x - P)$ is known as the *influence function* of the function ϕ. It can be computed as the ordinary derivative

$$\phi_P'(\delta_x - P) = \frac{d}{dt}\Big|_{t=0}\phi\big((1 - t)P + t\delta_x\big).$$

The name "influence function" originated in developing robust statistics. The function measures the change in the value $\phi(P)$ if an infinitesimally small part of P is replaced by a pointmass at x. In robust statistics, functions and estimators with an unbounded influence function are suspect, because a small fraction of the observations would have too much influence on the estimator if their values were equal to an x where the influence function is large.

In many examples the derivative takes the form of an "expectation operator" $\phi_P'(H) = \int \tilde{\phi}_P\, dH$, for some function $\tilde{\phi}_P$ with $\int \tilde{\phi}_P\, dP = 0$, at least for a subset of H. Then the influence function is precisely the function $\tilde{\phi}_P$.

20.2 Example (Mean). The sample mean is obtained as $\phi(\mathbb{P}_n)$ from the mean function $\phi(P) = \int s \, dP(s)$. The influence function is

$$\phi'_P(\delta_x - P) = \frac{d}{dt}\Big|_{t=0} \int s \, d[(1-t)P + t\delta_x](s) = x - \int s \, dP(s).$$

In this case, the approximation (20.1) is an identity, because the function is linear already. If the sample space is a Euclidean space, then the influence function is unbounded and hence the sample mean is not robust. □

20.3 Example (Wilcoxon). Let $(X_1, Y_1), \ldots, (X_n, Y_n)$ be a random sample from a bivariate distribution. Write $\mathbb{F}_n$ and $\mathbb{G}_n$ for the empirical distribution functions of the X_i and Y_j, respectively, and consider the Mann-Whitney statistic

$$T_n = \int \mathbb{F}_n \, d\mathbb{G}_n = \frac{1}{n^2} \sum_{i=1}^{n} \sum_{j=1}^{n} 1\{X_i \le Y_j\}.$$

This statistic corresponds to the function $\phi(F, G) = \int F \, dG$, which can be viewed as a function of two distribution functions, or also as a function of a bivariate distribution function with marginals F and G. (We have assumed that the sample sizes of the two samples are equal, to fit the example into the previous discussion, which, for simplicity, is restricted to i.i.d. observations.) The influence function is

$$\phi'_{(F,G)}(\delta_{x,y} - P) = \frac{d}{dt}\Big|_{t=0} \int \big[(1-t)F + t\delta_x\big] d\big[(1-t)G + t\delta_y\big]$$
$$= F(y) + 1 - G_-(x) - 2\int F \, dG.$$

The last step follows on multiplying out the two terms between square brackets: The function that is to be differentiated is simply a parabola in t. For this case (20.1) reads

$$\int \mathbb{F}_n \, d\mathbb{G}_n - \int F \, dG \approx \frac{1}{n} \sum_{i=1}^{n} \left(F(Y_i) + 1 - G_-(X_i) - 2\int F \, dG \right).$$

From the two-sample U-statistic theorem, Theorem 12.6, it is known that the difference between the two sides of the approximation sign is actually $o_P(1/\sqrt{n})$. Thus, the heuristic calculus leads to the correct answer. In the next section an alternative proof of the asymptotic normality of the Mann-Whitney statistic is obtained by making this heuristic approach rigorous. □

20.4 Example (Z-functions). For every θ in an open subset of $\mathbb{R}^k$, let $x \mapsto \psi_\theta(x)$ be a given, measurable map into $\mathbb{R}^k$. The corresponding Z-function assigns to a probability measure P a zero $\phi(P)$ of the map $\theta \mapsto P\psi_\theta$. (Consider only P for which a unique zero exists.) If applied to the empirical distribution, this yields a Z-estimator $\phi(\mathbb{P}_n)$.

Differentiating with respect to t across the identity

$$0 = (P + t\delta_x)\psi_{\phi(P+t\delta_x)} = P\psi_{\phi(P+t\delta_x)} + t\psi_{\phi(P+t\delta_x)}(x),$$

and assuming that the derivatives exist and that $\theta \mapsto \psi_\theta$ is continuous, we find

$$0 = \left(\frac{\partial}{\partial\theta} P\psi_\theta\right)_{\theta=\phi(P)} \left[\frac{d}{dt}\phi(P + t\delta_x)\right]_{t=0} + \psi_{\phi(P)}(x).$$

The expression enclosed by squared brackets is the influence function of the Z-function. Informally, this is seen to be equal to

$$-\left(\frac{\partial}{\partial\theta}P\psi_\theta\right)^{-1}_{\theta=\phi(P)}\psi_{\phi(P)}(x).$$

In robust statistics we look for estimators with bounded influence functions. Because the influence function is, up to a constant, equal to $\psi_{\phi(P)}(x)$, this is easy to achieve with Z-estimators!

The Z-estimators are discussed at length in Chapter 5. The theorems discussed there give sufficient conditions for the asymptotic normality, and an asymptotic expansion for $\sqrt{n}(\phi(\mathbb{P}_n)-\phi(P))$. This is of the type (20.1) with the influence function as in the preceding display. □

20.5   Example (Quantiles).   The pth quantile of a distribution function F is, roughly, the number $\phi(F) = F^{-1}(p)$ such that $FF^{-1}(p) = p$. We set $F_t = (1 - t)F + t\delta_x$, and differentiate with respect to t the identity

$$p = F_t F_t^{-1}(p) = (1 - t)F(F_t^{-1}(p)) + t\delta_x(F_t^{-1}(p)).$$

This "identity" may actually be only an inequality for certain values of p, t, and x, but we do not worry about this. We find that

$$0 \equiv -F(F^{-1}(p)) + f(F^{-1}(p))\left[\frac{d}{dt}F_t^{-1}(p)\right]_{|t=0} + \delta_x(F^{-1}(p)).$$

The derivative within square brackets is the influence function of the quantile function and can be solved from the equation as

$$\phi'_F(\delta_x - F) = -\frac{1_{[x,\infty)}(F^{-1}(p)) - p}{f(F^{-1}(p))}.$$

The graph of this function is given in Figure 20.1 and has the following interpretation. Suppose the pth quantile has been computed for a large sample, but an additional observation x is obtained. If x is to the left of the pth quantile, then the pth quantile decreases; if x is to the right, then the quantile increases. In both cases the rate of change is constant, irrespective of the location of x. Addition of an observation x at the pth quantile has an unstable effect.

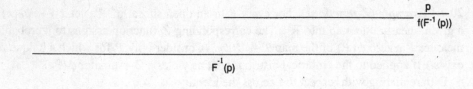

Figure 20.1.   Influence function of the pth quantile.

The von Mises calculus suggests that the sequence of empirical quantiles $\sqrt{n}\big(\mathbb{F}_n^{-1}(t) - F^{-1}(t)\big)$ is asymptotically normal with variance $\text{var}_F\, \phi_F'(\delta_{X_1}) = p(1-p)/f \circ F^{-1}(p)^2$. In Chapter 21 this is proved rigorously by the delta method of the following section. Alternatively, a pth quantile may be viewed as an M-estimator, and we can apply the results of Chapter 5. $\square$

20.1.1   *Higher-Order Expansions*

In most examples the analysis of the first derivative suffices. This statement is roughly equivalent to the statement that most limiting distributions are normal. However, in some important examples the quadratic term dominates the von Mises expansion.

The second derivative $\phi_P''(H)$ ought to correspond to a *bilinear map*. Thus, it is better to write it as $\phi_P''(H, H)$. If the first derivative in the von Mises expansion vanishes, then we expect that

$$\phi(\mathbb{P}_n) - \phi(P) \approx \frac{1}{2}\frac{1}{n}\phi_P''(\mathbb{G}_n, \mathbb{G}_n) = \frac{1}{2}\frac{1}{n^2}\sum_{i=1}^{n}\sum_{j=1}^{n}\phi_P''(\delta_{X_i} - P, \delta_{X_j} - P).$$

The right side is a *V-statistic* of degree 2 with kernel function equal to $h_P(x, y) = \frac{1}{2}\phi_P''(\delta_x - P, \delta_y - P)$. The kernel ought to be symmetric and degenerate in that $Ph_P(X, y) = 0$ for every y, because, by linearity and continuity,

$$\int \phi_P''(\delta_x - P, \delta_y - P)\, dP(x) = \phi_P''\left(\int (\delta_x - P)\, dP(x), \delta_y - P\right)$$
$$= \phi_P''(0, \delta_y - P) = 0.$$

If we delete the diagonal, then a V-statistic turns into a U-statistic and hence we can apply Theorem 12.10 to find the limit distribution of $n\big(\phi(\mathbb{P}_n) - \phi(P)\big)$. We expect that

$$n\big(\phi(\mathbb{P}_n) - \phi(P)\big) = \frac{2}{n}\sum_{i<j}\sum h_P(X_i, X_j) + \frac{1}{n}\sum_{i=1}^{n}h_P(X_i, X_i) + o_P(1).$$

If the function $x \mapsto h_P(x, x)$ is P-integrable, then the second term on the right only contributes a constant to the limit distribution. If the function $(x, y) \mapsto h_P^2(x, y)$ is $(P \times P)$-integrable, then the first term on the right converges to an infinite linear combination of independent χ_1^2-variables, according to Example 12.12.

20.6   *Example (Cramér–von Mises).* The Cramér–von Mises statistic is the function $\phi(\mathbb{F}_n)$ for $\phi(F) = \int (F - F_0)^2\, dF_0$ and a fixed cumulative distribution function F_0. By direct calculation,

$$\phi(F + tH) = \phi(F) + 2t \int (F - F_0) H\, dF_0 + t^2 \int H^2\, dF_0.$$

Consequently, the first derivative vanishes at $F = F_0$ and the second derivative is equal to $\phi_{F_0}''(H) = 2 \int H^2\, dF_0$. The von Mises calculus suggests the approximation

$$\phi(\mathbb{F}_n) - \phi(F_0) \approx \frac{1}{2}\frac{1}{n}\phi_{F_0}''(\mathbb{G}_n) = \frac{1}{n}\int \mathbb{G}_n^2\, dF_0.$$

This is certainly correct, because it is just the definition of the statistic. The preceding discussion is still of some interest in that it suggests that the limit distribution is nonnormal and can be obtained using the theory of V-statistics. Indeed, by squaring the sum that is hidden in $\mathbb{G}_n^2$, we see that

$$n\phi(\mathbb{F}_n) = \frac{1}{n} \sum_{i=1}^n \sum_{j=1}^n \int \left(1_{X_i \leq x} - F_0(x)\right)\left(1_{X_j \leq x} - F_0(x)\right) dF_0(x).$$

In Example 12.13 we used this representation to find that the sequence $n\phi(\mathbb{F}_n) \rightsquigarrow (1/6) + \sum_{j=1}^\infty j^{-2}\pi^{-2}(Z_j^2 - 1)$ for an i.i.d. sequence of standard normal variables $Z_1, Z_2, \ldots$, if the true distribution F_0 is continuous. $\square$

20.2 Hadamard-Differentiable Functions

Let T_n be a sequence of statistics with values in a normed space $\mathbb{D}$ such that $r_n(T_n - \theta)$ converges in distribution to a limit T, for a given, nonrandom θ, and given numbers $r_n \to \infty$. In the previous section the role of T_n was played by the empirical distribution $\mathbb{P}_n$, which might, for instance, be viewed as an element of the normed space $D[-\infty, \infty]$. We wish to prove that $r_n(\phi(T_n) - \phi(\theta))$ converges to a limit, for every appropriately differentiable map ϕ, which we shall assume to take its values in another normed space $\mathbb{E}$.

There are several possibilities for defining differentiability of a map $\phi : \mathbb{D} \mapsto \mathbb{E}$ between normed spaces. A map ϕ is said to be *Gateaux differentiable* at $\theta \in \mathbb{D}$ if for every fixed h there exists an element $\phi'_\theta(h) \in \mathbb{E}$ such that

$$\phi(\theta + th) - \phi(\theta) = t\phi'_\theta(h) + o(t), \qquad \text{as } t \downarrow 0.$$

For $\mathbb{E}$ the real line, this is precisely the differentiability as introduced in the preceding section. Gateaux differentiability is also called "directional differentiability," because for every possible direction h in the domain the derivative value $\phi'_\theta(h)$ measures the direction of the infinitesimal change in the value of the function ϕ. More formally, the $o(t)$ term in the previous displayed equation means that

$$\left\| \frac{\phi(\theta + th) - \phi(\theta)}{t} - \phi'_\theta(h) \right\|_{\mathbb{E}} \to 0, \qquad \text{as } t \downarrow 0. \tag{20.7}$$

The suggestive notation $\phi'_\theta(h)$ for the "tangent vectors" encourages one to think of the directional derivative as a map $\phi'_\theta : \mathbb{D} \mapsto \mathbb{E}$, which approximates the difference map $\phi(\theta + h) - \phi(\theta) : \mathbb{D} \mapsto \mathbb{E}$. It is usually included in the definition of Gateaux differentiability that this map $\phi'_\theta : \mathbb{D} \mapsto \mathbb{E}$ be linear and continuous.

However, Gateaux differentiability is too weak for the present purposes, and we need a stronger concept. A map $\phi : \mathbb{D}_\phi \mapsto \mathbb{E}$, defined on a subset $\mathbb{D}_\phi$ of a normed space $\mathbb{D}$ that contains θ, is called *Hadamard differentiable* at θ if there exists a continuous, linear map $\phi'_\theta : \mathbb{D} \mapsto \mathbb{E}$ such that

$$\left\| \frac{\phi(\theta + th_t) - \phi(\theta)}{t} - \phi'_\theta(h) \right\|_{\mathbb{E}} \to 0, \qquad \text{as } t \downarrow 0, \text{ every } h_t \to h.$$

(More precisely, for every $h_t \to h$ such that $\theta + th_t$ is contained in the domain of ϕ for all small $t > 0$.) The values $\phi'_\theta(h)$ of the derivative are the same for the two types of differentiability. The difference is that for Hadamard-differentiability the directions h_t are allowed to change with t (although they have to settle down eventually), whereas for Gateaux differentiability they are fixed. The definition as given requires that $\phi'_\theta : \mathbb{D} \mapsto \mathbb{E}$ exists as a map on the whole of $\mathbb{D}$. If this is not the case, but ϕ'_θ exists on a subset $\mathbb{D}_0$ and the sequences $h_t \to h$ are restricted to converge to limits $h \in \mathbb{D}_0$, then ϕ is called Hadamard differentiable *tangentially* to this subset.

It can be shown that Hadamard differentiability is equivalent to the difference in (20.7) tending to zero uniformly for h in compact subsets of $\mathbb{D}$. For this reason, it is also called *compact differentiability*. Because weak convergence of random elements in metric spaces is intimately connected with compact sets, through Prohorov's theorem, Hadamard differentiability is the right type of differentiability in connection with the delta method.

The derivative map $\phi'_\theta : \mathbb{D} \mapsto \mathbb{E}$ is assumed to be linear and continuous. In the case of finite-dimensional spaces a linear map can be represented by matrix multiplication and is automatically continuous. In general, linearity does not imply continuity.

Continuity of the map $\phi'_\theta : \mathbb{D} \mapsto \mathbb{E}$ should not be confused with continuity of the dependence $\theta \mapsto \phi'_\theta$ (if ϕ has derivatives in a neighborhood of θ-values). If the latter continuity holds, then ϕ is called *continuously differentiable*. This concept requires a norm on the set of derivative maps but need not concern us here.

For completeness we discuss a third, stronger form of differentiability. The map $\phi : \mathbb{D}_\phi \mapsto \mathbb{E}$ is called *Fréchet differentiable* at θ if there exists a continuous, linear map $\phi'_\theta : \mathbb{D} \mapsto \mathbb{E}$ such that

$$\left\| \phi(\theta + h) - \phi(\theta) - \phi'_\theta(h) \right\|_{\mathbb{E}} = o(\|h\|), \qquad \text{as } \|h\| \downarrow 0.$$

Because sequences of the type th_t, as employed in the definition of Hadamard differentiability, have norms satisfying $\|th_t\| = O(t)$, Fréchet differentiability is the most restrictive of the three concepts. In statistical applications, Fréchet differentiability may not hold, whereas Hadamard differentiability does. We did not have this problem in Section 3.1, because Hadamard and Fréchet differentiability are equivalent when $\mathbb{D} = \mathbb{R}^k$.

20.8 **Theorem (Delta method).** *Let $\mathbb{D}$ and $\mathbb{E}$ be normed linear spaces. Let $\phi : \mathbb{D}_\phi \subset \mathbb{D} \mapsto \mathbb{E}$ be Hadamard differentiable at θ tangentially to $\mathbb{D}_0$. Let $T_n : \Omega_n \mapsto \mathbb{D}_\phi$ be maps such that $r_n(T_n - \theta) \rightsquigarrow T$ for some sequence of numbers $r_n \to \infty$ and a random element T that takes its values in $\mathbb{D}_0$. Then $r_n(\phi(T_n) - \phi(\theta)) \rightsquigarrow \phi'_\theta(T)$. If ϕ'_θ is defined and continuous on the whole space $\mathbb{D}$, then we also have $r_n(\phi(T_n) - \phi(\theta)) = \phi'_\theta(r_n(T_n - \theta)) + o_P(1)$.*

Proof. To prove that $r_n(\phi(T_n) - \phi(\theta)) \rightsquigarrow \phi'_\theta(T)$, define for each n a map $g_n(h) = r_n(\phi(\theta + r_n^{-1}h) - \phi(\theta))$ on the domain $\mathbb{D}_n = \{h : \theta + r_n^{-1}h \in \mathbb{D}_\phi\}$. By Hadamard differentiability, this sequence of maps satisfies $g_{n'}(h_{n'}) \to \phi'_\theta(h)$ for every subsequence $h_{n'} \to h \in \mathbb{D}_0$. Therefore, $g_n(r_n(T_n - \theta)) \rightsquigarrow \phi'_\theta(T)$ by the extended continuous-mapping theorem, Theorem 18.11, which is the first assertion.

The seemingly stronger last assertion of the theorem actually follows from this, if applied to the function $\psi = (\phi, \phi'_\theta) : \mathbb{D} \mapsto \mathbb{E} \times \mathbb{E}$. This is Hadamard-differentiable at (θ, θ) with derivative $\psi'_\theta = (\phi'_\theta, \phi'_\theta)$. Thus, by the preceding paragraph, $r_n(\psi(T_n) - \psi(\theta))$ converges weakly to $(\phi'_\theta(T), \phi'_\theta(T))$ in $\mathbb{E} \times \mathbb{E}$. By the continuous-mapping theorem, the difference $r_n(\phi(T_n) - \phi(\theta)) - \phi'_\theta(r_n(T_n - \theta))$ converges weakly to $\phi'_\theta(T) - \phi'_\theta(T) = 0$. Weak convergence to a constant is equivalent to convergence in probability. $\blacksquare$

Without the *chain rule*, Hadamard differentiability would not be as interesting. Consider maps $\phi : \mathbb{D} \mapsto \mathbb{E}$ and $\psi : \mathbb{E} \mapsto \mathbb{F}$ that are Hadamard-differentiable at θ and $\phi(\theta)$, respectively. Then the composed map $\psi \circ \phi : \mathbb{D} \mapsto \mathbb{F}$ is Hadamard-differentiable at θ, and the derivative is the map obtained by composing the two derivative maps. (For Euclidean spaces this means that the derivative can be found through matrix multiplication of the two derivative matrices.) The attraction of the chain rule is that it allows a calculus of Hadamard-differentiable maps, in which differentiability of a complicated map can be established by decomposing this into a sequence of basic maps, of which Hadamard differentiability is known or can be proven easily. This is analogous to the chain rule for real functions, which allows, for instance, to see the differentiability of the map $x \mapsto \exp\cos\log(1 + x^2)$ in a glance.

20.9   Theorem (Chain rule). *Let $\phi : \mathbb{D}_\phi \mapsto \mathbb{E}_\psi$ and $\psi : \mathbb{E}_\psi \mapsto \mathbb{F}$ be maps defined on subsets $\mathbb{D}_\phi$ and $\mathbb{E}_\psi$ of normed spaces $\mathbb{D}$ and $\mathbb{E}$, respectively. Let ϕ be Hadamard-differentiable at θ tangentially to $\mathbb{D}_0$ and let ψ be Hadamard-differentiable at $\phi(\theta)$ tangentially to $\phi_\theta'(\mathbb{D}_0)$. Then $\psi \circ \phi : \mathbb{D}_\phi \mapsto \mathbb{F}$ is Hadamard-differentiable at θ tangentially to $\mathbb{D}_0$ with derivative $\psi_{\phi(\theta)}' \circ \phi_\theta'$.*

Proof.   Take an arbitrary converging path $h_t \to h$ in $\mathbb{D}$. With the notation $g_t = t^{-1}\big(\phi(\theta + th_t) - \phi(\theta)\big)$, we have

$$\frac{\psi \circ \phi(\theta + th_t) - \psi \circ \phi(\theta)}{t} = \frac{\psi\big(\phi(\theta) + tg_t\big) - \psi\big(\phi(\theta)\big)}{t}.$$

By Hadamard differentiability of ϕ, $g_t \to \phi_\theta'(h)$. Thus, by Hadamard differentiability of ψ, the whole expression goes to $\psi_{\phi(\theta)}'\big(\phi_\theta'(h)\big)$.   ∎

20.3   Some Examples

In this section we give examples of Hadamard-differentiable functions and applications of the delta method. Further examples, such as quantiles and trimmed means, are discussed in separate chapters.

The Mann-Whitney statistic can be obtained by substituting the empirical distribution functions of two samples of observations into the function $(F, G) \mapsto \int F \, dG$. This function also plays a role in the construction of other estimators. The following lemma shows that it is Hadamard-differentiable. The set $BV_M[a, b]$ is the set of all cadlag functions $z : [a, b] \mapsto [-M, M] \subset \mathbb{R}$ of variation bounded by M (the set of differences of $z_1 - z_2$ of two monotonely increasing functions that together increase no more than M).

20.10   Lemma. *Let $\phi : [0, 1] \mapsto \mathbb{R}$ be twice continuously differentiable. Then the function $(F_1, F_2) \mapsto \int \phi(F_1) \, dF_2$ is Hadamard-differentiable from the domain $D[-\infty, \infty] \times BV_1[-\infty, \infty] \subset D[-\infty, \infty] \times D[-\infty, \infty]$ into $\mathbb{R}$ at every pair of functions of bounded variation (F_1, F_2). The derivative is given by[†]*

$$(h_1, h_2) \mapsto h_2 \, \phi \circ F_1 \big|_{-\infty}^{\infty} - \int h_{2-} \, d\phi \circ F_1 + \int \phi'(F_1) h_1 \, dF_2.$$

[†] We denote by h_- the left-continuous version of a cadlag function h and abbreviate $h\big|_a^b = h(b) - h(a)$.

Furthermore, the function $(F_1, F_2) \mapsto \int_{(-\infty, \cdot]} \phi(F_1) \, dF_2$ *is Hamamard-differentiable as a map into* $D[-\infty, \infty]$.

Proof. Let $h_{1t} \to h_1$ and $h_{2t} \to h_2$ in $D[-\infty, \infty]$ be such that $F_{2t} = F_2 + t h_{2t}$ is a function of variation bounded by 1 for each t. Because F_2 is of bounded variation, it follows that h_{2t} is of bounded variation for every t. Now, with $F_{1t} = F_1 + t h_{1t}$,

$$\frac{1}{t} \left(\int \phi(F_{1t}) \, dF_{2t} - \int \phi(F_1) \, dF_2 \right)$$
$$= \int \left(\frac{\phi(F_{1t}) - \phi(F_1)}{t} - \phi'(F_1) h_1 \right) dF_{2t} + \int \phi(F_1) \, dh_{2t} + \int \phi'(F_1) h_1 \, dF_{2t}.$$

By partial integration, the second term on the right can be rewritten as $\phi \circ F_1 h_{2t} \big|_{-\infty}^{\infty} - \int h_{2t-} \, d\phi \circ F_1$. Under the assumption on h_{2t}, this converges to the first part of the derivative as given in the lemma. The first term is bounded above by $\left(\|\phi''\|_\infty t \|h_{1t}\|_\infty + \|\phi'\|_\infty \|h_{1t} - h_1\|_\infty \right) \int d|F_{2t}|$. Because the measures F_{2t} are of total variation at most 1 by assumption, this expression converges to zero. To analyze the third term on the right, take a grid $u_0 = -\infty < u_1 < \cdots < u_m = \infty$ such that the function $\phi' \circ F_1 h_1$ varies less than a prescribed value $\varepsilon > 0$ on each interval $[u_{i-1}, u_i)$. Such a grid exists for every element of $D[-\infty, \infty]$ (problem 18.6). Then

$$\left| \int \phi'(F_1) h_1 \, d(F_{2t} - F_2) \right| \le \varepsilon \left(\int d|F_{2t}| + d|F_2| \right)$$
$$+ \sum_{i=1}^{m+1} \left| (\phi' \circ F_1 h_1)(u_{i-1}) \right| \left| F_{2t}[u_{i-1}, u_i) - F_2[u_{i-1}, u_i) \right|.$$

The first term is bounded by $\varepsilon O(1)$, in which the ε can be made arbitrarily small by the choice of the partition. For each fixed partition, the second term converges to zero as $t \downarrow 0$. Hence the left side converges to zero as $t \downarrow 0$.

This proves the first assertion. The second assertion follows similarly. ∎

20.11 *Example (Wilcoxon).* Let $\mathbb{F}_m$ and $\mathbb{G}_n$ be the empirical distribution functions of two independent random samples $X_1, \ldots, X_m$ and $Y_1, \ldots, Y_n$ from distribution functions F and G, respectively. As usual, consider both m and n as indexed by a parameter ν, let $N = m + n$, and assume that $m/N \to \lambda \in (0, 1)$ as $\nu \to \infty$. By Donsker's theorem and Slutsky's lemma,

$$\sqrt{N}(\mathbb{F}_m - F, \mathbb{G}_n - G) \rightsquigarrow \left(\frac{\mathbb{G}_F}{\sqrt{\lambda}}, \frac{\mathbb{G}_G}{\sqrt{1 - \lambda}} \right),$$

in the space $D[-\infty, \infty] \times D[-\infty, \infty]$, for a pair of independent Brownian bridges $\mathbb{G}_F$ and $\mathbb{G}_G$. The preceding lemma together with the delta method imply that

$$\sqrt{N} \left(\int \mathbb{F}_m \, d\mathbb{G}_n - \int F \, dG \right) \rightsquigarrow - \int \frac{\mathbb{G}_{G-}}{\sqrt{1 - \lambda}} \, dF + \int \frac{\mathbb{G}_F}{\sqrt{\lambda}} \, dG.$$

The random variable on the right is a continuous, linear function applied to Gaussian processes. In analogy to the theorem that a linear transformation of a multivariate Gaussian vector has a Gaussian distribution, it can be shown that a continuous, linear transformation of a tight Gaussian process is normally distributed. That the present variable is normally distributed can be more easily seen by applying the delta method in its stronger form, which implies that the limit variable is the limit in distribution of the sequence

$$-\int \sqrt{N}(\mathbb{G}_n - G)_- \, dF + \int \sqrt{N}(\mathbb{F}_m - F) \, dG.$$

This can be rewritten as the difference of two sums of independent random variables, and next we can apply the central limit theorem for real variables. □

20.12 *Example (Two-sample rank statistics).* Let $\mathbb{H}_N$ be the empirical distribution function of a sample $X_1, \ldots, X_m, Y_1, \ldots, Y_n$ obtained by "pooling" two independent random samples from distributions F and G, respectively. Let $R_{N1}, \ldots, R_{NN}$ be the ranks of the pooled sample and let $\mathbb{G}_n$ be the empirical distribution function of the second sample. If no observations are tied, then $N\mathbb{H}_N(Y_j)$ is the rank of Y_j in the pooled sample. Thus,

$$\int \phi(\mathbb{H}_N) \, d\mathbb{G}_n = \frac{1}{n} \sum_{j=m+1}^{N} \phi\left(\frac{R_{Nj}}{N}\right)$$

is a two-sample rank statistic. This can be shown to be asymptotically normal by the preceding lemma. Because $N\mathbb{H}_N = m\mathbb{F}_m + n\mathbb{G}_n$, the asymptotic normality of the pair $(\mathbb{H}_N, \mathbb{G}_n)$ can be obtained from the asymptotic normality of the pair $(\mathbb{F}_m, \mathbb{G}_n)$, which is discussed in the preceding example. □

The *cumulative hazard function* corresponding to a cumulative distribution function F on $[0, \infty]$ is defined as

$$\Lambda_F(t) = \int_{[0,t]} \frac{dF}{1 - F_-}.$$

In particular, if F has a density f, then Λ_F has a density $\lambda_F = f/(1-F)$. If $F(t)$ gives the probability of "survival" of a person or object until time t, then $d\Lambda_F(t)$ can be interpreted as the probability of "instant death at time t given survival until t." The hazard function is an important modeling tool in survival analysis.

The correspondence between distribution functions and hazard functions is one-to-one. The cumulative distribution function can be explicitly recovered from the cumulative hazard function as the *product integral* of $-\Lambda$ (see the proof of Lemma 25.74),

$$1 - F_\Lambda(t) = \prod_{0 < s \leq t} \left(1 - \Lambda\{s\}\right)e^{-\Lambda^c(t)}. \tag{20.13}$$

Here $\Lambda\{s\}$ is the jump of Λ at s and $\Lambda^c(s)$ is the continuous part of Λ.

Under some restrictions the maps $F \leftrightarrow \Lambda_F$ are Hadamard differentiable. Thus, from an asymptotic-statistical point of view, estimating a distribution function and estimating a cumulative hazard function are the same problem.

20.14 *Lemma.* *Let $\mathbb{D}_\phi$ be the set of all nondecreasing cadlag functions $F : [0, \tau] \mapsto \mathbb{R}$ with $F(0) = 0$ and $1 - F(\tau) \geq \varepsilon > 0$ for some $\varepsilon > 0$, and let $\mathbb{E}_\psi$ be the set of all nondecreasing cadlag functions $\Lambda : [0, \tau] \mapsto \mathbb{R}$ with $\Lambda(0) = 0$ and $\Lambda(\tau) \leq M$ for some $M \in \mathbb{R}$.*

(i) *The map $\phi : \mathbb{D}_\phi \subset D[0, \tau] \mapsto D[0, \tau]$ defined by $\phi(F) = \Lambda_F$ is Hadamard differentiable.*

(ii) *The map $\psi : \mathbb{E}_\psi \subset D[0, \tau] \mapsto D[0, \tau]$ defined by $\psi(\Lambda) = F_\Lambda$ is Hadamard differentiable.*

Proof. Part (i) follows from the chain rule and the Hadamard differentiability of each of the three maps in the decomposition

$$F \mapsto (F, 1 - F_-) \mapsto \left(F, \frac{1}{1 - F_-} \right) \mapsto \int_{[0,t]} \frac{dF}{1 - F_-}.$$

The differentiability of the first two maps is easy to see. The differentiability of the last one follows from Lemma 20.10. The proof of (ii) is longer; see, for example, [54] or [55]. ∎

20.15 *Example (Nelson-Aalen estimator).* Consider estimating a distribution function based on *right-censored data*. We wish to estimate the distribution function F (or the corresponding cumulative hazard function Λ) of a random sample of "failure times" $T_1, \ldots, T_n$. Unfortunately, instead of T_i we only observe the pair (X_i, Δ_i), in which $X_i = T_i \wedge C_i$ is the minimum of T_i and a "censoring time" C_i, and $\Delta_i = 1\{T_i \leq C_i\}$ records whether T_i is censored ($\Delta_i = 0$) or not ($\Delta_i = 1$). The censoring time could be the closing date of the study or a time that a patient is lost for further observation. The cumulative hazard function of interest can be written

$$\Lambda(t) = \int_{[0,t]} \frac{1}{1 - F_-} dF = \int_{[0,t]} \frac{1}{1 - H_-} dH_1,$$

for $1 - H = (1 - F)(1 - G)$ and $dH_1 = (1 - G_-)dF$, and every choice of distribution function G. If we assume that the censoring times $C_1, \ldots, C_n$ are a random sample from G and are independent of the failure times T_i, then H is precisely the distribution function of X_i and H_1 is a "subdistribution function,"

$$1 - H(x) = \mathrm{P}(X_i > x), \qquad H_1(x) = \mathrm{P}(X_i \leq x, \Delta_i = 1).$$

An estimator for Λ is obtained by estimating these functions by the empirical distributions of the data, given by $\mathbb{H}_n(x) = n^{-1} \sum_{i=1}^n 1\{X_i \leq x\}$ and $\mathbb{H}_{1n}(x) = n^{-1} \sum_{i=1}^n 1\{X_i \leq x, \Delta_i = 1\}$, and next substituting these estimators in the formula for Λ. This yields the *Nelson-Aalen estimator*

$$\hat{\Lambda}_n(t) = \int_{[0,t]} \frac{1}{1 - \mathbb{H}_{n-}} d\mathbb{H}_{1n}.$$

Because they are empirical distribution functions, the pair $(\mathbb{H}_n, \mathbb{H}_{1n})$ is asymptotically normal in the space $D[-\infty, \infty] \times D[-\infty, \infty]$. The easiest way to see this is to consider them as continuous transformations of the (bivariate) empirical distribution function of the pairs (X_i, Δ_i). The Nelson-Aalen estimator is constructed through the maps

$$(A, B) \mapsto (1 - A, B) \mapsto \left(\frac{1}{1 - A}, B \right) \mapsto \int_{[0,t]} \frac{1}{1 - A_-} dB.$$

These are Hadamard differentiable on appropriate domains, the main restrictions being that $1 - A$ should be bounded away from zero and B of uniformly bounded variation. The asymptotic normality of the Nelson-Aalen estimator $\hat{\Lambda}_n(t)$ follows for every t such that $H(t) < 1$, and even as a process in $D[0, \tau]$ for every τ such that $H(\tau) < 1$.

If we apply the product integral given in (20.13) to the Nelson-Aalen estimator, then we obtain an estimator $1 - \hat{F}_n$ for the distribution function, known as the *product limit estimator* or *Kaplan-Meier estimator*. For a discrete hazard function the product integral is an ordinary product over the jumps, by definition, and it can be seen that

$$1 - \hat{F}_n(t) = \prod_{i\,:\,X_i \leq t} \frac{\#(j : X_j \geq X_i) - \Delta_i}{\#(j : X_j \geq X_i)} = \prod_{i\,:\,X_{(i)} \leq t} \left(\frac{n-i}{n-i+1} \right)^{\Delta_{(i)}}.$$

This estimator sequence is asymptotically normal by the Hadamard differentiability of the product integral. $\Box$

Notes

A calculus of "differentiable statistical functions" was proposed by von Mises [104]. Von Mises considered functions $\phi(\mathbb{F}_n)$ of the empirical distribution function (which he calls the "repartition of the real quantities $x_1, \ldots, x_n$") as in the first section of this chapter. Following Volterra he calls ϕ m times differentiable at F if the first m derivatives of the map $t \mapsto \phi(F + tH)$ at $t = 0$ exist and have representations of the form

$$\phi_F^{(k)}(H) = \int \cdots \int \psi(x_1, \ldots, x_k) \, dH(x_1) \cdots dH(x_k).$$

This representation is motivated in analogy with the finite-dimensional case, in which H would be a vector and the integrals sums. From the perspective of our section on Hadamard-differentiable functions, the representation is somewhat arbitrary, because it is required that a derivative be continuous, whence its general form depends on the norm that we use on the domain of ϕ. Furthermore, the Volterra representation cannot be directly applied to, for instance, a limiting Brownian bridge, which is not of bounded variation.

Von Mises' treatment is not at all informal, as is the first section of this chapter. After developing moment bounds on the derivatives, he shows that $n^{m/2}\big(\phi(\mathbb{F}_n) - \phi(F)\big)$ is asymptotically equivalent to $\phi_F^{(m)}(\mathbb{G}_n)$ if the first $m - 1$ derivatives vanish at F and the $(m+1)$th derivative is sufficiently regular. He refers to the approximating variables $\phi_F^{(m)}(\mathbb{G}_n)$, degenerate V-statistics, as "quantics" and derives the asymptotic distribution of quantics of degree 2, first for discrete observations and next in general by discrete approximation. Hoeffding's work on U-statistics, which was published one year later, had a similar aim of approximating complicated statistics by simpler ones but did not consider degenerate U-statistics.

The systematic application of Hamadard differentiability in statistics appears to have first been put forward in the (unpublished) thesis [125] of J Reeds and had a main focus on robust functions. It was revived by Gill [53] with applications in survival analysis in mind. With a growing number of functional estimators available (beyond the empirical distribution and product-limit estimator), the delta method is a simple but useful tool to standardize asymptotic normality proofs.

Our treatment allows the domain $\mathbb{D}_\phi$ of the map ϕ to be arbitrary. In particular, we do not assume that it is open, as we did, for simplicity, when discussing the Delta method for

Euclidean spaces. This is convenient, because many functions of statistical interest, such as zeros, inverses or integrals, are defined only on irregularly shaped subsets of a normed space, which, besides a linear space, should be chosen big enough to support the limit distribution of T_n.

PROBLEMS

1. Let $\phi(P) = \int \int h(u, v) \, dP(u) \, dP(v)$ for a fixed given function h. The corresponding estimator $\phi(\mathbb{P}_n)$ is known as a *V-statistic*. Find the influence function.

2. Find the influence function of the function $\phi(F) = \int a(F_1 + F_2) \, dF_2$ if F_1 and F_2 are the marginals of the bivariate distribution function F, and a is a fixed, smooth function. Write out $\phi(\mathbb{F}_n)$. What asymptotic variance do you expect?

3. Find the influence function of the map $F \mapsto \int_{[0,t]} (1 - F_-)^{-1} \, dF$ (the cumulative hazard function).

4. Show that a map $\phi : \mathbb{D} \mapsto \mathbb{E}$ is Hadamard differentiable at a point θ if and only if for every compact set $K \subset \mathbb{D}$ the expression in (20.7) converges to zero uniformly in $h \in K$ as $t \to 0$.

5. Show that the symmetrization map $(\theta, F) \mapsto \frac{1}{2}\big(F(t) + 1 - F(2\theta - t)\big)$ is (tangentially) Hadamard differentiable under appropriate conditions.

6. Let $g : [a, b] \mapsto \mathbb{R}$ be a continuously differentiable function. Show that the map $z \mapsto g \circ z$ with domain the functions $z : T \mapsto [a, b]$ contained in $\ell^\infty(T)$ is Hadamard differentiable. What does this imply for the function $z \mapsto 1/z$?

7. Show that the map $F \mapsto \int_{[a,b]} s \, dF(s)$ is Hadamard differentiable from the domain of all distribution functions to $\mathbb{R}$, for each pair of finite numbers a and b. View the distribution functions as a subset of $D[-\infty, \infty]$ equipped with supremum norm. What if a or b are infinite?

8. Find the first- and second-order derivative of the function $\phi(F) = \int (F - F_0)^2 \, dF$ at $F = F_0$. What limit distribution do you expect for $\phi(\mathbb{F}_n)$?

21

Quantiles and Order Statistics

In this chapter we derive the asymptotic distribution of estimators of quantiles from the asymptotic distribution of the corresponding estimators of a distribution function. Empirical quantiles are an example, and hence we also discuss some results concerning order statistics. Furthermore, we discuss the asymptotics of the median absolute deviation, which is the empirical 1/2-quantile of the observations centered at their 1/2-quantile.

21.1 Weak Consistency

The *quantile function* of a cumulative distribution function F is the generalized inverse $F^{-1} : (0, 1) \mapsto \mathbb{R}$ given by

$$F^{-1}(p) = \inf\{x : F(x) \geq p\}.$$

It is a left-continuous function with range equal to the support of F and hence is often unbounded. The following lemma records some useful properties.

21.1 Lemma. *For every $0 < p < 1$ and $x \in \mathbb{R}$,*
  (i) $F^{-1}(p) \leq x$ *iff* $p \leq F(x)$;
  (ii) $F \circ F^{-1}(p) \geq p$ *with equality iff p is in the range of F; equality can fail only if F is discontinuous at $F^{-1}(p)$;*
  (iii) $F_- \circ F^{-1}(p) \leq p$;
  (iv) $F^{-1} \circ F(x) \leq x$; *equality fails iff x is in the interior or at the right end of a "flat" of F;*
  (v) $F^{-1} \circ F \circ F^{-1} = F^{-1}$; $F \circ F^{-1} \circ F = F$;
  (vi) $(F \circ G)^{-1} = G^{-1} \circ F^{-1}$.

Proof. The proofs of the inequalities in (i) through (iv) are best given by a picture. The equalities (v) follow from (ii) and (iv) and the monotonicity of F and F^{-1}. If $p = F(x)$ for some x, then, by (ii) $p \leq F \circ F^{-1}(p) = F \circ F^{-1} \circ F(x) = F(x) = p$, by (iv). This proves the first statement in (ii); the second is immediate from the inequalities in (ii) and (iii). Statement (vi) follows from (i) and the definition of $(F \circ G)^{-1}$. ∎

Consequences of (ii) and (iv) are that $F \circ F^{-1}(p) \equiv p$ on $(0, 1)$ if and only if F is continuous (i.e., has range $[0, 1]$), and $F^{-1} \circ F(x) \equiv x$ on $\mathbb{R}$ if and only if F is strictly increasing (i.e., has no "flats"). Thus F^{-1} is a proper inverse if and only if F is both continuous and strictly increasing, as one would expect.

By (i) the random variable $F^{-1}(U)$ has distribution function F if U is uniformly distributed on $[0, 1]$. This is called the *quantile transformation*. On the other hand, by (i) and (ii) the variable $F(X)$ is uniformly distributed on $[0, 1]$ if and only if X has a continuous distribution function F. This is called the *probability integral transformation*.

A sequence of quantile functions is defined to *converge weakly* to a limit quantile function, denoted $F_n^{-1} \rightsquigarrow F^{-1}$, if and only if $F_n^{-1}(t) \to F^{-1}(t)$ at every t where F^{-1} is continuous. This type of convergence is not only analogous in form to the weak convergence of distribution functions, it is the same.

21.2 Lemma. *For any sequence of cumulative distribution functions, $F_n^{-1} \rightsquigarrow F^{-1}$ if and only if $F_n \rightsquigarrow F$.*

Proof. Let U be uniformly distributed on $[0, 1]$. Because F^{-1} has at most countably many discontinuity points, $F_n^{-1} \rightsquigarrow F^{-1}$ implies that $F_n^{-1}(U) \to F^{-1}(U)$ almost surely. Consequently, $F_n^{-1}(U)$ converges in law to $F^{-1}(U)$, which is exactly $F_n \rightsquigarrow F$ by the quantile transformation.

For a proof the converse, let V be a normally distributed random variable. If $F_n \rightsquigarrow F$, then $F_n(V) \overset{as}{\to} F(V)$, because convergence can fail only at discontinuity points of F. Thus $\Phi\big(F_n^{-1}(t)\big) = P\big(F_n(V) < t\big)$ (by (i) of the preceding lemma) converges to $P\big(F(V) < t\big) = \Phi\big(F^{-1}(t)\big)$ at every t at which the limit function is continuous. This includes every t at which F^{-1} is continuous. By the continuity of Φ^{-1}, $F_n^{-1}(t) \to F^{-1}(t)$ for every such t. ∎

A statistical application of the preceding lemma is as follows. If a sequence of estimators $\hat{F}_n$ of a distribution function F is weakly consistent, then the sequence of estimators $\hat{F}_n^{-1}$ is weakly consistent for the quantile function F^{-1}.

21.2 Asymptotic Normality

In the absence of information concerning the underlying distribution function F of a sample, the empirical distribution function $\mathbb{F}_n$ and empirical quantile function $\mathbb{F}_n^{-1}$ are reasonable estimators for F and F^{-1}, respectively. The empirical quantile function is related to the order statistics $X_{n(1)}, \ldots, X_{n(n)}$ of the sample through

$$\mathbb{F}_n^{-1}(p) = X_{n(i)}, \qquad \text{for } p \in \left(\frac{i-1}{n}, \frac{i}{n} \right].$$

One method to prove the asymptotic normality of empirical quantiles is to view them as M-estimators and apply the theorems given in Chapter 5. Another possibility is to express the distribution function $P(X_{n(i)} \le x)$ into binomial probabilities and apply approximations to these. The method that we follow in this chapter is to deduce the asymptotic normality of quantiles from the asymptotic normality of the distribution function, using the delta method.

An advantage of this method is that it is not restricted to empirical quantiles but applies to the quantiles of any estimator of the distribution function.

For a nondecreasing function $F \in D[a, b]$, $[a, b] \subset [-\infty, \infty]$, and a fixed $p \in \mathbb{R}$, let $\phi(F) \in [a, b]$ be an arbitrary point in $[a, b]$ such that

$$F(\phi(F)-) \le p \le F(\phi(F)).$$

The natural domain $\mathbb{D}_\phi$ of the resulting map ϕ is the set of all nondecreasing F such that there exists a solution to the pair of inequalities. If there exists more than one solution, then the precise choice of $\phi(F)$ is irrelevant. In particular, $\phi(F)$ may be taken equal to the pth quantile $F^{-1}(p)$.

21.3   Lemma. *Let $F \in \mathbb{D}_\phi$ be differentiable at a point $\xi_p \in (a, b)$ such that $F(\xi_p) = p$, with positive derivative. Then $\phi : \mathbb{D}_\phi \subset D[a, b] \mapsto \mathbb{R}$ is Hadamard-differentiable at F tangentially to the set of functions $h \in D[a, b]$ that are continuous at ξ_p, with derivative $\phi'_F(h) = -h(\xi_p)/F'(\xi_p)$.*

Proof.   Let $h_t \to h$ uniformly on $[a, b]$ for a function h that is continuous at ξ_p. Write ξ_{pt} for $\phi(F + th_t)$. By the definition of ϕ, for every $\varepsilon_t > 0$,

$$(F + th_t)(\xi_{pt} - \varepsilon_t) \le p \le (F + th_t)(\xi_{pt}).$$

Choose ε_t positive and such that $\varepsilon_t = o(t)$. Because the sequence h_t converges uniformly to a bounded function, it is uniformly bounded. Conclude that $F(\xi_{pt} - \varepsilon_t) + O(t) \le p \le F(\xi_{pt}) + O(t)$. By assumption, the function F is monotone and bounded away from p outside any interval $(\xi_p - \varepsilon, \xi_p + \varepsilon)$ around ξ_p. To satisfy the preceding inequalities the numbers ξ_{pt} must be to the right of $\xi_p - \varepsilon$ eventually, and the numbers $\xi_{pt} - \varepsilon_t$ must be to the left of $\xi_p + \varepsilon$ eventually. In other words, $\xi_{pt} \to \xi_p$.

By the uniform convergence of h_t and the continuity of the limit, $h_t(\xi_{pt} - \varepsilon_t) \to h(\xi_p)$ for every $\varepsilon_t \to 0$. Using this and Taylor's formula on the preceding display yields

$$p + (\xi_{pt} - \xi_p)F'(\xi_p) - o(\xi_{pt} - \xi_p) + O(\varepsilon_t) + th(\xi_p) - o(t)$$
$$\le p \le p + (\xi_{pt} - \xi_p)F'(\xi_p) + o(\xi_{pt} - \xi_p) + O(\varepsilon_t) + th(\xi_p) + o(t).$$

Conclude first that $\xi_{pt} - \xi_p = O(t)$. Next, use this to replace the $o(\xi_{pt} - \xi_p)$ terms in the display by $o(t)$ terms and conclude that $(\xi_{pt} - \xi_p)/t \to -(h/F')(\xi_p)$.   ∎

Instead of a single quantile we can consider the *quantile function* $F \mapsto (F^{-1}(p))_{p_1 < p < p_2}$, for fixed numbers $0 \le p_1 < p_2 \le 1$. Because any quantile function is bounded on an interval $[p_1, p_2]$ strictly contained in $(0, 1)$, we may hope that a quantile estimator converges in distribution in $\ell^\infty(p_1, p_2)$ for such an interval. The quantile function of a distribution with compact support is bounded on the whole interval $(0, 1)$, and then we may hope to strengthen the result to weak convergence in $\ell^\infty(0, 1)$.

Given an interval $[a, b] \subset \mathbb{R}$, let $\mathbb{D}_1$ be the set of all restrictions of distribution functions on $\mathbb{R}$ to $[a, b]$, and let $\mathbb{D}_2$ be the subset of $\mathbb{D}_1$ of distribution functions of measures that give mass 1 to (a, b).

21.4   Lemma.

(i) *Let $0 < p_1 < p_2 < 1$, and let F be continuously differentiable on the interval $[a, b] = \left[F^{-1}(p_1) - \varepsilon, F^{-1}(p_2) + \varepsilon \right]$ for some $\varepsilon > 0$, with strictly positive derivative f. Then the inverse map $G \mapsto G^{-1}$ as a map $\mathbb{D}_1 \subset D[a, b] \mapsto \ell^{\infty}[p_1, p_2]$ is Hadamard differentiable at F tangentially to $C[a, b]$.*

(ii) *Let F have compact support $[a, b]$ and be continuously differentiable on its support with strictly positive derivative f. Then the inverse map $G \mapsto G^{-1}$ as a map $\mathbb{D}_2 \subset D[a, b] \mapsto \ell^{\infty}(0, 1)$ is Hadamard differentiable at F tangentially to $C[a, b]$.*

In both cases the derivative is the map $h \mapsto -(h/f) \circ F^{-1}$.

Proof.   It suffices to make the proof of the preceding lemma uniform in p. We use the same notation.

(i). Because the function F has a positive density, it is strictly increasing on an interval $[\xi_{p'_1}, \xi_{p'_2}]$ that strictly contains $[\xi_{p_1}, \xi_{p_2}]$. Then on $[p'_1, p'_2]$ the quantile function F^{-1} is the ordinary inverse of F and is (uniformly) continuous and strictly increasing. Let $h_t \to h$ uniformly on $[\xi_{p'_1}, \xi_{p'_2}]$ for a continuous function h. By the proof of the preceding lemma, $\xi_{p_i t} \to \xi_{p_i}$ and hence every ξ_{pt} for $p_1 \le p \le p_2$ is contained in $[\xi_{p'_1}, \xi_{p'_2}]$ eventually. The remainder of the proof is the same as the proof of the preceding lemma.

(ii). Let $h_t \to h$ uniformly in $D[a, b]$, where h is continuous and $F + th_t$ is contained in $\mathbb{D}_2$ for all t. Abbreviate $F^{-1}(p)$ and $(F + th_t)^{-1}(p)$ to ξ_p and ξ_{pt}, respectively. Because F and $F + th_t$ are concentrated on (a, b) by assumption, we have $a < \xi_{pt}, \xi_p \le b$ for all $0 < p < 1$. Thus the numbers $\varepsilon_{pt} = t^2 \wedge (\xi_{pt} - a)$ are positive, whence, by definition,

$$(F + th_t)(\xi_{pt} - \varepsilon_{pt}) \le p \le (F + th_t)(\xi_{pt}).$$

By the smoothness of F we have $F(\xi_p) = p$ and $F(\xi_{pt} - \varepsilon_{pt}) = F(\xi_{pt}) + O(\varepsilon_{pt})$, uniformly in $0 < p < 1$. It follows that

$$-th(\xi_{pt}) + o(t) \le F(\xi_{pt}) - F(\xi_p) \le -th(\xi_{pt} - \varepsilon_{pt}) + o(t).$$

The $o(t)$ terms are uniform in $0 < p < 1$. The far left side and the far right side are $O(t)$; the expression in the middle is bounded above and below by a constant times $|\xi_{pt} - \xi_p|$. Conclude that $|\xi_{pt} - \xi_p| = O(t)$, uniformly in p. Next, the lemma follows by the uniform differentiability of F.   ∎

Thus, the asymptotic normality of an estimator of a distribution function (or another nondecreasing function) automatically entails the asymptotic normality of the corresponding quantile estimators. More precisely, to derive the asymptotic normality of even a single quantile estimator $\hat{F}_n^{-1}(p)$, we need to know that the estimators $\hat{F}_n$ are asymptotically normal as a process, in a neighborhood of $F^{-1}(p)$. The standardized empirical distribution function is asymptotically normal as a process indexed by $\mathbb{R}$, and hence the empirical quantiles are asymptotically normal.

21.5   Corollary.   *Fix $0 < p < 1$. If F is differentiable at $F^{-1}(p)$ with positive derivative $f\left(F^{-1}(p)\right)$, then*

$$\sqrt{n}\left(\mathbb{F}_n^{-1}(p) - F^{-1}(p)\right) = -\frac{1}{\sqrt{n}} \sum_{i=1}^{n} \frac{1\left\{X_i \le F^{-1}(p)\right\} - p}{f\left(F^{-1}(p)\right)} + o_P(1).$$

Consequently, the sequence $\sqrt{n}\big(\mathbb{F}_n^{-1}(p) - F^{-1}(p)\big)$ is asymptotically normal with mean 0 and variance $p(1 - p)/f^2\big(F^{-1}(p)\big)$. Furthermore, if F satisfies the conditions (i) or (ii) of the preceding lemma, then $\sqrt{n}(\mathbb{F}_n^{-1} - F^{-1})$ converges in distribution in $\ell^\infty[p_1, p_2]$ or $\ell^\infty(0, 1)$, respectively, to the process $\mathbb{G}_\lambda / f\big(F^{-1}(p)\big)$, where $\mathbb{G}_\lambda$ is a standard Brownian bridge.

Proof. By Theorem 19.3, the empirical process $\mathbb{G}_{n,F} = \sqrt{n}(\mathbb{F}_n - F)$ converges in distribution in $D[-\infty, \infty]$ to an F-Brownian bridge process $\mathbb{G}_F = \mathbb{G}_\lambda \circ F$. The sample paths of the limit process are continuous at the points at which F is continuous. By Lemma 21.3, the quantile function $F \mapsto F^{-1}(p)$ is Hadamard-differentiable tangentially to the range of the limit process. By the functional delta method, the sequence $\sqrt{n}\big(\mathbb{F}_n^{-1}(p) - F^{-1}(p)\big)$ is asymptotically equivalent to the derivative of the quantile function evaluated at $\mathbb{G}_{n,F}$, that is, to $-\mathbb{G}_{n,F}\big(F^{-1}(p)\big)/f\big(F^{-1}(p)\big)$. This is the first assertion. Next, the asymptotic normality of the sequence $\sqrt{n}\big(\mathbb{F}_n^{-1}(p) - F^{-1}(p)\big)$ follows by the central limit theorem.

The convergence of the quantile process follows similarly, this time using Lemma 21.4. ∎

21.6 Example. The uniform distribution function has derivative 1 on its compact support. Thus, the uniform empirical quantile process converges weakly in $\ell^\infty(0, 1)$. The limiting process is a standard Brownian bridge.

The normal and Cauchy distribution functions have continuous derivatives that are bounded away from zero on any compact interval. Thus, the normal and Cauchy empirical quantile processes converge in $\ell^\infty[p_1, p_2]$, for every $0 < p_1 < p_2 < 1$. □

The empirical quantile function at a point is equal to an order statistic of the sample. In estimating a quantile, we could also use the order statistics directly, not necessarily in the way that $\mathbb{F}_n^{-1}$ picks them. For the k_n-th order statistic $X_{n(k_n)}$ to be a consistent estimator for $F^{-1}(p)$, we need minimally that $k_n/n \to p$ as $n \to \infty$. For mean-zero asymptotic normality, we also need that $k_n/n \to p$ faster than $1/\sqrt{n}$, which is necessary to ensure that $X_{n(k_n)}$ and $\mathbb{F}_n^{-1}(p)$ are asymptotically equivalent. This still allows considerable freedom for choosing k_n.

21.7 Lemma. *Let F be differentiable at $F^{-1}(p)$ with positive derivative and let $k_n/n = p + c/\sqrt{n} + o(1/\sqrt{n})$. Then*

$$\sqrt{n}\big(X_{n(k_n)} - \mathbb{F}_n^{-1}(p)\big) \overset{\mathrm{P}}{\to} \frac{c}{f\big(F^{-1}(p)\big)}.$$

Proof. First assume that F is the uniform distribution function. Denote the observations by U_i, rather than X_i. Define a function $g_n : \ell^\infty(0, 1) \mapsto \mathbb{R}$ by $g_n(z) = z(k_n/n) - z(p)$. Then $g_n(z_n) \to z(p) - z(p) = 0$, whenever $z_n \to z$ for a function z that is continuous at p. Because the uniform quantile process $\sqrt{n}(\mathbb{G}_n^{-1} - G^{-1})$ converges in distribution in $\ell^\infty(0, 1)$, the extended continuous-mapping theorem, Theorem 18.11, yields $g_n\big(\sqrt{n}(\mathbb{G}_n^{-1} - G^{-1})\big) = \sqrt{n}\big(U_{n(k_n)} - \mathbb{G}_n^{-1}(p)\big) - \sqrt{n}(k_n/n - p) \rightsquigarrow 0$. This is the result in the uniform case.

A sample from a general distribution function F can be generated as $F^{-1}(U_i)$, by the quantile transformation. Then $\sqrt{n}\big(X_{n(k_n)} - \mathbb{F}_n^{-1}(p)\big)$ is equal to

$$\sqrt{n}\big[F^{-1}(U_{n(k_n)}) - F^{-1}(p)\big] - \sqrt{n}\big[F^{-1}\big(\mathbb{G}_n^{-1}(p)\big) - F^{-1}(p)\big].$$

Apply the delta method to the two terms to see that $f(F^{-1}(p))$ times their difference is asymptotically equivalent to $\sqrt{n}(U_{n(k_n)} - p) - \sqrt{n}(\mathbb{G}_n^{-1}(p) - p)$. ∎

21.8 Example (Confidence intervals for quantiles). If $X_1, \ldots, X_n$ is a random sample from a continuous distribution function F, then $U_1 = F(X_1), \ldots, U_n = F(X_n)$ are a random sample from the uniform distribution, by the probability integral transformation. This can be used to construct confidence intervals for quantiles that are distribution-free over the class of continuous distribution functions. For any given natural numbers k and l, the interval $(X_{n(k)}, X_{n(l)}]$ has coverage probability

$$P_F(X_{n(k)} < F^{-1}(p) \le X_{n(l)}) = P(U_{n(k)} < p \le U_{n(l)}).$$

Because this is independent of F, it is possible to obtain an exact confidence interval for every fixed n, by determining k and l to achieve the desired confidence level. (Here we have some freedom in choosing k and l but can obtain only finitely many confidence levels.) For large n, the values k and l can be chosen equal to

$$\frac{k, l}{n} = p \pm z_\alpha \sqrt{\frac{p(1-p)}{n}}.$$

To see this, note that, by the preceding lemma,

$$U_{n(k)}, U_{n(l)} = \frac{\mathbb{G}_n^{-1}(p)}{\sqrt{n}} \pm z_\alpha \sqrt{\frac{p(1-p)}{n}} + o_P\left(\frac{1}{\sqrt{n}}\right).$$

Thus the event $U_{n(k)} < p \le U_{n(l)}$ is asymptotically equivalent to the event $\sqrt{n}|\mathbb{G}_n^{-1}(p) - p| \le z_\alpha \sqrt{p(1-p)}$. Its probability converges to $1 - 2\alpha$.

An alternative is to use the asymptotic normality of the empirical quantiles $\mathbb{F}_n^{-1}$, but this has the unattractive feature of having to estimate the density $f(F^{-1}(p))$, because this appears in the denominator of the asymptotic variance. If using the distribution-free method, we do not even have to assume that the density exists. □

The application of the Hadamard differentiability of the quantile transformation is not limited to empirical quantiles. For instance, we can also immediately obtain the asymptotic normality of the quantiles of the product limit estimator, or any other estimator of a distribution function in semiparametric models. On the other hand, the results on empirical quantiles can be considerably strengthened by taking special properties of the empirical distribution into account. We discuss a few extensions, mostly for curiosity value.[†]

Corollary 21.5 asserts that $R_n(p) \xrightarrow{P} 0$, for, with $\xi_p = F^{-1}(p)$,

$$R_n(p) = f(\xi_p)\sqrt{n}(\mathbb{F}_n^{-1}(p) - F^{-1}(p)) + \sqrt{n}(\mathbb{F}_n(\xi_p) - F(\xi_p)).$$

The expression on the left is known as the standardized *empirical difference process*. "Standardized" refers to the leading factor $f(\xi_p)$. That a sum is called a difference is curious but stems from the fact that minus the second term is approximately equal to the first term. The identity shows an interesting symmetry between the empirical distribution and quantile

[†] See [134, pp. 586–587] for further information.

processes, particularly in the case that F is uniform, if $f(\xi_p) \equiv 1$ and $\xi_p \equiv p$. The result that $R_n(p) \overset{P}{\to} 0$ can be refined considerably. If F is twice-differentiable at ξ_p with positive first derivative, then, by the *Bahadur-Kiefer theorems*,

$$\limsup_{n\to\infty} \frac{n^{1/4}}{(\log\log n)^{3/4}} |R_n(p)| = \left[\frac{32}{27} p(1-p) \right]^{1/4}, \qquad \text{a.s.},$$

$$n^{1/4} R_n(p) \rightsquigarrow \frac{2}{\sqrt{p(1-p)}} \int_0^\infty \Phi\left(\frac{x}{\sqrt{y}} \right) \phi\left(\frac{y}{\sqrt{p(1-p)}} \right) dy.$$

The right side in the last display is a distribution function as a function of the argument x. Thus, the magnitude of the empirical difference process is $O_P(n^{-1/4})$, with the rate of its fluctuations being equal to $n^{-1/4}(\log\log n)^{3/4}$. Under some regularity conditions on F, which are satisfied by, for instance, the uniform, the normal, the exponential, and the logistic distribution, versions of the preceding results are also valid in supremum norm,

$$\limsup_{n\to\infty} \frac{n^{1/4}}{(\log n)^{1/2}(2\log\log n)^{1/4}} \|R_n\|_\infty = \frac{1}{\sqrt{2}}, \qquad \text{a.s.},$$

$$\frac{n^{1/4}}{(\log n)^{1/2}} \|R_n\|_\infty \rightsquigarrow \sqrt{\|\mathbb{Z}_\lambda\|_\infty}.$$

Here $\mathbb{Z}_\lambda$ is a standard Brownian motion indexed by the interval $[0, 1]$.

21.3   Median Absolute Deviation

The *median absolute deviation* of a sample $X_1, \ldots, X_n$ is the robust estimator of scale defined by

$$\text{MAD}_n = \underset{1 \le i \le n}{\text{med}} \left| X_i - \underset{1 \le i \le n}{\text{med}} X_i \right|.$$

It is the median of the deviations of the observations from their median and is often recommended for reducing the observations to a standard scale as a first step in a robust procedure. Because the median is a quantile, we can prove the asymptotic normality of the median absolute deviation by the delta method for quantiles, applied twice.

If a variable X has distribution function F, then the variable $|X - \theta|$ has the distribution function $x \mapsto F(\theta + x) - F_-(\theta - x)$. Let $(\theta, F) \mapsto \phi_2(\theta, F)$ be the map that assigns to a given number θ and a given distribution function F the distribution function $F(\theta + x) - F_-(\theta - x)$, and consider the function $\phi = \phi_3 \circ \phi_2 \circ \phi_1$ defined by

$$F \overset{\phi_1}{\mapsto} \left(\theta := F^{-1}(1/2), F \right) \overset{\phi_2}{\mapsto} G := F(\theta + \cdot) - F_-(\theta - \cdot) \overset{\phi_3}{\mapsto} G^{-1}(1/2).$$

If we identify the median with the $1/2$-quantile, then the median absolute deviation is exactly $\phi(\mathbb{F}_n)$. Its asymptotic normality follows by the delta method under a regularity condition on the underlying distribution.

21.9   Lemma. *Let the numbers m_F and m_G satisfy $F(m_F) = \frac{1}{2} = F(m_F + m_G) - F(m_F - m_G)$. Suppose that F is differentiable at m_F with positive derivative and is continuously differentiable on neighborhoods of $m_F + m_G$ and $m_F - m_G$ with positive derivative at*

$m_F + m_G$ and/or $m_F - m_G$. *Then the map $\phi: D[-\infty, \infty] \mapsto \mathbb{R}$, with as domain the distribution functions, is Hadamard-differentiable at F, tangentially to the set of functions that are continuous both at m_F and on neighborhoods of $m_F + m_G$ and $m_F - m_G$. The derivative $\phi'_F(H)$ is given by*

$$\frac{H(m_F)}{f(m_F)} \frac{f(m_F + m_G) - f(m_F - m_G)}{f(m_F + m_G) + f(m_F - m_G)} - \frac{H(m_F + m_G) - H(m_F - m_G)}{f(m_F + m_G) + f(m_F - m_G)}.$$

Proof. Define the maps ϕ_i as indicated previously.

By Lemma 21.3, the map $\phi_1: D[-\infty, \infty] \mapsto \mathbb{R} \times D[-\infty, \infty]$ is Hadamard-differentiable at F tangentially to the set of functions H that are continuous at m_F.

The map $\phi_2: \mathbb{R} \times D[-\infty, \infty] \mapsto D[m_G - \varepsilon, m_G + \varepsilon]$ is Hadamard-differentiable at the point (m_F, F) tangentially to the set of points (g, H) such that H is continuous on the intervals $[m_F \pm m_G - 2\varepsilon, m_F \pm m_G + 2\varepsilon]$, for sufficiently small $\varepsilon > 0$. This follows because, if $a_t \to a$ and $H_t \to H$ uniformly,

$$\frac{(F + t H_t)(m_F + t a_t + x) - F(m_F + x)}{t} \to a f(m_F + x) + H(m_F + x),$$

uniformly in $x \approx m_G$, and because a similar statement is valid for the differences $(F + t H_t)_-(m_F + t a_t - x) - F_-(m_F - x)$. The range of the derivative is contained in $C[m_G - \varepsilon, m_G + \varepsilon]$.

Finally, by Lemma 21.3, the map $\phi_3: D[m_G - \varepsilon, m_G + \varepsilon] \mapsto \mathbb{R}$ is Hadamard-differentiable at $G = \phi_2(m_F, F)$, tangentially to the set of functions that are continuous at m_G, because G has a positive derivative at its median, by assumption.

The lemma follows by the chain rule, where we ascertain that the tangent spaces match up properly. ∎

The F-Brownian bridge process $\mathbb{G}_F$ has sample paths that are continuous everywhere that F is continuous. Under the conditions of the lemma, they are continuous at the point m_F and in neighborhoods of the points $m_F + m_G$ and $m_F - m_G$. Thus, in view of the lemma and the delta method, the sequence $\sqrt{n}(\phi(\mathbb{F}_n) - \phi(F))$ converges in distribution to the variable $\phi'_F(\mathbb{G}_F)$.

21.10 *Example (Symmetric F).* If F has a density that is symmetric about 0, then its median m_F is 0 and the median absolute deviation m_G is equal to $F^{-1}(3/4)$. Then the first term in the definition of the derivative vanishes, and the derivative $\phi'_F(\mathbb{G}_F)$ at the F-Brownian bridge reduces to $-(\mathbb{G}_\lambda(3/4) - \mathbb{G}_\lambda(1/4))/2f(F^{-1}(3/4))$ for a standard Brownian bridge $\mathbb{G}_\lambda$. Then the asymptotic variance of $\sqrt{n}(\text{MAD}_n - m_G)$ is equal to $(1/16)/f \circ F^{-1}(3/4)^2$. □

21.11 *Example (Normal distribution).* If F is equal to the normal distribution with mean zero and variance σ^2, then $m_F = 0$ and $m_G = \sigma \Phi^{-1}(3/4)$. We find an asymptotic variance $(\sigma^2/16)\phi \circ \Phi^{-1}(3/4)^{-2}$. As an estimator for the standard deviation σ, we use the estimator $\text{MAD}_n/\Phi^{-1}(3/4)$, and as an estimator for σ^2 the square of this. By the delta method, the latter estimator has asymptotic variance equal to $(1/4)\sigma^4 \phi \circ \Phi^{-1}(3/4)^{-2}\Phi^{-1}(3/4)^{-2}$, which is approximately equal to $5.44\sigma^4$. The relative efficiency, relative to the sample variance, is approximately equal to 37%, and hence we should not use this estimator without a good reason. □

21.4   Extreme Values

The asymptotic behavior of order statistics $X_{n(k_n)}$ such that $k_n/n \to 0$ or 1 is, of course, different from that of central-order statistics. Because $X_{n(k_n)} \le x_n$ means that at most $n - k_n$ of the X_i can be bigger than x_n, it follows that, with $p_n = P(X_i > x_n)$,

$$\Pr(X_{n(k_n)} \le x_n) = P\big(\mathrm{bin}(n, p_n) \le n - k_n\big).$$

Therefore, limit distributions of general-order statistics can be derived from approximations to the binomial distribution. In this section we consider the most extreme cases, in which $k_n = n - k$ for a fixed number k, starting with the maximum $X_{n(n)}$. We write $\overline{F}(t) = P(X_i > t)$ for the survival distribution of the observations, a random sample of size n from F.

The distribution function of the maximum can be derived from the preceding display, or directly, and satisfies

$$P(X_{n(n)} \le x_n) = F(x_n)^n = \left(1 - \frac{n\,\overline{F}(x_n)}{n}\right)^n.$$

This representation readily yields the following lemma.

21.12   Lemma.   *For any sequence of numbers x_n and any $\tau \in [0, \infty]$, we have $P(X_{n(n)} \le x_n) \to e^{-\tau}$ if and only if $n\overline{F}(x_n) \to \tau$.*

In view of the lemma we can find "interesting limits" for the probabilities $P(X_{n(n)} \le x_n)$ only for sequences x_n such that $\overline{F}(x_n) = O(1/n)$. Depending on F this may mean that x_n is bounded or converges to infinity.

Suppose that we wish to find constants a_n and $b_n > 0$ such that $b_n^{-1}(X_{n(n)} - a_n)$ converges in distribution to a nontrivial limit. Then we must choose a_n and b_n such that $\overline{F}(a_n + b_n x) = O(1/n)$ for a nontrivial set of x. Depending on F such constants may or may not exist. It is a bit surprising that the set of possible limit distributions is extremely small.[†]

21.13   Theorem (Extremal types).   *If $b_n^{-1}(X_{n(n)} - a_n) \rightsquigarrow G$ for a nondegenerate cumulative distribution function G, then G belongs to the location-scale family of a distribution of one of the following forms:*
  (i) $e^{-e^{-x}}$ with support $\mathbb{R}$;
  (ii) $e^{-(1/x^{\alpha})}$ with support $[0, \infty)$ and $\alpha > 0$;
  (iii) $e^{-(-x)^{\alpha}}$ with support $(-\infty, 0]$ and $\alpha > 0$.

21.14   Example (Uniform).   If the distribution has finite support $[0, 1]$ with $\overline{F}(t) = (1 - t)^{\alpha}$, then $n\overline{F}(1 + n^{-1/\alpha}x) \to (-x)^{\alpha}$ for every $x \le 0$. In view of Lemma 21.12, the sequence $n^{1/\alpha}(X_{n(n)} - 1)$ converges in distribution to a limit of type (iii). The uniform distribution is the special case with $\alpha = 1$, for which the limit distribution is the negative of an exponential distribution. $\square$

[†] For a proof of the following theorem, see [66] or Theorem 1.4.2 in [90].

21.15 *Example (Pareto).* The survival distribution of the Pareto distribution satisfies $\overline{F}(t) = (\mu/t)^\alpha$ for $t \geq \mu$. Thus $n\overline{F}(n^{1/\alpha}\mu x) \to 1/x^\alpha$ for every $x > 0$. In view of Lemma 21.12, the sequence $n^{-1/\alpha} X_{n(n)}/\mu$ converges in distribution to a limit of type (ii). $\square$

21.16 *Example (Normal).* For the normal distribution the calculations are similar, but more delicate. We choose

$$a_n = \sqrt{2\log n} - \frac{1}{2}\frac{\log\log n + \log 4\pi}{\sqrt{2\log n}}, \qquad b_n = 1/\sqrt{2\log n}.$$

Using *Mill's ratio*, which asserts that $\overline{\Phi}(t) \sim \phi(t)/t$ as $t \to \infty$, it is straightforward to see that $n\overline{\Phi}(a_n + b_n x) \to e^{-x}$ for every x. In view of Lemma 21.12, the sequence $\sqrt{2\log n}(X_{n(n)} - a_n)$ converges in distribution to a limit of type (i). $\square$

The problem of convergence in distribution of suitably normalized maxima is solved in general by the following theorem. Let $\tau_F = \sup\{t: F(t) < 1\}$ be the right endpoint of F (possibly ∞).

21.17 *Theorem.* *There exist constants a_n and b_n such that the sequence $b_n^{-1}(X_{n(n)} - a_n)$ converges in distribution if and only if, as $t \to \tau_F$,*

(i) *There exists a strictly positive function g on $\mathbb{R}$ such that $\overline{F}(t + g(t)x)/\overline{F}(t) \to e^{-x}$, for every $x \in \mathbb{R}$;*

(ii) *$\tau_F = \infty$ and $\overline{F}(tx)/\overline{F}(t) \to 1/x^\alpha$, for every $x > 0$;*

(iii) *$\tau_F < \infty$ and $\overline{F}(\tau_F - (\tau_F - t)x)/\overline{F}(t) \to x^\alpha$, for every $x > 0$.*

The constants (a_n, b_n) can be taken equal to $(u_n, g(u_n))$, $(0, u_n)$ and $(\tau_F, \tau_F - u_n)$, respectively, for $u_n = F^{-1}(1 - 1/n)$.

Proof. We only give the proof for the "only if" part, which follows the same lines as the preceding examples. In every of the three cases, $n\overline{F}(u_n) \to 1$. To see this it suffices to show that the jump $F(u_n) - F(u_n-) = o(1/n)$. In case (i) this follows because, for every $x < 0$, the jump is smaller than $\overline{F}(u_n + g(u_n)x) - \overline{F}(u_n)$, which is of the order $\overline{F}(u_n)(e^{-x} - 1) \leq (1/n)(e^{-x} - 1)$. The right side can be made smaller than $\varepsilon(1/n)$ for any $\varepsilon > 0$, by choosing x close to 0. In case (ii), we can bound the jump at u_n by $\overline{F}(xu_n) - \overline{F}(u_n)$ for every $x < 1$, which is of the order $\overline{F}(u_n)(1/x^\alpha - 1) \leq (1/n)(1/x^\alpha - 1)$. In case (iii) we argue similarly.

We conclude the proof by applying Lemma 21.12. For instance, in case (i) we have $n\overline{F}(u_n + g(u_n)x) \sim n\overline{F}(u_n)e^{-x} \to e^{-x}$ for every x, and the result follows. The argument under the assumptions (ii) or (iii) is similar. ∎

If the maximum converges in distribution, then the $(k + 1)$-th largest-order statistics $X_{n(n-k)}$ converge in distribution as well, with the same centering and scaling, but a different limit distribution. This follows by combining the preceding results and the Poisson approximation to the binomial distribution.

21.18 *Theorem.* *If $b_n^{-1}(X_{n(n)} - a_n) \rightsquigarrow G$, then $b_n^{-1}(X_{n(n-k)} - a_n) \rightsquigarrow H$ for the distribution function $H(x) = G(x)\sum_{i=0}^{k}\left(-\log G(x)\right)^i/i!.$*

Proof. If $p_n = \overline{F}(a_n + b_n x)$, then $np_n \to -\log G(x)$ for every x where G is continuous (all x), by Lemma 21.12. Furthermore,

$$P\big(b_n^{-1}(X_{n(n-k)} - a_n) \le x\big) = P\big(\text{bin}(n, p_n) \le k\big).$$

This converges to the probability that a Poisson variable with mean $-\log G(x)$ is less than or equal to k. (See problem 2.21.) ∎

By the same, but more complicated, arguments, the sample extremes can be seen to converge jointly in distribution also, but we omit a discussion.

Any order statistic depends, by its definition, on all observations. However, asymptotically central and extreme order statistics depend on the observations in orthogonal ways and become stochastically independent. One way to prove this is to note that central-order statistics are asymptotically equivalent to means, and averages and extreme order statistics are asymptotically independent, which is a result of interest on its own.

21.19 Lemma. *Let g be a measurable function with $Fg = 0$ and $Fg^2 = 1$, and suppose that $b_n^{-1}(X_{n(n)} - a_n) \rightsquigarrow G$ for a nondegenerate distribution G. Then $\big(n^{-1/2}\sum_{i=1}^n g(X_i),$ $b_n^{-1}(X_{n(n)} - a_n)\big) \rightsquigarrow (U, V)$ for independent random variables U and V with distributions $N(0, 1)$ and G.*

Proof. Let $U_n = n^{-1/2} \sum_{i=1}^{n-1} g(X_{n(i)})$ and $V_n = b_n^{-1}(X_{n(n)} - a_n)$. Because $Fg^2 < \infty$, it follows that $\max_{1 \le i \le n} |g(X_i)| = o_P(\sqrt{n})$. Hence $n^{-1/2}|g(X_{n(n)})| \xrightarrow{P} 0$, whence the distance between $(\mathbb{G}_n g, V_n)$ and (U_n, V_n) converges to zero in probability. It suffices to show that $(U_n, V_n) \rightsquigarrow (U, V)$. Suppose that we can show that, for every u,

$$F_n(u \mid V_n) := P(U_n \le u \mid V_n) \xrightarrow{P} \Phi(u).$$

Then, by the dominated-convergence theorem, $EF_n(u \mid V_n)1\{V_n \le v\} = \Phi(u)E1\{V_n \le v\} + o(1)$, and hence the cumulative distribution function $EF_n(u \mid V_n)1\{V_n \le v\}$ of (U_n, V_n) converges to $\Phi(u)G(v)$.

The conditional distribution of U_n given that $V_n = v_n$ is the same as the distribution of $n^{-1/2} \sum X_{ni}$ for i.i.d. random variables $X_{n,1}, \ldots, X_{n,n-1}$ distributed according to the conditional distribution of $g(X_1)$ given that $X_1 \le x_n := a_n + b_n v_n$. These variables have absolute mean

$$|EX_{n1}| = \frac{\left|\int_{(-\infty, x_n]} g \, dF\right|}{F(x_n)} = \frac{\left|\int_{(x_n, \infty)} g \, dF\right|}{F(x_n)} \le \frac{\left(\int_{(x_n, \infty)} g^2 \, dF \, \overline{F}(x_n)\right)^{1/2}}{F(x_n)}.$$

If $v_n \to v$, then $P(V_n \le v_n) \to G(v)$ by the continuity of G, and, by Lemma 21.12, $n\overline{F}(x_n) = O(1)$ whenever $G(v) > 0$. We conclude that $\sqrt{n}EX_{n1} \to 0$. Because we also have that $EX_{n1}^2 \to Fg^2$ and $EX_{n1}^2 1\{|X_{n1}| \ge \varepsilon\sqrt{n}\} \to 0$ for every $\varepsilon > 0$, the Lindeberg-Feller theorem yields that $F_n(u \mid v_n) \to \Phi(u)$. This implies $F_n(u \mid V_n) \rightsquigarrow \Phi(u)$ by Theorem 18.11 or a direct argument. ∎

By taking linear combinations, we readily see from the preceding lemma that the empirical process $\mathbb{G}_n$ and $b_n^{-1}(X_{n(n)} - a_n)$, if they converge, are asymptotically independent as well. This independence carries over onto statistics whose asymptotic distribution can be derived from the empirical process by the delta method, including central order statistics $X_{n(k_n/n)}$ with $k_n/n = p + O(1/\sqrt{n})$, because these are asymptotically equivalent to averages.

Notes

For more results concerning the empirical quantile function, the books [28] and [134] are good starting points. For results on extreme order statistics, see [66] or the book [90].

PROBLEMS

1. Suppose that $F_n \to F$ uniformly. Does this imply that $F_n^{-1} \to F^{-1}$ uniformly or pointwise? Give a counterexample.

2. Show that the asymptotic lengths of the two types of asymptotic confidence intervals for a quantile, discussed in Example 21.8, are within $o_P(1/\sqrt{n})$. Assume that the asymptotic variance of the sample quantile (involving $1/f \circ F^{-1}(p)$) can be estimated consistently.

3. Find the limit distribution of the median absolute deviation from the mean, $\text{med}_{1 \le i \le n} |X_i - \overline{X}_n|$.

4. Find the limit distribution of the pth quantile of the absolute deviation from the median.

5. Prove that $\overline{X}_n$ and $X_{n(n-1)}$ are asymptotically independent.

22

L-Statistics

In this chapter we prove the asymptotic normality of linear combinations of order statistics, particularly those used for robust estimation or testing, such as trimmed means. We present two methods: The projection method presumes knowledge of Chapter 11 only; the second method is based on the functional delta method of Chapter 20.

22.1 Introduction

Let $X_{n(1)}, \ldots, X_{n(n)}$ be the order statistics of a sample of real-valued random variables. A linear combination of (transformed) order statistics, or *L-statistic*, is a statistic of the form

$$\sum_{i=1}^{n} c_{ni} a(X_{n(i)}).$$

The coefficients c_{ni} are a triangular array of constants and a is some fixed function. This "score function" can without much loss of generality be taken equal to the identity function, for an L-statistic with monotone function a can be viewed as a linear combination of the order statistics of the variables $a(X_1), \ldots, a(X_n)$, and an L-statistic with a function a of bounded variation can be dealt with similarly, by splitting the L-statistic into two parts.

22.1 *Example (Trimmed and Winsorized means).* The simplest example of an L-statistic is the sample mean. More interesting are the α-*trimmed means*[†]

$$\frac{1}{n - 2\lfloor \alpha n \rfloor} \sum_{i=\lfloor \alpha n \rfloor + 1}^{n - \lfloor \alpha n \rfloor} X_{n(i)},$$

and the α-*Winsorized means*

$$\frac{1}{n} \left[\lfloor \alpha n \rfloor X_{n(\lfloor \alpha n \rfloor)} + \sum_{i=\lfloor \alpha n \rfloor + 1}^{n - \lfloor \alpha n \rfloor} X_{n(i)} + \lfloor \alpha n \rfloor X_{n(n - \lfloor \alpha n \rfloor + 1)} \right].$$

[†] The notation $\lfloor x \rfloor$ is used for the greatest integer that is less than or equal to x. Also $\lceil x \rceil$ denotes the smallest integer greater than or equal to x. For a natural number n and a real number $0 \le x \le n$ one has $\lfloor n - x \rfloor = n - \lceil x \rceil$ and $\lceil n - x \rceil = n - \lfloor x \rfloor$.

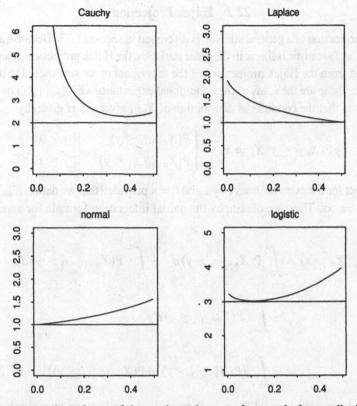

Figure 22.1. Asymptotic variance of the α-trimmed mean of a sample from a distribution F as function of α for four distributions F.

The α-trimmed mean is the average of the middle $(1 - 2\alpha)$-th fraction of the observations, the α-Winsorized mean replaces the αth fractions of smallest and largest data by $X_{n(\lfloor \alpha n \rfloor)}$ and $X_{n(n-\lfloor \alpha n \rfloor+1)}$, respectively, and next takes the average. Both estimators were already used in the early days of statistics as location estimators in situations in which the data were suspected to contain outliers. Their properties were studied systematically in the context of robust estimation in the 1960s and 1970s. The estimators were shown to have good properties in situations in which the data follows a heavier tailed distribution than the normal one. Figure 22.1 shows the asymptotic variances of the trimmed means as a function of α for four distributions. (A formula for the asymptotic variance is given in Example 22.11.) The four graphs suggest that 10% to 15% trimming may give an improvement over the sample mean in some cases and does not cost much even for the normal distribution. □

22.2 Example (Ranges). Two estimators of dispersion are the *interquartile range* $X_{n(\lceil 3n/4 \rceil)} - X_{n(\lceil n/4 \rceil)}$ and the *range* $X_{n(n)} - X_{n(1)}$. Of these, the range does not have a normal limit distribution and is not within the scope of the results of this chapter. □

We present two methods to prove the asymptotic normality of L-statistics. The first method is based on the Hájek projection; the second uses the delta method. The second method is preferable in that it applies to more general statistics, but it necessitates the study of empirical processes and does not cover the simplest L-statistic: the sample mean.

22.2 Hájek Projection

The Hájek projection of a general statistic is discussed in section 11.3. Because a projection is linear and an *L*-statistic is linear in the order statistics, the Hájek projection of an *L*-statistic can be found from the Hájek projections of the individual order statistics. Up to centering at mean zero, these are the sums of the conditional expectations $E(X_{n(i)} \mid X_k)$ over k. Some thought shows that the conditional distribution of $X_{n(i)}$ given X_k is given by

$$P(X_{n(i)} \leq y \mid X_k = x) = \begin{cases} P(X_{n-1(i)} \leq y) & \text{if } y < x, \\ P(X_{n-1(i-1)} \leq y) & \text{if } y \geq x. \end{cases}$$

This is correct for the extreme cases $i = 1$ and $i = n$ provided that we define $X_{n-1(0)} = -\infty$ and $X_{n-1(n)} = \infty$. Thus, we obtain, by the partial integration formula for an expectation, for $x \geq 0$,

$$E(X_{n(i)} \mid X_k = x) = \int_0^x P(X_{n-1(i)} > y) \, dy + \int_x^\infty P(X_{n-1(i-1)} > y) \, dy$$

$$- \int_{-\infty}^0 P(X_{n-1(i)} \leq y) \, dy$$

$$= - \int_x^\infty \big(P(X_{n-1(i)} > y) - P(X_{n-1(i-1)} > y) \big) \, dy + E X_{n-1(i)}.$$

The second expression is valid for $x < 0$ as well, as can be seen by a similar argument. Because $X_{n-1(i-1)} \leq X_{n-1(i)}$, the difference between the two probabilities in the last integral is equal to the probability of the event $\{X_{n-1(i-1)} \leq y < X_{n-1(i)}\}$. This is precisely the probability that a binomial $(n-1, F(y))$-variable is equal to $i - 1$. If we write this probability as $B_{n-1,F(y)}(i-1)$, then the Hájek projection $\hat{X}_{n(i)}$ of $X_{n(i)}$ satisfies, with $\mathbb{F}_n$ the empirical distribution function of $X_1, \ldots, X_n$,

$$\hat{X}_{n(i)} - E\hat{X}_{n(i)} = - \sum_{k=1}^n \int_{X_k}^\infty B_{n-1,F(y)}(i-1) \, dy + C_n$$

$$= - \int n(\mathbb{F}_n - F)(y) \, B_{n-1,F(y)}(i-1) \, dy.$$

For the projection of the *L*-statistic $T_n = \sum_{i=1}^n c_{ni} X_{n(i)}$ we find

$$\hat{T}_n - E\hat{T}_n = - \int n(\mathbb{F}_n - F)(y) \sum_{i=1}^n c_{ni} B_{n-1,F(y)}(i-1) \, dy.$$

Under some conditions on the coefficients c_{ni}, this sum (divided by $\sqrt{n}$) is asymptotically normal by the central limit theorem. Furthermore, the projection $\hat{T}_n$ can be shown to be asymptotically equivalent to the *L*-statistic T_n by Theorem 11.2. Sufficient conditions on the c_{ni} can take a simple appearance for coefficients that are "generated" by a function ϕ as in (13.4).

22.3 Theorem. *Suppose that* $EX_1^2 < \infty$ *and that* $c_{ni} = \phi(i/(n+1))$ *for a bounded function ϕ that is continuous at $F(y)$ for Lebesgue almost-every y. Then the sequence $n^{-1/2}(T_n - ET_n)$*

converges in distribution to a normal distribution with mean zero and variance

$$\sigma^2(\phi, F) = \iint \phi(F(x))\phi(F(y))(F(x \wedge y) - F(x)F(y)) \, dx \, dy.$$

Proof. Define functions $e(y) = \phi(F(y))$ and

$$e_n(y) = \sum_{i=1}^{n} c_{ni} B_{n-1, F(y)}(i - 1) = E\phi\left(\frac{B_n + 1}{n + 1}\right),$$

for B_n binomially distributed with parameters $(n - 1, F(y))$. By the law of large numbers $(B_n + 1)/(n + 1) \xrightarrow{P} F(y)$. Because ϕ is bounded, $e_n(y) \to e(y)$ for every y such that ϕ is continuous at $F(y)$, by the dominated-convergence theorem. By assumption, this includes almost every y.

By Theorem 11.2, the sequence $n^{-1/2}(T_n - \hat{T}_n)$ converges in second mean to zero if the variances of $n^{-1/2}T_n$ and $n^{-1/2}\hat{T}_n$ converge to the same number. Because $n^{-1/2}(\hat{T}_n - E\hat{T}_n) = -\int \mathbb{G}_n(y) e_n(y) \, dy$, the second variance is easily computed to be

$$\frac{1}{n}\text{var}\,\hat{T}_n = \iint (F(x \wedge y) - F(x)F(y)) e_n(x)e_n(y) \, dx \, dy.$$

This converges to $\sigma^2(\phi, F)$ by the dominated-convergence theorem. The variance of $n^{-1/2}T_n$ can be written in the form

$$\frac{1}{n}\text{var}\,T_n = \frac{1}{n}\sum_{i=1}^{n}\sum_{j=1}^{n} c_{ni} c_{nj} \text{cov}(X_{n(i)}, X_{n(j)}) = \iint R_n(x, y) \, dx \, dy,$$

where, because $\text{cov}(X, Y) = \iint \text{cov}(\{X \leq x\}, \{Y \leq y\}) \, dx \, dy$ for any pair of variables (X, Y),

$$R_n(x, y) = \frac{1}{n}\sum_{i=1}^{n}\sum_{j=1}^{n} \phi\left(\frac{i}{n+1}\right)\phi\left(\frac{j}{n+1}\right) \text{cov}(\{X_{n(i)} \leq x\}, \{X_{n(j)} \leq y\}).$$

Because the order statistics are positively correlated, all covariances in the double sum are nonnegative. Furthermore,

$$\frac{1}{n}\sum_{i=1}^{n}\sum_{j=1}^{n} \text{cov}(\{X_{n(i)} \leq x\}, \{X_{n(j)} \leq y\}) = \text{cov}(\mathbb{G}_n(x), \mathbb{G}_n(y))$$
$$= (F(x \wedge y) - F(x)F(y)).$$

For pairs (i, j) such that $i \approx nF(x)$ and $j \approx nF(y)$, the coefficient of the covariance is approximately $e(x)e(y)$ by the continuity of ϕ. The covariances corresponding to other pairs (i, j) are negligible. Indeed, for $i \geq nF(x) + n\varepsilon_n$,

$$0 \leq \text{cov}(\{X_{n(i)} \leq x\}, \{X_{n(j)} \leq y\}) \leq 2P(X_{n(i)} \leq x)$$
$$\leq 2P(\text{bin}(n, F(x)) \geq nF(x) + n\varepsilon_n)$$
$$\leq 2\exp{-2n\varepsilon_n^2},$$

by Hoeffding's inequality.[†] Thus, because ϕ is bounded, the terms with $i \geq nF(x) + n\varepsilon_n$ contribute exponentially little as $\varepsilon_n \to 0$ not too fast (e.g., $\varepsilon_n^2 = n^{-1/2}$). A similar argument applies to the terms with $i \leq nF(x) - n\varepsilon_n$ or $|j - nF(y)| \geq n\varepsilon_n$. Conclude that, for every (x, y) such that ϕ is continuous at both $F(x)$ and $F(y)$,

$$R_n(x, y) \to e(x)e(y)\big(F(x \wedge y) - F(x)F(y)\big).$$

Finally, we apply the dominated convergence theorem to see that the double integral of this expression, which is equal to $n^{-1}\operatorname{var} T_n$, converges to $\sigma^2(\phi, F)$.

This concludes the proof that T_n and $\hat{T}_n$ are asymptotically equivalent. To show that the sequence $n^{-1/2}(\hat{T}_n - E\hat{T}_n)$ is asymptotically normal, define $S_n = -\int \mathbb{G}_n(y)\, e(y)\, dy$. Then, by the same arguments as before, $n^{-1}\operatorname{var}(S_n - \hat{T}_n) \to 0$. Furthermore, the sequence $n^{-1/2}S_n$ is asymptotically normal by the central limit theorem. ∎

22.3   Delta Method

The order statistics of a sample $X_1, \ldots, X_n$ can be expressed in their empirical distribution $\mathbb{F}_n$, or rather the empirical quantile function, through

$$\mathbb{F}_n^{-1}(s) = X_{n(\lceil sn \rceil)} = X_{n(i)}, \qquad \text{for } \frac{i-1}{n} < s \leq \frac{i}{n}.$$

Consequently, an L-statistic can be expressed in the empirical distribution function as well. Given a fixed function a and a fixed signed measure K on $(0, 1)$[‡], consider the function

$$\phi(F) = \int_0^1 a(F^{-1})\, dK.$$

View ϕ as a map from the set of distribution functions into $\mathbb{R}$. Clearly,

$$\phi(\mathbb{F}_n) = \sum_{i=1}^n K\left(\frac{i-1}{n}, \frac{i}{n}\right] a(X_{n(i)}). \tag{22.4}$$

The right side is an L-statistic with coefficients $c_{ni} = K\big((i-1)/n, i/n\big]$. Not all possible arrays of coefficients c_{ni} can be "generated" through a measure K in this manner. However, most L-statistics of interest are almost of the form (22.4), so that not much generality is lost by assuming this structure. An advantage is simplicity in the formulation of the asymptotic properties of the statistics, which can be derived with the help of the von Mises method. More importantly, the function $\phi(F)$ can also be applied to other estimators besides $\mathbb{F}_n$. The results of this section yield their asymptotic normality in general.

22.5   *Example.* The α-trimmed mean corresponds to the uniform distribution K on the interval $(\alpha, 1 - \alpha)$ and a the identity function. More precisely, the L-statistic generated by

---

[†] See for example, the appendix of [117]. This inequality gives more than needed. For instance, it also works to apply Markov's inequality for fourth moments.
[‡] A signed measure is a difference $K = K_1 - K_2$ of two finite measures K_1 and K_2.

this measure is

$$\frac{1}{1-2\alpha} \int_\alpha^{1-\alpha} \mathbb{F}_n^{-1}(s)\, ds = \frac{1}{n-2\alpha n} \left[(\lceil \alpha n \rceil - \alpha n) X_{n(\lceil \alpha n \rceil)} \right.$$
$$\left. + \sum_{i=\lceil \alpha n \rceil + 1}^{n - \lceil \alpha n \rceil} X_{n(i)} + (\lceil \alpha n \rceil - \alpha n) X_{n(n - \lceil \alpha n \rceil + 1)} \right].$$

Except for the slightly different weight factor and the treatment of the two extremes in the averages, this agrees with the α-trimmed mean as introduced before. Because $X_{n(k_n)}$ converges in probability to $F^{-1}(p)$ if $k_n/n \to p$ and $(n - 2\lfloor \alpha n \rfloor)/(n - 2\alpha n) = 1 + O(1/n)$, the difference between the two versions of the trimmed mean can be seen to be $O_P(1/n)$. For the purpose of this chapter this is negligible.

The α-Winsorized mean corresponds to the measure K that is the sum of Lebesgue measure on $(\alpha, 1 - \alpha)$ and the discrete measure with pointmasses of size α at each of the points α and $1 - \alpha$. Again, the difference between the estimator generated by this K and the Winsorized mean is negligible.

The interquartile range corresponds to the discrete, signed measure K that has pointmasses of sizes 1 and -1 at the points $1/4$ and $3/4$, respectively. $\square$

Before giving a proof of asymptotic normality, we derive the influence function of an (empirical) L-statistic in an informal way. If $F_t = (1 - t)F + t\delta_x$, then, by definition, the influence function is the derivative of the map $t \mapsto \phi(F_t)$ at $t = 0$. Provided a and K are sufficiently regular,

$$\frac{d}{dt} \int_0^1 a(F_t^{-1})\, dK = \int_0^1 a'(F_t^{-1}) \left[\frac{d}{dt} F_t^{-1} \right] dK.$$

Here the expression within square brackets if evaluated at $t = 0$ is the influence function of the quantile function and is derived in Example 20.5. Substituting the representation given there, we see that the influence function of the L-function $\phi(F) = \int a(F^{-1})\, dK$ takes the form

$$\phi_F'(\delta_x - F) = -\int_0^1 a'(F^{-1}(u)) \frac{1_{[x,\infty)}(F^{-1}(u)) - u}{f(F^{-1}(u))}\, dK(u) \tag{22.6}$$
$$= -\int a'(y) \frac{1_{[x,\infty)}(y) - F(y)}{f(y)}\, dK \circ F(y).$$

The second equality follows by (a generalization of) the quantile transformation.

An alternative derivation of the influence function starts with rewriting $\phi(F)$ in the form

$$\phi(F) = \int a\, dK \circ F = \int_{(0,\infty)} \overline{(K \circ F)_-}\, da - \int_{(-\infty,0]} (K \circ F)_-\, da. \tag{22.7}$$

Here $\overline{K \circ F}(x) = K \circ F(\infty) - K \circ F(x)$ and the partial integration can be justified for a a function of bounded variation with $a(0) = 0$ (see problem 22.6; the assumption that $a(0) = 0$ simplifies the formula, and is made for convenience). This formula for $\phi(F)$ suggests as influence function

$$\phi_F'(\delta_x - F) = -\int K'(F(y)) \left(1_{[x,\infty)}(y) - F(y) \right) da(y). \tag{22.8}$$

Under appropriate conditions each of the two formulas (22.6) and (22.8) for the influence function is valid. However, already for the defining expressions to make sense very different conditions are needed. Informally, for equation (22.6) it is necessary that a and F be differentiable with a positive derivative for F, (22.8) requires that K be differentiable. For this reason both expressions are valuable, and they yield nonoverlapping results.

Corresponding to the two derivations of the influence function, there are two basic approaches towards proving asymptotic normality of L-statistics by the delta method, valid under different sets of conditions. Roughly, one approach requires that F and a be smooth, and the other that K be smooth.

The simplest method is to view the L-statistic as a function of the empirical quantile function, through the map $\mathbb{F}_n^{-1} \mapsto \int a \circ \mathbb{F}_n^{-1} dK$, and next apply the functional delta method to the map $Q \mapsto \int a \circ Q \, dK$. The asymptotic normality of the empirical quantile function is obtained in Chapter 21.

22.9 Lemma. *Let $a : \mathbb{R} \mapsto \mathbb{R}$ be continuously differentiable with a bounded derivative. Let K be a signed measure on the interval $(\alpha, \beta) \subset (0, 1)$. Then the map $Q \mapsto \int a(Q) \, dK$ from $\ell^\infty(\alpha, \beta)$ to $\mathbb{R}$ is Hadamard-differentiable at every Q. The derivative is the map $H \mapsto \int a'(Q) H \, dK$.*

Proof. Let $H_t \to H$ in the uniform norm. Consider the difference

$$\int \left| \frac{a(Q + tH_t) - a(Q)}{t} - a'(Q) H \right| dK.$$

The integrand converges to zero everywhere and is bounded uniformly by $\|a'\|_\infty \left(\|H_t\|_\infty + \|H\|_\infty \right)$. Thus the integral converges to zero by the dominated-convergence theorem. ∎

If the underlying distribution has unbounded support, then its quantile function is unbounded on the domain $(0, 1)$, and no estimator can converge in $\ell^\infty(0, 1)$. Then the preceding lemma can apply only to generating measures K with support (α, β) strictly within $(0, 1)$. Fortunately, such generating measures are the most interesting ones, as they yield bounded influence functions and hence robust L-statistics.

A more serious limitation of using the preceding lemma is that it could require unnecessary smoothness conditions on the distribution of the observations. For instance, the empirical quantile process converges in distribution in $\ell^\infty(\alpha, \beta)$ only if the underlying distribution has a positive density between its α- and β-quantiles. This is true for most standard distributions, but unnecessary for the asymptotic normality of empirical L-statistics generated by smooth measures K. Thus we present a second lemma that applies to smooth measures K and does not require that F be smooth. Let $DF[-\infty, \infty]$ be the set of all distribution functions.

22.10 Lemma. *Let $a : \mathbb{R} \mapsto \mathbb{R}$ be of bounded variation on bounded intervals with $\int (a^+ + a^-) \, d|K \circ F| < \infty$ and $a(0) = 0$. Let K be a signed measure on $(0, 1)$ whose distribution function K is differentiable at $F(x)$ for a almost-every x and satisfies $\left| K(u + h) - K(u) \right| \leq M(u)h$ for every sufficiently small $|h|$, and some function M such that $\int M(F_-) \, d|a| < \infty$. Then the map $F \mapsto \int a \circ F^{-1} dK$ from $DF[-\infty, \infty] \subset D[-\infty, \infty]$ to $\mathbb{R}$ is Hadamard-differentiable at F, with derivative $H \mapsto -\int (K' \circ F_-) H \, da$.*

Proof. First rewrite the function in the form (22.7). Suppose that $H_t \to H$ uniformly and set $F_t = F + t H_t$. By continuity of K, $(K \circ F)_- = K(F_-)$. Because $K \circ F(\infty) = K(1)$ for all F, the difference $\phi(F_t) - \phi(F)$ can be rewritten as $- \int (K \circ F_{t-} - K \circ F_-) \, da$. Consider the integral

$$\int \left| \frac{K(F_- + t H_{t-}) - K(F_-)}{t} - K'(F_-) H \right| d|a|.$$

The integrand converges a-almost everywhere to zero and is bounded by $M(F_-)\big(\|H_t\|_\infty + \|H\|_\infty \big) \le M(F_-)\big(2 \|H\|_\infty + 1\big)$, for small t. Thus, the lemma follows by the dominated-convergence theorem. ∎

Because the two lemmas apply to nonoverlapping situations, it is worthwhile to combine the two approaches. A given generating measure K can be split in its discrete and continuous part. The corresponding two parts of the L-statistic can next be shown to be asymptotically linear by application of the two lemmas. Their sum is asymptotically linear as well and hence asymptotically normal.

22.11 **Example (Trimmed mean).** The cumulative distribution function K of the uniform distribution on $(\alpha, 1 - \alpha)$ is uniformly Lipschitz and fails to be differentiable only at the points α and $1 - \alpha$. Thus, the trimmed-mean function is Hadamard-differentiable at every F such that the set $\{x : F(x) = \alpha, \text{ or } 1 - \alpha\}$ has Lebesgue measure zero. (We assume that $\alpha > 0$.) In other words, F should not have flats at height α or $1 - \alpha$. For such F the trimmed mean is asymptotically normal with asymptotic influence function $- \int_\alpha^{1-\alpha} \big(1_{x \le y} - F(y) \big) \, dy$ (see (22.8)), and asymptotic variance

$$\int_{F^{-1}(\alpha)}^{F^{-1}(1-\alpha)} \int_{F^{-1}(\alpha)}^{F^{-1}(1-\alpha)} \big(F(x \wedge y) - F(x) \, F(y) \big) \, dx \, dy.$$

Figure 22.1 shows this number as a function of α for a number of distributions. □

22.12 **Example (Winsorized mean).** The generating measure of the Winsorized mean is the sum of a discrete measure on the two points α and $1 - \alpha$, and Lebesgue measure on the interval $(\alpha, 1 - \alpha)$. The Winsorized mean itself can be decomposed correspondingly. Suppose that the underlying distribution function F has a positive derivative at the points $F^{-1}(\alpha)$ and $F^{-1}(1 - \alpha)$. Then the first part of the decomposition is asymptotically linear in view of Lemma 22.9 and Lemma 21.3, the second part is asymptotically linear by Lemma 22.10 and Theorem 19.3. Combined, this yields the asymptotic linearity of the Winsorized mean and hence its asymptotic normality. □

22.4 L-Estimators for Location

The α-trimmed mean and the α-Winsorized mean were invented as estimators for location. The question in this section is whether there are still other attractive location estimators within the class of L-statistics.

One possible method of generating L-estimators for location is to find the best L-estimators for given location families $\{ f(x - \theta) : \theta \in \mathbb{R} \}$, in which f is some fixed density. For instance, for the f equal to the normal shape this leads to the sample mean.

According to Chapter 8, an estimator sequence T_n is asymptotically optimal for estimating the location of a density with finite Fisher information I_f if

$$\sqrt{n}(T_n - \theta) = -\frac{1}{\sqrt{n}} \sum_{i=1}^{n} \frac{1}{I_f} \frac{f'}{f}(X_i - \theta) + o_P(1).$$

Comparison with equation (22.8) for the influence function of an L-statistic shows that the choices of generating measure K and transformation a such that

$$K'\big(F(x - \theta)\big)\, a'(x) = -\left(\frac{1}{I_f} \frac{f'}{f}(x - \theta) \right)'$$

lead to an L-statistic with the optimal asymptotic influence function. This can be accommodated by setting $a(x) \equiv x$ and

$$K'(u) = -\left(\frac{1}{I_f} \frac{f'}{f} \right)' \big(F^{-1}(u)\big).$$

The class of L-statistics is apparently large enough to contain an asymptotically efficient estimator sequence for the location parameter of any smooth shape. The L-statistics are not as simplistic as they may seem at first.

Notes

This chapter gives only a few of the many results available on L-statistics. For instance, the results on Hadamard differentiability can be refined by using a weighted uniform norm combined with convergence of the weighted empirical process. This allows greater weights for the extreme-order statistics. For further results and references, see [74], [134], and [136].

PROBLEMS

1. Find a formula for the asymptotic variance of the Winsorized mean.
2. Let $T(F) = \int F^{-1}(u)\, k(u)\, du$.
   (i) Show that $T(F) = 0$ for every distribution F that is symmetric about zero if and only if k is symmetric about $1/2$.
   (ii) Show that $T(F)$ is location equivariant if and only if $\int k(u)\, du = 1$.
   (iii) Show that "efficient" L-statistics obtained from symmetric densities possess both properties (i) and (ii).
3. Let $X_1, \ldots, X_n$ be a random sample from a continuous distribution function. Show that conditionally on $(X_{n(k)}, X_{n(l)}) = (x, y)$, the variables $X_{n(k+1)}, \ldots, X_{n(l-1)}$ are distributed as the order statistics of a random sample of size $l - k - 1$ from the conditional distribution of X_1 given that $x \leq X_1 \leq y$. How can you use this to study the properties of trimmed means?
4. Find an optimal L-statistic for estimating the location in the logistic and Laplace location families.
5. Does there exist a distribution for which the trimmed mean is asymptotically optimal for estimating location?

6. **(Partial Integration.)** If $a : \mathbb{R} \mapsto \mathbb{R}$ is right-continuous and nondecreasing with $a(0) = 0$, and $b : \mathbb{R} \mapsto \mathbb{R}$ is right-continuous, nondecreasing and bounded, then

$$\int a \, db = \int_{(0,\infty)} \big(b(\infty) - b_- \big) \, da + \int_{(-\infty,0]} \big(b(-\infty) - b_- \big) \, da.$$

Prove this. If a is also bounded, then the righthand side can be written more succinctly as $ab\big|_{-\infty}^{\infty} - \int b_- \, da$. (Substitute $a(x) = \int_{(0,x]} da$ for $x > 0$ and $a(x) = -\int_{(x,0]} da$ for $x \leq 0$ into the left side of the equation, and use Fubini's theorem separately on the integral over the positive and negative part of the real line.)

23

Bootstrap

This chapter investigates the asymptotic properties of bootstrap estimators for distributions and confidence intervals. The consistency of the bootstrap for the sample mean implies the consistency for many other statistics by the delta method. A similar result is valid with the empirical process.

23.1 Introduction

In most estimation problems it is important to give an indication of the precision of a given estimate. A simple method is to provide an estimate of the bias and variance of the estimator; more accurate is a confidence interval for the parameter. In this chapter we concentrate on bootstrap confidence intervals and, more generally, discuss the bootstrap as a method of estimating the distribution of a given statistic.

Let $\hat{\theta}$ be an estimator of some parameter θ attached to the distribution P of the observations. The distribution of the difference $\hat{\theta} - \theta$ contains all the information needed for assessing the precision of $\hat{\theta}$. In particular, if ξ_α is the upper α-quantile of the distribution of $(\hat{\theta} - \theta)/\hat{\sigma}$, then

$$P(\hat{\theta} - \xi_\beta \hat{\sigma} \leq \theta \leq \hat{\theta} - \xi_{1-\alpha} \hat{\sigma} \mid P) \geq 1 - \beta - \alpha.$$

Here $\hat{\sigma}$ may be arbitrary, but it is typically an estimate of the standard deviation of $\hat{\theta}$. It follows that the interval $[\hat{\theta} - \xi_\beta \hat{\sigma}, \hat{\theta} - \xi_{1-\alpha} \hat{\sigma}]$ is a confidence interval of level $1 - \beta - \alpha$. Unfortunately, in most situations the quantiles and the distribution of $\hat{\theta} - \theta$ depend on the unknown distribution P of the observations and cannot be used to assess the performance of $\hat{\theta}$. They must be replaced by estimators.

If the sequence $(\hat{\theta} - \theta)/\hat{\sigma}$ tends in distribution to a standard normal variable, then the normal $N(0, \hat{\sigma}^2)$-distribution can be used as an estimator of the distribution of $\hat{\theta} - \theta$, and we can substitute the standard normal quantiles z_α for the quantiles ξ_α. The weak convergence implies that the interval $[\hat{\theta} - z_\beta \hat{\sigma}, \hat{\theta} - z_{1-\alpha} \hat{\sigma}]$ is a confidence interval of asymptotic level $1 - \alpha - \beta$.

Bootstrap procedures yield an alternative. They are based on an estimate $\hat{P}$ of the underlying distribution P of the observations. The distribution of $(\hat{\theta} - \theta)/\hat{\sigma}$ under P can, in principle, be written as a function of P. The bootstrap estimator for this distribution is the "plug-in" estimator obtained by substituting $\hat{P}$ for P in this function. Bootstrap estimators for quantiles, and next confidence intervals, are obtained from the bootstrap estimator for the distribution.

The following type of notation is customary. Let $\hat{\theta}^*$ and $\hat{\sigma}^*$ be computed from (hypothetic) observations obtained according to $\hat{P}$ in the same way $\hat{\theta}$ and $\hat{\sigma}$ are computed from the true observations with distribution P. If $\hat{\theta}$ is related to $\hat{P}$ in the same way θ is related to P, then the bootstrap estimator for the distribution of $(\hat{\theta} - \theta)/\hat{\sigma}$ under P is the distribution of $(\hat{\theta}^* - \hat{\theta})/\hat{\sigma}^*$ under $\hat{P}$. The latter is evaluated given the original observations, that is, for a fixed realization of $\hat{P}$.

A bootstrap estimator for a quantile ξ_α of $(\hat{\theta} - \theta)/\hat{\sigma}$ is a quantile of the distribution of $(\hat{\theta}^* - \hat{\theta})/\hat{\sigma}^*$ under $\hat{P}$. This is the smallest value $x = \hat{\xi}_\alpha$ that satisfies the inequality

$$\mathrm{P}\left(\frac{\hat{\theta}^* - \hat{\theta}}{\hat{\sigma}^*} \leq x \,|\, \hat{P}\right) \geq 1 - \alpha. \tag{23.1}$$

The notation $\mathrm{P}(\cdot \,|\, \hat{P})$ indicates that the distribution of $(\hat{\theta}^*, \hat{\sigma}^*)$ must be evaluated assuming that the observations are sampled according to $\hat{P}$ *given* the original observations. In particular, in the preceding display $\hat{\theta}$ is to be considered nonrandom. The left side of the preceding display is a function of the original observations, whence the same is true for $\hat{\xi}_\alpha$.

If $\hat{P}$ is close to the true underlying distribution P, then the bootstrap quantiles should be close to the true quantiles, whence it should be true that

$$\mathrm{P}\left(\frac{\hat{\theta} - \theta}{\hat{\sigma}} \leq \hat{\xi}_\alpha \,|\, P\right) \approx 1 - \alpha.$$

In this chapter we show that this approximation is valid in an asymptotic sense: The probability on the left converges to $1 - \alpha$ as the number of observations tends to infinity. Thus, the bootstrap confidence interval

$$[\hat{\theta} - \hat{\xi}_\beta \hat{\sigma}, \hat{\theta} - \hat{\xi}_{1-\alpha}\hat{\sigma}] = \left\{\theta : \hat{\xi}_{1-\alpha} \leq \frac{\hat{\theta} - \theta}{\hat{\sigma}} \leq \hat{\xi}_\beta\right\}$$

possesses asymptotic confidence level $1 - \alpha - \beta$.

The statistic $\hat{\sigma}$ is typically chosen equal to an estimator of the (asymptotic) standard deviation of $\hat{\theta}$. The resulting bootstrap method is known as the *percentile t-method*, in view of the fact that it is based on estimating quantiles of the "studentized" statistic $(\hat{\theta} - \theta)/\hat{\sigma}$. (The notion of a t-statistic is used here in an abstract manner to denote a centered statistic divided by a scale estimate; in general, there is no relationship with Student's t-distribution from normal theory.) A simpler method is to choose $\hat{\sigma}$ independent of the data. If we choose $\hat{\sigma} = \hat{\sigma}^* = 1$, then the bootstrap quantiles $\hat{\xi}_\alpha$ are the quantiles of the centered statistic $\hat{\theta}^* - \hat{\theta}$. This is known as the *percentile method*. Both methods yield asymptotically correct confidence levels, although the percentile t-method is generally more accurate.

A third method, *Efron's percentile method*, proposes the confidence interval $[\hat{\zeta}_{1-\beta}, \hat{\zeta}_\alpha]$ for $\hat{\zeta}_\alpha$ equal to the upper α-quantile of $\hat{\theta}^*$: the smallest value $x = \hat{\zeta}_\alpha$ such that

$$\mathrm{P}(\hat{\theta}^* \leq x \,|\, \hat{P}) \geq 1 - \alpha.$$

Thus, $\hat{\zeta}_\alpha$ results from "bootstrapping" $\hat{\theta}$, while $\hat{\xi}_\alpha$ is the product of bootstrapping $(\hat{\theta} - \theta)/\hat{\sigma}$. These quantiles are related, and Efron's percentile interval can be reexpressed in the quantiles $\hat{\xi}_\alpha$ of $\hat{\theta}^* - \hat{\theta}$ (employed by the percentile method with $\hat{\sigma} = 1$) as

$$[\hat{\zeta}_{1-\beta}, \hat{\zeta}_\alpha] = [\hat{\theta} + \hat{\xi}_{1-\beta}, \hat{\theta} + \hat{\xi}_\alpha].$$

The logical justification for this interval is less strong than for the intervals based on boot-strapping $\hat{\theta} - \theta$, but it appears to work well. The two types of intervals coincide in the case that the conditional distribution of $\hat{\theta}^* - \hat{\theta}$ is symmetric about zero. We shall see that the difference is asymptotically negligible if $\hat{\theta}^* - \hat{\theta}$ converges to a normal distribution.

Efron's percentile interval is the only one among the three intervals that is invariant under monotone transformations. For instance, if setting a confidence interval for the correlation coefficient, the sample correlation coefficient might be transformed by Fisher's transformation before carrying out the bootstrap scheme. Next, the confidence interval for the transformed correlation can be transformed back into a confidence interval for the correlation coefficient. This operation would have no effect on Efron's percentile interval but can improve the other intervals considerably, in view of the skewness of the statistic. In this sense Efron's method automatically "finds" useful (stabilizing) transformations. The fact that it does not become better through transformations of course does not imply that it is good, but the invariance appears desirable.

Several of the elements of the bootstrap scheme are still unspecified. The missing probability $\alpha + \beta$ can be distributed over the two tails of the confidence interval in several ways. In many situations equal-tailed confidence intervals, corresponding to the choice $\alpha = \beta$, are reasonable. In general, these do not have $\hat{\theta}$ exactly as the midpoint of the interval. An alternative is the interval

$$\left[\hat{\theta} - \hat{\xi}^s_{\alpha+\beta}\hat{\sigma}, \hat{\theta} + \hat{\xi}^s_{\alpha+\beta}\hat{\sigma}\right],$$

with $\hat{\xi}^s_\alpha$ equal to the upper α-quantile of $|\hat{\theta}^* - \hat{\theta}|/\hat{\sigma}^*$. A further possibility is to choose α and β under the side condition that the difference $\hat{\xi}_\beta - \hat{\xi}_{1-\alpha}$, which is proportional to the length of the confidence interval, is minimal.

More interesting is the choice of the estimator $\hat{P}$ for the underlying distribution. If the original observations are a random sample $X_1, \ldots, X_n$ from a probability distribution P, then one candidate is the empirical distribution $\mathbb{P}_n = n^{-1} \sum \delta_{X_i}$ of the observations, leading to the *empirical bootstrap*. Generating a random sample from the empirical distribution amounts to resampling with replacement from the set $\{X_1, \ldots, X_n\}$ of original observations. The name "bootstrap" derives from this resampling procedure, which might be surprising at first, because the observations are "sampled twice." If we view the bootstrap as a nonparametric plug-in estimator, we see that there is nothing peculiar about resampling.

We shall be mostly concerned with the empirical bootstrap, even though there are many other possibilities. If the observations are thought to follow a specified parametric model, then it is more reasonable to set $\hat{P}$ equal to $P_{\hat{\theta}}$ for a given estimator $\hat{\theta}$. This is what one would have done in the first place, but it is called the *parametric bootstrap* within the present context. That the bootstrapping methodology is far from obvious is clear from the fact that the literature also considers the exchangeable, the Bayesian, the smoothed, and the wild bootstrap, as well as several schemes for bootstrap corrections. Even "resampling" can be carried out differently, for instance, by sampling fewer than n variables, or without replacement.

It is almost never possible to calculate the bootstrap quantiles $\hat{\xi}_\alpha$ numerically. In practice, these estimators are approximated by a simulation procedure. A large number of independent bootstrap samples $X_1^*, \ldots, X_n^*$ are generated according to the estimated distribution $\hat{P}$. Each sample gives rise to a bootstrap value $(\hat{\theta}^* - \hat{\theta})/\hat{\sigma}^*$ of the standardized statistic. Finally, the bootstrap quantiles $\hat{\xi}_\alpha$ are estimated by the empirical quantiles of these bootstrap values. This simulation scheme always produces an additional (random) error in the coverage probability of the resulting confidence interval. In principle, by using a sufficiently large number of bootstrap samples, possibly combined with an efficient method of simulation, this error can be made arbitrarily small. Therefore the additional error is usually ignored in the theory of the bootstrap procedure. This chapter follows this custom and concerns the "exact" distribution and quantiles of $(\hat{\theta}^* - \hat{\theta})/\hat{\sigma}^*$, without taking a simulation error into account.

23.2 Consistency

A confidence interval $[\hat{\theta}_{n,1}, \hat{\theta}_{n,2}]$ is (conservatively) *asymptotically consistent* at level $1 - \alpha - \beta$ if, for every possible P,

$$\liminf_{n \to \infty} P(\hat{\theta}_{n,1} \le \theta \le \hat{\theta}_{n,2} \mid P) \ge 1 - \alpha - \beta.$$

The consistency of a bootstrap confidence interval is closely related to the consistency of the bootstrap estimator of the distribution of $(\hat{\theta}_n - \theta)/\hat{\sigma}_n$. The latter is best defined relative to a metric on the collection of possible laws of the estimator. Call the bootstrap estimator for the distribution *consistent* relative to the Kolmogorov-Smirnov distance if

$$\sup_x \left| P\left(\frac{\hat{\theta}_n - \theta}{\hat{\sigma}_n} \le x \mid P \right) - P\left(\frac{\hat{\theta}_n^* - \hat{\theta}_n}{\hat{\sigma}_n^*} \le x \mid \hat{P}_n \right) \right| \xrightarrow{P} 0.$$

It is not a great loss of generality to assume that the sequence $(\hat{\theta}_n - \theta)/\hat{\sigma}_n$ converges in distribution to a continuous distribution function F (in our examples Φ). Then consistency relative to the Kolmogorov-Smirnov distance is equivalent to the requirements, for every x,

$$P\left(\frac{\hat{\theta}_n - \theta}{\hat{\sigma}_n} \le x \mid P \right) \to F(x), \qquad P\left(\frac{\hat{\theta}_n^* - \hat{\theta}_n}{\hat{\sigma}_n^*} \le x \mid \hat{P}_n \right) \xrightarrow{P} F(x). \qquad (23.2)$$

(See Problem 23.1.) This type of consistency implies the asymptotic consistency of confidence intervals.

23.3 Lemma. *Suppose that $(\hat{\theta}_n - \theta)/\hat{\sigma}_n \rightsquigarrow T$, and that $(\hat{\theta}_n^* - \hat{\theta}_n)/\hat{\sigma}_n^* \rightsquigarrow T$ given the original observations, in probability, for a random variable T with a continuous distribution function. Then the bootstrap confidence intervals $[\hat{\theta}_n - \hat{\xi}_{n,\beta}\hat{\sigma}_n, \hat{\theta}_n - \hat{\xi}_{n,1-\alpha}\hat{\sigma}_n]$ are asymptotically consistent at level $1 - \alpha - \beta$. If the conditions hold for nonrandom $\hat{\sigma}_n = \hat{\sigma}_n^*$, and T is symmetrically distributed about zero, then the same is true for Efron's percentile intervals.*

Proof. Every subsequence has a further subsequence along which the sequence $(\hat{\theta}_n^* - \hat{\theta}_n)/\hat{\sigma}_n^*$ converges weakly to T, conditionally, almost surely. For simplicity, assume that the whole sequence converges almost surely; otherwise, argue along subsequences.

If a sequence of distribution functions F_n converges weakly to a distribution function F, then the corresponding quantile functions F_n^{-1} converge to the quantile function F^{-1} at every continuity point (see Lemma 21.2). Apply this to the (random) distribution functions $\hat{F}_n$ of $(\hat{\theta}_n^* - \hat{\theta}_n)/\hat{\sigma}_n^*$ and a continuity point $1 - \alpha$ of the quantile function F^{-1} of T to conclude that $\hat{\xi}_{n,\alpha} = \hat{F}_n^{-1}(1 - \alpha)$ converges almost surely to $F^{-1}(1 - \alpha)$. By Slutsky's lemma, the sequence $(\hat{\theta}_n - \theta)/\hat{\sigma}_n - \hat{\xi}_{n,\alpha}$ converges weakly to $T - F^{-1}(1 - \alpha)$. Thus

$$P(\theta \geq \hat{\theta}_n - \hat{\sigma}_n \hat{\xi}_{n,\alpha}) = P\left(\frac{\hat{\theta}_n - \theta}{\hat{\sigma}_n} \leq \hat{\xi}_{n,\alpha} \mid P\right) \to P\big(T \leq F^{-1}(1 - \alpha)\big) = 1 - \alpha.$$

This argument applies to all except at most countably many α. Because both the left and the right sides of the preceding display are monotone functions of α and the right side is continuous, it must be valid for every α. The consistency of the bootstrap confidence interval follows.

Efron's percentile interval is the interval $[\hat{\zeta}_{n,1-\beta}, \hat{\zeta}_{n,\alpha}]$, where $\hat{\zeta}_{n,\alpha} = \hat{\theta}_n + \hat{\xi}_{n,\alpha}$. By the preceding argument,

$$P(\theta \geq \hat{\zeta}_{n,1-\beta}) = P(\hat{\theta}_n - \theta \leq -\hat{\xi}_{n,1-\beta} \mid P) \to P\big(T \leq -F^{-1}(\beta)\big) = 1 - \beta.$$

The last equality follows by the symmetry of T. The consistency follows. ∎

From now on we consider the empirical bootstrap; that is, $\hat{P}_n = \mathbb{P}_n$ is the empirical distribution of a random sample $X_1, \ldots, X_n$. We shall establish (23.2) for a large class of statistics, with F the normal distribution. Our method is first to prove the consistency for $\hat{\theta}_n$ equal to the sample mean and next to show that the consistency is retained under application of the delta method. Combining these results, we obtain the consistency of many bootstrap procedures, for instance for setting confidence intervals for the correlation coefficient.

In view of Slutsky's lemma, weak convergence of the centered sequence $\sqrt{n}(\hat{\theta}_n - \theta)$ combined with convergence in probability of $\hat{\sigma}_n/\sqrt{n}$ yields the weak convergence of the studentized statistics $(\hat{\theta}_n - \theta)/\hat{\sigma}_n$. An analogous statement is true for the bootstrap statistic, for which the convergence in probability of $\hat{\sigma}_n^*/\sqrt{n}$ must be shown conditionally on the original observations. Establishing (conditional) consistency of $\hat{\sigma}_n/\sqrt{n}$ and $\hat{\sigma}_n^*/\sqrt{n}$ is usually not hard. Therefore, we restrict ourselves to studying the nonstudentized statistics.

Let $\overline{X}_n$ be the mean of a sample of n random vectors from a distribution with finite mean vector μ and covariance matrix Σ. According to the multivariate central limit theorem, the sequence $\sqrt{n}(\overline{X}_n - \mu)$ is asymptotically normal $N(0, \Sigma)$-distributed. We wish to show the same for $\sqrt{n}(\overline{X}_n^* - \overline{X}_n)$, in which $\overline{X}_n^*$ is the average of n observations from $\mathbb{P}_n$, that is, of n values resampled from the set of original observations $\{X_1, \ldots, X_n\}$ with replacement.

23.4   Theorem (Sample mean).   *Let $X_1, X_2, \ldots$ be i.i.d. random vectors with mean μ and covariance matrix Σ. Then conditionally on $X_1, X_2, \ldots$, for almost every sequence $X_1, X_2, \ldots,$*

$$\sqrt{n}(\overline{X}_n^* - \overline{X}_n) \rightsquigarrow N(0, \Sigma).$$

Proof.   For a fixed sequence $X_1, X_2, \ldots$, the variable $\overline{X}_n^*$ is the average of n observations $X_1^*, \ldots, X_n^*$ sampled from the empirical distribution $\mathbb{P}_n$. The (conditional) mean and covariance matrix of these observations are

$$E(X_i^* \mid \mathbb{P}_n) = \sum_{i=1}^n \frac{1}{n} X_i = \overline{X}_n,$$

$$E\big((X_i^* - \overline{X}_n)(X_i^* - \overline{X}_n)^T \mid \mathbb{P}_n\big) = \sum_{i=1}^n \frac{1}{n}(X_i - \overline{X}_n)(X_i - \overline{X}_n)^T$$

$$= \overline{X_n X_n^T} - \overline{X}_n \overline{X}_n^T.$$

By the strong law of large numbers, the conditional covariance converges to Σ for almost every sequence $X_1, X_2, \ldots$.

The asymptotic distribution of $\overline{X}_n^*$ can be established by the central limit theorem. Because the observations $X_1^*, \ldots, X_n^*$ are sampled from a different distribution $\mathbb{P}_n$ for every n, a central limit theorem for a triangular array is necessary. The Lindeberg central limit theorem, Theorem 2.27, is appropriate. It suffices to show that, for every $\varepsilon > 0$,

$$\mathrm{E}\|X_i^*\|^2 1\{\|X_i^*\| > \varepsilon\sqrt{n}\} = \frac{1}{n}\sum_{i=1}^{n} \|X_i\|^2 1\{\|X_i\| > \varepsilon\sqrt{n}\} \overset{as}{\to} 0.$$

The left side is smaller than $n^{-1}\sum_{i=1}^{n}\|X_i\|^2 1\{\|X_i\| > M\}$ as soon as $\varepsilon\sqrt{n} \geq M$. By the strong law of large numbers, the latter average converges to $\mathrm{E}\|X_i\|^2 1\{\|X_i\| > M\}$ for almost every sequence $X_1, X_2, \ldots$. For sufficiently large M, this expression is arbitrarily small. Conclude that the limit superior of the left side of the preceding display is smaller than any number $\eta > 0$ almost surely and hence the left side converges to zero for almost every sequence $X_1, X_2, \ldots$. ∎

Assume that $\hat{\theta}_n$ is a statistic, and that ϕ is a given differentiable map. If the sequence $\sqrt{n}(\hat{\theta}_n - \theta)$ converges in distribution, then so does the sequence $\sqrt{n}(\phi(\hat{\theta}_n) - \phi(\theta))$, by the delta method. The bootstrap estimator for the distribution of $\phi(\hat{\theta}_n) - \phi(\theta)$ is $\phi(\hat{\theta}_n^*) - \phi(\hat{\theta}_n)$. If the bootstrap is consistent for estimating the distribution of $\hat{\theta}_n - \theta$, then it is also consistent for estimating the distribution of $\phi(\hat{\theta}_n) - \phi(\theta)$.

23.5 Theorem (Delta method for bootstrap). *Let $\phi : \mathbb{R}^k \mapsto \mathbb{R}^m$ be a measurable map defined and continuously differentiable in a neighborhood of θ. Let $\hat{\theta}_n$ be random vectors taking their values in the domain of ϕ that converge almost surely to θ. If $\sqrt{n}(\hat{\theta}_n - \theta) \rightsquigarrow T$, and $\sqrt{n}(\hat{\theta}_n^* - \hat{\theta}_n) \rightsquigarrow T$ conditionally almost surely, then both $\sqrt{n}(\phi(\hat{\theta}_n) - \phi(\theta)) \rightsquigarrow \phi_\theta'(T)$ and $\sqrt{n}(\phi(\hat{\theta}_n^*) - \phi(\hat{\theta}_n)) \rightsquigarrow \phi_\theta'(T)$, conditionally almost surely.*

Proof. By the mean value theorem, the difference $\phi(\hat{\theta}_n^*) - \phi(\hat{\theta}_n)$ can be written as $\phi'_{\tilde{\theta}_n}(\hat{\theta}_n^* - \hat{\theta}_n)$ for a point $\tilde{\theta}_n$ between $\hat{\theta}_n^*$ and $\hat{\theta}_n$, if the latter two points are in the ball around θ in which ϕ is continuously differentiable. By the continuity of the derivative, there exists for every $\eta > 0$ a constant $\delta > 0$ such that $\|\phi'_{\theta'}h - \phi'_\theta h\| < \eta\|h\|$ for every h and every $\|\theta' - \theta\| \leq \delta$. If n is sufficiently large, δ sufficiently small, $\sqrt{n}\|\hat{\theta}_n^* - \hat{\theta}_n\| \leq M$, and $\|\hat{\theta}_n - \theta\| \leq \delta$, then

$$R_n := \left\| \sqrt{n}(\phi(\hat{\theta}_n^*) - \phi(\hat{\theta}_n)) - \phi'_\theta \sqrt{n}(\hat{\theta}_n^* - \hat{\theta}_n) \right\|$$

$$= \left| (\phi'_{\tilde{\theta}_n} - \phi'_\theta)\sqrt{n}(\hat{\theta}_n^* - \hat{\theta}_n) \right| \leq \eta M.$$

Fix a number $\varepsilon > 0$ and a large number M. For η sufficiently small to ensure that $\eta M < \varepsilon$,

$$\mathrm{P}(R_n > \varepsilon \mid \hat{P}_n) \leq \mathrm{P}(\sqrt{n}\|\hat{\theta}_n^* - \hat{\theta}_n\| > M \text{ or } \|\hat{\theta}_n - \theta\| > \delta \mid \hat{P}_n).$$

Because $\hat{\theta}_n$ converges almost surely to θ, the right side converges almost surely to $\mathrm{P}(\|T\| \geq M)$ for every continuity point M of $\|T\|$. This can be made arbitrarily small by choice of M. Conclude that the left side converges to 0 almost surely. The theorem follows by an application of Slutsky's lemma. ∎

23.6 Example (Sample variance). The (biased) sample variance $S_n^2 = n^{-1}\sum_{i=1}^{n}(X_i - \overline{X}_n)^2$ equals $\phi(\overline{X}_n, \overline{X_n^2})$ for the map $\phi(x, y) = y - x^2$. The empirical bootstrap is consistent for estimation of the distribution of $(\overline{X}_n, \overline{X_n^2}) - (\alpha_1, \alpha_2)$, by Theorem 23.4, provided that the fourth moment of the underlying distribution is finite. The delta method shows that the empirical bootstrap is consistent for estimating the distribution of $S_n^2 - \sigma^2$ in that

$$\sup_x \left| P\left(\sqrt{n}\left(S_n^2 - \sigma^2\right) \le x | P\right) - P\left(\sqrt{n}\left(S_n^{*2} - S_n^2\right) \le x | \mathbb{P}_n\right) \right| \overset{\text{as}}{\to} 0.$$

The asymptotic variance of S_n^2 can be estimated by $S_n^4(k_n + 2)$, in which k_n is the sample kurtosis. The law of large numbers shows that this estimator is asymptotically consistent. The bootstrap version of this estimator can be shown to be consistent given almost every sequence of the original observations. Thus, the consistency of the empirical bootstrap extends to the studentized statistic $(S_n^2 - \sigma^2)/S_n^2\sqrt{k_n + 1}$. $\square$

*23.2.1 *Empirical Bootstrap*

In this section we follow the same method as previously, but we replace the sample mean by the empirical distribution and the delta method by the functional delta method. This is more involved, but more flexible, and yields, for instance, the consistency of the bootstrap of the sample median.

Let $\mathbb{P}_n$ be the empirical distribution of a random sample $X_1, \ldots, X_n$ from a distribution P on a measurable space $(\mathcal{X}, \mathcal{A})$, and let $\mathcal{F}$ be a Donsker class of measurable functions $f : \mathcal{X} \mapsto \mathbb{R}$, as defined in Chapter 19. Given the sample values, let $X_1^*, \ldots, X_n^*$ be a random sample from $\mathbb{P}_n$. The *bootstrap empirical distribution* is the empirical measure $\mathbb{P}_n^* = n^{-1}\sum_{i=1}^n \delta_{X_i^*}$, and the *bootstrap empirical process* $\mathbb{G}_n^*$ is

$$\mathbb{G}_n^* = \sqrt{n}(\mathbb{P}_n^* - \mathbb{P}_n) = \frac{1}{\sqrt{n}} \sum_{i=1}^n (M_{ni} - 1)\delta_{X_i},$$

in which M_{ni} is the number of times that X_i is "redrawn" from $\{X_1, \ldots, X_n\}$ to form $X_1^*, \ldots, X_n^*$. By construction, the vector of counts $(M_{n1}, \ldots, M_{nn})$ is independent of $X_1, \ldots, X_n$ and multinomially distributed with parameters n and (probabilities) $1/n, \ldots, 1/n$.

If the class $\mathcal{F}$ has a finite envelope function F, then both the empirical process $\mathbb{G}_n$ and the bootstrap process $\mathbb{G}_n^*$ can be viewed as maps into the space $\ell^\infty(\mathcal{F})$. The analogue of Theorem 23.4 is that the sequence $\mathbb{G}_n^*$ converges in $\ell^\infty(\mathcal{F})$ conditionally in distribution to the same limit as the sequence $\mathbb{G}_n$, a tight Brownian bridge process $\mathbb{G}_P$. To give a precise meaning to "conditional weak convergence" in $\ell^\infty(\mathcal{F})$, we use the *bounded Lipschitz metric*. It can be shown that a sequence of random elements in $\ell^\infty(\mathcal{F})$ converges in distribution to a tight limit in $\ell^\infty(\mathcal{F})$ if and only if[†]

$$\sup_{h \in \mathrm{BL}_1(\ell^\infty(\mathcal{F}))} \left| \mathrm{E}^* h(G_n) - \mathrm{E}h(G) \right| \to 0.$$

We use the notation E_M to denote "taking the expectation conditionally on $X_1, \ldots, X_n$," or the expectation with respect to the multinomial vectors M_n only.[‡]

[†] For a metric space $\mathbb{D}$, the set $\mathrm{BL}_1(\mathbb{D})$ consists of all functions $h : \mathbb{D} \mapsto [-1, 1]$ that are uniformly Lipschitz: $\left|h(z_1) - h(z_2)\right| \le d(z_1, z_2)$ for every pair (z_1, z_2). See, for example, Chapter 1.12 of [146].

[‡] For a proof of Theorem 23.7, see the original paper [58], or, for example, Chapter 3.6 of [146].

23.7   Theorem (Empirical bootstrap). *For every Donsker class $\mathcal{F}$ of measurable functions with finite envelope function F,*

$$\sup_{h \in \mathrm{BL}_1(\ell^\infty(\mathcal{F}))} \left| E_M h(\mathbb{G}_n^*) - Eh(\mathbb{G}_P) \right| \xrightarrow{P} 0.$$

Furthermore, the sequence $\mathbb{G}_n^*$ is asymptotically measurable. If $P^* F^2 < \infty$, then the convergence is outer almost surely as well.

Next, consider an analogue of Theorem 23.5, using the functional delta method. Theorem 23.5 goes through without too many changes. However, for many infinite-dimensional applications of the delta method the condition of *continuous* differentiability imposed in Theorem 23.5 fails. This problem may be overcome in several ways. In particular, continuous differentiability is not necessary for the consistency of the bootstrap "in probability" (rather than "almost surely"). Because this appears to be sufficient for statistical applications, we shall limit ourselves to this case.

Consider sequences of maps $\hat{\theta}_n$ and $\hat{\theta}_n^*$ with values in a normed space $\mathbb{D}$ (e.g., $\ell^\infty(\mathcal{F})$) such that the sequence $\sqrt{n}(\hat{\theta}_n - \theta)$ converges unconditionally in distribution to a tight random element T, and the sequence $\sqrt{n}(\hat{\theta}_n^* - \hat{\theta}_n)$ converges conditionally given $X_1, X_2, \ldots$ in distribution to the same random element T. A precise formulation of the second is that

$$\sup_{h \in \mathrm{BL}_1(\mathbb{D})} \left| E_M h(\sqrt{n}(\hat{\theta}_n^* - \hat{\theta}_n)) - Eh(T) \right| \xrightarrow{P} 0. \tag{23.8}$$

Here the notation E_M means the conditional expectation given the original data $X_1, X_2, \ldots$ and is motivated by the application to the bootstrap empirical distribution.[†] By the preceding theorem, the empirical distribution $\hat{\theta}_n = \mathbb{P}_n$ satisfies condition (23.8) if viewed as a map in $\ell^\infty(\mathcal{F})$ for a Donsker class $\mathcal{F}$.

23.9   Theorem (Delta method for bootstrap). *Let $\mathbb{D}$ be a normed space and let $\phi : \mathbb{D}_\phi \subset \mathbb{D} \mapsto \mathbb{R}^k$ be Hadamard differentiable at θ tangentially to a subspace $\mathbb{D}_0$. Let $\hat{\theta}_n$ and $\hat{\theta}_n^*$ be maps with values in $\mathbb{D}_\phi$ such that $\sqrt{n}(\hat{\theta}_n - \theta) \rightsquigarrow T$ and such that (23.8) holds, in which $\sqrt{n}(\hat{\theta}_n^* - \hat{\theta}_n)$ is asymptotically measurable and T is tight and takes its values in $\mathbb{D}_0$. Then the sequence $\sqrt{n}\big(\phi(\hat{\theta}_n^*) - \phi(\hat{\theta}_n)\big)$ converges conditionally in distribution to $\phi_\theta'(T)$, given $X_1, X_2, \ldots$, in probability.*

Proof. By the Hahn-Banach theorem it is not a loss of generality to assume that the derivative $\phi_\theta' : \mathbb{D} \mapsto \mathbb{R}^k$ is defined and continuous on the whole space. For every $h \in \mathrm{BL}_1(\mathbb{R}^k)$, the function $h \circ \phi_\theta'$ is contained in $\mathrm{BL}_{\|\phi_\theta'\|}(\mathbb{D})$. Thus (23.8) implies

$$\sup_{h \in \mathrm{BL}_1(\mathbb{R}^k)} \left| E_M h\big(\phi_\theta'(\sqrt{n}(\hat{\theta}_n^* - \hat{\theta}_n))\big) - Eh(\phi_\theta'(T)) \right| \xrightarrow{P} 0.$$

Because $|h(x) - h(y)|$ is bounded by $2 \wedge d(x, y)$ for every $h \in \mathrm{BL}_1(\mathbb{R}^k)$,

$$\sup_{h \in \mathrm{BL}_1(\mathbb{R}^k)} \left| E_M h\big(\sqrt{n}(\phi(\hat{\theta}_n^*) - \phi(\hat{\theta}_n))\big) - E_M h\big(\phi_\theta'(\sqrt{n}(\hat{\theta}_n^* - \hat{\theta}_n))\big) \right|$$

$$\le \varepsilon + 2 P_M \left(\left\| \sqrt{n}(\phi(\hat{\theta}_n^*) - \phi(\hat{\theta}_n)) - \phi_\theta'(\sqrt{n}(\hat{\theta}_n^* - \hat{\theta}_n)) \right\| > \varepsilon \right). \tag{23.10}$$

[†] It is assumed that $h(\sqrt{n}(\hat{\theta}_n^* - \hat{\theta}_n))$ is a measurable function of M.

The theorem is proved once it has been shown that the conditional probability on the right converges to zero in outer probability.

The sequence $\sqrt{n}(\hat{\theta}_n^* - \hat{\theta}_n, \hat{\theta}_n - \theta)$ converges (unconditionally) in distribution to a pair of two independent copies of T. This follows, because conditionally given $X_1, X_2, \ldots$, the second component is deterministic, and the first component converges in distribution to T, which is the same for every sequence $X_1, X_2, \ldots$. Therefore, by the continuous-mapping theorem both sequences $\sqrt{n}(\hat{\theta}_n - \theta)$ and $\sqrt{n}(\hat{\theta}_n^* - \theta)$ converge (unconditionally) in distribution to separable random elements that concentrate on the linear space $\mathbb{D}_0$. By Theorem 20.8,

$$\sqrt{n}\big(\phi(\hat{\theta}_n^*) - \phi(\theta)\big) = \phi'_\theta\big(\sqrt{n}(\hat{\theta}_n^* - \theta)\big) + o_P^*(1),$$
$$\sqrt{n}\big(\phi(\hat{\theta}_n) - \phi(\theta)\big) = \phi'_\theta\big(\sqrt{n}(\hat{\theta}_n - \theta)\big) + o_P^*(1).$$

Subtract the second from the first equation to conclude that the sequence $\sqrt{n}\big(\phi(\hat{\theta}_n^*) - \phi(\hat{\theta}_n)\big) - \phi'_\theta\big(\sqrt{n}(\hat{\theta}_n^* - \hat{\theta}_n)\big)$ converges (unconditionally) to 0 in outer probability. Thus, the conditional probability on the right in (23.10) converges to zero in outer mean. This concludes the proof. ∎

23.11 *Example (Empirical distribution function).* Because the cells $(-\infty, t] \subset \mathbb{R}$ form a Donsker class, the empirical distribution function $\mathbb{F}_n$ of a random sample of real-valued variables satisfies the condition of the preceding theorem. Thus, conditionally on $X_1, X_2, \ldots$, the sequence $\sqrt{n}\big(\phi(\mathbb{F}_n^*) - \phi(\mathbb{F}_n)\big)$ converges in distribution to the same limit as $\sqrt{n}\big(\phi(\mathbb{F}_n) - \phi(F)\big)$, for every Hadamard-differentiable function ϕ.

This includes, among others, quantiles and trimmed means, under the same conditions on the underlying measure F that ensure that empirical quantiles and trimmed means are asymptotically normal. See Lemmas 21.3, 22.9, and 22.10. □

23.3 Higher-Order Correctness

The investigation of the performance of a bootstrap confidence interval can be refined by taking into account the order at which the true level converges to the desired level. A confidence interval is (conservatively) *correct at level* $1 - \alpha - \beta$ *up to order* $O(n^{-k})$ if

$$P\big(\hat{\theta}_{n,1} \le \theta \le \hat{\theta}_{n,2} \mid P\big) \ge 1 - \alpha - \beta - O\left(\frac{1}{n^k}\right).$$

Similarly, the quality of the bootstrap estimator for distributions can be assessed more precisely by the rate at which the Kolmogorov-Smirnov distance between the distribution function of $(\hat{\theta}_n - \theta)/\hat{\sigma}_n$ and the conditional distribution function of $(\hat{\theta}_n^* - \hat{\theta}_n)/\hat{\sigma}_n^*$ converges to zero. We shall see that the percentile t-method usually performs better than the percentile method. For the percentile t-method, the Kolmogorov-Smirnov distance typically converges to zero at the rate $O_P(n^{-1})$, whereas the percentile method attains "only" an $O_P(n^{-1/2})$ rate of correctness. The latter is comparable to the error of the normal approximation.

Rates for the Kolmogorov-Smirnov distance translate directly into orders of correctness of one-tailed confidence intervals. The correctness of two-tailed or symmetric confidence intervals may be higher, because of the cancellation of the coverage errors contributed by the left and right tails. In many cases the percentile method, the percentile t-method, and the normal approximation all yield correct two-tailed confidence intervals up to order $O(n^{-1})$. Their relative qualities may be studied by a more refined analysis. This must also take into account the length of the confidence intervals, for an increase in length of order $O_P(n^{-3/2})$ may easily reduce the coverage error to the order $O(n^{-k})$ for any k.

The technical tool to obtain these results are *Edgeworth expansions*. Edgeworth's classical expansion is a refinement of the central limit theorem that shows the magnitude of the difference between the distribution function of a sample mean and its normal approximation. Edgeworth expansions have subsequently been obtained for many other statistics as well.

An Edgeworth expansion for the distribution function of a statistic $(\hat{\theta}_n - \theta)/\hat{\sigma}_n$ is typically an expansion in increasing powers of $1/\sqrt{n}$ of the form

$$\mathrm{P}\left(\frac{\hat{\theta}_n - \theta}{\hat{\sigma}_n} \le x \mid P\right) = \Phi(x) + \frac{p_1(x \mid P)}{\sqrt{n}} \phi(x) + \frac{p_2(x \mid P)}{n} \phi(x) + \cdots \qquad (23.12)$$

The remainder is of lower order than the last included term, uniformly in the argument x. Thus, in the present case the remainder is $o(n^{-1})$ (or even $O(n^{-3/2})$). The functions p_i are polynomials in x, whose coefficients depend on the underlying distribution, typically through (asymptotic) moments of the pair $(\hat{\theta}_n, \hat{\sigma}_n)$.

23.13 *Example (Sample mean).* Let $\overline{X}_n$ be the mean of a random sample of size n, and let $S_n^2 = n^{-1} \sum_{i=1}^n (X_i - \overline{X}_n)^2$ be the (biased) sample variance. If μ, σ^2, λ and κ are the mean, variance, skewness and kurtosis of the underlying distribution, then

$$\mathrm{P}\left(\frac{\overline{X}_n - \mu}{\sigma/\sqrt{n}} \le x \mid P\right) = \Phi(x) - \frac{\lambda(x^2 - 1)}{6\sqrt{n}} \phi(x)$$

$$- \frac{3\kappa(x^3 - 3x) + \lambda^2(x^5 - 10x^3 + 15x)}{72n} \phi(x) + O\left(\frac{1}{n\sqrt{n}}\right).$$

These are the first two terms of the classical expansion of Edgeworth. If the standard deviation of the observations is unknown, an Edgeworth expansion of the t-statistic is of more interest. This takes the form (see [72, pp. 71–73])

$$\mathrm{P}\left(\frac{\overline{X}_n - \mu}{S_n/\sqrt{n}} \le x \mid P\right) = \Phi(x) + \frac{\lambda(2x^2 + 1)}{6\sqrt{n}} \phi(x)$$

$$+ \frac{3\kappa(x^3 - 3x) - 2\lambda^2(x^5 + 2x^3 - 3x) - 9(x^3 + 3x)}{36n} \phi(x) + O\left(\frac{1}{n\sqrt{n}}\right).$$

Although the polynomials are different, these expansions are of the same form. Note that the polynomial appearing in the $1/\sqrt{n}$ term is even in both cases.

These expansions generally fail if the underlying distribution of the observations is discrete. *Cramér's condition* requires that the modulus of the characteristic function of the observations be bounded away from unity on closed intervals that do not contain the origin. This condition is satisfied if the observations possess a density with respect to Lebesgue measure. Next to Cramér's condition a sufficient number of moments of the observations must exist. $\square$

23.14   *Example (Studentized quantiles).* The pth quantile $F^{-1}(p)$ of a distribution function F may be estimated by the empirical pth quantile $\mathbb{F}_n^{-1}(p)$. This is the rth order statistic of the sample for r equal to the largest integer not greater than np. Its mean square error can be computed as

$$E\big(\mathbb{F}_n^{-1}(p) - F^{-1}(p)\big)^2 = r\binom{n}{r}\int_0^1 \big(F^{-1}(u) - F^{-1}(p)\big)^2 u^{r-1}(1-u)^{n-r}\,du.$$

An empirical estimator $\hat{\sigma}_n$ for the mean square error of $\mathbb{F}_n^{-1}(p)$ is obtained by replacing F by the empirical distribution function. If the distribution has a differentiable density f, then

$$P\left(\frac{\mathbb{F}_n^{-1}(p) - F^{-1}(p)}{\hat{\sigma}_n} \le x \mid F\right) = \Phi(x) + \frac{p_1(x \mid F)}{\sqrt{n}}\phi(x) + O\left(\frac{1}{n^{3/4}}\right),$$

where $p_1(x \mid F)$ is the polynomial of degree 3 given by (see [72, pp. 318–321])

$$p_1(x \mid F)\,12\sqrt{p(1-p)} = \frac{3}{\sqrt{\pi}}x^3 + \left[2 - 10p - 12p(1-p)\frac{f'}{f^2}\big(F^{-1}(p)\big)\right]x^2$$

$$+ \frac{3 + 6\sqrt{2}}{\sqrt{\pi}}x - 8 + 4p - 12(r - np).$$

This expansion is unusual in two respects. First, the remainder is of the order $O(n^{-3/4})$ rather than of the order $O(n^{-1})$. Second, the polynomial appearing in the first term is not even. For this reason several of the conclusions of this section are not valid for sample quantiles. In particular, the order of correctness of all empirical bootstrap procedures is $O_P(n^{-1/2})$, not greater. In this case, a "smoothed bootstrap" based on "resampling" from a density estimator (as in Chapter 24) may be preferable, depending on the underlying distribution. □

   If the distribution function of $(\hat{\theta}_n - \theta)/\hat{\sigma}_n$ admits an Edgeworth expansion (23.12), then it is immediate that the normal approximation is correct up to order $O(1/\sqrt{n})$. Evaluation of the expansion at the normal quantiles z_β and $z_{1-\alpha}$ yields

$$P(\hat{\theta}_n - z_\beta\,\hat{\sigma}_n \le \theta \le \hat{\theta}_n - z_{1-\alpha}\,\hat{\sigma}_n \mid P) = 1 - \alpha - \beta$$

$$+ \frac{p_1(z_\beta \mid P)\phi(z_\beta) - p_1(z_{1-\alpha} \mid P)\phi(z_{1-\alpha})}{\sqrt{n}} + O\left(\frac{1}{n}\right).$$

Thus, the level of the confidence interval $[\hat{\theta}_n - z_\beta\,\hat{\sigma}_n, \hat{\theta}_n - z_{1-\alpha}\,\hat{\sigma}_n]$ is $1 - \alpha - \beta$ up to order $O(1/\sqrt{n})$. For a two-tailed, symmetric interval, α and β are chosen equal. Inserting $z_\beta = z_\alpha = -z_{1-\alpha}$ in the preceding display, we see that the errors of order $1/\sqrt{n}$ resulting from the left and right tails cancel each other if p_1 is an even function. In this common situation the order of correctness improves to $O(n^{-1})$.

   It is of theoretical interest that the coverage probability can be corrected up to any order by making the normal confidence interval slightly wider than first-order asymptotics would suggest. The interval may be widened by using quantiles z_{α_n} with $\alpha_n < \alpha$, rather than z_α. In view of the preceding display, for any α_n,

$$P(\hat{\theta}_n - z_{\alpha_n}\,\hat{\sigma}_n \le \theta \le \hat{\theta}_n - z_{1-\alpha_n}\,\hat{\sigma}_n \mid P) = 1 - 2\alpha_n + O\left(\frac{1}{n}\right).$$

The $O(n^{-1})$ term results from the Edgeworth expansion (23.12) and is universal, independent of the sequence α_n. For $\alpha_n = \alpha - M/n$ and a sufficiently large constant M, the right side becomes

$$1 - 2\alpha + \frac{2M}{n} + O\left(\frac{1}{n}\right) \geq 1 - 2\alpha - O\left(\frac{1}{n^k}\right).$$

Thus, a slight widening of the normal confidence interval yields asymptotically correct (conservative) coverage probabilities up to any order $O(n^{-k})$. If $\hat{\sigma}_n = O_P(n^{-1/2})$, then the widened interval is $2(z_{\alpha_n} - z_\alpha)\hat{\sigma}_n = O_P(n^{-3/2})$ wider than the normal confidence interval. This difference is small relatively to the absolute length of the interval, which is $O_P(n^{-1/2})$. Also, the choice of the scale estimator $\hat{\sigma}_n$ (which depends on $\hat{\theta}_n$) influences the width of the interval stronger than replacing ξ_α by ξ_{α_n}.

An Edgeworth expansion usually remains valid in a conditional sense if a good estimator $\hat{P}_n$ is substituted for the true underlying distribution P. The bootstrap version of expansion (23.12) is

$$P\left(\frac{\hat{\theta}_n^* - \hat{\theta}_n}{\hat{\sigma}_n^*} \leq x \mid \hat{P}_n\right) = \Phi(x) + \frac{p_1(x \mid \hat{P}_n)}{\sqrt{n}}\phi(x) + \frac{p_2(x \mid \hat{P}_n)}{n}\phi(x) + \cdots.$$

In this expansion the remainder term is a random variable, which ought to be of smaller order in probability than the last term. In the given expansion the remainder ought to be $o_P(n^{-1})$ uniformly in x. Subtract the bootstrap expansion from the unconditional expansion (23.12) to obtain that

$$\sup_x \left| P\left(\frac{\hat{\theta}_n - \theta}{\hat{\sigma}_n} \leq x \mid P\right) - P\left(\frac{\hat{\theta}_n^* - \hat{\theta}_n}{\hat{\sigma}_n^*} \leq x \mid \hat{P}_n\right) \right|$$

$$\leq \sup_x \left| \frac{p_1(x \mid P) - p_1(x \mid \hat{P}_n)}{\sqrt{n}} + \frac{p_2(x \mid P) - p_2(x \mid \hat{P}_n)}{n} \right| \phi(x) + o_P\left(\frac{1}{n}\right).$$

The polynomials p_i typically depend on P in a smooth way, and the difference $\hat{P}_n - \hat{P}$ is typically of the order $O_P(n^{-1/2})$. Then the Kolmogorov-Smirnov distance between the true distribution function of $(\hat{\theta}_n - \theta)/\hat{\sigma}_n$ and its percentile t-bootstrap estimator is of the order $O_P(n^{-1})$.

The analysis of the percentile method starts from an Edgeworth expansion of the distribution function of the unstudentized statistic $\hat{\theta}_n - \theta$. This has as leading term the normal distribution with variance σ_n^2, the asymptotic variance of $\hat{\theta}_n - \theta$, rather than the standard normal distribution. Typically it is of the form

$$P(\hat{\theta}_n - \theta \leq x \mid P) = \Phi\left(\frac{x}{\sigma_n}\right) + \frac{1}{\sqrt{n}}q_1\left(\frac{x}{\sigma_n} \middle| P\right)\phi\left(\frac{x}{\sigma_n}\right)$$

$$+ \frac{1}{n}q_2\left(\frac{x}{\sigma_n} \middle| P\right)\phi\left(\frac{x}{\sigma_n}\right) + \cdots.$$

The functions q_i are polynomials, which are generally different from the polynomials occurring in the Edgeworth expansion for the studentized statistic. The bootstrap version of this expansion is

$$P(\hat{\theta}_n^* - \hat{\theta}_n \leq x \mid \hat{P}_n) = \Phi\left(\frac{x}{\hat{\sigma}_n}\right) + \frac{1}{\sqrt{n}}q_1\left(\frac{x}{\hat{\sigma}_n} \middle| \hat{P}_n\right)\phi\left(\frac{x}{\hat{\sigma}_n}\right)$$

$$+ \frac{1}{n}q_2\left(\frac{x}{\hat{\sigma}_n} \middle| \hat{P}_n\right)\phi\left(\frac{x}{\hat{\sigma}_n}\right) + \cdots.$$

The Kolmogorov-Smirnov distance between the distribution functions on the left in the preceding displays is of the same order as the difference between the leading terms $\Phi(x/\sigma_n) - \Phi(x/\hat{\sigma}_n)$ on the right. Because the estimator $\hat{\sigma}_n$ is typically not closer than $O_P(n^{-1/2})$ to σ, this difference may be expected to be at best of the order $O_P(n^{-1/2})$. Thus, the percentile method for estimating a distribution is correct only up to the order $O_P(n^{-1/2})$, whereas the percentile t-method is seen to be correct up to the order $O_P(n^{-1})$.

One-sided bootstrap percentile t and percentile confidence intervals attain orders of correctness that are equal to the orders of correctness of the bootstrap estimators of the distribution functions: $O_P(n^{-1})$ and $O_P(n^{-1/2})$, respectively. For equal-tailed confidence intervals both methods typically have coverage error of the order $O_P(n^{-1})$. The decrease in coverage error is due to the cancellation of the errors contributed by the left and right tails, just as in the case of normal confidence intervals. The proofs of these assertions are somewhat technical. The coverage probabilities can be expressed in probabilities of the type

$$P\left(\frac{\hat{\theta}_n - \theta}{\hat{\sigma}_n} \le \hat{\xi}_{n,\alpha} \mid P\right). \tag{23.15}$$

Thus we need an Edgeworth expansion of the distribution of $(\hat{\theta}_n - \theta)/\hat{\sigma}_n - \hat{\xi}_{n,\alpha}$, or a related quantity. A technical complication is that the random variables $\hat{\xi}_{n,\alpha}$ are only implicitly defined, as the solution of (23.1).

To find the expansions, first evaluate the Edgeworth expansion for $(\hat{\theta}_n^* - \hat{\theta}_n)/\hat{\sigma}_n^*$ at its the upper quantile $\hat{\xi}_{n,\alpha}$ to find that

$$1 - \alpha = \Phi(\hat{\xi}_{n,\alpha}) + \frac{p_1(\hat{\xi}_{n,\alpha} \mid \hat{P}_n)\phi(\hat{\xi}_{n,\alpha})}{\sqrt{n}} + O_P\left(\frac{1}{n}\right).$$

After expanding Φ, p_1 and ϕ in Taylor series around z_α, we can invert this equation to obtain the (conditional) *Cornish-Fisher expansion*

$$\hat{\xi}_{n,\alpha} = z_\alpha - \frac{p_1(z_\alpha \mid P)}{\sqrt{n}} + O_P\left(\frac{1}{n}\right).$$

In general, Cornish-Fisher expansions are asymptotic expansions of quantile functions, much in the same spirit as Edgeworth expansions are expansions of distribution functions. The probability (23.15) can be rewritten

$$P\left(\frac{\hat{\theta}_n - \theta}{\hat{\sigma}_n} - O_P\left(\frac{1}{n}\right) \le z_\alpha - \frac{p_1(z_\alpha \mid P)}{\sqrt{n}} \mid P\right).$$

For a rigorous derivation it is necessary to characterize the $O_P(n^{-1})$ term. Informally, this term should only contribute to terms of order $O(n^{-1})$ in an Edgeworth expansion. If we just ignore it, then the probability in the preceding display can be expanded with the help of (23.12) as

$$\Phi\left(z_\alpha - \frac{p_1(z_\alpha \mid P)}{\sqrt{n}}\right) + \frac{p_1(z_\alpha - n^{-1/2}p_1(z_\alpha \mid P) \mid P)}{\sqrt{n}}\phi\left(z_\alpha - \frac{p_1(z_\alpha \mid P)}{\sqrt{n}}\right) + O\left(\frac{1}{n}\right).$$

The linear term of the Taylor expansion of Φ cancels the leading term of the Taylor expansion of the middle term. Thus the expression in the last display is equal to $1 - \alpha$ up to the order

$O(n^{-1})$, whence the coverage error of a percentile t–confidence interval is of the order $O(n^{-1})$.

For percentile intervals we proceed in the same manner, this time inverting the Edgeworth expansion of the unstudentized statistic. The (conditional) Cornish-Fisher expansion for the quantile $\hat{\xi}_{n,\alpha}$ of $\hat{\theta}_n^* - \hat{\theta}_n$ takes the form

$$\frac{\hat{\xi}_{n,\alpha}}{\hat{\sigma}_n} = z_\alpha - \frac{q_1(z_\alpha \mid \hat{P}_n)}{\sqrt{n}} + O_P\left(\frac{1}{n}\right).$$

The coverage probabilities of percentile confidence intervals can be expressed in probabilities of the type

$$P(\hat{\theta}_n - \theta \le \hat{\xi}_{n,\alpha} \mid P) = P\left(\frac{\hat{\theta}_n - \theta}{\hat{\sigma}_n} \le \frac{\hat{\xi}_{n,\alpha}}{\hat{\sigma}_n} \mid P\right).$$

Insert the Cornish-Fisher expansion, again neglect the $O_P(n^{-1})$ term, and use the Edgeworth expansion (23.12) to rewrite this as

$$\Phi\left(z_\alpha - \frac{q_1(z_\alpha \mid P)}{\sqrt{n}}\right) + \frac{p_1\left(z_\alpha - n^{-1/2}q_1(z_\alpha \mid P) \mid P\right)}{\sqrt{n}}\phi\left(z_\alpha - \frac{q_1(z_\alpha \mid P)}{\sqrt{n}}\right) + O\left(\frac{1}{n}\right).$$

Because p_1 and q_1 are different, the cancellation that was found for the percentile t-method does not occur, and this is generally equal to $1 - \alpha$ up to the order $O(n^{-1/2})$. Consequently, asymmetric percentile intervals have coverage error of the order $O(n^{-1/2})$. On the other hand, the coverage probability of the symmetric confidence interval $[\hat{\theta}_n - \hat{\xi}_{n,\alpha}, \hat{\theta}_n - \hat{\xi}_{n,1-\alpha}]$ is equal to the expression in the preceding display minus this expression evaluated for $1 - \alpha$ instead of α. In the common situation that both polynomials p_1 and q_1 are even, the terms of order $O(n^{-1/2})$ cancel, and the difference is equal to $1 - 2\alpha$ up to the order $O(n^{-1})$. Then the percentile two-tailed confidence interval has the same order of correctness as the symmetric normal interval and the percentile t-intervals.

Notes

For a wider scope on the applications of the bootstrap, see the book [44], whose first author Efron is the inventor of the bootstrap. Hall [72] gives a detailed treatment of higher-order expansions of a number of bootstrap schemes. For more information concerning the consistency of the empirical bootstrap, and the consistency of the bootstrap under the application of the delta method, see Chapter 3.6 and Section 3.9.3 of [146], or the paper by Giné and Zinn [58].

PROBLEMS

1. Let $\hat{F}_n$ be a sequence of random distribution functions and F a continuous, fixed-distribution function. Show that the following statements are equivalent:
   (i) $\hat{F}_n(x) \xrightarrow{P} F(x)$ for every x.
   (ii) $\sup_x \left| \hat{F}_n(x) - F(x) \right| \xrightarrow{P} 0$.

2. Compare in a simulation study Efron's percentile method, the normal approximation in combination with Fisher's transformation, and the percentile method to set a confidence interval for the correlation coefficient.

3. Let $X_{(n)}$ be the maximum of a sample of size n from the uniform distribution on [0, 1], and let $X^*_{(n)}$ be the maximum of a sample of size n from the empirical distribution $\mathbb{P}_n$ of the first sample. Show that $P(X^*_{(n)} = X_{(n)} \mid \mathbb{P}_n) \to 1 - e^{-1}$. What does this mean regarding the consistency of the empirical bootstrap estimator of the distribution of the maximum?

4. Devise a bootstrap scheme for setting confidence intervals for β in the linear regression model $Y_i = \alpha + \beta x_i + e_i$. Show consistency.

5. (Parametric bootstrap.) Let $\hat{\theta}_n$ be an estimator based on observations from a parametric model P_θ such that $\sqrt{n}(\hat{\theta}_n - \theta - h_n/\sqrt{n})$ converges under $\theta + h_n/\sqrt{n}$ to a continuous distribution L_θ for every converging sequence h_n and every θ. (This is slightly stronger than *regularity* as defined in the chapter on asymptotic efficiency.) Show that the *parametric bootstrap* is consistent: If $\hat{\theta}^*_n$ is $\hat{\theta}_n$ computed from observations obtained from $P_{\hat{\theta}_n}$, then $\sqrt{n}(\hat{\theta}^*_n - \hat{\theta}_n) \rightsquigarrow L_\theta$ conditionally on the original observations, in probability. (The conditional law of $\sqrt{n}(\hat{\theta}^*_n - \hat{\theta}_n)$ is $L_{n,\hat{\theta}}$ if $L_{n,\theta}$ is the distribution of $\sqrt{n}(\hat{\theta}_n - \theta)$ under θ.)

6. Suppose that $\sqrt{n}(\hat{\theta}_n - \theta) \rightsquigarrow T$ and $\sqrt{n}(\hat{\theta}^*_n - \hat{\theta}_n) \rightsquigarrow T$ in probability given the original observations. Show that $\sqrt{n}\big(\phi(\hat{\theta}^*_n) - \phi(\hat{\theta}_n)\big) \rightsquigarrow \phi'_\theta(T)$ in probability for every map ϕ that is differentiable at θ.

7. Let U_n be a U-statistic based on a random sample $X_1, \ldots, X_n$ with kernel $h(x, y)$ such that both $Eh(X_1, X_1)$ and $Eh^2(X_1, X_2)$ are finite. Let $\hat{U}^*_n$ be the same U-statistic based on a sample $X^*_1, \ldots, X^*_n$ from the empirical distribution of $X_1, \ldots, X_n$. Show that $\sqrt{n}(\hat{U}^*_n - U_n)$ converges conditionally in distribution to the same limit as $\sqrt{n}(U_n - \theta)$, almost surely.

8. Suppose that $\sqrt{n}(\hat{\theta}_n - \theta) \rightsquigarrow T$ and $\sqrt{n}(\hat{\theta}^*_n - \hat{\theta}_n) \rightsquigarrow T$ in probability given the original observations. Show that, unconditionally, $\sqrt{n}(\hat{\theta}_n - \theta, \hat{\theta}^*_n - \hat{\theta}_n) \rightsquigarrow (S, T)$ for independent copies S and T of T. Deduce the unconditional limit distribution of $\sqrt{n}(\hat{\theta}^*_n - \theta)$.

24

Nonparametric Density Estimation

This chapter is an introduction to estimating densities if the underlying density of a sample of observations is considered completely unknown, up to existence of derivatives. We derive rates of convergence for the mean square error of kernel estimators and show that these cannot be improved. We also consider regularization by monotonicity.

24.1   Introduction

Statistical models are called *parametric models* if they are described by a Euclidean parameter (in a nice way). For instance, the binomial model is described by a single parameter p, and the normal model is given through two unknowns: the mean and the variance of the observations. In many situations there is insufficient motivation for using a particular parametric model, such as a normal model. An alternative at the other end of the scale is a *nonparametric model*, which leaves the underlying distribution of the observations essentially free. In this chapter we discuss one example of a problem of nonparametric estimation: estimating the density of a sample of observations if nothing is known a priori. From the many methods for this problem, we present two: kernel estimation and monotone estimation. Notwithstanding its simplicity, this method can be fully asymptotically efficient.

24.2   Kernel Estimators

The most popular nonparametric estimator of a distribution based on a sample of observations is the empirical distribution, whose properties are discussed at length in Chapter 19. This is a discrete probability distribution and possesses no density. The most popular method of nonparametric density estimation, the *kernel method*, can be viewed as a recipe to "smooth out" the pointmasses of sizes $1/n$ in order to turn the empirical distribution into a continuous distribution.

Let $X_1, \ldots, X_n$ be a random sample from a density f on the real line. If we would know that f belongs to the normal family of densities, then the natural estimate of f would be the normal density with mean $\overline{X}_n$ and variance S_n^2, or the function

$$x \mapsto \frac{1}{S_n \sqrt{2\pi}} e^{-\frac{1}{2}(x - \overline{X}_n)^2 / S_n^2}.$$

341

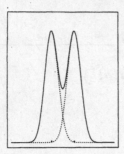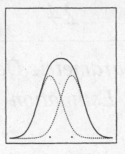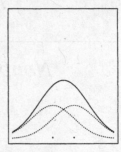

Figure 24.1. The kernel estimator with normal kernel and two observations for three bandwidths: small (*left*), intermediate (*center*) and large (*right*). The figures show both the contributions of the two observations separately (*dotted lines*) and the kernel estimator (*solid lines*), which is the sum of the two dotted lines.

In this section we suppose that we have no prior knowledge of the form of f and want to "let the data speak as much as possible for themselves."

Let K be a probability density with mean 0 and variance 1, for instance the standard normal density. A *kernel estimator* with *kernel* or *window K* is defined as

$$\hat{f}(x) = \frac{1}{n}\sum_{i=1}^{n}\frac{1}{h}K\left(\frac{x-X_i}{h}\right).$$

Here h is a positive number, still to be chosen, called the *bandwidth* of the estimator. It turns out that the choice of the kernel K is far less crucial for the quality of $\hat{f}$ as an estimator of f than the choice of the bandwidth. To obtain the best convergence rate the requirement that $K \geq 0$ may have to be dropped.

A kernel estimator is an example of a *smoothing method*. The construction of a density estimator can be viewed as a recipe for smoothing out the total mass 1 over the real line. Given a random sample of n observations it is reasonable to start with allocating the total mass in packages of size $1/n$ to the observations. Next a kernel estimator distributes the mass that is allocated to X_i smoothly around X_i, not homogenously, but according to the kernel and bandwidth.

More formally, we can view a kernel estimator as the sum of n small "mountains" given by the functions

$$x \mapsto \frac{1}{nh}K\left(\frac{x-X_i}{h}\right).$$

Every small mountain is centred around an observation X_i and has area $1/n$ under it, for any bandwidth h. For a small bandwidth the mountain is very concentrated (a peak), while for a large bandwidth the mountain is low and flat. Figure 24.1 shows how the mountains add up to a single estimator. If the bandwidth is small, then the mountains remain separated and their sum is peaky. On the other hand, if the bandwidth is large, then the sum of the individual mountains is too flat. Intermediate values of the bandwidth should give the best results.

Figure 24.2 shows the kernel method in action on a sample from the normal distribution. The solid and dotted lines are the estimator and the true density, respectively. The three pictures give the kernel estimates using three different bandwidths – small, intermediate, and large – each time with the standard normal kernel.

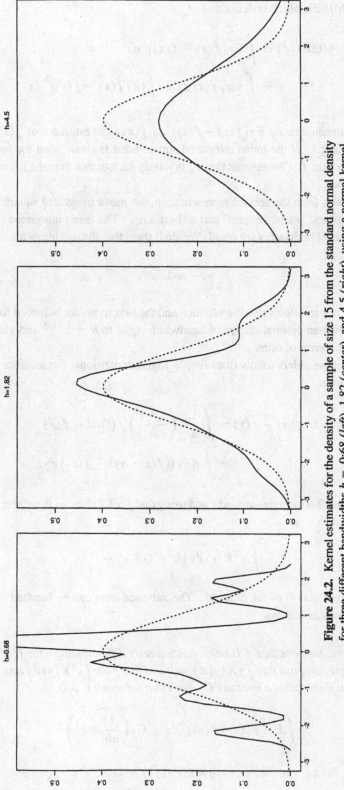

Figure 24.2. Kernel estimates for the density of a sample of size 15 from the standard normal density for three different bandwidths $h = 0.68$ (*left*), 1.82 (*center*), and 4.5 (*right*), using a normal kernel. The dotted line gives the true density.

A popular criterion to judge the quality of density estimators is the *mean integrated square error* (MISE), which is defined as

$$\text{MISE}_f(\hat{f}) = \int \text{E}_f\big(\hat{f}(x) - f(x)\big)^2 dx$$

$$= \int \text{var}_f \hat{f}(x)\, dx + \int \big(\text{E}_f \hat{f}(x) - f(x)\big)^2 dx.$$

This is the mean square error $\text{E}_f\big(\hat{f}(x) - f(x)\big)^2$ of $\hat{f}(x)$ as an estimator of $f(x)$ integrated over the argument x. If the mean integrated square error is small, then the function $\hat{f}$ is close to the function f. (We assume that $\hat{f}_n$ is jointly measurable to make the mean square error well defined.)

As can be seen from the second representation, the mean integrated square error is the sum of an integrated "variance term" and a "bias term." The mean integrated square error can be small only if both terms are small. We shall show that the two terms are of the orders

$$\frac{1}{nh}, \quad \text{and} \quad h^4,$$

respectively. Then it follows that the variance and the bias terms are balanced for $(nh)^{-1} \sim h^4$, which implies an optimal choice of bandwidth equal to $h \sim n^{-1/5}$ and yields a mean integrated square error of order $n^{-4/5}$.

Informally, these orders follow from simple Taylor expansions. For instance, the bias of $\hat{f}(x)$ can be written

$$\text{E}_f \hat{f}(x) - f(x) = \int \frac{1}{h} K\left(\frac{x-t}{h}\right) f(t)\, dt - f(x)$$

$$= \int K(y)\big(f(x-hy) - f(x)\big)\, dy.$$

Developing f in a Taylor series around x and using that $\int y K(y)\, dy = 0$, we see, informally, that this is equal to

$$\int y^2 K(y)\, dy \tfrac{1}{2} h^2 f''(x) + \cdots.$$

Thus, the squared bias is of the order h^4. The variance term can be handled similarly. A precise theorem is as follows.

24.1 Theorem. *Suppose that f is twice continuously differentiable with $\int |f''(x)|^2 dx < \infty$. Furthermore, suppose that $\int y K(y)\, dy = 0$ and that both $\int y^2 K(y)\, dy$ and $\int K^2(y)\, dy$ are finite. Then there exists a constant C_f such that for small $h > 0$*

$$\int \text{E}_f\big(\hat{f}(x) - f(x)\big)^2 dx \le C_f\left(\frac{1}{nh} + h^4\right).$$

Consequently, for $h_n \sim n^{-1/5}$, we have $\text{MISE}_f(\hat{f}_n) = O(n^{-4/5})$.

Proof. Because a kernel estimator is an average of n independent random variables, the variance of $\hat{f}(x)$ is $(1/n)$ times the variance of one term. Hence

$$\mathrm{var}_f\,\hat{f}(x) = \frac{1}{n}\mathrm{var}_f\,\frac{1}{h}K\!\left(\frac{x-X_1}{h}\right) \le \frac{1}{nh^2}\mathrm{E}_f K^2\!\left(\frac{x-X_1}{h}\right)$$

$$= \frac{1}{nh}\int K^2(y)f(x-hy)\,dy.$$

Take the integral with repect to x on both left and right sides. Because $\int f(x-hy)\,dx = 1$ is the same for every value of hy, the right side reduces to $(nh)^{-1}\int K^2(y)\,dy$, by Fubini's theorem. This concludes the proof for the variance term.

To upper bound the bias term we first write the bias $\mathrm{E}_f\hat{f}(x) - f(x)$ in the form as given preceding the statement of the theorem. Next we insert the formula

$$f(x+h) - f(x) = hf'(x) + h^2\int_0^1 f''(x+sh)(1-s)\,ds.$$

This is a Taylor expansion with the Laplacian representation of the remainder. We obtain

$$\mathrm{E}_f\hat{f}(x) - f(x) = \iint_0^1 K(y)\left[-hyf'(x) + (hy)^2 f''(x-shy)(1-s)\right]ds\,dy.$$

Because the kernel K has mean zero by assumption, the first term inside the square brackets can be deleted. Using the Cauchy-Schwarz inequality $(EUV)^2 \le EU^2 EV^2$ on the variables $U = Y$ and $V = Yf''(x - ShY)(1-S)$ for Y distributed with density K and S uniformly distributed on $[0, 1]$ independent of Y, we see that the square of the bias is bounded above by

$$h^4\int K(y)y^2\,dy \iint_0^1 K(y)y^2 f''(x-shy)^2 (1-s)^2\,ds\,dy.$$

The integral of this with respect to x is bounded above by

$$h^4\left(\int K(y)y^2\,dy\right)^2 \int f''(x)^2\,dx\,\frac{1}{3}.$$

This concludes the derivation for the bias term.

The last assertion of the theorem is trivial. ∎

The rate $O(n^{-4/5})$ for the mean integrated square error is not impressive if we compare it to the rate that could be achieved if we knew a priori that f belonged to some parametric family of densities f_θ. Then, likely, we would be able to estimate θ by an estimator such that $\hat{\theta} = \theta + O_P(n^{-1/2})$, and we would expect

$$\mathrm{MISE}_\theta(f_{\hat\theta}) = \int \mathrm{E}_\theta\big(f_{\hat\theta}(x) - f_\theta(x)\big)^2\,dx \sim \mathrm{E}_\theta(\hat{\theta}-\theta)^2 = O\!\left(\frac{1}{n}\right).$$

This is a factor $n^{-1/5}$ smaller than the mean square error of a kernel estimator.

This loss in efficiency is only a modest price. After all, the kernel estimator works for every density that is twice continuously differentiable whereas the parametric estimator presumably fails miserably if the true density does not belong to the postulated parametric model.

Moreover, the lost factor $n^{-1/5}$ can be (almost) covered if we assume that f has sufficiently many derivatives. Suppose that f is m times continuously differentiable. Drop the condition that the kernel K is a probability density, but use a kernel K such that

$$\int K(y)\,dy = 1, \qquad \int yK(y)\,dy = 0, \dots, \qquad \int y^{m-1}K(y)\,dy = 0,$$

$$\int |y|^m K(y)\,dy < \infty, \qquad \int K^2(y)\,dy < \infty.$$

Then, by the same arguments as before, the bias term can be expanded in the form

$$\mathrm{E}_f \hat{f}(x) - f(x) = \int K(y)\big(f(x - hy) - f(x)\big)\,dy$$

$$= \int K(y)\frac{1}{m!}(-1)^m h^m y^m f^{(m)}(x)\,dy + \cdots$$

Thus the squared bias is of the order h^{2m} and the bias-variance trade-off $(nh)^{-1} \sim h^{2m}$ is solved for $h \sim n^{1/(2m+1)}$. This leads to a mean square error of the order $n^{-2m/(2m+1)}$, which approaches the "parametric rate" n^{-1} as $m \to \infty$. This claim is made precise in the following theorem, whose proof proceeds as before.

24.2　Theorem. *Suppose that f is m times continuously differentiable with $\int |f^{(m)}(x)|^2 dx < \infty$. Then there exists a constant C_f such that for small $h > 0$*

$$\int \mathrm{E}_f\big(\hat{f}(x) - f(x)\big)^2 dx \le C_f\left(\frac{1}{nh} + h^{2m}\right).$$

Consequently, for $h_n \sim n^{-1/(2m+1)}$, we have $\mathrm{MISE}_f(\hat{f}_n) = O(n^{-2m/(2m+1)})$.

In practice, the number of derivatives of f is usually unknown. In order to choose a proper bandwidth, we can use *cross-validation* procedures. These yield a data-dependent bandwidth and also solve the problem of choosing the constant preceding $h^{-1/(2m+1)}$. The combined procedure of density estimator and bandwidth selection is called *rate-adaptive* if the procedure attains the upper bound $n^{-2m/(2m+1)}$ for the mean integrated square error for every m.

24.3　Rate Optimality

In this section we show that the rate $n^{-2m/(2m+1)}$ of a kernel estimator, obtained in Theorem 24.2, is the best possible. More precisely, we prove the following. Inspection of the proof of Theorem 24.2 reveals that the constants C_f in the upper bound are uniformly bounded in f such that $\int |f^{(m)}(x)|^2 dx$ is uniformly bounded. Thus, letting $\mathcal{F}_{m,M}$ be the class of all probability densities such that this quantity is bounded by M, there is a constant $C_{m,M}$ such that the kernel estimator with bandwidth $h_n = n^{-1/(2m+1)}$ satisfies

$$\sup_{f \in \mathcal{F}_{m,M}} \mathrm{E}_f \int \big(\hat{f}_n(x) - f(x)\big)^2 dx \le C_{m,M}\left(\frac{1}{n}\right)^{2m/(2m+1)}.$$

In this section we show that this upper bound is sharp, and the kernel estimator rate optimal, in that the maximum risk on the left side is bounded below by a similar expression for *every* density estimator $\hat{f}_n$, for every fixed m and M.

The proof is based on a construction of subsets $\mathcal{F}_n \subset \mathcal{F}_{m,M}$, consisting of 2^{r_n} functions, with $r_n = \lfloor n^{1/(2m+1)} \rfloor$, and on bounding the supremum over $\mathcal{F}_{m,M}$ by the average over $\mathcal{F}_n$. Thus the number of elements in the average grows fairly rapidly with n. An approach, such as in section 14.5, based on the comparison of $\hat{f}_n$ at only two elements of $\mathcal{F}_{m,M}$ does not seem to work for the integrated risk, although such an approach readily yields a lower bound for the maximum risk $\sup_f \mathrm{E}_f \big(\hat{f}_n(x) - f(x) \big)^2$ at a fixed x.

The subset $\mathcal{F}_n$ is indexed by the set of all vectors $\theta \in \{0, 1\}^{r_n}$ consisting of sequences of r_n zeros or ones. For $h_n = n^{-1/(2m+1)}$, let $x_{n,1} < x_{n,2} < \cdots < x_{n,n}$ be a regular grid of meshwidth $2h_n$. For a fixed probability density f and a fixed function K with support $(-1, 1)$, define, for every $\theta \in \{0, 1\}^{r_n}$,

$$f_{n,\theta}(x) = f(x) + h_n^m \sum_{j=1}^{r_n} \theta_j K \left(\frac{x - x_{n,j}}{h_n} \right).$$

If f is bounded away from zero on an interval containing the grid, $|K|$ is bounded, and $\int K(x)\, dx = 0$, then $f_{n,\theta}$ is a probability density, at least for large n. Furthermore,

$$\int \big| f_{n,\theta}^{(m)}(x) \big|^2 \, dx \leq 2 \int \big| f^{(m)}(x) \big|^2 \, dx + 2 h_n r_n \int \big| K^{(m)}(x) \big|^2 \, dx.$$

It follows that there exist many choices of f and K such that $f_{n,\theta} \in \mathcal{F}_{m,M}$ for every θ.

The following lemma gives a lower bound for the maximum risk over the parameter set $\{0, 1\}^r$, in an abstract form, applicable to the problem of estimating an arbitrary quantity $\psi(\theta)$ belonging to a metric space (with metric d). Let $H(\theta, \theta') = \sum_{i=1}^r |\theta_i - \theta_i'|$ be the *Hamming distance* on $\{0, 1\}^r$, which counts the number of positions at which θ and θ' differ. For two probability measures P and Q with densities p and q, write $\| P \wedge Q \|$ for $\int p \wedge q \, d\mu$.

24.3 Lemma (Assouad). *For any estimator T based on an observation in the experiment $\big(P_\theta : \theta \in \{0, 1\}^r \big)$, and any $p > 0$,*

$$\max_\theta 2^p \mathrm{E}_\theta d^p \big(T, \psi(\theta) \big) \geq \min_{H(\theta, \theta') \geq 1} \frac{d^p \big(\psi(\theta), \psi(\theta') \big)}{H(\theta, \theta')} \frac{r}{2} \min_{H(\theta, \theta')=1} \| P_\theta \wedge P_{\theta'} \|.$$

Proof. Define an estimator S, taking its values in $\Theta = \{0, 1\}^r$, by letting $S = \theta$ if $\theta' \mapsto d \big(T, \psi(\theta') \big)$ is minimal over Θ at $\theta' = \theta$. (If the minimum is not unique, choose a point of minimum in any consistent way.) By the triangle inequality, for any θ, $d \big(\psi(S), \psi(\theta) \big) \leq d \big(\psi(S), T \big) + d \big(\psi(\theta), T \big)$, which is bounded by $2 d \big(\psi(\theta), T \big)$, by the definition of S. If $d^p \big(\psi(\theta), \psi(\theta') \big) \geq \alpha H(\theta, \theta')$ for all pairs $\theta \neq \theta'$, then

$$2^p \mathrm{E}_\theta d^p \big(T, \psi(\theta) \big) \geq \mathrm{E}_\theta d^p \big(\psi(S), \psi(\theta) \big) \geq \alpha \mathrm{E}_\theta H(S, \theta).$$

The maximum of this expression over Θ is bounded below by the average, which, apart from the factor α, can be written

$$\frac{1}{2^r} \sum_\theta \sum_{j=1}^r E_\theta |S_j - \theta_j| = \frac{1}{2} \sum_{j=1}^r \left(\frac{1}{2^{r-1}} \sum_{\theta \,:\, \theta_j=0} \int S_j \, dP_\theta + \frac{1}{2^{r-1}} \sum_{\theta \,:\, \theta_j=1} \int (1 - S_j) \, dP_\theta \right).$$

This is minimized over S by choosing S_j for each j separately to minimize the jth term in the sum. The expression within brackets is the sum of the error probabilities of a test of

$$\overline{P}_{0,j} = \frac{1}{2^{r-1}} \sum_{\theta \,:\, \theta_j=0} P_\theta, \quad \text{versus} \quad \overline{P}_{1,j} = \frac{1}{2^{r-1}} \sum_{\theta \,:\, \theta_j=1} P_\theta.$$

Equivalently, it is equal to 1 minus the difference of power and level. In Lemma 14.30 this was seen to be at least $1 - \frac{1}{2}\|\overline{P}_{0,j} - \overline{P}_{1,j}\| = \|\overline{P}_{0,j} \wedge \overline{P}_{1,j}\|$. Hence the preceding display is bounded below by

$$\frac{1}{2} \sum_{j=1}^r \|\overline{P}_{0,j} \wedge \overline{P}_{1,j}\|.$$

Because the minimum $\overline{p}_m \wedge \overline{q}_m$ of two averages of numbers is bounded below by the average $m^{-1} \sum p_i \wedge q_i$ of the minima, the same is true for the total variation norm of a minimum: $\|\overline{P}_m \wedge \overline{Q}_m\| \geq m^{-1} \sum \|P_i \wedge Q_i\|$. The 2^{r-1} terms P_θ and $P_{\theta'}$ in the averages $\overline{P}_{0,j}$ and $\overline{P}_{1,j}$ can be ordered and matched such that each pair θ and θ' differ only in their jth coordinate. Conclude that the preceding display is bounded below by $\frac{1}{2} \sum_{j=1}^r \min \|P_\theta \wedge P_{\theta'}\|$, in which the minimum is taken over all pairs θ and θ' that differ by exactly one coordinate. ∎

We wish to apply Assouad's lemma to the product measures resulting from the densities $f_{n,\theta}$. Then the following inequality, obtained in the proof of Lemma 14.31, is useful. It relates the total variation, affinity, and Hellinger distance of product measures:

$$\|P^n \wedge Q^n\| \geq \tfrac{1}{2} A^2(P^n, Q^n) = \tfrac{1}{2}\left(1 - \tfrac{1}{2} H^2(P, Q)\right)^{2n}.$$

24.4   Theorem. *There exists a constant $D_{m,M}$ such that for any density estimator $\hat{f}_n$*

$$\sup_{f \in \mathcal{F}_{m,M}} E_f \int \left(\hat{f}_n(x) - f(x)\right)^2 dx \geq D_{m,M} \left(\frac{1}{n}\right)^{2m/(2m+1)}.$$

Proof. Because the functions $f_{n,\theta}$ are bounded away from zero and infinity, uniformly in θ, the squared Hellinger distance

$$\int \left(f_{n,\theta}^{1/2} - f_{n,\theta'}^{1/2}\right)^2 dx = \int \left(\frac{f_{n,\theta} - f_{n,\theta'}}{f_{n,\theta}^{1/2} + f_{n,\theta'}^{1/2}}\right)^2 dx$$

is up to constants equal to the squared L_2-distance between $f_{n,\theta}$ and $f_{n,\theta'}$. Because the functions $K\big((x - x_{n,j})/h_n\big)$ have disjoint supports, the latter is equal to

$$h_n^{2m} \sum_{j=1}^{r_n} |\theta_j - \theta_j'|^2 \int K^2\left(\frac{x - x_{n,j}}{h_n}\right) dx = h_n^{2m+1} H(\theta, \theta') \int K^2(x)\, dx.$$

This is of the order $1/n$. Inserting this in the lower bound given by Assouad's lemma, with $\psi(\theta) = f_{n,\theta}$ and $d\big(\psi(\theta), \psi(\theta')\big)$ the L_2-distance, we find up to a constant the lower bound $h_n^{2m+1}\,(r_n/2)\,\big(1 - O(1/n)\big)^{2n}$. ∎

24.4 Estimating a Unimodal Density

In the preceding sections the analysis of nonparametric density estimators is based on the assumption that the true density is smooth. This is appropriate for kernel-density estimation, because this is a smoothing method. It is also sensible to place some a priori restriction on the true density, because otherwise we cannot hope to achieve much beyond consistency. However, smoothness is not the only possible restriction. In this section we assume that the true density is monotone, or unimodal. We start with monotone densities and next view a unimodal density as a combination of two monotone pieces.

It is interesting that with monotone densities we can use maximum likelihood as the estimating principle. Suppose that $X_1, \ldots, X_n$ is a random sample from a Lebesgue density f on $[0, \infty)$ that is known to be nonincreasing. Then the maximum likelihood estimator $\hat{f}_n$ is defined as the nonincreasing probability density that maximizes the likelihood

$$f \mapsto \prod_{i=1}^{n} f(X_i).$$

This optimization problem would not have a solution if f were only restricted by possessing a certain number of derivatives, because very high peaks at the observations would yield an arbitrarily large likelihood. However, under monotonicity there is a unique solution.

The solution must necessarily be a left-continuous step function, with steps only at the observations. Indeed, if for a given f the limit from the right at $X_{(i-1)}$ is bigger than the limit from the left at $X_{(i)}$, then we can redistribute the mass on the interval $(X_{(i-1)}, X_{(i)}]$ by raising the value $f(X_{(i)})$ and lowering $f(X_{(i-1)}+)$, for instance by setting f equal to the constant value $(X_{(i)} - X_{(i-1)})^{-1} \int_{X_{(i-1)}}^{X_{(i)}} f(t)\, dt$ on the whole interval, resulting in an increase of the likelihood. By the same reasoning we see that the maximum likelihood estimator must be zero on $(X_{(n)}, \infty)$ (and $(-\infty, 0)$). Thus, with $f_i = \hat{f}_n(X_{(i)})$, finding the maximum likelihood estimator reduces to maximizing $\prod_{i=1}^{n} f_i$ under the side conditions (with $X_{(0)} = 0$)

$$f_1 \geq f_2 \geq \cdots \geq f_n \geq 0,$$

$$\sum_{i=1}^{n} f_i\big(X_{(i)} - X_{(i-1)}\big) = 1.$$

This problem has a nice graphical solution. The *least concave majorant* of the empirical distribution function $\mathbb{F}_n$ is defined as the smallest concave function $\hat{F}_n$ with $\hat{F}_n(x) \geq \mathbb{F}_n(x)$ for every x. This can be found by attaching a rope at the origin $(0, 0)$ and winding this (from above) around the empirical distribution function $\mathbb{F}_n$ (Figure 24.3). Because $\hat{F}_n$ is

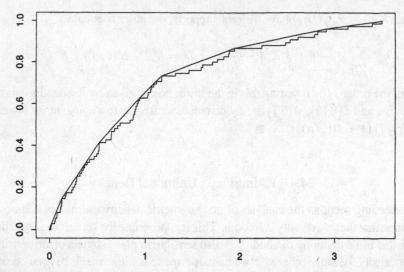

Figure 24.3. The empirical distribution and its concave majorant of a sample of size 75 from the exponential distribution.

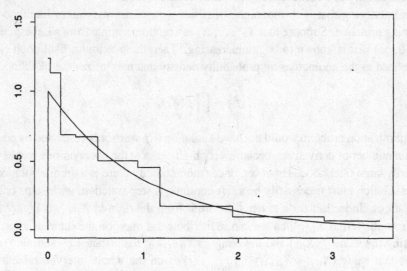

Figure 24.4. The derivative of the concave majorant of the empirical distribution and the true density of a sample of size 75 from the exponential distribution.

concave, its derivative is nonincreasing. Figure 24.4 shows the derivative of the concave majorant in Figure 24.3.

24.5   Lemma. *The maximum likelihood estimator $\hat{f}_n$ is the left derivative of the least concave majorant $\hat{F}_n$ of the empirical distribution $\mathbb{F}_n$, that is, on each of the intervals $(X_{(i-1)}, X_{(i)}]$ it is equal to the slope of $\hat{F}_n$ on this interval.*

Proof.   In this proof, let $\hat{f}_n$ denote the left derivative of the least concave majorant. We shall show that this maximizes the likelihood. Because the maximum likelihood estimator is necessarily constant on each interval $(X_{(i-1)}, X_{(i)}]$, we may restrict ourselves to densities f with this property. For such an f we can write $\log f = \sum a_i 1_{[0, x_{(i)}]}$ for the constants $a_i = \log f_i / f_{i+1}$ (with $f_{n+1} = 1$), and we obtain

$$\int \log f \, d\hat{F}_n = \sum_{i=1}^n a_i \, \hat{F}_n(X_{(i)}) \geq \sum_{i=1}^n a_i \, \mathbb{F}_n(X_{(i)}) = \int \log f \, d\mathbb{F}_n.$$

For $f = \hat{f}_n$ this becomes an equality. To see this, let $y_1 \leq y_2 \leq \cdots$ be the points where $\hat{F}_n$ touches $\mathbb{F}_n$. Then $\hat{f}_n$ is constant on each of the intervals $(y_{i-1}, y_i]$, so that we can write $\log \hat{f}_n = \sum b_i 1_{[0, y_i]}$, and obtain

$$\int \log \hat{f}_n \, d\hat{F}_n = \sum b_i \hat{F}_n(y_i) = \sum b_i \mathbb{F}_n(y_i) = \int \log \hat{f}_n \, d\mathbb{F}_n.$$

Third, by the identifiability property of the Kullback-Leibler divergence (see Lemma 5.35), for any probability density f,

$$\int \log \hat{f}_n \, d\hat{F}_n \geq \int \log f \, d\hat{F}_n,$$

with strict inequality unless $\hat{f}_n = f$. Combining the three displays, we see that $\hat{f}_n$ is the unique maximizer of $f \mapsto \int \log f \, d\mathbb{F}_n$. ∎

Maximizing the likelihood is an important motivation for taking the derivative of the concave majorant, but this operation also has independent value. Taking the concave majorant (or convex minorant) of the primitive function of an estimator and next differentiating the result may be viewed as a "smoothing" device, which is useful if the target function is known to be monotone. The estimator $\hat{f}_n$ can be viewed as the result of this procedure applied to the "naive" density estimator

$$\tilde{f}_n(x) = \frac{1}{n(X_{(i)} - X_{(i-1)})}, \qquad x \in (X_{(i-1)}, X_{(i)}].$$

This function is very rough and certainly not suitable as an estimator. Its primitive function is the polygon that linearly interpolates the extreme points of the empirical distribution function $\mathbb{F}_n$, and its smallest concave majorant coincides with the one of $\mathbb{F}_n$. Thus the derivative of the concave majorant of $\tilde{F}_n$ is exactly $\hat{f}_n$.

Consider the rate of convergence of the maximum likelihood estimator. Is the assumption of monotonicity sufficient to obtain a reasonable performance? The answer is affirmative if a rate of convergence of $n^{1/3}$ is considered reasonable. This rate is slower than the rate $n^{m/(2m+1)}$ of a kernel estimator if $m > 1$ derivatives exist and is comparable to this rate given one bounded derivative (even though we have not established a rate under $m = 1$). The rate of convergence $n^{1/3}$ can be shown to be best possible if only monotonicity is assumed. It is achieved by the maximum likelihood estimator.

24.6   Theorem. *If the observations are sampled from a compactly supported, bounded, monotone density f, then*

$$\int (\hat{f}_n(x) - f(x))^2 \, dx = O_P(n^{-2/3}).$$

Proof. This result is a consequence of a general result on maximum likelihood estimators of densities (e.g., Theorem 3.4.4 in [146].) We shall give a more direct proof using the convexity of the class of monotone densities.

The sequence $\| \hat{f}_n \|_\infty = \hat{f}_n(0)$ is bounded in probability. Indeed, by the characterization of $\hat{f}_n$ as the slope of the concave majorant of $\mathbb{F}_n$, we see that $\hat{f}_n(0) > M$ if and only if there exists $t > 0$ such that $\mathbb{F}_n(t) > Mt$. The claim follows, because, by concavity, $F(t) \le f(0)t$ for every t, and, by Daniel's theorem ([134, p. 642]),

$$P\left(\sup_{t > 0} \frac{\mathbb{F}_n(t)}{F(t)} > M \right) = \frac{1}{M}.$$

It follows that the rate of convergence of $\hat{f}_n$ is the same as the rate of the maximum likelihood estimator under the restriction that f is bounded by a (large) constant. In the remainder of the proof, we redefine $\hat{f}_n$ by the latter estimator.

Denote the true density by f_0. By the definition of $\hat{f}_n$ and the inequality $\log x \le 2(\sqrt{x} - 1)$,

$$0 \le \mathbb{F}_n \log \frac{\hat{f}_n}{\frac{1}{2}\hat{f}_n + \frac{1}{2}f_0} \le 2\mathbb{F}_n\left(\sqrt{\frac{2\hat{f}_n}{\hat{f}_n + f_0}} - 1 \right).$$

Therefore, we can obtain the rate of convergence of $\hat{f}_n$ by an application of Theorem 5.52 or 5.55 with $m_f = \sqrt{2f/(f + f_0)}$.

Because $(m_f - m_{f_0})(f_0 - f) \le 0$ for every f and f_0 it follows that $F_0(m_f - m_{f_0}) \le F(m_f - m_{f_0})$ and hence

$$F_0(m_f - m_{f_0}) \le \frac{1}{2}(F_0 + F)(m_f - m_{f_0}) = -\frac{1}{2}h^2\left(f, \frac{1}{2}f + \frac{1}{2}f_0\right) \lesssim -h^2(f, f_0),$$

in which the last inequality is elementary calculus. Thus the first condition of Theorem 5.52 is satisfied relative to the Hellinger distance h, with $\alpha = 2$.

The map $f \mapsto m_f$ is increasing. Therefore, it turns brackets $[f_1, f_2]$ for the functions $x \mapsto f(x)$ into brackets $[m_{f_1}, m_{f_2}]$ for the functions $x \mapsto m_f(x)$. The squared $L_2(F_0)$-size of these brackets satisfies

$$F_0(m_{f_1} - m_{f_2})^2 \le 4h^2(f_1, f_2).$$

It follows that the $L_2(F_0)$-bracketing numbers of the class of functions m_f can be bounded by the h-bracketing numbers of the functions f. The latter are the $L_2(\lambda)$-bracketing numbers of the functions $\sqrt{f}$, which are monotone and bounded by assumption. In view of Example 19.11,

$$\log N_{[]}\left(2\varepsilon, \{m_f : f \in \mathcal{F}\}, L_2(F_0)\right) \le \log N_{[]}\left(\varepsilon, \sqrt{\mathcal{F}}, L_2(\lambda)\right) \lesssim \frac{1}{\varepsilon}.$$

Because the functions m_f are uniformly bounded, the maximal inequality Lemma 19.36 gives, with $J(\delta) = \int_0^\delta \sqrt{1/\varepsilon}\, d\varepsilon = 2\sqrt{\delta}$,

$$\mathrm{E}_{f_0} \sup_{h(f, f_0) < \delta} \left| \mathbb{G}_n(f - f_0) \right| \lesssim \sqrt{\delta}\left(1 + \frac{\sqrt{\delta}}{\delta^2 \sqrt{n}} \right).$$

Therefore, Theorem 5.55 applies with $\phi_n(\delta)$ equal to the right side, and the Hellinger distance, and we conclude that $h(\hat{f}_n, f_0) = O_P(n^{-1/3})$.

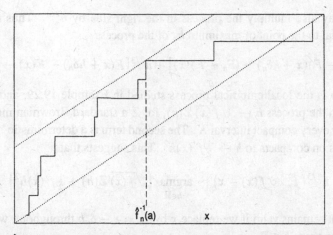

Figure 24.5. If $\hat{f}_n(x) \leq a$, then a line of slope a moved down vertically from $+\infty$ first hits $\mathbb{F}_n$ to the left of x. The point where the line hits is the point at which $\mathbb{F}_n$ is farthest above the line of slope a through the origin.

The $L_2(\lambda)$-distance between uniformly bounded densities is bounded up to a constant by the Hellinger distance, and the theorem follows. ∎

The most striking known results about estimating a monotone density concern limit distributions of the maximum likelihood estimator, for instance at a point.

24.7   Theorem.   *If f is differentiable at $x > 0$ with derivative $f'(x) < 0$, then, with $\{Z(h) : h \in \mathbb{R}\}$ a standard Brownian motion process (two-sided with $Z(0) = 0$),*

$$n^{1/3}\big(\hat{f}_n(x) - f(x)\big) \rightsquigarrow \big|4 f'(x) f(x)\big|^{1/3} \operatorname*{argmax}_{h \in \mathbb{R}}\{Z(h) - h^2\}.$$

Proof.   For simplicity we assume that f is continuously differentiable at x. Define a stochastic process $\{\hat{f}_n^{-1}(a) : a > 0\}$ by

$$\hat{f}_n^{-1}(a) = \operatorname*{argmax}_{s \geq 0}\{\mathbb{F}_n(s) - as\},$$

in which the largest value is chosen when multiple maximizers exist. The suggestive notation is justified, as the function $\hat{f}_n^{-1}$ is the inverse of the maximum likelihood estimator $\hat{f}_n$ in that $\hat{f}_n(x) \leq a$ if and only if $\hat{f}_n^{-1}(a) \leq x$, for every x and a. This is explained in Figure 24.5. We first derive the limit distribution of $\hat{f}_n^{-1}$. Let $\delta_n = n^{-1/3}$.

By the change of variable $s \mapsto x + h\delta_n$ in the definition of $\hat{f}_n^{-1}$, we have

$$n^{1/3}\big(\hat{f}_n^{-1} \circ f(x) - x\big) = \operatorname*{argmax}_{h \geq -n^{1/3}x}\{\mathbb{F}_n(x + h\delta_n) - f(x)(x + h\delta_n)\}.$$

Because the location of a maximum does not change by a vertical shift of the whole function, we can drop the term $f(x)x$ in the right side, and we may add a term $\mathbb{F}_n(x)$. For the same reason we may also multiply the process in the right side by $n^{2/3}$. Thus the preceding display is equal to the point of maximum $\hat{h}_n$ of the process

$$n^{2/3}\big[(\mathbb{F}_n - F)(x + h\delta_n) - (\mathbb{F}_n - F)(x)\big] + n^{2/3}\big[F(x + h\delta_n) - F(x) - f(x)h\delta_n\big].$$

The first term is the local empirical process studied in Example 19.29, and converges in distribution to the process $h \mapsto \sqrt{f(x)}\,\mathbb{Z}(h)$, for $\mathbb{Z}$ a standard Brownian motion process, in $\ell^\infty(K)$, for every compact interval K. The second term is a deterministic "drift" process and converges on compacta to $h \mapsto \frac{1}{2}f'(x)h^2$. This suggests that

$$n^{1/3}\big(\hat{f}_n^{-1} \circ f(x) - x\big) \rightsquigarrow \underset{h \in \mathbb{R}}{\mathrm{argmax}}\big\{\sqrt{f(x)}\,\mathbb{Z}(h) + \tfrac{1}{2}f'(x)h^2\big\}.$$

This argument remains valid if we replace x by $x_n = x - \delta_n b$ throughout, where the limit is the same for every $b \in \mathbb{R}$.

We can write the limit in a more attractive form by using the fact that the processes $h \mapsto \mathbb{Z}(\sigma h)$ and $h \mapsto \sqrt{\sigma}\mathbb{Z}(h)$ are equal in distribution for every $\sigma > 0$. First, apply the change of variables $h \mapsto \sigma h$, next pull σ out of $\mathbb{Z}(\sigma h)$, then divide the process by $\sqrt{f(x)\sigma}$, and finally choose σ such that the quadratic term reduces to $-h^2$, that is $\sqrt{f(x)\sigma} = -\frac{1}{2}f'(x)\sigma^2$. Then we obtain, for every $b \in \mathbb{R}$,

$$n^{1/3}\big(\hat{f}_n^{-1} \circ f(x - \delta_n b) - (x - \delta_n b)\big) \rightsquigarrow \left(\frac{\sqrt{f(x)}}{-\frac{1}{2}f'(x)}\right)^{2/3} \underset{h \in \mathbb{R}}{\mathrm{argmax}}\big\{\mathbb{Z}(h) - h^2\big\}.$$

The connection with the limit distribution of $\hat{f}_n(x)$ is that

$$\mathrm{P}\big(n^{1/3}\big(\hat{f}_n(x) - f(x)\big) \le -bf'(x)\big) = \mathrm{P}\big(\hat{f}_n(x) \le f(x - \delta_n b) + o(1)\big)$$

$$= \mathrm{P}\big(n^{1/3}\big(\hat{f}_n^{-1} \circ f(x - \delta_n b) - (x - \delta_n b)\big) \le b\big) + o(1).$$

Combined with the preceding display and simple algebra, this yields the theorem.

The preceding argument can be made rigorous by application of the argmax continuous-mapping theorem, Corollary 5.58. The limiting Brownian motion has continuous sample paths, and maxima of Gaussian processes are automatically unique (see, e.g., Lemma 2.6 in [87]). Therefore, we need only check that $\hat{h}_n = O_P(1)$, for which we apply Theorem 5.52 with

$$m_g = 1_{[0, x_n + g]} - 1_{[0, x_n]} - f(x_n)g.$$

(In Theorem 5.52 the function m_g can be allowed to depend on n, as is clear from its generalization, Theorem 5.55.) By its definition, $\hat{g}_n = \delta_n \hat{h}_n$ maximizes $g \mapsto \mathbb{F}_n m_g$, whence we wish to show that $\hat{g}_n = O_P(\delta_n)$. By Example 19.6 the bracketing numbers of the class of functions $\big\{1_{[0, x_n + g]} - 1_{[0, x_n]} : |g| < \delta\big\}$ are of the order δ/ε^2; the envelope function $|1_{[0, x_n + \delta]} - 1_{[0, x_n]}|$ has $L_2(F)$-norm of the order $\sqrt{f(x)\delta}$. By Corollary 19.35,

$$\mathrm{E}\sup_{|g| < \delta} |\mathbb{G}_n m_g| \lesssim \int_0^{\sqrt{\delta}} \sqrt{\log\frac{\delta}{\varepsilon^2}}\, d\varepsilon \lesssim \sqrt{\delta}.$$

By the concavity of F, the function $g \mapsto F(x_n + g) - F(x_n) - f(x_n)g$ is nonpositive and nonincreasing as g moves away from 0 in either direction (draw a picture.) Because

$f'(x_n) \to f'(x) < 0$, there exists a constant C such that, for sufficiently large n,

$$Fm_g = F(x_n + g) - F(x_n) - f(x_n)g \leq -C(g^2 \wedge |g|).$$

If we would know already that $\hat{g}_n \overset{P}{\to} 0$, then Theorem 5.52, applied with $\alpha = 2$ and $\beta = \frac{1}{2}$, yields that $\hat{g}_n = O_P(\delta_n)$.

The consistency of $\hat{g}_n$ can be shown by a direct argument. By the Glivenko-Cantelli theorem, for every $\varepsilon > 0$,

$$\sup_{|g| \geq \varepsilon} \mathbb{F}_n m_g \leq \sup_{|g| \geq \varepsilon} Fm_g + o_P(1) \leq -C \inf_{|g| \geq \varepsilon} (g^2 \wedge |g|) + o_P(1).$$

Because the right side is strictly smaller than $0 = \mathbb{F}_n m_0$, the maximizer $\hat{g}_n$ must be contained in $[-\varepsilon, \varepsilon]$ eventually. ∎

Results on density estimators at a point are perhaps not of greatest interest, because it is the overall shape of a density that counts. Hence it is interesting that the preceding theorem is also true in an L_1-sense, in that

$$n^{1/3} \int |\hat{f}_n(x) - f(x)| \, dx \rightsquigarrow \int |4f'(x)f(x)|^{1/3} \, dx \, \mathrm{E} \underset{h \in \mathbb{R}}{\operatorname{argmax}} \{\mathbb{Z}(h) - h^2\}.$$

This is true for every *strictly* decreasing, compactly supported, twice continuously differentiable true density f. For boundary cases, such as the uniform distribution, the behavior of $\hat{f}_n$ is very different. Note that the right side of the preceding display is degenerate. This is explained by the fact that the random variables $n^{1/3}(\hat{f}_n(x) - f(x))$ for different values of x are asymptotically independent, because they are only dependent on the observations X_i very close to x, so that the integral aggregates a large number of approximately independent variables. It is also known that $n^{1/6}$ times the difference between the left side and the right sides converges in distribution to a normal distribution with mean zero and variance not depending on f. For uniformly distributed observations, the estimator $\hat{f}_n(x)$ remains dependent on all n observations, even asymptotically, and attains a $\sqrt{n}$-rate of convergence (see [62]).

We define a density f on the real line to be *unimodal* if there exists a number M_f, such that f is nondecreasing on the interval $(-\infty, M_f]$ and nondecreasing on $[M_f, \infty)$. The *mode* M_f need not be unique. Suppose that we observe a random sample from a unimodal density.

If the true mode M_f is known a priori, then a natural extension of the preceding discussion is to estimate the distribution function F of the observations by the distribution function $\hat{F}_n$ that is the least concave majorant of $\mathbb{F}_n$ on the interval $[M_f, \infty)$ and the greatest convex minorant on $(-\infty, M_f]$. Next we estimate f by the derivative $\hat{f}_n$ of $\hat{F}_n$. Provided that none of the observations takes the value M_f, this estimator maximizes the likelihood, as can be shown by arguments as before. The limit results on monotone densities can also be extended to the present case. In particular, because the key in the proof of Theorem 24.7 is the characterization of $\hat{f}_n$ as the derivative of the concave majorant of $\mathbb{F}_n$, this theorem remains true in the unimodal case, with the same limit distribution.

If the mode is not known a priori, then the maximum likelihood estimator does not exist: The likelihood can be maximized to infinity by placing an arbitrary large mode at some fixed observation. It has been proposed to remedy this problem by restricting the likelihood to densities that have a modal interval of a given length (in which f must be constant and maximal). Alternatively, we could estimate the mode by an independent method and next apply the procedure for a known mode. Both of these possibilities break down unless f possesses some additional properties. A third possibility is to try every possible value M as a mode, calculate the estimator $\hat{f}_n^M$ for known mode M, and select the best fitting one. Here "best" could be operationalized as (nearly) minimizing the Kolmogorov-Smirnov distance $\|\hat{F}_n^M - \mathbb{F}_n\|_\infty$. It can be shown (see [13]) that this procedure renders the effect of the mode being unknown asymptotically negligible, in that

$$\int \left| \hat{f}_n^{\hat{M}}(x) - \hat{f}_n^{M_f}(x) \right| dx \leq 4\|\mathbb{F}_n - F\|_\infty = O_P\left(\frac{1}{\sqrt{n}}\right),$$

up to an arbitrarily small tolerance parameter if $\hat{M}$ only approximately achieves the minimum of $M \mapsto \|\hat{F}_n^M - \mathbb{F}_n\|_\infty$. This extra "error" is of lower order than the rate of convergence $n^{1/3}$ of the estimator with a known mode.

Notes

The literature on nonparametric density estimation, or "smoothing," is large, and there is an equally large literature concerning the parallel problem of nonparametric regression. Next to kernel estimation popular methods are based on classical series approximations, spline functions, and, most recently, wavelet approximation. Besides different methods, a good deal is known concerning other loss functions, for instance L_1-loss and automatic methods to choose a bandwidth. Most recently, there is a revived interest in obtaining exact constants in minimax bounds, rather than just rates of convergence. See, for instance, [14], [15], [36], [121], [135], [137], and [148] for introductions and further references. The kernel estimator is often named after its pioneers in the 1960s, Parzen and Rosenblatt, and was originally developed for smoothing the periodogram in spectral density estimation.

A lower bound for the maximum risk over Hölder classes for estimating a density at a single point was obtained in [46]. The lower bound for the L_2-risk is more recent. Birgé [12] gives a systematic study of upper and lower bounds and their relationship to the metric entropy of the model. An alternative for Assouad's lemma is Fano's lemma, which uses the Kullback-Leibler distance and can be found in, for example, [80].

The maximum likelihood estimator for a monotone density is often called the *Grenander estimator*, after the author who first characterized it in 1956. The very short proof of Lemma 24.5 is taken from [64]. The limit distribution of the Grenander estimator at a point was first obtained by Prakasa Rao in 1969 see [121]. Groeneboom [63] gives a characterization of the limit distribution and other interesting related results.

PROBLEMS

1. Show, informally, that under sufficient regularity conditions

$$\mathrm{MISE}_f(\hat{f}) \sim \frac{1}{nh} \int K^2(y)\,dy + \frac{1}{4}h^4 \int f''(x)^2\,dx \left(\int y^2 K(y)\,dy \right)^2.$$

What does this imply for an optimal choice of the bandwidth?

2. Let $X_1, \ldots, X_n$ be a random sample from the normal distribution with variance 1. Calculate the mean square error of the estimator $\phi(x - \overline{X}_n)$ of the common density.

3. Using the argument of section 14.5 and a submodel as in section 24.3, but with $r_n = 1$, show that the best rate for estimating a density at a fixed point is also $n^{-m/(2m+1)}$.

4. Using the argument of section 14.5, show that the rate of convergence $n^{1/3}$ of the maximum likelihood estimator for a monotone density is best possible.

5. **(Marshall's lemma.)** Suppose that F is concave on $[0, \infty)$ with $F(0) = 0$. Show that the least concave majorant $\hat{F}_n$ of $\mathbb{F}_n$ satisfies the inequality $\|\hat{F}_n - F\|_\infty \le \|\mathbb{F}_n - F\|_\infty$. What does this imply about the limiting behavior of $\hat{F}_n$?

25

Semiparametric Models

This chapter is concerned with statistical models that are indexed by infinite-dimensional parameters. It gives an introduction to the theory of asymptotic efficiency, and discusses methods of estimation and testing.

25.1 Introduction

Semiparametric models are statistical models in which the parameter is not a Euclidean vector but ranges over an "infinite-dimensional" parameter set. A different name is "model with a large parameter space." In the situation in which the observations consist of a random sample from a common distribution P, the *model* is simply the set $\mathcal{P}$ of all possible values of P: a collection of probability measures on the sample space. The simplest type of infinite-dimensional model is the *nonparametric model*, in which we observe a random sample from a completely unknown distribution. Then $\mathcal{P}$ is the collection of all probability measures on the sample space, and, as we shall see and as is intuitively clear, the empirical distribution is an asymptotically efficient estimator for the underlying distribution. More interesting are the intermediate models, which are not "nicely" parametrized by a Euclidean parameter, as are the standard classical models, but do restrict the distribution in an important way. Such models are often parametrized by infinite-dimensional parameters, such as distribution functions or densities, that express the structure under study. Many aspects of these parameters are estimable by the same order of accuracy as classical parameters, and efficient estimators are asymptotically normal. In particular, the model may have a natural parametrization $(\theta, \eta) \mapsto P_{\theta,\eta}$, where θ is a Euclidean parameter and η runs through a nonparametric class of distributions, or some other infinite-dimensional set. This gives a semiparametric model in the strict sense, in which we aim at estimating θ and consider η as a *nuisance parameter*. More generally, we focus on estimating the value $\psi(P)$ of some function $\psi : \mathcal{P} \mapsto \mathbb{R}^k$ on the model.

In this chapter we extend the theory of asymptotic efficiency, as developed in Chapters 8 and 15, from parametric to semiparametric models and discuss some methods of estimation and testing. Although the efficiency theory (lower bounds) is fairly complete, there are still important holes in the estimation theory. In particular, the extent to which the lower bounds are sharp is unclear. We limit ourselves to parameters that are $\sqrt{n}$-estimable, although in most semiparametric models there are many "irregular" parameters of interest that are outside the scope of "asymptotically normal" theory. Semiparametric testing theory has little more to offer than the comforting conclusion that tests based on efficient estimators are efficient. Thus, we shall be brief about it.

We conclude this introduction with a list of examples that shows the scope of semiparametric theory. In this description, X denotes a typical observation. Random vectors Y, Z, e, and f are used to describe the model but are not necessarily observed. The parameters θ and ν are always Euclidean.

25.1 Example (Regression). Let Z and e be independent random vectors and suppose that $Y = \mu_\theta(Z) + \sigma_\theta(Z)e$ for functions μ_θ and σ_θ that are known up to θ. The observation is the pair $X = (Y, Z)$. If the distribution of e is known to belong to a certain parametric family, such as the family of $N(0, \sigma^2)$-distributions, and the independent variables Z are modeled as constants, then this is just a classical regression model, allowing for heteroscedasticity. Semiparametric versions are obtained by letting the distribution of e range over all distributions on the real line with mean zero, or, alternatively, over all distributions that are symmetric about zero. $\square$

25.2 Example (Projection pursuit regression). Let Z and e be independent random vectors and let $Y = \eta(\theta^T Z) + e$ for a function η ranging over a set of (smooth) functions, and e having an $N(0, \sigma^2)$-distribution. In this model θ and η are confounded, but the direction of θ is estimable up to its sign. This type of regression model is also known as a *single-index* model and is intermediate between the classical regression model in which η is known and the nonparametric regression model $Y = \eta(Z) + e$ with η an unknown smooth function. An extension is to let the error distribution range over an infinite-dimensional set as well. $\square$

25.3 Example (Logistic regression). Given a vector Z, let the random variable Y take the value 1 with probability $1/(1 + e^{-r(Z)})$ and be 0 otherwise. Let $Z = (Z_1, Z_2)$, and let the function r be of the form $r(z_1, z_2) = \eta(z_1) + \theta^T z_2$. Observed is the pair $X = (Y, Z)$. This is a semiparametric version of the logistic regression model, in which the response is allowed to be nonlinear in part of the covariate. $\square$

25.4 Example (Paired exponential). Given an unobservable variable Z with completely unknown distribution, let $X = (X_1, X_2)$ be a vector of independent exponentially distributed random variables with parameters Z and $Z\theta$. The interest is in the ratio θ of the conditional hazard rates of X_1 and X_2. Modeling the "baseline hazard" Z as a random variable rather than as an unknown constant allows for heterogeneity in the population of all pairs (X_1, X_2), and hence ensures a much better fit than the two-dimensional parametric model in which the value z is a parameter that is the same for every observation. $\square$

25.5 Example (Errors-in-variables). The observation is a pair $X = (X_1, X_2)$, where $X_1 = Z + e$ and $X_2 = \alpha + \beta Z + f$ for a bivariate normal vector (e, f) with mean zero and unknown covariance matrix. Thus X_2 is a linear regression on a variable Z that is observed with error. The distribution of Z is unknown. $\square$

25.6 Example (Transformation regression). Suppose that $X = (Y, Z)$, where the random vectors Y and Z are known to satisfy $\eta(Y) = \theta^T Z + e$ for an unknown map η and independent random vectors e and Z with known or parametrically specified distributions.

The transformation η ranges over an infinite-dimensional set, for instance the set of all monotone functions. $\square$

25.7   Example (Cox). The observation is a pair $X = (T, Z)$ of a "survival time" T and a covariate Z. The distribution of Z is unknown and the conditional hazard function of T given Z is of the form $e^{\theta^T Z} \lambda(t)$ for λ being a completely unknown hazard function. The parameter θ has an interesting interpretation in terms of a ratio of hazards. For instance, if the ith coordinate Z_i of the covariate is a 0-1 variable then e^{θ_i} can be interpreted as the ratio of the hazards of two individuals whose covariates are $Z_i = 1$ and $Z_i = 0$, respectively, but who are identical otherwise. $\square$

25.8   Example (Copula). The observation X is two-dimensional with cumulative distribution function of the form $C_\theta\big(G_1(x_1), G_2(x_2)\big)$, for a parametric family of cumulative distribution functions C_θ on the unit square with uniform marginals. The marginal distribution functions G_i may both be completely unknown or one may be known. $\square$

25.9   Example (Frailty). Two survival times Y_1 and Y_2 are conditionally independent given variables (Z, W) with hazard function of the form $W e^{\theta^T Z} \lambda(y)$. The random variable W is not observed, possesses a gamma(ν, ν) distribution, and is independent of the variable Z which possesses a completely unknown distribution. The observation is $X = (Y_1, Y_2, Z)$. The variable W can be considered an unobserved regression variable in a Cox model. $\square$

25.10   Example (Random censoring). A "time of death" T is observed only if death occurs before the time C of a "censoring event" that is independent of T; otherwise C is observed. A typical observation X is a pair of a survival time and a 0-1 variable and is distributed as $\big(T \wedge C, 1\{T \le C\}\big)$. The distributions of T and C may be completely unknown. $\square$

25.11   Example (Interval censoring). A "death" that occurs at time T is only observed to have taken place or not at a known "check-up time" C. The observation is $X = \big(C, 1\{T \le C\}\big)$, and T and C are assumed independent with completely unknown or partially specified distributions. $\square$

25.12   Example (Truncation). A variable of interest Y is observed only if it is larger than a censoring variable C that is independent of Y; otherwise, nothing is observed. A typical observation $X = (X_1, X_2)$ is distributed according to the conditional distribution of (Y, C) given that $Y > C$. The distributions of Y and C may be completely unknown. $\square$

25.2   Banach and Hilbert Spaces

In this section we recall some facts concerning Banach spaces and, in particular, Hilbert spaces, which play an important role in this chapter.

Given a probality space $(\mathcal{X}, \mathcal{A}, P)$, we denote by $L_2(P)$ the set of measurable functions $g : \mathcal{X} \mapsto \mathbb{R}$ with $Pg^2 = \int g^2 \, dP < \infty$, where almost surely equal functions are identified. This is a *Hilbert space*, a complete inner-product space, relative to the inner product and norm

$$\langle g_1, g_2 \rangle = P g_1 g_2, \qquad \|g\| = \sqrt{P g^2}.$$

Given a Hilbert space $\mathbb{H}$, the *projection lemma* asserts that for every $g \in \mathbb{H}$ and convex, closed subset $C \subset \mathbb{H}$, there exists a unique element $\Pi g \in C$ that minimizes $c \mapsto \|g - c\|$ over C. If C is a closed, linear subspace, then the projection Πg can be characterized by the orthogonality relationship

$$\langle g - \Pi g, c \rangle = 0, \qquad \text{every } c \in C.$$

The proof is the same as in Chapter 11. If $C_1 \subset C_2$ are two nested, closed subspaces, then the projection onto C_1 can be found by first projecting onto C_2 and next onto C_1. Two subsets C_1 and C_2 are *orthogonal*, notation $C_1 \perp C_2$, if $\langle c_1, c_2 \rangle = 0$ for every pair of $c_i \in C_i$. The projection onto the sum of two orthogonal closed subspaces is the sum of the projections. The *orthocomplement* $C^\perp$ of a set C is the set of all $g \perp C$.

A *Banach space* is a complete, normed space. The *dual space* $\mathbb{B}^*$ of a Banach space $\mathbb{B}$ is the set of all continuous, linear maps $b^* : \mathbb{B} \mapsto \mathbb{R}$. Equivalently, all linear maps such that $|b^*(b)| \le \|b^*\| \|b\|$ for every $b \in \mathbb{B}$ and some number $\|b^*\|$. The smallest number with this property is denoted by $\|b^*\|$ and defines a norm on the dual space. According to the *Riesz representation theorem* for Hilbert spaces, the dual of a Hilbert space $\mathbb{H}$ consists of all maps

$$h \mapsto \langle h, h^* \rangle,$$

where h^* ranges over $\mathbb{H}$. Thus, in this case the dual space $\mathbb{H}^*$ can be identified with the space $\mathbb{H}$ itself. This identification is an isometry by the Cauchy-Schwarz inequality $|\langle h, h^* \rangle| \le \|h\| \|h^*\|$.

A linear map $A : \mathbb{B}_1 \mapsto \mathbb{B}_2$ from one Banach space into another is continuous if and only if $\|A b_1\|_2 \le \|A\| \|b_1\|$ for every $b_1 \in \mathbb{B}_1$ and some number $\|A\|$. The smallest number with this property is denoted by $\|A\|$ and defines a norm on the set of all continuous, linear maps, also called *operators*, from $\mathbb{B}_1$ into $\mathbb{B}_2$. Continuous, linear operators are also called "bounded," even though they are only bounded on bounded sets. To every continuous, linear operator $A : \mathbb{B}_1 \mapsto \mathbb{B}_2$ exists an *adjoint map* $A^* : \mathbb{B}_2^* \mapsto \mathbb{B}_1^*$ defined by $(A^* b_2^*) b_1 = b_2^* A b_1$. This is a continuous, linear operator of the same norm $\|A^*\| = \|A\|$. For Hilbert spaces the dual space can be identified with the original space and then the adjoint of $A : \mathbb{H}_1 \mapsto \mathbb{H}_2$ is a map $A^* : \mathbb{H}_2 \mapsto \mathbb{H}_1$. It is characterized by the property

$$\langle A h_1, h_2 \rangle_2 = \langle h_1, A^* h_2 \rangle_1, \qquad \text{every } h_1 \in \mathbb{H}_1, h_2 \in \mathbb{H}_2.$$

An operator between Euclidean spaces can be identified with a matrix, and its adjoint with the transpose. The adjoint of a restriction $A_0 : \mathbb{H}_{1,0} \subset \mathbb{H}_1 \mapsto \mathbb{H}_2$ of A is the composition $\Pi \circ A^*$ of the projection $\Pi : \mathbb{H}_1 \mapsto \mathbb{H}_{1,0}$ and the adjoint of the original A.

The range $R(A) = \{A b_1 : b_1 \in \mathbb{B}_1\}$ of a continuous, linear operator is not necessarily closed. By the "bounded inverse theorem" the range of a 1-1 continuous, linear operator between Banach spaces is closed if and only if its inverse is continuous. In contrast the kernel $N(A) = \{b_1 : A b_1 = 0\}$ of a continuous, linear operator is always closed. For an operator between Hilbert spaces the relationship $R(A)^\perp = N(A^*)$ follows readily from the characterization of the adjoint. The range of A is closed if and only if the range of A^* is closed if and only if the range of A^*A is closed. In that case $R(A^*) = R(A^*A)$.

If $A^*A: \mathbb{H}_1 \mapsto \mathbb{H}_1$ is continuously invertible (i.e., is 1-1 and onto with a continuous inverse), then $A(A^*A)^{-1}A^*: \mathbb{H}_2 \mapsto R(A)$ is the orthogonal projection onto the range of A, as follows easily by checking the orthogonality relationship.

25.3 Tangent Spaces and Information

Suppose that we observe a random sample $X_1, \ldots, X_n$ from a distribution P that is known to belong to a set $\mathcal{P}$ of probability measures on the sample space $(\mathcal{X}, \mathcal{A})$. It is required to estimate the value $\psi(P)$ of a functional $\psi: \mathcal{P} \mapsto \mathbb{R}^k$. In this section we develop a notion of information for estimating $\psi(P)$ given the model $\mathcal{P}$, which extends the notion of Fisher information for parametric models.

To estimate the parameter $\psi(P)$ given the model $\mathcal{P}$ is certainly harder than to estimate this parameter given that P belongs to a submodel $\mathcal{P}_0 \subset \mathcal{P}$. For every smooth parametric submodel $\mathcal{P}_0 = \{P_\theta: \theta \in \Theta\} \subset \mathcal{P}$, we can calculate the Fisher information for estimating $\psi(P_\theta)$. Then the information for estimating $\psi(P)$ in the whole model is certainly not bigger than the infimum of the informations over all submodels. We shall simply define the information for the whole model as this infimum. A submodel for which the infimum is taken (if there is one) is called *least favorable* or a "hardest" submodel.

In most situations it suffices to consider one-dimensional submodels $\mathcal{P}_0$. These should pass through the "true" distribution P of the observations and be differentiable at P in the sense of Chapter 7 on local asymptotic normality. Thus, we consider maps $t \mapsto P_t$ from a neighborhood of $0 \in [0, \infty)$ to $\mathcal{P}$ such that, for some measurable function $g: \mathcal{X} \mapsto \mathbb{R}$,[†]

$$\int \left[\frac{dP_t^{1/2} - dP^{1/2}}{t} - \frac{1}{2} g \, dP^{1/2} \right]^2 \to 0. \tag{25.13}$$

In other words, the parametric submodel $\{P_t: 0 < t < \varepsilon\}$ is differentiable in quadratic mean at $t = 0$ with *score function* g. Letting $t \mapsto P_t$ range over a collection of submodels, we obtain a collection of score functions, which we call a *tangent set* of the model $\mathcal{P}$ at P and denote by $\dot{\mathcal{P}}_P$. Because Pg^2 is automatically finite, the tangent space can be identified with a subset of $L_2(P)$, up to equivalence classes. The tangent set is often a linear space, in which case we speak of a *tangent space*.

Geometrically, we may visualize the model $\mathcal{P}$, or rather the corresponding set of "square roots of measures" $dP^{1/2}$, as a subset of the unit ball of $L_2(P)$, and $\dot{\mathcal{P}}_P$, or rather the set of all objects $\frac{1}{2} g \, dP^{1/2}$, as its tangent set.

Usually, we construct the submodels $t \mapsto P_t$ such that, for every x,

$$g(x) = \frac{\partial}{\partial t}\Big|_{t=0} \log dP_t(x).$$

[†] If P and every one of the measures P_t possess densities p and p_t with respect to a measure μ_t, then the expressions dP and dP_t can be replaced by p and p_t and the integral can be understood relative to μ_t (add $d\mu_t$ on the right). We use the notations dP_t and dP, because some models $\mathcal{P}$ of interest are not dominated, and the choice of μ_t is irrelevant. However, the model could be taken dominated for simplicity, and then dP_t and dP are just the densities of P_t and P.

However, the differentiability (25.13) is the correct definition for defining information, because it ensures a type of local asymptotic normality. The following lemma is proved in the same way as Theorem 7.2.

25.14 Lemma. *If the path* $t \mapsto P_t$ *in* $\mathcal{P}$ *satisfies* (25.13), *then* $Pg = 0$, $Pg^2 < \infty$, *and*

$$\log \prod_{i=1}^{n} \frac{dP_{1/\sqrt{n}}}{dP}(X_i) = \frac{1}{\sqrt{n}} \sum_{i=1}^{n} g(X_i) - \frac{1}{2} Pg^2 + o_P(1).$$

For defining the information for estimating $\psi(P)$, only those submodels $t \mapsto P_t$ along which the parameter $t \mapsto \psi(P_t)$ is differentiable are of interest. Thus, we consider only submodels $t \mapsto P_t$ such that $t \mapsto \psi(P_t)$ is differentiable at $t = 0$. More precisely, we define $\psi : \mathcal{P} \mapsto \mathbb{R}^k$ to be *differentiable* at P relative to a given tangent set $\dot{\mathcal{P}}_P$ if there exists a continuous linear map $\dot{\psi}_P : L_2(P) \mapsto \mathbb{R}^k$ such that for every $g \in \dot{\mathcal{P}}_P$ and a submodel $t \mapsto P_t$ with score function g,

$$\frac{\psi(P_t) - \psi(P)}{t} \to \dot{\psi}_P g.$$

This requires that the derivative of the map $t \mapsto \psi(P_t)$ exists in the ordinary sense, and also that it has a special representation. (The map $\dot{\psi}_P$ is much like a Hadamard derivative of ψ viewed as a map on the space of "square roots of measures.") Our definition is also relative to the submodels $t \mapsto P_t$, but we speak of "relative to $\dot{\mathcal{P}}_P$" for simplicity.

By the Riesz representation theorem for Hilbert spaces, the map $\dot{\psi}_P$ can always be written in the form of an inner product with a fixed vector-valued, measurable function $\tilde{\psi}_P : \mathcal{X} \mapsto \mathbb{R}^k$,

$$\dot{\psi}_P g = \langle \tilde{\psi}_P, g \rangle_P = \int \tilde{\psi}_P \, g \, dP.$$

Here the function $\tilde{\psi}_P$ is not uniquely defined by the functional ψ and the model $\mathcal{P}$, because only inner products of $\tilde{\psi}_P$ with elements of the tangent set are specified, and the tangent set does not span all of $L_2(P)$. However, it is always possible to find a candidate $\tilde{\psi}_P$ whose coordinate functions are contained in $\overline{\lin}\,\dot{\mathcal{P}}_P$, the closure of the linear span of the tangent set. This function is unique and is called the *efficient influence function*. It can be found as the projection of any other "influence function" onto the closed linear span of the tangent set.

In the preceding set-up the tangent sets $\dot{\mathcal{P}}_P$ are made to depend both on the model $\mathcal{P}$ and the functional ψ. We do not always want to use the "maximal tangent set," which is the set of all score functions of differentiable submodels $t \mapsto P_t$, because the parameter ψ may not be differentiable relative to it. We consider every subset of a tangent set a tangent set itself.

The maximal tangent set is a cone: If $g \in \dot{\mathcal{P}}_P$ and $a \geq 0$, then $ag \in \dot{\mathcal{P}}_P$, because the path $t \mapsto P_{at}$ has score function ag when $t \mapsto P_t$ has score function g. It is rarely a loss of generality to assume that the tangent set we work with is a cone as well.

25.15 Example (Parametric model). Consider a parametric model with parameter θ ranging over an open subset Θ of $\mathbb{R}^m$ given by densities p_θ with respect to some measure μ. Suppose that there exists a vector-valued measurable map $\dot{\ell}_\theta$ such that, as $h \to 0$,

$$\int \left[p_{\theta+h}^{1/2} - p_\theta^{1/2} - \frac{1}{2} h^T \dot{\ell}_\theta \, p_\theta^{1/2} \right]^2 d\mu = o(\|h\|^2).$$

Then a tangent set at P_θ is given by the linear space $\{h^T \dot\ell_\theta : h \in \mathbb{R}^m\}$ spanned by the score functions for the coordinates of the parameter θ.

If the Fisher information matrix $I_\theta = P_\theta \dot\ell_\theta \dot\ell_\theta^T$ is invertible, then every map $\chi : \Theta \mapsto \mathbb{R}^k$ that is differentiable in the ordinary sense as a map between Euclidean spaces is differentiable as a map $\psi(P_\theta) = \chi(\theta)$ on the model relative to the given tangent space. This follows because the submodel $t \mapsto P_{\theta + th}$ has score $h^T \dot\ell_\theta$ and

$$\frac{\partial}{\partial t}_{|t=0} \chi(\theta + th) = \dot\chi_\theta h = P_\theta \left[\left(\dot\chi_\theta I_\theta^{-1} \dot\ell_\theta \right) h^T \dot\ell_\theta \right].$$

This equation shows that the function $\tilde\psi_{P_\theta} = \dot\chi_\theta I_\theta^{-1} \dot\ell_\theta$ is the efficient influence function. In view of the results of Chapter 8, asymptotically efficient estimator sequences for $\chi(\theta)$ are asymptotically linear in this function, which justifies the name "efficient influence function." □

25.16 Example (Nonparametric models). Suppose that $\mathcal{P}$ consists of all probability laws on the sample space. Then a tangent set at P consists of all measurable functions g satisfying $\int g\, dP = 0$ and $\int g^2\, dP < \infty$. Because a score function necessarily has mean zero, this is the maximal tangent set.

It suffices to exhibit suitable one-dimensional submodels. For a bounded function g, consider for instance the exponential family $p_t(x) = c(t) \exp(tg(x)) p_0(x)$ or, alternatively, the model $p_t(x) = (1 + tg(x)) p_0(x)$. Both models have the property that, for every x,

$$g(x) = \frac{\partial}{\partial t}_{|t=0} \log p_t(x).$$

By a direct calculation or by using Lemma 7.6, we see that both models also have score function g at $t = 0$ in the L_2-sense (25.13). For an unbounded function g, these submodels are not necessarily well-defined. However, the models have the common structure $p_t(x) = c(t) k(tg(x)) p_0(x)$ for a nonnegative function k with $k(0) = k'(0) = 1$. The function $k(x) = 2(1 + e^{-2x})^{-1}$ is bounded and can be used with any g. □

25.17 Example (Cox model). The density of an observation in the Cox model takes the form

$$(t, z) \mapsto e^{-e^{\theta^T z} \Lambda(t)} \lambda(t) e^{\theta^T z} p_Z(z).$$

Differentiating the logarithm of this expression with respect to θ gives the score function for θ,

$$z - z e^{\theta^T z} \Lambda(t).$$

We can also insert appropriate parametric models $s \mapsto \lambda_s$ and differentiate with respect to s. If a is the derivative of $\log \lambda_s$ at $s = 0$, then the corresponding score for the model for the observation is

$$a(t) - e^{\theta^T z} \int_{[0,t]} a\, d\Lambda.$$

Finally, scores for the density p_Z are functions $b(z)$. The tangent space contains the linear span of all these functions. Note that the scores for Λ can be found as an "operator" working on functions a. □

25.18 *Example (Transformation regression model).* If the transformation η is increasing and e has density ϕ, then the density of the observation can be written $\phi\big(\eta(y) - \theta^T z\big)\,\eta'(y)\,p_Z(z)$. Scores for θ and η take the forms

$$-z\frac{\phi'}{\phi}\big(\eta(y) - \theta^T z\big), \qquad \frac{\phi'}{\phi}\big(\eta(y) - \theta^T z\big)a(y) + \frac{a'}{\eta'}(y),$$

where a is the derivative for η. If the distributions of e and Z are (partly) unknown, then there are additional score functions corresponding to their distributions. Again scores take the form of an operator acting on a set of functions. $\square$

To motivate the definition of information, assume for simplicity that the parameter $\psi(P)$ is one-dimensional. The Fisher information about t in a submodel $t \mapsto P_t$ with score function g at $t = 0$ is Pg^2. Thus, the "optimal asymptotic variance" for estimating the function $t \mapsto \psi(P_t)$, evaluated at $t = 0$, is the Cramér-Rao bound

$$\frac{\big(d\psi(P_t)/dt\big)^2}{Pg^2} = \frac{\langle \tilde{\psi}_P, g\rangle_P^2}{\langle g, g\rangle_P}.$$

The supremum of this expression over all submodels, equivalently over all elements of the tangent set, is a lower bound for estimating $\psi(P)$ given the model $\mathcal{P}$, if the "true measure" is P. This supremum can be expressed in the norm of the efficient influence function $\tilde{\psi}_P$.

25.19 *Lemma.* Suppose that the functional $\psi : \mathcal{P} \mapsto \mathbb{R}$ is differentiable at P relative to the tangent set $\dot{\mathcal{P}}_P$. Then

$$\sup_{g \in \mathrm{lin}\,\dot{\mathcal{P}}_P} \frac{\langle \tilde{\psi}_P, g\rangle_P^2}{\langle g, g\rangle_P} = P\tilde{\psi}_P^2.$$

Proof. This is a consequence of the Cauchy-Schwarz inequality $(P\tilde{\psi}_P g)^2 \le P\tilde{\psi}_P^2 Pg^2$ and the fact that, by definition, the efficient influence function $\tilde{\psi}_P$ is contained in the closure of $\mathrm{lin}\,\dot{\mathcal{P}}_P$. $\blacksquare$

Thus, the squared norm $P\tilde{\psi}_P^2$ of the efficient influence function plays the role of an "optimal asymptotic variance," just as does the expression $\dot{\psi}_\theta I_\theta^{-1}\dot{\psi}_\theta^T$ in Chapter 8. Similar considerations (take linear combinations) show that the "optimal asymptotic covariance" for estimating a higher-dimensional parameter $\psi : \mathcal{P} \mapsto \mathbb{R}^k$ is given by the covariance matrix $P\tilde{\psi}_P\tilde{\psi}_P^T$ of the efficient influence function.

In Chapter 8, we developed three ways to give a precise meaning to optimal asymptotic covariance: the convolution theorem, the almost-everywhere convolution theorem, and the minimax theorem. The almost-everywhere theorem uses the Lebesgue measure on the Euclidean parameter set, and does not appear to have an easy parallel for semiparametric models. On the other hand, the two other results can be generalized.

For every g in a given tangent set $\dot{\mathcal{P}}_P$, write $P_{t,g}$ for a submodel with score function g along which the function ψ is differentiable. As usual, an estimator T_n is a measurable function $T_n(X_1, \ldots, X_n)$ of the observations. An estimator sequence T_n is called *regular* at P for estimating $\psi(P)$ (relative to $\dot{\mathcal{P}}_P$) if there exists a probability measure L such that

$$\sqrt{n}\big(T_n - \psi(P_{1/\sqrt{n},g})\big) \overset{P_{1/\sqrt{n},g}}{\rightsquigarrow} L, \qquad \text{every } g \in \dot{\mathcal{P}}_P.$$

25.20 **Theorem (Convolution).** *Let the function $\psi : \mathcal{P} \mapsto \mathbb{R}^k$ be differentiable at P relative to the tangent cone $\dot{\mathcal{P}}_P$ with efficient influence function $\tilde{\psi}_P$. Then the asymptotic covariance matrix of every regular sequence of estimators is bounded below by $P\tilde{\psi}_P\tilde{\psi}_P^T$. Furthermore, if $\dot{\mathcal{P}}_P$ is a convex cone, then every limit distribution L of a regular sequence of estimators can be written $L = N\big(0, P\tilde{\psi}_P\tilde{\psi}_P^T\big) * M$ for some probability distribution M.*

25.21 **Theorem (LAM).** *Let the function $\psi : \mathcal{P} \mapsto \mathbb{R}^k$ be differentiable at P relative to the tangent cone $\dot{\mathcal{P}}_P$ with efficient influence function $\tilde{\psi}_P$. If $\dot{\mathcal{P}}_P$ is a convex cone, then, for any estimator sequence $\{T_n\}$ and subconvex function $\ell : \mathbb{R}^k \mapsto [0, \infty)$,*

$$\sup_I \liminf_{n \to \infty} \sup_{g \in I} \mathrm{E}_{P_{1/\sqrt{n},g}} \ell\Big(\sqrt{n}\big(T_n - \psi(P_{1/\sqrt{n},g})\big)\Big) \geq \int \ell \, dN\big(0, P\tilde{\psi}_P\tilde{\psi}_P^T\big).$$

Here the first supremum is taken over all finite subsets I of the tangent set.

Proofs. These results follow essentially by applying the corresponding theorems for parametric models to sufficiently rich finite-dimensional submodels. However, because we have defined the tangent set using one-dimensional submodels $t \mapsto P_{t,g}$, it is necessary to rework the proofs a little.

Assume first that the tangent set is a linear space, and fix an orthonormal base $g_P = (g_1, \ldots, g_m)^T$ of an arbitrary finite-dimensional subspace. For every $g \in \mathrm{lin}\, g_P$ select a submodel $t \mapsto P_{t,g}$ as used in the statement of the theorems. Each of the submodels $t \mapsto P_{t,g}$ is locally asymptotically normal at $t = 0$ by Lemma 25.14. Therefore, because the covariance matrix of g_P is the identity matrix,

$$\big(P_{1/\sqrt{n},h^T g_P}^n : h \in \mathbb{R}^m\big) \rightsquigarrow \big(N_m(h, I) : h \in \mathbb{R}^m\big)$$

in the sense of convergence of experiments. The function $\psi_n(h) = \psi(P_{1/\sqrt{n},h^T g_P})$ satisfies

$$\sqrt{n}\big(\psi_n(h) - \psi_n(0)\big) \to \dot{\psi}_P h^T g_P = \big(P\tilde{\psi}_P g_P^T\big)h =: Ah.$$

For the same $(k \times m)$ matrix the function Ag_P is the orthogonal projection of $\tilde{\psi}_P$ onto $\mathrm{lin}\, g_P$, and it has covariance matrix AA^T. Because $\tilde{\psi}_P$ is, by definition, contained in the closed linear span of the tangent set, we can choose g_P such that $\tilde{\psi}_P$ is arbitrarily close to its projection and hence AA^T is arbitrarily close to $P\tilde{\psi}_P\tilde{\psi}_P^T$.

Under the assumption of the convolution theorem, the limit distribution of the sequence $\sqrt{n}\big(T_n - \psi_n(h)\big)$ under $P_{1/\sqrt{n},h^T g_P}$ is the same for every $h \in \mathbb{R}^m$. By the asymptotic representation theorem, Proposition 7.10, there exists a randomized statistic T in the limit experiment such that the distribution of $T - Ah$ under h does not depend on h. By Proposition 8.4, the null distribution of T contains a normal $N(0, AA^T)$-distribution as a convolution factor. The proof of the convolution theorem is complete upon letting AA^T tend to $P\tilde{\psi}_P\tilde{\psi}_P^T$.

Under the assumption that the sequence $\sqrt{n}\big(T_n - \psi(P)\big)$ is tight, the minimax theorem is proved similarly, by first bounding the left side by the minimax risk relative to the submodel corresponding to g_P, and next applying Proposition 8.6. The tightness assumption can be dropped by a compactification argument. (see, e.g., [139], or [146]).

If the tangent set is a convex cone but not a linear space, then the submodel constructed previously can only be used for h ranging over a convex cone in $\mathbb{R}^m$. The argument can remain the same, except that we need to replace Propositions 8.4 and 8.6 by stronger results that refer to convex cones. These extensions exist and can be proved by the same Bayesian argument, now choosing priors that flatten out inside the cone (see, e.g., [139]).

If the tangent set is a cone that is not convex, but the estimator sequence is regular, then we use the fact that the matching randomized estimator T in the limit experiment satisfies $E_h T = Ah + E_0 T$ for every eligible h, that is, every h such that $h^T g_P \in \dot{\mathcal{P}}_P$. Because the tangent set is a cone, the latter set includes parameters $h = th_i$ for $t \geq 0$ and directions h_i spanning $\mathbb{R}^m$. The estimator T is unbiased for estimating $Ah + E_0 T$ on this parameter set, whence the covariance matrix of T is bounded below by AA^T, by the Cramér-Rao inequality. ■

Both theorems have the interpretation that the matrix $P\tilde{\psi}_P \tilde{\psi}_P^T$ is an optimal asymptotic covariance matrix for estimating $\psi(P)$ given the model $\mathcal{P}$. We might wish that this could be formulated in a simpler fashion, but this is precluded by the problem of superefficiency, as is already the case for the parametric analogues, discussed in Chapter 8. That the notion of asymptotic efficiency used in the present interpretation should not be taken absolutely is shown by the shrinkage phenomena discussed in section 8.8, but we use it in this chapter. We shall say that an estimator sequence is *asymptotically efficient* at P, if it is regular at P with limit distribution $L = N(0, P\tilde{\psi}_P \tilde{\psi}_P^T)$.[†]

The efficient influence function $\tilde{\psi}_P$ plays the same role as the normalized score function $I_\theta^{-1}\dot{\ell}_\theta$ in parametric models. In particular, a sequence of estimators T_n is asymptotically efficient at P if

$$\sqrt{n}(T_n - \psi(P)) = \frac{1}{\sqrt{n}}\sum_{i=1}^{n} \tilde{\psi}_P(X_i) + o_P(1). \qquad (25.22)$$

This justifies the name "efficient influence function."

25.23 Lemma. *Let the function $\psi : \mathcal{P} \mapsto \mathbb{R}^k$ be differentiable at P relative to the tangent cone $\dot{\mathcal{P}}_P$ with efficient influence function $\tilde{\psi}_P$. A sequence of estimators T_n is regular at P with limiting distribution $N(0, P\tilde{\psi}_P \tilde{\psi}_P^T)$ if and only if it satisfies (25.22).*

Proof. Because the submodels $t \mapsto P_{t,g}$ are locally asymptotically normal at $t = 0$, "if" follows with the help of Le Cam's third lemma, by the same arguments as for the analogous result for parametric models in Lemma 8.14.

To prove the necessity of (25.22), we adopt the notation of the proof of Theorem 25.20. The statistics $S_n = \psi(P) + n^{-1}\sum_{i=1}^{n} \tilde{\psi}_P(X_i)$ depend on P but can be considered a true estimator sequence in the local subexperiments. The sequence S_n trivially satisfies (25.22) and hence is another asymptotically efficient estimator sequence. We may assume for simplicity that the sequence $\sqrt{n}(S_n - \psi(P_{1/\sqrt{n},g}), T_n - \psi(P_{1/\sqrt{n},g}))$ converges under every local parameter g in distribution. Otherwise, we argue along subsequences, which can be

---

[†] If the tangent set is not a linear space, then the situation becomes even more complicated. If the tangent set is a convex cone, then the minimax risk in the left side of Theorem 25.21 cannot fall below the normal risk on the right side, but there may be nonregular estimator sequences for which there is equality. If the tangent set is not convex, then the assertion of Theorem 25.21 may fail. Convex tangent cones arise frequently; fortunately, nonconvex tangent cones are rare.

selected with the help of Le Cam's third lemma. By Theorem 9.3, there exists a matching randomized estimator $(S, T) = (S, T)(X, U)$ in the normal limit experiment. By the efficiency of both sequences S_n and T_n, the variables $S - Ah$ and $T - Ah$ are, under h, marginally normally distributed with mean zero and covariance matrix $P\tilde{\psi}_P\tilde{\psi}_P^T$. In particular, the expectations $E_h S = E_h T$ are identically equal to Ah. Differentiate with respect to h at $h = 0$ to find that

$$E_0 S X^T = E_0 T X^T = A.$$

It follows that the orthogonal projections of S and T onto the linear space spanned by the coordinates of X are identical and given by $\Pi S = \Pi T = AX$, and hence

$$\text{Cov}_0(S - T) = \text{Cov}_0(\Pi^\perp S - \Pi^\perp T) \le 2\,\text{Cov}_0 \Pi^\perp S + 2\,\text{Cov}_0 \Pi^\perp T.$$

(The inequality means that the difference of the matrices on the right and the left is nonnegative-definite.) We have obtained this for a fixed orthonormal set $g_P = (g_1, \ldots, g_m)$. If we choose g_P such that AA^T is arbitrarily close to $P\tilde{\psi}_P\tilde{\psi}_P^T$, equivalently $\text{Cov}_0 \Pi T = AA^T = \text{Cov}_0 \Pi S$ is arbitrarily close to $\text{Cov}_0 T = P\tilde{\psi}_P\tilde{\psi}_P^T = \text{Cov}_0 S$, and then the right side of the preceding display is arbitrarily close to zero, whence $S - T \approx 0$. The proof is complete on noting that $\sqrt{n}(S_n - T_n) \overset{0}{\rightsquigarrow} S - T$. ∎

25.24   *Example (Empirical distribution).* The empirical distribution is an asymptotically efficient estimator if the underlying distribution P of the sample is completely unknown. To give a rigorous expression to this intuitively obvious statement, fix a measurable function $f : \mathcal{X} \mapsto \mathbb{R}$ with $Pf^2 < \infty$, for instance an indicator function $f = 1_A$, and consider $\mathbb{P}_n f = n^{-1}\sum_{i=1}^n f(X_i)$ as an estimator for the function $\psi(P) = Pf$.

In Example 25.16 it is seen that the maximal tangent space for the nonparametric model is equal to the set of all $g \in L_2(P)$ such that $Pg = 0$. For a general function f, the parameter ψ may not be differentiable relative to the maximal tangent set, but it is differentiable relative to the tangent space consisting of all bounded, measurable functions g with $Pg = 0$. The closure of this tangent space is the maximal tangent set, and hence working with this smaller set does not change the efficient influence functions. For a bounded function g with $Pf = 0$ we can use the submodel defined by $dP_t = (1 + tg)\,dP$, for which $\psi(P_t) = Pf + tPfg$. Hence the derivative of ψ is the map $g \mapsto \dot{\psi}_P g = Pfg$, and the efficient influence function relative to the maximum tangent set is the function $\tilde{\psi}_P = f - Pf$. (The function f is an influence function; its projection onto the mean zero functions is $f - Pf$.)

The optimal asymptotic variance for estimating $P \mapsto Pf$ is equal to $P\tilde{\psi}_P^2 = P(f - Pf)^2$. The sequence of empirical estimators $\mathbb{P}_n f$ is asymptotically efficient, because it satisfies (25.22), with the $o_P(1)$-remainder term identically zero.   □

25.4   Efficient Score Functions

A function $\psi(P)$ of particular interest is the parameter θ in a semiparametric model $\{P_{\theta,\eta} : \theta \in \Theta, \eta \in H\}$. Here Θ is an open subset of $\mathbb{R}^k$ and H is an arbitrary set, typically of infinite dimension. The information bound for the functional of interest $\psi(P_{\theta,\eta}) = \theta$ can be conveniently expressed in an "efficient score function."

As submodels, we use paths of the form $t \mapsto P_{\theta+ta,\eta_t}$, for given paths $t \mapsto \eta_t$ in the parameter set H. The score functions for such submodels (if they exist) typically have the form of a sum of "partial derivatives" with respect to θ and η. If $\dot{\ell}_{\theta,\eta}$ is the ordinary score function for θ in the model in which η is fixed, then we expect

$$\frac{\partial}{\partial t}_{|t=0} \log d P_{\theta+ta,\eta_t} = a^T \dot{\ell}_{\theta,\eta} + g.$$

The function g has the interpretation of a score function for η if θ is fixed and runs through an infinite-dimensional set if we are concerned with a "true" semiparametric model. We refer to this set as the *tangent set for* η, and denote it by $_\eta\dot{\mathcal{P}}_{P_{\theta,\eta}}$.

The parameter $\psi(P_{\theta+ta,\eta_t}) = \theta + ta$ is certainly differentiable with respect to t in the ordinary sense but is, by definition, differentiable as a parameter on the model if and only if there exists a function $\tilde{\psi}_{\theta,\eta}$ such that

$$a = \frac{\partial}{\partial t}_{|t=0} \psi(P_{\theta+ta,\eta_t}) = \langle \tilde{\psi}_{\theta,\eta}, a^T \dot{\ell}_{\theta,\eta} + g \rangle_{P_{\theta,\eta}}, \qquad a \in \mathbb{R}^k, g \in {}_\eta\dot{\mathcal{P}}_{P_{\theta,\eta}}.$$

Setting $a = 0$, we see that $\tilde{\psi}_{\theta,\eta}$ must be orthogonal to the tangent set $_\eta\dot{\mathcal{P}}_{P_{\theta,\eta}}$ for the nuisance parameter. Define $\Pi_{\theta,\eta}$ as the orthogonal projection onto the closure of the linear span of $_\eta\dot{\mathcal{P}}_{P_{\theta,\eta}}$ in $L_2(P_{\theta,\eta})$.

The function defined by

$$\tilde{\ell}_{\theta,\eta} = \dot{\ell}_{\theta,\eta} - \Pi_{\theta,\eta}\dot{\ell}_{\theta,\eta}$$

is called the *efficient score function* for θ, and its covariance matrix $\tilde{I}_{\theta,\eta} = P_{\theta,\eta}\tilde{\ell}_{\theta,\eta}\tilde{\ell}_{\theta,\eta}^T$ is the *efficient information matrix*.

25.25 Lemma. *Suppose that for every $a \in \mathbb{R}^k$ and every $g \in {}_\eta\dot{\mathcal{P}}_{P_{\theta,\eta}}$ there exists a path $t \mapsto \eta_t$ in H such that*

$$\int \left[\frac{dP_{\theta+ta,\eta_t}^{1/2} - dP_{\theta,\eta}^{1/2}}{t} - \frac{1}{2}(a^T \dot{\ell}_{\theta,\eta} + g) dP_{\theta,\eta}^{1/2} \right]^2 \to 0. \qquad (25.26)$$

If $\tilde{I}_{\theta,\eta}$ is nonsingular, then the functional $\psi(P_{\theta,\eta}) = \theta$ is differentiable at $P_{\theta,\eta}$ relative to the tangent set $\dot{\mathcal{P}}_{P_{\theta,\eta}} = \lim \dot{\ell}_{\theta,\eta} + {}_\eta\dot{\mathcal{P}}_{P_{\theta,\eta}}$ with efficient influence function $\tilde{\psi}_{\theta,\eta} = \tilde{I}_{\theta,\eta}^{-1}\tilde{\ell}_{\theta,\eta}$.

Proof. The given set $\dot{\mathcal{P}}_{P_{\theta,\eta}}$ is a tangent set by assumption. The function ψ is differentiable with respect to this tangent set because

$$\langle \tilde{I}_{\theta,\eta}^{-1}\tilde{\ell}_{\theta,\eta}, a^T \dot{\ell}_{\theta,\eta} + g \rangle_{P_{\theta,\eta}} = \tilde{I}_{\theta,\eta}^{-1}\langle \tilde{\ell}_{\theta,\eta}, \dot{\ell}_{\theta,\eta}^T \rangle_{P_{\theta,\eta}} a = a.$$

The last equality follows, because the inner product of a function and its orthogonal projection is equal to the square length of the projection. Thus, we may replace $\dot{\ell}_{\theta,\eta}$ by $\tilde{\ell}_{\theta,\eta}$. ∎

Consequently, an estimator sequence is asymptotically efficient for estimating θ if

$$\sqrt{n}(T_n - \theta) = \frac{1}{\sqrt{n}} \sum_{i=1}^n \tilde{I}_{\theta,\eta}^{-1}\tilde{\ell}_{\theta,\eta}(X_i) + o_{P_{\theta,\eta}}(1).$$

This equation is very similar to the equation derived for efficient estimators in parametric models in Chapter 8. It differs only in that the ordinary score function $\tilde{\ell}_{\theta,\eta}$ has been replaced by the efficient score function (and similarly for the information). The intuitive explanation is that a part of the score function for θ can also be accounted for by score functions for the nuisance parameter η. If the nuisance parameter is unknown, a part of the information for θ is "lost," and this corresponds to a loss of a part of the score function.

25.27　*Example (Symmetric location).* Suppose that the model consists of all densities $x \mapsto \eta(x - \theta)$ with $\theta \in \mathbb{R}$ and the "shape" η symmetric about 0 with finite Fisher information for location I_η. Thus, the observations are sampled from a density that is symmetric about θ.

By the symmetry, the density can equivalently be written as $\eta(|x - \theta|)$. It follows that any score function for the nuisance parameter η is necessarily a function of $|x - \theta|$. This suggests a tangent set containing functions of the form $a(\eta'/\eta)(x - \theta) + b(|x - \theta|)$. It is not hard to show that all square-integrable functions of this type with mean zero occur as score functions in the sense of (25.26).[†]

A symmetric density has an asymmetric derivative and hence an asymmetric score function for location. Therefore, for every b,

$$\mathrm{E}_{\theta,\eta} \frac{\eta'}{\eta}(X - \theta)\, b(|X - \theta|) = 0.$$

Thus, the projection of the θ-score onto the set of nuisance scores is zero and hence the efficient score function coincides with the ordinary score function. This means that there is no difference in information about θ whether the form of the density is known or not known, as long as it is known to be symmetric. This surprising fact was discovered by Stein in 1956 and has been an important motivation in the early work on semiparametric models.

Even more surprising is that the information calculation is not misleading. There exist estimator sequences for θ whose definition does not depend on η that have asymptotic variance I_η^{-1} under any true η. See section 25.8. Thus a symmetry point can be estimated as well if the shape is known as if it is not, at least asymptotically. □

25.28　*Example (Regression).* Let g_θ be a given set of functions indexed by a parameter $\theta \in \mathbb{R}^k$, and suppose that a typical observation (X, Y) follows the regression model

$$Y = g_\theta(X) + e, \qquad \mathrm{E}(e \mid X) = 0.$$

This model includes the logistic regression model, for $g_\theta(x) = 1/(1 + e^{-\theta^T x})$. It is also a version of the ordinary linear regression model. However, in this example we do not assume that X and e are independent, but only the relations in the preceding display, apart from qualitative smoothness conditions that ensure existence of score functions, and the existence of moments. We shall write the formulas assuming that (X, e) possesses a density η. Thus, the observation (X, Y) has a density $\eta(x, y - g_\theta(x))$, in which η is (essentially) only restricted by the relations $\int e\eta(x, e)\, de \equiv 0$.

Because any perturbation η_t of η within the model must satisfy this same relation $\int e\eta_t(x, e)\, de = 0$, it is clear that score functions for the nuisance parameter η are functions

---

[†] That no other functions can occur is shown in, for example, [8, p. 56–57] but need not concern us here.

$a\big(x, y - g_\theta(x)\big)$ that satisfy

$$\mathrm{E}\big(ea(X, e) \mid X\big) = \frac{\int ea(X, e)\, \eta(X, e)\, de}{\int \eta(X, e)\, de} = 0.$$

By the same argument as for nonparametric models all bounded square-integrable functions of this type that have mean zero are score functions. Because the relation $\mathrm{E}\big(ea(X, e) \mid X\big) = 0$ is equivalent to the orthogonality in $L_2(\eta)$ of $a(x, e)$ to all functions of the form $eh(x)$, it follows that the set of score functions for η is the orthocomplement of the set $e\mathcal{H}$, of all functions of the form $(x, y) \mapsto \big(y - g_\theta(x)\big)h(x)$ within $L_2(P_{\theta, \eta})$, up to centering at mean zero.

Thus, we obtain the efficient score function for θ by projecting the ordinary score function $\dot{\ell}_{\theta, \eta}(x, y) = -\eta_2/\eta(x, e)\dot{g}_\theta(x)$ onto $e\mathcal{H}$. The projection of an arbitrary function $b(x, e)$ onto the functions $e\mathcal{H}$ is a function $eh_0(x)$ such that $\mathrm{E}b(X, e)eh(X) = \mathrm{E}eh_0(X)eh(X)$ for all measurable functions h. This can be solved for h_0 to find that the projection operator takes the form

$$\Pi_{e\mathcal{H}} b(X, e) = e\, \frac{\mathrm{E}\big(b(X, e)e \mid X\big)}{\mathrm{E}(e^2 \mid X)}.$$

This readily yields the efficient score function

$$\tilde{\ell}_{\theta, \eta}(X, Y) = -\frac{e\dot{g}_\theta(X)}{\mathrm{E}(e^2 \mid X)} \frac{\int \eta_2(X, e)e\, de}{\int \eta(X, e)\, de} = \frac{\big(Y - g_\theta(X)\big)\dot{g}_\theta(X)}{\mathrm{E}(e^2 \mid X)}.$$

The efficient information takes the form $\tilde{I}_{\theta, \eta} = \mathrm{E}\big(\dot{g}_\theta \dot{g}_\theta^T(X)/\mathrm{E}(e^2 \mid X)\big)$.   □

25.5   Score and Information Operators

The method to find the efficient influence function of a parameter given in the preceding section is the most convenient method if the model can be naturally partitioned in the parameter of interest and a nuisance parameter. For many parameters such a partition is impossible or, at least, unnatural. Furthermore, even in semiparametric models it can be worthwhile to derive a more concrete description of the tangent set for the nuisance parameter, in terms of a "score operator."

Consider first the situation that the model $\mathcal{P} = \{P_\eta : \eta \in H\}$ is indexed by a parameter η that is itself a probability measure on some measurable space. We are interested in estimating a parameter of the type $\psi(P_\eta) = \chi(\eta)$ for a given function $\chi : H \mapsto \mathbb{R}^k$ on the model H.

The model H gives rise to a tangent set $\dot{H}_\eta$ at η. If the map $\eta \mapsto P_\eta$ is differentiable in an appropriate sense, then its derivative maps every score $b \in \dot{H}_\eta$ into a score g for the model $\mathcal{P}$. To make this precise, we assume that a smooth parametric submodel $t \mapsto \eta_t$ induces a smooth parametric submodel $t \mapsto P_{\eta_t}$, and that the score functions b of the submodel $t \mapsto \eta_t$ and g of the submodel $t \mapsto P_{\eta_t}$ are related by

$$g = A_\eta b.$$

Then $A_\eta \dot{H}_\eta$ is a tangent set for the model $\mathcal{P}$ at P_η. Because A_η turns scores for the model H into scores for the model $\mathcal{P}$ it is called a *score operator*. It is seen subsequently here that if η

and P_η are the distributions of an unobservable Y and an observable $X = m(Y)$, respectively, then the score operator is a conditional expectation. More generally, it can be viewed as a derivative of the map $\eta \mapsto P_\eta$. We assume that A_η, as a map $A_\eta : \text{lin } \dot{H}_\eta \subset L_2(\eta) \mapsto L_2(P_\eta)$, is continuous and linear.

Next, assume that the function $\eta \mapsto \chi(\eta)$ is differentiable with influence function $\tilde{\chi}_\eta$ relative to the tangent set $\dot{H}_\eta$. Then, by definition, the function $\psi(P_\eta) = \chi(\eta)$ is pathwise differentiable relative to the tangent set $\dot{\mathcal{P}}_{P_\eta} = A_\eta \dot{H}_\eta$ if and only if there exists a vector-valued function $\tilde{\psi}_{P_\eta}$ such that

$$\langle \tilde{\psi}_{P_\eta}, A_\eta b \rangle_{P_\eta} = \frac{\partial}{\partial t}_{|t=0} \psi(P_{\eta_t}) = \frac{\partial}{\partial t}_{|t=0} \chi(\eta_t) = \langle \tilde{\chi}_\eta, b \rangle_\eta, \qquad b \in \dot{H}_\eta.$$

This equation can be rewritten in terms of the *adjoint score operator* $A_\eta^* : L_2(P_\eta) \mapsto \overline{\text{lin }} \dot{H}_\eta$. By definition this satisfies $\langle h, A_\eta b \rangle_{P_\eta} = \langle A_\eta^* h, b \rangle_\eta$ for every $h \in L_2(P_\eta)$ and $b \in \dot{H}_\eta$.[†] The preceding display is equivalent to

$$A_\eta^* \tilde{\psi}_{P_\eta} = \tilde{\chi}_\eta. \tag{25.29}$$

We conclude that the function $\psi(P_\eta) = \chi(\eta)$ is differentiable relative to the tangent set $\dot{\mathcal{P}}_{P_\eta} = A_\eta \dot{H}_\eta$ if and only if this equation can be solved for $\tilde{\psi}_{P_\eta}$; equivalently, if and only if $\tilde{\chi}_\eta$ is contained in the range of the adjoint A_η^*. Because A_η^* is not necessarily onto $\overline{\text{lin }} \dot{H}_\eta$, not even if it is one-to-one, this is a condition.

For multivariate functionals (25.29) is to be understood coordinate-wise. Two solutions $\tilde{\psi}_{P_\eta}$ of (25.29) can differ only by an element of the kernel $N(A_\eta^*)$ of A_η^*, which is the orthocomplement $R(A_\eta)^\perp$ of the range of $A_\eta : \text{lin } \dot{H}_\eta \mapsto L_2(P_\eta)$. Thus, there is at most one solution $\tilde{\psi}_{P_\eta}$ that is contained in $\overline{R}(A_\eta) = \overline{\text{lin }} A_\eta \dot{H}_\eta$, the closure of the range of A_η, as required.

If $\tilde{\chi}_\eta$ is contained in the smaller range of $A_\eta^* A_\eta$, then (25.29) can be solved, of course, and the solution can be written in the attractive form

$$\tilde{\psi}_{P_\eta} = A_\eta (A_\eta^* A_\eta)^- \tilde{\chi}_\eta. \tag{25.30}$$

Here $A_\eta^* A_\eta$ is called the *information operator*, and $(A_\eta^* A_\eta)^-$ is a "generalized inverse." (Here this will not mean more than that $b = (A_\eta^* A_\eta)^- \tilde{\chi}_\eta$ is a solution to the equation $A_\eta^* A_\eta b = \tilde{\chi}_\eta$.) In the preceding equation the operator $A_\eta^* A_\eta$ performs a similar role as the matrix $X^T X$ in the least squares solution of a linear regression model. The operator $A_\eta (A_\eta^* A_\eta)^{-1} A_\eta^*$ (if it exists) is the orthogonal projection onto the range space of A_η.

So far we have assumed that the parameter η is a probability distribution, but this is not necessary. Consider the more general situation of a model $\mathcal{P} = \{P_\eta : \eta \in H\}$ indexed by a parameter η running through an arbitrary set H. Let $\mathbb{H}_\eta$ be a subset of a Hilbert space that indexes "directions" b in which η can be approximated within H. Suppose that there exist continuous, linear operators $A_\eta : \text{lin } \mathbb{H}_\eta \mapsto L_2(P_\eta)$ and $\dot{\chi}_\eta : \text{lin } \mathbb{H}_\eta \mapsto \mathbb{R}^k$, and for every $b \in \mathbb{H}_\eta$ a path $t \mapsto \eta_t$ such that, as $t \downarrow 0$,

$$\int \left[\frac{dP_{\eta_t}^{1/2} - dP_\eta^{1/2}}{t} - \frac{1}{2} A_\eta b \, dP_\eta^{1/2} \right]^2 \to 0, \qquad \frac{\chi(\eta_t) - \chi(\eta)}{t} \to \dot{\chi}_\eta b.$$

[†] Note that we define A_η^* to have range $\overline{\text{lin }} \dot{H}_\eta$, so that it is the adjoint of $A_\eta : \dot{H}_\eta \mapsto L_2(P_\eta)$. This is the adjoint of an extension $A_\eta : L_2(\eta) \mapsto L_2(P_\eta)$ followed by the orthogonal projection onto $\overline{\text{lin }} \dot{H}_\eta$.

By the Riesz representation theorem for Hilbert spaces, the "derivative" $\dot{\chi}_\eta$ has a representation as an inner product $\dot{\chi}_\eta b = \langle \tilde{\chi}_\eta, b \rangle_{\mathbb{H}_\eta}$ for an element $\tilde{\chi}_\eta \in \overline{\text{lin}}\,\mathbb{H}_\eta^k$. The preceding discussion can be extended to this abstract set-up.

25.31 Theorem. *The map $\psi : \mathcal{P} \mapsto \mathbb{R}^k$ given by $\psi(P_\eta) = \chi(\eta)$ is differentiable at P_η relative to the tangent set $A_\eta \mathbb{H}_\eta$ if and only if each coordinate function of $\tilde{\chi}_\eta$ is contained in the range of $A_\eta^* : L_2(P_\eta) \mapsto \overline{\text{lin}}\,\mathbb{H}_\eta$. The efficient influence function $\tilde{\psi}_{P_\eta}$ satisfies (25.29). If each coordinate function of $\tilde{\chi}_\eta$ is contained in the range of $A_\eta^* A_\eta : \overline{\text{lin}}\,\mathbb{H}_\eta \mapsto \overline{\text{lin}}\,\mathbb{H}_\eta$, then it also satisfies (25.30).*

Proof. By assumption, the set $A_\eta \mathbb{H}_\eta$ is a tangent set. The map ψ is differentiable relative to this tangent set (and the corresponding submodels $t \mapsto P_{\eta_t}$) by the argument leading up to (25.29). ∎

The condition (25.29) is odd. By definition, the influence function $\tilde{\chi}_\eta$ is contained in the closed linear span of $\mathbb{H}_\eta$ and the operator A_η^* maps $L_2(P_\eta)$ into $\overline{\text{lin}}\,\mathbb{H}_\eta$. Therefore, the condition is certainly satisfied if A_η^* is onto. There are two reasons why it may fail to be onto. First, its range $\text{R}(A_\eta^*)$ may be a proper subspace of $\overline{\text{lin}}\,\mathbb{H}_\eta$. Because $b \perp \text{R}(A_\eta^*)$ if and only if $b \in \text{N}(A_\eta)$, this can happen only if A_η is not one-to-one. This means that two different directions b may lead to the same score function $A_\eta b$, so that the information matrix for the corresponding two-dimensional submodel is singular. A rough interpretation is that the parameter is not locally identifiable. Second, the range space $\text{R}(A_\eta^*)$ may be dense but not closed. Then for any $\tilde{\chi}_\eta$ there exist elements in $\text{R}(A_\eta^*)$ that are arbitrarily close to $\tilde{\chi}_\eta$, but (25.29) may still fail. This happens quite often. The following theorem shows that failure has serious consequences.[†]

25.32 Theorem. *Suppose that $\eta \mapsto \chi(\eta)$ is differentiable with influence function $\tilde{\chi}_\eta \notin \text{R}(A_\eta^*)$. Then there exists no estimator sequence for $\chi(\eta)$ that is regular at P_η.*

25.5.1 Semiparametric Models

In a semiparametric model $\{P_{\theta,\eta} : \theta \in \Theta, \eta \in H\}$, the pair (θ, η) plays the role of the single η in the preceding general discussion. The two parameters can be perturbed independently, and the score operator can be expected to take the form

$$A_{\theta,\eta}(a, b) = a^T \dot{\ell}_{\theta,\eta} + B_{\theta,\eta} b.$$

Here $B_{\theta,\eta} : \mathbb{H}_\eta \mapsto L_2(P_{\theta,\eta})$ is the score operator for the nuisance parameter. The domain of the operator $A_{\theta,\eta} : \mathbb{R}^k \times \text{lin}\,\mathbb{H}_\eta \mapsto L_2(P_{\theta,\eta})$ is a Hilbert space relative to the inner product

$$\big\langle (a, b), (\alpha, \beta) \big\rangle_\eta = a^T \alpha + \langle b, \beta \rangle_{\mathbb{H}_\eta}.$$

Thus this example fits in the general set-up, with $\mathbb{R}^k \times \mathbb{H}_\eta$ playing the role of the earlier $\mathbb{H}_\eta$. We shall derive expressions for the efficient influence functions of θ and η.

The efficient influence function for estimating θ is expressed in the *efficient score function* for θ in Lemma 25.25, which is defined as the ordinary score function minus its projection

---

[†] For a proof, see [140].

onto the score-space for η. Presently, the latter space is the range of the operator $B_{\theta,\eta}$. If the operator $B^*_{\theta,\eta} B_{\theta,\eta}$ is continuously invertible (but in many examples it is not), then the operator $B_{\theta,\eta}(B^*_{\theta,\eta} B_{\theta,\eta})^{-1} B^*_{\theta,\eta}$ is the orthogonal projection onto the nuisance score space, and

$$\tilde{\ell}_{\theta,\eta} = \left(I - B_{\theta,\eta}(B^*_{\theta,\eta} B_{\theta,\eta})^{-1} B^*_{\theta,\eta}\right)\dot{\ell}_{\theta,\eta}. \tag{25.33}$$

This means that $b = -(B^*_{\theta,\eta} B_{\theta,\eta})^{-1} B^*_{\theta,\eta} \dot{\ell}_{\theta,\eta}$ is a "least favorable direction" in H, for estimating θ. If θ is one-dimensional, then the submodel $t \mapsto P_{\theta+t,\eta_t}$ where η_t approaches η in this direction, has the least information for estimating t and score function $\tilde{\ell}_{\theta,\eta}$, at $t = 0$.

A function $\chi(\eta)$ of the nuisance parameter can, despite the name, also be of interest. The efficient influence function for this parameter can be found from (25.29). The adjoint of $A_{\theta,\eta} : \mathbb{R}^k \times \mathbb{H}_\eta \mapsto L_2(P_{\theta,\eta})$, and the corresponding information operator $A^*_{\theta,\eta} A_{\theta,\eta} : \mathbb{R}^k \times \mathbb{H}_\eta \mapsto \mathbb{R}^k \times \overline{\text{lin}}\,\mathbb{H}_\eta$ are given by, with $B^*_{\theta,\eta} : L_2(P_{\theta,\eta} \mapsto \overline{\text{lin}}\,\mathbb{H}_\eta$ the adjoint of $B_{\theta,\eta}$,

$$A^*_{\theta,\eta} g = \left(P_{\theta,\eta} g \dot{\ell}_{\theta,\eta}, B^*_{\theta,\eta} g\right),$$

$$A^*_{\theta,\eta} A_{\theta,\eta}(a, b) = \begin{pmatrix} I_{\theta,\eta} & P_{\theta,\eta}\dot{\ell}_{\theta,\eta} B_{\theta,\eta} \cdot \\ B^*_{\theta,\eta}\dot{\ell}^T_{\theta,\eta} & B^*_{\theta,\eta} B_{\theta,\eta} \end{pmatrix} \begin{pmatrix} a \\ b \end{pmatrix}.$$

The diagonal elements in the matrix are the information operators for the parameters θ and η, respectively, the former being just the ordinary Fisher information matrix $I_{\theta,\eta}$ for θ. If $\eta \mapsto \chi(\eta)$ is differentiable as before, then the function $(\theta, \eta) \mapsto \chi(\eta)$ is differentiable with influence function $(0, \tilde{\chi}_\eta)$. Thus, for a real parameter $\chi(\eta)$, equation (25.29) becomes

$$P_{\theta,\eta}\tilde{\psi}_{P_{\theta,\eta}}\dot{\ell}_{\theta,\eta} = 0, \qquad B^*_{\theta,\eta}\tilde{\psi}_{P_{\theta,\eta}} = \tilde{\chi}_\eta.$$

If $\tilde{I}_{\theta,\eta}$ is invertible and $\tilde{\chi}_\eta$ is contained in the range of $B^*_{\theta,\eta} B_{\theta,\eta}$, then the solution $\tilde{\psi}_{P_{\theta,\eta}}$ of these equations is

$$B_{\theta,\eta}(B^*_{\theta,\eta} B_{\theta,\eta})^- \tilde{\chi}_\eta - \left\langle B_{\theta,\eta}(B^*_{\theta,\eta} B_{\theta,\eta})^- \tilde{\chi}_\eta, \dot{\ell}_{\theta,\eta}\right\rangle^T_{P_{\theta,\eta}} \tilde{I}^{-1}_{\theta,\eta}\tilde{\ell}_{\theta,\eta}.$$

The second part of this function is the part of the efficient score function for $\chi(\eta)$ that is "lost" due to the fact that θ is unknown. Because it is orthogonal to the first part, it adds a positive contribution to the variance.

25.5.2 *Information Loss Models*

Suppose that a typical observation is distributed as a measurable transformation $X = m(Y)$ of an unobservable variable Y. Assume that the form of m is known and that the distribution η of Y is known to belong to a class H. This yields a natural parametrization of the distribution P_η of X. A nice property of differentiability in quadratic mean is that it is preserved under "censoring" mechanisms of this type: If $t \mapsto \eta_t$ is a differentiable submodel of H, then the induced submodel $t \mapsto P_{\eta_t}$ is a differentiable submodel of $\{P_\eta : \eta \in H\}$. Furthermore, the score function $g = A_\eta b$ (at $t = 0$) for the induced model $t \mapsto P_{\eta_t}$ can be obtained from the score function b (at $t = 0$) of the model $t \mapsto \eta_t$ by taking a conditional expectation:

$$A_\eta b(x) = \mathrm{E}_\eta\big(b(Y) \mid X = x\big).$$

If we consider the scores b and g as the carriers of information about t in the variables Y with law η_t and X with law P_{η_t}, respectively, then the intuitive meaning of the conditional expectation operator is clear. The information contained in the observation X is the information contained in Y diluted (and reduced) through conditioning.[†]

25.34   Lemma. *Suppose that* $\{\eta_t : 0 < t < 1\}$ *is a collection of probability measures on a measurable space* $(\mathcal{Y}, \mathcal{B})$ *such that for some measurable function* $b : \mathcal{Y} \mapsto \mathbb{R}$

$$\int \left[\frac{d\eta_t^{1/2} - d\eta^{1/2}}{t} - \frac{1}{2} b \, d\eta^{1/2} \right]^2 \to 0.$$

For a measurable map $m : \mathcal{Y} \mapsto \mathcal{X}$ *let* P_η *be the distribution of* $m(Y)$ *if* Y *has law* η *and let* $A_\eta b(x)$ *be the conditional expectation of* $b(Y)$ *given* $m(Y) = x$. *Then*

$$\int \left[\frac{dP_{\eta_t}^{1/2} - dP_\eta^{1/2}}{t} - \frac{1}{2} A_\eta b \, dP_\eta^{1/2} \right]^2 \to 0.$$

If we consider A_η as an operator $A_\eta : L_2(\eta) \mapsto L_2(P_\eta)$, then its adjoint $A_\eta^* : L_2(P_\eta) \mapsto L_2(\eta)$ is a conditional expectation operator also, reversing the roles of X and Y,

$$A_\eta^* g(y) = E_\eta \big(g(X) \mid Y = y \big).$$

This follows because, by the usual rules for conditional expectations, $EE\big(g(X) \mid Y\big) b(Y) = Eg(X)b(Y) = Eg(X)E\big(b(Y) \mid X\big)$. In the "calculus of scores" of Theorem 25.31 the adjoint is understood to be the adjoint of $A_\eta : \mathbb{H}_\eta \mapsto L_2(P_\eta)$ and hence to have range $\overline{\lin} \, \mathbb{H}_\eta \subset L_2(\eta)$. Then the conditional expectation in the preceding display needs to be followed by the orthogonal projection onto $\overline{\lin} \, \mathbb{H}_\eta$.

25.35   Example (Mixtures). Suppose that a typical observation X possesses a conditional density $p(x \mid z)$ given an unobservable variable $Z = z$. If the unobservable Z possesses an unknown probability distribution η, then the observations are a random sample from the mixture density

$$p_\eta(x) = \int p(x \mid z) \, d\eta(z).$$

This is a missing data problem if we think of X as a function of the pair $Y = (X, Z)$. A score for the mixing distribution η in the model for Y is a function $b(z)$. Thus, a score space for the mixing distribution in the model for X consists of the functions

$$A_\eta b(x) = E_\eta \big(b(Z) \mid X = x \big) = \frac{\int b(z) \, p(x \mid z) \, d\eta(z)}{\int p(x \mid z) \, d\eta(z)}.$$

If the mixing distribution is completely unknown, which we assume, then the tangent set $\dot{H}_\eta$ for η can be taken equal to the maximal tangent set $\{b \in L_2(\eta) : \eta b = 0\}$.

In particular, consider the situation that the kernel $p(x \mid z)$ belongs to an exponential family, $p(x \mid z) = c(z)d(x) \exp\big(z^T x\big)$. We shall show that the tangent set $A_\eta \dot{H}_\eta$ is dense

---

[†] For a proof of the following lemma, see, for example, [139, pp. 188–193].

in the maximal tangent set $\{g \in L_2(P_\eta) : P_\eta g = 0\}$, for every η whose support contains an interval. This has as a consequence that empirical estimators $\mathbb{P}_n g$, for a fixed squared-integrable function g, are efficient estimators for the functional $\psi(\eta) = P_\eta g$. For instance, the sample mean is asymptotically efficient for estimating the mean of the observations.

Thus nonparametric mixtures over an exponential family form very large models, which are only slightly smaller than the nonparametric model. For estimating a functional such as the mean of the observations, it is of relatively little use to know that the underlying distribution is a mixture. More precisely, the additional information does not decrease the asymptotic variance, although there may be an advantage for finite n. On the other hand, the mixture structure may express a structure in reality and the mixing distribution η may define the functional of interest.

The closure of the range of the operator A_η is the orthocomplement of the kernel $N(A_\eta^*)$ of its adjoint. Hence our claim is proved if this kernel is zero. The equation

$$0 = A_\eta^* g(z) = E\big(g(X) \mid Z = z\big) = \int g(x) \, p(x \mid z) \, d\nu(x)$$

says exactly that $g(X)$ is a zero-estimator under $p(x \mid z)$. Because the adjoint is defined on $L_2(\eta)$, the equation $0 = A_\eta^* g$ should be taken to mean $A_\eta^* g(Z) = 0$ almost surely under η. In other words, the display is valid for every z in a set of η-measure 1. If the support of η contains a limit point, then this set is rich enough to conclude that $g = 0$, by the completeness of the exponential family.

If the support of η does not contain a limit point, then the preceding approach fails. However, we may reach almost the same conclusion by using a different type of scores. The paths $\eta_t = (1 - ta)\eta + ta\eta_1$ are well-defined for $0 \le at \le 1$, for any fixed $a \ge 0$ and η_1, and lead to scores

$$\frac{\partial}{\partial t}_{\mid t=0} \log p_{\eta_t}(x) = a\left(\frac{p_{\eta_1}}{p_\eta}(x) - 1\right).$$

This is certainly a score in a pointwise sense and can be shown to be a score in the L_2-sense provided that it is in $L_2(P_\eta)$. If $g \in L_2(P_\eta)$ has $P_\eta g = 0$ and is orthogonal to all scores of this type, then

$$0 = P_{\eta_1} g = P_\eta g\left(\frac{p_{\eta_1}}{p_\eta} - 1\right), \quad \text{every } \eta_1.$$

If the set of distributions $\{P_\eta : \eta \in H\}$ is complete, then we can typically conclude that $g = 0$ almost surely. Then the closed linear span of the tangent set is equal to the nonparametric, maximal tangent set. Because this set of scores is also a convex cone, Theorems 25.20 and 25.21 next show that nonparametric estimators are asymptotically efficient. $\square$

25.36 *Example (Semiparametric mixtures).* In the preceding example, replace the density $p(x \mid z)$ by a parametric family $p_\theta(x \mid z)$. Then the model $p_\theta(x \mid z) \, d\eta(z)$ for the unobserved data $Y = (X, Z)$ has scores for both θ and η. Suppose that the model $t \mapsto \eta_t$ is differentiable with score b, and that

$$\iint \left[p_{\theta+a}^{1/2}(x \mid z) - p_\theta^{1/2}(x \mid z) - \frac{1}{2} a^T \dot\ell_\theta(x \mid z) \, p_\theta^{1/2}(x \mid z) \right]^2 d\mu(x) \, d\eta(z) = o(\|a\|^2).$$

Then the function $a^T \dot{\ell}_\theta(x \mid z) + b(z)$ can be shown to be a score function corresponding to the model $t \mapsto p_{\theta+ta}(x \mid z) \, d\eta_t(z)$. Next, by Lemma 25.34, the function

$$\mathrm{E}_{\theta,\eta}\big(a^T \dot{\ell}_\theta(X \mid Z) + b(Z) \mid X = x\big) = \frac{\int \big(\dot{\ell}_\theta(x \mid z) + b(z)\big) \, p_\theta(x \mid z) \, d\eta(z)}{\int p_\theta(x \mid z) \, d\eta(z)}$$

is a score for the model corresponding to observing X only. □

25.37 *Example (Random censoring).* Suppose that the time T of an event is only observed if the event takes place before a censoring time C that is generated independently of T; otherwise we observe C. Thus the observation $X = (Y, \Delta)$ is the pair of transformations $Y = T \wedge C$ and $\Delta = 1\{T \leq C\}$ of the "full data" (T, C). If T has a distribution function F and $t \mapsto F_t$ is a differentiable path with score function a, then the submodel $t \mapsto P_{F_t, G}$ for X has score function

$$A_{F,G} a(x) = \mathrm{E}_F\big(a(T) \mid X = (y, \delta)\big) = (1 - \delta) \frac{\int_{(y,\infty)} a \, dF}{1 - F(y)} + \delta a(y).$$

A score operator for the distribution of C can be defined similarly, and takes the form, with G the distribution of C,

$$B_{F,G} b(x) = (1 - \delta) b(y) + \delta \frac{\int_{[y,\infty)} b \, dG}{1 - G_-(y)}.$$

The scores $A_{F,G} a$ and $B_{F,G} b$ form orthogonal spaces, as can be checked directly from the formulas, because $\mathrm{E} A_F a(X) B_G b(X) = FaGb$. (This is also explained by the product structure in the likelihood.) A consequence is that knowing G does not help for estimating F in the sense that the information for estimating parameters of the form $\psi(P_{F,G}) = \chi(F)$ is the same in the models in which G is known or completely unknown, respectively. To see this, note first that the influence function of such a parameter must be orthogonal to every score function for G, because $d/dt \, \psi(P_{F,G_t}) = 0$. Thus, due to the orthogonality of the two score spaces, an influence function of this parameter that is contained in the closed linear span of $\mathrm{R}(A_{F,G}) + \mathrm{R}(B_{F,G})$ is automatically contained in $\mathrm{R}(A_{F,G})$. □

25.38 *Example (Current status censoring).* Suppose that we only observe whether an event at time T has happened or not at an observation time C. Then we observe the transformation $X = \big(C, 1\{T \leq C\}\big) = (C, \Delta)$ of the pair (C, T). If T and C are independent with distribution functions F and G, respectively, then the score operators for F and G are given by, with $x = (c, \delta)$,

$$A_{F,G} a(x) = \mathrm{E}_F\big(a(T) \mid C = c, \Delta = \delta\big) = (1 - \delta) \frac{\int_{(c,\infty)} a \, dF}{1 - F(c)} + \delta \frac{\int_{[0,c]} a \, dF}{F(c)},$$

$$B_{F,G} b(x) = \mathrm{E}\big(b(C) \mid C = c, \Delta = \delta\big) = b(c).$$

These score functions can be seen to be orthogonal with the help of Fubini's theorem. If we take F to be completely unknown, then the set of a can be taken all functions in $L_2(F)$ with $Fa = 0$, and the adjoint operator $A_{F,G}^*$ restricted to the set of mean-zero functions in $L_2(P_{F,G})$ is given by

$$A_{F,G}^* h(c) = \int_{[c,\infty)} h(u, 1) \, dG(u) + \int_{[0,c)} h(u, 0) \, dG(u).$$

For simplicity assume that the true F and G possess continuous Lebesgue densities, which are positive on their supports. The range of $A^*_{F,G}$ consists of functions as in the preceding display for functions h that are contained in $L_2(P_{F,G})$, or equivalently

$$\int h^2(u, 0)(1 - F)(u)\,dG(u) < \infty \quad \text{and} \quad \int h^2(u, 1)F(u)\,dG(u) < \infty.$$

Thus the functions $h(u, 1)$ and $h(u, 0)$ are square-integrable with respect to G on any interval inside the support of F. Consequently, the range of the adjoint $A^*_{F,G}$ contains only absolutely continuous functions, and hence (25.29) fails for every parameter $\chi(F)$ with an influence function $\tilde{\chi}_F$ that is discontinuous. More precisely, parameters $\chi(F)$ with influence functions that are not almost surely equal under F to an absolutely continuous function. Because this includes the functions $1_{[0,t]} - F(t)$, the distribution function $F \mapsto \chi(F) = F(t)$ at a point is not a differentiable functional of the model. In view of Theorem 25.32 this means that this parameter is not estimable at $\sqrt{n}$-rate, and the usual normal theory does not apply to it.

On the other hand, parameters with a smooth influence function $\tilde{\chi}_F$ may be differentiable. The score operator for the model $P_{F,G}$ is the sum $(a, b) \mapsto A_{F,G}a + B_{F,G}b$ of the score operators for F and G separately. Its adjoint is the map $h \mapsto (A^*_{F,G}h, B^*_{F,G}h)$. A parameter of the form $(F, G) \mapsto \chi(F)$ has an influence function of the form $(\tilde{\chi}_F, 0)$. Thus, for a parameter of this type equation (25.29) takes the form

$$A^*_{F,G}\tilde{\psi}_{P_{F,G}} = \tilde{\chi}_F, \qquad B^*_{F,G}\tilde{\psi}_{P_{F,G}} = 0.$$

The kernel $N(A^*_{F,G})$ consists of the functions $h \in L_2(P_{F,G})$ such that $h(u, 0) = h(u, 1)$ almost surely under F and G. This is precisely the range of $B_{F,G}$, and we can conclude that

$$R(A_{F,G})^\perp = N(A^*_{F,G}) = R(B_{F,G}) = N(B^*_{F,G})^\perp.$$

Therefore, we can solve the preceding display by first solving $A^*_{F,G}h = \tilde{\chi}_F$ and next projecting a solution h onto the closure of the range of $A_{F,G}$. By the orthogonality of the ranges of $A_{F,G}$ and $B_{F,G}$, the latter projection is the identity minus the projection onto $R(B_{F,G})$. This is convenient, because the projection onto $R(B_{F,G})$ is the conditional expectation relative to C.

For example, consider a function $\chi(F) = Fa$ for some fixed known, continuously differentiable function a. Differentiating the equation $a = A^*_{F,G}h$, we find $a'(c) = \big(h(c, 0) - h(c, 1)\big)g(c)$. This can happen for some $h \in L_2(P_{F,G})$ only if, for any τ such that $0 < F(\tau) < 1$,

$$\int_\tau^\infty \left(\frac{a'}{g}\right)^2 (1 - F)\,dG = \int_\tau^\infty \big(h(u, 0) - h(u, 1)\big)^2 (1 - F)(u)\,dG(u) < \infty,$$

$$\int_0^\tau \left(\frac{a'}{g}\right)^2 F\,dG = \int_0^\tau \big(h(u, 0) - h(u, 1)\big)^2 F(u)\,dG(u) < \infty.$$

If the left sides of these equations are finite, then the parameter $P_{F,G} \mapsto Fa$ is differentiable. An influence function is given by the function h defined by

$$h(c, 0) = \frac{a'(c)1_{[\tau,\infty)}(c)}{g(c)}, \quad \text{and} \quad h(c, 1) = -\frac{a'(c)1_{[0,\tau)}(c)}{g(c)}.$$

The efficient influence function is found by projecting this onto $\overline{R}(A_{F,G})$, and is given by

$$h(c, \delta) - E_{F,G}\big(h(C, \Delta) \mid C = c\big) = \big(h(c, 1) - h(c, 0)\big)\big(\delta - F(c)\big)$$
$$= -\delta \frac{1 - F(c)}{g(c)} a'(c) + (1 - \delta) \frac{F(c)}{g(c)} a'(c).$$

For example, for the mean $\chi(F) = \int u \, dF(u)$, the influence function certainly exists if the density g is bounded away from zero on the compact support of F. $\quad\square$

*25.5.3 *Missing and Coarsening at Random*

Suppose that from a given vector (Y_1, Y_2) we sometimes observe only the first coordinate Y_1 and at other times both Y_1 and Y_2. Then Y_2 is said to be *missing at random (MAR)* if the conditional probability that Y_2 is observed depends only on Y_1, which is always observed. We can formalize this definition by introducing an indicator variable Δ that indicates whether Y_2 is missing ($\Delta = 0$) or observed ($\Delta = 1$). Then Y_2 is missing at random if $P(\Delta = 0 \mid Y)$ is a function of Y_1 only.

If next to $P(\Delta = 0 \mid Y)$ we also specify the marginal distribution of Y, then the distribution of (Y, Δ) is fixed, and the observed data are the function $X = \big(\phi(Y, \Delta), \Delta\big)$ defined by (for instance)

$$\phi(y, 0) = y_1, \qquad \phi(y, 1) = y.$$

The tangent set for the model for X can be derived from the tangent set for the model for (Y, Δ) by taking conditional expectations. If the distribution of (Y, Δ) is completely unspecified, then so is the distribution of X, and both tangent spaces are the maximal "nonparametric tangent space". If we restrict the model by requiring MAR, then the tangent set for (Y, Δ) is smaller than nonparametric. Interestingly, provided that we make no further restrictions, the tangent set for X remains the nonparametric tangent set.

We shall show this in somewhat greater generality. Let Y be an arbitrary unobservable "full observation" (not necessarily a vector) and let Δ be an arbitrary random variable. The distribution of (Y, Δ) can be determined by specifying a distribution Q for Y and a conditional density $r(\delta \mid y)$ for the conditional distribution of Δ given Y.[†] As before, we observe $X = \big(\phi(Y, \Delta), \Delta\big)$, but now ϕ may be an arbitrary measurable map. The observation X is said to be *coarsening at random (CAR)* if the conditional densities $r(\delta \mid y)$ depend on $x = \big(\phi(y, \delta), \delta\big)$ only, for every possible value (y, δ). More precisely, $r(\delta \mid y)$ is a measurable function of x.

25.39 Example (Missing at random). If $\Delta \in \{0, 1\}$ the requirements are both that $P(\Delta = 0 \mid Y = y)$ depends only on $\phi(y, 0)$ and 0 and that $P(\Delta = 1 \mid Y = y)$ depends only on $\phi(y, 1)$ and 1. Thus the two functions $y \mapsto P(\Delta = 0 \mid Y = y)$ and $y \mapsto P(\Delta = 1 \mid Y = y)$ may be different (fortunately) but may depend on y only through $\phi(y, 0)$ and $\phi(y, 1)$, respectively.

If $\phi(y, 1) = y$, then $\delta = 1$ corresponds to observing y completely. Then the requirement reduces to $P(\Delta = 0 \mid Y = y)$ being a function of $\phi(y, 0)$ only. If $Y = (Y_1, Y_2)$ and $\phi(y, 0) = y_1$, then CAR reduces to MAR as defined in the introduction. $\quad\square$

[†] The density is relative to a dominating measure ν on the sample space for Δ, and we suppose that $(\delta, y) \mapsto r(\delta \mid y)$ is a Markov kernel.

Denote by $\mathcal{Q}$ and $\mathcal{R}$ the parameter spaces for the distribution Q of Y and the kernels $r(\delta \mid y)$ giving the conditional distribution of Δ given Y, respectively. Let $\mathcal{Q} \times \mathcal{R} = (Q \times R : Q \in \mathcal{Q}, R \in \mathcal{R})$ and $\mathcal{P} = (P_{Q,R} : Q \in \mathcal{Q}, R \in \mathcal{R})$ be the models for (Y, Δ) and X, respectively.

25.40   Theorem. *Suppose that the distribution Q of Y is completely unspecified and the Markov kernel $r(\delta \mid y)$ is restricted by CAR, and only by CAR. Then there exists a tangent set $\dot{\mathcal{P}}_{P_{Q,R}}$ for the model $\mathcal{P} = (P_{Q,R} : Q \in \mathcal{Q}, R \in \mathcal{R})$ whose closure consists of all mean-zero functions in $L_2(P_{Q,R})$. Furthermore, any element of $\dot{\mathcal{P}}_{P_{Q,R}}$ can be orthogonally decomposed as*

$$\mathrm{E}_{Q,R}\big(a(Y) \mid X = x\big) + b(x),$$

where $a \in \dot{\mathcal{Q}}_Q$ and $b \in \dot{\mathcal{R}}_R$. The functions a and b range exactly over the functions $a \in L_2(Q)$ with $Qa = 0$ and $b \in L_2(P_{Q,R})$ with $\mathrm{E}_R\big(b(X) \mid Y\big) = 0$ almost surely, respectively.

Proof.   Fix a differentiable submodel $t \mapsto Q_t$ with score a. Furthermore, for every fixed y fix a differentiable submodel $t \mapsto r_t(\cdot \mid y)$ for the conditional density of Δ given $Y = y$ with score $b_0(\delta \mid y)$ such that

$$\iint \left[\frac{r_t^{1/2}(\delta \mid y) - r^{1/2}(\delta \mid y)}{t} - \frac{1}{2} b_0(\delta \mid y) r^{1/2}(\delta \mid y) \right]^2 \, d\nu(\delta) \, dQ(y) \to 0.$$

Because the conditional densities satisfy CAR, the function $b_0(\delta \mid y)$ must actually be a function $b(x)$ of x only. Because it corresponds to a score for the conditional model, it is further restricted by the equations $\int b_0(\delta \mid y) r(\delta \mid y) \, d\nu(\delta) = \mathrm{E}_R\big(b(X) \mid Y = y\big) = 0$ for every y. Apart from this and square integrability, b_0 can be chosen freely, for instance bounded.

By a standard argument, with $Q \times R$ denoting the law of (Y, Δ) under Q and r,

$$\int \left[\frac{(dQ_t \times R_t)^{1/2} - (dQ \times R)^{1/2}}{t} - \frac{1}{2}\big(a(y) + b(x)\big)(dQ \times R)^{1/2} \right]^2 \to 0.$$

Thus $a(y) + b(x)$ is a score function for the model of (Y, Δ), at $Q \times R$. By Lemma 25.34 its conditional expectation $\mathrm{E}_{Q,R}\big(a(Y) + b(X) \mid X = x\big)$ is a score function for the model of X.

This proves that the functions as given in the theorem arise as scores. To show that the set of all functions of this type is dense in the nonparametric tangent set, suppose that some function $g \in L_2(P_{Q,R})$ is orthogonal to all functions $\mathrm{E}_{Q,R}\big(a(Y) \mid X = x\big) + b(x)$. Then $\mathrm{E}_{Q,R} g(X) a(Y) = \mathrm{E}_{Q,R} g(X) \mathrm{E}_{Q,R}\big(a(Y) \mid X\big) = 0$ for all a. Hence g is orthogonal to all functions of Y and hence is a function of the type b. If it is also orthogonal to all b, then it must be 0.   ∎

The interest of the representation of scores given in the preceding theorem goes beyond the case that the models $\mathcal{Q}$ and $\mathcal{R}$ are restricted by CAR only, as is assumed in the theorem. It shows that, under CAR, any tangent space for $\mathcal{P}$ can be decomposed into two orthogonal pieces, the first part consisting of the conditional expectations $\mathrm{E}_{Q,R}\big(a(Y) \mid X\big)$ of scores a for the model of Y (and their limits) and the second part being scores b for the model $\mathcal{R}$

describing the "missingness pattern." CAR ensures that the latter are functions of x already and need not be projected, and also that the two sets of scores are orthogonal. By the product structure of the likelihood $q(y)r(\delta \mid y)$, scores a and b for q and r in the model $Q \times R$ are always orthogonal. This orthogonality may be lost by projecting them on the functions of x, but not so under CAR, because b is equal to its projection.

In models in which there is a positive probability of observing the complete data, there is an interesting way to obtain all influence functions of a given parameter $P_{Q,R} \mapsto \chi(Q)$. Let C be a set of possible values of Δ leading to a complete observation, that is, $\phi(y, \delta) = y$ whenever $\delta \in C$, and suppose that $R(C \mid y) = P_R(\Delta \in C \mid Y = y)$ is positive almost surely. Suppose for the moment that R is known, so that the tangent space for X consists only of functions of the form $E_{Q,R}(a(Y) \mid X)$. If $\dot{\chi}_Q(y)$ is an influence function of the parameter $Q \mapsto \chi(Q)$ on the model Q, then

$$\dot{\psi}_{P_{Q,R}}(x) = \frac{1\{\delta \in C\}}{R(C \mid y)} \dot{\chi}_Q(y)$$

is an influence function for the parameter $\psi(P_{F,G}) = \chi(Q)$ on the model $\mathcal{P}$. To see this, first note that, indeed, it is a function of x, as the indicator $1\{\delta \in C\}$ is nonzero only if $(y, \delta) = x$. Second,

$$E_{Q,R} \dot{\psi}_{P_{Q,R}}(X) E_{Q,R}(a(Y) \mid X) = E_{Q,R} \frac{1\{\Delta \in C\}}{R(C \mid Y)} \dot{\chi}_Q(Y) a(Y)$$
$$= E_{Q,R} \dot{\chi}_Q(Y) a(Y).$$

The influence function we have found is just one of many influence functions, the other ones being obtainable by adding the orthocomplement of the tangent set. This particular influence function corresponds to ignoring incomplete observations altogether but reweighting the influence function for the full model to eliminate the bias caused by such neglect. Usually, ignoring all partial observations does not yield an efficient procedure, and correspondingly this influence function is usually not the efficient influence function.

All other influence functions, including the efficient influence function, can be found by adding the orthocomplement of the tangent set. An attractive way of doing this is:
– by varying $\dot{\chi}_Q$ over all possible influence functions for $Q \mapsto \chi(Q)$, combined with
– by adding all functions $b(x)$ with $E_R(b(X) \mid Y) = 0$.
This is proved in the following lemma. We still assume that R is known; if it is not, then the resulting functions need not even be influence functions.

25.41 Lemma. *Suppose that the parameter $Q \mapsto \chi(Q)$ on the model Q is differentiable at Q, and that the conditional probability $R(C \mid Y) = P(\Delta \in C \mid Y)$ of having a complete observation is bounded away from zero. Then the parameter $P_{Q,R} \mapsto \chi(Q)$ on the model $(P_{Q,R} : Q \in Q)$ is differentiable at $P_{Q,R}$ and any of its influence functions can be written in the form*

$$\frac{1\{\delta \in C\}}{R(C \mid y)} \dot{\chi}_Q(y) + b(x),$$

for $\dot{\chi}_Q$ an influence function of the parameter $Q \mapsto \chi(Q)$ on the model Q and a function $b \in L_2(P_{Q,R})$ satisfying $E_R(b(X) \mid Y) = 0$. This decomposition is unique. Conversely, every function of this form is an influence function.

Proof. The function in the display with $b = 0$ has already been seen to be an influence function. (Note that it is square-integrable, as required.) Any function $b(X)$ such that $E_R(b(X) \mid Y) = 0$ satisfies $E_{Q,R} b(X) E_{Q,R}(a(Y) \mid X) = 0$ and hence is orthogonal to the tangent set, whence it can be added to any influence function.

To see that the decomposition is unique, it suffices to show that the function as given in the lemma can be identically zero only if $\dot{\chi}_Q = 0$ and $b = 0$. If it is zero, then its conditional expectation with respect to Y, which is $\dot{\chi}_Q$, is zero, and reinserting this we find that $b = 0$ as well.

Conversely, an arbitrary influence function $\dot{\psi}_{P_{Q,R}}$ of $P_{Q,R} \mapsto \chi(Q)$ can be written in the form

$$\dot{\psi}_{P_{Q,R}}(x) = \frac{1\{\delta \in C\}}{R(C \mid y)} \dot{\chi}(y) + \left[\dot{\psi}_{P_{Q,R}}(x) - \frac{1\{\delta \in C\}}{R(C \mid y)} \dot{\chi}(y) \right].$$

For $\dot{\chi}(Y) = E_R(\dot{\psi}_{P_{Q,R}}(X) \mid Y)$, the conditional expectation of the part within square brackets with respect to Y is zero and hence this part qualifies as a function b. This function $\dot{\chi}$ is an influence function for $Q \mapsto \chi(Q)$, as follows from the equality $E_{Q,R} E_R(\dot{\psi}_{P_{Q,R}}(X) \mid Y) a(Y) = E_{Q,R} \dot{\psi}_{P_{Q,R}}(X) E_{Q,R}(a(Y) \mid X)$ for every a. ∎

Even though the functions $\dot{\chi}_Q$ and b in the decomposition given in the lemma are uniquely determined, the decomposition is not orthogonal, and (even under CAR) the decomposition does not agree with the decomposition of the (nonparametric) tangent space given in Theorem 25.40. The second term is as the functions b in this theorem, but the leading term is not in the maximal tangent set for Q.

The preceding lemma is valid without assuming CAR. Under CAR it obtains an interesting interpretation, because in that case the functions b range exactly over all scores for the parameter r that we would have had if R were completely unknown. If R is known, then these scores are in the orthocomplement of the tangent set and can be added to any influence function to find other influence functions.

A second special feature of CAR is that a similar representation becomes available in the case that R is (partially) unknown. Because the tangent set for the model $(P_{Q,R} : Q \in \mathcal{Q}, R \in \mathcal{R})$ contains the tangent set for the model $(P_{Q,R} : Q \in \mathcal{Q})$ in which R is known, the influence functions for the bigger model are a subset of the influence functions of the smaller model. Because our parameter $\chi(Q)$ depends on Q only, they are exactly those influence functions in the smaller model that are orthogonal to the set $_R\dot{\mathcal{P}}_{P_{Q,R}}$ of all score functions for R. This is true in general, also without CAR. Under CAR they can be found by subtracting the projections onto the set of scores for R.

25.42 Corollary. *Suppose that the conditions of the preceding lemma hold and that the tangent space $\dot{\mathcal{P}}_{P_{Q,R}}$ for the model $(P_{Q,R} : Q \in \mathcal{Q}, R \in \mathcal{R})$ is taken to be the sum $_Q\dot{\mathcal{P}}_{P_{Q,R}} + _R\dot{\mathcal{P}}_{P_{Q,R}}$ of tangent spaces of scores for Q and R separately. If $_Q\dot{\mathcal{P}}_{P_{Q,R}}$ and $_R\dot{\mathcal{P}}_{P_{Q,R}}$ are orthogonal, in particular under CAR, any influence function of $P_{Q,R} \mapsto \chi(Q)$ for the model $(P_{Q,R} : Q \in \mathcal{Q}, R \in \mathcal{R})$ can be obtained by taking the functions given by the preceding lemma and subtracting their projection onto $\overline{\lin}\, _R\dot{\mathcal{P}}_{P_{Q,R}}$.*

Proof. The influence functions for the bigger model are exactly those influence functions for the model in which R is known that are orthogonal to $_R\dot{\mathcal{P}}_{P_{Q,R}}$. These do not change by subtracting their projection onto this space. Thus we can find all influence functions as claimed.

If the score spaces for Q and R are orthogonal, then the projection of an influence function onto $\overline{\operatorname{lin}}\,_R\dot{\mathcal{P}}_{P_{Q,R}}$ is orthogonal to $_Q\dot{\mathcal{P}}_{P_{Q,R}}$, and hence the inner products with elements of this set are unaffected by subtracting it. Thus we necessarily obtain an influence function.   ∎

The efficient influence function $\tilde{\psi}_{P_{Q,R}}$ is an influence function and hence can be written in the form of Lemma 25.41 for some $\dot{\chi}_Q$ and b. By definition it is the unique influence function that is contained in the closed linear span of the tangent set. Because the parameter of interest depends on Q only, the efficient influence function is the same (under CAR or, more generally, if $_Q\dot{\mathcal{P}}_{P_{Q,R}} \perp \,_R\dot{\mathcal{P}}_{P_{Q,R}}$), whether we assume R known or not. One way of finding the efficient influence function is to minimize the variance of an arbitrary influence function as given in Lemma 25.41 over $\dot{\chi}_Q$ and b.

25.43   *Example (Missing at random).* In the case of MAR models there is a simple representation for the functions $b(x)$ in Lemma 25.41. Because MAR is a special case of CAR, these functions can be obtained by computing all the scores for R in the model for (Y, Δ) under the assumption that R is completely unknown, by Theorem 25.40. Suppose that Δ takes only the values 0 and 1, where 1 indicates a full observation, as in Example 25.39, and set $\pi(y) := \mathrm{P}(\Delta = 1 \mid Y = y)$. Under MAR $\pi(y)$ is actually a function of $\phi(y, 0)$ only. The likelihood for (Y, Δ) takes the form

$$q(y)r(\delta \mid y) = q(y)\pi(y)^\delta\big(1 - \pi(y)\big)^{1-\delta}.$$

Insert a path $\pi_t = \pi + tc$, and differentiate the log likelihood with respect to t at $t = 0$ to obtain a score for R of the form

$$\frac{\delta}{\pi(y)}c(y) - \frac{1 - \delta}{1 - \pi(y)}c(y) = \frac{\delta - \pi(y)}{\pi(y)(1 - \pi)(y)}c(y).$$

To remain within the model the functions π_t and π, whence c, may depend on y only through $\phi(y, 0)$. Apart from this restriction, the preceding display gives a candidate for b in Lemma 25.41 for any c, and it gives all such b.

Thus, with a slight change of notation any influence function can be written in the form

$$\frac{\delta}{\pi(y)}\dot{\chi}_Q(y) - \frac{\delta - \pi(y)}{\pi(y)}c(y).$$

One approach to finding the efficient influence function in this case is first to minimize the variance of this influence function with respect to c and next to optimize over $\dot{\chi}_Q$. The first step of this plan can be carried out in general. Minimizing with respect to c is a weighted least-squares problem, whose solution is given by

$$\tilde{c}(Y) = \mathrm{E}_{Q,R}\big(\dot{\chi}_Q(Y) \mid \phi(Y, 0)\big).$$

To see this it suffices to verify the orthogonality relation, for all c,

$$\frac{\delta}{\pi(y)}\dot{\chi}_Q(y) - \frac{\delta - \pi(y)}{\pi(y)}\tilde{c}(y) \perp \frac{\delta - \pi(y)}{\pi(y)}c(y).$$

Splitting the inner product of these functions on the first minus sign, we obtain two terms, both of which reduce to $\mathrm{E}_{Q,R}\dot{\chi}_Q(Y)c(Y)(1 - \pi)(Y)/\pi(Y)$.   □

25.6 Testing

The problem of testing a null hypothesis $H_0 : \psi(P) \leq 0$ versus the alternative $H_1 : \psi(P) > 0$ is closely connected to the problem of estimating the function $\psi(P)$. It ought to be true that a test based on an asymptotically efficient estimator of $\psi(P)$ is, in an appropriate sense, asymptotically optimal. For real-valued parameters $\psi(P)$ this optimality can be taken in the absolute sense of an asymptotically (locally) uniformly most powerful test. With higher-dimensional parameters we run into the same problem of defining a satisfactory notion of asymptotic optimality as encountered for parametric models in Chapter 15. We leave the latter case undiscussed and concentrate on real-valued functionals $\psi : \mathcal{P} \mapsto \mathbb{R}$.

Given a model $\mathcal{P}$ and a measure P on the boundary of the hypotheses, that is, $\psi(P) = 0$, we want to study the "local asymptotic power" in a neighborhood of P. Defining a local power function in the present infinite-dimensional case is somewhat awkward, because there is no natural "rescaling" of the parameter set, such as in the Euclidean case. We shall utilize submodels corresponding to a tangent set. Given an element g in a tangent set $\dot{\mathcal{P}}_P$, let $t \mapsto P_{t,g}$ be a differentiable submodel with score function g along which ψ is differentiable. For every such g for which $\dot{\psi}_P g = P \tilde{\psi}_P g > 0$, the submodel $P_{t,g}$ belongs to the alternative hypothesis H_1 for (at least) every sufficiently small, positive t, because $\psi(P_{t,g}) = t P \tilde{\psi}_P g + o(t)$ if $\psi(P) = 0$. We shall study the power at the alternatives $P_{h/\sqrt{n},g}$.

25.44 Theorem. *Let the functional $\psi : \mathcal{P} \mapsto \mathbb{R}$ be differentiable at P relative to the tangent space $\dot{\mathcal{P}}_P$ with efficient influence function $\tilde{\psi}_P$. Suppose that $\psi(P) = 0$. Then for every sequence of power functions $P \mapsto \pi_n(P)$ of level-α tests for $H_0 : \psi(P) \leq 0$, and every $g \in \dot{\mathcal{P}}_P$ with $P \tilde{\psi}_P g > 0$ and every $h > 0$,*

$$\limsup_{n \to \infty} \pi_n(P_{h/\sqrt{n},g}) \leq 1 - \Phi \left(z_\alpha - h \frac{P \tilde{\psi}_P g}{(P \tilde{\psi}_P^2)^{1/2}} \right).$$

Proof. This theorem is essentially Theorem 15.4 applied to sufficiently rich submodels. Because the present situation does not fit exactly in the framework of Chapter 15, we rework the proof. Fix arbitrary h_1 and g_1 for which we desire to prove the upper bound. For notational convenience assume that $P g_1^2 = 1$.

Fix an orthonormal base $g_P = (g_1, \ldots, g_m)^T$ of an arbitrary finite-dimensional subspace of $\dot{\mathcal{P}}_P$ (containing the fixed g_1). For every $g \in \text{lin} \, g_P$, let $t \mapsto P_{t,g}$ be a submodel with score g along which the parameter ψ is differentiable. Each of the submodels $t \mapsto P_{t,g}$ is locally asymptotically normal at $t = 0$ by Lemma 25.14. Therefore, with S^{m-1} the unit sphere of $\mathbb{R}^m$,

$$\left(P_{h/\sqrt{n}, a^T g_P}^n : h > 0, a \in S^{m-1} \right) \rightsquigarrow \left(N_m(ha, I) : h > 0, a \in S^{m-1} \right),$$

in the sense of convergence of experiments. Fix a subsequence along which the limsup in the statement of the theorem is taken for $h = h_1$ and $g = g_1$. By contiguity arguments, we can extract a further subsequence along which the functions $\pi_n(P_{h/\sqrt{n}, a^T g})$ converge pointwise to a limit $\pi(h, a)$ for every (h, a). By Theorem 15.1, the function $\pi(h, a)$ is the power function of a test in the normal limit experiment. If it can be shown that this test is of level α for testing $H_0 : a^T P \tilde{\psi}_P g_P = 0$, then Proposition 15.2 shows that, for every (a, h)

with $a^T P\tilde{\psi}_P g_P > 0$,

$$\pi(h, a) \le 1 - \Phi\left(z_\alpha - h \frac{a^T P\tilde{\psi}_P g_P}{\left(P\tilde{\psi}_P g_P^T P\tilde{\psi}_P g_P\right)^{1/2}}\right).$$

The orthogonal projection of $\tilde{\psi}_P$ onto lin g_P is equal to $(P\tilde{\psi}_P g_P^T)g_P$, and has length $P\tilde{\psi}_P g_P^T P\tilde{\psi}_P g_P$. By choosing lin g_P large enough, we can ensure that this length is arbitrarily close to $P\tilde{\psi}_P^2$. Choosing $(h, a) = (h_1, e_1)$ completes the proof, because $\limsup \pi_n(P_{h_1/\sqrt{n}, g_1}) \le \pi(h_1, e_1)$, by construction.

To complete the proof, we show that π is of level α. Fix any $h > 0$ and an $a \in S^{m-1}$ such that $a^T P\tilde{\psi}_P g_P < 0$. Then

$$\psi(P_{h/\sqrt{n}, a^T g}) = \psi(P) + \frac{h}{\sqrt{n}}\left(a^T P\tilde{\psi}_P g_P + o(1)\right)$$

is negative for sufficiently large n. Hence $P_{h/\sqrt{n}, a^T g}$ belongs to H_0 and

$$\pi(h, a) = \lim \pi_n(P_{h/\sqrt{n}, a^T g}) \le \alpha.$$

Thus, the test with power function π is of level α for testing $H_0 : a^T P\tilde{\psi}_P g_P < 0$. By continuity it is of level α for testing $H_0 : a^T P\tilde{\psi}_P g_P \le 0$. ∎

As a consequence of the preceding theorem, a test based on an efficient estimator for $\psi(P)$ is automatically "locally uniformly most powerful": Its power function attains the upper bound given by the theorem. More precisely, suppose that the sequence of estimators T_n is asymptotically efficient at P and that S_n is a consistent sequence of estimators of its asymptotic variance. Then the test that rejects $H_0 : \psi(P) = 0$ for $\sqrt{n}T_n/S_n \ge z_\alpha$ attains the upper bound of the theorem.

25.45 Lemma. *Let the functional $\psi : \mathcal{P} \mapsto \mathbb{R}$ be differentiable at P with $\psi(P) = 0$. Suppose that the sequence T_n is regular at P with a $N(0, P\tilde{\psi}_P^2)$-limit distribution. Furthermore, suppose that $S_n^2 \xrightarrow{P} P\tilde{\psi}_P^2$. Then, for every $h \ge 0$ and $g \in \dot{\mathcal{P}}_P$,*

$$\lim_{n\to\infty} P_{h/\sqrt{n}, g}\left(\frac{\sqrt{n}T_n}{S_n} \ge z_\alpha\right) = 1 - \Phi\left(z_\alpha - h\frac{P\tilde{\psi}_P g}{(P\tilde{\psi}_P^2)^{1/2}}\right).$$

Proof. By the efficiency of T_n and the differentiability of ψ, the sequence $\sqrt{n}T_n$ converges under $P_{h/\sqrt{n}, g}$ to a normal distribution with mean $hP\tilde{\psi}_P g$ and variance $P\tilde{\psi}_P^2$. ∎

25.46 Example (Wilcoxon test). Suppose that the observations are two independent random samples $X_1, \ldots, X_n$ and $Y_1, \ldots, Y_n$ from distribution functions F and G, respectively. To fit this two-sample problem in the present i.i.d. set-up, we pair the two samples and think of (X_i, Y_i) as a single observation from the product measure $F \times G$ on $\mathbb{R}^2$. We wish to test the null hypothesis $H_0 : \int F\, dG \le \frac{1}{2}$ versus the alternative $H_1 : \int F\, dG > \frac{1}{2}$. The Wilcoxon test, which rejects for large values of $\int \mathbb{F}_n\, d\mathbb{G}_n$, is asymptotically efficient, relative to the model in which F and G are completely unknown. This gives a different perspective on this test, which in Chapters 14 and 15 was seen to be asymptotically efficient for testing location in the logistic location-scale family. Actually, this finding is an example of the general principle that, in the situation that the underlying distribution of the observations is completely unknown, empirical-type statistics are asymptotically efficient for whatever they naturally estimate or test (also see Example 25.24 and Section 25.7). The present conclusion concerning the Wilcoxon test extends to most other test statistics.

By the preceding lemma, the efficiency of the test follows from the efficiency of the Wilcoxon statistic as an estimator for the function $\psi(F \times G) = \int F \, dG$. This may be proved by Theorem 25.47, or by the following direct argument.

The model $\mathcal{P}$ is the set of all product measures $F \times G$. To generate a tangent set, we can perturb both F and G. If $t \mapsto F_t$ and $t \mapsto G_t$ are differentiable submodels (of the collection of all probability distributions on $\mathbb{R}$) with score functions a and b at $t = 0$, respectively, then the submodel $t \mapsto F_t \times G_t$ has score function $a(x) + b(y)$. Thus, as a tangent space we may take the set of all square-integrable functions with mean zero of this type. For simplicity, we could restrict ourselves to bounded functions a and b and use the paths $d\,F_t = (1 + ta)d\,F$ and $dG_t = (1 + tb)dG$. The closed linear span of the resulting tangent set is the same as before. Then, by simple algebra,

$$\dot{\psi}_{F \times G}(a, b) = \frac{\partial}{\partial t} \psi(F_t \times G_t)_{|t=0} = \int (1 - G_-) a \, dF + \int F b \, dG.$$

We conclude that the function $(x, y) \mapsto (1 - G_-)(x) + F(y)$ is an influence function of ψ. This is of the form $a(x) + b(y)$ but does not have mean zero; the efficient influence function is found by subtracting the mean.

The efficiency of the Wilcoxon statistic is now clear from Lemma 25.23 and the asymptotic linearity of the Wilcoxon statistic, which is proved by various methods in Chapters 12, 13, and 20.   □

*25.7   Efficiency and the Delta Method

Many estimators can be written as functions $\phi(T_n)$ of other estimators. By the delta method asymptotic normality of T_n carries over into the asymptotic normality of $\phi(T_n)$, for every differentiable map ϕ. Does efficiency of T_n carry over into efficiency of $\phi(T_n)$ as well? With the right definitions, the answer ought to be affirmative. The matter is sufficiently useful to deserve a discussion and turns out to be nontrivial. Because the result is true for the functional delta method, applications include the efficiency of the product-limit estimator in the random censoring model and the sample median in the nonparametric model, among many others.

If T_n is an estimator of a Euclidean parameter $\psi(P)$ and both ϕ and ψ are differentiable, then the question can be answered by a direct calculation of the normal limit distributions. In view of Lemma 25.23, efficiency of T_n can be defined by the asymptotic linearity (25.22). By the delta method,

$$\sqrt{n} \left(\phi(T_n) - \phi \circ \psi(P) \right) = \phi'_{\psi(P)} \sqrt{n} \left(T_n - \psi(P) \right) + o_P(1)$$

$$= \frac{1}{\sqrt{n}} \sum_{i=1}^{n} \phi'_{\psi(P)} \tilde{\psi}_P(X_i) + o_P(1).$$

The asymptotic efficiency of $\phi(T_n)$ follows, provided that the function $x \mapsto \phi'_{\psi(P)} \tilde{\psi}_P(x)$ is the efficient influence function of the parameter $P \mapsto \phi \circ \psi(P)$. If the coordinates of

$\tilde{\psi}_P$ are contained in the closed linear span of the tangent set, then so are the coordinates of $\phi'_{\psi(P)}\tilde{\psi}_P$, because the matrix multiplication by $\phi'_{\psi(P)}$ means taking linear combinations. Furthermore, if ψ is differentiable at P (as a statistical parameter on the model $\mathcal{P}$) and ϕ is differentiable at $\psi(P)$ (in the ordinary sense of calculus), then

$$\frac{\phi \circ \psi(P_t) - \phi \circ \psi(P)}{t} \to \phi'_{\psi(P)}\dot{\psi}_P g = P\phi'_{\psi(P)}\tilde{\psi}_P g.$$

Thus the function $\phi'_{\psi(P)}\tilde{\psi}_P$ is an influence function and hence the efficient influence function.

More involved is the same question, but with T_n an estimator of a parameter in a Banach space, for instance a distribution in the space $D[-\infty, \infty]$ or in a space $\ell^\infty(\mathcal{F})$. The question is empty until we have defined efficiency for this situation. A definition of asymptotic efficiency of Banach-valued estimators can be based on generalizations of the convolution and minimax theorems to general Banach spaces.[†] We shall avoid this route and take a more naive approach.

The *dual space* $\mathbb{D}^*$ of a Banach space $\mathbb{D}$ is defined as the collection of all continuous, linear maps $d^* : \mathbb{D} \mapsto \mathbb{R}$. If T_n is a $\mathbb{D}$-valued estimator for a parameter $\psi(P) \in \mathbb{D}$, then $d^* T_n$ is a real-valued estimator for the parameter $d^* \psi(P) \in \mathbb{R}$. This suggests to defining T_n to be *asymptotically efficient* at $P \in \mathcal{P}$ if $\sqrt{n}(T_n - \psi(P))$ converges under P in distribution to a tight limit and $d^* T_n$ is asymptotically efficient at P for estimating $d^* \psi(P)$, for every $d^* \in \mathbb{D}^*$.

This definition presumes that the parameters $d^*\psi$ are differentiable at P in the sense of section 25.3. We shall require a bit more. Say that $\psi : \mathcal{P} \mapsto \mathbb{D}$ is *differentiable* at P relative to a given tangent set $\dot{\mathcal{P}}_P$ if there exists a continuous linear map $\dot{\psi}_P : L_2(P) \mapsto \mathbb{D}$ such that, for every $g \in \dot{\mathcal{P}}_P$ and a submodel $t \mapsto P_t$ with score function g,

$$\frac{\psi(P_t) - \psi(P)}{t} \to \dot{\psi}_P g.$$

This implies that every parameter $d^*\psi : \mathcal{P} \mapsto \mathbb{R}$ is differentiable at P, whence, for every $d^* \in \mathbb{D}^*$, there exists a function $\tilde{\psi}_{P,d^*} : \mathcal{X} \mapsto \mathbb{R}$ in the closed linear span of $\dot{\mathcal{P}}_P$ such that $d^*\dot{\psi}_P(g) = P\tilde{\psi}_{P,d^*}g$ for every $g \in \dot{\mathcal{P}}_P$. The efficiency of $d^* T_n$ for $d^*\psi$ can next be understood in terms of asymptotic linearity of $d^*\sqrt{n}(T_n - \psi(P))$, as in (25.22), with influence function $\tilde{\psi}_{P,d^*}$.

To avoid measurability issues, we also allow nonmeasurable functions $T_n = T_n(X_1, \ldots, X_n)$ of the data as estimators in this section. Let both $\mathbb{D}$ and $\mathbb{E}$ be Banach spaces.

25.47 Theorem. *Suppose that $\psi : \mathcal{P} \mapsto \mathbb{D}$ is differentiable at P and takes its values in a subset $\mathbb{D}_\phi \subset \mathbb{D}$, and suppose that $\phi : \mathbb{D}_\phi \subset \mathbb{D} \mapsto \mathbb{E}$ is Hadamard-differentiable at $\psi(P)$ tangentially to $\overline{\lin}\,\dot{\psi}_P(\dot{\mathcal{P}}_P)$. Then $\phi \circ \psi : \mathcal{P} \mapsto \mathbb{E}$ is differentiable at P. If T_n is a sequence of estimators with values in $\mathbb{D}_\phi$ that is asymptotically efficient at P for estimating $\psi(P)$, then $\phi(T_n)$ is asymptotically efficient at P for estimating $\phi \circ \psi(P)$.*

Proof. The differentiability of $\phi \circ \psi$ is essentially a consequence of the chain rule for Hadamard-differentiable functions (see Theorem 20.9) and is proved in the same way. The derivative is the composition $\phi'_{\psi(P)} \circ \dot{\psi}_P$.

[†] See for example, Chapter 3.11 in [146] for some possibilities and references.

First, we show that the limit distribution L of the sequence $\sqrt{n}(T_n - \psi(P))$ concentrates on the subspace $\overline{\operatorname{lin}}\,\dot{\psi}_P(\dot{\mathcal{P}}_P)$. By the Hahn-Banach theorem, for any $S \subset \mathbb{D}$,

$$\overline{\operatorname{lin}}\,\dot{\psi}_P(\dot{\mathcal{P}}_P) \cap S = \cap_{d^* \in \mathbb{D}^* : d^*\dot{\psi}_P = 0} \{d \in S : d^*d = 0\}.$$

For a separable set S, we can replace the intersection by a countable subintersection. Because L is tight, it concentrates on a separable set S, and hence L gives mass 1 to the left side provided $L(d : d^*d = 0) = 1$ for every d^* as on the right side. This probability is equal to $N(0, \|\tilde{\psi}_{d^*P}\|_P^2)\{0\} = 1$.

Now we can conclude that under the assumptions the sequence $\sqrt{n}(\phi(T_n) - \phi \circ \psi(P))$ converges in distribution to a tight limit, by the functional delta method, Theorem 20.8. Furthermore, for every $e^* \in \mathbb{E}^*$

$$\sqrt{n}(e^*\phi(T_n) - e^*\phi \circ \psi(P)) = e^*\phi'_{\psi(P)}\sqrt{n}(T_n - \psi(P)) + o_P(1),$$

where, if necessary, we can extend the definition of $d^* = e^*\phi'_{\psi(P)}$ to all of $\mathbb{D}$ in view of the Hahn-Banach theorem. Because $d^* \in \mathbb{D}^*$, the asymptotic efficiency of the sequence T_n implies that the latter sequence is asymptotically linear in the influence function $\tilde{\psi}_{P,d^*}$. This is also the influence function of the real-valued map $e^*\phi \circ \psi$, because

$$e^*\phi'_{\psi(P)} \circ \dot{\psi}_P g = d^*\dot{\psi}_P g = P\tilde{\psi}_{P,d^*}g, \qquad g \in \dot{\mathcal{P}}_P.$$

Thus, $e^*\phi(T_n)$ is asymptotically efficient at P for estimating $e^*\phi \circ \psi(P)$, for every $e^* \in \mathbb{E}^*$. ∎

The proof of the preceding theorem is relatively simple, because our definition of an efficient estimator sequence, although not unnatural, is relatively involved.

Consider, for instance, the case that $\mathbb{D} = \ell^\infty(S)$ for some set S. This corresponds to estimating a (bounded) function $s \mapsto \psi(P)(s)$ by a random function $s \mapsto T_n(s)$. Then the "marginal estimators" d^*T_n include the estimators $\pi_s T_n = T_n(s)$ for every fixed s – the coordinate projections $\pi_s : d \mapsto d(s)$ are elements of the dual space $\ell^\infty(S)^*$–, but include many other, more complicated functions of T_n as well. Checking the efficiency of every marginal of the general type d^*T_n may be cumbersome.

The deeper result of this section is that this is not necessary. Under the conditions of Theorem 17.14, the limit distribution of the sequence $\sqrt{n}(T_n - \psi(P))$ in $\ell^\infty(S)$ is determined by the limit distributions of these processes evaluated at finite sets of "times" $s_1, \ldots, s_k$. Thus, we may hope that the asymptotic efficiency of T_n can also be characterized by the behavior of the marginals $T_n(s)$ only. Our definition of a differentiable parameter $\psi : \mathcal{P} \mapsto \mathbb{D}$ is exactly right for this purpose.

25.48   Theorem (Efficiency in $\ell^\infty(S)$). *Suppose that $\psi : \mathcal{P} \mapsto \ell^\infty(S)$ is differentiable at P, and suppose that $T_n(s)$ is asymptotically efficient at P for estimating $\psi(P)(s)$, for every $s \in S$. Then T_n is asymptotically efficient at P provided that the sequence $\sqrt{n}(T_n - \psi(P))$ converges under P in distribution to a tight limit in $\ell^\infty(S)$.*

The theorem is a consequence of a more general principle that obtains the efficiency of T_n from the efficiency of d^*T_n for a sufficient number of elements $d^* \in \mathbb{D}^*$. By definition, efficiency of T_n means efficiency of d^*T_n for all $d^* \in \mathbb{D}^*$. In the preceding theorem the efficiency is deduced from efficiency of the estimators $\pi_s T_n$ for all coordinate projections π_s

on $\ell^\infty(S)$. The coordinate projections are a fairly small subset of the dual space of $\ell^\infty(S)$. What makes them work is the fact that they are of norm 1 and satisfy $\|z\|_S = \sup_s |\pi_s z|$.

25.49   Lemma. *Suppose that $\psi : \mathcal{P} \mapsto \mathbb{D}$ is differentiable at P, and suppose that $d'T_n$ is asymptotically efficient at P for estimating $d'\psi(P)$ for every d' in a subset $\mathbb{D}' \subset \mathbb{D}^*$ such that, for some constant C,*

$$\|d\| \le C \sup_{d' \in \mathbb{D}', \|d'\| \le 1} |d'(d)|.$$

Then T_n is asymptotically efficient at P provided that the sequence $\sqrt{n}(T_n - \psi(P))$ is asymptotically tight under P.

Proof. The efficiency of all estimators $d'T_n$ for every $d' \in \mathbb{D}'$ implies their asymptotic linearity. This shows that $d'T_n$ is also asymptotically linear and efficient for every $d' \in \text{lin } \mathbb{D}'$. Thus, it is no loss of generality to assume that $\mathbb{D}'$ is a linear space.

By Prohorov's theorem, every subsequence of $\sqrt{n}(T_n - \psi(P))$ has a further subsequence that converges weakly under P to a tight limit T. For simplicity, assume that the whole sequence converges; otherwise argue along subsequences. By the continuous-mapping theorem, $d^* \sqrt{n}(T_n - \psi(P))$ converges in distribution to d^*T for every $d^* \in \mathbb{D}^*$. By the assumption of efficiency, the sequence $d^* \sqrt{n}(T_n - \psi(P))$ is asymptotically linear in the influence function $\tilde\psi_{P,d^*}$ for every $d^* \in \mathbb{D}'$. Thus, the variable d^*T is normally distributed with mean zero and variance $P\tilde\psi_{P,d^*}^2$ for every $d^* \in \mathbb{D}'$. We show below that this is then automatically true for every $d^* \in \mathbb{D}^*$.

By Le Cam's third lemma (which by inspection of its proof can be seen to be valid for general metric spaces), the sequence $\sqrt{n}(T_n - \psi(P))$ is asymptotically tight under $P_{1/\sqrt{n}}$ as well, for every differentiable path $t \mapsto P_t$. By the differentiability of ψ, the sequence $\sqrt{n}(T_n - \psi(P_{1/\sqrt{n}}))$ is tight also. Then, exactly as in the preceding paragraph, we can conclude that the sequence $d^* \sqrt{n}(T_n - \psi(P_{1/\sqrt{n}}))$ converges in distribution to a normal distribution with mean zero and variance $P\tilde\psi_{P,d^*}^2$, for every $d^* \in \mathbb{D}^*$. Thus, d^*T_n is asymptotically efficient for estimating $d^*\psi(P)$ for every $d^* \in \mathbb{D}^*$ and hence T_n is asymptotically efficient for estimating $\psi(P)$, by definition.

It remains to prove that a tight, random element T in $\mathbb{D}$ such that d^*T has law $N(0, \|d^*\tilde\psi_P\|^2)$ for every $d^* \in \mathbb{D}'$ necessarily verifies this same relation for every $d^* \in \mathbb{D}^*$.[†] First assume that $\mathbb{D} = \ell^\infty(S)$ and that $\mathbb{D}'$ is the linear space spanned by all coordinate projections.

Because T is tight, there exists a semimetric ρ on S such that S is totally bounded and almost all sample paths of T are contained in $UC(S, \rho)$ (see Lemma 18.15). Then automatically the range of $\tilde\psi_P$ is contained in $UC(S, \rho)$ as well.

To see the latter, we note first that the map $s \mapsto ET(s)T(u)$ is contained in $UC(S, \rho)$ for every fixed u: If $\rho(s_m, t_m) \to 0$, then $T(s_m) - T(t_m) \to 0$ almost surely and hence in second mean, in view of the zero-mean normality of $T(s_m) - T(t_m)$ for every m, whence $|ET(s_m)T(u) - ET(t_m)T(u)| \to 0$ by the Cauchy-Schwarz inequality. Thus, the map

$$s \mapsto \tilde\psi_P(\tilde\psi_{P,\pi_u})(s) = \pi_s \tilde\psi_P(\tilde\psi_{P,\pi_u}) = \langle \tilde\psi_{P,\pi_u}, \tilde\psi_{P,\pi_s} \rangle_P = ET(u)T(s)$$

[†] The proof of this lemma would be considerably shorter if we knew already that there exists a tight random element T with values in $\mathbb{D}$ such that d^*T has a $N(0, \|d^*\tilde\psi_P\|_{P,2}^2)$-distribution for every $d^* \in \mathbb{D}^*$. Then it suffices to show that the distribution of T is uniquely determined by the distributions of d^*T for $d^* \in \mathbb{D}'$.

is contained in the space $UC(S, \rho)$ for every u. By the linearity and continuity of the derivative $\dot{\psi}_P$, the same is then true for the map $s \mapsto \dot{\psi}_P(g)(s)$ for every g in the closed linear span of the gradients $\tilde{\psi}_{P,\pi_u}$ as u ranges over S. It is even true for every g in the tangent set, because $\dot{\psi}_P(g)(s) = \dot{\psi}_P(\Pi g)(s)$ for every g and s, and Π the projection onto the closure of lin $\tilde{\psi}_{P,\pi_u}$.

By a minor extension of the Riesz representation theorem for the dual space of $C(\overline{S}, \rho)$, the restriction of a fixed $d^* \in \mathbb{D}^*$ to $UC(S, \rho)$ takes the form

$$d^* z = \int_{\overline{S}} \overline{z}(s) \, d\overline{\mu}(s),$$

for $\overline{\mu}$ a signed Borel measure on the completion $\overline{S}$ of S, and $\overline{z}$ the unique continuous extension of z to $\overline{S}$. By discretizing $\overline{\mu}$, using the total boundedness of S, we can construct a sequence d_m^* in $\text{lin}\{\pi_s : s \in S\}$ such that $d_m^* \to d^*$ pointwise on $UC(S, \rho)$. Then $d_m^* \dot{\psi}_P \to d^* \dot{\psi}_P$ pointwise on $\dot{\mathcal{P}}_P$. Furthermore, $d_m^* T \to d^* T$ almost surely, whence in distribution, so that $d^* T$ is normally distributed with mean zero. Because $d_m^* T - d_n^* T \to 0$ almost surely, we also have that

$$E(d_m^* T - d_n^* T)^2 = \|d_m^* \dot{\psi}_P - d_n^* \dot{\psi}_P\|_{P,2}^2 \to 0,$$

whence $d_m^* \dot{\psi}_P$ is a Cauchy sequence in $L_2(P)$. We conclude that $d_m^* \dot{\psi}_P \to d^* \dot{\psi}_P$ also in norm and $E(d_m^* T)^2 = \|d_m^* \dot{\psi}_P\|_{P,2}^2 \to \|d^* \dot{\psi}_P\|_{P,2}^2$. Thus, $d^* T$ is normally distributed with mean zero and variance $\|d^* \dot{\psi}_P\|_{P,2}^2$.

This concludes the proof for $\mathbb{D}$ equal to $\ell^\infty(S)$. A general Banach space $\mathbb{D}$ can be embedded in $\ell^\infty(\mathbb{D}_1')$, for $\mathbb{D}_1' = \{d' \in \mathbb{D}', \|d'\| \leq 1\}$, by the map $d \to z_d$ defined as $z_d(d') = d'(d)$. By assumption, this map is a norm homeomorphism, whence T can be considered to be a tight random element in $\ell^\infty(\mathbb{D}_1')$. Next, the preceding argument applies. ∎

Another useful application of the lemma concerns the estimation of functionals $\psi(P) = (\psi_1(P), \psi_2(P))$ with values in a product $\mathbb{D}_1 \times \mathbb{D}_2$ of two Banach spaces. Even though marginal weak convergence does not imply joint weak convergence, marginal efficiency implies joint efficiency!

25.50 **Theorem (Efficiency in product spaces).** *Suppose that $\psi_i : \mathcal{P} \mapsto \mathbb{D}_i$ is differentiable at P, and suppose that $T_{n,i}$ is asymptotically efficient at P for estimating $\psi_i(P)$, for $i = 1, 2$. Then $(T_{n,1}, T_{n,2})$ is asymptotically efficient at P for estimating $(\psi_1(P), \psi_2(P))$ provided that the sequences $\sqrt{n}(T_{n,i} - \psi_i(P))$ are asymptotically tight in $\mathbb{D}_i$ under P, for $i = 1, 2$.*

Proof. Let $\mathbb{D}'$ be the set of all maps $(d_1, d_2) \mapsto d_i^*(d_i)$ for d_i^* ranging over $\mathbb{D}_i^*$, and $i = 1, 2$. By the Hahn-Banach theorem, $\|d_i\| = \sup\{|d_i^*(d_i)| : \|d_i^*\| = 1, d_i^* \in \mathbb{D}_i^*\}$. Thus, the product norm $\|(d_1, d_2)\| = \|d_1\| \vee \|d_2\|$ satisfies the condition of the preceding lemma (with $C = 1$ and equality). ∎

25.51 **Example (Random censoring).** In section 25.10.1 it is seen that the distribution of $X = (C \wedge T, 1\{T \leq C\})$ in the random censoring model can be any distribution on the sample space. It follows by Example 20.16 that the empirical subdistribution functions $\mathbb{H}_{0n}$ and $\mathbb{H}_{1n}$ are asymptotically efficient. By Example 20.15 the product limit estimator is a Hadamard-differentiable functional of the empirical subdistribution functions. Thus, the product limit-estimator is asymptotically efficient. □

25.8   Efficient Score Equations

The most important method of estimating the parameter in a parametric model is the method of maximum likelihood, and it can usually be reduced to solving the score equations $\sum_{i=1}^{n} \dot{\ell}_\theta(X_i) = 0$, if necessary in a neighborhood of an initial estimate. A natural generalization to estimating the parameter θ in a semiparametric model $\{P_{\theta,\eta} : \theta \in \Theta, \eta \in H\}$ is to solve θ from the *efficient score equations*

$$\sum_{i=1}^{n} \tilde{\ell}_{\theta,\hat{\eta}_n}(X_i) = 0.$$

Here we use the efficient score function instead of the ordinary score function, and we substitute an estimator $\hat{\eta}_n$ for the unknown nuisance parameter. A refinement of this method has been applied successfully to a number of examples, and the method is likely to work in many other examples. A disadvantage is that the method requires an explicit form of the efficient score function, or an efficient algorithm to compute it. Because, in general, the efficient score function is defined only implicitly as an orthogonal projection, this may preclude practical implementation.

A variation on this approach is to obtain an estimator $\hat{\eta}_n(\theta)$ of η for each given value of θ, and next to solve θ from the equation

$$\sum_{i=1}^{n} \tilde{\ell}_{\theta,\hat{\eta}_n(\theta)}(X_i) = 0.$$

If $\hat{\theta}_n$ is a solution, then it is also a solution of the estimating equation in the preceding display, for $\hat{\eta}_n = \hat{\eta}_n(\hat{\theta}_n)$. The asymptotic normality of $\hat{\theta}_n$ can therefore be proved by the same methods as applying to this estimating equation. Due to our special choice of estimating function, the nature of the dependence of $\hat{\eta}_n(\theta)$ on θ should be irrelevant for the limiting distribution of $\sqrt{n}(\hat{\theta}_n - \theta)$. Informally, this is because the partial derivative of the estimating equation relative to the θ inside $\hat{\eta}_n(\theta)$ should converge to zero, as is clear from our subsequent discussion of the "no-bias" condition (25.52). The dependence of $\hat{\eta}_n(\theta)$ on θ does play a role for the consistency of $\hat{\theta}_n$, but we do not discuss this in this chapter, because the general methods of Chapter 5 apply. For simplicity we adopt the notation as in the first estimating equation, even though for the construction of $\hat{\theta}_n$ the two-step procedure, which "profiles out" the nuisance parameter, may be necessary.

In a number of applications the nuisance parameter η, which is infinite-dimensional, cannot be estimated within the usual order $O(n^{-1/2})$ for parametric models. Then the classical approach to derive the asymptotic behavior of Z-estimators – linearization of the equation in both parameters – is impossible. Instead, we utilize the notion of a Donsker class, as developed in Chapter 19. The auxiliary estimator for the nuisance parameter should satisfy[†]

$$P_{\hat{\theta}_n,\eta} \tilde{\ell}_{\hat{\theta}_n,\hat{\eta}_n} = o_P\big(n^{-1/2} + \|\hat{\theta}_n - \theta\|\big), \tag{25.52}$$

$$P_{\theta,\eta} \big\| \tilde{\ell}_{\hat{\theta}_n,\hat{\eta}_n} - \tilde{\ell}_{\theta,\eta} \big\|^2 \xrightarrow{P} 0, \qquad P_{\hat{\theta}_n,\eta} \big\| \tilde{\ell}_{\hat{\theta}_n,\hat{\eta}_n} \big\|^2 = O_P(1). \tag{25.53}$$

[†] The notation $P\ell_{\hat{\eta}}$ is an abbreviation for the integral $\int \ell_{\hat{\eta}}(x)\, dP(x)$. Thus the expectation is taken with respect to x only and not with respect to $\hat{\eta}$.

The second condition (25.53) merely requires that the "plug-in" estimator $\tilde{\ell}_{\theta, \hat{\eta}_n}$ is a consistent estimator for the true efficient influence function. Because $P_{\hat{\theta}_n, \eta} \tilde{\ell}_{\hat{\theta}_n, \eta} = 0$, the first condition (25.52) requires that the "bias" of the plug-in estimator, due to estimating the nuisance parameter, converge to zero faster than $1/\sqrt{n}$. Such a condition comes out naturally of the proofs. A partial motivation is that the efficient score function is orthogonal to the score functions for the nuisance parameter, so that its expectation should be insensitive to changes in η.

25.54 Theorem. *Suppose that the model $\{P_{\theta, \eta} : \theta \in \Theta\}$ is differentiable in quadratic mean with respect to θ at (θ, η) and let the efficient information matrix $\tilde{I}_{\theta, \eta}$ be nonsingular. Assume that (25.52) and (25.53) hold. Let $\hat{\theta}_n$ satisfy $\sqrt{n} \, \mathbb{P}_n \tilde{\ell}_{\hat{\theta}_n, \hat{\eta}_n} = o_{\hat{P}}(1)$ and be consistent for θ. Furthermore, suppose that there exists a Donsker class with square-integrable envelope function that contains every function $\tilde{\ell}_{\hat{\theta}_n, \hat{\eta}_n}$ with probability tending to 1. Then the sequence $\hat{\theta}_n$ is asymptotically efficient at (θ, η).*

Proof. Let $G_n(\theta', \eta') = \sqrt{n}(\mathbb{P}_n - P_{\theta, \eta}) \tilde{\ell}_{\theta', \eta'}$ be the empirical process indexed by the functions $\tilde{\ell}_{\theta', \eta'}$. By the assumption that the functions $\tilde{\ell}_{\hat{\theta}, \hat{\eta}}$ are contained in a Donsker class, together with (25.53),

$$G_n(\hat{\theta}_n, \hat{\eta}_n) = G_n(\theta, \eta) + o_P(1).$$

(see Lemma 19.24.) By the defining relationship of $\hat{\theta}_n$ and the "no-bias" condition (25.52), this is equivalent to

$$\sqrt{n}(P_{\hat{\theta}_n, \eta} - P_{\theta, \eta}) \tilde{\ell}_{\hat{\theta}_n, \hat{\eta}_n} = G_n(\theta, \eta) + o_P\left(1 + \sqrt{n} \|\hat{\theta}_n - \theta_0\|\right).$$

The remainder of the proof consists of showing that the left side is asymptotically equivalent to $\left(\tilde{I}_{\theta, \eta} + o_P(1)\right) \sqrt{n}(\hat{\theta}_n - \theta)$, from which the theorem follows. Because $\tilde{I}_{\theta, \eta} = P_{\theta, \eta} \tilde{\ell}_{\theta, \eta} \dot{\ell}_{\theta, \eta}^T$, the difference of the left side of the preceding display and $\tilde{I}_{\theta, \eta} \sqrt{n}(\hat{\theta}_n - \theta)$ can be written as the sum of three terms:

$$\sqrt{n} \int \tilde{\ell}_{\hat{\theta}_n, \hat{\eta}_n} \left(p_{\hat{\theta}_n, \eta}^{1/2} + p_{\theta, \eta}^{1/2}\right) \left[\left(p_{\hat{\theta}_n, \eta}^{1/2} - p_{\theta, \eta}^{1/2}\right) - \frac{1}{2}(\hat{\theta}_n - \theta)^T \dot{\ell}_{\theta, \eta} \, p_{\theta, \eta}^{1/2}\right] d\mu$$

$$+ \int \tilde{\ell}_{\hat{\theta}_n, \hat{\eta}_n} \left(p_{\hat{\theta}_n, \eta}^{1/2} - p_{\theta, \eta}^{1/2}\right) \frac{1}{2} \dot{\ell}_{\theta, \eta}^T \, p_{\theta, \eta}^{1/2} \, d\mu \, \sqrt{n}(\hat{\theta}_n - \theta)$$

$$- \int \left(\tilde{\ell}_{\hat{\theta}_n, \hat{\eta}_n} - \tilde{\ell}_{\theta, \eta}\right) \dot{\ell}_{\theta, \eta}^T \, p_{\theta, \eta} \, d\mu \, \sqrt{n}(\hat{\theta}_n - \theta).$$

The first and third term can easily be seen to be $o_P\left(\sqrt{n} \|\hat{\theta}_n - \theta\|\right)$ by applying the Cauchy-Schwarz inequality together with the differentiability of the model and (25.53). The square of the norm of the integral in the middle term can for every sequence of constants $m_n \to \infty$ be bounded by a multiple of

$$m_n^2 \int \left\|\tilde{\ell}_{\hat{\theta}_n, \hat{\eta}_n}\right\| \, p_{\theta, \eta}^{1/2} \left|p_{\hat{\theta}_n, \eta}^{1/2} - p_{\theta, \eta}^{1/2}\right| d\mu^2$$

$$+ \int \left\|\tilde{\ell}_{\hat{\theta}_n, \hat{\eta}_n}\right\|^2 (p_{\hat{\theta}_n, \eta} + p_{\theta, \eta}) \, d\mu \int_{\|\dot{\ell}_{\theta, \eta}\| > m_n} \|\dot{\ell}_{\theta, \eta}\|^2 \, p_{\theta, \eta} \, d\mu.$$

In view of (25.53), the differentiability of the model in θ, and the Cauchy-Schwarz inequality, the first term converges to zero in probability provided $m_n \to \infty$ sufficiently slowly to ensure that $m_n \|\hat{\theta}_n - \theta\| \xrightarrow{P} 0$. (Such a sequence exists. If $Z_n \xrightarrow{P} 0$, then there exists a sequence $\varepsilon_n \downarrow 0$ such that $P(|Z_n| > \varepsilon_n) \to 0$. Then $\varepsilon_n^{-1/2} Z_n \xrightarrow{P} 0$.) In view of the last part of (25.53), the second term converges to zero in probability for every $m_n \to \infty$. This concludes the proof of the theorem. ∎

The preceding theorem is best understood as applying to the efficient score functions $\tilde{\ell}_{\theta,\eta}$. However, its proof only uses this to ensure that, at the true value (θ, η),

$$\tilde{I}_{\theta,\eta} = P_{\theta,\eta} \tilde{\ell}_{\theta,\eta} \dot{\ell}_{\theta\eta}^T.$$

The theorem remains true for arbitrary, mean-zero functions $\tilde{\ell}_{\theta,\eta}$ provided that this identity holds. Thus, if an estimator $(\hat{\theta}, \hat{\eta})$ only approximately satisfies the efficient score equation, then the latter can be replaced by an approximation.

The theorem applies to many examples, but its conditions may be too stringent. A modification that can be theoretically carried through under minimal conditions is based on the one-step method. Suppose that we are given a sequence of initial estimators $\tilde{\theta}_n$ that is $\sqrt{n}$-consistent for θ. We can assume without loss of generality that the estimators are discretized on a grid of meshwidth $n^{-1/2}$, which simplifies the constructions and proof. Then the one-step estimator is defined as

$$\hat{\theta}_n = \tilde{\theta}_n + \left(\sum_{i=1}^n \tilde{\ell}_{\tilde{\theta}_n, \hat{\eta}_{n,i}} \tilde{\ell}_{\tilde{\theta}_n, \hat{\eta}_{n,i}}^T (X_i) \right)^{-1} \sum_{i=1}^n \tilde{\ell}_{\tilde{\theta}_n, \hat{\eta}_{n,i}} (X_i).$$

The estimator $\hat{\theta}_n$ can be considered a one-step iteration of the Newton-Raphson algorithm for solving the equation $\sum \tilde{\ell}_{\theta,\hat{\eta}}(X_i) = 0$ with respect to θ, starting at the initial guess $\tilde{\theta}_n$. For the benefit of the simple proof, we have made the estimators $\hat{\eta}_{n,i}$ for η dependent on the index i. In fact, we shall use only two different values for $\hat{\eta}_{n,i}$, one for the first half of the sample and another for the second half. Given estimators $\hat{\eta}_n = \hat{\eta}_n(X_1, \ldots, X_n)$ define $\hat{\eta}_{n,i}$ by, with $m = \lfloor n/2 \rfloor$,

$$\hat{\eta}_{n,i} = \begin{cases} \hat{\eta}_m(X_1, \ldots, X_m) & \text{if } i > m \\ \hat{\eta}_{n-m}(X_{m+1}, \ldots, X_n) & \text{if } i \le m. \end{cases}$$

Thus, for X_i belonging to the first half of the sample, we use an estimator $\hat{\eta}_{n,i}$ based on the second half of the sample, and vice versa. This sample-splitting trick is convenient in the proof, because the estimator of η used in $\tilde{\ell}_{\theta,\eta}(X_i)$ is always independent of X_i, simultaneously for X_i running through each of the two halves of the sample.

The discretization of $\tilde{\theta}_n$ and the sample-splitting are mathematical devices that rarely are useful in practice. However, the conditions of the preceding theorem can now be relaxed to, for every deterministic sequence $\theta_n = \theta + O(n^{-1/2})$,

$$\sqrt{n} P_{\theta_n,\eta} \tilde{\ell}_{\theta_n,\hat{\eta}_n} \xrightarrow{P} 0, \qquad P_{\theta_n,\eta} \|\tilde{\ell}_{\theta_n,\hat{\eta}_n} - \tilde{\ell}_{\theta_n,\eta}\|^2 \xrightarrow{P} 0. \tag{25.55}$$

$$\int \left\| \tilde{\ell}_{\theta_n,\eta} d P_{\theta_n,\eta}^{1/2} - \tilde{\ell}_{\theta,\eta} d P_{\theta,\eta}^{1/2} \right\|^2 \to 0. \tag{25.56}$$

25.57   Theorem. *Suppose that the model $\{P_{\theta,\eta} : \theta \in \Theta\}$ is differentiable in quadratic mean with respect to θ at (θ, η), and let the efficient information matrix $\tilde{I}_{\theta,\eta}$ be nonsingular.*

Assume that (25.55) and (25.56) hold. Then the sequence $\hat{\theta}_n$ is asymptotically efficient at (θ, η).

Proof. Fix a deterministic sequence of vectors $\theta_n = \theta + O(n^{-1/2})$. By the sample-splitting, the first half of the sum $\sum \tilde{\ell}_{\theta_n, \tilde{\eta}_{n,i}}(X_i)$ is a sum of conditionally independent terms, given the second half of the sample. Thus,

$$E_{\theta_n, \eta}\left(\sqrt{m}\mathbb{P}_m\left(\tilde{\ell}_{\theta_n, \hat{\eta}_{n,i}} - \tilde{\ell}_{\theta_n, \eta}\right) | X_{m+1}, \ldots, X_n\right) = \sqrt{m}P_{\theta_n, \eta}\tilde{\ell}_{\theta_n, \hat{\eta}_{n,i}},$$

$$\mathrm{var}_{\theta_n, \eta}\left(\sqrt{m}\mathbb{P}_m\left(\tilde{\ell}_{\theta_n, \hat{\eta}_{n,i}} - \tilde{\ell}_{\theta_n, \eta}\right) | X_{m+1}, \ldots, X_n\right) \le P_{\theta_n, \eta}\|\tilde{\ell}_{\theta_n, \hat{\eta}_{n,i}} - \tilde{\ell}_{\theta_n, \eta}\|^2.$$

Both expressions converge to zero in probability by assumption (25.55). We conclude that the sum inside the conditional expectations converges conditionally, and hence also unconditionally, to zero in probability. By symmetry, the same is true for the second half of the sample, whence

$$\sqrt{n}\mathbb{P}_n\left(\tilde{\ell}_{\theta_n, \hat{\eta}_{n,i}} - \tilde{\ell}_{\theta_n, \eta}\right) \xrightarrow{P} 0.$$

We have proved this for the probability under (θ_n, η), but by contiguity the convergence is also under (θ, η).

The second part of the proof is technical, and we only report the result. The condition of differentiabily of the model and (25.56) imply that

$$\sqrt{n}\mathbb{P}_n\left(\tilde{\ell}_{\theta_n, \eta} - \tilde{\ell}_{\theta, \eta}\right) + \sqrt{n}P_{\theta, \eta}\,\tilde{\ell}_{\theta, \eta}\dot{\ell}_{\theta, \eta}^T(\theta_n - \theta) \xrightarrow{P} 0$$

(see [139], p. 185. Under stronger regularity conditions, this can also be proved by a Taylor expansion of $\tilde{\ell}_{\theta, \eta}$ in θ.) By the definition of the efficient score function as an orthogonal projection, $P_{\theta, \eta}\tilde{\ell}_{\theta, \eta}\dot{\ell}_{\theta, \eta}^T = \tilde{I}_{\theta, \eta}$. Combining the preceding displays, we find that

$$\sqrt{n}\mathbb{P}_n\left(\tilde{\ell}_{\theta_n, \hat{\eta}_{n,i}} - \tilde{\ell}_{\theta, \eta}\right) + \tilde{I}_{\theta, \eta}\sqrt{n}(\theta_n - \theta) \xrightarrow{P} 0.$$

In view of the discretized nature of $\tilde{\theta}_n$, this remains true if the deterministic sequence θ_n is replaced by $\tilde{\theta}_n$; see the argument in the proof of Theorem 5.48.

Next we study the estimator for the information matrix. For any vector $h \in \mathbb{R}^k$, the triangle inequality yields

$$\left|\sqrt{\mathbb{P}_m(h^T\tilde{\ell}_{\theta_n, \hat{\eta}_{n,i}})^2} - \sqrt{\mathbb{P}_m(h^T\tilde{\ell}_{\theta_n, \eta})^2}\right|^2 \le \mathbb{P}_m\left(h^T\tilde{\ell}_{\theta_n, \hat{\eta}_{n,i}} - h^T\tilde{\ell}_{\theta_n, \eta}\right)^2.$$

By (25.55), the conditional expectation under (θ_n, η) of the right side given $X_{m+1}, \ldots, X_n$ converges in probability to zero. A similar statement is valid for the second half of the observations. Combining this with (25.56) and the law of large numbers, we see that

$$\mathbb{P}_n\tilde{\ell}_{\theta_n, \hat{\eta}_{n,i}}\tilde{\ell}_{\theta_n, \hat{\eta}_{n,i}}^T \xrightarrow{P} \tilde{I}_{\theta, \eta}.$$

In view of the discretized nature of $\tilde{\theta}_n$, this remains true if the deterministic sequence θ_n is replaced by $\tilde{\theta}_n$.

The theorem follows combining the results of the last two paragraphs with the definition of $\hat{\theta}_n$. ∎

A further refinement is not to restrict the estimator for the efficient score function to be a plug-in type estimator. Both theorems go through if $\tilde{\ell}_{\theta,\hat{\eta}}$ is replaced by a general estimator $\hat{\ell}_{n,\theta} = \hat{\ell}_{n,\theta}(\cdot \mid X_1, \ldots, X_n)$, provided that this satisfies the appropriately modified conditions of the theorems, and in the second theorem we use the sample-splitting scheme. In the generalization of Theorem 25.57, condition (25.55) must be replaced by

$$\sqrt{n} P_{\theta_n,\eta} \hat{\ell}_{n,\theta_n} \overset{P}{\to} 0, \qquad P_{\theta_n,\eta} \| \hat{\ell}_{n,\theta_n} - \tilde{\ell}_{\theta_n,\eta} \|^2 \overset{P}{\to} 0. \qquad (25.58)$$

The proofs are the same. This opens the door to more tricks and further relaxation of the regularity conditions. An intermediate theorem concerning one-step estimators, but without discretization or sample-splitting, can also be proved under the conditions of Theorem 25.54. This removes the conditions of existence and consistency of solutions to the efficient score equation.

The theorems reduce the problem of efficient estimation of θ to estimation of the efficient score function. The estimator of the efficient score function must satisfy a "no-bias" and a consistency conditions. The consistency is usually easy to arrange, but the no-bias condition, such as (25.52) or the first part of (25.58), is connected to the structure and the size of the model, as the bias of the efficient score equations must converge to zero at a rate faster than $1/\sqrt{n}$. Within the context of Theorem 25.54 condition (25.52) is necessary. If it fails, then the sequence $\hat{\theta}_n$ is not asymptotically efficient and may even converge at a slower rate than $\sqrt{n}$. This follows by inspection of the proof, which reveals the following adaptation of the theorem. We assume that $\tilde{\ell}_{\theta,\eta}$ is the efficient score function for the true parameter (θ, η) but allow it to be arbitrary (mean-zero) for other parameters.

25.59 Theorem. *Suppose that the conditions of Theorem 25.54 hold except possibly condition (25.52). Then*

$$\sqrt{n}(\hat{\theta}_n - \theta) = \frac{1}{\sqrt{n}} \sum_{i=1}^{n} \tilde{I}_{\theta,\eta}^{-1} \tilde{\ell}_{\theta,\eta}(X_i) + \sqrt{n} P_{\theta_n,\eta} \tilde{\ell}_{\hat{\theta}_n,\hat{\eta}_n} + o_P(1).$$

Because by Lemma 25.23 the sequence $\hat{\theta}_n$ can be asymptotically efficient (regular with $N(0, \tilde{I}_{\theta,\eta}^{-1})$-limit distribution) only if it is asymptotically equivalent to the sum on the right, condition (25.52) is seen to be necessary for efficiency.

The verification of the no-bias condition may be easy due to special properties of the model but may also require considerable effort. The derivative of $P_{\theta,\eta} \tilde{\ell}_{\theta,\hat{\eta}}$ with respect to θ ought to converge to $\partial/\partial\theta \, P_{\theta,\eta} \tilde{\ell}_{\theta,\eta} = 0$. Therefore, condition (25.52) can usually be simplified to

$$\sqrt{n} P_{\theta,\eta} \tilde{\ell}_{\theta,\hat{\eta}_n} \overset{P}{\to} 0.$$

The dependence on $\hat{\eta}$ is more interesting and complicated. The verification may boil down to a type of Taylor expansion of $P_{\theta,\eta} \tilde{\ell}_{\theta,\hat{\eta}}$ in $\hat{\eta}$ combined with establishing a rate of convergence for $\hat{\eta}$. Because η is infinite-dimensional, a Taylor series may be nontrivial. If $\hat{\eta} - \eta$ can occur as a direction of approach to η that leads to a score function $B_{\theta,\eta}(\hat{\eta} - \eta)$, then we can write

$$P_{\theta,\eta}\tilde{\ell}_{\theta,\hat{\eta}} = (P_{\theta,\eta} - P_{\theta,\hat{\eta}})(\tilde{\ell}_{\theta,\hat{\eta}} - \tilde{\ell}_{\theta,\eta})$$

$$- P_{\theta,\eta}\tilde{\ell}_{\theta,\eta}\left[\frac{p_{\theta,\hat{\eta}} - p_{\theta,\eta}}{p_{\theta,\eta}} - B_{\theta,\eta}(\hat{\eta} - \eta)\right]. \tag{25.60}$$

We have used the fact that $P_{\theta,\eta}\tilde{\ell}_{\theta,\eta}B_{\theta,\eta}h = 0$ for every h, by the orthogonality property of the efficient score function. (The use of $B_{\theta,\eta}(\hat{\eta} - \eta)$ corresponds to a score operator that yields scores $B_{\theta,\eta}h$ from paths of the form $\eta_t = \eta + th$. If we use paths $d\eta_t = (1 + th)\,d\eta$, then $B_{\theta,\eta}(d\hat{\eta}/d\eta - 1)$ is appropriate.) The display suggests that the no-bias condition (25.52) is certainly satisfied if $\|\hat{\eta} - \eta\| = O_P(n^{-1/2})$, for $\|\cdot\|$ a norm relative to which the two terms on the right are both of the order $o_P(\|\hat{\eta} - \eta\|)$. In cases in which the nuisance parameter is not estimable at $\sqrt{n}$-rate the Taylor expansion must be carried into its second-order term. If the two terms on the right are both $O_P(\|\hat{\eta} - \eta\|^2)$, then it is still sufficient to have $\|\hat{\eta} - \eta\| = o_P(n^{-1/4})$. This observation is based on a crude bound on the bias, an integral in which cancellation could occur, by norms and can therefore be too pessimistic (See [35] for an example.) Special properties of the model may also allow one to take the Taylor expansion even further, with the lower order derivatives vanishing, and then a slower rate of convergence of the nuisance parameter may be sufficient, but no examples of this appear to be known. However, the extreme case that the expression in (25.52) is identically zero occurs in the important class of models that are convex-linear in the parameter.

25.61 Example (Convex-linear models). Suppose that for every fixed θ the model $\{P_{\theta,\eta} : \eta \in H\}$ is convex-linear: H is a convex subset of a linear space, and the dependence $\eta \mapsto P_{\theta,\eta}$ is linear. Then for every pair (η_1, η) and number $0 \le t \le 1$, the convex combination $\eta_t = t\eta_1 + (1-t)\eta$ is a parameter and the distribution $t P_{\theta,\eta_1} + (1-t)P_{\theta,\eta} = P_{\theta,\eta_t}$ belongs to the model. The score function at $t = 0$ of the submodel $t \mapsto P_{\theta,\eta_t}$ is

$$\frac{\partial}{\partial t}_{|t=0} \log dP_{\theta,t\eta_1+(1-t)\eta} = \frac{dP_{\theta,\eta_1}}{dP_{\theta,\eta}} - 1.$$

Because the efficient score function for θ is orthogonal to the tangent set for the nuisance parameter, it should satisfy

$$0 = P_{\theta,\eta}\tilde{\ell}_{\theta,\eta}\left(\frac{dP_{\theta,\eta_1}}{dP_{\theta,\eta}} - 1\right) = P_{\theta,\eta_1}\tilde{\ell}_{\theta,\eta}.$$

This means that the unbiasedness conditions in (25.52) and (25.55) are trivially satisfied, with the expectations $P_{\theta,\eta}\tilde{\ell}_{\theta,\hat{\eta}}$ even equal to 0.

A particular case in which this convex structure arises is the case of estimating a linear functional in an information-loss model. Suppose we observe $X = m(Y)$ for a known function m and an unobservable variable Y that has an unknown distribution η on a measurable space $(\mathcal{Y}, \mathcal{A})$. The distribution $P_\eta = \eta \circ m^{-1}$ of X depends linearly on η. Furthermore, if we are interested in a linear function $\theta = \chi(\eta)$, then the nuisance-parameter space $H_\theta = \{\eta : \chi(\eta) = \theta\}$ is a convex subset of the set of probability measures on $(\mathcal{Y}, \mathcal{A})$. $\square$

25.8.1 *Symmetric Location*

Suppose that we observe a random sample from a density $\eta(x - \theta)$ that is symmetric about θ. In Example 25.27 it was seen that the efficient score function for θ is the ordinary score function,

$$\tilde{\ell}_{\theta,\eta}(x) = -\frac{\eta'}{\eta}(x - \theta).$$

We can apply Theorem 25.57 to construct an asymptotically efficient estimator sequence for θ under the minimal condition that the density η has finite Fisher information for location.

First, as an initial estimator $\tilde{\theta}_n$, we may use a discretized Z-estimator, solving $\mathbb{P}_n \psi(x - \theta) = 0$ for a well-behaved, symmetric function ψ. For instance, the score function of the logistic density. The $\sqrt{n}$-consistency can be established by Theorem 5.21.

Second, it suffices to construct estimators $\hat{\ell}_{n,\theta}$ that satisfy (25.58). By symmetry, the variables $T_i = |X_i - \theta|$ are, for a fixed θ, sampled from the density $g(s) = 2\eta(s)1\{s > 0\}$. We use these variables to construct an estimator $\hat{k}_n$ for the function g'/g, and next we set

$$\hat{\ell}_{n,\theta}(x; X_1, \ldots, X_n) = -\hat{k}_n(|x - \theta|; T_1, \ldots, T_n)\,\mathrm{sign}(x - \theta).$$

Because this function is skew-symmetric about the point θ, the bias condition in (25.58) is satisfied, with a bias of zero. Because the efficient score function can be written in the form

$$\tilde{\ell}_{\theta,\eta}(x) = -\frac{g'}{g}(|x - \theta|)\,\mathrm{sign}(x - \theta),$$

the consistency condition in (25.58) reduces to consistency of $\hat{k}_n$ for the function g'/g in that

$$\int \left(\hat{k}_n - \frac{g'}{g}\right)^2 (s)\, g(s)\, ds \overset{\mathrm{P}}{\to} 0. \tag{25.62}$$

Estimators $\hat{k}_n$ can be constructed by several methods, a simple one being the kernel method of density estimation. For a fixed twice continuously differentiable probability density ω with compact support, a bandwidth parameter σ_n, and further positive tuning parameters α_n, β_n, and γ_n, set

$$\hat{g}_n(s) = \frac{1}{\sigma_n} \sum_{i=1}^{n} \omega\left(\frac{s - T_i}{\sigma_n}\right),$$

$$\hat{k}_n(s) = \frac{\hat{g}_n'}{\hat{g}_n}(s) 1_{\hat{B}_n}(s), \tag{25.63}$$

$$\hat{B}_n = \left\{s : |\hat{g}_n'(s)| \le \alpha_n,\, \hat{g}_n(s) \ge \beta_n,\, s \ge \gamma_n\right\}.$$

Then (25.58) is satisfied provided $\alpha_n \uparrow \infty$, $\beta_n \downarrow 0$, $\gamma_n \downarrow 0$, and $\sigma_n \downarrow 0$ at appropriate speeds. The proof is technical and is given in the next lemma.

This particular construction shows that efficient estimators for θ exist under minimal conditions. It is not necessarily recommended for use in practice. However, any good initial estimator $\tilde{\theta}_n$ and any method of density or curve estimation may be substituted and will lead to a reasonable estimator for θ, which is theoretically efficient under some regularity conditions.

25.64 **Lemma.** *Let $T_1, \ldots, T_n$ be a random sample from a density g that is supported and absolutely continuous on $[0, \infty)$ and satisfies $\int (g'/\sqrt{g})^2(s)\, ds < \infty$. Then $\hat{k}_n$ given by (25.63) for a probability density ω that is twice continuously differentiable and supported on $[-1, 1]$ satisfies (25.62), if $\alpha_n \uparrow \infty$, $\gamma_n \downarrow 0$, $\beta_n \downarrow 0$, and $\sigma_n \downarrow 0$ in such a way that $\sigma_n \leq \gamma_n$, $\alpha_n^2 \sigma_n / \beta_n^2 \to 0$, $n\sigma_n^4 \beta_n^2 \to \infty$.*

Proof. Start by noting that $\|g\|_\infty \leq \int |g'(s)|\, ds \leq \sqrt{I_g}$, by the Cauchy-Schwarz inequality. The expectations and variances of $\hat{g}_n$ and its derivative are given by

$$g_n(s) := \mathrm{E}\hat{g}_n(s) = \mathrm{E}\frac{1}{\sigma}\omega\left(\frac{s - T_1}{\sigma}\right) = \int g(s - \sigma y)\,\omega(y)\, dy,$$

$$\mathrm{var}\,\hat{g}_n(s) = \frac{1}{n\sigma^2}\mathrm{var}\,\omega\left(\frac{s - T_1}{\sigma}\right) \leq \frac{1}{n\sigma^2}\|\omega\|_\infty^2,$$

$$\mathrm{E}\hat{g}_n'(s) = g_n'(s) = \int g'(s - \sigma y)\omega(y)\, dy, \qquad (s \geq \gamma),$$

$$\mathrm{var}\,\hat{g}_n'(s) \leq \frac{1}{n\sigma^4}\|\omega'\|_\infty^2.$$

By the dominated-convergence theorem, $g_n(s) \to g(s)$, for every $s > 0$. Combining this with the preceding display, we conclude that $\hat{g}_n(s) \overset{\mathrm{P}}{\to} g(s)$. If g' is sufficiently smooth, then the analogous statement is true for $\hat{g}_n'(s)$. Under only the condition of finite Fisher information for location, this may fail, but we still have that $\hat{g}_n'(s) - g_n'(s) \overset{\mathrm{P}}{\to} 0$ for every s; furthermore, $g_n' 1_{[\sigma, \infty)} \to g'$ in L_1, because

$$\int_\sigma^\infty |g_n' - g'|(s)\, ds \leq \iint |g'(s - \sigma y) - g'(s)|\, ds\,\omega(y)\, dy \to 0,$$

by the L_1-continuity theorem on the inner integral, and next the dominated-convergence theorem on the outer integral.

The expectation of the integral in (25.62) restricted to the complement of the set $\hat{B}_n$ is equal to

$$\int \left(\frac{g'}{g}\right)^2 (s)\, g(s)\, \mathrm{P}\left(|\hat{g}_n'|(s) > \alpha \text{ or } \hat{g}_n(s) < \beta \text{ or } s < \gamma\right) ds.$$

This converges to zero by the dominated-convergence theorem. To see this, note first that $\mathrm{P}\left(\hat{g}_n(s) < \beta\right)$ converges to zero for all s such that $g(s) > 0$. Second, the probability $\mathrm{P}\left(|\hat{g}_n'|(s) > \alpha\right)$ is bounded above by $1\{|g_n'|(s) > \alpha/2\} + o(1)$, and the Lebesgue measure of the set $\{s : |g_n'|(s) > \alpha/2\}$ converges to zero, because $g_n' \to g'$ in L_1.

On the set $\hat{B}_n$ the integrand in (25.62) is the square of the function $(\hat{g}_n'/\hat{g}_n - g'/g)g^{1/2}$. This function can be decomposed as

$$\frac{\hat{g}_n'}{\hat{g}_n}\left(g^{1/2} - g_n^{1/2}\right) + \frac{(\hat{g}_n' - g_n')g_n^{1/2}}{\hat{g}_n} - \frac{g_n'(\hat{g}_n - g_n)}{g_n^{1/2}\hat{g}_n} + \left(\frac{g_n'}{g_n^{1/2}} - \frac{g'}{g^{1/2}}\right).$$

On $\hat{B}_n$ the sum of the squares of the four terms on the right is bounded above by

$$\frac{\alpha^2}{\beta^2}|g_n - g| + \frac{1}{\beta^2}(\hat{g}_n' - g_n')^2 g_n + \frac{1}{\beta^2}\left(\frac{g_n'}{g_n^{1/2}}\right)^2(\hat{g}_n - g_n)^2 + \left(\frac{g_n'}{g_n^{1/2}} - \frac{g'}{g^{1/2}}\right)^2.$$

The expectations of the integrals over $\hat{B}_n$ of these four terms converge to zero. First, the integral over the first term is bounded above by

$$\frac{\alpha^2}{\beta^2} \iint_{s>\gamma} |g(s - \sigma t) - g(s)| \, \omega(t) \, dt \, ds \leq \frac{\alpha^2 \sigma}{\beta^2} \int |g'(t)| \, dt \int |t| \omega(t) \, dt.$$

Next, the sum of the second and third terms gives the contribution

$$\frac{1}{n\sigma^4 \beta^2} \|\omega'\|_\infty^2 \int g_n(s) \, ds + \frac{1}{n\sigma^2 \beta^2} \|\omega\|_\infty^2 \int \left(\frac{g_n'}{g_n^{1/2}}\right)^2 ds.$$

The first term in this last display converges to zero, and the second as well, provided the integral remains finite. The latter is certainly the case if the fourth term converges to zero. By the Cauchy-Schwarz inequality,

$$\frac{\left(\int g'(s - \sigma y) \, \omega(y) \, dy\right)^2}{\int g(s - \sigma y) \, \omega(y) \, dy} \leq \int \left(\frac{g'}{g^{1/2}}\right)^2 (s - \sigma y) \, \omega(y) \, dy.$$

Using Fubini's theorem, we see that, for any set B, and B^σ its σ-enlargement,

$$\int_B \left(\frac{g_n'}{g_n^{1/2}}\right)^2 (s) \, ds \leq \int_{B^\sigma} \left(\frac{g'}{g^{1/2}}\right)^2 ds.$$

In particular, we have this for $B = B^\sigma = \mathbb{R}$, and $B = \{s : g(s) = 0\}$. For the second choice of B, the sets B^σ decrease to B, by the continuity of g. On the complement of B, $g_n'/g_n^{1/2} \to g'/g^{1/2}$ in Lebesgue measure. Thus, by Proposition 2.29, the integral of the fourth term converges to zero. ∎

25.8.2 *Errors-in-Variables*

Let the observations be a random sample of pairs (X_i, Y_i) with the same distribution as

$$X = Z + e$$
$$Y = \alpha + \beta Z + f,$$

for a bivariate normal vector (e, f) with mean zero and covariance matrix Σ and a random variable Z with distribution η, independent of (e, f). Thus Y is a linear regression on a variable Z which is observed with error. The parameter of interest is $\theta = (\alpha, \beta, \Sigma)$ and the nuisance parameter is η. To make the parameters identifiable one can put restrictions on either Σ or η. It suffices that η is not normal (if a degenerate distribution is considered normal with variance zero); alternatively it can be assumed that Σ is known up to a scalar.

Given (θ, Σ) the statistic $\psi_\theta(X, Y) = (1, \beta)\Sigma^{-1}(X, Y - \alpha)^T$ is sufficient (and complete) for η. This suggests to define estimators for (α, β, Σ) as the solution of the "conditional score equation" $\mathbb{P}_n \tilde{\ell}_{\theta, \hat{\eta}} = 0$, for

$$\tilde{\ell}_{\theta, \eta}(X, Y) = \dot{\ell}_{\theta, \eta}(X, Y) - \mathrm{E}_\theta\big(\dot{\ell}_{\theta, \eta}(X, Y) \mid \psi_\theta(X, Y)\big).$$

This estimating equation has the attractive property of being unbiased in the nuisance parameter, in that

$$P_{\theta, \eta} \tilde{\ell}_{\theta, \eta'} = 0, \qquad \text{every } \theta, \eta, \eta'.$$

Therefore, the no-bias condition is trivially satisfied, and the estimator $\hat{\eta}$ need only be consistent for η (in the sense of (25.53)). One possibility for $\hat{\eta}$ is the maximum likelihood estimator, which can be shown to be consistent by Wald's theorem, under some regularity conditions.

As the notation suggests, the function $\tilde{\ell}_{\theta,\eta}$ is equal to the efficient score function for θ. We can prove this by showing that the closed linear span of the set of nuisance scores contains all measurable, square-integrable functions of $\psi_\theta(x, y)$, because then projecting on the nuisance scores is identical to taking the conditional expectation.

As explained in Example 25.61, the functions $p_{\theta,\eta_1}/p_{\theta,\eta} - 1$ are score functions for the nuisance parameter (at (θ, η)). As is clear from the factorization theorem or direct calculation, they are functions of the sufficient statistic $\psi_\theta(X, Y)$. If some function $b(\psi_\theta(x, y))$ is orthogonal to all scores of this type and has mean zero, then

$$\mathrm{E}_{\theta,\eta_1} b(\psi_\theta(X, Y)) = \mathrm{E}_{\theta,\eta} b(\psi_\theta(X, Y)) \left(\frac{p_{\theta,\eta_1}}{p_{\theta,\eta}} - 1 \right) = 0.$$

Consequently, $b = 0$ almost surely by the completeness of $\psi_\theta(X, Y)$.

The regularity conditions of Theorem 25.54 can be shown to be satisfied under the condition that $\int |z|^9 \, d\eta(z) < \infty$. Because all coordinates of the conditional score function can be written in the form $Q_\theta(x, y) + P_\theta(x, y) \mathrm{E}_\eta(Z \mid \psi_\theta(X, Y))$ for polynomials Q_θ and P_θ of orders 2 and 1, respectively, the following lemma is the main part of the verification.[†]

25.65 Lemma. *For every $0 < \alpha \leq 1$ and every probability distribution η_0 on $\mathbb{R}$ and compact $K \subset (0, \infty)$, there exists an open neighborhood U of η_0 in the weak topology such that the class $\mathcal{F}$ of all functions*

$$(x, y) \mapsto (a_0 + a_1 x + a_2 y) \frac{\int z \, e^{z(b_0 + b_1 x + b_2 y)} \, e^{-cz^2} \, d\eta(z)}{\int e^{z(b_0 + b_1 x + b_2 y)} \, e^{-cz^2} \, d\eta(z)},$$

with η ranging over U, c ranging over K, and a and b ranging over compacta in $\mathbb{R}^3$, satisfies

$$\log N_{[]}(\varepsilon, \mathcal{F}, L_2(P)) \leq C \left(\frac{1}{\varepsilon} \right)^V \left(P(1 + |x| + |y|)^{5+2\alpha+4/V+\delta} \right)^{V/2},$$

for every $V \geq 1/\alpha$, every measure P on $\mathbb{R}^2$ and $\delta > 0$, and a constant C depending only on α, η_0, U, V, the compacta, and δ.

25.9 General Estimating Equations

Taking the efficient score equation as the basis for estimating a parameter is motivated by our wish to construct asymptotically efficient estimators. Perhaps, in certain situations, this is too much to ask, and it is better to aim at estimators that come close to attaining efficiency or are efficient only at the elements of a certain "ideal submodel." The pay off could be a gain in robustness, finite-sample properties, or computational simplicity. The information bounds then have the purpose of quantifying how much efficiency has possibly been lost.

[†] See [108] for a proof.

We retain the requirement that the estimator is $\sqrt{n}$-consistent and regular at every distribution P in the model. A somewhat stronger but still reasonable requirement is that it be *asymptotically linear* in that

$$\sqrt{n}\big(T_n - \psi(P)\big) = \frac{1}{\sqrt{n}} \sum_{i=1}^{n} \dot{\psi}_P(X_i) + o_P(1).$$

This type of expansion and regularity implies that $\dot{\psi}_P$ is an influence function of the parameter $\psi(P)$, and the difference $\dot{\psi}_P - \tilde{\psi}_P$ must be orthogonal to the tangent set $\dot{\mathcal{P}}_P$.

This suggests that we compute the set of all influence functions to obtain an indication of which estimators T_n might be possible. If there is a nice parametrization $\dot{\psi}_{\theta,\tau}$ of these sets of functions in terms of a parameter of interest θ and a nuisance parameter τ, then a possible estimation procedure is to solve θ from the estimating equation, for given τ,

$$\sum_{i=1}^{n} \dot{\psi}_{\theta,\tau}(X_i) = 0.$$

The choice of the parameter τ determines the efficiency of the estimator $\hat{\theta}$. Rather than fixing it at some value we also can make it data-dependent to obtain efficiency at every element of a given submodel, or perhaps even the whole model. The resulting estimator can be analyzed with the help of, for example, Theorem 5.31.

If the model is parametrized by a partitioned parameter (θ, η), then any influence function for θ must be orthogonal to the scores for the nuisance parameter η. The parameter τ might be indexing both the nuisance parameter η and "position" in the tangent set at a given (θ, η). Then the unknown η (or the aspect of it that plays a role in τ) must be replaced by an estimator. The same reasoning as for the "no-bias" condition discussed in (25.60) allows us to hope that the resulting estimator for θ behaves as if the true η had been used.

25.66   *Example (Regression).* In the regression model considered in Example 25.28, the set of nuisance scores is the orthocomplement of the set $e\mathcal{H}$ of all functions of the form $(x, y) \mapsto \big(y - g_\theta(x)\big)h(x)$, up to centering at mean zero. The efficient score function for θ is equal to the projection of the score for θ onto the set $e\mathcal{H}$, and an arbitrary influence function is obtained, up to a constant, by adding any element from $e\mathcal{H}$ to this. The estimating equation

$$\sum_{i=1}^{n} \big(Y_i - g_\theta(X_i)\big)h(X_i) = 0$$

leads to an estimator with influence function in the direction of $\big(y - g_\theta(x)\big)h(x)$. Because the equation is unbiased for any h, we easily obtain $\sqrt{n}$-consistent estimators, even for data-dependent h. The estimator is more efficient if h is closer to the function $\dot{g}_\theta(x)/\mathrm{E}_\eta(e^2 \mid X = x)$, which gives the efficient influence function. For full efficiency it is necessary to estimate the function $x \mapsto \mathrm{E}_\eta(e^2 \mid X = x)$ nonparametrically, where consistency (for the right norm) suffices. $\Box$

25.67   *Example (Missing at random).* In Lemma 25.41 and Example 25.43 the influence functions in a MAR model are characterized as the sums of reweighted influence functions in the original model and the influence functions obtained from the MAR specification. If the function π is known, then this leads to estimating equations of the form

$$\sum_{i=1}^{n} \frac{\Delta_i}{\pi(Y_i)} \dot{\psi}_{\theta,\tau}(X_i) - \sum_{i=1}^{n} \frac{\Delta_i - \pi(Y_i)}{\pi(Y_i)} c(Y_i) = 0.$$

For instance, if the original model is the regression model in the preceding example, then $\dot{\psi}_{\theta,\tau}(y)$ is $(y - g_\theta(x))h(x)$. The efficiency of the estimator is influenced by the choice of c (the optimal choice is given in Example 25.43) and the choice of $\dot{\psi}_{\theta,\tau}$. (The efficient influence function of the original model need not be efficient here.) If π is correctly specified, then the second part of the estimating equation is unbiased for any c, and the asymptotic variance when using a random c should be the same as when using the limiting value of c. □

25.10 Maximum Likelihood Estimators

Estimators for parameters in semiparametric models can be constructed by any method – for instance, M-estimation or Z-estimation. However, the most important method to obtain asymptotically efficient estimators may be the method of maximum likelihood, just as in the case of parametric models. In this section we discuss the definitions of likelihoods and give some examples in which maximum likelihood estimators can be analyzed by direct methods. In Sections 25.11 and 25.12 we discuss two general approaches for analyzing these estimators.

Because many semiparametric models are not dominated or are defined in terms of densities that maximize to infinity, the functions that are called the "likelihoods" of the models must be chosen with care. For some models a likelihood can be taken equal to a density with respect to a dominating measure, but for other models we use an "empirical likelihood." Mixtures of these situations occur as well, and sometimes it is fruitful to incorporate a "penalty" in the likelihood, yielding a "penalized likelihood estimator"; maximize the likelihood over a set of parameters that changes with n, yielding a "sieved likelihood estimator"; or group the data in some way before writing down a likelihood. To bring out this difference with the classical, parametric maximum likelihood estimators, our present estimators are sometimes referred to as "nonparametric maximum likelihood estimators" (NPMLE), although semiparametric rather than nonparametric seems more correct. Thus we do not give an abstract definition of "likelihood," but describe "likelihoods that work" for particular examples. We denote the likelihood for the parameter P given one observation x by $\text{lik}(P)(x)$.

Given a measure P, write $P\{x\}$ for the measure of the one-point set $\{x\}$. The function $x \mapsto P\{x\}$ may be considered the density of P, or its absolutely continuous part, with respect to counting measure. The *empirical likelihood* of a sample $X_1, \ldots, X_n$ is the function,

$$P \mapsto \prod_{i=1}^{n} P\{X_i\}.$$

Given a model $\mathcal{P}$, a maximum likelihood estimator could be defined as the distribution $\hat{P}$ that maximizes the empirical likelihood over $\mathcal{P}$. Such an estimator may or may not exist.

25.68   *Example (Empirical distribution).* Let $\mathcal{P}$ be the set of all probability distributions on the measurable space $(\mathcal{X}, \mathcal{A})$ (in which one-point sets are measurable). Then, for n

fixed different values $x_1, \ldots, x_n$, the vector $\left(P\{x_1\}, \ldots, P\{x_n\} \right)$ ranges over all vectors $p \geq 0$ such that $\sum p_i \leq 1$ when P ranges over $\mathcal{P}$. To maximize $p \mapsto \prod_i p_i$, it is clearly best to choose p maximal: $\sum_i p_i = 1$. Then, by symmetry, the maximizer must be $p = (1/n, \ldots, 1/n)$. Thus, the empirical distribution $\mathbb{P}_n = n^{-1} \sum \delta_{X_i}$ maximizes the empirical likelihood over the nonparametric model, whence it is referred to as the nonparametric maximum likelihood estimator.

If there are ties in the observations, this argument must be adapted, but the result is the same.

The empirical likelihood is appropriate for the nonparametric model. For instance, in the case of a Euclidean space, even if the model is restricted to distributions with a continuous Lebesgue density p, we still cannot use the map $p \mapsto \prod_{i=1}^{n} p(X_i)$ as a likelihood. The supremum of this likelihood is infinite, for we could choose p to have an arbitrarily high, very thin peak at some observation. $\quad\square$

Given a partitioned parameter (θ, η), it is sometimes helpful to consider the *profile likelihood*. Given a likelihood $\mathrm{lik}_n(\theta, \eta)(X_1, \ldots, X_n)$, the profile likelihood for θ is defined as the function

$$\theta \mapsto \sup_{\eta} \mathrm{lik}_n(\theta, \eta)(X_1, \ldots, X_n).$$

The supremum is taken over all possible values of η. The point of maximum of the profile likelihood is exactly the first coordinate of the maximum likelihood estimator $(\hat{\theta}, \hat{\eta})$. We are simply computing the maximum of the likelihood over (θ, η) in two steps.

It is rarely possible to compute a profile likelihood explicitly, but its numerical evaluation is often feasible. Then the profile likelihood may serve to reduce the dimension of the likelihood function. Profile likelihood functions are often used in the same way as (ordinary) likelihood functions of parametric models. Apart from taking their points of maximum as estimators $\hat{\theta}$, the second derivative at $\hat{\theta}$ is used as an estimate of minus the inverse of the asymptotic covariance matrix of $\hat{\theta}$. Recent research appears to validate this practice.

25.69 *Example (Cox model).* Suppose that we observe a random sample from the distribution of $X = (T, Z)$, where the conditional hazard function of the "survival time" T with covariate Z takes the form

$$\lambda_{T \mid Z}(t) = e^{\theta Z} \lambda(t).$$

The hazard function λ is completely unspecified. The density of the observation $X = (T, Z)$ is equal to

$$e^{\theta z} \lambda(t) e^{-e^{\theta z} \Lambda(t)},$$

where Λ is the primitive function of λ (with $\Lambda(0) = 0$). The usual estimator for (θ, Λ) based on a sample of size n from this model is the maximum likelihood estimator $(\hat{\theta}, \hat{\Lambda})$, where the likelihood is defined as, with $\Lambda\{t\}$ the jump of Λ at t,

$$(\theta, \Lambda) \mapsto \prod_{i=1}^{n} e^{\theta z_i} \Lambda\{t_i\} e^{-e^{\theta z_i} \Lambda(t_i)}.$$

This is the product of the density at the observations, but with the hazard function $\lambda(t)$ replaced by the jumps $\Lambda\{t\}$ of the cumulative hazard function. (This likelihood is close but not exactly equal to the empirical likelihood of the model.) The form of the likelihood forces the maximizer $\hat{\Lambda}$ to be a jump function with jumps at the observed "deaths" t_i, only and hence the likelihood can be reduced to a function of the unknowns $\Lambda\{t_1\}, \ldots, \Lambda\{t_n\}$. It appears to be impossible to derive the maximizers $(\hat{\theta}, \hat{\Lambda})$ in closed-form formulas, but we can make some headway in characterizing the maximum likelihood estimators by "profiling out" the nuisance parameter Λ. Elementary calculus shows that, for a fixed θ, the function

$$(\lambda_1, \ldots, \lambda_n) \mapsto \prod_{i=1}^{n} e^{\theta z_i} \lambda_i e^{-e^{\theta z_i} \sum_{j:t_j \leq t_i} \lambda_j}$$

is maximal for

$$\frac{1}{\lambda_k} = \sum_{i:t_i \geq t_k} e^{\theta z_i}.$$

The profile likelihood for θ is the supremum of the likelihood over Λ for fixed θ. In view of the preceding display this is given by

$$\theta \mapsto \prod_{i=1}^{n} \frac{e^{\theta z_i}}{\sum_{j:t_j \geq t_i} e^{\theta z_j}} e^{-1}.$$

The latter expression is known as the *Cox partial likelihood*. The original motivation for this criterion function is that the terms in the product are the conditional probabilities that the ith subject dies at time i given that one of the subjects at risk dies at that time. The maximum likelihood estimator for Λ is the step function with jumps

$$\hat{\Lambda}\{t_k\} = \frac{1}{\sum_{i:t_i \geq t_k} e^{\hat{\theta} z_i}}.$$

The estimators $\hat{\theta}$ and $\hat{\Lambda}$ are asymptotically efficient, under some restrictions. (See section 25.12.1.) We note that we have ignored the fact that jumps of hazard functions are smaller than 1 and have maximized over all measures Λ. □

25.70   *Example (Scale mixture).* Suppose we observe a sample from the distribution of $X = \theta + Z\varepsilon$, where the unobservable variables Z and ε are independent with completely unknown distribution η and a known density ϕ, respectively. Thus, the observation has a mixture density $\int p_\theta(x \mid z) \, d\eta(z)$ for the kernel

$$p_\theta(x \mid z) = \frac{1}{z}\phi\left(\frac{x-\theta}{z}\right).$$

If ϕ is symmetric about zero, then the mixture density is symmetric about θ, and we can estimate θ asymptotically efficiently with a fully adaptive estimator, as discussed in Section 25.8.1. Alternatively, we can take the mixture form of the underlying distribution into account and use, for instance, the maximum likelihood estimator, which maximizes the likelihood

$$(\theta, \eta) \mapsto \prod_{i=1}^{n} \int p_\theta(X_i \mid z) \, d\eta(z).$$

Under some conditions this estimator is asymptotically efficient.

Because the efficient score function for θ equals the ordinary score function for θ, the maximum likelihood estimator satisfies the efficient score equation $\mathbb{P}_n \tilde{\ell}_{\theta,\eta} = 0$. By the convexity of the model in η, this equation is unbiased in η. Thus, the asymptotic efficiency of the maximum likelihood estimator $\hat{\theta}$ follows under the regularity conditions of Theorem 25.54. Consistency of the sequence of maximum likelihood estimators $(\hat{\theta}_n, \hat{\eta}_n)$ for the product of the Euclidean and the weak topology can be proved by the method of Wald. The verification that the functions $\tilde{\ell}_{\theta,\eta}$ form a Donsker class is nontrivial but is possible using the techniques of Chapter 19. $\square$

25.71 **Example (Penalized logistic regression).** In this model we observe a random sample from the distribution of $X = (V, W, Y)$, for a 0-1 variable Y that follows the logistic regression model

$$P_{\theta,\eta}(Y = 1 \mid V, W) = \Psi(\theta V + \eta(W)),$$

where $\Psi(u) = 1/(1 + e^{-u})$ is the logistic distribution function. Thus, the usual linear regression of (V, W) has been replaced by the partial linear regression $\theta V + \eta(W)$, in which η ranges over a large set of "smooth functions." For instance, η is restricted to the Sobolev class of functions on $[0, 1]$ whose $(k - 1)$st derivative exists and is absolutely continuous with $J(\eta) < \infty$, where

$$J^2(\eta) = \int_0^1 \left(\eta^{(k)}(w)\right)^2 dw.$$

Here $k \geq 1$ is a fixed integer and $\eta^{(k)}$ is the kth derivative of η with respect to z.

The density of an observation is given by

$$p_{\theta,\eta}(x) = \Psi(\theta v + \eta(w))^y \left(1 - \Psi(\theta v + \eta(w))\right)^{1-y} f_{V,W}(v, w).$$

We cannot use this directly for defining a likelihood. The resulting maximizer $\hat{\eta}$ would be such that $\hat{\eta}(w_i) = \infty$ for every w_i with $y_i = 1$ and $\hat{\eta}(w_i) = -\infty$ when $y_i = 0$, or at least we could construct a sequence of finite, smooth η_m approaching this extreme choice. The problem is that qualitative smoothness assumptions such as $J(\eta) < \infty$ do not restrict η on a finite set of points $w_1, \ldots, w_n$ in any way.

To remedy this situation we can restrict the maximization to a smaller set of η, which we allow to grow as $n \to \infty$; for instance, the set of all η such that $J(\eta) \leq M_n$ for $M_n \uparrow \infty$ at a slow rate, or a sequence of spline approximations.

An alternative is to use a penalized likelihood, of the form

$$(\theta, \eta) \mapsto \mathbb{P}_n \log p_{\theta,\eta} - \hat{\lambda}_n^2 J^2(\eta).$$

Here $\hat{\lambda}_n$ is a "smoothing parameter" that determines the importance of the penalty $J^2(\eta)$. A large value of $\hat{\lambda}_n$ leads to smooth maximizers $\hat{\eta}$, for small values the maximizer is more like the unrestricted maximum likelihood estimator. Intermediate values are best and are often chosen by a data-dependent scheme, such as cross-validation. The penalized estimator $\hat{\theta}$ can be shown to be asymptotically efficient if the smoothing parameter is constructed to satisfy $\hat{\lambda}_n = o_P(n^{-1/2})$ and $\hat{\lambda}_n^{-1} = O_P(n^{k/(2k+1)})$ (see [102]). $\square$

25.72 **Example (Proportional odds).** Suppose that we observe a random sample from the distribution of the variable $X = (T \wedge C, 1\{T \leq C\}, Z)$, in which, given Z, the variables

T and C are independent, as in the random censoring model, but with the distribution function $F(t \mid z)$ of T given Z restricted by

$$\frac{F(t \mid z)}{1 - F(t \mid z)} = e^{z^T \theta} \eta(t).$$

In other words, the conditional *odds* given z of survival until t follows a Cox-type regression model. The unknown parameter η is a nondecreasing, cadlag function from $[0, \infty)$ into itself with $\eta(0) = 0$. It is the odds of survival if $\theta = 0$ and T is independent of Z.

If η is absolutely continuous, then the density of $X = (Y, \Delta, Z)$ is

$$\left(\frac{e^{-z^T\theta} \eta'(y) \left(1 - F_C(y - \mid z)\right)}{\left(\eta(y) + e^{-z^T\theta}\right)^2} \right)^\delta \left(\frac{e^{-z^T\theta} f_C(y \mid z)}{\eta(y) + e^{-z^T\theta}} \right)^{1-\delta} f_Z(z).$$

We cannot use this density as a likelihood, for the supremum is infinite unless we restrict η in an important way. Instead, we view η as the distribution function of a measure and use the empirical likelihood. The probability that $X = x$ is given by

$$\left(\frac{e^{-z^T\theta} \eta\{y\} \left(1 - F_C(y - \mid z)\right)}{\left(\eta(y) + e^{-z^T\theta}\right)\left(\eta(y-) + e^{-z^T\theta}\right)} \right)^\delta \left(\frac{e^{-z^T\theta} F_C(\{y\} \mid z)}{\eta(y) + e^{-z^T\theta}} \right)^{1-\delta} F_Z\{z\},$$

For likelihood inference concerning (θ, η) only, we may drop the terms involving F_C and F_Z and define the likelihood for one observation as

$$\text{lik}(\theta, \eta)(x) = \left(\frac{e^{-z^T\theta} \eta\{y\}}{\left(\eta(y) + e^{-z^T\theta}\right)\left(\eta(y-) + e^{-z^T\theta}\right)} \right)^\delta \left(\frac{e^{-z^T\theta}}{\eta(y) + e^{-z^T\theta}} \right)^{1-\delta}.$$

The presence of the jumps $\eta\{y\}$ causes the maximum likelihood estimator $\hat\eta$ to be a step function with support points at the observed survival times (the values y_i corresponding to $\delta_i = 1$). First, it is clear that each of these points must receive a positive mass. Second, mass to the right of the largest y_i such that $\delta_i = 1$ can be deleted, meanwhile increasing the likelihood. Third, mass assigned to other points can be moved to the closest y_i to the right such that $\delta_i = 1$, again increasing the likelihood. If the biggest observation y_i has $\delta_i = 1$, then $\hat\eta\{y_i\} = \infty$ and that observation gives a contribution 1 to the likelihood, because the function $p \mapsto p/(p + r)$ attains for $p \geq 0$ its maximal value 1 at $p = \infty$. On the other hand, if $\delta_i = 0$ for the largest y_i, then all jumps of $\hat\eta$ must be finite.

The maximum likelihood estimators have been shown to be asymptotically efficient under some conditions in [105]. $\square$

25.10.1 *Random Censoring*

Suppose that we observe a random sample $(X_1, \Delta_1), \ldots, (X_n, \Delta_n)$ from the distribution of $\left(T \wedge C, 1\{T \leq C\}\right)$, in which the "survival time" T and the "censoring time" C are independent with completely unknown distribution functions F and G, respectively. The distribution of a typical observation (X, Δ) satisfies

$$P_{F,G}(X \leq x, \Delta = 0) = \int_{[0,x]} (1 - F) \, dG,$$

$$P_{F,G}(X \leq x, \Delta = 1) = \int_{[0,x]} (1 - G_-) \, dF.$$

Consequently, if F and G have densities f and g (relative to some dominating measures), then (X, Δ) has density

$$(x, \delta) \mapsto \big((1 - F)(x)g(x)\big)^{\delta}\big((1 - G_-)(x)f(x)\big)^{1-\delta}.$$

For f and g interpreted as Lebesgue densities, we cannot use this expression as a factor in a likelihood, as the resulting criterion would have supremum infinity. (Simply choose f or g to have a very high, thin peak at an observation X_i with $\Delta_i = 1$ or $\Delta_i = 0$, respectively.) Instead, we may take f and g as densities relative to counting measure. This leads to the empirical likelihood

$$(F, G) \mapsto \prod_{i=1}^{n}\big((1 - F)(X_i)G\{X_i\}\big)^{1-\Delta_i}\prod_{i=1}^{n}\big((1 - G_-)(X_i)F\{X_i\}\big)^{\Delta_i}.$$

In view of the product form, this factorizes in likelihoods for F and G separately. The maximizer $\hat{F}$ of the likelihood $F \mapsto \prod_{i=1}^{n}(1 - F)(X_i)^{1-\Delta_i}F\{X_i\}^{\Delta_i}$ turns out to be the *product limit estimator*, given in Example 20.15.

That the product limit estimator maximizes the likelihood can be seen by direct arguments, but a slight detour is more insightful. The next lemma shows that under the present model the distribution $P_{F,G}$ of (X, Δ) can be any distribution on the sample space $[0, \infty) \times \{0, 1\}$. In other words, if F and G range over all possible probability distributions on $[0, \infty]$, then $P_{F,G}$ ranges over all distributions on $[0, \infty) \times \{0, 1\}$. Moreover, the relationship $(F, G) \leftrightarrow P_{F,G}$ is one-to-one on the interval where $(1-F)(1-G) > 0$. As a consequence, there exists a pair $(\hat{F}, \hat{G})$ such that $P_{\hat{F}, \hat{G}}$ is the empirical distribution $\mathbb{P}_n$ of the observations

$$P_{\hat{F}, \hat{G}}\{X_i, \Delta_i\} = \mathbb{P}_n\{X_i, \Delta_i\}, \qquad 1 \le i \le n.$$

Because the empirical distribution maximizes $P \mapsto \prod_{i=1}^{n}P\{X_i, \Delta_i\}$ over all distributions, it follows that $(\hat{F}, \hat{G})$ maximizes $(F, G) \mapsto \prod_{i=1}^{n}P_{F,G}\{X_i, \Delta_i\}$ over all (F, G). That $\hat{F}$ is the product limit estimator next follows from Example 20.15.

To complete the discussion, we study the map $(F, G) \leftrightarrow P_{F,G}$. A probability distribution on $[0, \infty) \times \{0, 1\}$ can be identified with a pair (H_0, H_1) of subdistribution functions on $[0, \infty)$ such that $H_0(\infty)+H_1(\infty) = 1$, by letting $H_i(x)$ be the mass of the set $[0, x] \times \{i\}$. A given pair of distribution functions (F_0, F_1) on $[0, \infty)$ yields such a pair of subdistribution functions (H_0, H_1), by

$$H_0(x) = \int_{[0,x]} (1 - F_1)\, dF_0, \qquad H_1(x) = \int_{[0,x]} (1 - F_{0-})\, dF_1. \qquad (25.73)$$

Conversely, the pair (F_0, F_1) can be recovered from a given pair (H_0, H_1) by, with ΔH_i the jump in H_i, $H = H_0 + H_1$ and Λ_i^c the continuous part of Λ_i,

$$\Lambda_0(x) = \int_{[0,x]} \frac{dH_0}{1 - H_- - \Delta H_1}, \qquad \Lambda_1(x) = \int_{[0,x]} \frac{dH_1}{1 - H_-},$$

$$1 - F_i(x) = \prod_{0 \le s \le x}\big(1 - \Lambda_i\{s\}\big)e^{-\Lambda_i^c(x)}.$$

25.74 Lemma. *Given any pair (H_0, H_1) of subdistribution functions on $[0, \infty)$ such that $H_0(\infty) + H_1(\infty) = 1$, the preceding display defines a pair (F_0, F_1) of subdistribution functions on $[0, \infty)$ such that (25.73) holds.*

Proof.   For any distribution function A and cumulative hazard function B on $[0, \infty)$, with B^c the continuous part of B,

$$1 - A(t) = \prod_{0 \le s \le t} \left(1 - B\{s\}\right) e^{-B^c(t)} \text{ iff } B(t) = \int_{[0,t]} \frac{dA}{1 - A_-}.$$

To see this, rewrite the second equality as $(1 - A_-) \, dB = dA$ and $B(0) = A(0)$, and integrate this to rewrite it again as the *Volterra equation*

$$(1 - A) = 1 + \int_{[0,\cdot]} (1 - A_-) \, d(-B).$$

It is well known that the Volterra equation has the first equation of the display as its unique solution.[†]

Combined with the definition of F_i, the equivalence in the preceding display implies immediately that $d\Lambda_i = dF_i/(1 - F_{i-})$. Secondly, as immediate consequences of the definitions,

$$(1 - F_0)(1 - F_1)(t) = \prod_{s \le t}(1 - \Delta\Lambda_0 - \Delta\Lambda_1 + \Delta\Lambda_0\Delta\Lambda_1)(s)e^{-(\Lambda_0 + \Lambda_1)^c(t)},$$

$$(\Lambda_0 + \Lambda_1)(t) - \sum_{s \le t} \Delta\Lambda_0(s)\Delta\Lambda_1(s) = \int_{[0,t]} \frac{dH}{1 - H_-}.$$

(Split $dH_0/(1 - H_- - \Delta H_1)$ into the parts corresponding to dH_0^c and ΔH_0 and note that ΔH_1 may be dropped in the first part.) Combining these equations with the Volterra equation, we obtain that $1 - H = (1 - F_0)(1 - F_1)$. Taken together with $dH_1 = (1 - H_-) \, d\Lambda_1$, we conclude that $dH_1 = (1 - F_{0-})(1 - F_{1-}) \, d\Lambda_1 = (1 - F_{0-}) \, dF_1$, and similarly $dH_0 = (1 - F_1) \, dF_0$. ∎

25.11 Approximately Least-Favorable Submodels

If the maximum likelihood estimator satisfies the efficient score equation $\mathbb{P}_n \tilde{\ell}_{\hat\theta, \hat\eta} = 0$, then Theorem 25.54 yields its asymptotic normality, provided that its conditions can be verified for the maximum likelihood estimator $\hat\eta$. Somewhat unexpectedly, the efficient score function may not be a "proper" score function and the maximum likelihood estimator may not satisfy the efficient score equation. This is because, by definition, the efficient score function is a projection, and nothing guarantees that this projection is the derivative of the log likelihood along some submodel. If there exists a "least favorable" path $t \mapsto \eta_t(\hat\theta, \hat\eta)$ such that $\eta_0(\hat\theta, \hat\eta) = \hat\eta$, and, for every x,

$$\tilde{\ell}_{\hat\theta, \hat\eta}(x) = \frac{\partial}{\partial t}\Big|_{t=0} \log \text{lik}\big(\hat\theta + t, \eta_t(\hat\theta, \hat\eta)\big)(x),$$

then the maximum likelihood estimator satisfies the efficient score equation; if not, then this is not clear. The existence of an exact least favorable submodel appears to be particularly uncertain at the maximum likelihood estimator $(\hat\theta, \hat\eta)$, as this tends to be on the "boundary" of the parameter set.

[†] See, for example, [133, p. 206] or [55] for an extended discussion.

A method around this difficulty is to replace the efficient score equation by an approximation. First, it suffices that $(\hat{\theta}, \hat{\eta})$ satisfies the efficient score equation approximately, for Theorem 25.54 goes through provided $\sqrt{n}\, \mathbb{P}_n \tilde{\ell}_{\hat{\theta}, \hat{\eta}} = o_P(1)$. Second, it was noted following the proof of Theorem 25.54 that this theorem is valid for estimating equations of the form $\mathbb{P}_n \tilde{\ell}_{\theta, \hat{\eta}} = 0$ for arbitrary mean-zero functions $\tilde{\ell}_{\theta, \eta}$; its assertion remains correct provided that at the true value of (θ, η) the function $\tilde{\ell}_{\theta, \eta}$ is the efficient score function. This suggests to replace, in our proof, the function $\tilde{\ell}_{\theta, \eta}$ by functions $\tilde{\kappa}_{\theta, \eta}$ that are proper score functions and are close to the efficient score function, at least for the true value of the parameter. These are derived from "approximately-least favorable submodels."

We define such submodels as maps $t \mapsto \eta_t(\theta, \eta)$ from a neighborhood of $0 \in \mathbb{R}^k$ to the parameter set for η with $\eta_0(\theta, \eta) = \eta$ (for every (θ, η)) such that

$$\tilde{\kappa}_{\theta, \eta}(x) = \frac{\partial}{\partial t}_{|t=0} \log \operatorname{lik}\big(\theta + t, \eta_t(\theta, \eta)\big)(x),$$

exists (for every x) and is equal to the efficient score function at $(\theta, \eta) = (\theta_0, \eta_0)$. Thus, the path $t \mapsto \eta_t(\theta, \eta)$ must pass through η at $t = 0$, and at the true parameter (θ_0, η_0) the submodel is truly least favorable in that its score is the efficient score for θ. We need such a submodel for every fixed (θ, η), or at least for the true value (θ_0, η_0) and every possible value of $(\hat{\theta}, \hat{\eta})$.

If $(\hat{\theta}, \hat{\eta})$ maximizes the likelihood, then the function $t \mapsto \mathbb{P}_n \log \operatorname{lik}\big(\theta + t, \eta_t(\hat{\theta}, \hat{\eta})\big)$ is maximal at $t = 0$ and hence $(\hat{\theta}, \hat{\eta})$ satisfies the stationary equation $\mathbb{P}_n \tilde{\kappa}_{\hat{\theta}, \hat{\eta}} = 0$. Now Theorem 25.54, with $\tilde{\ell}_{\theta, \eta}$ replaced by $\tilde{\kappa}_{\theta, \eta}$, yields the asymptotic efficiency of $\hat{\theta}_n$. For easy reference we reformulate the theorem.

$$P_{\hat{\theta}_n, \eta_0} \tilde{\kappa}_{\hat{\theta}_n, \hat{\eta}_n} = o_P\big(n^{-1/2} + \|\hat{\theta}_n - \theta_0\|\big) \tag{25.75}$$

$$P_{\theta_0, \eta_0} \big\| \tilde{\kappa}_{\hat{\theta}_n, \hat{\eta}_n} - \tilde{\kappa}_{\theta_0, \eta_0} \big\|^2 \xrightarrow{\text{P}} 0, \qquad P_{\hat{\theta}_n, \eta_0} \big\| \tilde{\kappa}_{\hat{\theta}_n, \hat{\eta}_n} \big\|^2 = O_P(1). \tag{25.76}$$

25.77 Theorem. *Suppose that the model* $\{P_{\theta, \eta} : \theta \in \Theta\}$, *is differentiable in quadratic mean with respect to* θ *at* (θ_0, η_0) *and let the efficient information matrix* $\tilde{I}_{\theta_0, \eta_0}$ *be nonsingular. Assume that* $\tilde{\kappa}_{\theta, \eta}$ *are the score functions of approximately least-favorable submodels (at* (θ_0, η_0)*), that the functions* $\tilde{\kappa}_{\hat{\theta}, \hat{\eta}}$ *belong to a* P_{θ_0, η_0}*-Donsker class with square-integrable envelope with probability tending to 1, and that* (25.75) *and* (25.76) *hold. Then the maximum likelihood estimator* $\hat{\theta}_n$ *is asymptotically efficient at* (θ_0, η_0) *provided that it is consistent.*

The no-bias condition (25.75) can be analyzed as in (25.60), with $\tilde{\ell}_{\theta, \hat{\eta}}$ replaced by $\tilde{\kappa}_{\theta, \hat{\eta}}$. Alternatively, it may be useful to avoid evaluating the efficient score function at $\hat{\theta}$ or $\hat{\eta}$, and (25.60) may be adapted to

$$P_{\hat{\theta}, \eta_0} \tilde{\kappa}_{\hat{\theta}, \hat{\eta}} = (P_{\hat{\theta}, \eta_0} - P_{\hat{\theta}, \hat{\eta}})(\tilde{\kappa}_{\hat{\theta}, \hat{\eta}} - \tilde{\kappa}_{\theta_0, \eta_0})$$
$$- \int \tilde{\kappa}_{\theta_0, \eta_0} \big[p_{\hat{\theta}, \hat{\eta}} - p_{\hat{\theta}, \eta_0} - B_{\theta_0, \eta_0}(\hat{\eta} - \eta_0)\, p_{\theta_0, \eta_0} \big]\, d\mu. \tag{25.78}$$

Replacing $\hat{\theta}$ by θ_0 should make at most a difference of $o_P\big(\|\hat{\theta} - \theta_0\|\big)$, which is negligible in the preceding display, but the presence of $\hat{\eta}$ may require a rate of convergence for $\hat{\eta}$. Theorem 5.55 yields such rates in some generality and can be translated to the present setting as follows.

Consider estimators $\hat{\tau}_n$ contained in a set H_n that, for a given $\hat{\lambda}_n$ contained in a set $\Lambda_n \subset \mathbb{R}$, maximize a criterion $\tau \mapsto \mathbb{P}_n m_{\tau,\lambda_n}$, or at least satisfy $\mathbb{P}_n m_{\tau,\lambda_n} \geq \mathbb{P}_n m_{\tau_0,\lambda_n}$. Assume that for every $\lambda \in \Lambda_n$, every $\tau \in H_n$ and every $\delta > 0$,

$$P(m_{\tau,\lambda} - m_{\tau_0,\lambda}) \lesssim -d_\lambda^2(\tau,\tau_0) + \lambda^2, \tag{25.79}$$

$$\mathrm{E}^* \sup_{\substack{d_\lambda(\tau,\tau_0)<\delta \\ \lambda \in \Lambda_n, \tau \in H_n}} \left| \mathbb{G}_n(m_{\tau,\lambda} - m_{\tau_0,\lambda}) \right| \lesssim \phi_n(\delta). \tag{25.80}$$

25.81 Theorem. *Suppose that (25.79) and (25.80) are valid for functions ϕ_n such that $\delta \mapsto \phi_n(\delta)/\delta^\alpha$ is decreasing for some $\alpha < 2$ and sets $\Lambda_n \times H_n$ such that $\mathrm{P}(\hat{\lambda}_n \in \Lambda_n, \hat{\tau}_n \in H_n) \to 1$. Then $d_\lambda(\hat{\tau}_n, \tau_0) \leq O_P^*(\delta_n + \hat{\lambda}_n)$ for any sequence of positive numbers δ_n such that $\phi_n(\delta_n) \leq \sqrt{n}\, \delta_n^2$ for every n.*

25.11.1 Cox Regression with Current Status Data

Suppose that we observe a random sample from the distribution of $X = (C, \Delta, Z)$, in which $\Delta = 1\{T \leq C\}$, that the "survival time" T and the observation time C are independent given Z, and that T follows a Cox model. The density of X relative to the product of $F_{C,Z}$ and counting measure on $\{0, 1\}$ is given by

$$p_{\theta,\Lambda}(x) = \left(1 - \exp\left(-e^{\theta^T z}\Lambda(c)\right)\right)^\delta \left(\exp\left(-e^{\theta^T z}\Lambda(c)\right)\right)^{1-\delta}.$$

We define this as the likelihood for one observation x. In maximizing the likelihood we restrict the parameter θ to a compact in $\mathbb{R}^k$ and restrict the parameter Λ to the set of all cumulative hazard functions with $\Lambda(\tau) \leq M$ for a fixed large constant M and τ the end of the study.

We make the following assumptions. The observation times C possess a Lebesgue density that is continuous and positive on an interval $[\sigma, \tau]$ and vanishes outside this interval. The true parameter Λ_0 is continuously differentiable on this interval, satisfies $0 < \Lambda_0(\sigma-) \leq \Lambda_0(\tau) < M$, and is continuously differentiable on $[\sigma, \tau]$. The covariate vector Z is bounded and $\mathrm{E}\,\mathrm{cov}(Z \mid C) > 0$. The function h_{θ_0,Λ_0} given by (25.82) has a version that is differentiable with a bounded derivative on $[\sigma, \tau]$. The true parameter θ_0 is an inner point of the parameter set for θ.

The score function for θ takes the form

$$\dot{\ell}_{\theta,\Lambda}(x) = z\Lambda(c)Q_{\theta,\Lambda}(x),$$

for the function $Q_{\theta,\Lambda}$ given by

$$Q_{\theta,\Lambda}(x) = e^{\theta^T z}\left[\delta \frac{e^{-e^{\theta^T z}\Lambda(c)}}{1 - e^{-e^{\theta^T z}\Lambda(c)}} - (1 - \delta)\right].$$

For every nondecreasing, nonnegative function h and positive number t, the submodel $\Lambda_t = \Lambda + th$ is well defined. Inserting this in the log likelihood and differentiating with respect to t at $t = 0$, we obtain a score function for Λ of the form

$$B_{\theta,\Lambda}h(x) = h(c)Q_{\theta,\Lambda}(x).$$

The linear span of these score functions contains $B_{\theta,\Lambda}h$ for all bounded functions h of bounded variation. In view of the similar structure of the scores for θ and Λ, projecting $\dot{\ell}_{\theta,\Lambda}$ onto the closed linear span of the nuisance scores is a weighted least-squares problem with weight function $Q_{\theta,\Lambda}$. The solution is given by the vector-valued function

$$h_{\theta,\Lambda}(c) = \Lambda(c)\frac{\mathrm{E}_{\theta,\Lambda}\big(Z Q_{\theta,\Lambda}^2(X)\mid C = c\big)}{\mathrm{E}_{\theta,\Lambda}\big(Q_{\theta,\Lambda}^2(X)\mid C = c\big)}. \tag{25.82}$$

The efficient score function for θ takes the form

$$\tilde{\ell}_{\theta,\Lambda}(x) = \big(z\Lambda(c) - h_{\theta,\Lambda}(c)\big)Q_{\theta,\Lambda}(x).$$

Formally, this function is the derivative at $t = 0$ of the log likelihood evaluated at $(\theta + t, \Lambda - t^T h_{\theta,\Lambda})$. However, the second coordinate of the latter path may not define a nondecreasing, nonnegative function for every t in a neighborhood of 0 and hence cannot be used to obtain a stationary equation for the maximum likelihood estimator. This is true in particular for discrete cumulative hazard functions Λ, for which $\Lambda + th$ is nondecreasing for both $t < 0$ and $t > 0$ only if h is constant between the jumps of Λ.

This suggests that the maximum likelihood estimator does not satisfy the efficient score equation. To prove the asymptotic normality of $\hat{\theta}$, we replace this equation by an approximation, obtained from an approximately least favorable submodel.

For fixed (θ, Λ), and a fixed bounded, Lipschitz function ϕ, define

$$\Lambda_t(\theta, \Lambda) = \Lambda - t^T \phi(\Lambda)\big(h_{\theta_0,\Lambda_0} \circ \Lambda_0^{-1}\big)(\Lambda).$$

Then $\Lambda_t(\theta, \Lambda)$ is a cumulative hazard function for every t that is sufficiently close to zero, because for every $u \leq v$,

$$\Lambda_t(\theta, \Lambda)(v) - \Lambda_t(\theta, \Lambda)(u) \geq \big(\Lambda(v) - \Lambda(u)\big)\Big(1 - \|t\|\,\big\|\phi\, h_{\theta_0,\Lambda_0} \circ \Lambda_0^{-1}\big\|_{\mathrm{Lip}}\Big).$$

Inserting $\big(\theta + t, \Lambda_t(\theta, \Lambda)\big)$ into the log likelihood, and differentiating with respect to t at $t = 0$, yields the score function

$$\tilde{\kappa}_{\theta,\Lambda}(x) = \Big(z\Lambda(c) - \phi\big(\Lambda(c)\big)\big(h_{\theta_0,\Lambda_0} \circ \Lambda_0^{-1}\big)\big(\Lambda(c)\big)\Big)Q_{\theta,\Lambda}(x).$$

If evaluated at (θ_0, Λ_0) this reduces to the efficient score function $\tilde{\ell}_{\theta_0,\Lambda_0}(x)$ provided $\phi(\Lambda_0) = 1$, whence the submodel is approximately least favorable. To prove the asymptotic efficiency of $\hat{\theta}_n$ it suffices to verify the conditions of Theorem 25.77.

The function ϕ is a technical device that has been introduced in order to ensure that $0 \leq \Lambda_t(\theta, \Lambda) \leq M$ for all t that are sufficiently close to 0. This is guaranteed if $0 \leq y\,\phi(y) \leq c\big(y \wedge (M - y)\big)$ for every $0 \leq y \leq M$, for a sufficiently large constant c. Because by assumption $\big[\Lambda_0(\sigma-),\,\Lambda_0(\tau)\big] \subset (0, M)$, there exists such a function ϕ that also fulfills $\phi(\Lambda_0) = 1$ on $[\sigma, \tau]$.

In order to verify the no-bias condition (25.52) we need a rate of convergence for $\hat{\Lambda}_n$.

25.83 Lemma. *Under the conditions listed previously, $\hat{\theta}_n$ is consistent and $\|\hat{\Lambda}_n - \Lambda_0\|_{P_0,2} = O_P(n^{-1/3})$.*

Proof. Denote the index (θ_0, Λ_0) by 0, and define functions

$$m_{\theta,\Lambda} = \log\big(p_{\theta,\Lambda} + p_0\big)/2.$$

The densities $p_{\theta,\Lambda}$ are bounded above by 1, and under our assumptions the density p_0 is bounded away from zero. It follows that the functions $m_{\theta,\Lambda}(x)$ are uniformly bounded in (θ, Λ) and x.

By the concavity of the logarithm and the definition of $(\hat{\theta}, \hat{\Lambda})$,

$$\mathbb{P}_n m_{\hat{\theta},\hat{\Lambda}} \geq \tfrac{1}{2} \mathbb{P}_n \log p_{\hat{\theta},\hat{\Lambda}} + \tfrac{1}{2} \mathbb{P}_n \log p_0 \geq \mathbb{P}_n \log p_0 = \mathbb{P}_n m_0.$$

Therefore, Theorem 25.81 is applicable with $\tau = (\theta, \Lambda)$ and without λ. For technical reasons it is preferable first to establish the consistency of $(\hat{\theta}, \hat{\Lambda})$ by a separate argument.

We apply Wald's proof, Theorem 5.14. The parameter set for θ is compact by assumption, and the parameter set for Λ is compact relative to the weak topology. Wald's theorem shows that the distance between $(\hat{\theta}, \hat{\Lambda})$ and the set of maximizers of the Kullback-Leibler divergence converges to zero. This set of maximizers contains (θ_0, Λ_0), but this parameter is not fully identifiable under our assumptions: The parameter Λ_0 is identifiable only on the interval (σ, τ). It follows that $\hat{\theta} \overset{P}{\to} \theta_0$ and $\hat{\Lambda}(t) \overset{P}{\to} \Lambda_0(t)$ for every $\sigma < t < \tau$. (The convergence of $\hat{\Lambda}$ at the points σ and τ does not appear to be guaranteed.)

By the proof of Lemma 5.35 and Lemma 25.85 below, condition (25.79) is satisfied with $d\big((\theta, \Lambda), (\theta_0, \Lambda_0)\big)$ equal to $\|\theta - \theta_0\| + \|\Lambda - \Lambda_0\|_2$. By Lemma 25.84 below, the bracketing entropy of the class of functions $m_{\theta,\Lambda}$ is of the order $(1/\varepsilon)$. By Lemma 19.36 condition (25.80) is satisfied for

$$\phi_n(\delta) = \sqrt{\delta}\left(1 + \frac{\sqrt{\delta}}{\delta^2 \sqrt{n}}\right).$$

This leads to a convergence rate of $n^{-1/3}$ for both $\|\hat{\theta} - \theta_0\|$ and $\|\hat{\Lambda} - \Lambda_0\|_2$. ∎

To verify the no-bias condition (25.75), we use the decomposition (25.78). The integrands in the two terms on the right can both be seen to be bounded, up to a constant, by $(\hat{\Lambda} - \Lambda_0)^2$, with probability tending to one. Thus the bias $P_{\hat{\theta},\eta_0}\tilde{\kappa}_{\hat{\theta},\hat{\eta}}$ is actually of the order $O_P(n^{-2/3})$.

The functions $x \mapsto \tilde{\kappa}_{\theta,\Lambda}(x)$ can be written in the form $\psi\big(z, e^{\theta^T z}, \Lambda(c), \delta\big)$ for a function ψ that is Lipschitz in its first three coordinates, for $\delta \in \{0, 1\}$ fixed. (Note that $\Lambda \mapsto \Lambda Q_{\theta,\Lambda}$ is Lipschitz, as $\Lambda \mapsto h_{\theta_0,\Lambda_0} \circ \Lambda_0^{-1}(\Lambda)/\Lambda = (h_{\theta_0,\Lambda_0}/\Lambda_0) \circ \Lambda_0^{-1}(\Lambda)$.) The functions $z \mapsto z$, $z \mapsto \exp\theta^T z, c \mapsto \Lambda(c)$ and $\delta \mapsto \delta$ form Donsker classes if θ and Λ range freely. Hence the functions $x \mapsto \Lambda(c)Q_{\theta,\Lambda}(x)$ form a Donsker class, by Example 19.20. The efficiency of $\hat{\theta}_n$ follows by Theorem 25.77.

25.84 Lemma. *Under the conditions listed previously, there exists a constant C such that, for every $\varepsilon > 0$,*

$$\log N_{[\,]}\Big(\varepsilon, \big\{m_{\theta,\Lambda}, (\theta, \Lambda)\big\}, L_2(P_0)\Big) \leq C\left(\frac{1}{\varepsilon}\right).$$

Proof. First consider the class of functions $m_{\theta,\Lambda}$ for a fixed θ. These functions depend on Λ monotonely if considered separately for $\delta = 0$ and $\delta = 1$. Thus a bracket $\Lambda_1 \leq \Lambda \leq \Lambda_2$ for Λ leads, by substitution, readily to a bracket for $m_{\theta,\Lambda}$. Furthermore, because this dependence is Lipschitz, there exists a constant D such that

$$\int (m_{\theta,\Lambda_1} - m_{\theta,\Lambda_2})^2 \, dF_{C,Z} \leq D \int_\sigma^\tau \big(\Lambda_1(c) - \Lambda_2(c)\big)^2 \, dc.$$

Thus, brackets for Λ of L_2-size ε translate into brackets for $m_{\theta,\Lambda}$ of $L_2(P_{\theta,\Lambda})$-size proportional to ε. By Example 19.11 we can cover the set of all Λ by $\exp C(1/\varepsilon)$ brackets of size ε.

Next, we allow θ to vary freely as well. Because θ is finite-dimensional and $\partial/\partial\theta\, m_{\theta,\Lambda}(x)$ is uniformly bounded in (θ, Λ, x), this increases the entropy only slightly. ∎

25.85 Lemma. *Under the conditions listed previously there exist constants $C, \varepsilon > 0$ such that, for all Λ and all $\|\theta - \theta_0\| < \varepsilon$,*

$$\int \left(p_{\theta,\Lambda}^{1/2} - p_{\theta_0,\Lambda_0}^{1/2}\right)^2 d\mu \geq C \int_\sigma^\tau (\Lambda - \Lambda_0)^2(c)\, dc + C\|\theta - \theta_0\|^2.$$

Proof. The left side of the lemma can be rewritten as

$$\int \frac{(p_{\theta,\Lambda} - p_{\theta_0,\Lambda_0})^2}{\left(p_{\theta,\Lambda}^{1/2} + p_{\theta_0,\Lambda_0}^{1/2}\right)^2} d\mu.$$

Because p_0 is bounded away from zero, and the densities $p_{\theta,\Lambda}$ are uniformly bounded, the denominator can be bounded above and below by positive constants. Thus the Hellinger distance (in the display) is equivalent to the L_2-distance between the densities, which can be rewritten

$$2\int \left[e^{-e^{\theta^T z}\Lambda(c)} - e^{-e^{\theta_0^T z}\Lambda_0(c)}\right]^2 dF^{Y,Z}(c, z).$$

Let $g(t)$ be the function $\exp\left(-e^{\theta^T z}\Lambda(c)\right)$ evaluated at $\theta_t = t\theta + (1 - t)\theta_0$ and $\Lambda_t = t\Lambda + (1 - t)\Lambda_0$, for fixed (c, z). Then the integrand is equal to $\left(g(1) - g(0)\right)^2$, and hence, by the mean value theorem, there exists $0 \leq t = t(c, z) \leq 1$ such that the preceding display is equal to

$$P_0\left(e^{-\Lambda_t(c)e^{\theta_t^T z}} e^{\theta_t^T z}\left[(\Lambda - \Lambda_0)(c)\left(1 + t(\theta - \theta_0)^T z\right) + (\theta - \theta_0)^T z\Lambda_0(c)\right]\right)^2.$$

Here the multiplicative factor $e^{-\Lambda_t(c)e^{\theta_t^T z}} e^{\theta_t^T z}$ is bounded away from zero. By dropping this term we obtain, up to a constant, a lower bound for the left side of the lemma. Next, because the function Q_{θ_0,Λ_0} is bounded away from zero and infinity, we may add a factor Q_{θ_0,Λ_0}^2, and obtain the lower bound, up to a constant,

$$P_0\left(\left(1 + t(\theta - \theta_0)^T z\right) B_{\theta_0,\Lambda_0}(\Lambda - \Lambda_0)(x) + (\theta - \theta_0)^T \ell_{\theta_0,\Lambda_0}(x)\right)^2.$$

Here the function $h = \left(1 + t(\theta - \theta_0)^T z\right)$ is uniformly close to 1 if θ is close to θ_0. Furthermore, for any function g and vector a,

$$\left(P_0(B_{\theta_0,\Lambda_0}g)a^T \ell_{\theta_0,\Lambda_0}\right)^2 = \left(P_0(B_{\theta_0,\Lambda_0}g)a^T (\dot\ell_{\theta_0,\Lambda_0} - \tilde\ell_0)\right)^2$$
$$\leq P_0(B_{\theta_0,\Lambda_0}g)^2 a^T (I_0 - \tilde I_0)a,$$

by the Cauchy-Schwarz inequality. Because the efficient information $\tilde I_0$ is positive-definite, the term $a^T (I_0 - \tilde I_0)a$ on the right can be written $a^T I_0 a c$ for a constant $0 < c < 1$. The lemma now follows by application of Lemma 25.86 ahead. ∎

25.86   Lemma. *Let h, g_1 and g_2 be measurable functions such that $c_1 \leq h \leq c_2$ and $(Pg_1g_2)^2 \leq cPg_1^2Pg_2^2$ for a constant $c < 1$ and constants $c_1 < 1 < c_2$ close to 1. Then*

$$P(hg_1 + g_2)^2 \geq C(Pg_1^2 + Pg_2^2),$$

for a constant C depending on c, c_1 and c_2 that approaches $1 - \sqrt{c}$ as $c_1 \uparrow 1$ and $c_2 \downarrow 1$.

Proof.   We may first use the inequalities

$$
\begin{aligned}
(hg_1 + g_2)^2 &\geq c_1 hg_1^2 + 2hg_1g_2 + c_2^{-1}hg_2^2 \\
&= h(g_1 + g_2)^2 + (c_1 - 1)hg_1^2 + \left(1 - c_2^{-1}\right)hg_2^2 \\
&\geq c_1\left(g_1^2 + 2g_1g_2 + g_2^2\right) + (c_1 - 1)c_2 g_1^2 + \left(c_2^{-1} - 1\right)g_2^2.
\end{aligned}
$$

Next, we integrate this with respect to P, and use the inequality for Pg_1g_2 on the second term to see that the left side of the lemma is bounded below by

$$c_1\left(Pg_1^2 - 2\sqrt{cPg_1^2Pg_2^2} + Pg_2^2\right) + (c_1 - 1)c_2 Pg_1^2 + \left(c_2^{-1} - 1\right)c_2 Pg_2^2.$$

Finally, we apply the inequality $2xy \leq x^2 + y^2$ on the second term.   ∎

25.11.2   *Exponential Frailty*

Suppose that the observations are a random sample from the density of $X = (U, V)$ given by

$$p_{\theta,\eta}(u, v) = \int ze^{-zu}\,\theta ze^{-\theta zv}\,d\eta(z).$$

This is a density with respect to Lebesgue measure on the positive quadrant of $\mathbb{R}^2$, and we may take the likelihood equal to just the joint density of the observations. Let $(\hat{\theta}_n, \hat{\eta}_n)$ maximize

$$(\theta, \eta) \mapsto \prod_{i=1}^{n} p_{\theta,\eta}(U_i, V_i).$$

This estimator can be shown to be consistent, under some conditions, for the Euclidean and weak topology, respectively, by, for instance, the method of Wald, Theorem 5.14.

The "statistic" $\psi_\theta(U, V) = U + \theta V$ is, for fixed and known θ, sufficient for the nuisance parameter. Because the likelihood depends on η only through this statistic, the tangent set $_\eta\dot{\mathcal{P}}_{P_{\theta,\eta}}$ for η consists of functions of $U + \theta V$ only. Furthermore, because $U + \theta V$ is distributed according to a mixture over an exponential family (a gamma-distribution with shape parameter 2), the closed linear span of $_\eta\dot{\mathcal{P}}_{P_{\theta,\eta}}$ consists of all mean-zero, square-integrable functions of $U + \theta V$, by Example 25.35. Thus, the projection onto the closed linear span of $_\eta\dot{\mathcal{P}}_{P_{\theta,\eta}}$ is the conditional expectation with respect to $U + \theta V$, and the efficient score function for θ is the "conditional score," given by

$$
\begin{aligned}
\tilde{\ell}_{\theta,\eta}(x) &= \dot{\ell}_{\theta,\eta}(x) - \mathrm{E}_\theta\left(\dot{\ell}_{\theta,\eta}(X) \mid \psi_\theta(X) = \psi_\theta(x)\right) \\
&= \frac{\int \frac{1}{2}(u - \theta v)z^3 e^{-z(u+\theta v)}\,d\eta(z)}{\int \theta z^2 e^{-z(u+\theta v)}\,d\eta(z)},
\end{aligned}
$$

where we may use that, given $U + \theta V = s$, the variables U and θV are uniformly distributed on the interval $[0, s]$. This function turns out to be also an actual score function, in that there exists an exact least favorable submodel, given by

$$\eta_t(\theta, \eta)(B) = \eta\left(B\left(1 - \frac{t}{2\theta}\right)\right).$$

Inserting $\eta_t(\theta, \eta)$ in the log likelihood, making the change of variables $z\big(1 - t/(2\theta)\big) \to z$, and computing the (ordinary) derivative with respect to t at $t = 0$, we obtain $\tilde{\ell}_{\theta,\eta}(x)$. It follows that the maximum likelihood estimator satisfies the efficient score equation, and its asymptotic normality can be proved with the help of Theorem 25.54.

The linearity of the model in η (or the formula involving the conditional expectation) implies that

$$P_{\theta,\eta_0} \tilde{\ell}_{\theta,\eta} = 0, \qquad \text{every } \theta, \eta, \eta_0.$$

Thus, the "no-bias" condition (25.52) is trivially satisfied. The verification that the functions $\tilde{\ell}_{\theta,\eta}$ form a Donsker class is more involved but is achieved in the following lemma.[†]

25.87   Lemma. *Suppose that $\int (z^2 + z^{-5})\, d\eta_0(z) < \infty$. Then there exists a neighborhood V of η_0 for the weak topology such that the class of functions*

$$(x, y) \mapsto \frac{\int (a_1 + a_2 z x + a_3 z y)\, z^2\, e^{-z(b_1 x + b_2 y)}\, d\eta(z)}{\int z^2\, e^{-z(b_1 x + b_2 y)}\, d\eta(z)},$$

where $(a_1, \ldots, a_3)$ ranges over a bounded subset of $\mathbb{R}^3$, (b_1, b_2) ranges over a compact subset of $(0, \infty)^2$, and η ranges over V, is P_{θ_0,η_0}-Donsker with square-integrable envelope.

25.11.3   Partially Linear Regression

Suppose that we observe a random sample from the distribution of $X = (V, W, Y)$, in which for some unobservable error e independent of (V, W),

$$Y = \theta V + \eta(W) + e.$$

Thus, the independent variable Y is a regression on (V, W) that is linear in V with slope θ but may depend on W in a nonlinear way. We assume that V and W take their values in the unit interval $[0, 1]$, and that η is twice differentiable with $J(\eta) < \infty$, for

$$J^2(\eta) = \int_0^1 \eta''(w)^2\, dw.$$

This smoothness assumption should help to ensure existence of efficient estimators of θ and will be used to define an estimator.

If the (unobservable) error is assumed to be normal, then the density of the observation $X = (V, W, Y)$ is given by

$$p_{\theta,\eta}(x) = \frac{1}{\sigma\sqrt{2\pi}} e^{-\frac{1}{2}(y - \theta v - \eta(w))^2/\sigma^2}\, p_{V,W}(v, w).$$

[†] For a proof see [106].

We cannot use this directly to define a maximum likelihood estimator for (θ, η), as a maximizer for η will interpolate the data exactly: A choice of η such that $\eta(w_i) = y_i - \theta v_i$ for every i maximizes $\prod p_{\theta,\eta}(x_i)$ but does not provide a useful estimator. The problem is that so far η has only been restricted to be differentiable, and this does not prevent it from being very wiggly. To remedy this we use a penalized log likelihood estimator, defined as the minimizer of

$$(\theta, \eta) \mapsto \mathbb{P}_n\big(y - \theta v - \eta(w)\big)^2 + \hat{\lambda}_n^2 J^2(\eta).$$

Here $\hat{\lambda}_n$ is a "smoothing parameter" that may depend on the data, and determines the weight of the "penalty" $J^2(\eta)$. A large value of $\hat{\lambda}_n$ gives much influence to the penalty term and hence leads to a smooth estimate of η, and conversely. Intermediate values are best. For the purpose of estimating θ we may use any values in the range

$$\hat{\lambda}_n^2 = o_P(n^{-1/2}), \qquad \hat{\lambda}_n^{-1} = O_P(n^{2/5}).$$

There are simple numerical schemes to compute the maximizer $(\hat{\theta}_n, \hat{\eta}_n)$, the function $\hat{\eta}_n$ being a natural cubic spline with knots at the values $w_1, \ldots, w_n$. The sequence $\hat{\theta}_n$ can be shown to be asymptotically efficient provided that the regression components involving V and W are not confounded or degenerate. More precisely, we assume that the conditional distribution of V given W is nondegenerate, that the distribution of W has at least two support points, and that $h_0(w) = \mathrm{E}(V \mid W = w)$ has a version with $J(h_0) < \infty$. Then, we have the following lemma on the behavior of $(\hat{\theta}_n, \hat{\eta}_n)$.

Let $\|\cdot\|_W$ denote the norm of $L_2(P_W)$.

25.88 Lemma. *Under the conditions listed previously, the sequence $\hat{\theta}_n$ is consistent for θ_0, $\|\hat{\eta}_n\|_\infty = O_P(1)$, $J(\hat{\eta}_n) = O_P(1)$, and $\|\hat{\eta}_n - \eta\|_W = O_P(\hat{\lambda}_n)$, under (θ_0, η_0).*

Proof. Write $g(v, w) = \theta v + \eta(w)$, let $\mathbb{P}_n$ and P_0 be the empirical and true distribution of the variables (e_i, V_i, W_i), and define functions

$$m_{g,\lambda}(e, v, w) = \big(y - g(v, w)\big)^2 + \lambda^2 \big(J^2(\eta) - J^2(\eta_0)\big).$$

Then $\hat{g}(v, w) = \hat{\theta} v + \hat{\eta}(w)$ minimizes $g \mapsto \mathbb{P}_n m_{g,\hat{\lambda}}$, and

$$m_{g,\lambda} - m_{g_0,\lambda} = 2e(g_0 - g) + (g_0 - g)^2 + \lambda^2 J^2(\eta) - \lambda^2 J^2(\eta_0).$$

By the orthogonality property of a conditional expectation and the Cauchy-Schwarz inequality, $(\mathrm{E}V\eta(W))^2 \leq \mathrm{E}\mathrm{E}(V \mid W)^2 \mathrm{E}\eta^2(W) < \mathrm{E}V^2\|\eta\|_W^2$. Therefore, by Lemma 25.86,

$$P_0(g - g_0)^2 \gtrsim |\theta - \theta_0|^2 + \|\eta - \eta_0\|_W^2.$$

Consequently, because $P_0 e = 0$ and e is independent of (V, W),

$$P_0(m_{g,\lambda} - m_{g_0,\lambda}) \gtrsim |\theta - \theta_0|^2 + \|\eta - \eta_0\|_W^2 + \lambda^2 J^2(\eta) - \lambda^2.$$

This suggests to apply Theorem 25.81 with $\tau = (\theta, \eta)$ and $d_\lambda^2(\tau, \tau_0)$ equal to the sum of the first three terms on the right.

Because $\hat{\lambda}_n^{-1} = O_P(1/\lambda_n)$ for $\lambda_n = n^{-2/5}$, it is not a real loss of generality to assume that $\hat{\lambda}_n \in \Lambda_n = [\lambda_n, \infty)$. Then $d_\lambda(\tau, \tau_0) < \delta$ and $\lambda \in \Lambda_n$ implies that $|\theta - \theta_0| < \delta$, that $\|\eta - \eta_0\|_W < \delta$ and that $J(\eta) \leq \delta/\lambda_n$. Assume first that it is known already that $|\hat{\theta}|$ and

$\|\hat{\eta}\|_\infty$ are bounded in probability, so that it is not a real loss of generality to assume that $|\hat{\theta}| \vee \|\hat{\eta}\|_\infty \leq 1$. Then

$$P_0\big(e^{|e\eta|} - 1 - |e\eta|\big) = \sum_{m \geq 2} P_0 \frac{|e\eta|^m}{m!} \leq P_0 \eta^2 \mathrm{E} e^{|e|}.$$

Thus a bound on the $\|\cdot\|_W$-norm of η yields a bound on the "Bernstein norm" of $e\eta$ (given on the left) of proportional magnitude. A bracket $[\eta_1, \eta_2]$ for η induces a bracket $[e^+\eta_1 - e^-\eta_2, e^+\eta_2 - e^-\eta_1]$ for the functions $e\eta$. In view of Lemma 19.37 and Example 19.10, we obtain

$$\mathrm{E} \sup_{d_\lambda(\tau, \tau_0) < \delta} \big|\mathbb{G}_n e(\eta - \eta_0)\big| \lesssim \phi_n(\delta) := J_n(\delta)\bigg(1 + \frac{J_n(\delta)}{\delta^2 \sqrt{n}}\bigg),$$

for

$$J_n(\delta) = \int_0^\delta \sqrt{\bigg(\frac{1 + \delta/\lambda_n}{\varepsilon}\bigg)^{1/2}}\, d\varepsilon \lesssim \delta^{3/4} + \frac{\delta}{\lambda_n^{1/4}}.$$

This bound remains valid if we replace $\eta - \eta_0$ by $g - g_0$, for the parametric part θv adds little to the entropy. We can obtain a similar maximal inequality for the process $\mathbb{G}_n(g - g_0)^2$, in view of the inequality $P_0(g - g_0)^4 \leq 4P_0(g - g_0)^2$, still under our assumption that $|\theta| \vee \|\eta\|_\infty \leq 1$. We conclude that Theorem 25.81 applies and yields the rate of convergence $|\hat{\theta} - \theta_0| + \|\hat{\eta} - \eta_0\|_W = O_P(n^{-2/5} + \hat{\lambda}_n) = O_P(\hat{\lambda}_n)$.

Finally, we must prove that $\hat{\theta}$ and $\|\hat{\eta}\|_\infty$ are bounded in probability. By the Cauchy-Schwarz inequality, for every w and η,

$$\big|\eta(w) - \eta(0) - \eta'(0)w\big| \leq \int_0^w \int_0^u |\eta''|(s)\, ds\, du \leq J(\eta).$$

This implies that $\|\eta\|_\infty \leq |\eta(0)| + |\eta'(0)| + J(\eta)$, whence it suffices to show that $\hat{\theta}, \hat{\eta}(0)$, $\hat{\eta}'(0)$, and $J(\hat{\eta})$ remain bounded. The preceding display implies that

$$\big|\theta v + \eta(0) + \eta'(0)w\big| \leq \big|g(v, w)\big| + J(\eta).$$

The empirical measure applied to the square of the left side is equal to $a^T A_n a$ for $a = \big(\theta, \eta(0), \eta'(0)\big)$ and $A_n = \mathbb{P}_n(v, 1, w)(v, 1, w)^T$ the sample second moment matrix of the variables $(V_i, 1, W_i)$. By the conditions on the distribution of (V, W), the corresponding population matrix is positive-definite, whence we can conclude that $\hat{a}$ is bounded in probability as soon as $\hat{a}^T A_n \hat{a}$ is bounded in probability, which is certainly the case if $\mathbb{P}_n \hat{g}^2$ and $J(\hat{\eta})$ are bounded in probability.

We can prove the latter by applying the preceding argument conditionally, given the sequence $V_1, W_1, V_2, W_2, \ldots$. Given these variables, the variables e_i are the only random part in $m_{g,\lambda} - m_{g_0,\lambda}$ and the parts $(g - g_0)^2$ only contribute to the centering function. We apply Theorem 25.81 with square distance equal to

$$d_\lambda^2(\tau, \tau_0) = \mathbb{P}_n(g - g_0)^2 + \lambda^2 J^2(\eta).$$

An appropriate maximal inequality can be derived from, for example Corollary 2.2.8 in [146], because the stochastic process $\mathbb{G}_n eg$ is sub-Gaussian relative to the $L_2(\mathbb{P}_n)$-metric on the set of g. Because $d_\lambda(\tau, \tau_0) < \delta$ implies that $\mathbb{P}_n(g - g_0)^2 < \delta^2$, $J(\eta) \leq \delta/\lambda_n$, and $|\theta|^2 \vee \|\eta\|_\infty^2 \leq C\big(\mathbb{P}_n(g - g_0)^2 + J^2(\eta)\big)$ for C dependent on the smallest eigenvalue of the second moment matrix A_n, the maximal inequality has a similar form as before, and we conclude that $\mathbb{P}_n(\hat{g} - g_0)^2 + \hat{\lambda}^2 J^2(\hat{\eta}) = O_P(\hat{\lambda}^2)$. This implies the desired result. ∎

The normality of the error e motivates the least squares criterion and is essential for the efficiency of $\hat{\theta}$. However, the penalized least-squares method makes sense also for nonnormal error distributions. The preceding lemma remains true under the more general condition of exponentially small error tails: $Ee^{c|e|} < \infty$ for some $c > 0$.

Under the normality assumption (with $\sigma = 1$ for simplicity) the score function for θ is given by

$$\dot{\ell}_{\theta,\eta}(x) = (y - \theta v - \eta(w))v.$$

Given a function h with $J(h) < \infty$, the path $\eta_t = \eta + th$ defines a submodel indexed by the nuisance parameter. This leads to the nuisance score function

$$B_{\theta,\eta}h(x) = (y - \theta v - \eta(w))h(w).$$

On comparing these expressions, we see that finding the projection of $\dot{\ell}_{\theta,\eta}$ onto the set of η-scores is a weighted least squares problem. By the independence of e and (V, W), it follows easily that the projection is equal to $B_{\theta,\eta}h_0$ for $h_0(w) = E(V \mid W = w)$, whence the efficient score function for θ is given by

$$\tilde{\ell}_{\theta,\eta}(x) = (y - \theta v - \eta(w))(v - h_0(w)).$$

Therefore, an exact least-favorable path is given by $\eta_t(\theta, \eta) = \eta - th_0$.

Because $(\hat{\theta}_n, \hat{\eta}_n)$ maximizes a penalized likelihood rather than an ordinary likelihood, it certainly does not satisfy the efficient score equation as considered in section 25.8. However, it satisfies this equation up to a term involving the penalty. Inserting $(\hat{\theta} + t, \eta_t(\hat{\theta}, \hat{\eta}))$ into the least-squares criterion, and differentiating at $t = 0$, we obtain the stationary equation

$$\mathbb{P}_n \tilde{\ell}_{\hat{\theta},\hat{\eta}} - 2\hat{\lambda}^2 \int_0^1 \hat{\eta}''(w)h_0''(w)\, dw = 0.$$

The second term is the derivative of $\hat{\lambda}^2 J^2(\eta_t(\hat{\theta}, \hat{\eta}))$ at $t = 0$. By the Cauchy-Schwarz inequality, it is bounded in absolute value by $2\hat{\lambda}^2 J(\hat{\eta})J(h_0) = o_P(n^{-1/2})$, by the first assumption on $\hat{\lambda}$ and because $J(\hat{\eta}) = O_P(1)$ by Lemma 25.88. We conclude that $(\hat{\theta}_n, \hat{\eta}_n)$ satisfies the efficient score equation up to a $o_P(n^{-1/2})$-term. Within the context of Theorem 25.54 a remainder term of this small order is negligible, and we may use the theorem to obtain the asymptotic normality of $\hat{\theta}_n$.

A formulation that also allows other estimators $\hat{\eta}$ is as follows.

25.89   Theorem. *Let $\hat{\eta}_n$ be any estimators such that $\|\hat{\eta}_n\|_\infty = O_P(1)$ and $J(\hat{\eta}_n) = O_P(1)$. Then any consistent sequence of estimators $\hat{\theta}_n$ such that $\sqrt{n}\ \mathbb{P}_n \tilde{\ell}_{\hat{\theta},\hat{\eta}} = o_P(1)$ is asymptotically efficient at (θ_0, η_0).*

Proof.   It suffices to check the conditions of Theorem 25.54. Since

$$P_{\theta,\eta}\tilde{\ell}_{\theta,\hat{\eta}} = P_{\theta,\eta}(\eta(w) - \hat{\eta}(w))(v - h_0(w)) = 0,$$

for every (θ, η), the no-bias condition (25.52) is satisfied.

That the functions $\tilde{\ell}_{\vartheta, \hat{\eta}}$ are contained in a Donsker class, with probability tending to 1, follows from Example 19.10 and Theorem 19.5.

The remaining regularity conditions of Theorem 25.54 can be seen to be satisfied by standard arguments. ∎

In this example we use the smoothness of η to define a penalized likelihood estimator for θ. This automatically yields a rate of convergence of $n^{-2/5}$ for $\hat{\eta}$. However, efficient estimators for θ exist under weaker smoothness assumptions on η, and the minimal smoothness of η can be traded against smoothness of the function $g(w) = E(V|W = w)$, which also appears in the formula for the efficient score function and is unknown in practice. The trade-off is a consequence of the bias $P_{\theta, \eta, g} \tilde{\ell}_{\theta, \hat{\eta}, \hat{g}}$ being equal to the cross product of the biases in $\hat{\eta}$ and $\hat{g}$. The square terms in the second order expansion (25.60), in which the derivative relative to (η, g) (instead of η) is a (2×2)-matrix, vanish. See [35] for a detailed study of this model.

25.12 Likelihood Equations

The "method of the efficient score equation" isolates the parameter θ of interest and characterizes an estimator $\hat{\theta}$ as the solution of a system of estimating equations. In this system the nuisance parameter has been replaced by an estimator $\hat{\eta}$. If the estimator $\hat{\eta}$ is the maximum likelihood estimator, then we may hope that a solution $\hat{\theta}$ of the efficient score equation is also the maximum likelihood estimator for θ, or that this is approximately true.

Another approach to proving the asymptotic normality of maximum likelihood estimators is to design a system of likelihood equations for the parameter of interest and the nuisance parameter jointly. For a semiparametric model, this necessarily is a system of infinitely many equations.

Such a system can be analyzed much in the same way as a finite-dimensional system. The system is linearized in the estimators by a Taylor expansion around the true parameter, and the limit distribution involves the inverse of the derivative applied to the system of equations. However, in most situations an ordinary pointwise Taylor expansion, the classical argument as employed in the introduction of section 5.3, is impossible, and the argument must involve some advanced tools, in particular empirical processes. A general scheme is given in Theorem 19.26, which is repeated in a different notation here. A limitation of this approach is that both $\hat{\theta}$ and $\hat{\eta}$ must converge at $\sqrt{n}$-rate. It is not clear that a model can always appropriately parametrized such that this is the case; it is certainly not always the case for the natural parametrization.

The system of estimating equations that we are looking for consists of stationary equations resulting from varying either the parameter θ or the nuisance parameter η. Suppose that our maximum likelihood estimator $(\hat{\theta}, \hat{\eta})$ maximizes the function

$$(\theta, \eta) \mapsto \prod \text{lik}(\theta, \eta)(X_i),$$

for $\text{lik}(\theta, \eta)(x)$ being the "likelihood" given one observation x.

The parameter θ can be varied in the usual way, and the resulting stationary equation takes the form

$$\mathbb{P}_n \dot{\ell}_{\hat{\theta}, \hat{\eta}} = 0.$$

This is the usual maximum likelihood equation, except that we evaluate the score function at the joint estimator $(\hat{\theta}, \hat{\eta})$, rather than at the single value $\hat{\theta}$. A precise condition for this equation to be valid is that the partial derivative of $\log \mathrm{lik}(\theta, \eta)(x)$ with respect to θ exists and is equal to $\dot{\ell}_{\theta, \eta}(x)$, for every x, (at least for $\eta = \hat{\eta}$ and at $\theta = \hat{\theta}$).

Varying the nuisance parameter η is conceptually more difficult. Typically, we can use a selection of the submodels $t \mapsto \eta_t$ used for defining the tangent set and the information in the model. If scores for η take the form of an "operator" $B_{\theta, \eta}$ working on a set of indices h, then a typical likelihood equation takes the form

$$\mathbb{P}_n B_{\hat{\theta}, \hat{\eta}} h = P_{\hat{\theta}, \hat{\eta}} B_{\hat{\theta}, \hat{\eta}} h.$$

Here we have made it explicit in our notation that a score function always has mean zero, by writing the score function as $x \mapsto B_{\theta, \eta} h(x) - P_{\theta, \eta} B_{\theta, \eta} h$ rather than as $x \mapsto B_{\theta, \eta} h(x)$. The preceding display is valid if, for every (θ, η), there exists some path $t \mapsto \eta_t(\theta, \eta)$ such that $\eta_0(\theta, \eta) = \eta$ and, for every x,

$$B_{\theta, \eta} h(x) - P_{\theta, \eta} B_{\theta, \eta} h = \frac{\partial}{\partial t}_{|t=0} \log \mathrm{lik}\big(\theta + t, \eta_t(\theta, \eta)\big).$$

Assume that this is the case for every h in some index set $\mathcal{H}$, and suppose that the latter is chosen in such a way that the map $h \mapsto B_{\theta, \eta} h(x) - P_{\theta, \eta} B_{\theta, \eta} h$ is uniformly bounded on $\mathcal{H}$, for every x and every (θ, η).

Then we can define random maps $\Psi_n : \mathbb{R}^k \times H \mapsto \mathbb{R}^k \times \ell^\infty(\mathcal{H})$ by $\Psi_n = (\Psi_{n1}, \Psi_{n2})$ with

$$\Psi_{n1}(\theta, \eta) = \mathbb{P}_n \dot{\ell}_{\theta, \eta},$$
$$\Psi_{n2}(\theta, \eta) h = \mathbb{P}_n B_{\theta, \eta} h - P_{\theta, \eta} B_{\theta, \eta} h, \qquad h \in \mathcal{H}.$$

The expectation of these maps under the parameter (θ_0, η_0) is the deterministic map $\Psi = (\Psi_1, \Psi_2)$ given by

$$\Psi_1(\theta, \eta) = P_{\theta_0, \eta_0} \dot{\ell}_{\theta, \eta},$$
$$\Psi_2(\theta, \eta) h = P_{\theta_0, \eta_0} B_{\theta, \eta} h - P_{\theta, \eta} B_{\theta, \eta} h, \qquad h \in \mathcal{H}.$$

By construction, the maximum likelihood estimators $(\hat{\theta}_n, \hat{\eta}_n)$ and the "true" parameter (θ_0, η_0) are zeros of these maps,

$$\Psi_n(\hat{\theta}_n, \hat{\eta}_n) = 0 = \Psi(\theta_0, \eta_0).$$

The argument next proceeds by linearizing these equations. Assume that the parameter set H for η can be identified with a subset of a Banach space. Then an adaptation of Theorem 19.26 is as follows.

25.90   Theorem. *Suppose that the functions $\dot{\ell}_{\theta, \eta}$ and $B_{\theta, \eta} h$, if h ranges over $\mathcal{H}$ and (θ, η) over a neighborhood of (θ_0, η_0), are contained in a P_{θ_0, η_0}-Donsker class, and that*

$$P_{\theta_0, \eta_0} \big\| \dot{\ell}_{\hat{\theta}, \hat{\eta}} - \dot{\ell}_{\theta_0, \eta_0} \big\|^2 \xrightarrow{\mathrm{P}} 0, \qquad \sup_{h \in \mathcal{H}} P_{\theta_0, \eta_0} \big| B_{\hat{\theta}, \hat{\eta}} h - B_{\theta_0, \eta_0} h \big|^2 \xrightarrow{\mathrm{P}} 0.$$

Furthermore, suppose that the map $\Psi : \Theta \times H \mapsto \mathbb{R}^k \times \ell^\infty(\mathcal{H})$ is Fréchet-differentiable at (θ_0, η_0), with a derivative $\dot{\Psi}_0 : \mathbb{R}^k \times \mathrm{lin}\, H \mapsto \mathbb{R}^k \times \ell^\infty(\mathcal{H})$ that has a continuous inverse *on its range. If the sequence $(\hat{\theta}_n, \hat{\eta}_n)$ is consistent for (θ_0, η_0) and satisfies $\Psi_n(\hat{\theta}_n, \hat{\eta}_n) = o_P(n^{-1/2})$, then*

$$\dot{\Psi}_0 \sqrt{n}(\hat{\theta}_n - \theta_0, \hat{\eta}_n - \eta_0) = -\sqrt{n}\Psi_n(\theta_0, \eta_0) + o_P(1).$$

The theorem gives the joint asymptotic distribution of $\hat{\theta}_n$ and $\hat{\eta}_n$. Because $\sqrt{n}\Psi_n(\theta_0, \eta_0)$ is the empirical process indexed by the Donsker class consisting of the functions $\dot{\ell}_{\theta_0, \eta_0}$ and $B_{\theta_0, \eta_0}h$, this process is asymptotically normally distributed. Because normality is retained under a continuous, linear map, such as $\dot{\Psi}_0^{-1}$, the limit distribution of the sequence $\sqrt{n}(\hat{\theta}_n - \theta_0, \hat{\eta}_n - \eta_0)$ is Gaussian as well.

The case of a partitioned parameter (θ, η) is an interesting one and illustrates most aspects of the application of the preceding theorem. Therefore, we continue to write the formulas in the corresponding partitioned form. However, the preceding theorem, applies more generally. In Example 25.5.1 we wrote the score operator for a semiparametric model in the form

$$A_{\theta, \eta}(a, b) = a^T \dot{\ell}_{\theta, \eta} + B_{\theta, \eta}b.$$

Corresponding to this, the system of likelihood equations can be written in the form

$$\mathbb{P}_n A_{\theta, \eta}(a, b) = P_{\theta, \eta}A_{\theta, \eta}(a, b), \qquad \text{every } (a, b).$$

If the partitioned parameter (θ, η) and the partitioned "directions" (a, b) are replaced by a general parameter τ and general direction c, then this formulation extends to general models. The maps Ψ_n and Ψ then take the forms

$$\Psi_n(\tau)c = \mathbb{P}_n A_\tau c - P_\tau A_\tau g, \qquad \Psi(\tau)c = P_{\tau_0} A_\tau c - P_\tau A_\tau c.$$

The theorem requires that these can be considered maps from the parameter set into a Banach space, for instance a space $\ell^\infty(C)$.

To gain more insight, consider the case that η is a measure on a measurable space $(\mathcal{Z}, \mathcal{C})$. Then the directions h can often be taken equal to bounded functions $h : \mathcal{Z} \mapsto \mathbb{R}$, corresponding to the paths $d\eta_t = (1 + th) d\eta$ if η is a completely unknown measure, or $d\eta_t = (1 + t(h - \eta h)) d\eta$ if the total mass of each η is fixed to one. In the remainder of the discussion, we assume the latter. Now the derivative map $\dot{\Psi}_0$ typically takes the form

$$(\theta - \theta_0, \eta - \eta_0) \mapsto \begin{pmatrix} \dot{\Psi}_{11} & \dot{\Psi}_{12} \\ \dot{\Psi}_{21} & \dot{\Psi}_{22} \end{pmatrix} \begin{pmatrix} \theta - \theta_0 \\ \eta - \eta_0 \end{pmatrix}$$

where

$$\dot{\Psi}_{11}(\theta - \theta_0) = -P_{\theta_0, \eta_0} \dot{\ell}_{\theta_0, \eta_0} \dot{\ell}_{\theta_0, \eta_0}^T (\theta - \theta_0),$$

$$\dot{\Psi}_{12}(\eta - \eta_0) = -\int B_{\theta_0, \eta_0}^* \dot{\ell}_{\theta_0, \eta_0} d(\eta - \eta_0),$$

$$\dot{\Psi}_{21}(\theta - \theta_0)h = -P_{\theta_0, \eta_0}(B_{\theta_0, \eta_0}h) \dot{\ell}_{\theta_0, \eta_0}^T (\theta - \theta_0),$$

$$\dot{\Psi}_{22}(\eta - \eta_0)h = -\int B_{\theta_0, \eta_0}^* B_{\theta_0, \eta_0} h \, d(\eta - \eta_0).$$

$$(25.91)$$

For instance, to find the last identity in an informal manner, consider a path η_t in the direction of g, so that $d\eta_t - d\eta_0 = tg\,d\eta_0 + o(t)$. Then by the definition of a derivative

$$\Psi_2(\theta_0, \eta_t) - \Psi_2(\theta_0, \eta_0) \approx \dot{\Psi}_{22}(\eta_t - \eta_0) + o(t).$$

On the other hand, by the definition of Ψ, for every h,

$$\begin{aligned}
\Psi_2(\theta_0, \eta_t)h - \Psi(\theta_0, \eta_0)h &= -(P_{\theta_0, \eta_t} - P_{\theta_0, \eta_0})B_{\theta_0, \eta_t}h \\
&\approx -t P_{\theta_0, \eta_0}(B_{\theta_0, \eta_0}g)(B_{\theta_0, \eta_0}h) + o(t) \\
&= -\int (B_{\theta_0, \eta_0}^* B_{\theta_0, \eta_0}h)\,tg\,d\eta_0 + o(t).
\end{aligned}$$

On comparing the preceding pair of displays, we obtain the last line of (25.91), at least for $d\eta - d\eta_0 = g\,d\eta_0$. These arguments are purely heuristic, and this form of the derivative must be established for every example. For instance, within the context of Theorem 25.90, we may need to apply $\dot{\Psi}_0$ to η that are not absolutely continuous with respect to η_0. Then the validity of (25.91) depends on the version that is used to define the adjoint operator B_{θ_0, η_0}^*. By definition, an adjoint is an operator between L_2-spaces and hence maps equivalence classes into equivalence classes.

The four partial derivatives $\dot{\Psi}_{ij}$ in (25.91) involve the four parts of the information operator $A_{\theta, \eta}^* A_{\theta, \eta}$, which was written in a partitioned form in Example 25.5.1. In particular, the map $\dot{\Psi}_{11}$ is exactly the Fisher information for θ, and the operator $\dot{\Psi}_{22}$ is defined in terms of the information operator for the nuisance parameter. This is no coincidence, because the formulas can be considered a version of the general identity "expectation of the second derivative is equal to minus the information." An abstract form of the preceding argument applied to the map $\Psi(\tau)c = P_{\tau_0}A_\tau c - P_\tau A_\tau c$ leads to the identity, with τ_t a path with derivative $\dot{\tau}_0$ at $t = 0$ and score function $A_{\tau_0}d$,

$$\dot{\Psi}_0(\dot{\tau}_0)c = \langle A_{\tau_0}^* A_{\tau_0}c, d\rangle_{\tau_0}.$$

In the case of a partitioned parameter $\tau = (\theta, \eta)$, the inner inner product on the right is defined as $\langle (a, b), (\alpha, \beta)\rangle_{\tau_0} = a^T\alpha + \int b\beta\,d\eta_0$, and the four formulas in (25.91) follow by Example 25.5.1 and some algebra. A difference with the finite-dimensional situation is that the derivatives $\dot{\tau}_0$ may not be dense in the domain of $\dot{\Psi}_0$, so that the formula determines $\dot{\Psi}_0$ only partly.

An important condition in Theorem 25.90 is the continuous invertibility of the derivative. Because a linear map between Euclidean spaces is automatically continuous, in the finite-dimensional set-up this condition reduces to the derivative being one-to-one. For infinite-dimensional systems of estimating equations, the continuity is far from automatic and may be the condition that is hardest to verify. Because it refers to the $\ell^\infty(\mathcal{H})$-norm, we have some control over it while setting up the system of estimating equations and choosing the set of functions $\mathcal{H}$. A bigger set $\mathcal{H}$ makes $\dot{\Psi}_0^{-1}$ more readily continuous but makes the differentiability of Ψ and the Donsker condition more stringent.

In the partitioned case, the continuous invertibility of $\dot{\Psi}_0$ can be verified by ascertaining the continuous invertibility of the two operators $\dot{\Psi}_{11}$ and $\dot{V} = \dot{\Psi}_{22} - \dot{\Psi}_{21}\dot{\Psi}_{11}^{-1}\dot{\Psi}_{12}$. In that case we have

$$\dot{\Psi}_0^{-1} = \begin{pmatrix} \dot{\Psi}_{11}^{-1} + \dot{\Psi}_{11}^{-1}\dot{\Psi}_{12}\dot{V}^{-1}\dot{\Psi}_{21}\dot{\Psi}_{11}^{-1} & -\dot{\Psi}_{11}^{-1}\dot{\Psi}_{12}\dot{V}^{-1} \\ -\dot{V}^{-1}\dot{\Psi}_{21}\dot{\Psi}_{11}^{-1} & \dot{V}^{-1} \end{pmatrix}$$

The operator $\dot\Psi_{11}$ is the Fisher information matrix for θ if η is known. If this would not be invertible, then there would be no hope of finding asymptotically normal estimators for θ. The operator $\dot V$ has the form

$$\dot V(\eta - \eta_0)h = -\int \left(B^*_{\theta_0,\eta_0} B_{\theta_0,\eta_0} + K\right)h \, d(\eta - \eta_0),$$

where the operator K is defined as

$$K h = -\left(P_{\theta_0,\eta_0}\left(B_{\theta_0,\eta_0}h\right)\dot\ell^T_{\theta_0,\eta_0}\right)I^{-1}_{\theta_0,\eta_0} B^*_{\theta_0,\eta_0}\dot\ell_{\theta_0,\eta_0}.$$

The operator $\dot V : \lim H \mapsto \ell^\infty(\mathcal H)$ is certainly continuously invertible if there exists a positive number ϵ such that

$$\sup_{h\in\mathcal H}\left|\dot V(\eta - \eta_0)h\right| \ge \varepsilon \parallel \eta - \eta_0 \parallel .$$

In the case that η is identified with the map $h \mapsto \eta h$ in $\ell^\infty(\mathcal H)$, the norm on the right is given by $\sup_{h\in\mathcal H}|(\eta - \eta_0)h|$. Then the display is certainly satisfied if, for some $\varepsilon > 0$,

$$\left\{\left(B^*_{\theta_0,\eta_0} B_{\theta_0,\eta_0} + K\right)h : h \in \mathcal H\right\} \supset \epsilon\mathcal H.$$

This condition has a nice interpretation if $\mathcal H$ is equal to the unit ball of a Banach space $\mathbb B$ of functions. Then the preceding display is equivalent to the operator $B^*_{\theta_0,\eta_0} B_{\theta_0,\eta_0} + K : \mathbb B \mapsto \mathbb B$ being continuously invertible. The first part of this operator is the information operator for the nuisance parameter. Typically, this is continuously invertible if the nuisance parameter is regularly estimable at a $\sqrt n$-rate (relatively to the norm used) if θ is known. The following lemma guarantees that the same is then true for the operator $B^*_{\theta_0,\eta_0} B_{\theta_0,\eta_0} + K$ if the efficient information matrix for θ is nonsingular, that is, the parameters θ and η are not locally confounded.

25.92 Lemma. *Let $\mathbb B$ be a Banach space contained in $\ell^\infty(\mathcal Z)$. If $\tilde I_{\theta_0,\eta_0}$ is nonsingular $B^*_{\theta_0,\eta_0} B_{\theta_0,\eta_0} : \mathbb B \mapsto \mathbb B$ is onto and continuously invertible and $B^*_{\theta_0,\eta_0}\dot\ell_{\theta_0,\eta_0} \in \mathbb B$, then $B^*_{\theta_0,\eta_0} B_{\theta_0,\eta_0} + K : \mathbb B \mapsto \mathbb B$ is onto and continuously invertible.*

Proof. Abbreviate the index (θ_0, η_0) to 0. The operator K is compact, because it has a finite-dimensional range. Therefore, by Lemma 25.93 below, the operator $B^*_0 B_0 + K$ is continuously invertible provided that it is one-to-one.

Suppose that $(B^*_0 B_0 + K)h = 0$ for some $h \in \mathbb B$. By assumption there exists a path $t \mapsto \eta_t$ with score function $\overline B_0 h = B_0 h - P_0 B_0 h$ at $t = 0$. Then the submodel indexed by $t \mapsto (\theta_0 + ta_0, \eta_t)$, for $a_0 = -I^{-1}_0 P_0(B_0 h)\dot\ell_0$, has score function $a^T_0\dot\ell_0 + \overline B_0 h$ at $t = 0$, and information

$$a^T_0 I_0 a_0 + P_0(\overline B_0 h)^2 + 2a^T_0 P_0\dot\ell_0(B_0 h) = P_0(\overline B_0 h)^2 - a^T_0 I_0 a_0.$$

Because the efficient information matrix is nonsingular, this information must be strictly positive, unless $a_0 = 0$. On the other hand,

$$0 = \eta_0 h(B^*_0 B_0 + K)h = P_0(B_0 h)^2 + a^T_0 P_0(B_0 h)\dot\ell_0.$$

This expression is at least the right side of the preceding display and is positive if $a_0 \neq 0$. Thus $a_0 = 0$, whence $Kh = 0$. Reinserting this in the equation $(B_0^* B_0 + K)h = 0$, we find that $B_0^* B_0 h = 0$ and hence $h = 0$. ∎

The proof of the preceding lemma is based on the Fredholm theory of linear operators. An operator $K : \mathbb{B} \mapsto \mathbb{B}$ is *compact* if it maps the unit ball into a totally bounded set. The following lemma shows that for certain operators continuous invertibility is a consequence of their being one-to-one, as is true for matrix operators on Euclidean space.[†] It is also useful to prove the invertibility of the information operator itself.

25.93   Lemma. *Let $\mathbb{B}$ be a Banach space, let the operator $A : \mathbb{B} \mapsto \mathbb{B}$ be continuous, onto and continuously invertible and let $K : \mathbb{B} \mapsto \mathbb{B}$ be a compact operator. Then $\mathrm{R}(A + K)$ is closed and has codimension equal to the dimension of $\mathrm{N}(A + K)$. In particular, if $A + K$ is one-to-one, then $A + K$ is onto and continuously invertible.*

The asymptotic covariance matrix of the sequence $\sqrt{n}(\hat{\theta}_n - \theta_0)$ can be computed from the expression for $\dot{\Psi}_0$ and the covariance function of the limiting process of the sequence $\sqrt{n}\Psi_n(\theta_0, \eta_0)$. However, it is easier to use an asymptotic representation of $\sqrt{n}(\hat{\theta}_n - \theta_0)$ as a sum. For a continuously invertible information operator $B_{\theta_0, \eta_0}^* B_{\theta_0, \eta_0}$ this can be obtained as follows.

In view of (25.91), the assertion of Theorem 25.90 can be rewritten as the system of equations, with a subscript 0 denoting (θ_0, η_0),

$$-I_0(\hat{\theta}_n - \theta_0) - (\hat{\eta}_n - \eta_0)B_0^* \dot{\ell}_0 = -(\mathbb{P}_n - P_0)\dot{\ell}_0 + o_P(1/\sqrt{n}),$$

$$-P_0(B_0 h)\dot{\ell}_0^T(\hat{\theta}_n - \theta_0) - (\hat{\eta}_n - \eta_0)B_0^* B_0 h = -(\mathbb{P}_n - P_0)B_0 h + o_P(1/\sqrt{n}).$$

The $o_P(1/\sqrt{n})$-term in the second line is valid for every $h \in \mathcal{H}$ (uniformly in h). If we can also choose $h = (B_0^* B_0)^{-1} B_0^* \dot{\ell}_0$, and subtract the first equation from the second, then we arrive at

$$\tilde{I}_{\theta_0, \eta_0} \sqrt{n}(\hat{\theta}_n - \theta_0) = \sqrt{n}(\mathbb{P}_n - P_0)\tilde{\ell}_{\theta_0, \eta_0} + o_P(1).$$

Here $\tilde{\ell}_{\theta_0, \eta_0}$ is the efficient score function for θ, as given by (25.33), and $\tilde{I}_{\theta_0, \eta_0}$ is the efficient information matrix. The representation shows that the sequence $\sqrt{n}(\hat{\theta}_n - \theta_0)$ is asymptotically linear in the efficient influence function for estimating θ. Hence the maximum likelihood estimator $\hat{\theta}$ is asymptotically efficient.[‡] The asymptotic efficiency of the estimator $\hat{\eta}h$ for ηh follows similarly.

We finish this section with a number of examples. For each example we describe the general structure and main points of the verification of the conditions of Theorem 25.90, but we refer to the original papers for some of the details.

25.12.1   *Cox Model*

Suppose that we observe a random sample from the distribution of the variable $X = (T \wedge C, 1\{T \leq C\}, Z)$, where, given Z, the variables T and C are independent, as in the

---

[†] For a proof see, for example, [132, pp. 99–103].

[‡] This conclusion also can be reached from general results on the asymptotic efficiency of the maximum likelihood estimator. See [56] and [143].

random censoring model, and T follows the Cox model. Thus, the density of $X = (Y, \Delta, Z)$ is given by

$$\left(e^{\theta z}\lambda(y)e^{-e^{\theta z}\Lambda(y)}\left(1 - F_{C|Z}(y - |z)\right)\right)^{\delta}\left(e^{-e^{\theta z}\Lambda(y)}f_{C|Z}(y|z)\right)^{1-\delta}p_z(z).$$

We define a likelihood for the parameters (θ, Λ) by dropping the factors involving the distribution of (C, Z), and replacing $\lambda(y)$ by the pointmass $\Lambda\{y\}$,

$$\text{lik}(\theta, \Lambda)(x) = \left(e^{\theta z}\Lambda\{y\}e^{-e^{\theta z}\Lambda(y)}\right)^{\delta}(e^{-e^{\theta z}\Lambda(y)})^{1-\delta}.$$

This likelihood is convenient in that the profile likelihood function for θ can be computed explicitly, exactly as in Example 25.69. Next, given the maximizer $\hat{\theta}$, which must be calculated numerically, the maximum likelihood estimator $\hat{\Lambda}$ is given by an explicit formula.

Given the general results put into place so far, proving the consistency of $(\hat{\theta}, \hat{\Lambda})$ is the hardest problem. The methods of section 5.2 do not apply directly, because of the empirical factor $\Lambda\{y\}$ in the likelihood. These methods can be adapted. Alternatively, the consistency can be proved using the explicit form of the profile likelihood function. We omit a discussion.

For simplicity we make a number of partly unnecessary assumptions. First, we assume that the covariate Z is bounded, and that the true conditional distribution of T given Z possesses a continuous Lebesgue density. Second, we assume that there exists a finite number $\tau > 0$ such that $P(C \geq \tau) = P(C = \tau) > 0$ and $P_{\theta_0, \Lambda_0}(T > \tau) > O$. The latter condition is not unnatural: It is satisfied if the survival study is stopped at some time τ at which a positive fraction of individuals is still "at risk" (alive). Third, we assume that, for any measurable function h, the probability that $Z \neq h(Y)$ is positive. The function Λ now matters only on $[0, \tau]$; we shall identify Λ with its restriction to this interval. Under these conditions the maximum likelihood estimator $(\hat{\theta}, \hat{\Lambda})$ can be shown to be consistent for the product of the Euclidean topology and the topology of uniform convergence on $[0, \tau]$.

The score function for θ takes the form

$$\dot{\ell}_{\theta, \Lambda}(x) = \delta_z - ze^{\theta z}\Lambda(y).$$

For any bounded, measurable function $h: [0, \tau] \mapsto \mathbb{R}$, the path defined by $d\Lambda_t = (1 + th)d\Lambda$ defines a submodel passing through Λ at $t = 0$. Its score function at $t = 0$ takes the form

$$B_{\theta, \Lambda}h(x) = \delta h(y) - e^{\theta z}\int_{[0,y]}hd\Lambda.$$

The function $h \mapsto B_{\theta, \Lambda}h(x)$ is bounded on every set of uniformly bounded functions h, for any finite measure Λ, and is even uniformly bounded in x and in (θ, Λ) ranging over a neighborhood of $(\theta_0, \Lambda 0)$.

It is not difficult to find a formula for the adjoint $B^*_{\theta, \Lambda}$ of $B_{\theta, \Lambda}: L_2(\Lambda) \mapsto L_2(P_{\theta, \Lambda})$, but this is tedious and not insightful. The information operator $B^*_{\theta, \Lambda}B_{\theta, \Lambda}: L_2(\Lambda) \mapsto L_2(\Lambda)$ can be calculated from the identity $P_{\theta, \Lambda}(B_{\theta, \Lambda}g)(B_{\theta, \Lambda}h) = \Lambda g(B^*_{\theta, \Lambda}B_{\theta, \Lambda}h)$. For continuous Λ it takes the surprisingly simple form

$$B^*_{\theta, \Lambda}B_{\theta, \Lambda}h(y) = h(y)\text{E}_{\theta, \Lambda}1_{Y \geq y}e^{\theta Z}.$$

To see this, write the product $B_{\theta,\Lambda} g B_{\theta,\Lambda} h$ as the sum of four terms

$$\delta h(y)g(y) - \delta h(y)e^{\theta z} \int_0^y g\, d\Lambda - \delta g(y)e^{\theta z}\int_0^y h\, d\Lambda + e^{2\theta z}\int_0^y g\, d\Lambda \int_0^y h\, d\Lambda.$$

Take the expectation under $P_{\theta,\Lambda}$ and interchange the order of the integrals to represent $B^*_{\theta,\Lambda}B_{\theta,\Lambda}h$ also as a sum of four terms. Partially integrate the fourth term to see that this cancels the second and third terms. We are left with the first term. The function $B^*_{\theta,\Lambda}\dot{\ell}_{\theta,\Lambda}$ can be obtained by a similar argument, starting from the identity $P_{\theta,\Lambda}\dot{\ell}_{\theta,\Lambda}B_{\theta,\Lambda}h = \Lambda(B^*_{\theta,\Lambda}\dot{\ell}_{\theta,\Lambda})h$. It is given by

$$B^*_{\theta,\Lambda}\dot{\ell}_{\theta,\Lambda} = E_{\theta,\Lambda} 1_{Y\geq y} Z e^{\theta Z}.$$

The calculation of the information operator in this way is instructive, but only to check (25.91) for this example. As in other examples a direct derivation of the derivative of the map $\Psi = (\Psi_1, \Psi_2)$ given by $\Psi_1(\theta, \Lambda) = P_0\dot{\ell}_{\theta,\Lambda}$ and $\Psi_2(\theta, \Lambda)h = P_0 B_{\theta,\Lambda}h$ requires less work. In the present case this is almost trivial, for the map Ψ is already linear in Λ. Writing $G_0(y\mid Z)$ for the distribution function of Y given Z, this map can be written as

$$\Psi_1(\theta, \Lambda) = EZe^{\theta_0 Z}\int \bar{G}_0(y\mid Z)\, d\Lambda_0(y) - EZe^{\theta Z}\int \Lambda(y)\, dG_0(y\mid Z),$$

$$\Psi_2(\theta, \Lambda)h = Ee^{\theta_0 Z}\int h(y)\bar{G}_0(y\mid Z)\, d\Lambda_0(y) - Ee^{\theta Z}\iint_{[0,y]} h\, d\Lambda\, dG_0(y\mid Z).$$

If we take $\mathcal{H}$ equal to the unit ball of the space $BV[0, \tau]$ of bounded functions of bounded variation, then the map $\Psi : \mathbb{R} \times \ell^\infty(\mathcal{H}) \mapsto \mathbb{R} \times \ell^\infty(\mathcal{H})$ is linear and continuous in Λ, and its partial derivatives with respect to θ can be found by differentiation under the expectation and are continuous in a neighborhood of (θ_0, Λ_0). Several applications of Fubini's theorem show that the derivative takes the form (25.91).

We can consider $B_0^* B_0$ as an operator of the space $BV[0, \tau]$ into itself. Then it is continuously invertible if the function $y \mapsto E_{\theta_0,\Lambda_0} 1_{Y\geq y} e^{\theta_0 Z}$ is bounded away from zero on $[0, \tau]$. This we have (indirectly) assumed. Thus, we can apply Lemma 25.92. The efficient score function takes the form (25.33), which, with $M_i(y) = E_{\theta_0,\Lambda_0} 1_{Y\geq y} Z^i e^{\theta_0 Z}$, reduces to

$$\tilde{\ell}_{\theta_0,\Lambda_0}(x) = \delta\left(z - \frac{M_1}{M_0}(y)\right) - e^{\theta_0 z}\int_{[0,y]}\left(z - \frac{M_1}{M_0}(t)\right) d\Lambda_0(t).$$

The efficient information for θ can be computed from this as

$$\tilde{I}_{\theta_0,\Lambda_0} = Ee^{\theta_0 Z}\int\left(Z - \frac{M_1}{M_0}(y)\right)^2 \bar{G}_0(y\mid Z)\, d\Lambda_0(y).$$

This is strictly positive by the assumption that Z is not equal to a function of Y.

The class $\mathcal{H}$ is a universal Donsker class, and hence the first parts $\delta h(y)$ of the functions $B_{\theta,\Lambda}h$ form a Donsker class. The functions of the form $\int_{[0,y]} h\, d\Lambda$ with h ranging over $\mathcal{H}$ and Λ ranging over a collection of measures of uniformly bounded variation are functions of uniformly bounded variation and hence also belong to a Donsker class. Thus the functions $B_{\theta,\Lambda}h$ form a Donsker class by Example 19.20.

25.12.2 *Partially Missing Data*

Suppose that the observations are a random sample from a density of the form

$$(x, y, z) \mapsto \int p_\theta(x \mid s) \, d\eta(s) \, p_\theta(y \mid z) \, d\eta(z) =: p_\theta(x \mid \eta) \, p_\theta(y \mid z) \, d\eta(z).$$

Here the parameter η is a completely unknown distribution, and the kernel $p_\theta(\cdot \mid s)$ is a given parametric model indexed by the parameters θ and s, relative to some density μ. Thus, we obtain equal numbers of bad and good (direct) observations concerning η. Typically, by themselves the bad observations do not contribute positive information concerning the cumulative distribution function η, but along with the good observations they help to cut the asymptotic variance of the maximum likelihood estimators.

25.94 Example. This model can arise if we are interested in the relationship between a response Y and a covariate Z, but because of the cost of measurement we do not observe Z for a fraction of the population. For instance, a full observation $(Y, Z) = (D, W, Z)$ could consist of
– a logistic regression D on $\exp Z$ with intercept and slope β_0 and β_1, respectively, and
– a linear regression W on Z with intercept and slope α_0 and α_1, respectively, and an $N(0, \sigma^2)$-error.
Given Z the variables D and W are assumed independent, and Z has a completely unspecified distribution η on an interval in $\mathbb{R}$. The kernel is equal to, with Ψ denoting the logistic distribution function and ϕ denoting the standard normal density,

$$p_\theta(d, w \mid z) = \Psi(\beta_0 + \beta_1 e^z)^d \big(1 - \Psi(\beta_0 + \beta_1 e^z)\big)^{1-d} \frac{1}{\sigma} \phi\Big(\frac{w - \alpha_0 - \alpha_1 z}{\sigma}\Big).$$

The precise form of this density does not play a major role in the following.

In this situation the covariate Z is a gold standard, but, in view of the costs of measurement, for a selection of observations only the "surrogate covariate" W is available. For instance, Z corresponds to the LDL cholesterol and W to total cholesterol, and we are interested in heart disease $D = 1$. For simplicity, each observation in our set-up consists of one full observation $(Y, Z) = (D, W, Z)$ and one reduced observation $X = (D, W)$. □

25.95 Example. If the kernel $p_\theta(y \mid z)$ is equal to the normal density with mean z and variance θ, then the observations are a random sample $Z_1, \ldots, Z_n$ from η, a random sample $X_1, \ldots, X_n$ from η perturbed by an additive (unobserved) normal error, and a sample $Y_1, \ldots, Y_n$ of random variables that given $Z_1, \ldots, Z_n$ are normally distributed with means Z_i and variance θ. In this case the interest is perhaps focused on estimating η, rather than θ. □

The distribution of an observation (X, Y, Z) is given by two densities and a nonparametric part. We choose as likelihood

$$\mathrm{lik}(\theta, \eta)(x, y, z) = p_\theta(x \mid \eta) \, p_\theta(y \mid z) \, \eta\{z\}.$$

Thus, for the completely unknown distribution η of Z we use the empirical likelihood for the other part of the observations we use the density, as usual. It is clear that the maximum likelihood estimator $\hat{\eta}$ charges all observed values $z_1, \ldots, z_n$, but the term $p_\theta(x \mid \eta)$ leads to some additional support points as well. In general, these are not equal to values of the observations.

The score function for θ is given by

$$\dot{\ell}_{\theta, \eta}(x, y, z) = \dot{\kappa}_{\theta, \eta}(x) + \dot{\kappa}_\theta(y \mid z) = \frac{\int \dot{\kappa}_\theta(x \mid s) \, p_\theta(x \mid s) \, d\eta(s)}{p_\theta(x \mid \eta)} + \dot{\kappa}_\theta(y \mid z).$$

Here $\dot{\kappa}_\theta(y \mid z) = \partial/\partial\theta \, \log p_\theta(y \mid z)$ is the score function for θ for the conditional density $p_\theta(y \mid z)$, and $\dot{\kappa}_{\theta, \eta}(x)$ is the score function for θ of the mixture density $p_\theta(x \mid \eta)$.

Paths of the form $d\eta_t = (1 + th) \, d\eta$ (with $\eta h = 0$) yield scores

$$B_{\theta, \eta}h(x, z) = C_{\theta, \eta}h(x) + h(z) = \frac{\int h(s) p_\theta(x \mid s) \, d\eta(s)}{p_\theta(x \mid \eta)} + h(z).$$

The operator $C_{\theta, \eta} : L_2(\eta) \mapsto L_2\big(p_\theta(\cdot \mid \eta)\big)$ is the score operator for the mixture part of the model. Its Hilbert-space adjoint is given by

$$C_{\theta, \eta}^* g(z) = \int g(x) \, p_\theta(x \mid z) \, d\mu(x).$$

The range of $B_{\theta, \eta}$ is contained in the subset G of $L_2\big(p_\theta(\cdot \mid \eta) \times \eta\big)$ consisting of functions of the form $(x, z) \mapsto g_1(x) + g_2(z) + c$. This representation of a function of this type is unique if both g_1 and g_2 are taken to be mean-zero functions. With $P_{\theta, \eta}$ the distribution of the observation (X, Y, Z),

$$P_{\theta, \eta}(g_1 \oplus g_2 \oplus c) B_{\theta, \eta}h = P_{\theta, \eta}g_1 C_{\theta, \eta}h + \eta g_2 h + 2\eta hc = \eta(C_{\theta, \eta}^* g_1 + g_2 + 2c)h.$$

Thus, the adjoint $B_{\theta, \eta}^* : G \mapsto L_2(\eta)$ of the operator $B_{\theta, \eta} : L_2(\eta) \mapsto G$ is given by

$$B_{\theta, \eta}^*(g_1 \oplus g_2 \oplus c) = C_{\theta, \eta}^* g_1 + g_2 + 2c.$$

Consequently, on the set of mean-zero functions in $L_2(\eta)$ we have the identity $B_{\theta, \eta}^* B_{\theta, \eta} = C_{\theta, \eta}^* C_{\theta, \eta} + I$. Because the operator $C_{\theta, \eta}^* C_{\theta, \eta}$ is nonnegative definite, the operator $B_{\theta, \eta}^* B_{\theta, \eta}$ is strictly positive definite and hence continuously invertible as an operator of $L_2(\eta)$ into itself. The following lemma gives a condition for continuous invertibility as an operator on the space $C^\alpha(\mathcal{Z})$ of all "α-smooth functions." For $\alpha_0 \leq \alpha$ the smallest integer strictly smaller than α, these consist of the functions $h : \mathcal{Z} \subset \mathbb{R}^d \mapsto \mathbb{R}$ whose partial derivatives up to order α_0 exist and are bounded and whose α_0-order partial derivatives are Lipschitz of order $\alpha - \alpha_0$. These are Banach spaces relative to the norm, with D^k a differential operator $\partial^{k_1} \cdots \partial^{k_d} / \partial z_1^{k_1} \cdots z_k^{k_d}$,

$$\|h\|_\alpha = \max_{|k| < \alpha} \sup_{z \in \mathcal{Z}} |D^k h(z)| \vee \max_{|k| = \alpha_0} \sup_{z_1 \neq z_2 \in \mathcal{Z}} \frac{|D^k(z_1) - D^k(z_2)|}{\|z_1 - z_2\|^{\alpha - \alpha_0}}.$$

The unit ball of one of these spaces is a good choice for the set $\mathcal{H}$ indexing the likelihood equations if the maps $z \mapsto p_{\theta_0}(x \mid z)$ are sufficiently smooth.

25.96   Lemma. *Let $\mathcal{Z}$ be a bounded, convex subset of $\mathbb{R}^d$ and assume that the maps $z \mapsto p_0(x \mid z)$ are continuously differentiable for each x with partial derivatives $\partial/\partial z_i \, p_{\theta_0}(x \mid z)$*

satisfying, for all z, z' in $\mathcal{Z}$ and fixed constants K and $\alpha > 0$,

$$\int \left| \frac{\partial}{\partial z_i} p_0(x \mid z) - \frac{\partial}{\partial z_i} p_0(x \mid z') \right| d\mu(x) \leq K \|z - z'\|^\alpha,$$

$$\int \left| \frac{\partial}{\partial z_i} p_0(x \mid z) \right| d\mu(x) \leq K.$$

Then $B_{\theta_0,\eta_0}^* B_{\theta_0,\eta_0} : C^\beta(\mathcal{Z}) \mapsto C^\beta(\mathcal{Z})$ is continuously invertible for every $\beta < \alpha$.

Proof. By its strict positive-definiteness in the Hilbert-space sense, the operator $B_0^* B_0 :$ $\ell^\infty(\mathcal{Z}) \mapsto \ell^\infty(\mathcal{Z})$ is certainly one-to-one in that $B_0^* B_0 h = 0$ implies that $h = 0$ almost surely under η_0. On reinserting this we find that $-h = C_0^* C_0 h = C_0^* 0 = 0$ everywhere. Thus $B_0^* B_0$ is also one-to-one in a pointwise sense. If it can be shown that $C_0^* C_0 : C^\beta(\mathcal{Z}) \mapsto$ $C^\beta(\mathcal{Z})$ is compact, then $B_0^* B_0$ is onto and continuously invertible, by Lemma 25.93.

It follows from the Lipschitz condition on the partial derivatives that $C_0^* h(z)$ is differentiable for every bounded function $h : \mathcal{X} \mapsto \mathbb{R}$ and its partial derivatives can be found by differentiating under the integral sign:

$$\frac{\partial}{\partial z_i} C_0^* h(z) = \int h(x) \frac{\partial}{\partial z_i} p_0(x \mid z) \, d\mu(x).$$

The two conditions of the lemma imply that this function has Lipschitz norm of order α bounded by $K \|h\|_\infty$. Let h_n be a uniformly bounded sequence in $\ell^\infty(\mathcal{X})$. Then the partial derivatives of the sequence $C_0^* h_n$ are uniformly bounded and have uniformly bounded Lipschitz norms of order α. Because $\mathcal{Z}$ is totally bounded, it follows by a strengthening of the Arzela-Ascoli theorem that the sequences of partial derivatives are precompact with respect to the Lipschitz norm of order β for every $\beta < \alpha$. Thus there exists a subsequence along which the partial derivatives converge in the Lipschitz norm of order β. By the Arzela-Ascoli theorem there exists a further subsequence such that the functions $C_0^* h_n(z)$ converge uniformly to a limit. If both a sequence of functions itself and their continuous partial derivatives converge uniformly to limits, then the limit of the functions must have the limits of the sequences of partial derivatives as its partial derivatives. We conclude that $C_0^* h_n$ converges in the $\|\cdot\|_{1+\beta}$-norm, whence $C_0^* : \ell^\infty(\mathcal{X}) \mapsto C^\beta(\mathcal{Z})$ is compact. Then the operator $C_0^* C_0$ is certainly compact as an operator from $C^\beta(\mathcal{Z})$ into itself. ∎

Because the efficient information for θ is bounded below by the information for θ in a "good" observation (Y, Z), it is typically positive. Then the preceding lemma together with Lemma 25.92 shows that the derivative $\dot{\Psi}_0$ is continuously invertible as a map from $\mathbb{R}^k \times \ell^\infty(\mathcal{H}) \times \mathbb{R}^k \times \ell^\infty(\mathcal{H})$ for $\mathcal{H}$ the unit ball of $C^\beta(\mathcal{Z})$. This is useful in the cases that the dimension of $\mathcal{Z}$ is not bigger than 3, for, in view of Example 19.9, we must have that $\beta > d/2$ in order that the functions $B_{\theta,\eta} h = C_{\theta,\eta} h \oplus h$ form a Donsker class, as required by Theorem 25.90. Thus $\alpha > 1/2, 2, 3/2$ suffice in dimensions 1, 2, 3, but we need $\beta > 2$ if $\mathcal{Z}$ is of dimension 4.

Sets $\mathcal{Z}$ of higher dimension can be treated by extending Lemma 25.96 to take into account higher-order derivatives, or alternatively, by not using a $C^\alpha(\mathcal{Z})$-unit ball for $\mathcal{H}$. The general requirements for a class $\mathcal{H}$ that is the unit ball of a Banach space $\mathbb{B}$ are that $\mathcal{H}$ is η_0-Donsker, that $C_0^* C_0 \mathbb{B} \subset \mathbb{B}$, and that $C_0^* C_0 : \mathbb{B} \mapsto \mathbb{B}$ is compact. For instance, if $p_\theta(x \mid z)$ corresponds to a linear regression on z, then the functions $z \mapsto C_0^* C_0 h(z)$ are of the form $z \mapsto g(\alpha^T z)$

for functions g with a one-dimensional domain. Then the dimensionality of $\mathcal{Z}$ does not really play an important role, and we can apply similar arguments, under weaker conditions than required by treating $\mathcal{Z}$ as general higher dimensional, with, for instance, $\mathbb{B}$ equal to the Banach space consisting of the linear span of the functions $z \mapsto g(\alpha^T z)$ in $C_1^1(\mathcal{Z})$ and $\mathcal{H}$ its unit ball.

The second main condition of Theorem 25.92 is that the functions $\tilde{\ell}_{\theta,\eta}$ and $B_{\theta,\eta}h$ form a Donsker class. Dependent on the kernel $p_\theta(x \mid z)$, a variety of methods may be used to verify this condition. One possibility is to employ smoothness of the kernel in x in combination with Example 19.9. If the map $x \mapsto p_\theta(x \mid z)$ is appropriately smooth, then so is the map $x \mapsto C_{\theta,\eta}h(x)$. Straightforward differentiation yields

$$\frac{\partial}{\partial x_i} C_{\theta,\eta} h(x) = \mathrm{cov}_x \left(h(Z), \frac{\partial}{\partial x_i} \log p_\theta(x \mid Z) \right),$$

where for each x the covariance is computed for the random variable Z having the (conditional) density $z \mapsto p_\theta(x \mid z)\, d\eta(z)/p_\theta(x \mid \eta)$. Thus, for a given bounded function h,

$$\left| \frac{\partial}{\partial x_i} C_{\theta,\eta} h(x) \right| \leq \|h\|_\infty \frac{\int \left| \frac{\partial}{\partial x_i} \log p_\theta(x \mid z) \right| p_\theta(x \mid z)\, d\eta(z)}{\int p_\theta(x \mid z)\, d\eta(z)}.$$

Depending on the function $\partial/\partial x_i \log p_\theta(x \mid z)$, this leads to a bound on the first derivative of the function $x \mapsto C_{\theta,\eta}h(x)$. If $\mathcal{X}$ is an interval in $\mathbb{R}$, then this is sufficient for applicability of Example 19.9. If $\mathcal{X}$ is higher dimensional, the we can bound higher-order partial derivatives in a similar manner.

If the main interest is in the estimation of η rather than θ, then there is also a nontechnical criterion for the choice of $\mathcal{H}$, because the final result gives the asymptotic distribution of $\hat{\eta}h$ for every $h \in \mathcal{H}$, but not necessarily for $h \notin \mathcal{H}$. Typically, a particular h of interest can be added to a set $\mathcal{H}$ that is chosen for technical reasons without violating the results as given previously. The addition of an infinite set would require additional arguments. Reference [107] gives more details concerning this example.

Notes

Most of the results in this chapter were obtained during the past 15 years, and the area is still in development. The monograph by Bickel, Klaassen, Ritov, and Wellner [8] gives many detailed information calculations, and heuristic discussions of methods to construct estimators. See [77], [101], [102], [113], [122], [145] for a number of other, also more recent, papers. For many applications in survival analysis, counting processes offer a flexible modeling tool, as shown in Andersen, Borgan, Gill, and Keiding [1], who also treat semiparametric models for survival analysis. The treatment of maximum likelihood estimators is motivated by (partially unpublished) joint work with Susan Murphy. Apparently, the present treatment of the Cox model is novel, although proofs using the profile likelihood function and martingales go back at least 15 years. In connection with estimating equations and CAR models we profited from discussions with James Robins, the representation in section 25.53 going back to [129]. The use of the empirical likelihood goes back a long way, in particular in survival analysis. More recently it has gained popularity as a basis for constructing likelihood ratio based confidence intervals. Limitations of the information bounds and the type of asymptotics discussed in this chapter are pointed out in [128]. For further information concerning this chapter consult recent journals, both in statistics and econometrics.

PROBLEMS

1. Suppose that the underlying distribution of a random sample of real-valued observations is known to have mean zero but is otherwise unknown.
   (i) Derive a tangent set for the model.
   (ii) Find the efficient influence function for estimating $\psi(P) = P(C)$ for a fixed set C.
   (iii) Find an asymptotically efficient sequence of estimators for $\psi(P)$.

2. Suppose that the model consists of densities $p(x - \theta)$ on $\mathbb{R}^k$, where p is a smooth density with $p(x) = p(-x)$. Find the efficient influence function for estimating θ.

3. In the regression model of Example 25.28, assume in addition that e and X are independent. Find the efficient score function for θ.

4. Find a tangent set for the set of mixture distributions $\int p(x \mid z) \, dF(z)$ for $x \mapsto p(x \mid z)$ the uniform distribution on $[z, z + 1]$. Is the linear span of this set equal to the nonparametric tangent set?

5. **(Neyman-Scott problem)** Suppose that a typical observation is a pair (X, Y) of variables that are conditionally independent and $N(Z, \theta)$-distributed given an unobservable variable Z with a completely unknown distribution η on $\mathbb{R}$. A natural approach to estimating θ is to "eliminate" the unobservable Z by taking the difference $X - Y$. The maximum likelihood estimator based on a sample of such differences is $T_n = \frac{1}{2} n^{-1} \sum_{i=1}^{n} (X_i - Y_i)^2$.
   (i) Show that the closed linear span of the tangent set for η contains all square-integrable, mean-zero functions of $X + Y$.
   (ii) Show that T_n is asymptotically efficient.
   (iii) Is T_n equal to the semiparametric maximum likelihood estimator?

6. In Example 25.72, calculate the score operator and the information operator for η.

7. In Example 25.12, express the density of an observation X in the marginal distributions F and G of Y and C and
   (i) Calculate the score operators for F and G.
   (ii) Show that the empirical distribution functions $\hat{F}^*$ and $\hat{G}^*$ of the Y_i and C_j are asymptotically efficient for estimating the marginal distributions F^* and G^* of Y and C, respectively;
   (iii) Prove the asymptotic normality of the estimator for F given by

$$\hat{F}(y) = 1 - \prod_{0 \leq s \leq y} \left(1 - \hat{\Lambda}\{s\}\right), \qquad \hat{\Lambda}(y) = \int_{[0,y]} \frac{d\hat{F}^*}{\hat{G}^* - \hat{F}^*};$$

   (iv) Show that this estimator is asymptotically efficient.

8. **(Star-shaped distributions)** Let $\mathcal{F}$ be the collection of all cumulative distribution functions on $[0, 1]$ such that $x \mapsto F(x)/x$ is nondecreasing. (This is a famous example in which the maximum likelihood estimator is inconsistent.)
   (i) Show that there exists a maximizer $\hat{F}_n$ (over $\mathcal{F}$) of the empirical likelihood $F \mapsto \prod_{i=1}^{n} F\{x_i\}$, and show that this satisfies $\hat{F}_n(x) \to xF(x)$ for every x.
   (ii) Show that at every $F \in \mathcal{F}$ there is a convex tangent cone whose closed linear span is the nonparametric tangent space. What does this mean for efficient estimation of F?

9. Show that a U-statistic is an asymptotically efficient estimator for its expectation if the model is nonparametric.

10. Suppose that the model consists of all probability distributions on the real line that are symmetric.
    (i) If the symmetry point is known to be 0, find the maximum likelihood estimator relative to the empirical likelihood.
    (ii) If the symmetry point is unknown, characterize the maximum likelihood estimators relative to the empirical likelihood; are they useful?

11. Find the profile likelihood function for the parameter θ in the Cox model with censoring discussed in Section 25.12.1.

12. Let $\mathcal{P}$ be the set of all probability distributions on $\mathbb{R}$ with a positive density and let $\psi(P)$ be the median of P.
    (i) Find the influence function of ψ.
    (ii) Prove that the sample median is asymptotically efficient.

References

[1] Andersen, P.K., Borgan, O., Gill, R.D., and Keiding, N. (1992). *Statistical Models Based on Counting Processes*. Springer, Berlin.

[2] Arcones, M.A., and Giné, E. (1993). Limit theorems for U-processes. *Annals of Probability* **21**, 1494–1542.

[3] Bahadur, R.R. (1967). An optimal property of the likelihood ratio statistic. *Proceedings of the Fifth Berkeley Symposium on Mathematical Statistics and Probability (1965/66)* **I**, 13–26. University of California Press, Berkeley.

[4] Bahadur, R.R. (1971). Some limit theorems in statistics. *Conference Board of the Mathematical Sciences Regional Conference Series in Applied Mathematics* **4**. Society for Industrial and Applied Mathematics, Philadelphia.

[5] Bamdoff-Nielsen, O.E., and Hall, P. (1988). On the level-error after Bartlett adjustment of the likelihood ratio statistic. *Biometrika* **75**, 378–378.

[6] Bauer, H. (1981). *Probability Theory and Elements of Measure Theory*. Holt, Rinehart, and Winston, New York.

[7] Bentkus, V., Gotze, E, van Zwet, W.R. (1997). An Edgeworth expansion for symmetric statistics. Annals of Statistics **25**, 851–896.

[8] Bickel, PJ., Klaassen, C.A.J., Ritov, Y., and Wellner, J.A. (1993). Efficient and Adaptive Estimationfor Semi parametric Models. Johns Hopkins University Press, Baltimore.

[9] Bickel, PJ., and Ghosh, J.K. (1990). A decomposition for the likelihood ratio statistic and the Bartlett correction-a Bayesian argument. Annals of Statistics 18, 1070-1090.

[10] Bickel, PJ., and Rosenblatt, M. (1973). On some global measures of the deviations of density function estimates. Annals of Statistics 1,1071-1095.

[11] Billingsley, P. (1968). Convergence of Probability Measures. John Wiley, New York.

[12] Birge, L. (1983). Approximation dans les espaces metriques et theorie de l'estimation. Zeitschriftfiir Wahrscheinlichkeitstheorie und Verwandte Gebiete 65, 181-238.

[13] Birge, L. (1997). Estimation of unimodal densities without smoothness assumptions. Annals of Statistics 25, 970-981.

[14] Birge, L., and Massart, P. (1993). Rates of convergence for minimum contrast estimators. Probability Theory and Related Fields 97, 113-150.

[15] Birge, L., and Massart, P. (1997). From model selection to adaptive estimation. Festschriftfor Lucien Le Cam. Springer, New York, 55-87.

[16] Birman, M.S., and Solomjak, M.Z. (1967). Piecewise-polynomial approximation of functions of the classes Wp. Mathematics of the USSR Sbomik 73,295-317.

[17] Brown, L. (1987). Fundamentals of Statistical Exponential Families with Applications in Statistical Decision Theory. Institute of Mathematical Statistics, California.

[18] Brown, L.D., and Fox, M. (1974). Admissibility of procedures in two-dimensional location parameter problems. Annals of Statistics 2, 248-266.

[19] Cantelli, EP. (1933). Sulla determinazione empirica delle leggi di probabilitA. Giomale dell'lstituto Italiano degli Attuari 4, 421-424.

[20] Chernoff, H. (1952). A measure of asymptotic efficiency for tests of a hypothesis based on the sum of observations. *Annals of Mathematical Statistics* **23**, 493–507.

[21] Chernoff, H. (1954). On the distribution of the likelihood ratio statistic. *Annals of Mathematical Statistics* **25**, 573–578.

[22] Chernoff, H., and Lehmann, E.L. (1954). The use of maximum likelihood estimates in χ^2 tests for goodness of fit. *Annals of Mathematical Statistics* **25**, 579–586.

[23] Chow, Y.S., and Teicher, H. (1978). *Probability Theory.* Springer-Verlag, New York.

[24] Cohn, D.L. (1980). *Measure Theory.* Birkhäuser, Boston.

[25] Copas, J. (1975). On the unimodality of the likelihood for the Cauchy distribution. *Biometrika* **62**, 701–704.

[26] Cramér, H. (1938). Sur un nouveau théoréme-limite de la théorie des probabilités. *Actualités Scientifiques et Industrielles* **736**, 5–23.

[27] Cramér, H. (1946). *Mathematical Methods of Statistics.* Princeton University Press, Princeton.

[28] Csörgő, M. (1983). *Quantile Processes with Statistical Applications. CBMS-NSF Regional Conference Series in Applied Mathematics* **42**. Society for Industrial and Applied Mathematics (SIAM), Philadelphia.

[29] Dacunha-Castelle, D., and Duflo, M. (1993). *Probabilités et Statistiques, tome II*. Masson, Paris.

[30] Davies, R.B. (1973). Asymptotic inference in stationary Gaussian time-series. *Advances in Applied Probability* **4**, 469–497.

[31] Dembo, A., and Zeitouni, O. (1993). *Large Deviation Techniques and Applications.* Jones and Bartlett Publishers, Boston.

[32] Deuschel, J.D., and Stroock, D.W. (1989). *Large Deviations.* Academic Press, New York.

[33] Devroye, L., and Gyorfi, L. (1985). *Nonparametric Density Estimation: The L_1-View.* John Wiley & Sons, New York.

[34] Diaconis, P., and Freedman, D. (1986). On the consistency of Bayes estimates. *Annals of Statistics* **14**, 1–26.

[35] Donald, S.G., and Newey, W.K. (1994). Series estimation of semilinear models. *Journal of Multivariate Analysis* **50**, 30–40.

[36] Donoho, D.L., and Johnstone, I.M. (1994). Idea spatial adaptation by wavelet shrinkage. *Biometrika* **81**, 425–455.

[37] Donoho, D.L., and Liu, R.C. (1991). Geometrizing rates of convergence II, III. *Annals of Statistics* **19**, 633–701.

[38] Donsker, M.D. (1952). Justification and extension of Doob's heuristic approach to the Kolmogorov-Smirnov theorems. *Annals of Mathematical Statistics* **23**, 277–281.

[39] Doob, J. (1948). Application of the theory of martingales. *Le Calcul des Probabilités et ses Applications. Colloques Internationales du CNRS Paris*, 22–28.

[40] Drost, F.C. (1988). *Asymptotics for Generalized Chi-Square Goodness-of-Fit Tests. CWI tract* **48**. Centrum voor Wiskunde en Informatica, Amsterdam.

[41] Dudley, R.M. (1976). *Probability and Metrics: Convergence of Laws on Metric Spaces. Mathematics Institute Lecture Notes Series* **45**. Aarhus University, Denmark.

[42] Dudley, R.M. (1989). *Real Analysis and Probability*, Wadsworth, Belmont, California.

[43] Dupač, V., and Hájek, J. (1969). Asymptotic normality of simple linear rank statistics under alternatives, *Annals of Mathematical Statistics* II **40**, 1992–2017.

[44] Efron, B., and Tibshirani, R.J. (1993). *An Introduction to the Bootstrap.* Chapman and Hall, London.

[45] Fahrmeir, L., and Kaufmann, H. (1985). Consistency and asymptotic normality of the maximum likelihood estimator in generalized linear models. *Annals of Statistics* **13**, 342–368. (Correction: *Annals of Statistics* **14**, 1643.)

[46] Farrell, R.H. (1972). On the best obtainable asymptotic rates of convergence in estimation of a density function at a point. *Annals of Mathematical Statistics* **43**, 170–180.

[47] Feller, W. (1971). *An Introduction to Probability Theory and Its Applications, vol. II.* John Wiley & Sons, New York.

[48] Fisher, R.A. (1922). On the mathematical foundations of theoretical statistics. *Philosophical Transactions of the Royal Society of London, Series A* **222**, 309–368.

[49] Fisher, R.A. (1924). The conditions under which χ^2 measures the discrepancy between observations and hypothesis. *Journal Royal Statist. Soc.* **87**, 442–450.

[50] Fisher, R.A. (1925). Theory of statistical estimation. *Proceedings of the Cambridge Philosophical Society* **22**, 700–725.

[51] van de Geer, S.A. (1988). *Regression Analysis and Empirical Processes. CWI Tract* **45**. Centrum voor Wiskunde en Informatica, Amsterdam.

[52] Ghosh, J.K. (1994). *Higher Order Asymptotics*. Institute of Mathematical Statistics, Hayward.

[53] Gill, R.D. (1989). Non- and semi-parametric maximum likelihood estimators and the von-Mises method (part I). *Scandinavian Journal of Statistics* **16**, 97–128.

[54] Gill, R.D. (1994). Lectures on survival analysis. *Lecture Notes in Mathematics* **1581**, 115–241.

[55] Gill, R.D., and Johansen, S. (1990). A survey of product-integration with a view towards application in survival analysis. *Annals of Statistics* **18**, 1501–1555.

[56] Gill, R.D., and van der Vaart, A.W. (1993). Non- and semi-parametric maximum likelihood estimators and the von Mises method (part II). *Scandinavian Journal of Statistics* **20**, 271–288.

[57] Giné, E., and Zinn, J. (1986). Lectures on the central limit theorem for empirical processes. *Lecture Notes in Mathematics* **1221**, 50–113.

[58] Giné, E., and Zinn, J. (1990). Bootstrapping general empirical measures. *Annals of Probability* **18**, 851–869.

[59] Glivenko, V. (1933). Sulla determinazione empirica della leggi di probabilità. *Giornale dell'Istituto Italiano degli Attuari* **4**, 92–99.

[60] Greenwood, P.E., and Nikulin, M.S. (1996). *A Guide to Chi-Squared Testing*. John Wiley & Sons, New York.

[61] Groeneboom, P. (1980). *Large Deviations and Bahadur Efficiencies. MC tract* **118**, Centrum voor Wiskunde en Informatica, Amsterdam.

[62] Groeneboom, P. (1985). Estimating a monotone density. *Proceedings of the Berkeley Conference in Honor of Jerzy Neyman and Jack Kiefer* **2**, 539–555. Wadsworth, Monterey, California.

[63] Groeneboom, P. (1988). Brownian Motion with a parabolic drift and Airy functions. *Probability Theory and Related Fields* **81**, 79–109.

[64] Groeneboom, P., Lopuhaä, H.P. (1993). Isotonic estimators of monotone densities and distribution functions: basic facts. *Statistica Neerlandica* **47**, 175–183.

[65] Groeneboom, P., Oosterhoff, J., and Ruymgaart, F. (1979). Large deviation theorems for empirical probability measures. *Annals of Probability* **7**, 553–586.

[66] de Haan, L. (1976). Sample extremes: An elementary introduction. *Statistica Neerlandica* **30**, 161–172.

[67] Hájek, J. (1961). Some extensions of the Wald-Wolfowitz-Noether theorem. *Annals of Mathematical Statistics* **32**, 506–523.

[68] Hájek, J. (1968). Asymptotic normality of simple linear rank statistics under alternatives. *Annals of Mathematical Statistics* **39**, 325–346.

[69] Hájek, J. (1970). A characterization of limiting distributions of regular estimates. *Zeitschrift für Wahrscheinlichkeitstheorie und Verwandte Gebiete* **14**, 323–330.

[70] Hájek, J. (1972). Local asymptotic minimax and admissibility in estimation. *Proceedings of the Sixth Berkeley Symposium on Mathematical Statistics and Probability* **1**, 175–194.

[71] Hájek, J., and Šidák, Z. (1967). *Theory of Rank Tests*. Academic Press, New York.

[72] Hall, P. (1992). *The Bootstrap and Edgeworth Expansion. Springer Series in Statistics*. Springer-Verlag, New York.

[73] Hampel, F.R., Ronchetti, E.M., Rousseeuw, P.J., and Stahel, W.A. (1986). *Robust Statistics: the Approach Based on Influence Functions*. Wiley, New York.

[74] Helmers, R. (1982). *Edgeworth Expansions for Linear Combinations of Order Statistics. Mathematical Centre Tracts* **105**. Mathematisch Centrum, Amsterdam.

[75] Hoeffding, W. (1948). A class of statistics with asymptotically normal distribution. *Annals of Mathematical Statistics* **19**, 293–325.

[76] Hoffmann-Jørgensen, J. (1991). *Stochastic Processes on Polish Spaces. Various Publication Series* **39**. Aarhus Universitet, Aarhus, Denmark.

[77] Huang, J. (1996). Efficient estimation for the Cox model with interval censoring. *Annals of Statistics* **24**, 540–568.

[78] Huber, P. (1967). The behavior of maximum likelihood estimates under nonstandard conditions. *Proceedings of the Fifth Berkeley Symposium on Mathematical Statistics and Probability* **1**, 221–233. University of California Press, Berkeley.

[79] Huber, P. (1974). *Robust Statistics*. Wiley, New York.

[80] Ibragimov, I.A., and Has'minskii, R.Z. (1981). *Statistical Estimation: Asymptotic Theory*. Springer-Verlag, New York.

[81] Jagers, P. (1975). *Branching Processes with Biological Applications*. John Wiley & Sons, London-New York-Sydney.

[82] Janssen, A., and Mason, D.M. (1990). *Nonstandard Rank Tests. Lecture Notes in Statistics* **65**. Springer-Verlag, New York.

[83] Jensen, J.L. (1993). A historical sketch and some new results on the improved log likelihood ratio statistic. *Scandinavian Journal of Statistics* **20**, 1–15.

[84] Kallenberg, W.C.M. (1983). Intermediate efficiency, theory and examples. *Annals of Statistics* **11**, 498–504.

[85] Kallenberg, W.C.M., and Ledwina, T. (1987). On local and nonlocal measures of efficiency. *Annals of Statistics* **15**, 1401–1420.

[86] Kallenberg, W.C.M., Oosterhoff, J., and Schriever, B.F. (1980). The number of classes in chi-squared goodness-of-fit tests. *Journal of the American Statistical Association* **80**, 959–968.

[87] Kim, J., and Pollard, D. (1990). Cube root asymptotics. *Annals of Statistics* **18**, 191–219.

[88] Kolčinskii, V.I. (1981). On the central limit theorem for empirical measures. *Theory of Probability and Mathematical Statistics* **24**, 71–82.

[89] Koul, H.L., and Pflug, G.C. (1990). Weakly adaptive estimators in explosive autoregression. *Annals of Statistics* **18**, 939–960.

[90] Leadbetter, M.R., Lindgren, G., and Rootzén, H. (1983). *Extremes and Related Properties of Random Sequences and Processes*. Springer-Verlag, New York.

[91] Le Cam, L. (1953). On some asymptotic properties of maximum likelihood estimates and related Bayes estimates. *University of California Publications in Statistics* **1**, 277–330.

[92] Le Cam, L. (1960). Locally asymptotically normal families of distributions. *University of California Publications in Statistics* **3**, 37–98.

[93] Le Cam, L. (1969). *Théorie Asymptotique de la Décision Statistique*. Les Presses de l'Université de Montréal, Montreal.

[94] Le Cam, L. (1970). On the assumptions used to prove asymptotic normality of maximum likelihood estimates. *Annals of Mathematical Statistics* **41**, 802–828.

[95] Le Cam, L. (1972). Limits of experiments. *Proceedings of the Sixth Berkeley Symposium on Mathematical Statistics and Probability* **1**, 245–261. University of California Press, Berkeley.

[96] Le Cam, L. (1986). *Asymptotic Methods in Statistical Decision Theory*. Springer-Verlag, New York.

[97] Le Cam, L.M., and Yang, G. (1990). *Asymptotics in Statistics: Some Basic Concepts*. Springer-Verlag, New York.

[98] Ledoux, M., and Talagrand, M. (1991). *Probability in Banach Spaces: Isoperimetry and Processes*. Springer-Verlag, Berlin.

[99] Lehmann, E.L. (1983). *Theory of Point Estimation*. Wiley, New York.

[100] Lehmann, E.L. (1991). *Testing Statistical Hypotheses, 2nd edition*. Wiley, New York.

[101] Levit, B.Y. (1978). Infinite-dimensional informational lower bounds. *Theory of Probability and its Applications* **23**, 388–394.

[102] Mammen, E., and van de Geer, S.A. (1997). Penalized quasi-likelihood estimation in partial linear models. *Annals of Statistics* **25**, 1014–1035.

[103] Massart, P. (1990). The tight constant in the Dvoretsky-Kiefer-Wolfowitz inequality. *Annals of Probability* **18**, 1269–1283.

[104] von Mises, R. (1947). On the asymptotic distribution of differentiable statistical functions. *Annals of Mathematical Statistics* **18**, 309–348.

[105] Murphy, S.A., Rossini, T.J., and van der Vaart, A.W. (1997). MLE in the proportional odds model. *Journal of the American Statistical Association* **92**, 968–976.

[106] Murphy, S.A., and van der Vaart, A.W. (1997). Semiparametric likelihood ratio inference. *Annals of Statistics* **25**, 1471–1509.

[107] Murphy, S.A., and van der Vaart, A.W. (1996). Semiparametric mixtures in case-control studies.

[108] Murphy, S.A., and van der Vaart, A.W. (1996). Likelihood ratio inference in the errors-in-variables model. *Journal of Multivariate Analysis* **59**, 81–108.

[109] Noether, G.E. (1955). On a theorem of Pitman. *Annals of Mathematical Statistics* **25**, 64–68.

[110] Nussbaum, M. (1996). Asymptotic equivalence of density estimation and Gaussian white noise. *Annals of Statistics* **24**, 2399–2430.

[111] Ossiander, M. (1987). A central limit theorem under metric entropy with L_2 bracketing. *Annals of Probability* **15**, 897–919.

[112] Pearson, K. (1900). On the criterion that a given system of deviations from the probable in the case of a correlated system of variables is such that it can be reasonably supposed to have arisen from random sampling. *Philosopical Magazine, Series 5* **50**, 157–175. (Reprinted in: *Karl Pearson's Early Statistical Papers*, Cambridge University Press, 1956.)

[113] Pfanzagl, J., and Wefelmeyer, W. (1982). *Contributions to a General Asymptotic Statistical Theory. Lecture Notes in Statistics* **13**. Springer-Verlag, New York.

[114] Pfanzagl, J., and Wefelmeyer, W. (1985). *Asymptotic Expansions for General Statistical Models. Lecture Notes in Statistics* **31**. Springer-Verlag, New York.

[115] Pflug, G.C. (1983). The limiting loglikelihood process for discontinuous density families. *Zeitschrift für Wahrscheinlichkeitstheorie und Verwandte Gebiete* **64**, 15–35.

[116] Pollard, D. (1982). A central limit theorem for empirical processes. *Journal of the Australian Mathematical Society A* **33**, 235–248.

[117] Pollard, D. (1984). *Convergence of Stochastic Processes*. Springer-Verlag, New York.

[118] Pollard, D. (1985). New ways to prove central limit theorems. *Econometric Theory* **1**, 295–314.

[119] Pollard, D. (1989). A maximal inequality for sums of independent processes under a bracketing condition.

[120] Pollard, D. (1990). *Empirical Processes: Theory and Applications. NSF-CBMS Regional Conference Series in Probability and Statistics* **2**. Institute of Mathematical Statistics and American Statistical Association. Hayward, California.

[121] Prakasa Rao, B.L.S. (1983). *Nonparametric Functional Estimation*. Academic Press, Orlando.

[122] Qin, J., and Lawless, J. (1994). Empirical likelihood and general estimating equations. *Annals of Statistics* **22**, 300–325.

[123] Rao, C.R. (1973). *Linear Statistical Inference and Its Applications*. Wiley, New York.

[124] Reed, M., and Simon, B. (1980). *Functional Analysis*. Academic Press, Orlando.

[125] Reeds, J.A. (1976). *On the Definition of von Mises Functionals*. Ph.D. dissertation, Department of Statistics, Harvard University, Cambridge, MA.

[126] Reeds, J.A. (1985). Asymptotic number of roots of Cauchy location likelihood equations. *Annals of Statistics* **13**, 775–784.

[127] Révész, P. (1968). *The Laws of Large Numbers*. Academic Press, New York.

[128] Robins, J.M., and Ritov, Y. (1997). Towards a curse of dimensionality appropriate (CODA) asymptotic theory for semi-parametric models. *Statistics in Medicine* **16**, 285–319.

[129] Robins, J.M., and Rotnitzky, A. (1992). Recovery of information and adjustment for dependent censoring using surrogate markers. In *AIDS Epidemiology–Methodological Issues*, 297–331, eds: N. Jewell, K. Dietz, and V. Farewell. Birkhäuser, Boston.

[130] Roussas, G.G. (1972). *Contiguity of Probability Measures*. Cambridge University Press, Cambridge.

[131] Rubin, H., and Vitale, R.A. (1980). Asymptotic distribution of symmetric statistics. *Annals of Statistics* **8**, 165–170.

[132] Rudin, W. (1973). *Functional Analysis*. McGraw-Hill, New York.

[133] Shiryayev, A.N. (1984). *Probability*. Springer-Verlag, New York-Berlin.

[134] Shorack, G.R., and Wellner, J.A. (1986). *Empirical Processes with Applications to Statistics*. Wiley, New York.

[135] Silverman, B.W. (1986). *Density Estimation for Statistics and Data Analysis*. Chapman and Hall, London.

[136] Stigler, S.M. (1974). Linear functions of order statistics with smooth weight functions. *Annals of Statistics* **2**, 676–693. (Correction: Annals of Statistics **7**, 466.)

[137] Stone, C.J. (1990). Large-sample inference for log-spline models. *Annals of Statistics*, **18**, 717–741.

[138] Strasser, H. (1985). *Mathematical Theory of Statistics*. Walter de Gruyter, Berlin.

[139] van der Vaart, A.W. (1988). *Statistical Estimation in Large Parameter Spaces. CWI Tracts* **44**. Centrum voor Wiskunde en Informatica, Amsterdam.

[140] van der Vaart, A.W. (1991). On differentiable functionals. *Annals of Statistics* **19**, 178–204.

[141] van der Vaart, A.W. (1991). An asymptotic representation theorem. *International Statistical Review* **59**, 97–121.

[142] van der Vaart, A.W. (1994). *Limits of Experiments*. Lecture Notes, Yale University.

[143] van der Vaart, A.W. (1995). Efficiency of infinite dimensional M-estimators. *Statistica Neerlandica* **49**, 9–30.

[144] van der Vaart, A.W. (1994). Maximum likelihood estimation with partially censored observations. *Annals of Statistics* **22**, 1896–1916.

[145] van der Vaart, A.W. (1996). Efficient estimation in semiparametric models. *Annals of Statistics* **24**, 862–878.

[146] van der Vaart, A.W., and Wellner, J.A. (1996). *Weak Convergence and Empirical Processes*. Springer, New York.

[147] Vapnik, V.N., and Červonenkis, A.Y. (1971). On the uniform convergence of relative frequencies of events to their probabilities. *Theory of Probability and Its Applications* **16**, 264–280.

[148] Wahba, G. (1990). *Spline models for observational data. CBMS-NSF Regional Conference Series in Applied Mathematics* **59**. Society for Industrial and Applied Mathematics, Philadelphia.

[149] Wald, A. (1943). Test of statistical hypotheses concerning several perameters when the number of observations is large. *Transactions of the American Mathematical Society* **54**, 426–482.

[150] Wilks, S.S. (1938). The large-sample distribution of the likelihood ratio for testing composite hypotheses. *Annals of Mathematical Statistics* **19**, 60–62.

[151] van Zwet, W.R. (1984). A Berry-Esseen bound for symmetric statistics. *Zeitschrift für Wahrscheinlichkeitstheorie und Verwandte Gebiete* **66**, 425–440.

Index

α-Winsorized means, 316
α-trimmed means, 316
absolute rank, 181
absolutely
    continuous, 85, 268
    continuous part, 85
accessible, 150
adaptation, 223
adjoint map, 361
adjoint score operator, 372
antiranks, 184
Assouad's lemma, 347
asymptotic
    consistent, 44, 329
    differentiable, 106
    distribution free, 164
    efficient, 64, 367, 387
    equicontinuity, 262
    influence function, 58
    of level α, 192
    linear, 401
    lower bound, 108
    measurable, 260
    relative efficiency, 195
    risk, 109
    tight, 260
    tightness, 262
    uniformly integrable, 17

Bahadur
    efficiency, 203
    relative efficiency, 203
    slope, 203, 239
Bahadur-Kiefer theorems, 310
Banach space, 361
bandwidth, 342
Bartlett correction, 238
Bayes
    estimator, 138
    risk, 138
Bernstein's inequality, 285
best regular, 115
bilinear map, 295

bootstrap
    empirical distribution, 332
    empirical process, 332
    parametric, 328, 340
Borel
    σ-field, 256
    measurable, 256
bounded
    Lipschitz metric, 332
    in probability, 8
bowl-shaped, 113
bracket, 270
bracketing
    integral, 270
    number, 270
Brownian
    bridge, 266
    bridge process, 168
    motion, 268

cadlag, 257
canonical link functions, 235
Cartesian product, 257
Cauchy sequence, 255
central
    limit theorem, 6
    moments, 27
chain rule, 298
chaining argument, 285
characteristic function, 13
chi-square distribution, 242
Chibisov-O'Reilly theorem, 273
closed, 255
closure, 255
coarsening at random (CAR), 379
coefficients, 173
compact, 255, 424
compact differentiability, 297
complete, 255
completion, 257
concordant, 164
conditional expectation, 155
consistent, 3, 149, 193, 329

439

Printed in the United States
by Baker & Taylor Publisher Services